Silk Tree, Guanacaste, Monkey's Earring

The MEMOIRS are published at irregular intervals in volumes of various sizes and are designed to include results of original botanical research by members of The New York Botanical Garden's staff, or by botanists who have collaborated in one or more of The New York Botanical Garden's research programs. Ordinarily only manuscripts of 100 or more typewritten pages will be considered for publication.

Manuscripts should be submitted to the Editor. For further information regarding editorial policy and instructions for the preparation of manuscripts, contact the Editor.

Orders for published and forthcoming issues and volumes should be placed with:

NYBG

Scientific Publications Department
THE NEW YORK BOTANICAL GARDEN
Bronx, New York 10458–5126 U.S.A.
(718) 817-8721
FAX (718) 817-8842
e-mail scipubs@nybg.org

Silk Tree, Guanacaste, Monkey's Earring

A Generic System for the Synandrous Mimosaceae of the Americas

Part I. Abarema, Albizia, and Allies

RUPERT C. BARNEBY AND JAMES W. GRIMES

Issued: 25 March 1996
Memoirs of THE NEW YORK BOTANICAL GARDEN
Volume 74, Part I

THE NEW YORK BOTANICAL GARDEN
Bronx, New York 10458-5126 U.S.A.

The New York Botanical Garden

Published by

The New York Botanical Garden
Bronx, New York 10458

Composition by Eisner/Martin Typographics

Manufactured by Braun-Brumfield, Inc.

International Standard Serial Number 0071-5794

The paper used in this publication meets the requirements of the American National Standard for Information Sciences—Permanence of Paper for Publications and Documents in Libraries and Archives, ANSI/NISO (Z39.48–1992).

Printed in the United States of America on recycled paper
Printed using soy-based ink

Library of Congress Cataloging in Publication Data

Barneby, Rupert C.
Silk tree, Guanacaste, Monkey's earring : a generic system for the synandrous Mimosaceae of the Americas / Rupert C. Barneby, James W. Grimes
p. cm.—(Memoirs of the New York Botanical Garden, ISSN 0071-5794 ; v. 74)
Includes bibliographical references (v. 1, p.) and index.
Contents: pt. 1. Abarema, Albizia, and allies
ISBN 0-89327-395-3 (alk. paper)
1. Mimosaceae—West Indies—Classification. 2. Mimosaceae—South America—Classification. I. Grimes, James W., 1953– . II. Title. III. Series
QK1.N526 Vol. 74
[QK495.M545]
581 s—dc20
[583'.321] 95-39082
CIP

MEMOIRS OF THE NEW YORK BOTANICAL GARDEN **74:** 1–292

Silk Tree, Guanacaste, Monkey's Earring: A Generic System for the Synandrous Mimosaceae of the Americas. Part I. *Abarema, Albizia,* and Allies

RUPERT BARNEBY AND JAMES GRIMES

Contents

Abstract

Barneby, R. C. and J. W. Grimes (The New York Botanical Garden, Bronx, NY 10458-5126, U.S.A.). Silk tree, guanacaste, monkey's earring: A generic system for the synandrous Mimosaceae of the Americas. Part I. *Abarema, Albizia,* and allies. Mem. New York Bot. Gard. **74:** 1–292. 1996.—Presented here are (a) definitions, new or confirmed, of the genera of Mimosaceae tribe Ingeae either native or naturalized in the Americas; (b) phylogenetic analyses and taxonomic revision of *Hydrochorea* Barneby & Grimes (4 spp.), *Balizia* Barneby & Grimes (3 spp.), *Abarema* Pittier (45 spp.), *Hesperalbizia* Barneby & Grimes (1 sp.), *Pseudosamanea* Harms (2 spp.), *Samanea* Merrill (3 spp.), *Blanchetiodendron* Barneby & Grimes (1 sp.), *Leucochloron* Barneby & Grimes (4 spp.), *Chloroleucon* (Bentham) Britton & Rose (10 spp.), *Sphinga* Barneby & Grimes (3 spp.), *Havardia* Small (5 spp.), *Ebenopsis* Britton & Rose (3 spp.), *Painteria* Britton & Rose (3 spp.), and

Macrosamanea Britton & Killip (11 spp.); (c) alpha-taxonomic revision of *Albizia* Durazzini (23 spp.), *Cedrelinga* Ducke (1 sp.), *Falcataria* (Nielsen) Barneby & Grimes (1 sp.), and *Paraserianthes* Nielsen (1 sp.); and (d) summaries of *Enterolobium* Martius (11 spp.) and *Lysiloma* Bentham (8 spp.). Cladistic analyses and revision of the genera *Calliandra* Bentham, *Cojoba* Britton & Rose, *Pithecellobium* Martius (s.str.), and *Zygia* P. Browne will appear in Part II.

Key words: Mimosaceae, Ingeae, *Abarema, Albizia, Balizia, Blanchetiodendron, Calliandra, Cedrelinga, Chloroleucon, Cojoba, Ebenopsis, Enterolobium, Falcataria, Havardia, Hesperalbizia, Hydrochorea, Leucochloron, Lysiloma, Macrosamanea, Painteria, Paraserianthes, Pithecellobium, Pseudosamanea, Samanea, Sphinga, Zygia,* Americas.

Introduction

This monograph presents revisions, at generic and specific levels, of all West Indian and South American, and some Central American, genera of tribe Ingeae (Mimosaceae) ever referred to *Pithecellobium* Martius. By definition it excludes the genera *Inga* Miller (including *Affonsea* St.-Hilaire), *Calliandra* Bentham, and *Zapoteca* H. Hernández, none of which have ever been confused with the core of *Pithecellobium*. As a precaution, however, we include a brief account of West Indian *Lysiloma* Bentham, a genus perhaps more closely related to tribe Acacieae but easily confounded with pithecellobioid elements of genus *Albizia* Durazzini. The genera *Enterolobium* Martius and *Lysiloma* are treated in less detail than the rest, due to excellent modern monographs (Mesquita, 1990; Thompson, 1980, respectively).

This statement of contents must be modified for the three genera *Pithecellobium*, *Zygia* P. Browne, and *Cojoba* Britton & Rose, of which only generic definitions are herein given. We have in manuscript full treatments of all West Indian and South American species of these genera, but defer publication until their Mexican and Central American relatives have been examined in detail and phylogenetic analyses performed.

The tribe Ingeae was originally so defined by Bentham (1875) as to include all Mimosaceae with stamens united into a filament tube. Recent phylogenetic analyses (Grimes, unpubl. data; Maslin & Chappill, pers. comm.) show that the tribe is probably paraphyletic with respect to Acacieae, some members of which also have stamens shortly united into a tube, though no formal realignments have been proposed. Bentham's concepts of *Inga* and *Calliandra* have been relatively stable, the former characterized by simply pinnate leaves and seed-testa modified into a sweet pulp, the latter by a fruit elastically dehiscent from apex but not coiling or twisted. The only modification of either of these groups with any prospect of permanence is the segregation of *Zapoteca* from *Calliandra* (Hernández, 1986). The remainder of the species have had a more tumultuous nomenclatural history, the majority of them at one time or another having been referred to *Pithecellobium* Martius. Historically, workers have adopted either a broad or a narrow definition of *Pithecellobium*. In the broad sense, the taxon has been construed in negative terms: Plants do not possess the fruits of *Calliandra* or the fruits or leaves of *Inga*. In a narrower sense, *Pithecellobium* is characterized by the syndrome of stipular spines and arillate seeds, though these characters are not found together in *all* species. A strictly defined *Pithecellobium* does not solve the problem of a genus defined only in exclusive terms but transfers the problem from *Pithecellobium* to *Albizia*. The nomenclatural history of the tribe has been reviewed by Nielsen (1981), who has provided the most durable alternative to Bentham's classification, so it will not be reviewed here.

The persistent problem of how best to define *Pithecellobium*, and consequently other genera in Ingeae, revolves around one issue: the importance attached to characters of the fruit in interpretation of affinities and in construction of classifications. Nielsen has made the first attempt to modify the emphasis on fruit-characters inherited from Bentham. His work may be criticized only, if at all, because of his lack of familiarity with the New World species. Likewise our work may suffer weakness because of lack of familiarity with the Old World species. However, Nielsen's many publications provide us with necessary comparative data.

A phylogenetic analysis of the whole tribe, containing the data garnered from this study, is to be included in the forthcoming volume of *Advances in Legume Systematics 7*. In that study (Grimes, 1995), monophyly of various generic segregates is tested and preliminary hypotheses of sister-group relationships are presented.

When we started this project, other workers were studying the Mexican and Central American members of the *Pithecellobium*-complex, so we undertook to revise at specific level only the West Indian and South American elements. Nonetheless, in order to ascertain generic distinctions at first hand we have included several genera primarily of Central American dispersal; phylogenetic analyses of all these genera are presented below. Because it is now apparent that work on the Mexican and Central American species of *Pithecellobium*, *Cojoba*, and *Zygia* will not come to monographic fruition, it is our intent to present a comprehensive revision of those genera in a second part of this work.

A note on authorship is appropriate. At the beginning of the project Barneby and Grimes independently started descriptive work on some of the genera treated herein. A year into the project Grimes traveled to visit major herbaria and to do fieldwork; Barneby continued with the descriptive work. After returning from the field, Grimes reviewed the work by Barneby and, at the same time, started putting together the data sets for phylogenetic analysis. Barneby, therefore, assumes first authorship on the descriptions presented herein, and Grimes first authorship on the phylogenetic analyses. Both authors contributed equally to the section on morphology and are jointly responsible for taxonomic decisions.

Anatomy and Morphology of the *Pithecellobium*-complex

Habit

The American Ingeae are mostly monopodial or sympodial woody plants with monomorphic or dimorphic branch-systems. The species vary in stature from the dwarf shrublets of *Painteria revoluta* to gigantic trees of *Cedrelinga*, emergent from the canopy of high rain forest, or of *Samanea saman*, with an umbrella-shaped crown attaining diameters of 40 m or more. A few South American species of *Calliandra* are functionally herbaceous subshrubs with stems arising annually from a xylopodium and are often fire-pruned. Ingeae are dispersed into most of the neotropical plant communities, from the Sonoran Desert in Mexico to ever-moist or intermittently flooded forest of Amazonia, and they are modified accordingly in stature and habit. In general, xeromorphic Ingeae are relatively slow-growing and have hard, dense wood, whereas the more mesophytic and more rapidly maturing ones have softer and less durable wood.

While preparing data sets for cladistic analysis, Grimes (1992) found that classical definitions of inflorescences did not reflect homologous organ-systems. Furthermore, it was found that inflorescence structure was intricately linked to branching patterns. In order to define homologues, Grimes introduced the term *Repeating Growth-Unit* (RGU; Fig. 1) as a morphological unit at a hierarchical level between the metamer and the module. A metamer is defined as the internode, node, and associated axillary buds; the RGU is defined as the smallest complete, or unitary, repeating sequence of metamers produced by a meristem. A module then is everything produced by a single apical meristem. The habit of a plant depends in large part on whether growth is monopodial or sympodial, and this has not been conclusively demonstrated for the majority of taxa. As far as is known, all axes in American Ingeae are blastotelic (Briggs & Johnson, 1979); that is, the growth of the axes terminates with the formation of a bud. Anthotelic axes (i.e., those in which growth of the meristem ends with the formation of a flower) are unknown in any Ingeae. Derstine and Tucker (1991) have shown that in *Acacia baileyana* the apparently terminal flower, rather than being formed directly from the apical meristem, is probably a lateral flower that moves into a terminal position after elongation of the axis has stopped. We make the assumption that this is also true of Ingeae and call such flowers pseudoterminal. Blastotelic meristems are further divided into auxotelic meristems and anauxotelic ones. Auxotelic axes continue growth after the formation of inflorescences; anauxotelic ones do not. It follows that growth in plants with auxotelic axes is monopodial, while that of anauxotelic ones is sympodial.

In taxa with sympodial growth, growth may or may not be articulate (Tomlinson & Gill, 1973); that is, there may or may not be a distinct morphological discontinuity or articulation between RGUs. When there is a discontinuity (e.g., *Calliandra*, *Chloroleucon*, *Leucochloron*), the articulations are demarked by scars of bud-scales.

Habit depends not only on whether growth is sympodial or monopodial but also on the formation and disposition of branch-systems. The formation of branch-systems in the American Ingeae varies depending on whether branch-systems arise from proleptic or sylleptic buds. Sylleptic buds are those that form and develop concomitant with the development of the RGU on which they are found. Proleptic

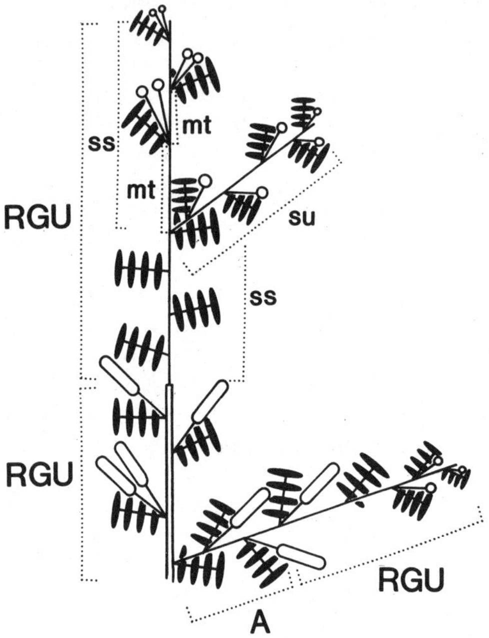

FIG. 1. The Repeating Growth Unit (RGU) is defined as the smallest complete, or unitary, repeating sequence of metamers produced by a meristem. mt = metamer. su = subunit, defined as any branch that repeats a sequence of metamers produced on the main axis. ss = subsequence, or any partial sequence of metamers on a RGU. "A" was a sylleptic branch of a penultimate RGU (hollow stem) and is a subunit that formed the base of a new branch system of RGUs.

buds form during the development of the RGU on which they are found but develop after the RGU is fully formed. Adventitious bud formation (i.e., the development of a bud independent of the formation of an RGU) has not been demonstrated in American Ingeae. In the American Ingeae with monomorphic branch-systems, the RGU always consists of a sequence of simple metamers and fertile metamers (those metamers producing the units of inflorescence, or partial inflorescence), though their order may differ (see below, Inflorescence). In these taxa, branch-systems apparently always develop from sylleptic branches, and the metamers that produce the branches develop before the simple metamers and those that produce inflorescences. In other taxa, only proleptic buds (those that form with the development of the RGU holding them) form vegetative branch-systems.

There are many forms of branch dimorphism in American Ingeae. Branch dimorphism is defined as the condition wherein the sequence and/or form of metamers on one branch-system differs from the sequence on another contemporaneous system on the same plant. The kinds of branch dimorphism are as follows. In some taxa brachyblasts, or short-shoots, form in an axillary position. Brachyblasts are shoot-systems in which development of internodes is suppressed. Often brachyblasts change into long-shoots and there is a proliferation of the vegetative branch-system. Such brachyblasts may have metamers that produce either both leaves and inflorescences or only inflorescences. This latter type leads to cauliflory, which is common in American Ingeae. In other taxa, sylleptic branches produce only annual, axillary, strictly reproductive branch systems in which leaves are suppressed. These systems are ephemeral and do not further participate in growth of the plant. Vegetative growth arises solely from proleptic buds, and growth is apparently always sympodial.

The majority of American Ingeae exhibit spiral phyllotaxy. However, *Calliandra* and *Zapoteca* are noteworthy for their distichous phyllotaxy. These character states remain to be determined in Old World Ingeae and in some American taxa.

INDUMENTUM

The indumentum of Ingeae is composed largely or entirely of simple unicellular hairs, in some genera white or gray, in others varying from yellow to brown or bronze; often mixed with these are smaller, reddish or brown, clavate or seemingly granular, several-celled trichomes. In the genera treated herein no gland-tipped excretory hairs have been observed. Pubescence that clothes the young developing stems and new leaves of the season may become dispersed or deciduous in age. Pubescence of the perianth is more stable within a species than pubescence of the foliage, which is often labile and unpredictably diverse. In many Ingeae an ovary that is glabrous at early anthesis becomes more or less densely puberulent after fertilization.

ARMAMENT

The majority of Ingeae are unarmed. The armament most commonly encountered in the tribe consists of stipules early modified, either throughout or at tip only, into straight or recurved, vulnerant spines that persist after fall of the associated leaf. Stipules of this sort characterize *Pithecellobium* (sens. lat., with few exceptions), several small xerophytic Mexican genera (*Havardia*, *Painteria*, *Ebenopsis*, *Sphinga*), one species of *Zapoteca*, a few Old World taxa debatedly referred to *Calliandra*, and they make an independent appearance in Asiatic *Thailandentadopsis* Kostermans. In the small group of Antillian *Calliandra* assembled by Britton and Rose (1928: 52) under the title *Anneslia* ser. *Haematostomae* (nom. nud.) the so-called stipular spines are of different origin. Here the stipule-blade is expanded but herbaceous and is itself armed, dorsally close above the point of insertion, with a long-persistent spine. The armament of *Chloroleucon* has been incautiously described (Nielsen, 1981: 182; Rico, 1991: 519; but not Bentham, 1875: 597) as of stipular origin. In reality, it consists of axillary inflorescence-axes modified into thorns, the true stipules being membranous and ephemeral, as often found in *Lysiloma*. The random thorns of the Asiatic *Cathormion umbellatum* (Vahl) Kostermans seem similarly derived. Genuine aculei, so widespread in tribes Mimoseae and Acacieae, are unknown in Ingeae, although the peculiar infrapetiolar hooked prickles of Asiatic *Albizia corniculata* (Loureiro) Druce and *A. myriophylla* Bentham associated with a scandent, machaerium-like habit of growth, are easily mistaken for such.

Buds and Bud-Scales

Striately nerved bud-scales reminiscent of stipules of *Calliandra* are a feature of the genera *Chloroleucon*, *Leucochloron*, and *Blanchetiodendron* and appear sporadically in some *Lysiloma*. These scales protect the precocious flowering shoots during the drought period preceding anthesis and are cast off promptly as the shoots come into active growth. True bud-scales are nearly always morphologically distinct from stipules. In some taxa (e.g., *Calliandra*, *Zygia* pro parte), the branches are apparently preformed. That is, most of the nodes are developed in bud, and growth takes place mostly through elongation of the internodes. In these taxa, the stipules are crowded on the bud. In most other taxa (e.g., *Abarema*, *Samanea*), the buds are not preformed; that is, the nodes form and develop sequentially. The stipules of these taxa are obscure or obsolete, and the adaxial surface of the leaf-rachis of developing leaves serves to protect the apical meristem.

Stipules

The stipules in the genera of Ingeae described herein are for the most part poorly differentiated, often ephemeral, or sometimes lacking. Dilated membranous stipules occur in native *Chloroleucon* and *Lysiloma* and in introduced *Albizia chinensis*, but they are quickly deciduous before maturity of the associated leaf and are never encountered on fruiting branchlets. In *Pithecellobium* (few exceptions), *Havardia*, *Painteria*, *Ebenopsis*, and *Sphinga*, the stipules are modified into spines (see Armament, above). Stipules of *Calliandra* and *Zapoteca*, on the contrary, are often persistent and prominently striate-nerved. In *Abarema*, the pallid scar left by a caducous stipule often becomes dilated and low-convex and can be mistaken for a nectary. As a general rule, stipules have no obvious function in the life-history of American Ingeae adapted to moist tropical conditions, but they assume a protective role, in the form of lignified and often recurved spines, in those adapted to monsoon and desert climates.

Leaves

The leaf of Ingeae, other than that of *Inga* (including *Affonsea*) itself and one anomalous species each of *Zygia*, *Cojoba*, and *Calliandra*, is the conventionally bipinnate leaf of the family. As in previous publications (e.g., Barneby, 1991), we present for each taxon a *leaf-formula*, calculated from the upper limits, not the full range, of numbers observed. Thus a leaf-formula expressed as iii–vii/9–16 indicates that in the material studied, the largest leaf of all specimens consisted of 3 to 7 pairs of pinnae and the longest pinna of 9 to 16 pairs of leaflets. The range in leaf-formula is extremely wide; pinnae of well-developed leaves of flowering branches vary from 1 to more than 20 pairs, and leaflets from exactly 2 per leaf to nearly 3000 (*Enterolobium glaziovii* (Bentham) Mesquita). In *Zygia* and *Cojoba*, the rachises of the single or furthest pair of pinnae tend to be of unequal length and vigor, which suggests that the simply paripinnate leaf of *Z. inundata* (Ducke) Lima and *C. rufescens* (Bentham) Britton & Rose arose through elimination of the weaker member of a pair. Following a pattern prevalent throughout the family, number and size of leaflets are reciprocally adjusted, the blades becoming larger when few, smaller when relatively many. The always pulvinate leaf-stalk varies from terete to ventrally canaliculate, the sulcus varying in width and depth, but outside *Inga* the leaf-stalk is only rarely green-marginate or winged. Secondary leaf-axes repeat the primary one and may be either shorter or longer than it; the pairs of pinnae, when more than one per leaf, are decrescent at each end, proximally decrescent and thence subequilong, or more often progressively longer upward. Pulvini of primary and secondary leaf-axes and pulvinules of the leaflets provide pivots that enable the leaf, pinnae, and leaflets, separately or in unison, to adjust their attitudes to sunlight and to fold at night or after shock. The leaflets are obligately opposite or nearly so in most genera of New World Ingeae, but in exceptional species (e.g., *Havardia albicans* (Kunth) Britton & Rose, *Falcataria moluccana* (Miquel) Barneby & Grimes) they are distinctly alternate along the rachis. The leaflet-pulvinules vary from linear to broader than long, and the leaflet-blades in consequence are either visibly stalked or seemingly sessile against the rachis.

We distinguish in the descriptive text five main types of leaflet venation: (1) pinnate (Fig. 2A, B), in which the inner posterior nerve arising from the pulvinule is weaker, or not stronger, than the first posterior secondary nerve arising from the midrib on the same side; (2) palmate (Fig. 2D), in which at least one primary nerve posterior to the midrib is clearly longer and stronger than any posterior secondary nerve; (3) palmate-pinnate (Fig. 2C), intermediate between the two preceding; (4) dimidiate-palmate (Fig. 2E), and (5) simple (not illustrated), when reduced to the midrib, as often in the shortest narrowest leaflets. These types conform to the following general types proposed by Hickey (1973). The pinnate types are camptodromous. In most species treated herein the pattern is brochidodromous, though eucamptodromous patterns are known in some species of *Macrosamanea.* The palmate-pinnate type is actinodromous, the palmate types acrodromous, and those reduced to midrib hyphodromous. Many members of Ingeae are notable in that the acroscopic portion of the lamina is much less wide than the basiscopic. This is independent of venation pattern. The midrib varies in position from centric to strongly displaced toward the anterior margin of the blade and becomes diagonal in blades of trapeziform outline. If arcuate, the midrib is nearly always either gently sigmoid or simply incurved beyond the middle; but it is arched backward in *Cedrelinga*. The basiscopic margin of the lamina is always convex in American Ingeae, though the acroscopic margin may be either convex (Fig. 2A–D) or concave (Fig. 2E).

Leaf-nectaries are almost ubiquitous in Ingeae, but vary in number from one, situated on the leaf-stalk either between or below the first pair of pinnae, to indefinitely numerous, when scattered both along the primary leaf-stalk and between pairs of leaflets, especially toward the tip of pinnae. The nectaries vary most commonly from patelliform to cupular or infundibular, and from sessile to stipitate; exceptionally they are mounded or verruciform with a small apical pore, or amorphous and more or less sunken into the petiolar tissue. In two species of *Abarema*, the proximal petiolar nectary is hypertrophied into a conspicuous colored funnel. Rudimentary leaf-stalks associated with an inflorescence-branch or peduncle are distinguished from bracts by a disproportionately large nectary on their adaxial face. In South American *Macrosamanea* and some species of Asiatic *Archidendron*, a morphologically similar nectary occurs on the ventral face of some or all floral bracts of the capitulum. There have been several hypotheses (e.g., Bentley, 1977; McKey, 1989) as to the role or function of extrafloral nectaries. Most have to do with either pollinator attraction or protection by ants.

INFLORESCENCE

Grimes (1992) defined the inflorescence as that sequence of metamers on any RGU participating in the production and presentation of flowers and fruit. In practice, this usually denotes the whole array of flowers borne during any season on a major branch or coherent branch-system. However, because growth in many tropical trees is independent of seasonal phemomena (Coster 1927–1928; Holttum, 1953; Koriba, 1958), and this has been observed in some Amazonian Ingeae, the more precise definition is presented. All inflorescences in American Ingeae are compound. Compound inflorescences are those in which flowers are produced on metamers that are themselves aggregated (the unit inflorescence) and produced on metamers of different morphology (the compound inflorescence). The primary inflorescence of Ingeae is composed of *unit inflorescences* in the form of racemes; spikes; and corymbiform, umbelliform, or globose capitula that arise, either singly or in fascicles of two to six (rarely more), from nodes subtended by a pair of caducous stipules, an expanded leaf, a rudimentary leafstalk, or a simple bract. Following a pattern ubiquitous in Mimosaceae, paired unit inflorescences are mutually divergent in orientation; when more than two at a node, the unit inflorescences are seen to arise from biseriate supranodal buds. The organization and aspect of the compound inflorescence are determined by modifications of spatial and temporal relations of these elements of the inflorescence to each other. For instance, the unit inflorescences may be basitonic (Carr & Carr, 1959; Briggs & Johnson, 1979), produced early in the development of the RGU, or acrotonic, produced at the

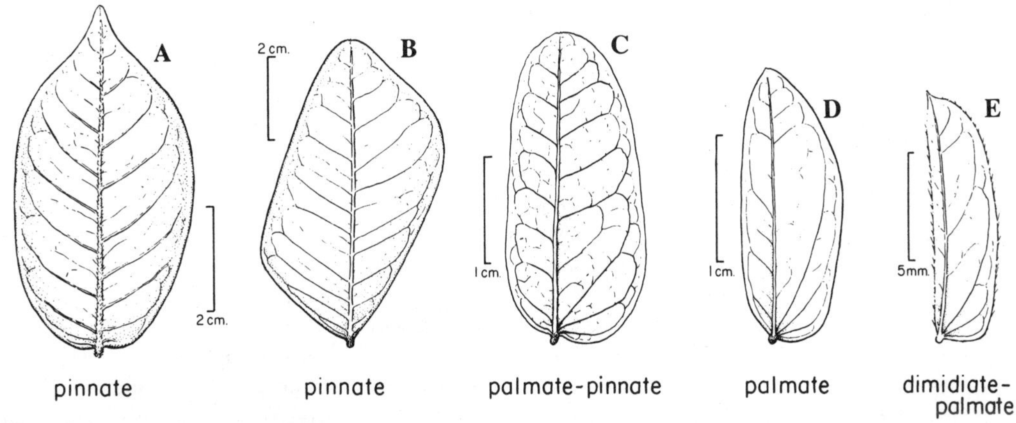

FIG. 2. Illustrations of types of venation in species of New World Ingeae. **A.** Pinnate venation, symmetrical leaflet. (Drawn from *Abarema lehmanii* (Britton & Killip) Barneby & Grimes.) **B.** Pinnate venation, symmetrically rhombic leaflet, in which the widest portion of the acroscopic half of the blade is not on the same transverse plane as the widest portion of the basiscopic half of the blade. Leaflets of some species are rhombic, but asymmetrically so, the acro- and basiscopic halves being inequally developed. (Drawn from *Abarema jupunba* (Willdenow) Britton & Killip.) **C.** Palmate-pinnate, venation in which 2–3 veins arise from the pulvinule, but secondary venation from the midrib is essentially pinnate. These secondary veins from the midrib are of approximately the same diameter as the subordinate veins arising from the pulvinule. (Drawn from *Albizzia lebbeck* (Linneaus) Bentham. **D.** Palmate venation, in which several veins arise from the pulvinule, but the secondary veins from midrib are not as large as those from the pulvinule, yet are more or less equal in diameter on both the acro- and basiscopic halves of the blade. The acroscopic margin of the lamina is convex. (Drawn from *Albizia subdimidiata* (Splitgerber) Barneby & Grimes.) **E.** Dimidiate-palmate venation, in which the secondary veins on the acroscopic half of the lamina are not as thick as those on the basiscopic half and the acroscopic margin of the lamina is straight or concave. (Drawn from *Albizia julibrissin* Durazzini.)

terminal nodes of the RGU. These positional relations may be further modified by temporal ones; there may be heterochronic development of the leaf subtending the primary inflorescences. The leaves may be suppressed, in which case the leaf-primordium develops into a small bract. Or the development of the leaf may be delayed with respect to the development of the primary inflorescence, so that the leaves complete development sometime near the end of anthesis or during development of the fruit.

No Ingeae are known in which the terminal meristem ends in the formation of a flower. So, the branch-systems of Ingeae fall into two primary categories: that in which the terminal meristem is continuous through more than one complete flowering cycle and that in which the terminal meristem perishes at the completion of a cycle and is replaced in the following cycle by regrowth from lateral buds. We have chosen to follow the terminology of Briggs and Johnson (1979) in calling these meristems *auxotelic* and *anauxotelic,* respectively. Because the distinction is ambiguous in many herbarium specimens, particularly those in young flower, its value as a practical taxonomic character is limited. We find it, nevertheless, a feature constant within species and a fundamental feature of higher taxa.

In the indeterminate group the unit inflorescences develop: (a) exclusively or in great part prior to the earliest leaves of the hornotinous branch (as in *Hesperalbizia* and *Chloroleucon*); (b) coevally with leaves of the season (*Abarema* in great part, *Cojoba*); (c) on brachyblasts lateral to annotinous long-shoots (*Ebenopsis*, *Sphinga*); (d) from the axil of some distal, developmentally delayed leaves (*Samanea saman*); or (e) collected into fascicles or compact efoliate panicles on knots of long-defoliate branches equivalent to leafless brachyblasts (*Zygia* in great part). This same range of form can be expressed in terms of asynchronous maturation or spatial separation of leaf and flower. Types a, b, and d demonstrate reciprocal adjustments in timing of leaf and flower maturity, whereas types (c) and (e) demonstrate displacement of reproductive organs from contemporarily developing stem-axes. Each of these adjustments seems plausibly related to annual climatic cycles and environmental factors, among which pollinators are doubtless crucial. Thus the inflorescence of *Abarema* and *Cojoba* is adapted to a relatively uniform climatic regime and abscence of pronounced dry season. The cauliflory almost but not quite charactersitic of *Zygia* is expressed under conditions of year-round groundwater in the understory of riparian forest. The precocious inflorescences of *Samanea* and some *Enterolobium* are associated with a rapid flush of new growth that follows a more or less prolonged resting period. The floriferous brachyblasts differentiated from barren long-shoots are related to long dry summers that prevail in the sclerophyllous chaparral communities in the West Indies and the shrub deserts of the Mexican plateau. Unfortunately, intrageneric lability of each type defies exact definitions, however unmistakable are the extremes.

The inflorescence of Ingeae with determinate meristems is less complex. It varies principally in the level at which leaves are supplanted by bracts and the degree of ramification. The extreme forms are *Albizia julibrissin*, in which the panicle is reduced to a simple, efoliate terminal raceme, and *Enterolobium contortisiliquum* and the African *Albizia ferruginea* Bentham, in which every peduncle or every partial inflorescence may be subtended by a fully developed leaf.

In American Ingeae, the unit of inflorescence is either a raceme or a spike, occasionally long and loose but more often than not variably condensed, becoming an umbelliform corymb when the individual flowers are pedicellate and a hemispherical or subspherical capitulum when they are sessile. The common peduncle is neither articulate nor bracteolate. Each flower is subtended by a bract, often small and caducous, but in one genus (*Macrosamanea*, with two exceptions), persistent, reflexed, and charged on the ventral face with a nectary resembling those of leaf-stalks. The pedicels vary from linear terete to thickened obconical, then sometimes simulating a solid hypanthium. By definition, the individual flowers of American Ingeae are rather conventional mimosoid flowers, notable only for the union of the more or less numerous filaments. The calyx and corolla are usually composed of five, sometimes four or six, segments united to some degree. The stamens and corolla are usually fused to some degree into a structure referred to as the stemonozone. Rarely this is absent. The corolla of *Cedrelinga cateniformis* is unusual in the tribe in that the petals are free to the stemonozone. The staminal tube is sometimes very short, and indeed, is found in some members of American Acacieae. The stamens are nearly always distinctly longer than the corolla. Throughout this treatment, we use the word *tassel* to refer to the collection of free filaments that surpass the corolla. Color in the flower may be contributed by the corolla, by the filaments, or by both. The anthers of most American Ingeae are elliptic in outline, that is, they are distinctly wider than long. In contrast, the anthers of American Acacieae are more or less round. The anthers of *Lysiloma* are somewhat intermediate between these two states. The ovary is either distinctly stipitate or (sub)sessile, though in the context of all species these states are continuous. The body of the ovary may be glabrous, pubescent in part, or glabrous before fertilization and pubescent after. The flowers of any unit may be homomorphic, as always in *Cojoba* and *Zygia*, or dimorphic, as in most species of *Albizia* and *Abarema*; the apparently terminal flower is enlarged, functionally staminate, and appears to arise from the apex of the receptacular axis, often elevated on a distinct pedestal and sometimes expanding before the smaller proximal flowers. The floral axis of dimorphic capitula appears determinate. However, the modified distal flower is subtended on one side by a bract like the flower next below it and is here interpreted as terminal in a topographical but not in a morphological sense. These are referred to as *pseudoterminal flowers* in following discussions. Not rarely, two or more flowers transitional in size or form are found between the enlarged central flower and the peripheral ones. The androecium of the pseudoterminal flower usually has a longer and more widely flaring tube than that of the potentially fertile peripheral flowers and at least in some cases provides more copious nectar from the interstaminal disc. Exceptionally, the pseudoterminal flower (*Lysiloma latisiliquum* (Linnaeus) Bentham) or all flowers of a capitulum (*Macrosamanea* spp.) have two ovaries.

In much of the literature on Ingeae the term *heteromorphic* is employed when the terminal flower(s) is morphologically different from the peripheral ones. We adopt a recommendation by J. Ramírez-Domenech that *heteromorphy* be used to refer to the various systems of andromonoecy found, for instance, in *Parkia* R. Brown and *Dichrostachys* Wight & Arnott, and that *dimorphy* be used to refer to the morphologically differentiated flowers of Ingeae.

Dehiscence

In most American Ingeae, dehiscence of the fruit is determined, or mediated, by two independently variable qualities: texture and curvature. The thicker the texture of the valves and sutures, the more tenacious are the vertical lines of potential fracture between the sutural keels, one or both of which may remain permanently sealed, when the seeds are released only by decay of the fruit or by animal agency. Curvature, proceeding always outward and downward, is brought about by differential growth of the two sutures, the adaxial (seminiferous) suture more or less rapidly outpacing the abaxial one in the process of maturation. Curvature sets up obliquely transverse tensions in the valves that are suddenly released at dehiscence, causing the valves either to gape, when separated along the ventral suture only, or to coil and twist upon themselves, when simultaneously separated along both sutures. Straight pods that become dry and papery or thinly coriaceous when ripe normally dehisce inertly along each suture, sometimes only tardily, after falling to the ground. Variation in texture and in degree of curvature, which at the extreme may wind the fruit into a compressed helix of several turns, is continuous in the tribe, each proceeding independently of the other, as evidenced by divergent syndromes of seed dispersal. The types of dehiscence, for the present ignoring special cases described below, can be analyzed into four main groups, each comprising variant subcategories:

1. Pod at once lignescent and straight: (a) indehiscent (*Samanea*); (b) dehiscent tardily through ventral suture only (*Balizia pedicellaris*); (c) indehiscent or dehiscent tardily through both sutures on the ground (*Ebenopsis*).
2. Pod at once chartaceous-coriaceous and straight: (a) inertly dehiscent through the length of both sutures (*Albizia* ex parte, *Pseudosamanea, Macrosamanea, Zygia* ex parte, *Paraserianthes, Leucochloron, Havardia*, and *Sphinga*); (b) technically indehiscent but shattering into lomentiform one-seeded segments, further described below.
3. Pod at once lignescent and retrofalcate or coiled: (a) indehiscent (*Enterolobium, Balizia leucocalyx*); (b) indehiscent only ventrally, hence follicular (*Abarema* spp., e.g., *A. curvicarpa, A. macradenia*; *Pithecellobium* ex parte; *Chloroleucon* spp.).
4. Pod at once coriaceous to chartaceous and retrofalcate or coiled: (a) fully dehiscent through both sutures (*Pithecellobium* ex parte, the majority of *Abarema*), the valves continuous (i.e., remaining intact); (b) early or tardily segmented in some fashion, the individual propagules either dehiscent or not.

In addition to these major categories of dehiscence, specialized conditions are observed in:

1. *Cojoba*: valves of the variously decurved or moniliform pod more or less fleshy and colored, dehiscent through one or both sutures, then shriveling and randomly contorted.
2. *Lysiloma*: sutures of straight, papery pod permanently sealed, sometimes separating from the valves in the form of a replum, the valves then falling away as entire panels from the sutural frame. In a few species, the pod at once straight, papery, and permanently indehiscent, the exocarp, however, peeling in flakes.
3. *Cedrelinga:* valves of straight pod papery, separating at interseminal constrictions into relatively very large, one-seeded, free-falling samaras.
4. *Albizia* ex parte, *Hydrochorea*: valves straight, chartaceous-coriaceous, not dehiscent through sutures, but separating at interseminal constrictions or along transverse interseminal fracture lines into one-seeded packets, each either (*Hydrochorea*) at length falling free from the sutures, or (*Albizia multiflora*) incorporating a segment of the sutures.
5. *Albizia inundata* (Martius) Barneby & Grimes: a *cryptoloment*, conventionally dehiscent with continuous valves but segmented endocarp, each segment free-falling monospermous.
6. *Calliandra, Zapoteca*: valves coriaceous or lignescent, straight, elastically separating from one another, from apex to base, along both sutures, and each recurved in the vertical plane but not twisted.

SEEDS

The seeds of American Ingeae defy easy description. They may be either overgrown sensu Corner (1951) or overgrown-like (Maumont, 1993), with corresponding modifications of the seed-coat (Maumont, 1993). The seed-coat may be thin and papery, tough-rubbery, or extremely hard and thickened. When the seed-coat is moistened, mucilage may appear. A pleurogram may be absent or, if present, either complete or incomplete. The areole may be of the same or of different color from the remainder of the seed-coat. The embryo itself is usually colorless, though both pink and blue embryos occur. Obese ("overgrown") seeds with papery testa and immediate germination may have a selective advantage in moist conditions. In contrast, seeds in seasonally dry habitats have a hard seed-coat with pleurogram and are potentially long-dormant. Seeds of xerophytic *Ebenopsis* are of obese type, but they are protected from drying out by exceptionally thick, tardily dehiscent pod-valves. Both types occur in *Cojoba* and *Pithecellobium*, genera with species adapted to both climatic regimes.

Characters, Phylogenetic Analysis, and Classification of Ingeae

Several classical, intuitive proposals for classification of the Ingeae have been reviewed by Nielsen (1981) and will not be recapitulated here. We present below phylogenetic analyses for all genera for which we have complete revisions. Outgroups were chosen from a tribal-level analysis of the genera using exemplars (Grimes, 1995; see below).

All cladistic analyses employed by us are based on parsimony methods. The exemplar analysis was done using PAUP (Swofford, 1990). The analyses included herein were done using HENNIG86 (version 1.5; Farris, 1988) and CLADOS (version 1.32; Nixon, 1992). In all analyses (with one exception) autapomorphies were excluded, and all characters were unordered, unless stated otherwise. The mhennig* and bb* options of HENNIG86 were chosen to generate trees. Even though the mhennig* option constructs several trees by adding the terminals in different sequences, it was found that the order of input of the taxa influenced tree length because of the phenomenon of tree islands. Therefore, in all analyses the taxa were randomly reordered 10 times to give 10 data sets, each of which was run. The analyses yielding the shortest trees were saved. In the exemplar analysis a majority rule consensus was generated using PAUP. Strict consensus trees were generated in HENNIG86.

Taxa at all ranks are statements of relationship recognized on the basis of shared derived characters. As such they are statements of monophyly. In practice, the genera herein recognized consist of the smallest monophyletic group of species. In some cases, monophyly is supported by a unique combination of character-states. We reject the alternative of recognizing paraphyla or larger genera so internally diverse so as to defy description.

Prerequisites of phylogenetic analysis are that the group under study be monophyletic and, for rigorous results, a strong hypothesis of sister-group relationships be available. The conflicting classifications proposed by earlier students of Ingeae, and the fact that our intuitive concepts of monophyletic groups were further in conflict with those of other workers, led us to realize that monophyly of any group was open to serious challenge. Furthermore, there existed no hypotheses of sister-group relationships whatsoever. In order to provide preliminary hypotheses of sister-group relationships, Grimes (1995) presented an exemplar study of Ingeae and Acacieae. A discussion of that analysis follows. From it a number of monophyletic groups (clades) were recognized. These clades are herein referred to as the *Abarema*-alliance, the *Samanea*-alliance, the *Chloroleucon*-alliance, and the *Pithecellobium*-alliance. Species-level analyses of these clades, and the genus *Macrosamanea*, are presented below, preceding the taxonomic treatments. A global analysis of all species is not yet possible because of the large number of missing data.

The generic-level study included exemplars (the types of described genera, species arbitrarily chosen for putative genera) of putative monophyletic groups ("genera") from our revision of the West Indian and South American species, plus the Old World genera of Ingeae recognized by Nielsen (1981), as well as *Obolinga* Barneby and *Zapoteca* H. Hernández, genera described since Nielsen's study. If a genus was polymorphic for any character(s) used in the analysis, species from the genus were added as terminals until all variation in that genus was included. If other genera were variable for the same characters they were split up at the same time. If the "genus" did not remain monophyletic in this exemplar analysis it was divided in subsequent analyses. Each analysis included a different number of terminal taxa, depending on which putative genus was being studied. Sister-groups, as determined on a consensus tree, were chosen from this analysis for the species-level phylogenetic analyses of genera treated below. Readers are referred to Grimes (1995), for a discussion of the characters and the data set used in the exemplar analysis.

The majority-rule consensus tree, shown in Figure 3A, B, was used to choose outgroups. A majority-rule consensus was chosen because a strict consensus would not show relationships that occurred, for example, 98% of the time. However, all groups subjected to analysis below also occur in the strict consensus. The majority-rule consensus was used only to generate hypotheses of likely outgroups. The numbers in boldface on the majority-rule tree (Fig. 3) refer to genera or groups of genera that are treated in this revision. These are: (1) the *Abarema*-alliance (Fig. 3B, no. 1), consisting of CORYMBOSA, PEDICELL, and ABAREMA, treated herein as the genera *Hydrochorea*, *Balizia*, and *Abarema* (genera I, II, and III); (2) the *Samanea*-alliance (Fig. 3B, no. 2), consisting of SAMANEA, PSEUDOSA and ALOCCID, treated herein as the genera *Hesperalbizia*, *Pseudosamanea*

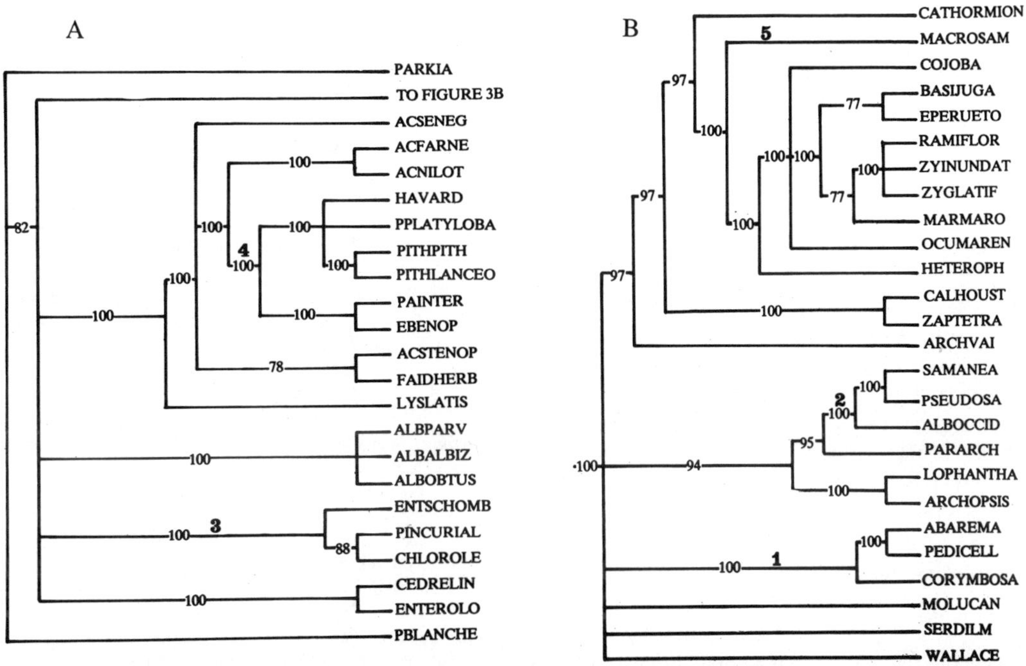

Fig. 3 A. The basal portion of a majority-rule consensus tree from an exemplar analysis of genera of Ingeae, used to establish sister-groups for phylogenetic analysis of New World members of the *Pithecellobium*-complex. Numbers in bold refer to genera or groups of genera for which phylogenetic analyses are presented. The numbers on the branches refer to the percent of the 8800 trees that have the topology shown. **B.** The upper portion of the majority rule consensus tree.

and *Samanea* (genera IV, V, VI); (3) the *Chloroleucon*-alliance (Fig. 3A, no. 3), including PINCURIAL and CHLOROLE, treated herein as the genera *Leucochloron*, and *Chloroleucon* (genera VIII, IX); BLANCHET (*Blanchetiodendron*, genus VII) was also included in this analysis; (4) the *Pithecellobium*-alliance (Fig. 3A, no. 4), containing the terminals PPLATYLOBA, HAVARD, EBENOP, PAINTER, PITHPITH, PITHLANCEO, PITHFURCAT, and PITHMACRAN, representing *Sphinga*, *Havardia*, *Ebenopsis*, *Painteria*, and *Pithecellobium* (genera X, XI, XII, XIII; a revision of *Pithecellobium* will be presented in a forthcoming treatment); and (5) the genus *Macrosamanea* (Fig. 3B, no. 5; genus XIV). Phylogenetic analyses are presented of all these save *Pithecellobium*. Following these genera we present a complete description of the West Indian and South American species of *Albizia* Durazzini (genus XV) and *Cedrelinga* Ducke (genus XVII), and abbreviated notes on *Enterolobium* Martius (genus XVI). These last three are obviously closely related, not only to each other, but also to the African and Australasian species of *Albizia* (including *Cathormion* of African authors, not Hasskarl). Phylogenetic analyses and resolution of generic limits of these genera await monography of the African species. Finally we provide treatments of the introduced *Falcataria* (Nielsen) Barneby & Grimes (genus XVIII), *Paraserianthes* Nielsen (genus XIX), and of the West Indian species of *Lysiloma* Bentham (genus XX). Revisions of *Cojoba*, *Pithecellobium*, and *Zygia* will be provided in a companion monograph.

The results of the exemplar analysis, as well as those presented below, lead to a classification of American Ingeae different from any others. Though older names are available for most genera we recognize, the circumscription of those genera is usually new.

Britton and coworkers (Britton & Rose, 1928; Britton & Killip, 1936) provided many of the generic names taken up in this treatment. However, their circumscription of taxa were arbitrarily based on characters of the fruit. They were following a tradition started by Bentham himself, but the result is that many of their genera are para- or polyphyletic. For example, *Samanea* sensu Britton & Rose includes species that are referred herein to the genera *Balizia*, *Samanea*, *Cojoba*, and *Pithecellobium*.

Our studies, we believe, provide a more thorough analysis and description of characters and character-states than antecedent ones. It follows, therefore, that we place more faith in the results of the exemplar analysis (Grimes, 1995) and the species-level revisions presented following. We have found that synapomorphic characters (i.e., diagnostic ones) are, for the most part, those that have not been used in previous classifications. Examples are the axillary thorns of *Chloroleucon*, the anthocyanic seeds of *Abarema* Pittier, monopodial versus sympodial growth, and the heterochronic development of leaves or inflorescences. Our revision also demonstrates that characters that are descriptive or even diagnostic in one group are variable in others. Polyad number provides one example: *Abarema*, *Balizia* Barneby & Grimes, and *Hydrochorea* Barneby & Grimes, consistently have polyads comprising 16 monads; *Zygia* P. Browne, *Macrosamanea* Britton & Rose, and *Inga* Miller, in contrast, are monophyletic groups that are polymorphic for polyad number. Floral dimorphy provides another example: *Macrosamanea*, *Inga*, and *Zygia* have monomorphic flowers; *Samanea* Merrill, *Pseudosamanea* Harms, and *Albizia* Durazzini sensu stricto, consistently have dimorphic flowers. *Chloroleucon*, *Enterolobium*, and *Abarema*, however, are polymorphic for that character, even on a single individual.

Because so many characters exhibit this type of variation, characters whose states are known in but a few species are not used in the following analyses. We make no assumptions about the distribution of their states. We recognize further that our hypotheses of phylogenetic relationships are preliminary. Whole classes of data, such as cytological, molecular, and chemical, are still unavailable.

A conspectus of species is provided for each genus immediately preceding each taxonomic treatment as an adjunct to the all-choice key. These conspectuses neither imply nor reflect hypotheses of phylogenetic relationships. Likewise, we present below a conspectus of the native American genera of Ingeae together with those Old World genera of which relationships are certain. Though this generic conspectus does reflect our understanding of relationships thus far, it is no more than a practical, heuristic summary intended to provide a framework into which Part II of this study, and subsequent monographic work, can be placed. Only numbered genera are revised in detail in this monograph.

Conspectus of native American genera of Ingeae

¶ The *Abarema*-alliance. Growth monopodial; branches unarmed monomorphic, sylleptic; buds protected by the adaxial side of the petiole; leaves coeval or late-suppressed relative to inflorescence; flowers monomorphic or dimorphic.

I. *Hydrochorea* Barneby & Grimes
II. *Balizia* Barneby & Grimes
III. *Abarema* Pittier

¶¶ The *Samanea*-alliance. As ¶, but branches proleptic; leaves developmentally delayed relative to inflorescence; flowers monomorphic or dimorphic.

IV. *Hesperalbizia* Barneby & Grimes
V. *Pseudosamanea* Harms
VI. *Samanea* Merrill

¶¶¶ The *Chloroleucon*-alliance. Growth sympodial; branches unarmed, proleptic, monomorphic or commonly dimorphic with ephemeral lateral-shoot-systems, or (*Blanchetiodendron*) with sylleptic, annual, axillary, strictly reproductive branch systems; buds protected by perulate scales, shoots perhaps preformed; leaves early suppressed to coeval; flowers monomorphic or dimorphic.

VII. *Blanchetiodendron* Barneby & Grimes
VIII. *Leucochloron* Barneby & Grimes
IX. *Chloroleucon* (Bentham) Britton & Rose

¶¶¶¶ The *Pithecellobium*-alliance. Growth sympodial or perhaps monopodial; branches armed with stipular spines, proleptic, dimorphic, with vegetative and reproductive short-shoot-systems; buds protected by adaxial side of petiole; leaves coeval or late-suppressed; flowers monomorphic.

X. *Sphinga* Barneby & Grimes
XI. *Havardia* Small
XII. *Ebenopsis* Britton & Rose
XIII. *Painteria* Britton & Rose
Pithecellobium Martius

¶¶¶¶¶ The *Inga*-alliance. Type of growth undetermined for most genera; branching proleptic or sylleptic; shoots unarmed dimorphic, with perennating, axillary, strictly reproductive meristems leading to ramiflory and cauliflory; buds apparently preformed, protected by stipules; leaves coeval or late-suppressed; flowers monomorphic.

XIV. *Macrosamanea* Britton & Killip
Cojoba Britton & Rose
Inga Miller
Zygia P. Browne, (including *Marmaroxylon* Record)
Calliandra Bentham
Zapoteca H. Hernández
Archidendron F. v. Mueller

¶¶¶¶¶¶ Genera incertae sedis. Until a comprehensive monograph of African, Asian, and Australian *Albizia* is available, the relationships and generic limits of the following are unknown. Treatments are provided herein for the New World taxa only. Growth sympodial, vegetative branches unarmed proleptic, but sylleptic, annual, axillary, strictly reproductive branch-systems present on most; buds with perulate scales present or not; hysteranthy early-suppressed or late-suppressed; flowers monomorphic or dimorphic.

XV. *Albizia* Durazzini

XVI. *Enterolobium* Martius

XVII. *Cedrelinga* Ducke

¶¶¶¶¶¶¶ *Lysiloma* Bentham (XX) is intermediate between tribes Ingeae and Acacieae.

A key to the genera of Ingeae native, naturalized, and planted in the Americas

[N. B.: This pragmatic key to native, naturalized, and planted species of Ingeae in the Americas is designed exclusively for purposes of identification and has only incidental phylogenetic overtones. It covers all taxa hitherto referred to the tribe with the exception of *Lysiloma*, which is ambiguous between Ingeae and Acacieae. *Lysiloma*, being native to Mexico, adjacent Arizona and Central America, and the West Indies, is summarily mentioned at p. 257 of this revision, but only the two West Indian species are described in detail. Genera not preceded by a number are not treated.]

1. Stems armed at all or at most nodes with either (a) a pair of spinescent stipules, or (b) a pair of spines at each node, one arising from dorsal face of each stipule, or (c) one or two axillary thorns [if misled hither by plant armed at nodes with a pair of epidermal prickles, see *Acacia*].
   2. Stipules not modified into spines, the stem-nodes armed with:
      3. A spine emergent from dorsal face of each stipule; plants with axillary short-shoots; perulate bud-scales lacking; Greater Antilles and Bahamas only .... *Calliandra* spp. affin. *C. haematommati* Bentham
      3. Axillary thorns (sometimes at only few scattered nodes); axillary short-shoots only rarely and poorly developed; perulate resting buds present; widespread in seasonally dry continental N and S America as well as the West Indies .... IX. *Chloroleucon* (p. 136)
   2. Armament of spinescent stipules.
      4. No petiolar nectary .... *Zapoteca* subgen. *Aculeata*
      4. At least one nectary on each lf-stk, either between or below the first pair of pinnae.
         5. Lf-axis broadly green-winged .... *Pithecellobium*
         5. Lf-axes terete or ventrally channeled, not wing-marginate.
            6. Specimen in flower:
               7. Petiolar nectary exactly interpinnal; venation of lfts pinnate.......... *Pithecellobium*
               7. Either the petiolar nectary borne below first (or only) pair of pinnae or venation of lfts palmate.
                  8. Fl-buds obovoid-pyriform from inception; calyx campanulate 1–3 mm; corolla 3–6.3 mm; androecium 0.65–1.45 cm.
                     9. First petiolar nectary situated below proximal pair of pinnae; lft venation pinnate or in smallest lfts simple; capitula not, or not exclusively, arising from brachyblasts .... XI. *Havardia* (p. 165)
                     9. First petiolar nectary situated exactly between proximal pinna-pair; lft-venation palmate, or in smallest lfts simple; capitula all arising from scaley brachyblasts.
                        10. Petiolar nectary sessile, shallowly cupular; ovary-stipe 1 mm or longer.......... XIII. *Painteria* (p. 178)
                        10. Petiolar nectary stoutly stipitate, more or less columnar; ovary-stipe very short .... XII. *Ebenopsis* (p. 173)
                  8. Fl-buds flask-shaped, until shortly before anthesis acute or acuminate; calyx cylindro-campanulate 5.5–12.5 mm; corolla 11.5–27 mm; androecium 4–9 cm .... X. *Sphinga* (p. 160)
            6. Specimen in fruit:
               11. Seeds cupped by a fleshy aril.......... *Pithecellobium*
               11. Seeds borne on a filiform or narrowly ribbonlike funicle.
                  12. Pod ligneous subterete, its cavity divided by interseminal septa; seeds obese, reddish brown .... XII. *Ebenopsis* (p. 173)
                  12. Pod leathery or papery, laterally compressed, unilocular; seeds otherwise.
                     13. All or most capitula racemose or paniculate on hornotinous branches .... XI. *Havardia* (p. 165)
                     13. All capitula arising from brachyblasts.
                        14. Pod leathery, strongly recurved; petiolar nectaries interpinnal .... XIII. *Painteria* (p. 178)
                        14. Pod stiffly papery, straight; petiolar nectaries below midpetiole .... X. *Sphinga* (p. 160)

1. Stems unarmed.
   15. Lvs simply paripinnate.
      16. Fruit variable in form, compression, and texture, but indehiscent; seeds covered with enlarged, sticky, sweet-tasting hairs; genus of many species, ubiquitous in neotropical lowland and low-montane habitat .... *Inga*
      16. Fruit dehiscent through one or both sutures.
         17. Inflorescence cauliflorous at least in part; pod plumply compressed, at dehiscence the valves narrowly gaping to release seeds; seeds 3–3.5 cm diam, the testa papery, fragile; E. Amazonian Brazil .... *Zygia*
         17. Units of the inflorescences axillary to lvs; pods compressed, subterete or moniliform, the valves elastically dehiscent and sometimes contorting; seeds (unknown in *Calliandra hymenaeodes*) 5.5–14 mm in greatest diameter; Panama, Costa Rica, NW Colombia, and the Guianas.
            18. Lvs lacking nectary; lft venation of 4 nerves from pulvinule; pod unknown, but presumably plano-compressed and elastically dehiscent though both sutures, the valves straight; known only from the Guianas .... *Calliandra hymenaeodes* Bentham
            18. Lvs with a nectary at insertion of every lft-pair; lft venation pinnate; pod subterete or moniliform, the fleshy red valves becoming wrinkled and fuscous after dehiscence; Costa Rica, Panama, NW Colombia .... *Cojoba rufescens* (Bentham) Britton & Rose
   15. Lvs bipinnate [caveat: one pair of long pinnae borne on very short primary lf-stk in some *Zygia* spp. are easily mistaken for two abruptly pinnate lvs].
      19. Petiolar nectaries wanting .... *Calliandra* and most species of *Zapoteca*
      19. Petiolar nectary between or below the proximal (or only) pinna-pair.
         20. Terminal and axillary resting-buds ovoid, composed of firm, brown, closely imbrictate, striately nerved perules.
            21. Specimen in flower:
               22. Capitula organized in an efoliate terminal panicle or into efoliate pseudoracemes axillary to coeval lvs, the individual peduncles not subtended by a lf; flowers dimorphic .... VII. *Blanchetiodendron* (p. 127)
               22. Capitula either solitary or fasciculate, either in coeval lf-axils or at efoliate nodes anterior to lvs.
                  23. Stem-growth differentiated into flowerless long-shoots and fertile lateral branchlets often contracted into stemless brachyblasts; fls of each capitulum dimorphic; thornless states of .... IX. *Chloroleucon* (p. 136)
                  23. Stem-growth monomorphic; capitula mostly axillary to a coeval lf; fls of each capitulum monomorphic .... VIII. *Leucochloron* (p. 130)
            21. Specimen in fruit:
               24. Pod papery plano-compressed; seeds discoid, the testa papery-membranous, translucent, exareolate.
                  25. Fruits terminally paniculate; stipe of pod 4–8 mm .... VII. *Blanchetiodendron* (p. 127)
                  25. Fruits either axillary, or lateral and below annotinous foliage; stipe of pod commonly 0, in *L. foederale* 1–5 mm .... VIII. *Leucochloron* (p. 130)
               24. Pod firm, compressed but plumply so, varying from straight to falcate to spirally coiled; seeds plump, the testa hard, areolate; thornless state of .... IX. *Chloroleucon* (p. 136)
         20. No resting-buds, or if such are present their scales either (a) not striately venulose or (b) linear-lanceolate and loosely imbricate.
            26. Plant obligately and exclusively cauliflorous, the units of inflorescence, whether spikes, or capitula, or congregations of either, arising from defoliate nodes of main trunk or of annotinous and older branchlets, clearly separated from and below all expanded coeval lvs.
               27. Calyx 14–20 mm; rare species of Surinam and French Guiana .... XIV. *Macrosamanea kegelii* (p. 202)
               27. Calyx much less than 14 mm; widespread .... *Zygia*
            26. Plant not, or only in small part, cauliflorous, all or most unit inflorescences borne on stems of the current year, most often subtended by a coeval or early hysteranthous lf, but sometimes at efoliate nodes of young stems, or not infrequently aggregated into lateral efoliate pseudoracemes or efoliate terminal panicles.
               28. Specimen in flower.
                  29. Units of inflorescence evidently racemose or spiciform, the individual fls dispersed along an axis at least 3 cm long.
                     30. Lvs bifoliolate, the pinnae 1 pair, each unifoliolate; Central American .... *Zygia confusa* Rico
                     30. Lvs plurifoliolate.
                        31. Units of inflorescence subtended by a fully expanded coeval lf; native, several species of .... III. *Abarema* (p. 41)

31. Units of inflorescence assembled into a compound panicle; 1 Asiatic species, cultivated .... XVIII. *Falcataria* (p. 256)

29. Units of inflorescence capituliform or cymose-umbelliform, or if spiciform relatively dense and the fl-axis <2.5 cm.

32. Fls of each unit inflorescence monomorphic.

33. Inflorescence a terminal panicle, efoliate throughout or with an expanded lf subtending only a few proximal secondary axes, the individual capitula never subtended by a leaf.

34. Lfts of longer pinnae 2–5-jug. and the larger blades 7–13 cm; midrib of lfts gently decurved; gigantic trees of equatorial forest, native .... XVII. *Cedrelinga* (p. 251)

34. Lfts of longer pinnae 8–13-jug. and the larger blades 2–5 cm; midrib of lfts straight or slightly curved forward; slender trees of Asiatic origin, planted but only weakly naturalized in neotropics .... XV. *Albizia procera* (p. 244)

33. Inflorescence composed wholly or in large part of axillary units, these consisting either of solitary/fasciculate capitula or of efoliate pseudoracemes of capitula, then either subtended by a coeval lf or arising at efoliate nodes below (not above) coeval foliage.

35. Bracts subtending individual fls of the capitulum charged on ventral face with a nectary resembling that of lf-stk .... XIV. *Macrosamanea* (p. 182)

35. Bracts without a nectary.

36. Calyx tumidly cylindro-campanulate 20–29 mm; ovaries 2–3 per fl .... XIV. *Macrosamanea macrocalyx* (p. 186)

36. Calyx much shorter, not tumid; ovary solitary.

37. Macrophyllidious, the lf-formula ii/1 and larger lfts 1–2 dm... *Zygia ocumarensis*

37. Mostly microphyllidious, the lfts commonly more numerous, always much smaller.

38. Receptacle of each unit of inflorescence capitate, globose, or pyriform.

39. Inflorescence precocious, the peduncles arising below and earlier than foliage of the current growing season, either at efoliate nodes of new branches or congregated in efoliate pseudoracemes (three genera best separated in fruit).

40. Branchlets thickened at the nodes by a prominent buttress under each lf-scar; fruit a craspedium or indehiscent legume .... XX. *Lysiloma* (p. 257)

40. Branchlets terete or nearly so at nodes and buttresses lacking at nodes; fruit either dehiscent, or if indehiscent thick-ligneous and strongly recurved.

41. Stipules cordate; lft-pulvinules 1–3 mm; pod long-stipitate, broad-linear plano-compressed, straight .... IV. *Hesperalbizia* (p. 112)

41. Stipules not cordate; lft-pulvinules <1 mm; pod sessile, broad-linear decurved through ½ to 2 circles .... XVI. *Enterolobium* (p. 245)

39. Inflorescence otherwise, the peduncles axillary to coevally expanding or fully expanded lvs; stipules lanceolate, subulate, linear, or 0.

42. Corolla campanulate.

43. Pubescence of new growth of white or sordid white, never brown or golden hairs; fruit either a craspedium or an indehiscent legume, the valves not colored internally .... XX. *Lysiloma* (p. 257)

43. Pubescence of new growth of brown or golden hairs; fruit a pod dehiscent through one or both sutures, the valves internally colored red or orange ....III. *Abarema* (p. 41)

42. Corolla tubular .... *Cojoba*

38. Receptacle of unit inflorescences linear or linear-clavate.

44. Pinnae 1-jug.; intrastaminal disc around ovary; local astride Brazil-Bolivia boundary ....*Zygia odoratissima* (Ducke) L. Rico

44. Pinnae pluri-jug.; no intrastaminal disc; cultivated and locally naturalized.

45. Fls pedicellate, the racemes borne in contemporary lf-axils ....XIX. *Paraserianthes lophantha* (p. 256)

45. Fls sessile, the spikes terminally paniculate ....XVIII. *Falcataria* (p. 254)

32. Fls of each unit inflorescence dimorphic.

46. Lf-formula i/1, all lvs amply 4-foliate; inflorescence paniculate, but most unit inflorescences subtended by a rudimentary lf-stk with nectary.

47. Corolla of peripheral fls of each capitulum 11.5–15.5 mm ....III. *Abarema levelii* (p. 106)

47. Corolla of peripheral fls of each capitulum 7 mm or less; few paucifoliate species of ....III. *Abarema* (p. 41)

46. Lf-formula higher in some respect, all or most lvs at least 8-foliolate.

48. Venation of lfts palmate, or the blade, if very narrow, 1-nerved.

49. Capitula all subtended by a bract and disposed along a simple, terminal, abruptly flexuous axis distal to all expanded lvs of current year ....XV. *Albizia julibrissin* (p. 243)

49. Capitula either subtended by a coeval lf, or borne in compound pseudoracemes or panicles, or arising precociously on annotinous branches proximal to foliage of current year.

50. Capitula (either solitary or fasciculate) or efoliate pseudoracemes of capitula arising from the axil of lvs of current year or from efoliate nodes below them.

51. Lfts of longer pinnae 40–80-jug., the longer ones 2.5–6 mm, 1-nerved; ovary truncate at apex ....XVI. *Enterolobium* sect. *Robrichia* (p. 249)

51. Lfts fewer and larger, palmately nerved; ovary tapering at apex ....XVI. *Enterolobium* sect. *Enterolobium* (p. 247)

50. Capitula either all axillary to coevally expanding lvs, or together paniculate above or at least beyond foliage of current year.

52. All capitula subtended by a coevally expanding or early hysteranthous lf (this sometimes reduced distally to a rudimentary lf-stk with nectary); peduncles 3–13 cm; lfts of longer pinnae 4–10-jug. and to 3 cm or more ....XV. *Albizia lebbeck* (p. 242)

52. Capitula all or most subtended by a bract and disposed in compound panicles above and beyond the foliage; peduncles in most species. <3 cm, if longer then the lfts at least 30-jug. and much smaller.

53. Stipules foliaceous 1–2.5 cm, caducous but at full anthesis some of them usually remaining at proximal nodes of panicle; Asiatic, planted in neotropics ....XV. *Albizia chinensis* (p. 241)

53. Stipules much smaller, or lacking; native ....XV. *Albizia* sect. *Arthrosamanea* (p. 206)

48. Venation of lfts pinnate.

54. Ovary truncate; three genera, abruptly differentiated by fruits but not technically separable at anthesis. III. *Abarema* ex parte (p. 41), II. *Balizia* (p. 34), I. *Hydrochorea* (p. 23)
54. Ovary tapering at apex.
 55. Corolla of peripheral fls at most 7 mm .... III. *Abarema* ex parte (p. 41)
 55. Corolla of peripheral fls 9–18 mm.
  56. Pedicel of outermost peripheral fls of capitulum 11–22 mm .... V. *Pseudosamanea* (p. 113)
  56. Pedicel of outermost peripheral fls of capitulum 1–9.5 mm .... VI. *Samanea* (p. 117)

28. Specimen in ripe fruit:
 57. Seeds exareolate.
  58. Fruit a long papery samariform loment 20–70 × 3.5–6 cm, pinched between the plano-compressed, indehiscent articles and at each isthmus twisted through ±90° .... XVII. *Cedrelinga* (p. 251)
  58. Fruit otherwise, not segmented, dehiscent.
   59. Pod diverse in curvature, texture, and compression, but the endocarp internally either red, maroon, or orange-brown, contrasting in color with the more or less translucent, either blue, gray-blue, or pale tan (sometimes partly opaque and bright white) seed-testa .... III. *Abarema* (p. 41)
   59. Pod and seeds otherwise.
    60. Fruit cylindroid, torulose or moniliform, the ripe valves before dehiscence red externally, after dehiscence leathery and shriveled ...... *Cojoba*
    60. Fruit somewhat laterally compressed, the valves either flat or low-convex, never red externally, not shriveled after dehiscence .... XIV. *Macrosamanea* (p. 182)
 57. Seeds areolate.
  61. Venation of lfts pinnate.
   62. Pod indehiscent, strongly biconvex or subcylindroid, the continuous valves fibro-ligneous or pulpy, the endocarp at least narrowly septiferous; macrophyllidious, the lfts of longer pinnae 3–9 pairs and (2–)2.5–6 cm.
    63. Sutures of pod externally differentiated into prominent broad keels; seeds plumply oblong-ellipsoid, less than 3 times as long as wide .... VI. *Samanea* (p. 117)
    63. Sutures of pod immersed; seeds narrowly oblong-ellipsoids about 3 times as long as wide .... II. *Balizia* sect. *Leucosamanea* (p. 36)
   62. Pod not as above, always strongly compressed laterally, but either (a) papery plano-compressed, (b) leathery and elastically dehiscent, (c) tardily follicular, or (d) indehiscent and incipiently lomentiform, and if the cavity divided by interseminal septa the lvs microphyllidious and the longest lfts <1.6 cm.
    64. Valves of pod continuous.
     65. Pod falcately recurved or coiled, the endocarp internally either red, maroon, or orange-brown, contrasting in color with the somewhat translucent, either blue, gray-blue, or pale tan (sometimes in part opaque and bright white) seed-testa .... III. *Abarema* (p. 41)
     65. Pod straight or almost so, not red inside; seeds mostly brown.
    64. Sutures of pod dilated, framing the transversely fibrous valves; seeds narrowly oblong, about 3 times as long as wide .... II. *Balizia* sect. *Balizia* (p. 37)
     66. Sutures of pod slender, not forming a raised frame around the valves, sometimes immersed; seeds in profile round or elliptic, nearly isodiametric.
      67. Stipe of pod 0–4 mm.
       68. Peduncles 2.5–8 cm; valves of pod densely pubescent; native in Central America, Cuba, and N South America .... V. *Pseudosamanea* (p. 113)
       68. Peduncles <1 cm; valves of pod glabrous; cultivated and perhaps sparingly naturalized in the neotropics .... XV. *Albizia procera* (p. 244)

67. Stipe of pod 10–40 mm ....IV. *Hesperalbizia* (p. 112)
64. Valves of pod tardily segmented between seeds, incipiently lomentiform ....I. *Hydrochorea* (p. 23)
61. Venation of lfts palmate, or the lfts very narrow and 1-nerved.
69. Pod at once lignescent and decurved through at least ¾ circle, often further into a compressed spiral, indehiscent; seeds separated by complete septa.
70. Lfts very numerous, small and narrow, 40–80-jug. in longer pinnae and only 2.5–6 mm long, 1-nerved ....XVI. *Enterolobium* sect. *Robrichia* (p. 249)
70. Lfts less numerous and larger, clearly palmate-veined ....XVI. *Enterolobium* sect. *Enterolobium* (p. 247)
69. Pod straight, not lignescent, commonly but not always inertly dehiscent; cavity continuous, septa lacking.
71. Receptacle of each unit inflorescence elongate, the pedicel-scars well separated along an axis 1.5–6 cm ....XVIII. *Falcataria mollucana* (p. 255)
71. Receptacle of each unit inflorescence globose, shortly oblongoid, or pyriform, the pedicel-scars contiguous.
72. Pod 25–52 mm wide; lfts of longer pinnae 4–10-jug. and the longer ones ±3–5 cm, obtuse ....XV. *Albizia lebbeck* (p. 242)
72. Pod 16–26 mm wide, in two species to 32 mm wide but if >25 mm wide then the leaflets much more numerous and smaller.
73. Lfts dimidiate, the midrib forwardly displaced to within 0.5 mm of anterior margin, the apex acute; Asiatic, planted, and naturalized in the Americas.
74. Lfts bicolored, pallid beneath; inflorescence of each leafy branch of the season terminal and simple, its primary axis abruptly flexuous; pod inertly dehiscent through both sutures; pleurogram occupying middle of each seed-face ....XV. *Albizia julibrissin* (p. 243)
74. Lfts pallid on both faces; inflorescence of each foliate branch of the season more or less branched; pod falling indehisced; pleurogram small, displaced well below middle of seed-face.. ....XV. *Albizia chinensis* (p. 241)
73. Lfts often inequilateral but not dimidiate, if the midrib much displaced then the apex obtuse; native ....XV. *Albizia* sect. *Arthrosamanea* (p. 206)

The *Abarema*-alliance

The exemplar study by Grimes (1995) identified a monophyletic clade (Fig. 3B, no. 1) of three exemplars recognized by the following homoplasious character-states: auxotelic meristems; vegetative growth from sylleptic and proleptic buds; leaves late-suppressed (see Grimes, 1992); pollen with areolate tectum; and seeds with resinoid tissue. The exemplar CORYMBOSA represented the following species: *Pithecellobium corymbosum* (L. C. Richard) Bentham, *P. gonggrijpii* Kleinhoonte, *P. marginatum* Spruce ex Bentham, and *P. acreanum* Macbride. This exemplar was sister-group to the two others: ABAREMA, representing approximately 42 species and including the types of *Abarema* Pittier, *Jupunba* Britton & Rose, *Punjuba* Britton & Rose, and *Klugiodendron* Britton & Killip; and PEDICELL, representing *Pithecellobium pedicellare* (de Candolle) Bentham, *P. elegans* Ducke, and *P. leucocalyx* (Britton & Rose) Standley. These species are included in a phylogenetic analysis following. The exemplar analysis (Grimes, 1995) showed that the *Abarema*-alliance occurred at an unresolved polychotomy. *Wallaceodendron celebicum* Koorders was chosen as the outgroup, in part arbitrarily, but also in part because in many other exemplar analyses it was consistently sister-group to the *Abarema*-clade.

Many of the species of Ingeae are distinguished by meristic differences such as number of pairs of pinnae and leaflets, as well as by the size of various organs. These differences are not easily amenable to coding, especially as they are variable on individuals. Five species belonging to *Abarema* still not known from both flowering and fruiting specimens have been omitted from the analysis.

In this and all following analyses the characters are numbered according to the conventions of HENNIG86. The terminal taxa are species (labels are self-evident) with the following exceptions. *Hydrochorea acreana* (Macbride) Barneby & Grimes remains unknown in fruit; otherwise, it resembles closely *Hydrochorea corymbosa* (L. C. Richard) Barneby & Grimes, *H. gonggrijpii* (Kleinhoonte) Barneby & Grimes, and *H. marginata* (Bentham) Barneby & Grimes. Some species are indistiguishable on the basis of the characters used in this analysis, so OPPNIPE includes both *Abarema oppositifolia* (Urban) Barneby & Grimes and *A. nipensis* (Britton) Barneby & Grimes; FLORIFER includes both *A. floribunda* (Bentham) Barneby & Grimes and *A. ferruginea* (Bentham) Pittier; and LEHMKIL includes *A. lehmannii* (Britton & Killip) Barneby & Grimes and *A. killipii* (Britton & Killip) Barneby & Grimes.

The data set is shown in Table I; the characters used in the analysis are as follows:

0. Stipules: 0 = absent; 1 = present.
1. Pinnae: 0 = distally accrescent; 1 = proximally decrescent, thence subequilong, but the penultimate or distalmost pair the longest; 2 = proximally decrescent thence all subequilong; 3 = decrescent at each end; 4 = 1 pair.
2. Nectaries: 0 = present between some leaflet-pairs; 1 = absent between leaflet-pairs.
3. Leaves: 0 = forming and developing coevally with all inflorescences; 1 = coeval with the oldest inflorescences, but suppressed below inflorescences formed late in the growing season. There is heterochronic development of the leaves with regard to the subtended peduncles. This heterochrony is discussed in detail in Grimes, 1992.
4. Paraphyllidia: 0 = present on some leaves; 1 = lacking on all leaves.
5. Leaflets: 0 = proximally decrescent, thence subequilong; 1 = distally accrescent.
6. Leaflets: 0 = not ciliolate; 1 = ciliolate.
7. Leaflets: 0 = broadest at or above midblade; 1 = broadest below midblade.
8. Leaflets: 0 = margins revolute; 1 = margins plane.
9. Leaflets: 0 = the widest portion of one half of the blade is on the same transverse plane (i.e., continous with) the widest portion of the other half of the blade (Fig. 2A, C). The leaflets in this case are (ob)elliptic or (ob)ovate, but not necessarily symmetrically so. 1 = the widest portion of one half of the blade is on a different transverse plane (i.e., is not continuous with) than the widest portion of the other half, the leaflets being rhombic (Fig. 2B).
10. Venation: 0 = palmate with several secondary nerves from pulvinule (Fig. 2D); 1 = pinnate, but sometimes with one secondary nerve from the pulvinule (palmate-pinnate, Fig. 2C); 2 = pinnate, never with secondary nerves from pulvinule (Fig. 2A).
11. Bracts: 0 = caducous; 1 = the distal ones deciduous, the proximal persistent; 2 = all persistent.
12. Inflorescence: 0 = with a pseudoterminal flower; 1 = without a pseudoterminal flower. As far as is known, all Ingeae have auxotelic inflorescence-axes; however, they appear to be blastotelic. Apparently the last axillary flower formed moves into a "pseudoterminal" position, displacing the vegetative growing tip (see Derstine & Tucker, 1991).
13. Inflorescence: 0 = of capitula; 1 = of capituliform racemes.
14. Pedicels: 0 = lacking or broader than long, in any case all about equal in length; 1 = distally decrescent. Unit-inflorescences (i.e., individual capitula) with distally decrescent pedicels have been described as subcorymbiform or umbelliform; the flowers tend to lie on a flat plane above the receptacle.
15. Terminal flower: 0 = not broader than peripheral ones; 1 = broader than peripheral ones.
16. Perianth of terminal flower: 0 = dimorphic; 1 = homomorphic. Dimorphic flowers, in which the terminal flower is enlarged relative to the lateral ones, are common in New World Ingeae. In *Abarema*, this heteromorphy may involve the calyx, corolla or stamens, or combinations of these. (Compare no. 18 below.)
17. Corolla: 0 = pink to red; 1 = greenish to white.
18. Stamens of terminal flower relative to those of peripheral flowers: 0 = dimorphic, that is usually more or less reduplicated in number and usually with stouter, longer tube; 1 = homomorphic.
19. Stamens: 0 = greenish to white; 1 = distally pink to red.
20. Filaments: 0 = not dilated and recurved at point of separation; 1 = dilated and recurved at point of separation.
21. Ovary: 0 = ellipsoid; 1 = truncate at apex.
22. Ovary: 0 = pubescent in part; 1 = glabrous.

Table 1

Data set used in the analysis of the *Abarema*-alliance

	0	5	10	15	20	25	30	35	40	45
WALLACEODENDRON	0000?	00000	000?0	??00?	?0000	00000	00000	000-0	0000?	?0
GONGGRIJPII	10100	0?010	00001	00100	01011	00000	02010	10000	0?200	00
MARGINATA	14100	10010	00001	00100	01011	00000	02010	10000	00200	00
CORYMBOSA	10100	?0010	00001	00100	01011	00000	02010	10000	00200	11
PEDICELLARIS	13110	00000	00001	00100	010-0	00000	01100	11000	10010	11
ELEGANS	12110	00000	00001	00100	010-0	00000	01100	11000	10010	11
LEUCOCALYX	12100	11010	00001	00100	01010	00000	01100	11000	10010	22
OPPNIPE	00011	00000	00001	?0110	01100	00100	00011	11111	11221	22
GLAUCA	00111	10010	10011	01110	01000	00101	00011	11111	10221	22
GANYMEDEA	10111	00010	10001	00100	00120	00110	0001?	?11?1	????1	22
JUPUNBA	10110	10000	00001	00100	00000	00100	00012	11111	10121	22
OBOVATA	12101	11000	10011	00100	01100	00100	00011	11111	10221	22
MICROCALYX	10?11	1?000	00011	00100	00120	?0?00	00012	11111	10??1	22
FLORFER	14110	11010	00001	00001	01?20	00100	00011	11111	10221	22
IDIOPODA	11110	0?010	00001	00100	01020	00100	00011	11111	1???1	22
MATAYBIFOLIA	00001	10101	00001	00100	01100	00100	00012	11111	10121	22
LANGSDORFII	10011	0?000	10011	001?0	01000	00100	0011	11111	1??21	22
OBOVALIS	02?11	10000	00011	01110	01000	00100	10012	11111	11221	22
CURVICARPA	04111	00010	00?11	11110	01120	11000	00012	11111	1?121	22
FILATUR	11111	10000	00011	00100	01100	00101	00011	11111	10221	22
CAMPESTRIS	12111	10111	0100?	??0?0	00100	00100	00012	11111	1?221	22
ASPLENIFOLIA	02111	10000	00011	00110	01000	00100	00011	11111	10221	22
BRACHYSTACHYS	11100	11000	10011	00100	01100	00100	00011	11111	10221	22
BARBOURIANA	11110	000?0	0000?	00100	00000	00100	0001?	11111	10?21	22
PIRESII	10111	01000	02000	00100	01120	01?10	00010	11111	10121	22
LONGIPEDUNCULATA	14110	10010	00000	??0?1	01120	00000	00011	11111	1?121	22
MACRADENIA	10101	000?0	00011	10100	01120	10100	00011	11111	10121	22
ZOLLERANA	11101	00000	00011	00100	00100	00?00	10012	11111	1??21	22
LAETA	14101	1?111	00000	10110	01100	00100	10012	11111	10121	22
GALLORUM	10111	10010	00001	00001	01000	00000	00011	11111	10??1	22
COCHLIACARPOS	11111	10000	00000	01100	01100	00101	00011	11111	10221	22
ABBOTTII	03111	00000	00001	??110	011?0	00?00	00012	11111	10221	22
CALLEJASII	04111	10111	00111	01110	00100	00000	00012	11111	1?121	22
AURICULATA	12111	01010	02000	00100	10100	00100	00012	11111	10121	22
LEHMKIL	04111	10111	00110	01110	00100	00100	00012	11111	1?121	22
COMMUTATA	10111	10000	00001	00100	00020	00100	00012	11111	1???1	22
LEUCOPHYLLA	04101	1?000	00001	00101	11100	00000	10012	11111	10121	22
ADENOPHORA	11100	10000	00011	00100	10000	00100	00011	11111	10121	22
COCHLEATA	14111	101?1	01001	00100	00100	?0100	00012	11111	10121	22
RACEMIFLORA	04111	10111	00110	01110	00100	00101	00011	11111	1?121	22
ALEXANDRI	00011	00010	00101	01110	01100	00100	00011	11111	10221	22

Note: ? = state not observed in this terminal; - = state inapplicable or of uncertain homology.

23. Fruit: 0 = dehiscent through both sutures; 1 = indehiscent; 2 = follicular. *Pithecellobium pedicellare* and *P. elegans* are scored as state inapplicable. Technically, the pods are either indehiscent or follicular. However, even when the pods are follicular, dehiscence nearly always takes place only after the fruit has fallen from the tree (compare no. 46.)
24. Fruit: 0 = not lomentiform; 1 = lomentiform. This character refers to shape of fruit only, as many lomentiform fruits in Ingeae do not actually break apart into loments.
25. Pod: 0 = not grossly lignified; 1 = grossly lignified.
26. Valves of fruit: 0 = brownish; 1 = externally yellow with red or orange splotches.
27. Valves of fruit: 0 = not spiraling with dehiscence; 1 = spiraling with dehiscence.
28. Fruit: 0 = not alternately twisting between every two or three seed chambers; 1 = alternately twisting between every two or three seed chambers.

29. Valves of fruit: 0 = not externally papillate; 1 = externally papillate.
30. Valves of fruit: 0 = not internally papillate; 1 = internally papillate.
31. Meso- and epicarp: 0 = not cracking between seeds; 1 = cracking between seeds but not through sutures; 2 = cracking between seeds and through sutures.
32. Valves: 0 = meso- and epicarp remaining contiguous; 1 = meso- and epicarp irregularly peeling or flaking away from one another.
33. Mesocarp: 0 = reticular-granular; 1 = not reticular-granular.
34. Endocarp: 0 = uniformly tan or brown; 1 = pigmented red or orange in seed cavities; 2 = internally pigmented overall.
35. Endocarp: 0 = becoming free from mesocarp; 1 = adherent to mesocarp.
36. Endocarp: 0 = adherent between sutures to endocarp of opposite valve; 1 = free from endocarp of opposite valves.
37. Endocarp: 0 = not cracking between seeds; 1 = cracking between seeds.
38. Seeds: 0 = narrowly elliptic in outline; 1 = lentiform.
39. Seeds: 0 = not persistent on the funicle; 1 = persistent on the funicle.
40. Testa: 0 = not translucent; 1 = translucent in part.
41. Funicle: 0 = ribbonlike and coiled near hilum; 1 = narrow and ligulate.
42. Pleurogram: 0 = incomplete hippocrepiform or narrowly U-shaped; 1 = lacking; 2 = complete.
43. Embryo: 0 = not colored; 1 = chlorophyllous or pink; 2 = blue.
44. Peduncles: 0 = held stiffly erect-ascending in fruit; 1 = not held stiffly erect-ascending in fruit.
45. Fruit: 0 = held stiffly erect-ascending; 1 = plagiotropic (i.e., held about perpendicular to the plane formed by the long-axis of the peduncle); 2 = of indeterminate attitude.
46. Seed-dispersal: 0 = hydrochorous; 1 = dispersed in pod; 2 = ornithochorous.

The analysis yielded eight equally parsimonious trees of 168 steps with a consistency index of 0.34, and a retention index of 0.66. One group of species accounted for the numerous trees. One tree from this analysis is shown in Figure 4, the alternative topologies of the group of five species is shown in Figure 5. Features common to all trees will be discussed first, using Figure 4 as an example; then the alternative topologies are discussed. The three members of the *corymbosum*-complex form an unresolved trichotomy sister-group to the remainder of the *Abarema*-alliance. Two synapomorphies support this clade; the lomentiform fruit (no. 24) that breaks apart into true loments (no. 31). They are further supported by homoplasious character-states of a non-reticular-granular mesocarp (no. 33) and a complete pleurogram (no. 42). The three species, as well as *Pithecellobium acreanum* (not included in the analysis) are distinguished inter se on the basis of autapomorphies. These species are treated below as genus *Hydrochorea* (no. I). The remainder of the species form a clade with the following synapomorphies: endocarp free from the endocarp of the opposite valve (no. 36), a seed-testa translucent in part (no. 40), and fruits not held stiffly erect-ascending (no. 45).

The three species in the *pedicellare*-alliance form a monophyletic group with the following synapomorphies: a meso- and epicarp that crack between the seeds but not through the sutures (no. 31); the layers of the valves (endo-, meso-, and epicarp) irregularly separating from each other with age (no. 32); a chlorophyllous or pink embryo (no. 43); and seeds dispersed in the pod (no. 46). These three species are treated in the genus *Balizia* (no. II) below. *Balizia leucocalyx* is the more primitive of the three, retaining the indehiscent fruit and leaf-architecture of the *corymbosum*-complex. *Balizia pedicellaris* and *B. elegans* share two homoplastic (in the context of the tribe) leaf-characters.

Sister-group to *Balizia* are the remainder of the species, treated herein in *Abarema*. Included within this clade are the type species of the genera *Jupunba* Britton & Rose, *Punjuba* Britton & Rose, and *Klugiodendron* Britton & Killip. The following synapomorphies support the clade: the endocarp not cracking between the seeds (no. 37); the lentiform seeds (no. 38) persistent on the funicles (no. 39); the pleurogram incomplete or lacking (no. 42); the blue-colored embryo (no. 43); neither peduncles (no. 44) nor fruit (no. 45) held stiffly erect-ascending or plagiotropic; and ornithochorous seed-dispersal (no. 46). It should be noted that all of these save the blue embryo are homoplastic in the context of the whole tribe.

As mentioned above, unresolved relationships among one group of species accounted for the 20 equally parsimonious trees. This group included the following taxa: *Abarema ganymedea*, *A. piresii*, *A. auriculata*, *A. curvicarpa*, and *A. macradenia*. Alternative trees are shown in Figure 5. The alternative topologies depend on two characters: the internal

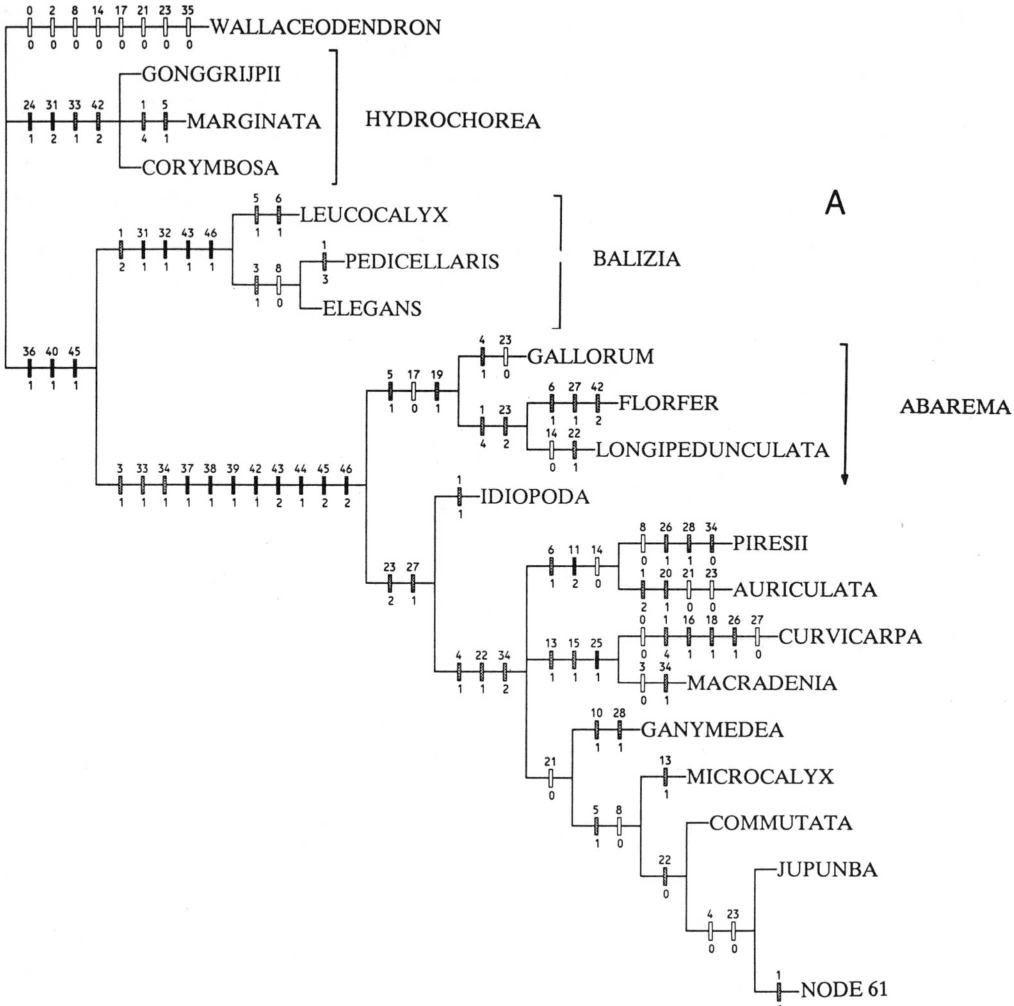

FIG. 4 [*above* (**A**) *and facing page* (**B**)]. One of 24 equally parsimonious trees from an analysis of the *Abarema*-alliance. **A.** Nodes 77 to 61.

coloration of the valves of the pod (no. 34), whether pigmented only in the seed-cavities (state 1) or overall (state 2); and the shape of the ovary (no. 21), whether truncate (state 1) or ellipsoid (state 0). In the tree shown in Figure 5A, valves internally colored overall are derived in parallel in *curvicarpa* and the clade including *piresii*, *auriculata*, and *ganymedea*. It must be noted that this character changes another seven times on the tree above these nodes. In this tree, a reversal to an ellipsoid ovary occurs below *ganymedea* and then reverses again to state 1 in *auriculata*. Three of the species—*A. ganymedea*, *A. microcalyx*, and *A. commutata*—are still unknown in seed.

In Figure 5B valves internally colored overall are derived in parallel in *curvicarpa* and the clade including *piresii*, *auriculata*, and *ganymedea*, and a reversal to the ellipsoid ovary occurs at the same time, with another reversal in *piresii*.

In Figure 5C valves internally colored overall is derived in all the taxa above *idiopoda*, with a reversal to state 1 in *macradenia*, and a reversal to the ellipsoid ovary occurs in parallel in *auriculata* and the clade including *ganymedea* and the remainder of *Abarema*. Resolution among taxa in *Abarema* depends mostly on homoplasious character states. However, several species-pairs or groups are united by synapomorphies: *A. piresii* and *A. auriculata* have persistent floral

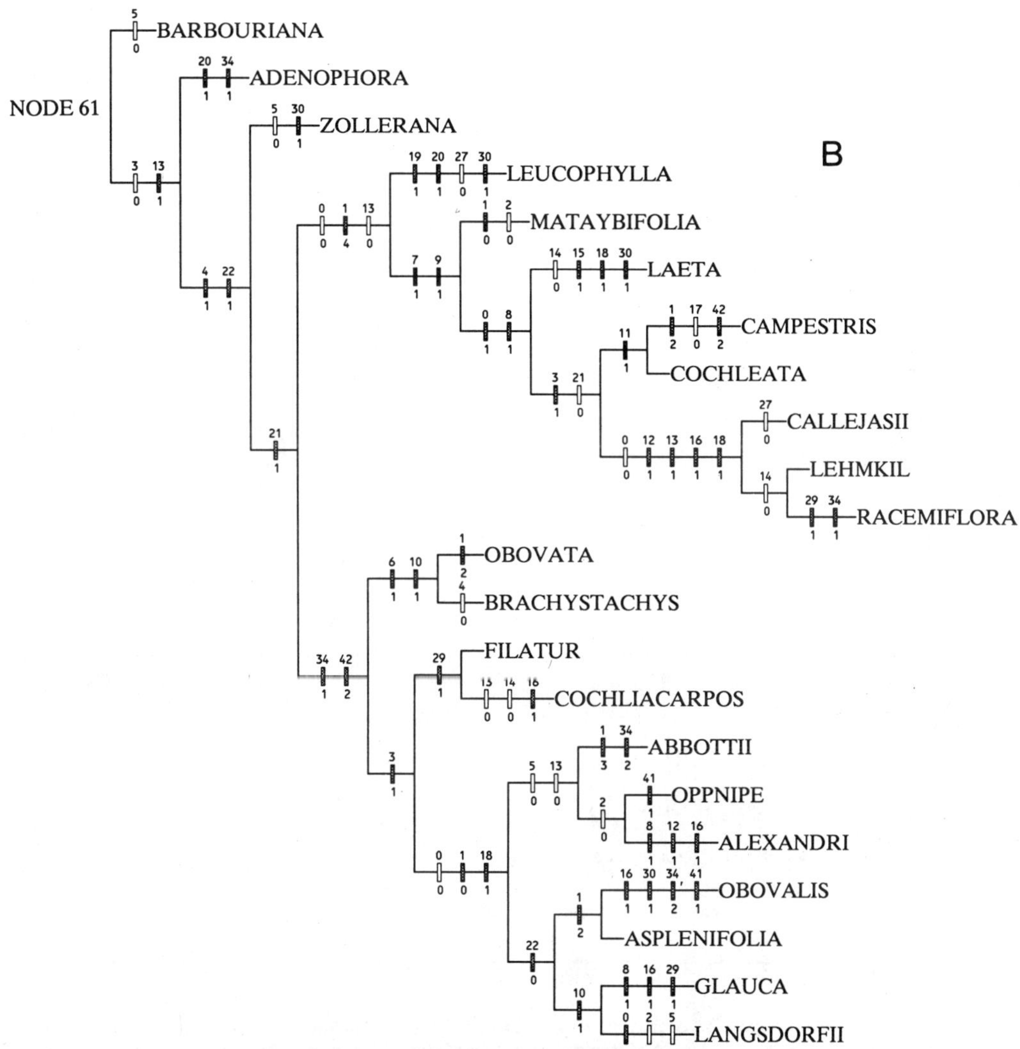

FIG. 4 (*continued*). **B.** Nodes 61 and above.

bracts (no. 11); *A. curivcarpa* and *A. macradenia* have grossly lignified fruits (no. 25). *Abarema microcalyx* is polymorphic for this character-state as there is a variety of that species with grossly lignified fruits. When divided into two taxa to reflect this variation, the state is homoplastic and secondarily derived within *A. microcalyx.* The clade including *A. mataybifolia, A. laeta, A. campestris, A. cochleata, A. callejasii, A. lehmanii,* and *A. racemiflora* is united by two synapomorphies: leaflets that are both broadest below mid-blade (no. 7) and have the broadest portions of each half of the leaflets on the same transverse plane (no. 9). These characters appear to be correlated; however, they occur independently outside *Abarema.*

I. HYDROCHOREA Barneby & Grimes

Hydrochorea Barneby & Grimes, gen. nov., *Baliziae* arcte affinis sed legumine coriaceo lomentiformi, maturo pluriarticulato, suturis continuis, articulis indehiscentibus demum liberis diversum. — Sp. typica: *H. corymbosa* (L. C. Richard) Barneby & Grimes ≡ *Mimosa corymbosa* L. C. Richard ≡ *Pithecolobium corymbosum* (L. C. Richard) ≡ *Arthrosamanea corymbosa* (L. C. Richard) Kleinhoonte ≡ *Cathormion corymbosum* (L. C. Richard) Burkart ≡ *Albizia corymbosa* (L. C. Richard) G. P. Lewis. — Etymology: Gr. *hydro,* water + *chorein,* to travel; in allusion to water-borne propagules.

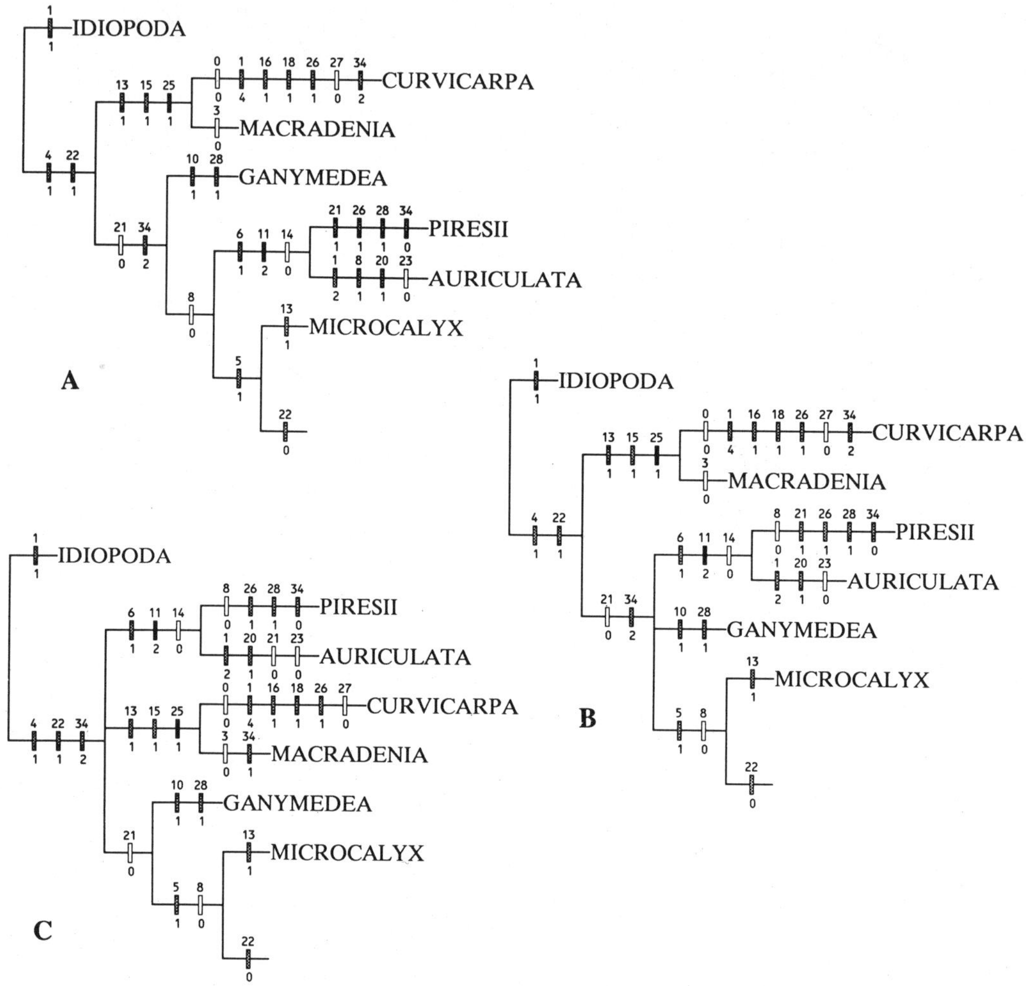

FIG. 5. Alternative topologies for equally parsimonious trees for the taxa *Abarema curvicarpa*, *A. macradenia*, *A. ganymedea*, *A. piresii*, and *A. auriculata*. **A.** A parallel derivation of states in characters 21 and 33. **B.** Reversal in character 21, a parallelism in 34. **C.** Reversal in character 34, a parallelism in 21.

Pithecolobium sect. *Samanea* ser. *Corymbosa* [*sic*] Bentham *in* Hooker, London J. Bot. **3:** 221. 1844. — Sp. typica (unica): *P. corymbosum* Bentham.

Pithecolobium sect. *Samanea* ser. *Subarticulatae* [*sic*] (*Americanae*) sensu Bentham, 1875: 586, magna ex parte, exclus. spp. no. 60, 62.

Unarmed *trees* and *arborescent shrubs* commonly 5–27 m, some precociously fertile when smaller, the young stems and foliage gray- or brown-pilosulous, the lfts nearly always conspicuously bicolored, the inflorescence of umbelliform units of dimorphic flowers mostly axillary to coeval or distally to hysteranthous lvs, the meristem continuous beyond them, in one species (of uncertain affinity) appearing terminally paniculate; phyllotaxy spiral. *Stipules* linear or subulate, caducous, perhaps sometimes 0. *Lf-formula* i–vii/3–35; petiolar nectary between or close below insertion of first or only pinna-pair, either sessile or stipitate (rarely suppressed), the head variably plane, convex or shallow-cupular; lft-pulvinules ±0.5–2.5 mm; venation of lfts pinnate. *Peduncles* 2–11(–14) cm; fls compactly racemose-capitulate, the axis including pedestal 1.5–6 mm, the peripheral fls slenderly pedicellate, mostly 5-merous, the 1–4 distal fls sessile or shortly stoutly pedicellate, 5–8-merous; calyx campanulate or turbinate-campanulate (1–)1.4–

7 mm; corolla (2.6–)3–11(–12) mm; androecium of peripheral fls 10–24(–36)-merous, 11–43 mm; ovary truncate at apex, 8–12-ovulate. *Pods* (that of *H. acreana* unknown) often several per capitulum, stiffly erect-ascending, sessile or shortly pseudostipitate, in profile linear or broad-linear, straight or almost so, compressed but low-biconvex over each seed and depressed between them, the valves leathery or sublignescent, the mature fruit lomentiform (4.5–)5–12.5 × 0.75–1.9 cm, readily breaking horizontally between seeds and less readily so through the shallowly undulate sutures into square or transversely oblong, 1-seeded articles closed at each end by half of the interseminal septum; seeds hard, brown, areolate, the pleurogram complete. — Spp. 4, widespread in riparian habitats in the Orinoco and Amazon basins and in the Guianas, in Brazil extending E to Maranhão and S to the Pantanal in Mato Grosso do Sul.

The relationships and diagnostic features of *Hydrochorea* are discussed under the genus *Balizia*, where *Hydrochorea* is shown to differ principally from *Balizia* and *Abarema* in its lomentiform fruit adapted to dispersal by water and brown (as opposed to partly translucent) seed-coat. The commoner species of *Hydrochorea* have been associated by students of Ingeae with those American albizias that have comparable jointed fruits (but not true loments). Consequently, they have been transferred successively to the Old World genus *Cathormion* Hasskarl, the neotropical genus *Arthrosamanea*, and latterly to a loosely defined and polymorphic *Albizia*. They form, nevertheless, a closely coherent group of species different from all the above in duration of meristem, features of inflorescence-architecture, leaf-venation, or shape of ovary, or in some combination of those. As the species of *Hydrochorea* differ mostly by autapomorphic characters, their relationships to each other remain unresolved (Fig. 4A).

Conspectus of genus *Hydrochorea*

¶ Peduncles all, or at least those at base of inflorescence, subtended at anthesis by an expanded lf; lf-formula ii–vii/12–35, the largest lfts 7.5–17.5 × 1.5–4.5 mm.
1. *H. gonggrijpii* (Kleinhoonte) Barneby & Grimes

¶¶ Peduncles of ¶; lf-formula ii–vii/4–11(–14), the pinnae in Venezuela not less than 3-jug., the largest lfts 12–35 × 4–14 mm.
2. *H. corymbosa* (L. C. Richard) Barneby & Grimes

¶¶¶ Peduncles of ¶; lf-formula i–ii/3–6, the pinnae 2-jug., only locally in Venezuela, elsewhere 1-jug., and the largest lfts ±30–60 × 13–32 mm.
3. *H. marginata* (Bentham) Barneby & Grimes
a. var. *marginata*
b. var. *panurensis* (Bentham) Barneby & Grimes
c. var. *scheryi* Barneby & Grimes

¶¶¶¶ Peduncles all bracteate, at anthesis forming a panicle above the expanded foliage; lf-formula iv–v/4–6, the largest lfts ±3.5–6 cm.
4. *H. acreana* (Macbride) Barneby & Grimes

Key to the species of genus *Hydrochorea*

1. Peduncles all, or all but a few distal ones, subtended at anthesis by a fully expanded lf, hence manifestly axillary.
  2. Lfts of longer pinnae 14–35-jug.; pods 7.5–9 mm wide ....1. *H. gonggrijpii* (p. 25)
  2. Lfts of longer pinnate (2–)3–11(–14)-jug.; pods (9–)10–19 mm wide.
    3. Pinnae of larger lvs 2–7-jug., but exceptionally less than 3-jug.; lfts of longer pinnae 4–11(–14)-jug. and the largest 1–3 cm ....2. *H. corymbosa* (p. 27)
    3. Pinnae mostly 1-jug., randomly 2-jug. only in var. *scheryi*, local on upper Orinoco river; lfts of longer pinnae (2–)3–6-jug. and the largest ±3–6 cm. ....3. *H. marginata* (p. 29)
1. Peduncles all subtended by a bract, together forming an efoliate panicle raised at anthesis above the foliage ....4. *H. acreana* (p. 33)

1. Hydrochorea gonggrijpii (Kleinhoonte) Barneby & Grimes, comb. nov. *Pithecolobium gonggrijpii* Kleinhoonte, Recueil Trav. Bot. Néerl. **22:** 414. 1925. — "Surinam: im Reservat der Zanderij I, die nummerierten Bäume n. 102 (Herb. [Acad. Rhenotraiect.] n. 1529, im Dez. 1915, und n. 4350 bl. im Juli 1919) und n. 141 (Herb. n. 4357, bl. im Juli 1919.)" — Syntypi, U!; isosyntypi (omnes 3), NY!. — *Arthrosamanea gonggrijpii* (Kleinhoonte) Kleinhoonte *in* Pulle, Fl. Suriname **2(2):** 328. 1940. According to J. C. Lindeman (pers. comm.), Kleinhoonte later marked BW 4357 (2 sheets) as types and the others as cotypes, and annotated the copy of Recueil Trav. Bot. Néerl. in pencil.

Pithecolobium pullei Kleinhoonte, Recueil Trav. Bot. Néerl. **22:** 415. 1925. — "Surinam: Paragebiet, Zanderij I: am Rande des Urwaldes, Pulle n. 65 im Juli 1920." — Holotypus, U (2 sheets)!; isotypus, K!. — *Arthrosamanea pullei* (Kleinhoonte) Kleinhoonte *in* Pulle, Fl. Suriname **2(2):** 328. 1940.

Pithecolobium sabanensis [*sic*] Schery, Fieldiana, Bot. **28:** 258. 1952. — "[Venezuela, estado Bolívar:] . . . in Gran Sabana south of Mount Roraima, alt. 1065 m, October 2, 1944, *Julian A. Steyermark 59146*." — Holotypus, F! = F Neg. *59146*; isotypi, MO!, VEN *31276*!.

Pithecollobium gonggrijpii sensu Sandwith, Kew Bull. **9:** 8. 1939, who suggested that *P. pullei* was based on juvenile material of the same; Lindeman & Mennega, Bomenboek Surinam 191. 1963.

Pithecellobium sabanense sensu Cárdenas, 1974: 122 (in nota), who reported a gland on the anthers, not confirmed in this study.

Microphyllidious arborescent *shrubs* (*trees* fide Lindeman, pers. comm.) (1)2–20(–27) m tall, the young stems, all lf-axes and peduncles puberulent-tomentulose with forwardly incurved, sordid or rusty hairs 0.1–0.2 mm, the lvs conspicuously bicolored, the lfts glossy dark green (when dry brunnescent) and (except for sometimes ciliolate midrib) glabrous above, beneath pale olivaceous (drying tan), dull, either glabrous or minutely thinly strigulose, ciliolate or not, the umbelliform capitula of small, pinkish or reddish, white-stamened fls axillary to coevally expanding lvs, immersed in foliage. *Stipules* linear-lanceolate or -attenuate 2–7.5 × 0.3–0.7 mm, early deciduous. *Lf-formula* (ii)iii–vi(in juvenile lvs occasionally –viii)/(12–)14–35(–"40"); lf-stks of lvs on most terminal fl stems (2.5–)3–12 cm, of lvs on vigorous long-shoots and sapling branches up to 16–23 cm, the petiole (0.8–)1–2.4(–5) cm, the longer interpinnal segments (7)9–23(–28) mm; nectary between or close below first pinna-pair most often shortly stoutly (sometimes cryptically) stipitate, (0.8–)1.1–2.1 mm diam, in profile 0.5–1.6 mm tall (often distorted by drying under pressure), the stipe as wide to less than half as wide as the head, this commonly convex pileiform, dimpled or not, but sometimes shallowly concave or cupular, a similar nectary at tip of lf-stk and often between 2 or more distal pinna-pairs, and much smaller trumpet-shaped ones on pinna-rachises between 2 to several further pairs of lfts; pinnae accrescent distally, the rachis of longer ones (3–)4.5–8.5(–10.5) cm, the longer interfoliolar segments 2–4(in sapling lvs –6.5) mm; minute paraphyllidia ascending from apex of pinna-pulvini; lft-pulvinules wrinkled, usually livid, 0.45–0.8(–1.3) × 0.4–0.7 mm; lfts except at far ends of rachis subequilong, in outline linear or narrowly oblong from inequilaterally rounded or on distal side cuneate base, broadly obtuse and sometimes minutely mucronulate at apex, those near and beyond midrachis (7.5–)8–16(–17.5) × (1.5–)2–4(–4.5) mm, 3.6–5(–5.5) times as long as wide; venation pinnate, the subcentric, nearly straight midrib commonly giving rise to 6–9 perceptible but always very slender major and often about as many intercalary secondary nerves brochidodrome just short of the plane margin, the tertiary venulation faint, irregular, or externally invisible, the whole venation, on one or both faces, either finely prominulous or immersed and discolored, less commonly immersed except for dorsally prominulous midrib. *Peduncles* solitary or geminate 2–5.5(–8) cm; capitula subhemispherical (11–)17–40-fld, the clavate axis, including short terminal pedestal, 1.5–4 mm; bracts linear-oblanceolate ±1.5–2.5 mm, early caducous; fls dimorphic, the peripheral ones slenderly pedicellate, the terminal 1–3 either sessile or stoutly short-pedicellate, with modified androecium; perianth 5-merous (or terminal fls randomly –8-merous), commonly strigulose overall but sometimes only distally, rarely glabrous except for ciliate corolla-lobes; PERIPHERAL FLS: pedicel of lowest fls (3–)4–9 mm, at middle 0.2–0.3 mm diam; calyx (1.7–)2.6–4(–4.8) × (1.1–)1.3–2.9 mm, openly turbinate-campanulate to narrowly campanulate, the teeth (0.3–)0.6–1.3(–1.8) mm; corolla 5.3–8.5(–9) mm, the lobes (1.5–)2–3 × 1.2–2 mm; androecium 10–12(–14)-merous, (13–)22–42 mm, the stemonozone 0.5–1.8 mm, the tube (1.5–)1.8–4.4 mm; ovary narrowly oblong-oblanceolate in profile, dilated and truncate at apex, usually densely puberulent overall, sometimes glabrous below middle, rarely puberulent only around the top; TERMINAL FL(S): calyx campanulate, sometimes deeply so, (2.8–)3–6 × (1.6–) 1.8–3.3 mm; corolla (7.5–)8–12(–13.5) mm; androecial tube (7–)9–16 mm, usually exserted at least shortly from corolla and somewhat dilated upward, at separation of filaments 1–2.4 mm diam. *Pods* 1–7 per capitulum, not seen fully ripe but apparently similar in structure and dehiscence to those of *H. corymbosa,* commonly 7.5–9 (in T. F. Amazonas, Venezuela to 11.5) mm wide, 8–13-seeded.

On river banks, at gallery margins, and in low-lying swamp forest, 40–1400 m, locally plentiful within and along NE periphery of the Guayana Highland in SE Colombia, S Venezuela, Guyana, Surinam, and E-centr. Amazonian Brazil: at 400 m upward on Gran Sabana in state of Bolívar, Venezuela, and adjacent Guyana and upland interior Surinam, N in Bolívar at lower elevations into T. F. Delta-Amacuro and NW Guyana; lowland N Guyana and Surinam; at 150 m at base of C. Duida in T. F. Amazonas, Venezuela; riverine forest of Igara-paraná in com. Amazonas, Colombia; in Brazil on Río Univini in Roraima, and near Maués in E Amazonas. — Map 1. — Fl. almost throughout the year, most abundantly X–IV. — *Manariballi, swamp manariballi* (Guyana); *manaliballi* (Suriname); *water-* or *boschtamarinde* (Suriname).

Hydrochorea gonggrijpii is closely related to *H. corymbosa,* differing most clearly in more numerous leaflets of narrower outline, but also in flowers on the average fewer per capitulum and a little larger, and in pods, as yet little known, at least on the average a little narrower. Like that of *H. corymbosa,* the perianth varies from densely strigulose overall to glabrous except for corolla-lobes minutely ciliolate at apex. We

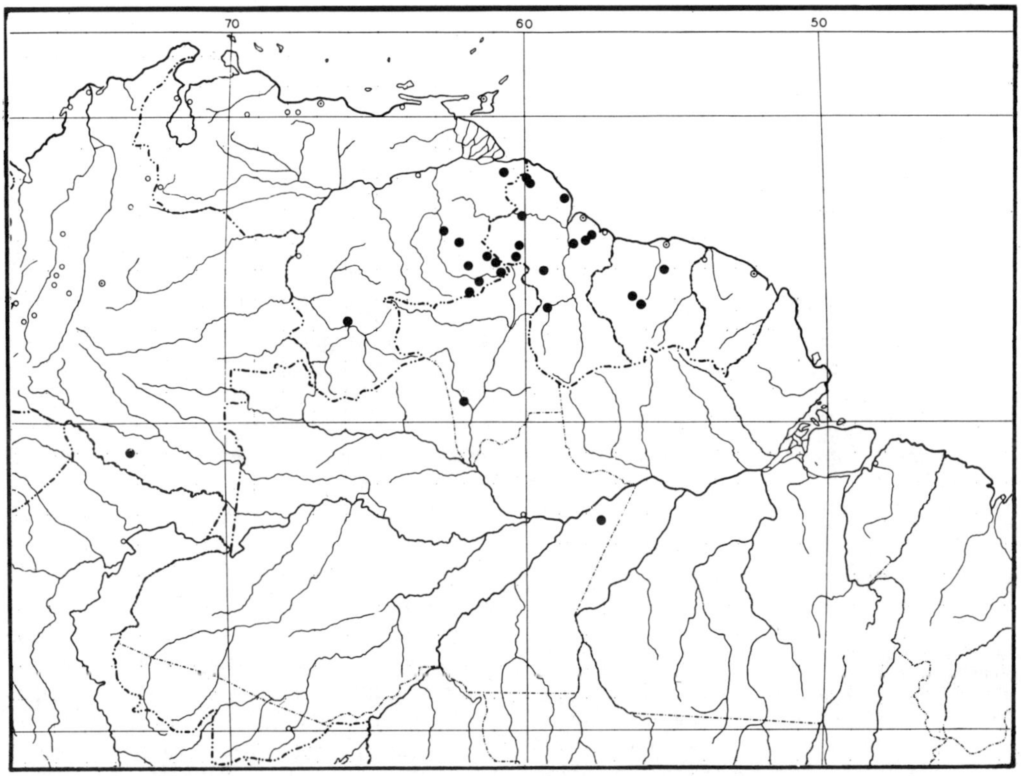

MAP 1. Northern South America. Range of *Hydrochorea gonggrijpii* (Kleinhoonte) Barneby & Grimes.

have found a narrow overlap between the two species in all measurements, even in leaflet number; but the overlaps are encountered at random and singly, and we have no really ambiguous specimens. The range of *H. gonggrijpii* lies within the northeastern quarter of that of *H. corymbosa* (compare Maps 1 and 2) and is there found in essentially the same riparian habitats.

Internal variation, other than in vesture of the perianth, is most marked in amplitude of the perianth, which varies from widely turbinate-campanulate, as in the type of *Pithecellobium gonggrijpii,* to narrowly and deeply campanulate, as in that of *P. sabanense*; the extremes are strikingly different, but intermediate forms are not uncommon, especially on the Gran Sabana. A contrast in leaf-formula between leaves of lush vegetative (or juvenile) branches and those of flowering branches of adult trees, which Kleinhoonte incautiously interpreted as indicating specific differentiation of the two precisely sympatric species *P. gonggrijpii* and *P. pullei,* is characteristic of the genus; in fact both types of foliage occur in some modern collections (e.g., *Davidse 4893A,* NY: lf-formulae iii/16 and vii/28!). This condition was foretold by Sandwith in 1939, when few collections had been made and no direct evidence of it was available. We have found further that the secondary and tertiary venulation of the leaflets varies in prominence (in the dry state), but this has no correlation with other characters or with dispersal. It may be due to edaphic influences, or to age of the plant, or to both. One of the two Brazilian specimens known to us (*Pires 14240,* NY) is notable for an extremely small peripheral flower, resembling that of *H. corymbosa,* combined with fully immersed venation of the leaflets. This may possibly represent a taxonomic variant. The other (*Ducke* in RB *23236,* from Maués, Amazonas), is also small-flowered, but seems otherwise typical.

2. Hydrochorea corymbosa (L. C. Richard) Barneby & Grimes, comb. nov. *Mimosa corymbosa* L. C. Richard, Actes Soc. Hist. Nat. Paris **1:** 113. 1792. — ". . . e Cayenna ineunte anno 1792 . . . a domino Le Blond." — Holotypus, "frequens in sylvis ripariis fluvii Kourou", P(herb. richard.)!. — *Pithecolobium corymbosum* (L. C. Richard) Bentham, Trans. Linn. Soc. London **30:** 587. 1875, nom. illeg. *Feuilleea corymbosa* (L. C. Richard) O. Kuntze, Revis. Gen. Pl. **1:** 187. 1891. *Arthrosamanea corymbosa* (L. C. Richard) Kleinhoonte *in* Pulle, Fl. Suriname **2:** 327. 1940. *Cathormion corymbosum* (L. C. Richard) Burkart, Darwiniana **13:** 446. 1964. *Albizia corymbosa* (L. C. Richard) G. P. Lewis & P. E. Owen, Leg. Ilha de Maracá 40, fig. 3J, pl. 4D. 1989. Non *Pithecellobium corymbosum* Gagnepain, Bull. Soc. Bot. France **99:** 49. 1952. (= *Albizia attopeuense* (Pierre) I. Nielsen).

Pithecolobium corymbosum Bentham, London J. Bot. **3:** 221. 1844. — "British Guiana, [Richard & Robert] *Schomburgk,* n. 114 of 1841 [2nd. coll.]; Cayenne, *Martin*; Surinam, *Hostmann,* n. 214, 262, and 1190; Borba on the Rio Negro [really lower Río Madeira, *Riedel,* commun.] *Langsdorff.*" — Lectotypus, *Schomburgk 114* of 1841, K!; paratypi, *Riedel* [no. 67], *Martin s.n., Hostmann 214, 1190,* all K!. — *Samanea corymbosa* (Bentham) Pittier, Bol. Ci. Técn. Mus. Com. Venez. **1:** 55. 1926, without full citation. — Equated with the preceding by Bentham, 1875: 587; 1876: 440, with reservation; Kleinhoonte, 1940: 327 (all excluding syn. Steudel).

Pithecolobium subcorymbosum Hoehne, Comiss. Linhas Telegr. Estratég, Mato Grosso Amazonas, Bot. **8:** 18, est. 133. 1919. — "Hoehne Nos.: 4582–4586 . . . em S. Luiz de Caceres, nas margens do rio Paraguay, perto da Campina [in SW Mato Grosso, Brazil, near 16°S, 57°40′W]. [16°S, 57°40′W] — Syntypi, SP, not seen, but the illustrated protologue unambiguous.

Pithecolobium corymbosum var. *longipes* Spruce *ex* Ducke, Bol. Técn. Inst. Agron. N. **18:** 37. 1949, nom. nud. [*Spruce 2429* e Panuré ad Río Uaupés, K!, NY!].

Broad-crowned riparian *trees* attaining 5–20 m with trunk 1–3 dm dbh, the annotinous epidermis cracking vertically and exfoliating, the young ones and all lf-axes (at least ventrally) and peduncles pilosulous with incurved, gray or sordid hairs up to 0.1–0.25 mm, the amply microphyllidious lvs bicolored, the lfts on upper face dark olivaceous brunnescent (overall, or in spots, rarely remaining green), either lustrous venulose or dull and nearly veinless, on lower face dull, pale olivaceous or pale tan and almost always minutely, often remotely strigulose, the umbelliform capitula consisting of many small crowded, slenderly pedicellate, whitish or reddish peripheral fls surmounted by a sometimes elongate pedestal bearing 1–4(–7) much longer and sessile or subsessile, white or white and basally pink sterile fls with modified androecium, the peduncles either 1–3(–4) together and axillary to fully developed lvs, or the upper ones, sometimes nearly all, axillary to hysteranthous lvs, the inflorescence then becoming terminally pseudoracemose and sometimes shortly exserted (but the terminal meristem potentially continuous). *Stipules* linear or linear-lanceolate 2–7(–8.5) × 0.3–0.7 mm, caducous with early expansion of associated lf, absent from mature spms. *Lf-formula* ii–v(–vii)/(4–)5–11(–14); lf-stk of most developed lvs 4–14(–18) cm, but that of depauperate lvs of some lateral flowering branchlets only 0.8–3.5 cm, the petiole (0.7–)1–4(–5) cm, at middle ±1–2 mm diam, the one or the longest interpetiolar segment (0.7–)1–3.5(–4.2) cm, the shallow ventral groove of each narrowly dilated upward; petiolar nectaries sessile or shortly stoutly stipitate, in vertical view round or obtusangulately scutiform ±1–2.5 mm diam, either mounded and dimpled, or almost plane, or cupular and coarsely marginate, the first one nearly always at or close below first pair of pinnae (exceptionally wanting) and similar ones between all, some, or no further pairs, and smaller, distinctly stipitate, cupular or trumpet-shaped nectaries on pinna-rachises between 2–7 distal pairs of lfts; pinnae usually subaccrescent distally, the rachis of distal or penultimate pair (3–)3.5–9.5(–11) cm, the interfoliolar segments dilated upward, the longer ones 5–13(–15) mm; minute paraphyllidia close above most pinna-pulvini; lft-pulvinules cross-wrinkled 0.5–1.2 × 0.6–0.9(–1) mm; lfts accrescent upward from base of rachis or subequiform (except for broader farthest pair) upward from midrachis, in outline broadly or narrowly rhombic-oblong from inequilateral, postically rounded, antically cuneate base, obtusangulate and at apex broadly obtuse or faintly emarginate, those near and above midrachis (11–)12–31(–35) × (4)4.5–14 mm, (1.7–)2–3.3(–4.5) times as long as wide; venation pinnate, the slender, straight or weakly incurved, subcentric or subdiagonal midrib giving rise to 7–12(–13) major and indefinite intercalary secondary nerves brochidodrome well within the plane margin and to an open sinuous reticulum, the whole venation most commonly finely prominulous on both faces, sometimes only beneath. *Peduncles* (1.5–)3–11(–13) cm, in fruit ascending stout and lignescent, long persistent; capitula (30–)35–75-fld, at full anthesis subhemispherical, the fls strongly dimorphic, the pedicellate peripheral ones crowded on a narrowly clavate receptacle 2.5–5.5 mm, the pedestal of the 1–4(–5) furthest sessile or subsessile fls drum-shaped ±1 mm diam to linear and to 5–8 mm; bracts dimorphic, those of peripheral fls linear or linear-oblanceolate 0.7–2 mm, those of (sub)terminal ones linear-elliptic 2–4.5 mm, all early dry caducous; PERIPHERAL FLS: pedicel of lower ones (4–)5–11.5(–13) × 0.1–0.2 mm; perianth 5(–6)-merous, usually finely minutely strigulose overall, but sometimes only the calyx-teeth and corolla-lobes puberulent, or the whole perianth glabrous except for ciliolate corolla-lobes; calyx campanulate or turbinate-campanulate (1–)1.4–2.6 × 0.6–1.2(–1.4) mm, the ovate-triangular, often unequal teeth 0.3–0.7 mm; corolla (2.6–)3.5–6(–6.5) mm, the lobes (1–)1.2–2.3 × 0.7–1.2 mm; androecium 10–18(–22, in occasional fls transitional to the terminal ones, –32)-merous, (11–)13–24 mm, the stemonozone 0.4–0.9 mm, the tube 1.3–2.6(–3) mm, the filaments white or greenish white; ovary a trifle dilated and truncate at apex, puberulent around the top or over the distal half, otherwise glabrous; style as long as or a trifle longer than longer filaments, at apex poriform or slightly dilated, less than 0.2 mm diam at the stigma; TERMINAL FL(S): pedicel 0 or to 1 mm, nearly as thick; calyx either broadly campanulate or cylindro-campanulate 3–6 × (0.8–)1.3–2.2 mm; corolla (6.5–)7.5–11.5(–15) mm; androecium 24–34-merous, the stemonozone 2–4 mm, the indurated tube 7–11 mm, ±1 mm shorter to 3 mm longer than corolla, the free part of the biseriate filaments more or less thickened and tortuous. *Pods* 1–7(–8) per capitulum, erect, either sessile or contracted at base into a pseudostipe 1–5 mm, in profile broad-linear (4.5–)5–10.5 × (0.9–)1–1.7(–1.9) cm, broadly rounded and apiculate at apex, straight or nearly so (sometimes bent sideways, but not decurved), the valves at first plano-compressed, brown-fuscous and often somewhat lustrous, minutely puberulent especially toward the base, becoming coriaceous or lignescent, dull brown or purplish black and often distally glabrate, low-convex over seeds and depressed between them, transversely venulose, the mature fruit lomentiform, the valves readily cracking between seeds and more reluctantly through the dorsally plane, shallowly undulate sutures (0.8–1.6 mm wide), the 1-seeded articles transversely oblong, as wide as the pod and (5–)5.5–9(–10) mm long, along the line of fracture 0.9–2 mm thick; seeds transverse, snugly fitting the cavity of each segment, basifixed on a filiform crumpled funicle, compressed but plump, in broad view 7.3–9.8 × 4–5.3(–5.6) mm, the firm crustaceous, tan or brown testa closely investing the hard greenish embryo, the complete, narrowly oblong pleurogram ±6–7.5 × 1.6–2.5 mm.

In periodically or permanently inundated riparian forest and (southward) in gallery forest, mostly below 200 m but attaining 480 m in Venezuelan Guayana, widespread through lowlands of the Orinoco and Amazon basins, in Venezuela from Apure and Guárico E to Delta-Amacuro and nearly throughout Venezuelan Guayana, thence W to the llanos of Arauca in NE Colombia and S in Brazil to Acre, adjacent Pando in Bolivia, and the Madre de Dios valley in SE Peru, to Rondônia, Mato Grosso, and N Goiás, in Mato Grosso extending into the Paraguay basin S to the Pantanal; from Brazil W through extreme SE Colombia to the upper Napo in Ecuador and the middle Ucayali valley in Peru; the three Guianas; and in Brazil E to Maranhão. — Map 2. — Fl. most prolifically V–IX, randomly in other months. — *Arepito, bizcochuelo, hueso de pescado rebalsero* (Venezuela); *paschaca, paschaca blanca* (Peru); *swamp manariballi* (Guyana); *bois macaque, préfontaine blanc, tamalin* (French Guiana); *angico branco, pau bixeiro, saboeira da várzea* (Amazonian Brazil); *tipa macho* (Bolivia).

Hydrochorea corymbosa varies in leaf-formula, flower-size, shape of terminal flowers, and width of fruit, but no pattern of correlation has been found among these or between any one of them and dispersal. Leaves on vigorous new long-shoots have relatively many and large leaflets, whereas leaves of short lateral branchlets, as exemplified by the typus of *Pithecellobium subcorymbosum,* tend to be shorter and fewer. The peripheral flowers of a capitulum in this species may be uniform or slightly accrescent distally, occasionally becoming transitional to the terminal flower in a combination of slender pedicel and duplex androecium. The modified terminal flowers may be exactly one per capitulum or two to three, when one it is either sessile or carried up beyond adjacent peripheral flowers on a stalklike pedestal. The calyx in the first case is campanulate, in the second more deeply campanulate or tubular. Collectors who have noted the color of flowers agree that the filaments are whitish and the corolla greenish white or pink-tinged. The flowers of *Prance 23591* (NY, from Río Urubú, Amazonas, Brazil) are described as *dark red,* but this may refer only to the corolla. Normally the perianth is gray-puberulent, conspicuously gray-tipped in bud, but a form with glabrous calyx and almost glabrous corolla has been encountered on upper Río Branco in Brazil and in neighboring Guyana (cf. *Black 51-13300, Maas 6362*; both NY).

3. Hydrochorea marginata (Bentham) Barneby & Grimes, comb. nov. *Pithecolobium marginatum* Spruce *ex* Bentham, Trans. Linn. Soc. London **30:** 586. 1875 & in Martius, Fl. Bras. **15(2):** 439. t. 114. 1876. — Typus infra sub var. *marginata* indicatur.

Macrophyllidious *trees* attaining 22(–27) m with trunk 1–3.5 dm dbh, sometimes precociously flowering at (1.5–)2–3 m, closely resembling *H. corymbosa*

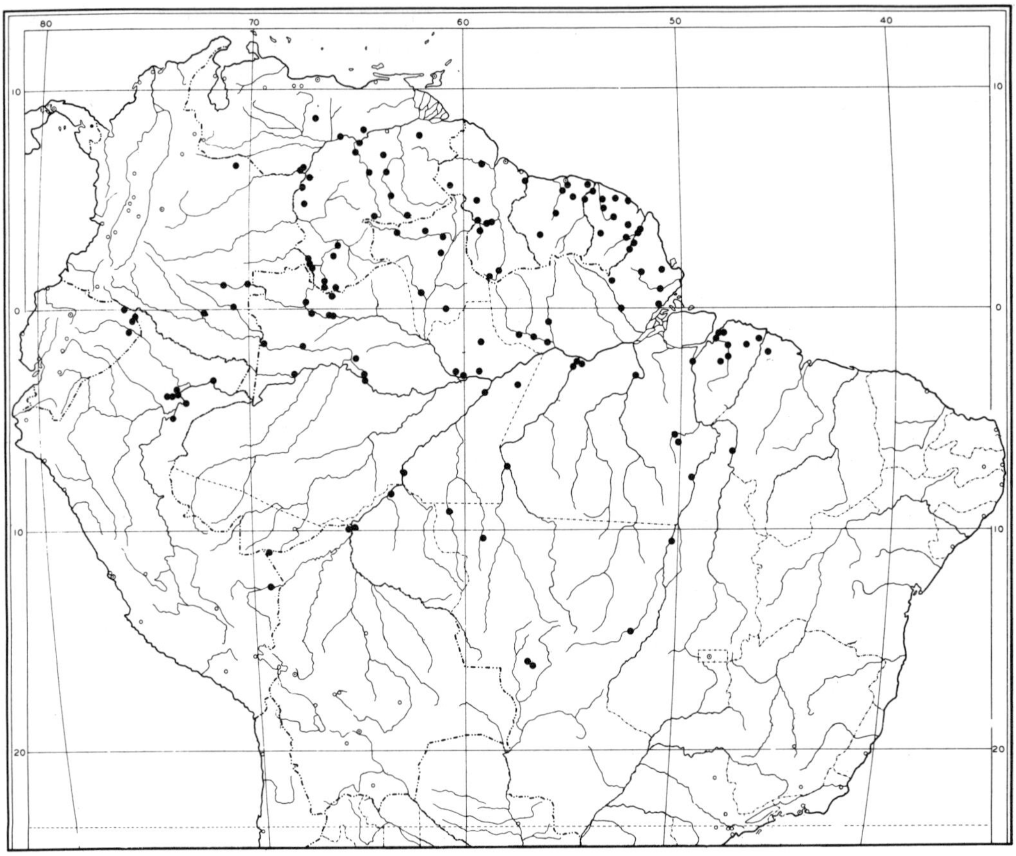

MAP 2. Central South America. Range of *Hydrochorea corymbosa* (L. C. Richard) Barneby & Grimes.

in habit, ecology, inflorescence, and fruit, but always different in lf-formula and vars. *panurensis* and *scheryi* further in greater size of peripheral fls, appearing glabrous but the very young stems and lf-axes minutely sordid-puberulent, the lvs conspicuously bicolored, on upper face lustrous dark-green (commonly brunnescent or brown-mottled when dry), on lower face dull pale olivaceous (fading tan) and sometimes microscopically thinly puberulent, the solitary or geminate, umbelliform capitula of greenish or reddish, ordinarily white-stamened fls axillary to coeval lvs, sometimes in early anthesis subpaniculate but in fruit immersed in foliage. *Stipules* linear-subulate 0.6–3.5 mm, caducous, absent from mature spms. *Lf-formula* i, in var. *scheryi* i–ii/(2–)3–6; lf-stks including somewhat wrinkled pulvinus (3–)6–20(–24) mm, at middle 0.8–1.8 mm diam, shallowly widely sulcate ventrally, the one interpinnal segment in var. *scheryi* 6–15 mm; nectary at tip of lf-stk (exceptionally wanting, in var. *scheryi* at insertion of proximal pinna-pair) either sessile or shortly stoutly stipitate, squatly funnel- or tube-shaped 0.9–2.2 mm diam, similar, sometimes larger nectaries 1–3.5 mm diam at tip of pinna-rachises and sometimes between 1–3 further lft-pairs; rachis of longer pinnae (2.5–)3.3–8.5 cm, bridged at insertion of lfts, each interpinnal segment dilated upward and often narrowly herbaceous-marginate, the longest (10–)12–23 mm; 1–2 minute subulate, deciduous paraphyllidia next to pinna-pulvinus; lft-pulvinules wrinkled (1.1–)1.3–2.6 × (0.8–)0.9–1.4 mm; lfts strongly or moderately accrescent distally, either obliquely oblanceolate, or rhombic-oblanceolate, -oblong, or -obovate from inequilateral, cuneate, or proximally shallowly semicordate base, broadly obtuse or shallowly emarginate and minutely mucronulate, the furthest pair (27–)30–62 × (11–)13–32 mm, 1.5–2.3(–2.5) times as long as wide; venation pinnate, the subcentric, straight or gently incurved midrib giving rise on each side to (6–)7–11(–12) major and random intercalary secondary nerves, brochidodrome well within the plane or proximally recurved, minutely ciliolate margin, and these in turn to a sinuous reticulum of venules, the whole venation often more sharply prominulous above than beneath, but sometimes equally defined on both faces, or almost immersed on both. *Peduncles* 3–11(–14) cm; capitula subhemispherical 17–38-fld, the fls dimorphic, the peripheral ones slenderly pedicellate, the 1–3(–4) much longer (sub)terminal ones either sessile or raised on a grossly thickened pedicel to 2.4 mm; bracts submembranous, narrowly ovate or linear-oblanceolate 0.3–3.5 mm, caducous long before anthesis of associated fl; perianth of peripheral fls 5-merous, of some terminal ones to 8-merous, the calyx and corolla either appressed-puberulent overall or glabrous except at apex; PERIPHERAL FLS: pedicel of lower ones commonly 6.5–13(–17.5) mm, in var. *scheryi* only 4–5 mm, at middle 0.2–0.4 mm diam, the distal ones progressively a little shorter; calyx campanulate (1.5–) 1.7–7 mm, further described under the vars.; corolla greenish or reddish, 4.5–11(–12) mm; androecium (10–)12–24(in var. *scheryi* to 36)-merous, 2.2–4.3 cm, the stemonozone 1–2.6 mm, the tube 1.8–6.5 mm, filaments white or yellowish; ovary subsessile, at apex a trifle dilated, obliquely truncate, and puberulent, glabrous proximally at anthesis, but after fertilization becoming strigulose; style of most fls exserted, the stigma either poriform or a little dilated, to 0.2 mm diam; TERMINAL FL(S): calyx 2.5–10 × 2–5 mm; corolla 7–14(–23) mm; androecial tube 8–28 mm, dilated upward and 1.8–3 mm diam at separation of filaments, these thickened and spreading at base. *Pods* ascending 1–5(–7) per capitulum, sessile or contracted at base into a short pseudostipe, in profile broad-linear, straight or almost so (but often sinuously bent sideways), 6–11(–12.5) × 1.4–1.8 cm, essentially like that of *H. corymbosa* in other respects; seeds (few seen) 9–10.5 × 5–6 × 1.5–2 mm, the testa hard, lustrous brown, the pleurogram oblong-elliptic complete.

Hydrochorea marginata is very closely related to *H. corymbosa,* of which it may be a macrophyllidious derivative. The two species are alike in ecology, have precisely the same fruit, and, where sympatric, especially when past flowering, are reliably distinguished only by leaf-formula. While the range of *H. marginata* nests within that of its commoner relative, and the two species have been collected close together on the upper forks of the Amazon, there is no evidence of intergradation in these circumstances. Typical var. *marginata* has essentially the same flower as *H. corymbosa,* but vars. *panurensis* and *scheryi* differ in fewer, larger flowers with a characteristic 5-nerved calyx; the latter simulates *H. corymbosa* to the extent that random leaves have two pairs of pinnae.

Ducke considered *Pithecolobium panurense* a mere form of *P. marginatum,* but it is probable that he knew the former only from Bentham's description. Modern collections demonstrate its morphological differences and discrete dispersal. These are such that a case could be made for maintaining *P. panurense* at a coordinate level with *H. marginata,* but the multiple resemblances suggest that they are better assessed as geographic segregates of one species. *Pithecolobium bijugatum,* which has precisely the flower and fruit of *H. marginata* var. *panurensis,* differs only in somewhat shorter pedicels, random leaves with a second pair of pinnae, and a narrow dispersal along the Orinoco, perhaps in more specialized environment. It falls without strain into our enlarged concept of *H. marginata.*

Key to the varieties of *H. marginata*

1. Calyx of peripheral fls commonly puberulent, rarely glabrous, (1.5–)1.7–3 × 1.2–1.8 mm, bluntly 5-angulate, the 5 ribs simple, immersed; corolla of peripheral fls 4.5–7 mm; filaments of peripheral fls (10–)12–16; calyx of terminal fl(s) campanulate 2.5–4 × 2–3 mm; centr. Amazonian Brazil in states of Amazonas and W Pará..3a. var. *marginata*
1. Calyx of peripheral fls consistently glabrous, 4–7 × 2–3 mm, sharply 5-ribbed, the ribs finely prominulous, often 1–2-branched distally; corolla of peripheral fls 7–11(–12) mm; filaments of peripheral fls (14–)16–24; calyx of terminal fl(s) cylindro-campanulate 7–10 × 2–5 mm; middle and upper Negro-Vaupés basin in state of Amazonas, Brazil, adjacent SE Colombia, and thence N across the Negro-Orinoco watershed in T. F. Amazonas, Venezuela, in Brazil E just into the lower Branco basin in SW state of Roraima.
   2. Pedicel of lower peripheral fls 6.5–13 mm; pinnae consistently 1-jug. in all lvs; range as given above, but in Venezuela N only to the mouth of Río Atabapo in lat. 4°N...........3b. var. *panurensis*
   2. Pedicel of lower peripheral fls 4–5 mm; pinnae of random upper lvs 2-jug.; local on upper Río Orinoco between 5° and 6°N.............3c. var. *scheryi*

3a. Hydrochorea marginata (Bentham) Barneby & Grimes var. **marginata.** *Pithecolobium marginatum* Bentham, 1875, 1876, ll. cc., sens. str., exclus. *Spruce 3320* (quae = var. *panurensis*). — ". . . in silvis ad Santarem prov. Para: Martius; ad ripas Rio Negro prope Manáos: Spruce n. 1658; in fluminis Vasivae ripis [T. F. Amazonas, Venezuela]: Spruce n. 3320." — Lectotypus, *Spruce 1658,* labeled "Barra [do rio Negro], July 1851" (fl), K(herb. bentham.)! = NY

Neg. *2012*; isotypus, NY!; paratypus (*Martius*), not seen. — *Feuilleea dubia* O. Kuntze, Revis. Gen. Pl. **1:** 185. 1891. *Samanea marginata* (Bentham) Pittier, Bol. Ci. Técn. Mus. Com. Venez. **1:** 55. 1925, lacking explicit citation of protologue. *Arthrosamanea marginata* (Bentham) Pittier, 3rd Conf. Interam. Agric. Caracas 359. 1945.

Pithecolobium marginatum sensu Ducke, Notizbl. Bot. Gart. Berlin-Dahlem **11:** 583. 1932, exclus. syn. *panurense*; 1949: 37.

Pinnae of all lvs 1-jug.; *capitula* 20–38-fld; floral bracts less than 1.5 mm; pedicel of lower peripheral fls 6.5–12(–17.5) × 0.2–0.3 mm; calyx usually strigulose overall, sometimes externally glabrous, (1.5–) 1.7–3.4 × 1.2–1.8 mm; stamens (10–)12–16, the stemonozone 1–1.6 mm, the tube 1.8–3.7 mm; calyx of central fl(s) 2.5–4 × 2–3 mm, the corolla 7–10.5 mm, the androecial tube 8–14 × 1.8–3 mm.

In igapó woodland, on river banks and lake shores, below 250 m, always within the floodline, preferring clearwater and blackwater streams, scattered along the Amazon and its affluents S of the equator from the middle Río Negro, middle Japurá and lower Juruá rivers downstream to the mouth of the Tapajós, S to the middle Madeira basin, between 55° and 77°W longitude in the Brazilian states of Amazonas and W Pará; to be expected in adjacent Loreto, Peru. — Map 3. — Fl. IV–XI. — Wood hard, heavy.

The *Pithecolobium marginatum* reported by Bentham from Venezuela was based on a fruiting specimen of var. *panurensis.* Typical var. *marginata* is unknown from Venezuela and not to be expected there.

3b. Hydrochorea marginata var. **panurensis** (Bentham) Barneby & Grimes, comb. & stat. nov. *Pithecolobium panurense* Bentham, Trans. Linn. Soc. London **30:** 586. 1875 & *in* Martius, Fl. Bras. **15(2):** 440. 1876. — ". . . in silvis 'Gapó' ad flumen Uaupés prope Panuré, prov. do Alto Amazonas: Spruce n. 2425." — Holotypus, K(herb. bentham.)! = NY Neg. *2014*; isotypi, †B = F Neg. *1214*!, NY!, P!. — *Feuilleea panurensis* (Bentham) O. Kuntze, Revis. Gen. Pl. **1:** 188. 1891. *Arthrosamanea panurensis* (Bentham) Pittier, 3rd Conf. Interam. Agric. Caracas 359. 1945.

Pithecolobium marginatum sensu Ducke, 1949: 37, quoad syn.; auct. venez. recentior.; non Bentham, s. str.

Pinnae of all lvs 1-jug.; *capitula* 17–26-fld; bracts 2–3.5 mm; pedicel of lower peripheral fls a little stouter than in var. *marginata,* 6.5–13 × 0.25–0.4 mm; calyx glabrous overall, 4–7(–8) × 2–3(–3.7) mm, the teeth 0.8–2.6 mm; corolla 7–11(–12) mm, glabrous or puberulent at tip of lobes, these often unequal, 0.8–4.5 × 1.7–2.3 mm; stamens (14–)16–24, the stemonozone 1.4–2.5 mm, the tube 3.5–6.5 mm; calyx of central fl(s) 2.5–4 × 2–3 mm, the corolla 8–23 mm, the androecial tube (9)11–28 × 2–3 mm.

In seasonally flooded habitats of var. *marginata,* and like it mostly, perhaps exclusively, on blackwater, below 250 m, nearly confined to the Río Negro basin in T. F. Amazonas, Venezuela, and state of Amazonas, Brazil, downstream in typical form to Barcelós and intergradient to var. *marginata* as far as Río Cueiras NW of Manaus (not mapped), E just into S state of Roraima (Río Cheriuini) and W to the upper Vaupés and Apaporis in Vaupés, Colombia, in Venezuela not known from N of the mouth of Río Atabapo. — Map 3. — Fl. V–X.

One leaf of *Colella 2161* (NY) from Río Casiquiare has two pairs of pinnae suggesting passage into var. *scheryi.*

3c. Hydrochorea marginata var. **scheryi** Barneby & Grimes, nom. & stat. nov. *Pithecolobium bijugatum* Schery, Fieldiana, Bot. **28:** 258. 1952. — "[Venezuela.] . . . at Sanariapo, Territorio Federal Amazonas, alt. 124 m, 1942, *Llewellyn Williams 15953.*" — Holotypus, F!; isotypi, NY!, US *1834936*!, VEN *2237*!.

Shrubs and *treelets* 1.5–6 m; *lf-formula* i–ii/3–5, the pinnae in some specimens 1 pair in all lvs, in others 2 pairs in some or all distal lvs, the true petiole then very short and the one interpinnal segment of lf-stk 6–15 mm, the lfts as in var. *panurensis*; peripheral *fls* of var. *panurensis* except for shorter pedicels, the lower ones ±4–5 mm.

In laja-gallery ecotone, and sometimes on rocky stream banks, below 200 m, localized along the Orinoco river between Pto Ayacucho and Sanariapo, T. F. Amazonas, Venezuela, and to be expected across river in adjacent Vichada, Colombia. — Map 3. — Fl. V, VII–VIII, the full season not known.

Although Schery possessed poor material lacking expanded flowers and pods he correctly (even though tentatively) assigned *P. bijugatum* to *Samanea* ser. *Subarticulatae.* Subsequent collections show it to have flowers and pods indistinguishable from those of *H. marginata* var. *panurensis,* except insofar as the pedicel of the peripheral flowers is decidedly shorter. As the foliage is otherwise not different from that of *H. marginata* sens. lat., we interpret *P. bijugatum* as a variety of that species, apparently representing it to

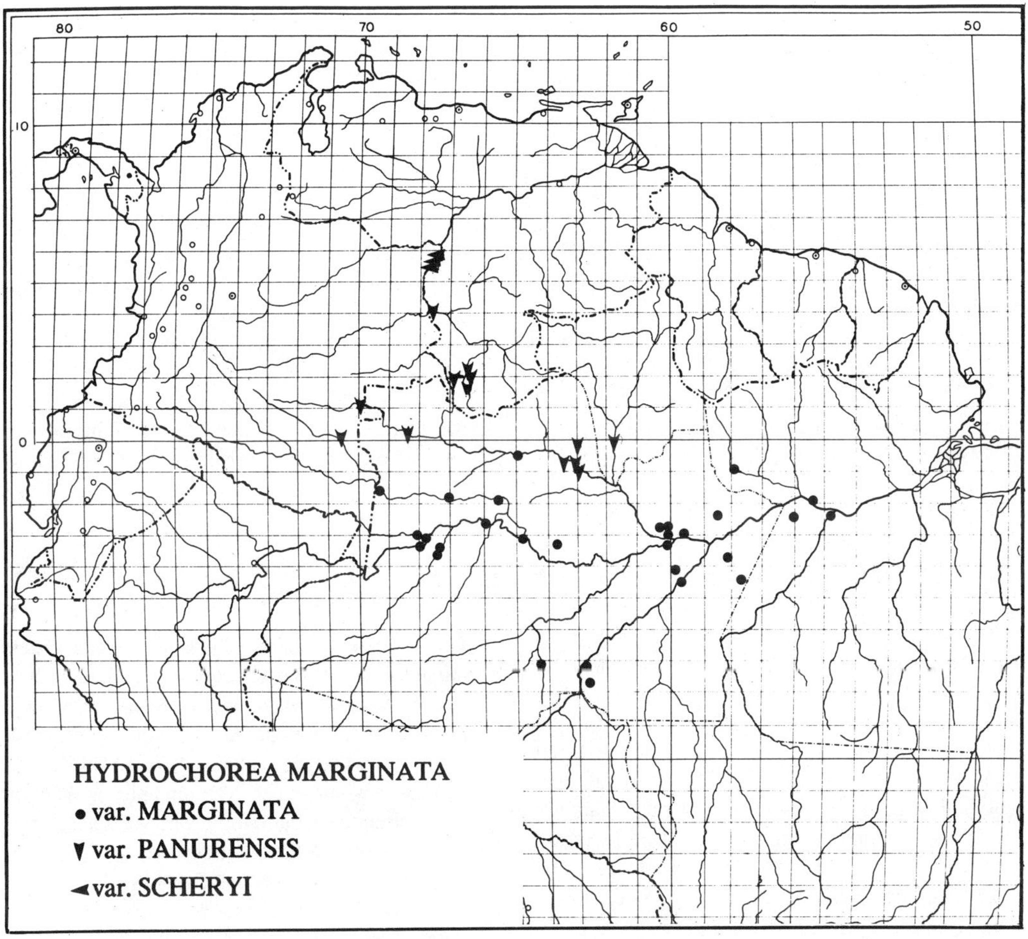

MAP 3. Upper Amazon and Orinoco basins. Range of *Hydrochorea marginata* (Bentham) Barneby & Grimes.

the exclusion of other forms on the upper Orinoco river between 5° and 6°N latitude. From scant information recorded by collectors it appears that var. *scheryi* is not found in flooded forest but in the woodland-laja interface, or on rocky riverbanks, perhaps above the floodline; its ecology needs observation.

Because the epithet *bijugatum* is misleadingly inappropriate to much of the material examined, we have exercised the privilege of altering the epithet along with the rank.

4. Hydrochorea (?) acreana (Macbride) Barneby & Grimes, comb. nov. *Pithecolobium acreanum* Macbride, Publ. Field Mus. Nat. Hist., Bot. Ser. **13** (Fl. Peru), pt. **3(1):** 51, quoad typum, pl. Klugiana excepta. 1943. — "[Brazil.] Rio [properly Terr. do] Acre: On terra firma, mouth of Río Macahuán [±9°20′S, 69°W], *Krukoff 5631*." — Holotypus, F *810951*! = F Neg. *55091*; isotypi, AA!, IAN!, MO!, NY!, U!, US!.

Macrophyllidious unarmed *tree* ±25–30 m, with smooth terete nonlenticellate hornotinous branches, the lf-axes and the terminal, efoliate, broadly corymbose-paniculate inflorescence of umbelliform racemes finely densely brown-puberulent with incurved hairs not over 0.15 mm, the foliage strongly bicolored, the chartaceous, finely venulose lfts when dry lustrous brown above, olivaceous beneath, glabrous facially except for very sparsely pubescent midrib, microscopically ciliolate. *Stipules* 0 (or perhaps very early caducous, but no scars observed). *Lf-formula* iv–v/4–6; lf-stks stout 10–21 cm, the petiole 4–6.5

cm, at middle 2–3 mm diam, the interpinnal segments 2–5.5 cm, the wide shallow ventral groove interrupted at each pinna-pair by a nectary, the one near or above midpetiole sessile cupular elliptic 2–3 mm diam, 0.8–1 mm tall, distal ones on lf-stk and between 1–several distal lft-pairs progressively smaller; pinnae accrescent distally, the rachis of penultimate or furthest pair up to 9–15 cm, the interfoliolar segments 1.4–3 cm; pulvinule of lfts in dorsal view 1.3–2 × 0.7–1 mm, strongly wrinkled, puberulent; lfts accrescent distally, obtusely rhombic-ovate from inequilateral, broadly rounded base, very shortly or obscurely acuminate, obtuse at very apex, the largest ±3.5–6.2 × 1.6–3.7 cm, 1.7–1.8 times as long as wide; venation pinnate, the nearly centric, forwardly arched midrib giving rise to ±10–17 pairs of major and various minor, straight, widely ascending secondary nerves brochidodrome well within the revolute margin, and these in turn to a reticulum of venules, the whole venation prominulous on both faces but more sharply so beneath. *Primary axis* of inflorescence 7–18 cm, that of the lowest branches ±5–10 cm, the upper progressively shorter, all widely ascending, the peduncles crowded distally, together forming a broad-topped panicle; bracts subtending inflorescence-branches 1–3 mm, caducous; peduncles fasciculate by 2–3, 20–33 mm; racemes ±25–30-fld, the axis including pedestal of terminal fl 3–4 mm, shallowly pitted; bracts minute caducous; fls either homo- or dimorphic, the peripheral ones pedicellate, the terminal one, when present, sessile, with larger calyx, the perianth of all 5-merous, minutely puberulent externally, brown when dry; PERIPHERAL FLS: pedicels 2.5–5 × 0.3–0.5 mm; calyx narrowly campanulate ±2 × 0.8–1.2 mm, the deltate-triangular obtuse teeth 0.4 mm; corolla narrowly trumpet-shaped ±4 mm, the ovate, weakly 1-nerved lobes 1–1.2 × 0.7–0.8 mm; androecium 10–12-merous, the stemonozone ±1 mm, the tube 1.7–2 mm, the longer stamens 13–15.5 mm; ovary short-stipitate (±0.5 mm), glabrous, truncate at apex; style about as long as stamens, the slightly dilated stigma 0.1 mm wide at apex; TERMINAL FL: calyx including solid base 2.8–4.5 × 1.2–1.5 mm, the ovate-triangular teeth 0.7–0.8 mm; corolla tubular 7 mm, the ovate lobes 13–14 × 0.8–0.9 mm; androecium 10-merous, ±12 mm long, the leathery tube ±6.5 mm, the filaments grossly thickened at point of separation, attenuate distally, apparently ascending (not sigmoid). *Pod* unknown.

In non-inundated Amazonian forest, known certainly only from the Brazilian state of Acre (9°18′–20′S, 69°–72°40′W). The paratype (*Klug 3079*) is here referred to *Abarema adenophora.* Suggestively similar is sterile material from Bolívar, Venezuela (*C. Blanco 751* from Reserva Forestal La Paragua (VEN) and *E. L. Little 17653* (VEN) from between El Palmar and Río Grande, 200 m [samán montañero acc. Blanco]), though slightly different in narrower ventral groove of lf-axes and lfts minutely silky-strigulose beneath. — *Ingá fava.*

Hydrochorea acreana is known with certainty from only two flowering collections, which differs from other species of *Hydrochorea* in the panicle of capitula efoliate at anthesis; pending discovery of the fruit its affinity cannot be assessed with finality. The small peripheral flower with androecium reduced to 10 or 12 members is like that of some forms of the variable *H. corymbosa,* but the foliage, except for smaller petiolar nectaries, is more reminiscent of *Abarema adenophora.*

II. BALIZIA BARNEBY & GRIMES

Balizia Barneby & Grimes, gen. nov., inter affinia floribus uniuscujusque capituli more *Albiziae* heteromorphis instructa his notulis praestans: rami indeterminati, ultra flores producti; foliola penninervia; capituli flores peripherici pedicellati, umbelliformes vel quasi corymbiformes; ovarium apice truncatum; legumen lignosum, tarde folliculare vel indehiscens, intus septatum sed haud lomentiforme. — Sp. typica: *Balizia pedicellaris* (de Candolle) Barneby & Grimes ≡ *Inga pedicellaris* de Candolle ≡ *Pithecolobium pedicellare* (de Candolle) Bentham. — Etymology: anagram of *Albizia.*

Pithecolobium sect. *Samanea* ser. *Carnosae* Bentham, 1875: 587, minima ex parte (sp. no. 64).

Unarmed *trees*, either macro- or microphyllidious, attaining 30–45 m but flowering at 10 m upward, the fertile branches indeterminate, the inflorescence composed of umbelliform capitula or corymbiform racemes arising singly or fasciculate in the axil of either coeval or hysteranthous lvs. Indumentum of young stems and lf-axes composed of short simple, usually sordid or brown hairs. *Stipules* linear-lanceolate or ligulate, externally nerveless, caducous (? sometimes obsolete). *Lf-formula* iii–xv/5–33; petiolar nectaries sessile or shortly stipitate, the first situated either near midpetiole or between pinnae of first pinna-pair; lft-pulvinules <2 mm; largest lfts 6–45 mm; venation of lfts pinnate, the midrib not or scarcely excentric. *Receptacle* of capitula, including terminal pedestal, 1.5–4.5 mm; bracts small caducous; fls dimorphic, the peripheral ones slenderly pedicellate, mostly bisexual, the longer pedicels 2.5–10 mm, the terminal fl larger than the rest and functionally sterile; perianth of peripheral fls 5-merous (random anomalies), the calyx (2–)2.3–5 mm, the corolla 5.5–10 mm; androecium of peripheral fls 12–26-merous, 17–36 mm, the filaments white, greenish white, or roseate; ovary truncate at apex; ovules 10–20. *Pods* of two types (further described in conspectus of species): (a) woody, plano-compressed, transversely fibrous, tardily follicular, and (b) ligneous, plumply compressed, indehiscent; seeds horizontal on subfiliform or ribbonlike, somewhat contorted funicle, narrowly oblong, the testa either putty-colored and rather discolored within the pleurogram or partly white and partly translucent, pleurogram narrowly U-shaped; the embryo green or anthocyanic (pink or orange). — Spp. 3, one widespread in the Orinoco and Amazon basins, the Guianas, and extending S in Atlantic forest of Brazil to Paraná, one bicentric in lower Amazonia and Central America, one localized in Gulf-Caribbean forest of Belize and adjoining republics.

Balizia and *Hydrochorea*, next above, are newly defined segregates from *Pithecellobium* sensu Bentham or from *Albizia* sensu Nielsen; because of their evident close kinship they can profitably be discussed together. In flower, the two genera closely resemble *Albizia*, but are distinguished from *Albizia*, as we understand it, by indeterminate inflorescence-axis and vegetative branches arising from sylleptic and proleptic buds (as opposed to proleptic buds only), and further from the native American species of *Albizia* (sect. *Arthrosamanea*) by pinnate leaflet-venation and truncate ovary. In gross aspect and almost all technical features, flowering specimens of *Balizia* and *Hydrochorea* are essentially indistinguishable from *Abarema*, or from each other for that matter, but their fruits are fundamentally different, so much so as to have prompted contradictory alignments within tribe Ingeae and repeated shifts of particular species from genus to genus. Thus *Balizia pedicellaris*, first described as an *Inga*, has been transferred successively to *Pithecellobium*, *Samanea*, and *Macrosamanea*; *B. leucocalyx*, described (because of its thick-walled indehiscent fruit) as a *Samanea*, has been removed to *Pithecellobium* and, inferentially, to *Albizia* sensu Nielsen; and *Hydrochorea corymbosa* has been classified in turn in the genera *Mimosa*, *Pithecellobium*, *Arthrosamanea*, and *Cathormion*, emphasis being laid always on the lomentiform fruit as the generic signature.

Balizia and *Hydrochorea* are conjectured to have arisen from common ancestry and to have diverged principally by evolution of different dispersal strategies and pari passu of distinctive pod morphologies. The ligneous fruits of genus *Balizia* are indehiscent or tardily dehiscent, their seeds being released sometimes only after decay of the valves, on the floor of terra firme forest, whereas the lomentiform fruits of riparian *Hydrochorea* are adapted to dispersal by water. Although fruits of *Hydrochorea* recall those of some species of *Albizia* sect. *Arthrosamanea* adapted to similar riparian habitats, the plants, as already stated, are significantly different in form of inflorescence, leaflet-venation, and shape of ovary.

Conspectus of genus *Balizia*

¶ Macrophyllidious, the lf-formula iii–vi/5–8, the longer lfts 24–45 mm; pod gently decurved, the sutures immersed, the body compressed but biconvex; dehiscence 0; terminal fl with intrastaminal disc; 1 species of Belize and contiguous states; sect. *Leucosamanea*.
1. *B. leucocalyx* (Britton & Rose) Barneby & Grimes

¶¶ Microphyllidious, the lf-formula vi–xv/16–33 and the longer lfts 6–14(–16) mm; pod straight, the sutures dilated, forming a raised rim around the nearly plane or low-umbonate valves; dehiscence follicular; terminal fl of capitula lacking intrastaminal disc; 2 species of equatorial and E. South America, one of them also disjunct in Nicaragua and Costa Rica; sect. *Balizia*.
2. *B. pedicellaris* (de Candolle) Barneby & Grimes
3. *B. elegans* (Ducke) Barneby & Grimes

Key to the species of genus *Balizia*

1. Lf-formula iii–vi/5–8, and largest lfts 24–45 mm; pod gently decruved, compressed but biconvex, not framed by the sutures, indehiscent, the epidermis of valves not cracking; Belize and countries immediately contiguous to W and S; sect. *Leucosamanea*. . . . . 1. *B. leucocalyx* (p. 36)
1. Lf-formula vi–xv/16–33, and largest lfts 6–14(–16) mm; pod straight, plano-compressed, framed by dilated sutures, tardily dehiscent, the epidermis of valves cracking transversely to reveal strong horizontal fibers of mesocarp; South Amerca, NW into Nicaragua and Costa Rica; sect. *Balizia*.
   2. Calyx 2–3 mm and corolla 5.3–7.4 mm; transverse fibers in mesocarp of pod-valves contiguous or almost so; widespread in tropical South America . . . . 2. *B. pedicellaris* (p. 37)
   2. Calyx 4–5 mm and corolla 8–10 mm; transverse fibers in mesocarp separated by intervals wider than their diam; of bicentric dispersal in lower Amazonian Brazil and SE Central America (Nicaragua, Costa Rica). . . . . 3. *B. elegans* (p. 40)

IIA. BALIZIA SECT. LEUCOSAMANEA BARNEBY & GRIMES

Balizia sect. **Leucosamanea** Barneby & Grimes, sect. nov. monotypica, a sect. *Balizia* legumine leviter recurvo indehiscenti, suturis haud prominentibus, florisque terminalis uniuscujusque capituli disco intrastamineo ovarii basin cingenti diversa. — Sp. unica: *B. leucocalyx* (Britton & Rose) Barneby & Grimes ≡ *Samanea leucocalyx* Britton & Rose ≡ *Pithecolobium leucocalyx* (Britton & Rose) Standley. — Etymology: Gr. *leucos*, white + Lat. *Samanea*, a generic name in Ingeae.

Macrophyllidious, the *lf-formula* iii–iv/5–8; *pods* oblong, slightly decurved, compressed but plump, indehiscent, the sutures immersed in woody valves; ovary of terminal fl of each capitulum surrounded at base by a nectarial disc. — Sp. 1, of SE Mexico, Belize, adjacent Guatemala, and Honduras.

1. **Balizia leucocalyx** (Britton & Rose) Barneby & Grimes, comb. nov. *Samanea leucocalyx* Britton & Rose, N. Amer. Fl. **23:** 34. 1928. — "El Limon, Tabasco [Mexico], July 12, 1891, *J. N. Rovirosa 976*." — Holotypus, US *1319837*!" clastotypus (fragm. + photo), NY! — *Pithecellobium leucocalyx* (Britton & Rose) Standley, Publ. Field Mus. Nat. Hist., Bot. Ser. **4:** 308. 1929. *Albizia leucocalyx* (Britton & Rose) L. Rico, J. Linn. Soc., Bot. **108:** 272. 1992.

Pithecellobium leucocalyx sensu Pennington & Sarukhán, 1968: 186 + fig.

Macrophyllidious *trees* 12–33 m with smooth or scaly gray trunk 4–12 dm dbh and rounded crown, the annotinous branches lenticellate, the young ones together with lf- and inflorescence-axes at first densely sordid-pilosulous with forwardly incurved hairs to 0.1–0.15 mm, the lvs bicolored, the papery lfts lustrous dark green (brunnescent) above, dull pale olivaceous beneath, either glabrous facially except for minutely ciliolate midrib or minutely remotely strigulose beneath, the margin ciliolate, the inflorescence a terminal efoliate pseudoraceme or few-branched panicle of long-pedunculate, rather large and strikingly white-silky-pilosulous capitula, the thick-walled indehiscent fruits commonly persistent for a year or more on annotinous branchlets. *Stipules* linear 2–5.5 mm, deciduous long before maturity of associated lf, absent from fruiting spms. *Lf-formula* iii-v(–vi)/5–8; lf-stks 5.5–13 cm, the petiole including firm pulvinus (1.3)2–4 cm, at middle 1.2–2 mm diam, the longer interpinnal segments 15–32 mm; nectary at or near midpetiole (exceptionally 0 or 2) subsessile but borne on a wide stout pediment 0.3–0.5 mm tall in profile, the shallowly concave or almost flat but crumpled head (0.6–)1–1.6 mm diam in vertical view, similar but smaller nectaries between each pair of pinnae and yet smaller ones between furthest 1–2 pairs of lfts; pinnae accrescent distally, the rachis of furthest pair (5–)6–11 cm, the longer interfoliolar segments 10–21 mm; lft-pulvinules in dorsal view 0.8–1.8 × 0.8–1.1 mm, deeply wrinkled; lfts strongly accrescent upward but otherwise (except for more broadly rhombic-obovate furthest pair) subequiform, in outline bluntly rhombic-oblong around a diagonal, gently incurved or almost straight midrib, inequilaterally flabellate at base, shallowly emarginate at apex, the penultimate pair 24–45 × 10–21 mm, (1.5–)1.9–2.5 times as long as wide; venation essentially pinnate, the 1–2 posterior primary nerves (if present) shorter and weaker than the first secondary ones, the major secondary nerves 6–10 from each side of midrib, widely ascending to anastomosis well within the plane, finely corneous margin and giving rise to a random tertiary venulation, the whole venation finely sharply prominulous on both faces of blade and commonly pallid on upper one. *Primary axis* of pseudoraceme 2–8 cm; peduncles solitary or 2–3 per node, the lower ones 4–6 cm, the distal ones somewhat shorter; capitula umbelliform 9–17-fld, the receptacle including pedestal of terminal fl 1.5–2.5 mm, the fls dimorphic, the peripheral ones pedicellate, the pedicel of lowest ones 2.5–6 × 0.4–0.5 mm (or upper somewhat shorter), the terminal fl sessile and with modified androecium, the perianth of all 5-merous, densely gray-silky-pilosulous externally, glabrous within; bracts narrowly oblanceolate 1.4–4 mm, caducous as the young pyriform fl-buds begin to swell, the mature perianth greenish beneath the indumentum, the filaments white drying pinkish brown; PERIPHERAL FLS: calyx turbinate-campanulate 3.8–4.5 × 2.4–3.5 mm, the deltate teeth (often unequal) 1–1.4 mm; corolla 7–9.3 mm, the ovate lobes 2–3.5 × 1.8–2.5 mm; androecium 28–36-merous, the longest stamens 25–36 mm, the stemonozone 1–1.8 mm, the tube 2–4.6 mm, disc 0; ovary subtruncate, distally puberulent; style a little longer than longest filaments, the narrowly dilated stigma 0.15–0.2 mm diam; TERMINAL FL: calyx squatly campanulate ± 4.5 mm; corolla 7.5–11 mm; filament-tube 8–10.5 mm, flaring at or shortly beyond mouth of corolla; ovary surrounded at base by a disc 0.4 mm tall. *Pods* solitary, sessile, in profile narrowly oblong, gently arched downward, when well fertilized (6.5–)7–10 × 2.1–2.7 cm and 10–14-seeded, rounded or broadly cuneate at base, rounded and abruptly apiculate at apex, strongly compressed but ±4–5 mm thick in section, not or scarcely elevated over seeds, girdled by obscurely 3-angulate, evenly arched sutures 2.5–3.5 mm wide, the rigid, early glabrate valves consisting when ripe of: (1) thin black epidermis cracking both transversely and lengthwise and randomly flaking to reveal the transversely fibrous exocarp, (2) a thin fibrous mesocarp, and (3) woody endocarp produced between seeds into complete septa; the individual seed-chambers ±4.5–6 mm long and as wide as the pod itself; dehiscence 0, the method of seed dispersal unknown; funicle ribbonlike, folded into a triple-S; seeds basally attached, horizontal to the pod's long axis, in broad view narrowly elliptic, 13–16 × 4–5 mm, only a little compressed, the testa in lower ⅓–⅔ membranous translucent and only loosely investing the embryo but distally white bony (±0.4 mm thick) and adherent to the thick horny cotyledons, these sagittate at base and fully concealing the radicle, the cotyledons either pale orange or magenta-pink, the narrowly U-shaped pleurogram ±6 × 1 mm.

In wet tropical forest, often on river banks, sometimes surviving disturbance as a pasture shade-tree, below 400 m, local on the Gulf-Caribbean lowlands in far SE Mexico (Tabasco and N Chiapas), E Guatemala (Petén, Izabal), Belize (Stann Creek, El Cayo, and Toledo), and NW Honduras (Atlántida). — Map 4. — Fl. V–VII. — *Caracolillo* (Tabasco); *cenicero* (Petén, elsewhere applied to *Samanea saman*); *guacibán, guacibe* (Chiapas); *wild tamarind* (Belize, also applied to *Abarema idiopoda*).

The cladogram suggests that *B. leucocalyx* is the most primitive of the three species of the genus; the other two species share a fruit in which the valves become convex at dehiscence. *Balizia leucocalyx*, on the other hand, retains an indehiscent fruit, as does *Hydrochorea*, though the form of the fruits are decidedly different.

IIB. BALIZIA BARNEBY & GRIMES SECT. BALIZIA

Balizia Barneby & Grimes sect. **Balizia.** *Balizia* Barneby & Grimes, ut supra descripta, sens. str. — Typus sub genere indicatur.

Pithecolobium sect. *Samanea* ser. *Carnosae* Bentham, 1875: 587, minima ex parte (sp. no. 64).

Macrosamanea sensu Kleinhoonte, 1940: 329, minore ex parte (*M. pedicellaris*); non Britton & Rose.

Microphyllidious, the *lf-formula* vi–xv/16–33; *pods* broad-linear or oblong, straight, with dilated sutures framing plane or almost plane, woody valves, these traversed horizontally by crowded tough mesocarpic fibers; dehiscence tardy, follicular, through the seminiferous suture, the valves folding out flat, like covers of a book, hinged on the abaxial suture. — Spp. 2, one widely dispersed in South America, one local in lower Amazonian Brazil and disjunct in Central America.

2. Balizia pedicellaris (de Candolle) Barneby & Grimes, comb. nov. *Inga pedicellaris* de Candolle, Prodr. **2:** 441. 1825. — "... in Cayenna.," the collector not given. — Holotypus, G-DC! = F Neg. *6972.* — *Pithecolobium pedicellare* (de Candolle) Bentham *in* Hooker, London J. Bot. **3:** 219. 1844. *Feuilleea pedicellaris* (de Candolle) O. Kuntze, Revis. Gen. Pl. **1:** 188. 1891. *Samanea pedicellaris* (de Candolle) Killip *ex* Record, Trop. Woods **63:** 4. 1940 (Sept). *Macrosamanea* (?) *pedicellaris* (de Candolle) Kleinhoonte *in* Pulle, Fl. Suriname **2(2):** 329. 1940 (Dec).

Mimosa terminalis Vellozo., Fl. Flumin **11:** t. 30. 1829 & Arq. Mus. Nac. Rio de Janeiro **5:** 437. 1881. — "Habitat silvis maritimis." — Holotypus, the cited plate (the terminal pinna, however, anomalous)!.

Pithecolobium pedicellare sensu Bentham, 1875: 588; 1876: 442; Ducke, 1949: 39; Lindeman & Mennega, Bomenboek Surinam 191. 1963.

Macrosamanea pedicellaris sensu Nielsen, 1981: 181, fig. 1(4); Lewis, Leg. Bahia 165, fig. 11 (O–R), pl. 9(F). 1987.

Microphyllidious *trees* attaining 40 m in height with trunk to 1 m diam dbh but often fertile as slender treelets as little as 3 m tall, the foliage crowded toward the end of hornotinous branchlets, these like all axes of lvs and inflorescence sordid-pilosulous or -strigulose with ascending-incurved or appressed hairs to 0.1–0.3 mm, the foliage bicolored, the firm, marginally revolute lfts on upper face convex but depressed along midrib, rich green, glabrous and often lustrous, beneath pale brown dull, thinly pilosulous and densely pallid-papillate (the papillae forming a rosette at base of each hair), the umbelliform racemes of small reddish, white-stamened fls single or 2–3 together in the axil of coeval lvs and immersed in foliage, the apical meristem of each branch indeterminate. *Stipules* caducous from the expanding lf and therefore present only about the young shoot-apex, linear-lanceolate or ligulate, (2–)2.5–8 × (0.3–)0.5–1.3 mm, puberulent like the stem, externally nerveless. *Lf-formula* vi–x(–xiv)/(–16)18–29(–33), locally in upland Bahia xi–xvii/22–31; lf-stk

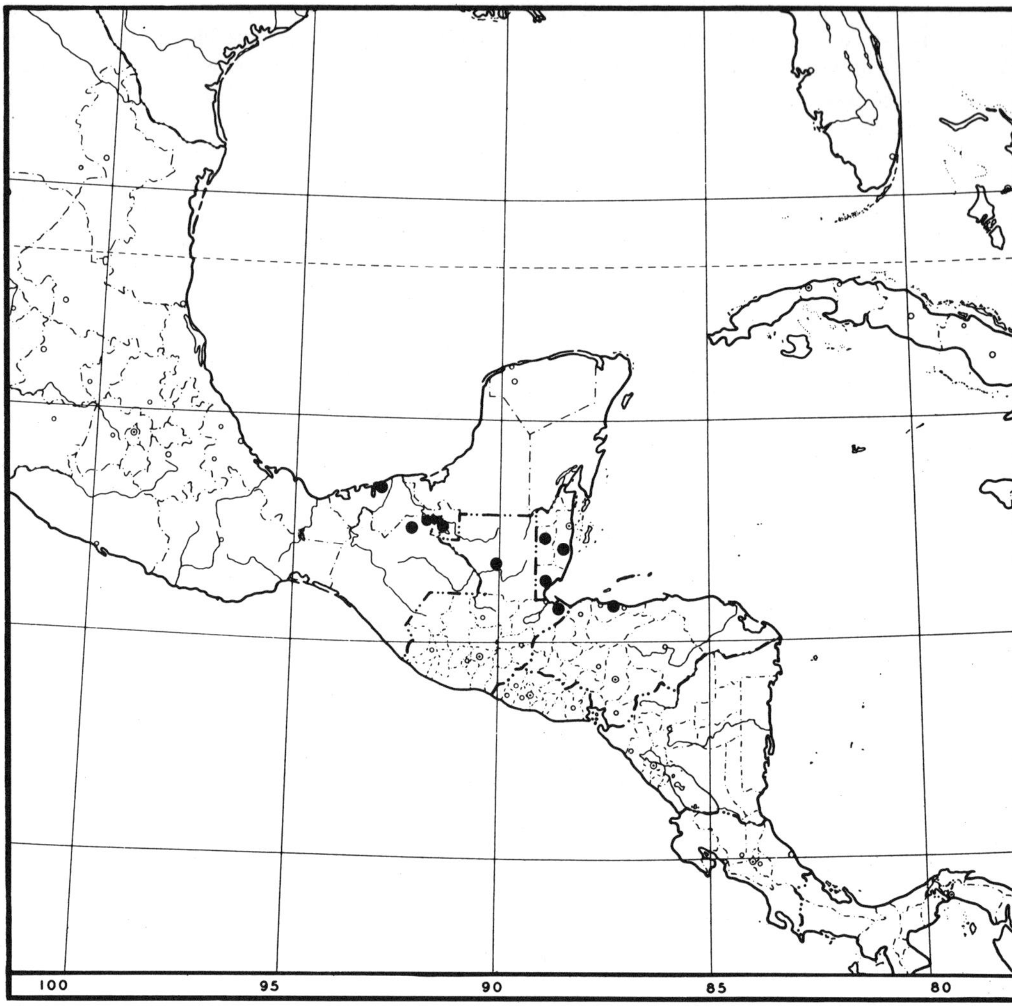

MAP 4. Belize, Honduras, Nicaragua, and southern Mexico. Range of *Balizia leucocalyx* (Britton & Rose) Barneby & Grimes.

of major lvs 7–18(–25) cm, the petiole 2–4.2(–5) cm, the longer interpinnal segments (7–)9–17(–21) mm, the ventral sulcus continuous except where interrupted by a nectary; petiolar nectaries variable in number, position, and form: almost always one near midpetiole, often a second between first and others between 1–2 furthest pairs of pinnae, exceptionally one between first pair only (inland Bahia), the proximal ones either sessile or very shortly stipitate and pileiform, convex or less often plane, in profile 0.5–1.2(–1.7) mm tall and 1.3–2.7 mm diam, and nearly always small goblet-shaped nectaries on pinnae at furthest 1–8 pairs of lfts; pinnae proximally decrescent, beyond midrachis subequilong, the rachis of distal ones (4–)5–9(–10) cm, the longer interfoliolar segments 2–4 mm; pulvinules 0.3–0.7 mm; lfts strongly decrescent proximally, less so distally, the first pair often represented by minute paraphyllidia, the blades narrowly oblong, narrowly lance-oblong, or oblong-elliptic from inequilateral (postically obtuse, antically angulate) base, broadly obtuse, those near midrachis 6–13.5(–16) × 1.8–3.3(–4) mm, (3–)3.2–4.2(–5) times as long as wide; venation of subcentric, dorsally cariniform midrib, this commonly simple at maturity of blade but when young, rarely later, pinnately branched, the secondary venules ±7–

12 on each side, ascending at wide angles and expiring just short of revolute margin. *Peduncles* (2.5–)3–6.5(–8) cm, becoming stout and persistent into a second year; capitula ±(15–)20–40-fld, the receptacle 2–4.5 mm, the 5-, randomly 6-merous fls dimorphic, the peripheral ones slenderly pedicellate, one or commonly 2–3 distal ones (sub)sessile, longer, and with exserted androecial tube; bracts minute, caducous long before anthesis; perianth pubescent externally, the calyx sordid-puberulent overall, the corolla silky dorsally, the lobes white-ciliolate; PERIPHERAL FLS: lower pedicels (4–)4.5–7.5 mm; calyx campanulate or turbinate-campanulate (2–)2.3–3 × 1.3–1.8 (–2) mm, the deltate-triangular teeth 0.3–0.8(–1) mm; corolla 5.3–7.4 mm, the lobes (1.3–)1.5–2.3(–2.5) mm; androecium 12–22(–28)-merous, 17–26(–36) mm long, the stemonozone ±1 mm, the tube (1.5–) 2.3–3.7 mm; ovary truncate, puberulent at and near apex, proximally glabrous or almost so; TERMINAL FL(S): calyx plumply campanulate from turbinate base (3–)3.4–4.6 × 2–2.8 mm; corolla (6.5)7–11 mm; androecial tube 7–12(–16) mm, as long as corolla or up to 2 mm longer. *Pods* 1 3 per capitulum, stiffly spreading-ascending from receptacle, sessile at oblique base, in profile oblong or broad-linear, straight or almost so, abruptly contracted at apex into a linear-subulate (deciduous) beak to 1 cm, the whole 7–12.5(–14) × (1.7–)1.8–3.2 cm, (12–)14–20-ovulate, the lignescent, never pulpy (but sometimes resinous) valves at first plane and dark reddish brown, becoming blackish brown and either low-convex or prominently umbonate over each seed, framed by almost straight, plane woody sutures 2.5–4 mm wide, the ripe valves consisting of: (a) thin blackish exocarp breaking into small tetragonal tesserae; (b) a midlayer of coarse transverse, parallel, and subcontiguous woody fibers; and (c) a crustaceous endocarp coherent between seeds but not septiferous; dehiscence follicular, through the seminiferous suture, the valves gaping to release the seeds but the pod often deciduous before this occurs, the valves also splitting transversely between seeds into linear-oblong panels 4–7(–8) mm wide, but these remaining permanently attached to the sutures; seeds transverse near middle of pod on slender tapelike sigmoid funicle, exactly basifixed, in broad outline narrowly oblong or oblong-elliptic (6–)7.5–9 × 3–4 mm, compressed but plump, either putty-colored overall or colored brown or green within the narrowly U-shaped pleurogram, the hard testa ±0.2–0.3 mm thick.

In non-inundated primary rain forest, mostly below 200 m but ascending on the sources of the Amazon between Ecuador and Bolivia to ±700–775 m, and southeasterly in lowland Atlantic forest and (locally in Bahia) in gallery forest near 800 m, discontinuously widespread in South America: relatively frequent from E Venezuela (Sucre, Bolívar, Delta-Amacuro) SE through the Guianas and lower Amazonian Brazil to Maranhão, thence sporadically W through the Amazon basin to SE Ecuador, SE Colombia (Vaupés), centr. Peru (Huánuco), and Bolivia (Nor-Yungas, Cochabamba); disjunct along the SE Brazilian coast between 13° and 24°30′S in Bahia, Rio de Janeiro, and São Paulo; represented in upland Bahia on the headwaters of Río Paraguaçú (near 12°30′S) by an aberrant form (see discussion following). — Map 5. — Fl. in equatorial latitudes ill-defined, apparently at intervals throughout the year, in SE Brazil XI–II. — *Samiknum, zamitmi* (Ecuador); *vilco colorado* (Peru); *arraigan, bayan* (Bolivia); *hueso de pescado amarillo, hueso de pescado colorado, kairemap-yek, samán montanero* (Venezuela); (red) *manaliballi, manariballi* (Guyana, Surinam); *aprobigi, tamarenprokoni, tamarinde plokonie* (Surinam); *assao, bois la morue, bois macaque* (French Guiana); *kalaipe* (Wayampi of French Guiana); *esponja, esponjeira, faveira* (Amazonian Brazil); *arapucumira* (Ka'apor of Maranhão); *juruena branca* (Bahia).

Populations of *B. pedicellaris* on the eastern slope of Chapada Diamantina in Bahia were mentioned by Lewis (1987: 165) as distinguished from the typical form of lowland Atlantic forest by more numerous petiolar nectaries and different indumentum of leaflets. These plants are indeed aberrant in position of nectaries and in very short true petiole of the leaves. It is as though a supernumerary pair of short pinnae were inserted at the lowest nectary, which elsewhere is situated well below the first pinna-pair. In context of the whole species, the indumentum is not really characteristic.

The specimens from prae-Andean Amazonia are indistinguishable morphologically from those of lower Amazonia and the Guianas.

We have no solution to the mystery surrounding *Mimosa chrysantha* Vahl (Eclog. Amer. **2:** 38. 1798), which Bentham (1875: 588) attributed with reservations to *Pithecolobium pedicellare*. The holotypus, collected at Cayenne by von Rohr, was seen by Bentham in Vahl's herbarium, but can no longer be located there (Mrs. Fox Maule, in litt.!). The leaf-formula given in the protologue is compatible with *B. pedicellaris*, but the abscence of petiolar nectary and yellow color of the corolla are not. De Candolle (Prodr. **2:** 471), who knew it only from description,

Map 5. South America. Range of *Balizia pedicellaris* (de Candolle) Barneby & Grimes.

hesitantly referred *M. chrysantha* to *Acacia*. The only bipinnate species of Ingeae with yellow flowers likely to have been found in French Guiana is *Zygia* (*Marmaroxylon*) *racemosa*, but this has nectaries that could hardly have been overlooked.

3. Balizia elegans (Ducke) Barneby & Grimes, comb. nov. *Pithecolobium elegans* Ducke, Arch. Jard. Bot. Rio de Janeiro **3:** 64. 1922. — "... in silvis non inundatis, l. A. Ducke prope Alcobaca (Tocantins) 17-7-1916 n. 16.271 (florif. et leguminibus junioribus), ad flumen Tapajoz prope Cachoeira do Mangabal 7-9-1916, n. 16.470 (fructif. et floribus siccis)." — Lectotypus, *Ducke* in MG *16271*, MG (2 sheets)!; isotypi, BM!, G!, K!, P!, US *1040853*!; paratypi, *Ducke 16470*, MG!, RB!.

Pithecolobium elegans sensu Ducke, 1949: 38.

Trees with smooth rusty-brown, yellowish, or pale orange trunk, attaining 30–45 m but flowering when 8 m or less, closely akin to and resembling *B. pedicellaris* in habit, indumentum, stipules, foliage, and lf-nectaries, but with larger fls and fruit different in venation of the valves. *Lf-formula* x–xv/20–31; lf-stks 8–18 cm, the petiole ±2.5–4.5 cm, the longer interpinnal segments ±8–14 mm; first petiolar nectary either near or well above midpetiole; rachis of longer pinnae 5–10 cm, the longer interfoliolar segments 1.7–3 mm;

longer lfts 6–9.5 × 1.7–3 mm, 2.4–3.4 times as long as wide, when dried dull on upper face and sometimes thinly puberulent along midrib, or thinly so overall, beneath less obviously papillate than those of *B. pedicellaris*. PERIPHERAL FLS: pedicels 6–10 (ex char.: –20) mm; calyx 4–5 × 2–3.5 mm, the teeth ±1–1.5 mm; corolla 8–10 mm, the lobes 2.5–3.5 × 1.6–2 mm; androecium 25–35-merous, ±3–3.5 cm, the stemonozone 1.3–2 mm, the tube 3–5.5 mm, filaments green or white distally; ovary truncate, densely minutely golden-strigulose overall; style long-exserted. *Pods* (not seen fully ripe) almost of *B. pedicellaris*, 9–15 × 2.4–3 cm, 12–23-ovulate, the glabrous valves coarsely transverse-venulose, the major fibers separated from each other by intervals much greater than their own diam, freely branched; seeds 7 × 4 mm in broad view, oblong, vivid green within, whitish or pink without, the pleurogram almost complete.

In primary rain forest on terra firme, below 350 m, discontinuously scattered between the equator and 11°S through the basins of ríos Madeira, Tapajoz, Jarí, and Tocantins in Hylaean Brazil, SW on the upper Río Acre just into Bolivia (Pando); distantly disjunct in Costa Rica (Limón) and Nicaragua. — Map 6. — Fl. V–VII(–?). — *Toco blanco* (Pando), *mapuchiquí, mapuxiqui vermelho* (Amazonas).

III. ABAREMA Pittier

Abarema Pittier, Arb. Legum. **1:** 56. 1927, based on *Pithecolobium* sect. *Abaremotemon* Bentham *in* Hooker, London J. Bot. **3:** 203. 1844. — Sp. lectotypica (Britton & Killip, Ann. New York Acad. Sci. **35:** 126. 1936): *P. auaremotemo* Martius = *Abarema cochliacarpos* (Gomes) Grimes & Barneby. — The subsequently proposed lectotypus (Cowan, Taxon **8:** 58. 1959) is rejected. Cowan believed the type of *Abarema* should be among the species listed by Pittier and chose *A. trapezifolia* as the better known. The genus was based, however, on sect. *Abaremotemon* Bentham, of which *P. auaremotemo* is the unavoidable lectotypus.

Pithecolobium sect. *Samanea* ser. *Axillares* Bentham *in* Hooker, London J. Bot. **3:** 216, ex parte (spp. no. 55–57 omissis). 1844. — Sp. lectotypica, one of few in which Bentham knew the dehiscent pod: *P. rhombeum* Bentham = *Abarema brachystachya* (de Candolle) Barneby & Grimes.

Pithecolobium sect. *Gyrolobium* Grisebach, Fl. Brit. W. I. 226. 1860, ex parte, exclus. *P. filicifolio*. — Sp. lectotypica: *P. micradenium* Bentham = *Abarema jupunba* (Willdenow) Britton & Killip var. *jupunba*.

Jupunba Britton & Rose, N. Amer. Fl. **23:** 25. 1928. — Sp. typica: *J. jupunba* (Willdenow) Britton & Rose ≡ *Acacia jupunba* Willdenow ≡ *Pithecellobium jupunba* (Willdenow) Urban ≡ *Abarema jupunba* (Willdenow) Barneby & Grimes.

Punjuba Britton & Rose, N. Amer. Fl. **23:** 28. 1928. — Sp. typica: *P. racemiflora* (Donnell Smith) Britton & Rose ≡ *Pithecellobium racemiflorum* Donnell Smith ≡ *Abarema racemiflora* (Donnell Smith) Barneby & Grimes.

Klugiodendron Britton & Killip, Ann. New York Acad. Sci. **35:** 125, 1936. — Sp. typica: *K. laetum* (Poeppig & Endlicher) Britton & Killip ≡ *Abarema laeta* (Poeppig & Endlicher) Barneby & Grimes.

Pithecolobium sect. *Abaremotemo* [*sic*] sensu Bentham, 1875: 581; 1876: 43 ("*Abaremotemon*").

Pithecolobium sect. *Clypearia* sensu Mohlenbrock, 1963: 446, ex parte, typo exclus. 1963; *Punjuba* sensu Nielsen, 1981: 190.

Unarmed *trees*, some flowering first as bushy treelets but few remaining shrubby when adult. Growth monopodial (not confirmed in all species), the shoots monomorphic; phyllotaxy spiral. *Indumentum* of simple hairs, usually brown, tawny, or bronze, dense on young branchlets and on axillary buds, often thin or deciduous in age, the lfts often glabrous on upper face, but the inflorescence always at least thinly puberulent. *Stipules* commonly small, narrow, caducous long before maturity of the associated lf, sometimes 0; no imbricate bud-scales. *Lvs* bipinnate, the pairs of pinnae and lfts one to many, the lft size and number reciprocally adjusted; one nectary on lf-stks almost always between or close below proximal pair of pinna-pulvinules, often one smaller one between further pairs, and nearly always between furthest pairs of lfts, none elsewhere on plant, the nectaries campanulate to cupular, patelliform, or verruciform, only rarely stipitate, sometimes immersed; venation of lfts primarily pinnate, the secondary veins always brochidodrome within the margin and often giving rise to a tertiary or quaternary reticulum. *Inflorescence* of spikes, racemes, or capitula, these axillary to coeval or rarely to hysteranthous or rudimentary lvs, the whole inflorescence when efoliate a pseudoraceme or short panicle (no cauliflory); fls bracteate in vernation, ebracteolate, the bracts nearly always caducous; the fls of any unit of inflorescence either homomorphic or dimorphic, the 1–few distal fls sessile and longer, or no longer but stouter, than the peripheral ones, and their androecium often thicker, with more numerous filaments united into a longer tube, this dimorphic fl expanding either before or with the rest; perianth with random exceptions 5-merous, the calyx campanulate or turbinate-campanulate, the often unequal

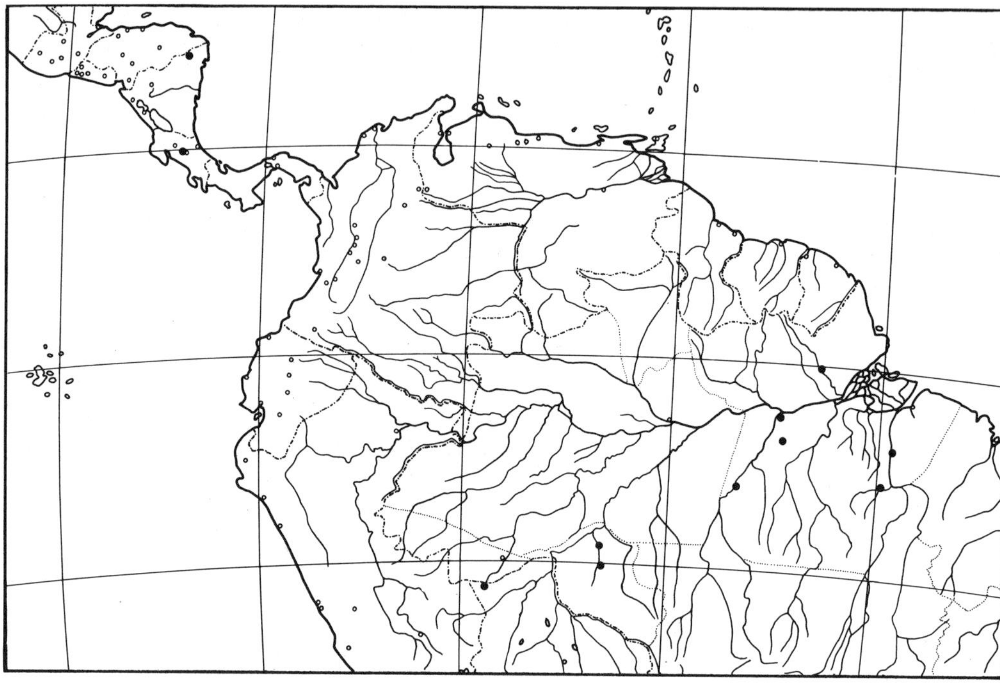

MAP 6. Central and South America. Range of *Balizia elegans* (Ducke) Barneby & Grimes.

teeth broader than long, the corolla narrowly funnelform, its lobes ascending; androecium 10–60-merous, the filaments united into a tube either shorter or longer than the corolla and the tube itself adherent to the corolla to form a short stemonozone; ovary either fusiform or truncate distally; style usually exserted from among longer stamens, not or scarcely dilated at apex; *ovules* (6–)8–14(–16). *Pods* sessile but sometimes backwardly attenuate at base, long-persistent after dehiscence, in profile broad-linear, falcately or spirally recurved (or equivalently contorted), the coarse sutures framing valves that are at first plane but in varying degree thickened at maturity and elevated over each seed, the ventral suture commonly undulate or in some species deeply recessed between seeds, the valves themselves varying from stiffly papery to leathery or ligneous, the always-thin exocarp either smooth or papillate and often coarsely venulose, the endocarp, as revealed on dehiscence, most often red, crimson, orange, or castaneous (especially in the cup-shaped seed-cavities) but often pale tan between seeds; dehiscence elastic, downward through both sutures but more readily and sometimes exclusively through the seminiferous (convex) outer one, the valves both vertically recurved and horizontally twisted but the torsion inhibited in some pods of extremely thick texture; seeds transverse or obliquely basipetal, following dehiscence suspended on a strap-shaped funicle sigmoidally bent distally, plumply lentiform or somewhat compressed, the testa thin, sometimes partly white opaque but otherwise translucent, colorless or becoming yellowish brownish in age, in texture resembling a grape-skin, girdled by a simple nerve, either closely or loosely investing the embryo, in some species charged on each face with finely engraved, elliptic or almost round pleurogram, but in others not so; endosperm 0; cotyledons horny, plano-convex, closely appressed by their flat side, convex externally and there either aniline-blue or fuscous, their color transmitted through the translucent parts of the seed-coat; plumule concealed by the cotyledons. Seedlings seen from few species, producing an onion- or garliclike odor upon germination, cotyledons epigeal, sessile, auriculate at base; eophylls opposite, bipinnate. — Spp. 44, endemic to the New World tropics, but one extending N to warm-temperate Bahamas and two S along the coast of E Brazil to Paraná and Sta Catarina, most numerous in the Amazon basin and on the Guayana Highland, but secondarily developed in the Greater Antilles, Central America, and Atlantic Brazil.

The genus *Abarema*, first formulated by Bentham as a section of *Pithecolobium* in 1844, has now grown so large that it can no longer be easily defined in exact terms. Its constant features are: (a) unarmed stems; (b) inflorescence composed of axillary peduncles, either solitary or fasciculate but never assembled into efoliate pseudoracemes, and borne on an auxotelic primary axis; (c) pinnately veined leaflets; and (d) a fruit that in dehiscence exposes seeds with lustrous, white or variably colored but never black seed-coat against a background of red or red-brown endocarp. Ornithochary is the established method of seed dispersal. The embryo is nearly always aniline-blue due to the presence of delphinidin (B. Meurer-Grimes, unpubl. data), and this may be synapomorphic. In a few instances, the embryo is described as gray or brown. However, it is not yet determined in these cases if this is the mature condition. Observations on various species (e.g., *A. jupunba*, *A. curvicarpa*, *A. mataybifolia*, and others) in the field show that the seeds do not attain the blue color until well near dehiscence; seeds in indehisced pods are often brown, and seeds that fail to germinate quickly lose the blue color. Likewise seeds damaged by predation or pathogens often lack coloration. In any case, even if the seeds are brown or gray at maturity, the cladistic analysis indicates that the brown condition represents a loss of delphinidin and not the pleisiomorphic condition.

The analysis further shows that *Abarema* is monophyletic on the basis of five synapomorphies: the lentiform seeds (no. 37) persistent on funicles (no. 38) against a background of red or red-brown endocarp (no. 33), the blue embryo (no. 43), and the geotropic pods (no. 45). All but the blue embryo are homoplastic in the context of the whole of Ingeae. The red or red-brown endocarp and seeds persistent on the funicle are found again in *Archidendron* and *Pithecellobium* sensu stricto; and the geotropic pods are found in almost all other members of the tribe.

The genus *Cojoba* resembles *Abarema* in organization of the inflorescence and auxotelic meristems, but has fleshy, red, not dry papery, leathery or lignescent valves of the pod, and seeds with black seed-coat. The slenderly tubular corolla of *Cojoba*, several times longer than the small campanulate calyx, is further distinctive, more like that of *Chloroleucon* than that of *Abarema*. *Pithecellobium*, as now defined, may have a fruit similar in texture and dehiscence to that of *Abarema*, but it differs not only in inflorescence architecture but also in arillate seeds and (with one exception) randomly armed stems. *Hydrochorea* and *Balizia* share with *Abarema* the indeterminate inflorescence of axillary peduncles and differ materially only in their fruits and seeds, as described in our key to major American taxa of Ingeae (p. 13).

Features common but not universal (except perhaps the anthocyanic embryo, not recorded in a few species) are: (a) brown or golden pubescence of nascent stems and leaf buds; (b) shortly racemose, cymose, or capitate units of inflorescence terminating in a dimorphic flower; (c) white filaments; and (d) pods strongly decurved or backwardly coiled into a compressed spiral and elastically dehiscent through both sutures. The leaf-formula and leaflet-size vary widely between species, the ovary may be either truncate or conical at apex, dehiscence of the pod, depending on texture of the valves, may be either bivalvar or follicular, and the seeds may be areolate or not. With the possible exception of *A. levelii*, the fruit of which is not yet known, there is no species of *Abarema* doubtfully referred to the genus or any that is in any real sense ambiguous between *Abarema* and another genus.

The genus *Klugiodendron* has lately been recognized by Nielsen (1981) as a genus of indefinite scope related to *Abarema* and *Cojoba*. Mohlenbrock (1963) subordinated it to *Pithecellobium*, with the false attribute of arillate seeds. The myth of an arillate seed can be traced back to Poeppig's descriptions of *Inga laeta* and *Pithecolobium polycarpum*, the latter based on fruiting specimens said to have seeds with smooth blue testa and a short white fleshy aril at the base, which we interpret as the crumpled funicle. Poeppig's two species were synonymized by Bentham and the so-called aril of *P. polycarpum* was carried over into the description of *P. laetum*, and thence into that of *Klugiodendron*.

Conspectus of genus *Abarema*

¶ Fls umbellate or racemose, homo- or dimorphic; filaments white; ovary truncate; dehiscence of pod elastic, through both sutures, the valves twisted; seeds areolate, the pleurogram complete, the testa partly or wholly translucent, the embryo blue, fuscescent, or rarely blackish. — West Indies and E Brazil; one in SE Mexico and Central America.

* West Indian.

32. *A. asplenifolia* (Grisebach) Barneby & Grimes
33. *A. glauca* (Urban) Barneby & Grimes
34. *A. obovalis* (A. Richard) Barneby & Grimes
35. *A. alexandri* (Urban) Barneby & Grimes
    a. var. *alexandri*
    b. var. *troyana* (Urban) Barneby & Grimes
36. *A. oppositifolia* (Urban) Barneby & Grimes
37. *A. nipensis* (Britton) Barneby & Grimes

38. *A. abbottii* (Rose & Leonard) Barneby & Grimes

** South Mexican, Central American.

5. *A. idiopoda* (Blake) Barneby & Grimes

*** East Brazilian.

26. *A. obovata* (Bentham) Barneby & Grimes
27. *A. brachystachya* (de Candolle) Barneby & Grimes
28. *A. filamentosa* (Bentham) Pittier
29. *A. turbinata* (Bentham) Barneby & Grimes
30. *A. cochliacarpos* (Gomes) Barneby & Grimes
31. *A. langsdorfii* (Bentham) Barneby & Grimes

¶¶ Characters of ¶ except for exareolate seeds. — Orinoco and Amazon basins, Venezuela and Brazil.

17. *A. leucophylla* (Bentham) Barneby & Grimes
    a. var. *leucophylla*
    b. var. *vaupesensis* Barneby & Grimes
18. *A. mataybifolia* (Sandwith) Barneby & Grimes
19. *A. laeta* (Bentham) Barneby & Grimes
39. *A. levelii* (Cowan) Barneby & Grimes

¶¶¶ Characters of ¶ except for follicular dehiscence of pod and exareolate seeds. — Brazilian Amazonia, Guayana, and Pacific Ecuador N into Panama.

4. *A. longipedunculata* (Irwin) Barneby & Grimes
6. *A. curvicarpa* (Irwin) Barneby & Grimes
    a. var. *curvicarpa*
    b. var. *rodriguesii* Barneby & Grimes
7. *A. macradenia* (Pittier) Barneby & Grimes
9. *A. piresii* Barneby & Grimes

¶¶¶¶ Characters of ¶ except for purplish corolla and rose-purple filaments. — Amazonian basin, the Guianas, Guayana Highland, N Colombia.

1. *A. floribunda* (Bentham) Barneby & Grimes
2. *A. ferruginea* (Bentham) Pittier
3. *A. gallorum* Barneby & Grimes

¶¶¶¶¶ Characters of ¶ except for ovary conical, not truncate at apex, and seeds of some species exareolate. — Amazonia, the Guianas, Guayana Highland, W to Pacific Ecuador and Colombia, NW into Central America and NE into Lesser Antilles.

8. *A. ganymedea* Barneby & Grimes
10. *A. auriculata* (Bentham) Barneby & Grimes
11. *A. microcalyx* (Bentham) Barneby & Grimes
    a. var. *microcalyx*
    b. var. *enterolobioides* Barneby & Grimes
    c. var. *parauaquarae* (Ducke) Barneby & Grimes
12. *A. commutata* Barneby & Grimes
13. *A. jupunba* (Willdenow) Britton & Killip
    a. var. *jupunba*
    b. var. *trapezifolia* (Vahl) Barneby & Grimes
14. *A. barbouriana* (Standley) Barneby & Grimes
    a. var. *barbouriana*
    b. var. *arenaria* (Ducke) Barneby & Grimes
15. *A. adenophora* (Ducke) Barneby & Grimes
16. *A. zollerana* (Standley & Steyermark) Barneby & Grimes
20. *A. campestris* (Bentham) Barneby & Grimes
21. *A. cochleata* (Willdenow) Barneby & Grimes
    a. var. *cochleata*
    b. var. *moniliformis* (Ducke) Barneby & Grimes
40. *A. oxyphyllidia* Barneby & Grimes
41. *A. villifera* (Ducke) Barneby & Grimes

¶¶¶¶¶¶ Fls openly racemose or spicate, homomorphic; ovary conical at apex; dehiscence of ¶; seeds exareolate. — Cordilleran, Bolivia to Costa Rica, most diverse in Andean Colombia.

22. *A. callejasii* Barneby & Grimes
23. *A. lehmannii* (Britton & Killip) Barneby & Grimes
24. *A. killipii* (Britton & Killip) Barneby & Grimes
25. *A. racemiflora* (Donnell Smith) Barneby & Grimes
42. *A. josephi* Barneby & Grimes
43. *A. centiflora* Barneby & Grimes

¶¶¶¶¶¶¶ Of unknown affinity.

44. *A. ricoae* nom. provis. — Nicaragua
45. *A. agropecuaria* nom. provis. — Brazil

Because speciation in *Abarema* has taken independent but often parallel paths in different quarters of the neotropics, there is hardly any repetition of taxa between Mexico with Central America, the Greater Antilles, equatorial and northern South America, and southeastern Brazil. The most efficient key to species, and to specimens of known provenance, is therefore based in first instance on distribution patterns.

Keys to the species of genus *Abarema*

A. Of Mexico and Central America . . . . Subkey I (p. 44)
A. Allopatric.
    B. Of the Bahamas and Greater Antilles (Cuba, Hispaniola, Jamaica) . . . . Subkey II (p. 45)
    B. Of continental South America and Lesser Antilles . . . . Subkey III (p. 45)

Subordinate key I: Mexico and Central America

1. First petiolar nectary funnelform-campanulate and 5–15 × 4–12 mm, if shorter than 8 mm, fls subhomomorphic.
    2. Lfts 2–4(–5) pairs per pinna; fls dimorphic; centr. Amazon basin westward to Peru and disjunct in NW Colombia and NE Nicaragua . . . . 15. *A. adenophora* (p. 74)
    2. Lfts 7–12 per pinna; fls subhomomorphic; Nicaragua and Panama . . . . 7. *A. macradenia* (p. 57)
1. First petiolar nectary 0.8–3 × (0.4–)0.6–2.4 mm; if longer than 2.4 mm fls dimorphic.
    3. Lf-formula i–ii/2–3, the larger lfts ±10–15 cm; fls openly racemose, the floral axis 20–30 cm; montane Costa Rica. . . . . 25. *A. racemiflora* (p. 90)
    3. Lf-formula iii–x/8–40, the larger lfts <5 cm; fls shortly racemose or pseudo-umbellate, the floral axis not over 1.5 cm.

4. Pinnae of larger lvs 3–5 pairs and lfts of longer pinnae 6-jug.; Nicaragua only . . . . 44. *A. ricoae* (p. 110)
4. Pinnae of larger lvs 4–13 pairs and lfts of longer pinnae more numerous and smaller, of longer pinnae 10–40 pairs and the largest ones not over 20 mm.
  5. Largest lfts 12–20 × 5–9 mm; S Mexico and Guatemala . . . . 16. *A. zollerana* (p. 75)
  5. Largest lfts 6.5–13 × 2–5 mm.
    6. Lfts of longer pinnae 17–40 pairs, plane, either glabrous or merely barbellate dorsally; lower fls of capitulum pedicellate; Panama NW to Belize and Oaxaca.
      7. Lfts acute; calyx ±5 mm and corolla 9–10 mm; ovary conic at apex . 40. *A. oxyphyllidia* (p. 107)
      7. Lfts obtuse; calyx 1.5–3 mm and corolla 5–7 mm; ovary truncate at apex . . . . 5. *A. idiopoda* (p. 53)
    6. Lfts of longer pinnae 10–16 pairs, very obtuse, convex ventrally, pilosulous dorsally; fls sessile or almost so; ovary conic at apex; centr. Panama southeastward . . . . 14. *A. barbouriana* (p. 70)

Subordinate key II: Bahamas and Greater Antilles

1. Fls openly spicate, the floral axis 2.5–8 cm long; Jamaica only . . . . 35. *A. alexandri* (p. 101)
1. Fls capitate, pseudo-umbellate, or shortly compactly racemose, the floral axis 2–22 mm; Bahamas, Cuba, Hispaniola.
  2. Largest lfts 12–65 mm long.
    3. Lf-formula i–ii(–iii)/3–5(–6), the largest lfts 40–65 × 30–55 mm. . . . . 34. *A. obovalis* (p. 99)
    3. Lf-formula otherwise, either the pinnae or the lfts more numerous, and the latter <35 × 20 mm.
      4. Lfts of longer pinnae 9–12 pairs and the largest of them 12–18 × 5–8(–10) mm; localized in NE Oriente, Cuba. . . . . 32. *A. asplenifolia* (p. 97)
      4. Lfts of longer pinnae mostly 6–8, seldom 10 pairs and the largest of them 20–32 × 9–19 mm; widespread in Bahamas, Cuba, and Hispaniola . . . . 33. *A. glauca* (p. 98)
  2. Largest lfts 3–10 mm long.
    5. Lfts to 5–8.5(10) × 2–5 mm.
      6. Lf-stks of longer lvs 4–13 cm, the petiole 6–20 mm, and the lfts of longer pinnae 16–25(–30) pairs; a nectary below first pinna-pair; calyx 3.4–4.2 mm; centr. and SW Dominican Republic W through S Haiti to centr. Cuba . . . . 36. *A. oppositifolia* (p. 103)
      6. Lf-stks of longer lvs 2–6 cm, the petiole almost 0, and the lfts of longer pinnae 8–12 pairs; no nectary below first pinna-pair; calyx 2.2–3 mm; NE Dominican Republic . . . . 38. *A. abbottii* (p. 105)
    5. Lfts to 3–5 × 1.5–1.8 mm; Oriente, Cuba. . . . . 37. *A. nipensis* (p. 104)

Subordinate key III: South America and Lesser Antilles

1. Fls racemose or spicate, the floral axis 3–13 cm.
  2. Lf-formula (v–)vi–xii/14–26, lfts 120 or more per leaf; centr. Amazonian Brazil and the Guianas . . . . 6. *A. curvicarpa* (p. 54)
  2. Lf-formula i–iii/2–7; lfts (12–)16–36(–44) per lf; Andes of Colombia and Ecuador.
    3. Pedicels all obsolete or almost so, when present broader than long; Colombia.
      4. Lf-formula i–ii/3–6, the lfts (12–)16–36(–44) per lf; Andes of Colombia and Ecuador, southward from state of Antioquia.
        5. Longer lfts (4.5–)5–9(–11.5) × 3–6.5 cm; blade of lfts dorsally pilosulous overall and the intervenium densely minutely pallid-papillate, consequently notably discolorous; lft-margin incipiently revolute; Antioquia, Colombia, near lat. 6°N. . . . . 23. *A. lehmannii* (p. 88)
        5. Longer lfts 9–15 × 4–5 cm; blade of lfts dorsally glabrous or only minutely puberulent along principal nerves, not notably discolorous; lft-margin plane; Caldas, Colombia, southward into Ecuador in lat. 4°40′N–3°S. . . . . 24. *A. killipii* (p. 89)
      4. Lf-formula ii–iii/6–7, the lfts 44–68 per lf; Cordillera Oriental near 1°50′N . . . . . 42. *A. josephi* (p. 109)
    3. Pedicels 1–2.5 mm, much longer than diameter.
      6. Racemes 80–100-fld; lower pedicels ±1 mm; rachis of the 1 pair of pinnae 13–15 cm; stamens 11 mm long; Bolivia (Cochabamba) . . . . 43. *A. centiflora* (p. 109)
      6. Racemes 25–45-fld; lower pedicels ±2–2.5 mm; rachis of the 1 or of the distal pair of pinnae 2.5–7 cm; stamens 15 mm long; Colombia (Antioquia) . . . . 22. *A. callejasii* (p. 85)
1. Fls capitate or densely shortly racemose or pseudo-umbellate, the floral axis much less than 3 cm.
  7. Lfts of longest pinnae 1–5 pairs and the larger ones at least 2.5 cm long.
    8. Plants distributed within the Amazon Basin and northward into Pacific Colombia, Venezuela, Trinidad, and the Guianas.
      9. Lfts ovate or lance-ovate from subequilateral base, distinctly even though only shortly acuminate, broadest below midblade.

10. Lfts of all pinnae exactly 1 pair.
    11. Lft-pulvinules 1.5–3 mm; secondary venation of lfts consisting of many parallel, widely ascending, nearly straight nerves of subequal strength; corolla of peripheral fls 4.5–7 mm, and their androecial tube ±2–4.5 mm; foliage retaining a bright and pale green color when dried; widespread through middle and upper Amazon basin, randomly N to Guyana and N Venezuela. . . . . . . 19. *A. laeta* (p. 79)
    11. Lft-pulvinules 4–8 mm; secondary venation of lfts consisting of 3–5 stronger, incurved-ascending as well as widely spreading, straight and weaker nerves; corolla of peripheral fls 11.5–15.5 mm, and their androecial tube 8–14 mm; rare species, known from centr. Amazonian Brazil and upper Orinoco basin in Venezuela . . . . 39. *A. levelii* (p. 106)
10. Lfts of longer pinnae 3–5 pairs.
    12. Primary lf-stk 7–23 cm, almost always longer than the rachis of the distal pinnae; lfts when dry commonly olivaceous on both faces, scarcely paler beneath; corolla of peripheral fls 6.5–8.5 mm and their stamens 28–46; valves of pod low-convex over each seed, this charged on each face with a faint pleurogram. . . . . 20. *A. campestris* (p. 82)
    12. Primary lf-stk 0.5–7(–9) cm, mostly much shorter than, rarely as long as, the rachis of the 1 or of the longer of 2 pairs of pinnae; lfts when dry almost always strongly bicolored, dark brown above, pale brown beneath; corolla of peripheral fls ±4.5–6.3 mm and their stamens 12–22; valves of pod hemi-spherically elevated over each seed, this lacking pleurogram . . . . 21. *A. cochleata* (p. 83)

9. Lfts elliptic, obovate, or rhombic, not acuminate, broadest at or above mid-blade, or if broadest below midblade then inequilateral at base.
    13. First petiolar nectary funnelform-campanulate and 10–13 × 8–12 mm; centr. Amazon basin W to Peru and disjunct in NW Colombia and NE Nicaragua . . . . . 15. *A. adenophora* (p. 74)
    13. First petiolar nectary either cupular-patelliform, or mounded, or immersed in petiolar groove, in any case <2 mm tall.
        14. Dorsal face of lfts pilosulous or tomentose with incurved-ascending hairs.
            15. Lf-formula i–ii/2–3; calyx of peripheral fls 3.5–4 mm and corolla 5.5–6 mm; seasonally inundated forest or campo below 200 m on upper río Negro in state of Amazonas, Brazil and T. F. Amazonas, Venezuela . . . . 41. *A. villifera* (p. 108)
            15. Lf-formula ii–iv/3–6; calyx of peripheral fls (unknown in "*agropecuaria*") 4.5–6 mm and corolla 6–9 mm.
                16. Petiole (0.5–)0.8–2.4 cm; distal lfts (3–)3.5–6.5 × 2.1–4.5 cm; upland savanna of Guayana Highland in SE Venezuela, Guyana, and Surinam . . . . 12. *A. commutata* (p. 65)
                16. Petiole 3–4 cm; distal lfts ±20 × 9 mm; lowland forest of Amazonas, Brazil . . . . 45. *A.* "*agropecuaria*" nom. prov. (p. 110)
        14. Dorsal face of lfts either glabrous or minutely appressed-puberulent.
            17. Calyx of peripheral fls not over 1.2 mm; nectary near first pair of pinnae verruciform with minute pore; lfts glabrous dorsally . . . . 11. *A. microcalyx* (p. 63)
            17. Calyx of peripheral fls larger, at least 1.7 mm; either (a) nectary cupular-patelliform, or immersed in petiolar groove, or (b) lfts puberulent beneath.
                18. Lfts of longer pinnae 1–3 pairs and their pulvinules 3–9 mm long.
                    19. Lfts 2–3 pairs in distal (or only) pair of pinnae and their pulvinule 5–9 mm; filaments of terminal flower not dilated, ascending, straight; pod 10–20 × 1.5–2.6 cm, 9–14-seeded; middle and lower Amazon valley and the Guianas . . . . 18. *A. mataybifolia* (p. 78)
                    19. Either lfts of almost all pinnae 1 pair, 2 pairs rare and exceptional on any plant, absent from most, or lft-pulvinules only 3-4 mm; filaments of terminal flower(s) dilated and sigmoidally arched out- and then upward; pod (of var. *leucophylla*, that of var. *vaupesensis* unknown) 2.7–5 × 0.9–1.4 cm, 6–8-seeded; middle Amazonian Brazil, SE Colombia, and SW Venezuela. . . . . 17. *A. leucophylla* (p. 76)
                18. Lfts of longer pinnae 4–6 pairs and their pulvinules <3 mm long.
                    20. Lfts bluntly rhombic in outline, finely minutely strigulose beneath; pod (so far as known) recurved through half to nearly a full circle, often further.
                        21. Lfts firm, lustrous above, reticulately venulose on both surfaces, the midrib ascending from the pinna-axis at wide angles; the Guianas and adjacent countries.
                            22. Filaments pink; ovary truncate; pods falcate, 12.5–16 × 2.2–2.4 cm; endocarp dark red-brown in seed-locules, tan between; larger lfts 4–8 cm; French Guiana . . . . 3. *A. gallorum* (p. 51)

22. Filaments white; ovary conical at apex; pods annulate, 6–9.5 × 1–1.6 cm; endocarp uniformly orange overall; larger lfts at most 3.5 cm; var. *trapezifolia* of . . . 13. *A. jupunba* (p. 66)
21. Lfts thin-textured, dull above, the tertiary venulation faint; lfts subvertically imbricated along rachis, the midrib ascending parallel to it; Uaupés valley in Amazonas, Brazil; undescribed variant of . . . . 1. *A. floribunda* (p. 48)
20. Lfts obovate in outline, glabrous beneath; pod gently decurved through ¼ circle or less; Gran Sabana in edo. Bolívar, Venezuela. . . . . 4. *A. longipedunculata* (p. 52)
8. Plants distributed in extra-Amazonian Brazil, from Paraíba to Paraná.
23. Adult lfts glabrous, even on margin.
24. Lfts relatively few, in largest lvs of any plant 4–24 in number and intricately reticulate on both faces; petiolar nectaries verruciform; androecium 21–36 mm; pod 14–24 mm wide at seeds; coastal Bahia in lat. 13°–18°S.
25. In longest pinnae lfts 2–3 pairs, obovate, and the largest 4–8.5(–10) × 2.2–5 cm; range as just given . . . . 28. *A. filamentosa* (p. 92)
25. In longest pinnae lfts 1–2 pairs, elliptic, and the largest 8–14 × 4.5–7 mm; occupying an enclave in range of the preceding, between 14° and 16°S. . . . . . . 29. *A. turbinata* (p. 94)
24. Lfts more numerous, in largest lvs of a plant 28–56 in number and their tertiary venulation weak and random; petiolar nectaries shallow-cupular, or plane, or immersed; androecium 13–16 mm; pod 9–14 mm wide at seeds; near the coast from Paraíba to Rio de Janeiro, inland to upland Bahia . . . . 30. *A. cochliacarpos* (p. 94)
23. Adult lfts ciliolate, sometimes also pilosulous on lower face.
26. Lfts of larger lvs 24–36 in number, obovate or obscurely rhombic-obovate, those of distal pinnae 4 pairs, and the largest to 3.5–6 × 2.1–3.5 cm; upland Minas Gerais . . . . 26. *A. obovata* (p. 91)
26. Lfts of larger lvs 48 or more, distinctly, often sharply rhombic, those of distal pinnae 4–6 pairs, the largest to 2–3.5 × 0.9–2.1 mm; coastal restinga, Rio de Janeiro to Paraná . . . . 27. *A. brachystachya* (p. 91)
7. Lfts of longer pinnae 7–many pairs.
27. Plants distributed within the Amazon Basin and northward to Pacific Colombia and Ecuador, Venezuela, the Guianas, and Lesser Antilles.
28. First petiolar nectary of lvs on fertile branches campanulate, greatly enlarged 5–15 × 4–11 mm; pod woody suberously thickened over each seed-locule; Pacific lowland Ecuador (?and Colombia) . . . . 7. *A. macradenia* (p. 57)
28. Petiolar nectaries cupular or patelliform; pod (so far as known) otherwise.
29. Lfts at maturity convex or bullate ventrally, dorsally pilosulous overall or at least in anterior basal angle of midrib . . . . 14. *A. barbouriana* (p. 70)
29. Lfts plane (except for sometimes narrowly revolute margin), dorsally glabrous or finely appressed-strigulose.
30. Peripheral fls of the capitulum relatively small, the calyx 1.7–3.5 mm, the corolla 4–6.5 mm; filaments whitish.
31. Ovary truncate glabrous; pod deeply strangulate between seeds by depression of seminiferous suture; dehiscence of pod follicular, through ventral suture only; endocarp of ripe pod brown; E Amazonian Brazil . . . . 9. *A. piresii* (p. 61)
31. Ovary conical at apex, either glabrous or pubescent; pod undulately constricted between seeds, or deeply strangulate only in *A. ganymedea* of NW Colombia and Pacific Ecuador; dehiscence through both sutures; endocarp of ripe pod orange overall; widely dispersed.
32. Lfts of longer pinnae 6–10(–12)-jug., finely minutely strigulose beneath, the largest at least 13 mm long; widespread in tropical South America and Lesser Antilles; var. *jupunba* of . . . . . 13. *A. jupunba* (p. 66)
32. Lfts of longer pinnae 9–18-jug., if <12-jug. glabrous beneath, the largest in any case only 6–12 mm long.
33. Lfts of longer pinnae 9–12-jug., glabrous dorsally; first petiolar nectary verruciform; floral bracts persistent; of Amazonian Peru and Brazil, mostly in white sand campirana . . . 10. *A. auriculata* (p. 62)
33. Lfts of longer pinnae 14–19-jug., minutely puberulent dorsally; first petiolar nectary infundibuliform, 1–2 mm tall; floral bracts caducous; NW Colombia (Antioquia), W Ecuador (Esmeraldas), apparently in rain forest . . . . 8. *A. ganymedea* (p. 59)
30. Peripheral fls of the capitulum relatively large, the calyx 4–9 mm, the corolla 8–13 mm; filaments rose or red.

34. Longer lf-stks (6–)8–23 cm and pinnae (3–)4–6-jug.; rachis of longer pinnae (7–)8–14(–20) cm; discontinuously widespread in Amazon basin, mostly below 400 m . . . . 1. *A. floribunda* (p. 48)
34. Longer lf-stks 5–7 cm and pinnae 3–4-jug.; rachis of longer pinnae 4–7 cm; SE Guayana Highland at ±1100–1500 m . . . . 2. *A. ferruginea* (p. 49)
27. Plants distributed in extra-Amazonian S Brazil, from Bahia to Sta Catarina, mostly along the coastal plain, but sparingly inland to submontane Bahia and Minas Gerais.
35. Lfts of longer lvs 6–10-jug., to 1–2 cm wide.
36. Lfts trapeziform, glabrous; restinga below 25 m, Rio de Janeiro to Paraná. . . . . 27. *A. brachystachya* (p. 91)
36. Lfts obliquely obovate, pilosulous dorsally; upland Minas Gerais . . . . 26. *A. obovata* (p. 91)
35. Lfts of longer lvs 14–28-jug., rhombic-elliptic, 2–4 mm wide; both in restinga and to 840 m in the coast ranges, SE Bahia to Sta Catarina. . . . . 31. *A. langsdorfii* (p. 95)

1. Abarema floribunda (Bentham) Barneby & Grimes, comb. nov. *Pithecolobium floribundum* Spruce *ex* Bentham, Trans. Linn. Soc. London **30:** 584. 1875 & *in* Martius, Fl. Bras. **15(2):** 437. 1876. — "Habitat in silvis Rio Uaupés prope Panuré, prov. do Alto Amazonas: Spruce n. 2471." — Holotypus, K(herb. bentham.) = NY Neg. *2809*!; isotypi, †B = F Neg. *1195*!, GH!, GOET!, K(herb. hooker.)!, NY!, P!, RB (2)!. — *Feuilleea floribunda* (Bentham) O. Kuntze, Revis. Gen. Pl. **1:** 187. 1891.

Pithecolobium floribundum sensu Ducke, 1949: 37.

Trees 10–20(–29) m with trunk 2–5 dm diam, but sometimes flowering precociously as a shrubby treelet, the young parts, except lfts, densely pubescent with sordid or golden-brown hairs ±0.1–0.3 mm but glabrate with age, the foliage bicolored, the young lfts subvertically imbricate along rachis but spreading in age, submembranous becoming chartaceous and sharply venulose, when dry brown above, pale brown beneath, commonly puberulent on both faces (or on upper one only along midrib) and thinly ciliolate, but sometimes glabrate on both faces, the large, many-fld capitula arising singly or paired in distal lf-axils on peduncle shorter than the associated lf-stk. *Stipules* linear-lanceolate, lanceolate or narrowly ovate (3–) 4–13 × 1.5–4 mm, quickly deciduous from the expanding lf. *Lf-formula* (iii–)iv–vi/8–17(–21); lf-stks (6–)8–15(–23) cm, the petiole (2–)3–7 cm, the longer interpinnal segments 1.3–4 cm; nectary immediately below first pinna-pair sessile low-convex or scutiform 1.5–4.5 mm diam, sometimes also one on petiole and/or between further pairs, similar or shortly stipitate, always smaller ones on pinna-rachis at 1–15 furthest pairs of lfts; pinnae either scarcely or strongly graduated, (7–)8–14(–20) cm, the longer interfoliolar segments 5–15 mm; paraphyllidia minute or 0; pulvinules wrinkled 0.7–1.4 × 0.5–1 mm; lfts a little distally accrescent, oblong from inequilateral, broadly cuneate base, broadly obtuse, the longer ones (10–) 17–28(–37) × (3.5–)5.5–12.5(–17) mm, 2.3–3.3 times as long as wide; the major secondary nerves 7–12(–15) on each side of midrib, the tertiary venules brochidodrome within the plane, somewhat thickened margin. *Peduncles* 3.5–8 cm; capitula ±40–60-fld, without filaments ±20–24 mm diam, the clavate receptacle 5–14 × 3–6 mm; lower bracts ovate-elliptic, the upper ones becoming spatulate, all ±1 mm, caducous; the 5(randomly 6)-merous fls dimorphic, the peripheral ones pedicellate, the 1–3 terminal ones sessile and larger, the perianth of all densely appressed-silky, the indumentum of corolla golden, of calyx brown, the corolla of peripheral ones purplish, of terminal ones (fide Spruce & Ducke) white, the filaments rose-purple; PERIPHERAL FLS: pedicel 1–5 × 0.4–0.55(–0.7) mm; calyx campanulate 4–9 × 2.1–2.4 mm, the teeth ±0.5–1.4 mm; corolla 8–13 mm, the lobes ±2.5 × 1.2–1.5 mm; androecium 18–26-merous, 3.3–4 cm, the stemonozone 1.6–2 mm, the tube 4–5 mm; ovary glabrous or rarely puberulent above middle, truncate and dilated at apex; TERMINAL FL (few seen): calyx 6–9.5 × 2.5–3.5 mm; corollas 15–17 mm; androecial tube shortly exserted. *Pods* 1–2 per capitulum, sessile or basally attenuate into a short pseudostipe, in profile broad-linear, evenly retrofalcate through ⅓–⅔-circle, 12–26(–33) × (1.4–)1.7–2.5 cm, the coriaceous, dark brown, glabrous valves framed by stout, shallowly undulate sutures 2–3 mm wide, becoming low-convex and wrinkled or verruculose over each of 9–15 seeds; dehiscence through both sutures, the valves coiling, the endocarp reddish in the seed-cavities and tan between them; seeds (one seen) transverse on tapelike, apically sigmoid funicle, plumply discoid 8–8.5 mm diam, the testa (when old) lustrous brown, but (fide Ducke) green and white when fresh, the pleurogram 6.5 mm diam.

In forest on terra firme, known in typical form from lowland centr. and W Amazonian Brazil, from Manaus and the Urubú basin NW to the Uaupés, W on the Solimões into Peru (Loreto, San Martín, E Huánuco) and on the Napo into NE Ecuador, and SW to Acre (Cruzeiro do Sul), N just into Venezuela (Amazonas); further (see discussion below) represented by morphologically, perhaps taxonomically, distinct forms, in (a) upland Guyana (Mazaruni basin), (b) N. Peru (Alto Amazonas, prov. Bagua), SE Colombia (Com. Vaupés), and N Colombia. — Map 7. — Fl. in Brazil VIII–IX. — *Pashaco* (Peru); *marupa* (Venezuela).

The foregoing description applies specifically to *A. floribunda* sensu stricto. Four collections, in some respects different both from typical *A. floribunda* and from each other, may represent distinct taxa, but these are at present too little known for certain definition or formal description:

variant 1 [*P. H. Allen 3379*. COLOMBIA. Vaupés: Guaracapuri cachoeira [Raudal Uaracapuri], Río Vaupés [69°55′W], 60 m, 21 Nov 1945 (NY, US).]: *Trees* ±10 m; *stipules* ±1 mm wide; *lf-formula* i–ii/5; lf-nectaries flat; larger lfts ±30 × 13 mm; *capitula* 15–20-fld; calyx of peripheral fls 4.5–5.5 mm and corolla 7.5–8.5 mm; ovary puberulent. — The lf-formula and relatively few-fld capitula are visually arresting.

variant 2 [*J. J. Wurdack 1844*. PERU. Amazonas, prov. Bagua: valley of Río Maranón above Cascadas de Mayasi, 450–600 m, 5 Sept. 1962 (NY, P, S, US)]: Bushy *trees* 5 m; *stipules* 1–1.5 mm wide; *lf-formula* iii–v/10–12; lf-nectaries low-convex; larger lfts ±35 × 12–15 mm; *capitula* ±50–70-fld; calyx of peripheral fls ±4.5 mm and corolla only 5.5–6.5 mm; ovary puberulent. — The ample lfts, small flowers, and puberulent ovary are peculiar.

variant 3 [*Tillett & Tillett 45554*. GUYANA. Mazaruni basin: Kako R., 950 m, 29 Sept 1960 (NY)]: *Trees* (stature not recorded); *stipules* ±1 mm wide; *lf-formula* iv–vi/12–13; lf-nectaries shallowly concave; larger lfts 16–22 × 6–9 mm; *capitula* ±25-fld; calyx of peripheral fls ±5.5 mm and corolla 10.5–11 mm; androecium 32-merous; ovary glabrous. — Congruent with *A. floribunda* in form and texture of leaves and in relatively short peduncles, but approaching the almost sympatric *A. ferruginea* in smaller flowers and thin-textured leaflets.

variant 4 [*Callejas et al. 8590*; *Fonnegra et al. 3057, 3115*. COLOMBIA. Antioquia, mun. San Carlos, near Vereda Miraflores (all HUA)]: *Trees* 10–20 m; *stipules* 2–4.5 mm; *lf-formula* ii–v/9–12; lf-nectaries almost flat, wrinkled; larger lfts 20–35 × 9–15 mm; only fragmentary withered peripheral *fls* seen, calyx 4–4.5 mm, corolla 7–7.5 mm; *seeds* white at base, translucent blue-gray and darker-mottled ("moradas") distally. — Very close to typical *A. floribunda*, but incompatible ecologically, and the androecium, so far as observed, only about 16 (not 33–40) mm long.

The differences may be expressed in the following key:

1. Lf-formula i–ii/5; E Colombia .... *Allen 3379*
1. Lf-formula iii–vi/10–21.
    2. Corolla of peripheral fls 5.5–6.5 mm, only 1–2-mm longer than calyx; NE Peru .... *Wurdack 1844*
    2. Corolla of peripheral fls 7–13 mm, 3–5 mm longer than calyx; allopatric.
        3. Capitula 40–60-fld; androecium 16–26-merous.
            4. Androecium of peripheral fls 33–40 mm; lowland Amazonian Brazil westward from Manaus to adjacent Colombia and Peru ....typical *A. floribunda*
            4. Androecium (poorly known) of peripheral fls ±16 mm; montane woodland in Antioquia, Colombia, at 750–900 m ....*Callejas et al. 8950* etc.
        3. Capitula ±25-fld, androecium ±32-merous; Guyana, 950 m.................. *Tillett & Tillett 45554*

2. Abarema ferruginea (Bentham) Pittier, 3rd Conf. Interam. Agric. Caracas 360. 1945. *Pithecolobium ferrugineum* Bentham, London J. Bot. **3:** 216. 1844. — "British Guiana [but very likely on the Venezuelan slope of Mt Roraima], Schomburgk, n. 663 of 1843." — Holotypus, *Rich. & Rob. Schomburgk 663 = 994*, K!; isotypus, †B = F Neg. *1193*; possible isotypus, U *38869*!. — *Feuilleea fulvescens* (Bentham) O. Kuntze, Revis. Gen. Pl. **1:** 185. 1891; non *F. ferruginea* (Guillemin & Perrotet) O. Kuntze, l. c.

Pithecolobium ferrugineum sensu Bentham, 1875: 583.

Arborescent *shrubs* attaining 8 m, the bark gray lenticellate, the stout, straight, rather distantly foliate hornotinous branches together with all lf- and inflorescence-axes densely pilosulous with golden-brown hairs to 0.1–0.3 mm, the thick-textured, sublustrous but sharply venulose lfts upwardly imbricate against the pinna-rachis, when dry brown above, pale brown beneath, puberulent along midrib on both faces, thinly microscopically ciliolate, the large, many-fld capitula arising singly in distal lf-axils on peduncles as long or commonly longer than the associated lf-stk; terminal meristem indeterminate. *Stipules* linear-lanceolate to -elliptic 4–7 × 0.75–1.5 mm, quickly deciduous from the expanding lf. *Lf-formula* iii–iv/8–11; lf-stks 5–7 cm, the petiole 1.2–2.0 cm, the longer interpinnal

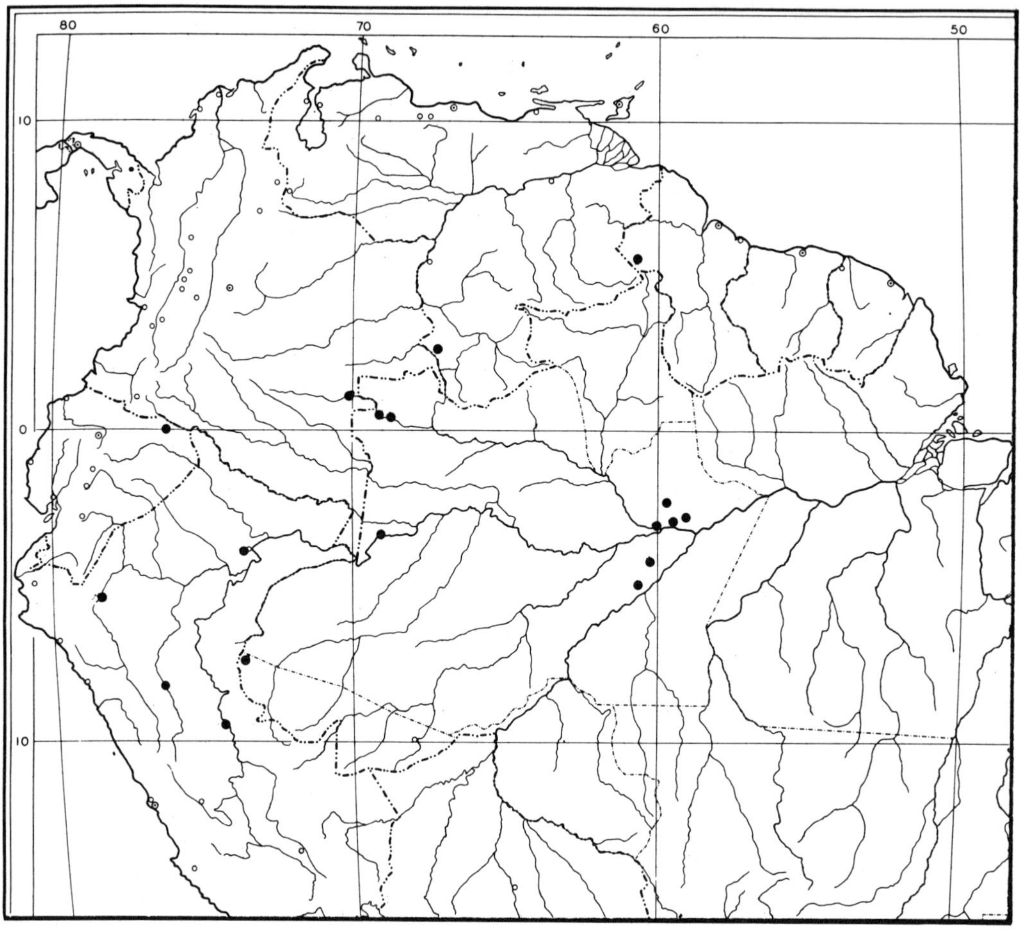

MAP 7. Northern South America. Range of *Abarema floribunda* (Bentham) Barneby & Grimes.

segments 1–2 cm; nectaries between or immediately below pinna-pairs concave-patelliform to scutiform (2–)2.5–4 mm diam, livid becoming pallid in age, similar but much smaller ones on pinna-rachis at 1–2 furthest pairs of lfts; pinnae little graduated, 4–7 cm, the longer interfoliolar segments 5–8 mm; paraphyllidia not seen, but scars apparent on some specimens; lft-pulvinules wrinkled ±1.5 × 1 mm; lfts a little accrescent distally, inequilaterally rhombic-oblong-elliptic from rounded base, broadly obtuse, the longer ones 18–26 × 9–14 mm, 1.8–2.2 times as long as wide; venation of 3 nerves from pulvinule and thence pinnate, the moderately displaced, dorsally prominulous midrib incurved distally, the inner posterior primary nerve produced much short of midblade and the outer one very short, the 7–8 major and numerous intercalary secondary nerves brochidodrome within the plane, somewhat thickened margin, these in turn giving rise to a coarse reticulum of veinlets prominulous on both faces. *Peduncles* 5–13 cm; capitula ±45–60-fld, without filaments ±15 mm diam, the 5(randomly 6)-merous fls seemingly homomorphic (few whole capitula available for study) and sessile or almost so, the stout pedicel <1.5 mm; bracts decrescent upward, the lowest oblanceolate or trullate 5–7 × 2–2.5 mm, the upper ones spatulate shorter, all caducous; perianth densely appressed-silky overall, the indumentum of corolla golden, of calyx brown; calyx deeply campanulate 6.5–7.5 × 3.5–4 mm, the ovate teeth unequal, 2–3 mm; corolla 12–13 mm, the lobes ±3 × 2.2–2.5 mm; androecium 32–36-merous, ±3 cm, biseriate; ovary glabrous, truncately dilated at apex. *Pods* straight to ever so slightly falcate 5.5–6.6 cm long, 1.75–2.0 cm broad over seed-cavities, somewhat constricted between them, the lustrous dark-brown finely venulose valves framed by dilated sutures to 2.0 mm broad; dehiscence primarlily follicular, though perhaps eventually through both sutures, the valves not twisting or contorting, internally light red or orange-red over seed-cavities; seeds somewhat obliquely inserted on a broad, strap-shaped, twisted funicle, the seed broadly ovate elliptic, in profile 8–9.5 × 6.5–7.5 mm, with a finely incised, U-shaped pleurogram 4.25–5 × 2.5–3 mm, the seed-coat bicolored, the basal ¼–⅓ white, the distal portion translucent, the embryo blue, but quickly fading to dark brown.

On wooded slopes and ridges, 1100–1525 m, local in the Pakaraima Mts and on the Gran Sabana in SE corner of state of Bolívar, Venezuela and adjacent state of Roraima (Sa Sabang), Brazil, in lat. ±4°30′–5°30′N. — Fl. XII–I(–?), VII, the full season conjectural.

Pithecolobium ferrugineum was described from a collection in flower, and this seems to be the only one seen by Bentham. In his monograph of Mimoseae (1875: 584) he nevertheless described the fruit as "legumen villosum, arcuatum v. circinatum, valvis 6–8 lin. latis contortis." We have seen only one fruit of *A. ferruginea*, and that fragmentary, but it is glabrous, as is the ovary of all known flowering specimens. The fruit described by Bentham, as noted by Sandwith in the protologue of *P. fanshawei*, was collected by Schomburgk and mistakenly associated with *P. ferrugineum* by Bentham. Sandwith thought it might belong to *P. fanshawei* or a close relative; we interpret the broken fragments of foliage as representing *A. barbouriana* var. *arenaria*.

This species is closely related to *A. floribunda* and perhaps not specifically distinct. *Abarema floribunda*, primarily a plant of lower elevations, has leaves of thinner texture, longer leaf-stalks (and consequently proportionately shorter peduncles), and somewhat larger peripheral flowers (as described in the key to species). The petiolar nectaries of *A. ferruginea* are flat or shallowly concave, whereas those of typical *A. floribunda* are convex. However, a single collection of the upper Mazaruni basin in Guyana, described above as a variant of *A. floribunda*, with which it agrees in length of peduncles, in texture of foliage, and in size of flowers, has the flat nectaries of almost sympatric *A. ferruginea*, flowers only a trifle smaller, and comes from a habitat at 950 m. Further collecting may show that *A. ferruginea* is only an upland variant of the already highly variable *A. floribunda*, but this problem must be postponed for lack of critical material.

3. Abarema gallorum Barneby & Grimes, sp. nov., *A. floribundae*, quacum legumine exacte quadrat, affinis, ab ea foliorum amplissimorum formula i–iii/4–5 (nec iii–vi/8–21) primo intuitu diversa, ab *A. jupunba* var. *trapezifolia*, sympatrica et pinnarum foliolorumque numeris congrua, foliolis maximis (5–8 cm usque longis) dorso subglabris necnon leguminis magni latioris 12.5–16 × 2.1–2.4 (nec 6–9.5 × 1–1.6) cm endocarpio tantum ad seminum loculos colorato, inter semina ochraceo, ut videtur distantior. — FRENCH GUIANA. SW de Sinnamary, piste de St. Elie (arbre n° E38), mars 1980 (fr), *Lescure 877*. — Holotypus, P *ex* ORSTOM; isotypus, CAY (2 sheets)!; Paratypi, s. loc., *H. Jacquemin 1553*, P (2 sheets, fr.).

Amply macrophyllidious, unarmed *trees* attaining 20(–?) m with trunk 2–3.5 dm dbh, the branchlets, lf-axes, and especially the axillary lf-buds densely tomentulose with sordid gray or brown hairs to 0.15–0.3 mm, the foliage bicolored, the lfts when dry stiffly papery, lustrous dark brown-olivaceous and glabrous above, beneath paler dull and remotely microscopically strigulose, the capitula solitary or fasciculate by 2–4 in coeval lf-axils. *Stipules* (few seen) lanceolate ±4–4.5 mm, caducous. *Lf formula* i–iii/4–5; lf-stks mostly 4–18 cm, the petiole 2–5 cm, at middle ±1.5–3 mm diam, shallowly narrowly sulcate ventrally, the one or the longer interpinnal segment 2.5–6.5 cm, but in some distal lvs with geminate pinnae the petiole only 1–2 cm; nectaries immediately below each pinna-pair sessile, round or scutiform, plane or shallowly concave, the first one 1.5–2.3 mm diam, the rest scarcely smaller, shallowly cupular, round or elliptic and below all but 1–2 proximal pair of lfts of each pinna-rachis; pinnae when more than one pair distally accrescent, the rachis of furthest pair 5.5–13.5 cm, the longest interfoliolar segment 1.5–4 cm; lft-pulvinules 1.6–2.8 × 0.9–1.4 mm, discolored, grossly cross-wrinkled; lateral lfts oblong around a diagonal, slightly porrect midrib, at base very inequilaterally rounded, at apex broadly obtuse and evidently or incipiently emarginate, the blades of the penultimate pair ±4–8 × 2–3.5 cm, 1.9–2.5 times as long as wide, the terminal pair of lfts broadly elliptic obtuse ±5–8.5 cm; the major secondary nerves ±9–15 on each side of midrib, nearly straight and widely ascending to anastomosis close within the plane corneous margin and connected by random 3- and 4-nary venules, the whole venation sharply finely prominulous on both faces. *Peduncles* 2.5–7.5 cm; receptacle obovoid, including short terminal pedestal 2.5–5 × 2–3.5 mm; bracts minute caducous; fls dimorphic, the pedicel of lower peripheral ones to ±1.2 × 0.6 mm, that of terminal fl or of several subterminal ones 0, the perianth of all densely minutely brownish silky-puberulent; PERIPHERAL FLS: calyx campanulate, narrowed into pedicel, 4.2–5 mm, the ovate teeth 0.4–0.9 mm; corolla pinkish, ±8 mm, the erect lobes 2–2.4 mm; androecium 26–32-merous, the stemonozone 1–1.2 mm, the tube 4.5–5.5 mm, the tassel pink, expanded at nightfall; nectarial disc 0; ovary oblong, a little dilated and abruptly truncate at puberulent apex; TERMINAL FL: calyx ±6 mm; corolla 11–12 mm, the lobes ±2.5 mm; androecial tube a little longer than corolla. *Pods* solitary, undulately broad-linear and falcately retroarcuate, 11–16 × 2–2.4 cm, contracted at base into a short pseudostipe and at apex into a short cusp, 10–14-seeded, the leathery, almost plane, fuscous glabrous evenulose valves framed by prominent sutures 1.5–2 mm wide, the endocarp smooth, brown in each seed-cavity, tan in between; dehiscence through both sutures, the valves randomly twisted; funicles linear compressed, sigmoid distally; seeds unknown.

In virgin forest at low elevations, known only from N French Guiana: in the Sinnamary basin at 53°W, 5°20′N; and in the Arataye basin at Station des Nouragues near 52°42′W, 4°3′N. — Map 8. — Fl. VIII–IX.

Etymology: of the French, in particular of the ORSTOM botanical collectors Jacquemin, Lescure, Prévost, and Sabatier, who obtained the plants described herein.

Abarema gallorum has the pink filaments, truncate ovary, and broad pod with endocarp mottled brown and tan that characterize *A. floribunda* and *A. ferruginea*, but is distinguished by the low number and great size of the leaflets. The almost sympatric *A. jupunba* var. *trapezifolia* is similar in leaf-formula, but differs in somewhat smaller flowers, white filaments, ovary tapering into the style, and a narrower pod (<1 cm, not 2–2.4 cm wide) uniformly orange inside. Gigantic shade-leaves (*Grimes 3338*, ex parte, NY) have leaf-stalks to 6.5 dm, pinna-rachises to 4 dm, and leaflets to 8 cm in length.

4. Abarema longipedunculata (Irwin) Barneby & Grimes, comb. nov. *Pithecellobium longipedunculatum* Irwin, Acta Bot. Venez. **2:** 223. 1967. — "Auyan-tepui, Estado Bolívar, Venezuela . . . a lo largo del Río Churún, vecindad del Campamento sur, sureste del 'Second Wall.' Alt. 1690 metros, 3 May 1964, *J. A. Steyermark 93316*." — Holotypus, NY! isotypi, F!, U!, US!, VEN!.

Macrophyllidious *shrubs* and *trees* 2–20 m, with densely rusty-tomentellous growing tips but early glabrate, the lvs bicolored, the coriaceous lfts dark green (when dry brown) and sublustrous above, paler dull (when dry tan) beneath, the capitula solitary and geminate in distal lf-axils, equaling or shorter than associated lvs. *Stipules* linear or linear-lanceolate 1.5–2.5 mm, early deciduous, absent from fruiting specimens. *Lf-formula* i–iii/3–5; lf-stks stout 0.7–5.5 cm, the petiole 0.7–2.2 cm, at middle 1.6–2.6 mm diam, the longer interpinnal segment(s) 0–1.8 cm; petiolar nectary between first pinna-pair sessile, round crateriform or shallow-cupular 2–3.5 mm diam, a similar one at tip of lf-stk and smaller, elliptic, shallowly concave ones at 1–3 further nodes of pinna-rachises; pinnae when more than one pair accrescent distally, the rachis of distal (or only) pair 4–7 cm, the longer interfoliolar segments 9–23 mm; paraphyllidia minute; pulvinules in dorsal view 1.6–3 × 1–1.5 mm, deeply cross-wrinkled; lfts distally accrescent, asymmetrically elliptic, oblong-, or obovate-elliptic from antically cuneate, postically rounded base, obtuse or obscurely emarginate, the larger ones 3.8–6.5 × 2-33.5 cm, 1.4–2 times as long as wide; the gently porrect midrib subcentric, the 9–12 pairs of major secondary nerves brochidodrome shortly within the thickened but scarcely revolute margin, the tertiary venulation weak and random, faintly raised on one or both faces of blade. *Peduncles* (3–)5–12.5 cm; capitula ±20–25-fld, the clavate receptacle 4–8 mm; bracts (caducous) not seen; fls (known only fallen) subsessile and obscurely pedicellate, probably dimorphic but the terminal one not observed, the perianth of peripheral ones 5-merous, densely brown-puberulent externally, the corolla more thinly so below the lobes; PERIPHERAL FLS: calyx narrowly campanulate 5–7.3 × 2–2.3 mm, the teeth 0.6–1.8 mm; corolla 8–9.5 mm, the lobes ±2 mm; androecium 30–36-merous, not seen complete and the color (probably white) uncertain, the stemonozone 1.2–1.8 mm, the tube 7–8.5 mm; ovary truncate and a trifle dilated at apex, glabrous; TERMINAL FL: unknown. *Pods* to 6 per capitulum, sessile or cuneately narrowed at base into an obscure pseudostipe, in profile linear, when well fertilized 7–10.5 × 1.3–1.7 cm, 10–13-seeded, gently evenly retrofalcate through less than ¼-circle, the plano-compressed leathery, glabrous reddish valves becoming low-convex over each developing seed, when dry fuscous, coarsely cross-nerved and reticulate, framed by evenly arched or (peripherally) shallowly undulate sutures 1.6–2 mm wide, the endocarp internally orange-red at seed-cavities and pallid between them; dehiscence through both sutures, the valves narrowly gaping to release

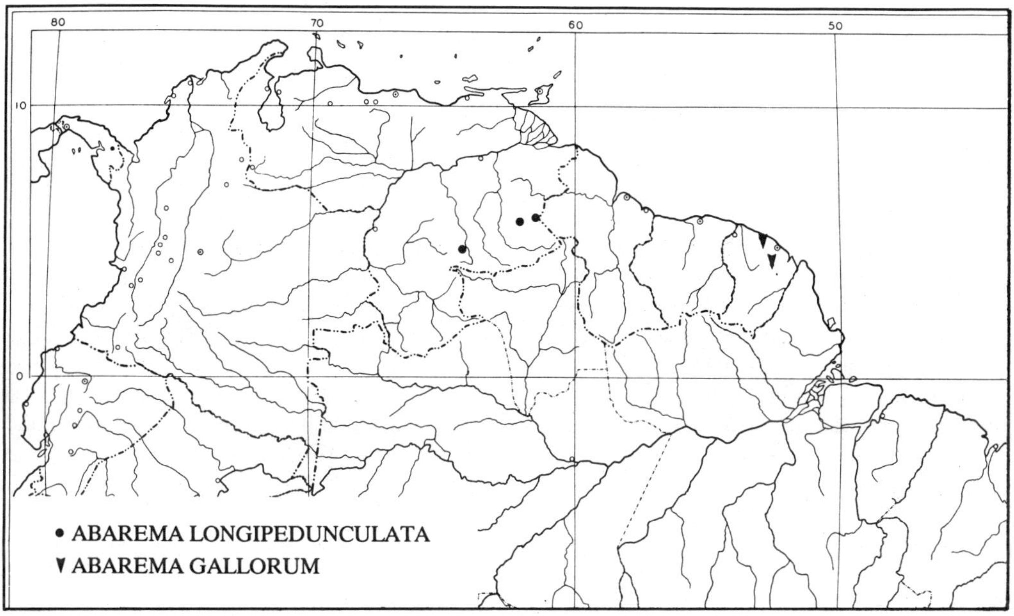

MAP 8. Venezuela and the Guianas. Range of *Abarema gallorum* Barneby & Grimes and *A. longipedunculata* (Irwin) Barneby & Grimes.

the seeds but apparently not elastically recurved or coiling; seeds plumply oblong-ellipsoid, in broad profile ±8.5–10.5 × 5.5–7 mm, the thin testa whitish at base and thence translucent pale tan, loosely vesting the gray-blue embryo, pleurogram 0.

On rocky summits, on cliff ledges, and in *zanjones* of sandstone table mountains, at (1100–)1400–2100 m, endemic to Venezuelan Guayana in state of Bolívar: Meseta Jáua, C. Auyántepui, and C. Venamo, to be sought on C. Guaiquinima and Chimantátepui. — Map 8.

The protologue of *P. longipedunculatum* pointed to *P. marginatum* and *P. villiferum* as its nearest relatives. The latter, also native but not endemic to sandstone table mountains of Guayana, differs obviously in softly pubescent hypophyllum of the leaflets and in pubescent ovary conical (not truncate) at apex. Because fruit and seed are still unknown in *A. villifera*, it was not included in the cladistic analysis, where *A. longipedunculata* is sister-group to the terminal containing *A. floribunda* and *A. ferruginea*. These together with *A. gallorum* form a clade united, in part, by the red or pinkish corolla and stamens.

Pithecellobium marginatum, superficially similar to *A. longipedunculata* in glabrous foliage and nearly straight pod, differs fundamentally in small long-pedicellate peripheral flowers and a pod that when ripe cracks transversely between seeds but does not dehisce conventionally through the sutures. Furthermore its pleurogrammic seeds are quite different from those of *A. longipedunculata*, which have a characteristic translucent, unmarked testa; and its lowland riparian habitat is further distinctive. It is referred herein to the genus *Hydrochorea*. Among the abaremas of the Guayana Highland *A. longipedunculata* is readily known by the combination of low leaf-formula, glabrous adult foliage, and gently falcate, not circinnate legume.

5. Abarema idiopoda (Blake) Barneby & Grimes, comb. nov. *Pithecolobium idiopodum* Blake, Contr. Gray Herb., n. ser. **52:** 70. 1917. — "BRITISH HONDURAS: . . . near Manatee Lagoon [Belize distr.], 12 May 1906, *M. E. Peck 437* . . . " — Holotypus, GH (not found in 1989); clastotypus (+ photo) NY!; isotypus, NY!.

Jupunba pseudo-tamarindus Britton *in* Britton & Rose, N. Amer. Fl. **23:** 193. 1928. — "Bocas Island, Bocas del Toro, Panama, January–March, 1928, *G. Proctor Cooper 461*." — Holotypus, NY!; isotypi, BM!, F!, US *1521844*!. — *Pithecolobium pseudo-tamarindus* (Britton) Standley, [Trop. Woods **16:** 17. 1928, nomen] Publ. Field Columbian Mus., Bot. Ser. **4:** 193. 1929.

Pithecolobium halogenes Standley, Publ. Carnegie Inst. Wash. **461:** 59. 1935. — "British Honduras: Punta Gorda,

mangrove swamp, July 3 1933, *William A. Schipp 1196*." — Holotypus, F!; isotypi, BM!, GH!, K!, NY!, UC *519115*!.
P. halogenes sensu Standley & Steyermark, 1946: 75.

Microphyllidious *trees* attaining 20 m with trunk to 2–8 dm dbh but flowering when smaller, the young stems, lf-axes and peduncles densely pilosulous with spreading-incurved or incumbent, pallid or commonly golden-brown hairs to 0.1–0.3 mm, the foliage bicolored, the lfts above dark green (brown when dry), sublustrous and glabrous except for sometimes microciliolate midrib, pale and either glabrous or remotely strigulose beneath, minutely ciliolate or not, the umbelliform capitula of white-stamened fls axillary to distal lvs, immersed in foliage. *Stipules* (few seen) narrowly lanceolate 2–3 mm, convex dorsally, caducous from small white scar. *Lf-formula* v–xiv (–xix)/18–40; lf-stk of major lvs 7–16 cm, the petiole including the enlarged pulvinus (1–)1.5–3.5(–4.3) cm, at middle 0.9–1.4 mm diam, the longer interpinnal segments 0.9–2 cm; petiolar-nectaries variable in position, size, and number, cupular on a broad pediment, the first situated at midpetiole, or close to first pinna-pair, or somewhere between these points, 0.8–2.4 mm diam, similar but smaller nectaries often inserted below some or all pinna-pairs (but sometimes wanting), and yet smaller ones on pinna-rachises at 1–6 furthest pairs of lfts; pinnae accrescent distally, often much so, the rachis of those at and beyond mid-lf 6–11 cm, the longer interfoliolar segments 2.5–5 mm; paraphyllidia immediately next to pinna-pulvini minute, caducous; pulvinules 0.5–0.7 × 0.4–0.6 mm; lfts decrescent at base of pinnae, thence subequilong, in outline narrowly oblong from inequilaterally cuneate or flabellate, rarely shallowly semicordate base, broadly obtuse, either straight or obscurely falcate, the larger ones 9–13 × 2.5–3.5 (or at antically dilated base sometimes –5) mm; the subcentric, straight or obscurely sigmoid midrib giving rise on its anterior side (or on both sides) to 5–8 weak, usually simple secondary veins faintly brochidodrome within the minutely thickened but not revolute margin, tertiary veinlets 0 or obscure, the whole venation immersed or almost so on upper face, finely prominulous beneath. *Peduncles* (3–)3.5–9.5 cm, lignescent in fruit and long persistent; capitula 25–55-fld, the axis including terminal pedestal (not always developed) 2.5–4.5 mm; bracts minute caducous; pedicels spreading and ascending 0.2–0.3 mm diam, the lowest (3)4–8.5 mm, the rest progressively shorter upward, that of the 1(–3), weakly dimorphic fls ±0.5–1.5 mm; perianth of all fls 5-merous (random exceptions), greenish white, thinly puberulent externally, the calyx-tube sometimes glabrate; PERIPHERAL FLS: calyx obconic-turbinate (1.5–)1.7–3 × 1.5–2.1 mm, the depressed-deltate teeth 0.1–0.6 mm; corolla 5–7.2 mm, the lobes 2–3 × 1.1–1.5 mm; androecium 24–36-merous, 20–24 mm, the stemonozone <1 mm or nearly obsolete, the tube 2–2.4 mm; ovary densely pubescent, obliquely truncate at apex, contracted at base into a short glabrate stipe; TERMINAL FL: calyx 2.2–3 × 1.6–2 mm; corolla 6–7.5 mm; stamen-tube 6–8.5 mm, as long as or shortly exserted from corolla. *Pods* 1(–2) per capitulum, sessile or shortly pseudostipitate, in profile undulately linear, curved through ±1–1.5 circles, ±7–10 × 1–1.4 cm, 8–12-seeded, the stiffly leathery, minutely puberulent, weakly venulose fuscous-brown valves framed by sutures 1.8–2.5 mm wide, at maturity low-convex over each seed, within dark reddish brown in seed-cavities, tan between; dehiscence through both sutures, the valves elastically recurving and coiling; seeds not seen, described by one collector (*Croat 15291*, MO) as bluish.

In rain forest, in mixed pine-broadleaf woodland, and in semideciduous forest, reported once from mangrove swamp, near sea level and to 1200 m, scattered on the Gulf and Caribbean slopes of S Mexico and Central America: E Oaxaca and N Chiapas E through N Guatemala and Belize to the Mosquito Coast of Nicaragua and central Panama, in Guatemala S to Suchitepéquez and in W Panama crossing the Cordillera to the Pacific slope in Veraguas. — Map 9. — Fl. III–V, X–XI.

Our description of *A. idiopoda* is based on 26 collections, but none of these has well-formed or well-preserved seeds. In the Central American flora, *A. idiopoda* is the only multifoliolate and microphyll abarema (leaflets of longer pinnae 18 pairs upward and none at midblade more than 3.5 mm wide) with truncate ovary and may be distinguished at anthesis from the similar but rarer *A. oxyphyllidia* by this feature combined with obtuse leaflets.

Pithecolobium idiopodum was overlooked by Britton & Rose in *North American Flora. Jupunba pseudotamarindus* and *Pithecolobium halogenes* were simply inadvertent duplications, not compared at the time of publication with any previously known species.

6. Abarema curvicarpa (Irwin) Barneby & Grimes, Acta Amaz. **14** (1/2, Supl.): 95. 1984 [1986]. *Pithecellobium curvicarpum* Irwin, Mem. New York Bot. Gard. **15:** 107. 1966. — Typus infra sub var. *curvicarpa* indicatur. Fig. 6

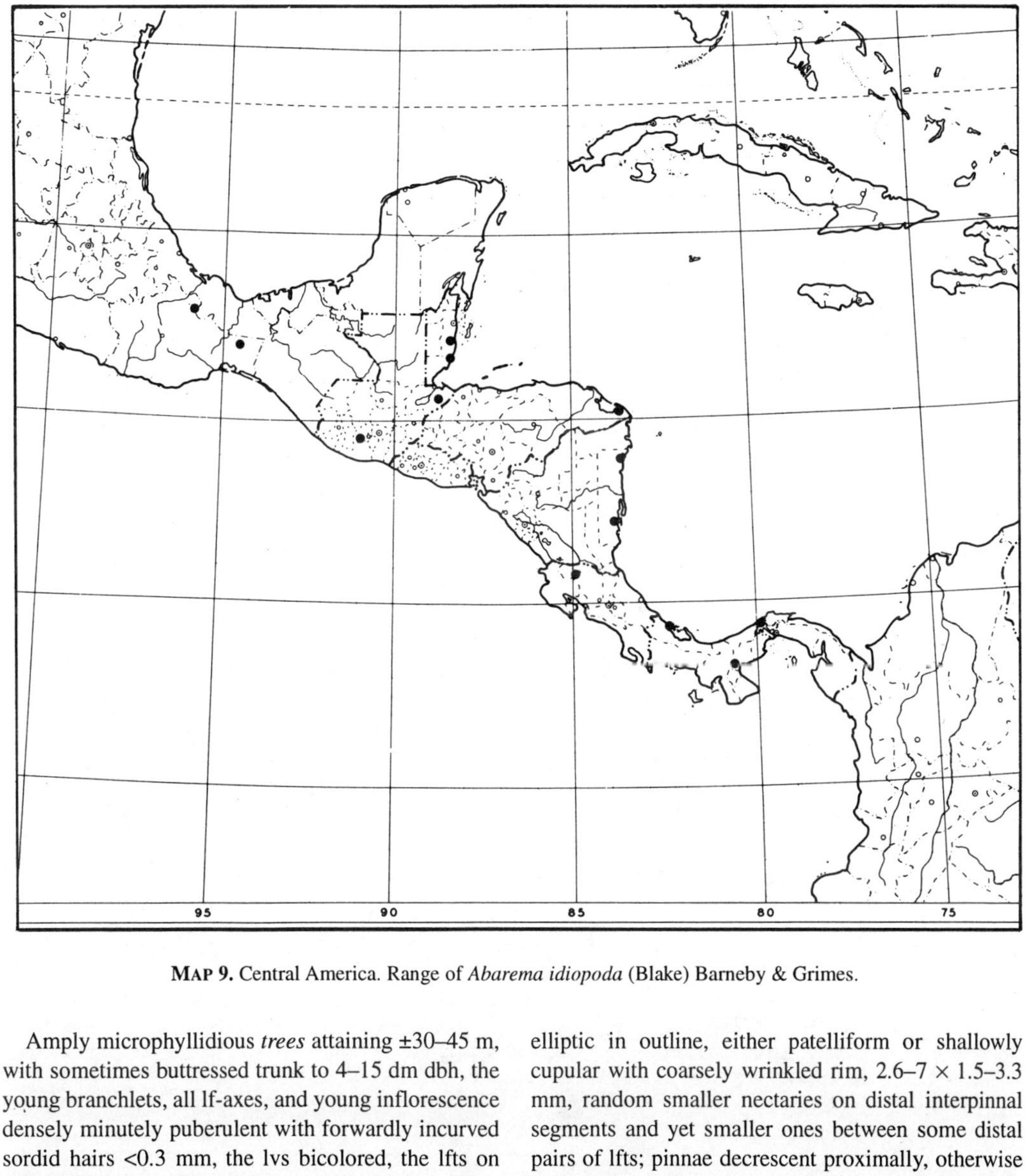

MAP 9. Central America. Range of *Abarema idiopoda* (Blake) Barneby & Grimes.

Amply microphyllidious *trees* attaining ±30–45 m, with sometimes buttressed trunk to 4–15 dm dbh, the young branchlets, all lf-axes, and young inflorescence densely minutely puberulent with forwardly incurved sordid hairs <0.3 mm, the lvs bicolored, the lfts on upper face brown-olivaceous smooth, glabrous except for a line of hairs along midrib, paler dull and remotely minutely strigulose or glabrate beneath, the spiciform racemes arising singly or geminate in the axil either of coeval or of new-fallen lvs, the ligneous fruits borne below current foliage. *Stipules* 0 (absent in vernation). *Lf-formula* (v–)vi–xii (on shade lvs –xiv)/14–26 (lfts 120 or more per lf); lf-stks (6–)9–15 cm, the ventrally flattened petiole 2–3.5 cm, the longer interpinnal segments ±12–17 mm; lf-nectaries sessile, the first, situated either near mid-petiole or immediately below insertion of proximal pinna-pair, elliptic in outline, either patelliform or shallowly cupular with coarsely wrinkled rim, 2.6–7 × 1.5–3.3 mm, random smaller nectaries on distal interpinnal segments and yet smaller ones between some distal pairs of lfts; pinnae decrescent proximally, otherwise subequilong, the rachis of those at and beyond mid-lf 4–9 cm, the longer interfoliolar segments 2–6.5 mm; lfts slightly decrescent at base or at each end of rachis, the rest (disregarding the slightly broader furthest pair) subequiform, in outline obliquely oblong, elliptic-oblong or linear-oblong from inequilateral base, broadly rounded or shallowly emarginate at apex, the longer ones 7–18 × 1.8–6 mm, 2.7–4.4 times as long as wide; venation weakly prominulous on lower face only, the slender subcentric midrib either straight or obscurely sigmoidally curved, either simple or pinnately branched, the 4–7 pairs of secondary nerves

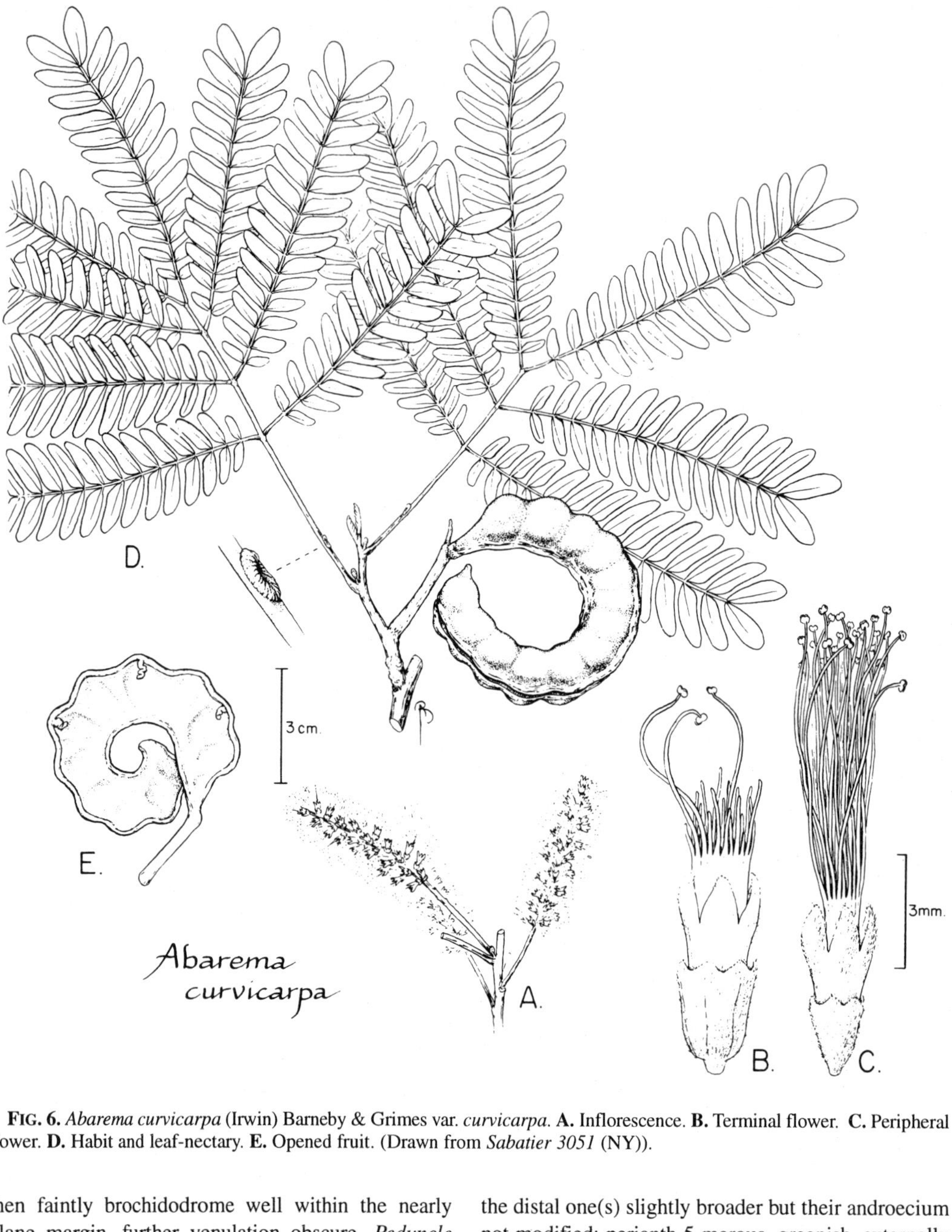

FIG. 6. *Abarema curvicarpa* (Irwin) Barneby & Grimes var. *curvicarpa*. **A.** Inflorescence. **B.** Terminal flower. **C.** Peripheral flower. **D.** Habit and leaf-nectary. **E.** Opened fruit. (Drawn from *Sabatier 3051* (NY)).

then faintly brochidodrome well within the nearly plane margin, further venulation obscure. *Peduncle* and raceme-axis together ±6–12 cm, the raceme ±30–50-fld, the floral axis in var. *curvicarpa* becoming 3.5–4.5 cm, in var. *rodriguesii* perhaps shorter; bracts (few seen) very early dry and caducous, linear-oblanceolate 1–1.6 mm; pedicel of proximal fls 0.3–0.9 mm, that of distal ones as broad as long or obsolete; fls (of var. *curvicarpa*) subhomomorphic, the distal one(s) slightly broader but their androecium not modified; perianth 5-merous, greenish, externally yellow-silky-puberulent, the corolla lobes tipped with white polyp-like trichomes, the filaments white; calyx submembranous campanulate, that of most fls 2–2.2 × 1.6 mm, the depressed-deltate teeth 0.2 mm, that of the furthest fl ±2.6–3 × 2 mm; corolla 4.5–5.2 mm, the ovate lobes ±1.3 × 0.9 mm; androecium ±28-merous, 11–12 mm, the stemonozone ±1.5 mm, the tube 5–5.5

× 1 mm; ovary glabrous; style scarcely dilated at apex. *Pods* turgid, moniliform and evenly incurved through 1–2 full spirals, 10–16-seeded, in profile ±10–16 × 1.7–2.3 cm, contracted at base into a cuneate neck, the prominent, undulate or serpentine ventral (seminiferous) and evenly recurved dorsal (interior) sutures 4–8 mm wide, the ventral one rather shallowly depressed or deeply recessed between seeds, the woody valves to 4–8 mm thick where grossly dilated and domed over each seed, externally coarsely rugose, bright red or red and yellow, glabrous, the endocarp orange-red, the seed cavities separated by solid isthmi; funicle ligulate ±1.2–2.3 mm wide, sigmoid distally; dehiscence tardy, primarily along the ventral suture, the valves widely gaping to disperse the seeds, their thick woody texture inhibiting lateral twist; seeds (of var. *curvicarpa*) transverse, compressed-obovoid, in broad view ±9 × 7 mm, the testa (at least when fresh) white on basal portion, translucent above, ±0.1 mm thick in section, loosely investing the fuscous, when fresh aniline-tinged, coarsely wrinkled embryo, pleurogram 0. Eophylls opposite, bipinnate, these and next few lvs of formula i/6–12.

Assuming that the grossly lignified pods of *A. curvicarpa* and *A. macradenia* are homologous, these species must be more closely related to each other than to any known species. They are further alike in the broader, but otherwise homomorphic flowers and lack of pleurogram on the seed. Because the two species differ in length of the floral receptacle they are not directly contrasted in the key to the species, as they deserve:

1. Lf-formula v–xii/14–26, the larger lfts 4–18 × 1.8–6 mm; first lf-nectary shallowly cupular or patelliform, in profile 1.5–3.3 mm tall; floral receptacle (unknown in var. *rodriguesii*) 3.5–4.5 cm; corolla of peripheral fls 4.5–5.2 mm; central Amazonian Brazil and the Guianas .... *A. curvicarpa*
1. Lf-formula ii–vi/7–12, the larger lfts 27–46 × 9–20 mm; first lf-nectary campanulate, in pro file 4–11 mm tall; floral receptacle 5–11 mm; corolla of peripheral fls 6.5–9.5 mm; SE Central America and Pacific NW South America (S to Ecuador) .... *A. macradenia*

The bicentric range of *A. curvicarpa*, which resembles that of *Monopteryx inpae* W. Rodrigues (Barneby & Grimes, 1984: 46), coincides with somewhat modified foliage and perhaps with variation in length of the floral receptacle, but this will remain uncertain until flowering material from central Amazonia is obtained. The fruits are essentially similar at the extremes of dispersal, but the differences in size and venation of the leaflets require formal recognition.

Key to the varieties of *A. curvicarpa*

1. Lfts oblong, the longer ones 11.5–18 × 3.5–6 mm; hypophyllum weakly but evidently pinnate-veined, the secondary venules brochidodrome; known only from the Guianas .... 6a. var. *curvicarpa*
1. Lfts linear-oblong, the longer ones 7–11 × 1.8–2.7 mm, dorsally 1-nerved; known from centr. Brazilian Amazonia (vic. of Manaus) .... 6b. var. *rodriguesii*

6a. Abarema curvicarpa (Irwin) Barneby & Grimes var. **curvicarpa.** *Pithecellobium curvicarpum* Irwin, 1966, l. c., fig. 1(G–L). — "Essequibo R., British Guiana, Kamuni Creek, Gorete (= Groete) Creek [±6°40′N, 58°40′W], *Bassett Maguire & D. B. Fanshawe 22950*." — Holotypus, collected 21.IV.1944 (fr) NY!; isotypi, G!, GH!, MO!, US *190957*!.

Lfts relatively ample and perceptibly pinnate-veined on lower face, as described in key to varieties; *inflorescence* and fls as described for the species.

In humid riverine and moist upland forest, reported in French Guiana from 200 to 400 m, known only from the lower Essequibo valley in Guyana (Groete Creek) and from French Guiana (Piste St. Elie; Saül). — Map 10. — Fr. IV–V(–VIII). — Foliage from a sterile tree (no. *1503*, collector unknown, herb. no *1847*, MO) from Watramiri, Surinam, may possibly represent *A. curvicarpa*.

6b. Abarema curvicarpa var. **rodriguesii** Barneby & Grimes, var. nov., a var. *curvicarpa* foliolis minoribus angustioribus (7–11 × 1.8–2.7, nec 11.5–18 × 3.5–6 mm usque), dorso 1-nerviis (nec manifeste pinnatim nervosis) patriaque remota abstans. — BRAZIL. Amazonas: Manaus, Reserva Florestal Ducke, ao lado da Q.11 no bosque, 6.5.65 (fr.), *W. Rodrigues* (with *Osmarino*) *6930*. Holotypus, INPA *15483* = NY Neg. *12665*.

Lfts relatively small and narrow, 1-nerved, as described in key to varieties; *inflorescence* not known at anthesis, the receptacle perhaps shorter than that of var. *curvicarpa*, the fruit essentially similar.

In forest on heavy clay soils, known only from centr. Amazonian Brazil (Reserva Ducke near Manaus, and 30 km N of Manaus on road to Caracaraí). — Map 10. — Fl. season unknown.

7. Abarema macradenia (Pittier) Barneby & Grimes, comb. nov. *Pithecollobium macradenium* Pittier, Contr. U.S. Natl. Herb. **20:** 465. 1922. — ". . . at Monte Lirio, Canal Zone, Panama, May 6, 1912, by E. D. Christopherson (no. 196)." — Holotypus, US

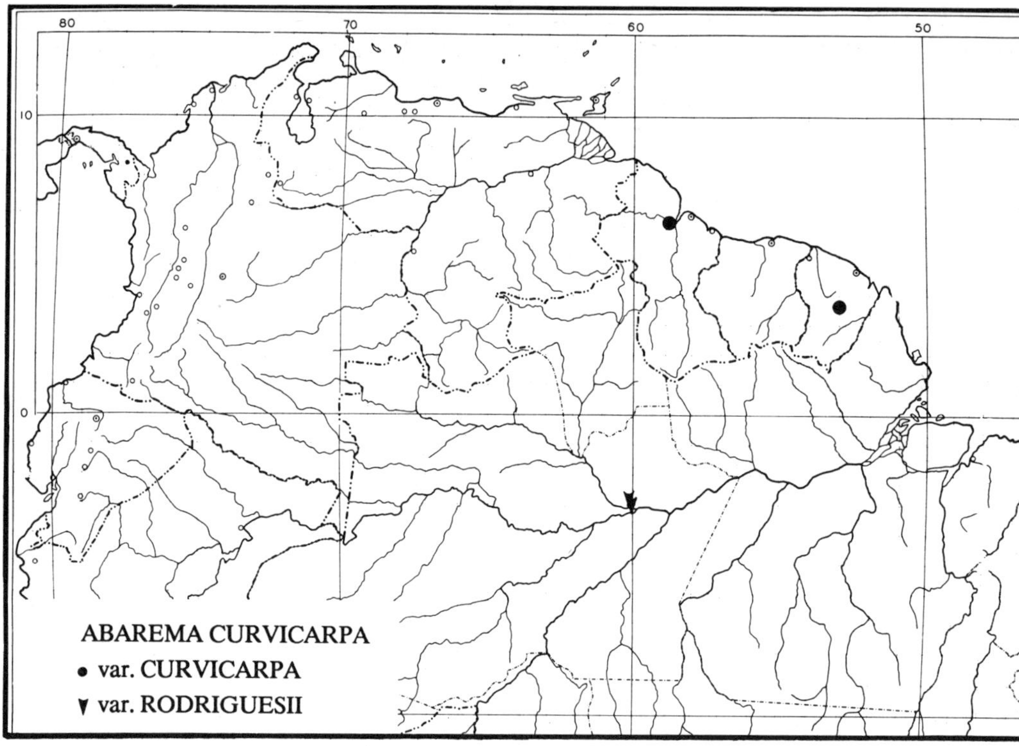

MAP 10. The Guianas and central Amazonia. Range of *Abarema curvicarpa* (Irwin) Barneby & Grimes var. *curvicarpa*, and var. *rodriguesii* Barneby & Grimes.

716842!; isotypi, NY! = NY Neg. *10365*, US!. — *Samanea macradenia* (Pittier) Britton & Rose, N. Amer. Fl. **23:** 35. 1928.

Pithecellobium macradenium sensu Elias, Bot. Gaz. 133: 38, figs. 1–7 (anatomy of lf-nectaries). 1972.

Amply macrophyllidious, unarmed, broad-crowned *trees* 9–36 m when fertile and attaining 12 dm dbh, the longitudinally sulcate young stems, all lf-axes, and inflorescence densely sordid-puberulent with incurved hairs to 0.1–0.2 mm, the lvs strongly bicolored, the thinly papery lfts on upper face olivaceous (brunnescent) and except for minutely ciliolate midrib glabrous, on lower face pale green and minutely thinly strigulose (along midrib sometimes villosulous), the capituliform racemes of pale green, white-stamened fls solitary or 2–3-nate in the axil of coevally expanding lvs or sometimes forming when young a subcorymbose panicle, the woody fruits long persistent below the new lvs. *Stipules* subulate to 2–5.5 mm, very early caducous from a small pallid scar. *Lf-formula* of lvs associated with fls or fruits (ii–)iii–vi/7–12, that of sapling lvs, described below in brackets only, vi–xii/11–20; lf-stks (7–)9–24 [25–60] cm, the petiole 2.5–6.5 [6–18] cm, the longer interpinnal segments (2–)2.5–4 [3.5–7.5] cm; petiolar nectaries of lvs associated with flowers dimorphic, the first one, situated shortly below insertion of first pair of pinnae, sessile campanulate 5–15 × 4–11 mm, its walls leathery and striate lengthwise, becoming subligneous in age, the nectaries between further pairs of pinnae either sessile or stipitate, cupular ±1 mm diam, and yet smaller cupular ones between some distal lft-pairs [nectaries of ample juvenile lvs all small or the first petiolar one lacking]; pinnae accrescent distally, the rachis of furthest pair 8–15 [15–27] cm, the longer interfoliolar segments 11–19 mm; pulvinules in dorsal view 0.6–1.2 × 0.6–0.8 mm, finely wrinkled; lfts proximally decrescent, those from mid-rachis upward (except for slightly broader distal pair no further mentioned) subequiform, in outline obliquely oblong-elliptic or very obtusely rhombic around a diagonal, straight or slightly incurved midrib, inequilaterally flabellate at base, broadly rounded and muticous at apex, the larger ones (27)30–46 × 13–20 [9–14] mm, 1.8–2.7(–3) times as long as wide;

the midrib giving rise on each side to 7–10 major (and random weaker intercalary) secondary nerves widely ascending to anastomosis well within the subrevolute margin, the tertiary venules weak and random, all venation subequally prominulous on each face. *Peduncles* 5–12 cm, in fruit lignescent 2–5 mm diam; racemes ±30–45-fld, the narrowly clavate receptacle 5–11 mm; bracts caducous, linear-spatulate ±2 mm (or the first one, lacking a subtended fl, broader); fls subhomomorphic except for distal ones slightly larger and sessile or very shortly pedicellate, their androecium hardly modified; pedicel of lower peripheral fls 0.7–3.6 × ±0.3 mm, that of distal ones progressively shorter; perianth 5-merous, finely yellowish silky-puberulent externally, the corolla-lobes ciliolate with white coralloid trichomes; calyx campanulate 2.5–4 × 1.6–2.2 mm, the ovate or depressed-deltate, sometimes very unequal teeth (0.2–) 0.4–1.2(–1.4) mm; corolla narrowly trumpet-shaped (6.5–)7–9.5 mm, the ovate acute lobes 1.4–3.2 × 1.3–1.8 mm; androecium 18–26-merous, to 18–25 mm long, the stemonozone 1.2–2 mm, the tube 3.5–7.5 mm; ovary glabrous truncate; style a little shorter or longer than longest stamens, the stigma punctiform. *Pods* usually 1 per capitulum, subsessile, in profile broad-linear but evenly recurved through ¾ to 1.5 circle, 9–14 × 1.3–2.4 cm, compressed but strongly turgid over each seed and depressed between them, the prominent sutures 3–5 mm wide, the exterior (seminiferous) one shallowly undulate, the interior one evenly curved, the woody valves externally coarsely rugulose, fuscous-castaneous glabrous, nearly 3 mm thick over each seed-locule, the transverse depressions between seeds 3–8 mm deep, the crustaceous endocarp ±0.5 mm thick in section, the seed-locules castaneous or dark crimson internally; funicles ligulate, coiled distally; dehiscence of the genus, the valves rigidly recurved and coiled; seeds 10–15, horizontal, in broad view 6.5–8 × 5–7 mm, the translucent testa white in proximal ⅓–½, pale tan distally, in section at most 0.2 mm thick, loosely investing the scarcely rugulose blue embryo, pleurogram 0.

In humid primary forest, sometimes surviving disturbance as a shade tree in pasture, apparently local in SE Central America and in N Pacific South America: Nicaragua (Zelaya, at 13°3′N); in hill-country at 550 m in S Costa Rica (Puntarenas); near sea level in the Canal Zone of Panama (Isla Barro Colorado; Mte Lirio; Fort S. Lorenzo); and at 70–220 m in prov. Los Ríos, Ecuador (Palenque Biological Station; Jauneche Forest; Pichilingue). — Map 11. — Fl. in Central America I–V, in Ecuador VIII–IX, the full season probably greater than recorded. — *Coralillo* (Nicaragua), *guavo, guavo de montana* (Costa Rica); *bantano* (Ecuador).

This is a species ordinarily recognized by tall arboreal stature and either by the enlarged leathery bell-shaped petiolar nectary on leaves associated with flowers or by the woody pod grossly enlarged over the seed-locules, or by both. It appears most closely related to the Guyanan *A. curvicarpa*, which has a comparable ligneous pod but patelliform leaf-nectaries, significantly different leaf-formula, and an extended raceme-axis.

There is incipient racial differentiation between the Central American and Ecuadorean populations of *A. macradenia*, but too little marked as yet to deserve taxonomic separation. The calyx in Ecuador is on the average smaller (2.5–3.3, not 3.3–4 mm long), and the pod rather narrower (mostly 13–17, not 19–24 mm wide), with shallower transverse depressions between seeds. A sterile specimen from Pacific Colombia (Valle: 10 km NW of Buenaventura; *A. Gentry 56959*, MO, NY), not certainly of this species, suggests a link between the at-present widely dissevered known localities. The enlarged petiolar nectary is sometimes eaten down to the base, perhaps by ants, and is not developed in the leaves of saplings or stump-sprouts, leaves twice or more the size of those on fertile branchlets of adult trees.

8. Abarema ganymedea Barneby & Grimes, sp. nov., omnium fere partium adspectu ovarioque glabro *Ab. auriculatae* proxima, attamen pinnarum longiorum foliolis 14–19 (nec 9–12)-jugis usque, infra parce puberulis (nec glabris), petiolorum nectario primo majusculo poculato (nec verruciformi), bracteis floralibus caducis (nec persistentibus), florum periphericorum androecio 10 (nec 14–18)-mero, necnon patria valde remota abstans. — COLOMBIA. Antioquia: Chigorodó, carretera a Turbo, 100–200 m, 20–23.xii.1962, *H. García-Barriga 17624*. — Holotypus, US *2558153* = NY Neg. *12544*; isotypi, COL, GH.

Amply microphyllidious *trees* attaining 20 m, with much-branched crown and mottled gray annotinous branches, the young stems, all lf-axes and peduncles densely pilosulous with sordid but pallescent hairs <0.3 mm, the lvs bicolored, the lfts when dry dark brown and except for exiguously ciliolate midrib glabrous above, beneath dull pale gray-green and minutely remotely strigulose, the capitula of white fls solitary and paired in the axil of few distal lvs,

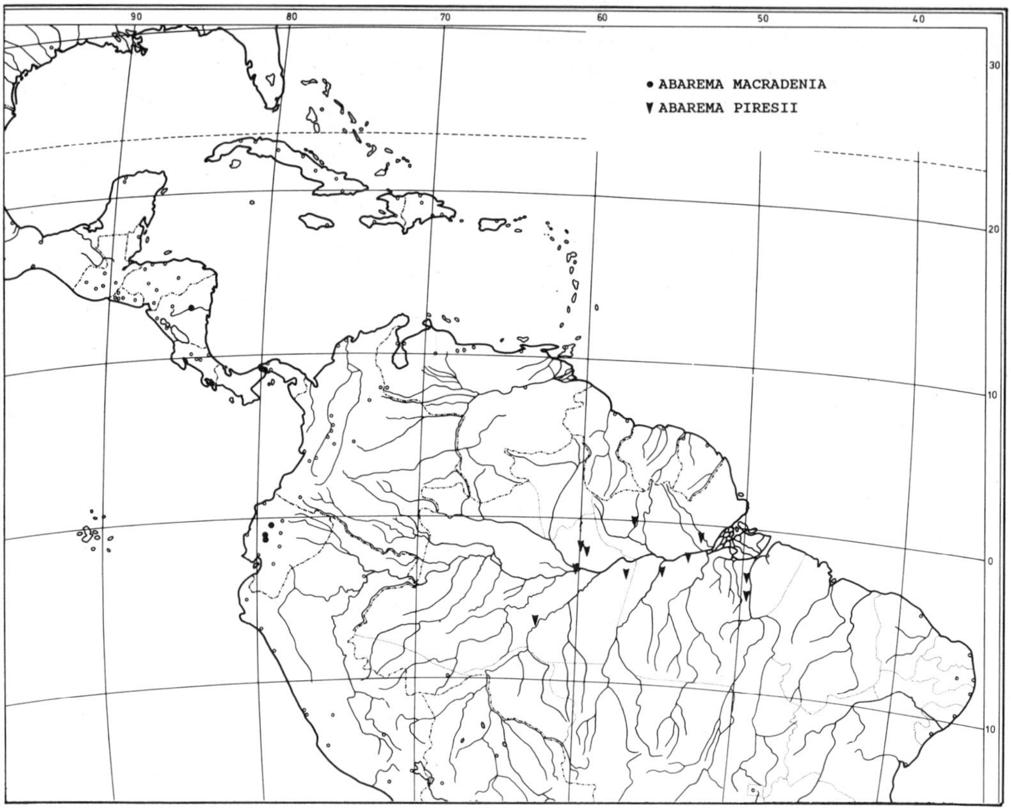

Map 11. Central America and northern South America. Range of *Abarema macradenia* (Pittier) Barneby & Grimes and *A. piresii* Barneby & Grimes.

immersed in foliage. *Stipules* (few seen) lanceolate ±2.5 mm, pubescent dorsally like stem, caducous. *Lf-formula* (vi–)viii–x/14–19; lf-stks 8–13.5 mm, the petiole 10–19 mm, at middle 1.2–1.5 mm diam, the longer interpinnal segments 10–15 mm; a coarsely coriaceous infundibuliform nectary 1–2 mm tall and 1.2–2.5 mm diam immediately below first pair of pinnae and similar but smaller ones between 2–4 distal pinna-pairs and at tip of each pinna-rachis; pinnae proximally decrescent, thence subequilong, the rachis of median and distal ones 4–7 mm, the first pair of lfts arising directly next to pinna-pulvinus, the longer interfoliolar segments 3–5.5 mm; lft-pulvinules ±0.2 × 0.4 mm; lfts (except for slightly broader furthest pair) subequilong, in outline oblong from inequilaterally flabellate base, broadly obtuse, the larger ones 6.5–8.5 × 2.7–3.5 mm, ±2.4–2.8 times as long as wide; venation of 1–2 nerves from pulvinule, the medially diagonal, nearly straight midrib and one fine (almost obsolete) posterior nerve incurved to anastomosis a little beyond midblade and close within the plane, minutely ciliolate margin, the midrib pinnately 6-branched on each side, tertiary venules random obscure, the whole venation discolored but immersed on upper face, finely prominulous beneath. *Peduncles* 1.5–4 cm; capitula ±12–25-fld, without filaments ±10 mm diam, the receptacle including short terminal pedestal 2–3 mm; bracts obovate 0.5–0.8 mm, sericeous dorsally, caducous; fls dimorphic, the peripheral ones raised on pedicel to 0.5–0.8 mm, the terminal one sessile, larger, and with modified androecium, the perianth of all 5-merous and sordid-strigulose externally, the corolla lobes densely so at tip; PERIPHERAL FLS: calyx ±1.8 × 1.1 mm, the deltate teeth 0.3–0.5 mm; corolla 4–4.2 mm, the ovate lobes 1.7–0.7 mm; androecium 10-merous, to 15.5 mm, the stemonozone almost 0, the tube 1.3 mm; ovary glabrous, conical at apex; style longer than longest filaments, the stigma poriform; TERMINAL FL: calyx 2.8 × 1 mm; corolla 6.5 mm; androecium 15-merous, the stemonozone 1.8 mm, the stout and firm, apically dilated tube as long as corolla, the free filaments thickened proximally;

gynoecium depauperate, probably infertile; style included. *Pods* geotropic, undulately linear ±14 × 1 cm, attenuate at base into a pseudostipe ±1 cm, cuspidate at apex, ±11-seeded, the body recurved through more than 360° and also randomly twisted, the seminiferous suture deeply recessed between seeds, the brownish puberulent leathery valves coarsely reticulate-venulose over each seed; ripe seeds not seen.

In wet forest, known only from between 100 and 200 m along the road between Chigorodó and Turbo, near 8°N in far N Antioquia, Colombia, and from 250 m in Cantón S. Lorenzo at 0°55′N in Esmeraldas, Ecuador. — Map 12. — Fl. XII–?.

Etymology: of Ganymedes, cup-bearer to Zeus, in allusion to the cupular petiolar nectaries.

Abarema ganymedea suggests a Pacific lowland counterpart of the Amazonian *A. auriculata*, from which it differs in more numerous leaflets (see Latin diagnosis for figures) puberulent on lower face, in the cupular (not mounded) petiolar nectary, and (in one known flowering specimen) an androecium reduced to 10 members. If we correctly associate with the flowering holotype the fruiting specimen (*Rubio & Quelal 722*, NY) from Ecuador, *A. ganymedea* differs further from *A. auriculata* in the deeply recessed ventral suture of the pod, which altogether is reminiscent of *A. piresii* in quasimoniliform outline.

9. Abarema piresii Barneby & Grimes, Acta Amaz. **14** (1/2, Supl.): 96, fig. 1. 1984 [1986]. — "Brazil, Fronteira Amazonas–Pará–Mato Grosso: Boca do rio Juruena, terra firme, 31.XII.1951, *J. Murça Pires 3704*."–Holotypus, NY!; isotypus, IAN!.

Microphyllidious *trees* (5–)8–25 m with trunk 1.2–4 dm dbh, the young branches, lf-axes and peduncles densely puberulent with sordid or rusty, forwardly incurved hairs to 0.15–0.3 mm, the lfts bicolored, lustrous olivaceous above, paler dull beneath, except for often minutely puberulent midrib glabrous on both faces but minutely remotely ciliolate, the loosely few-fld capitula solitary or geminate in distal lf-axils, immersed in foliage. *Stipules* erect, linear-attenuate 2–5 × 0.4–0.5 mm, caducous from small white callus. *Lf-formula* (i–)ii–iii(–iv)/8–12(–13); lf-stks (0.5–) 1.2–5.5 cm, the petiole including a scarcely differentiated pulvinus 3–10 mm, the longer interpinnal segments (0–)3–25 mm, the shallow ventral sulcus interrupted at each pair of pinnae by a globose, pyriform or fungiform, broadly obtuse or subtruncate nectary 0.6–1.4 × 0.6–2.5 mm, and smaller, stipitate or less often sessile nectaries at 4–8 further nodes of pinna-rachises; pinnae accrescent distally, the rachises of distal pair (3–)4–10.5 cm, the longer interfoliolar segments 5–9 mm; lft-pulvinules 0.5–0.8 × 0.5–0.9 mm; lfts decrescent proximally, thence (except for longer narrower furthest pair) subequilong, the opposite blades sessile against rachis, in outline obtusely rhombic from broadly cuneate base, those near and beyond midrachis 12–23 × 7–12 mm, 1.6–2.3 times as long as wide; the diagonal, nearly straight midrib giving rise to crowded secondary venules weakly brochidodrome just within the narrowly revolute margin, the whole venation bluntly prominulous above, more sharply so beneath. *Peduncles* (1.5–)2–5.5 cm; capitula ±12–22-fld, without androecia ±9–10 mm diam, the receptacle 3–5 mm; bracts herbaceous lanceolate 2–4.5 × 0.4–1 mm, 3–5-nerved, persistent at least through anthesis, often into fruiting; fls dimorphic, all sessile or almost so, the greenish perianth of all 5-merous, densely brown-puberulent on teeth or lobes but glabrescent proximally, the filaments white; PERIPHERAL FLS: calyx narrowly or turbinately campanulate, 3–3.5 mm, the obtuse teeth 0.8–1 mm; corolla cylindro-campanulate, 5–6 mm, the lanceolate lobes 1.7–2.5 × 0.8–1 mm; androecium 12–24-merous, 15–21 mm, the stemonozone ±1.5 mm, the tube 2–4 mm; ovary obliquely truncate at apex, glabrous; TERMINAL FL: calyx 2.7–4 × 2–2.5 mm; corolla 7–9.5 mm; staminal tube 9–10 mm. *Pods* 1–2 per capitulum, sessile or nearly so, persistent, in profile submoniliform, recurved or equivalently contorted through 1–1.5 circles, the dorsal suture evenly decurved, the ventral suture deeply impressed between seeds to form isthmi 1.5–4 mm wide, the (5–)8–14 biconvex 1-seeded segments 9–11 mm diam, as wide as or a little wider than long, framed by dorsally plane sutures, the ventral one 1.2–1.6 mm wide, the subligneous valves externally orange brunnescent, glabrous overall, densely coarsely sinuously venulose and reticulate, internally pale brown (not red), the endocarp adherent to mesocarp; dehiscence follicular, along the seminiferous suture only, the valves gaping to expose the seeds, these suspended on a crumpled but not dilated funicle, filling the cavity, when not quite ripe plumply discoid ±6.5 mm diam, the pale brown, translucent testa loosely investing the embryo, pleurogram 0.

In forest or at forest margin, on terra firme, preferring clayey or sandy clay soils, below 150 m, occasional in E Amazonian Brazil in lat. ±0–7°30′S, from E Amazonas downstream to Río Jarí in Pará, S to the lower Madeira and middle Tapajós rivers, at the mouth of Río Juruena perhaps entering extreme N Mato Grosso. — Map 11. — Fl. VI–IX. — *Angico branco* (a name applied also to allied species).

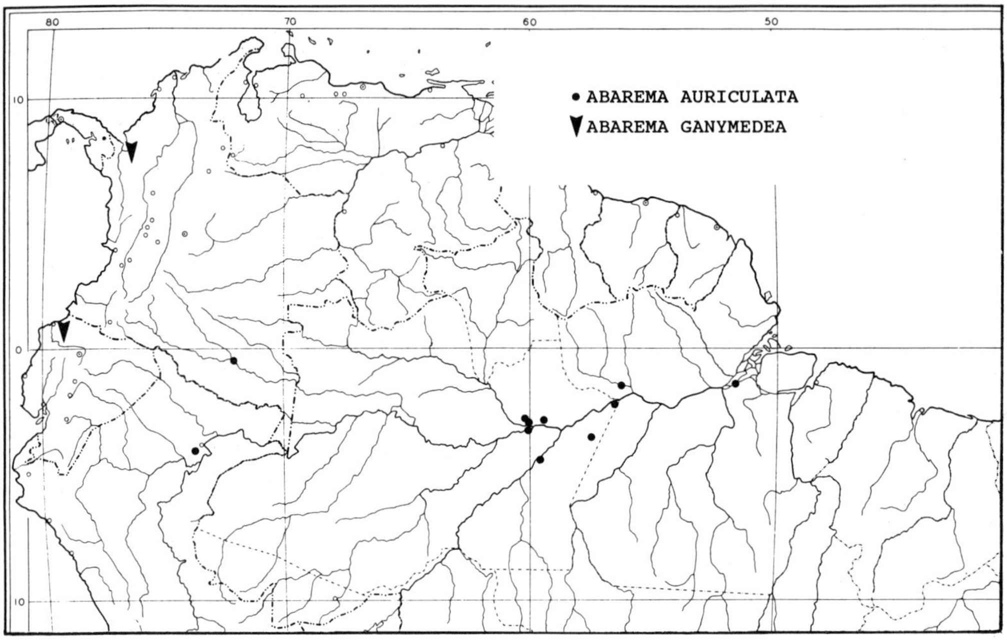

MAP 12. Northern South America. Range of *Abarema auriculata* (Bentham) Barneby & Grimes and *A. ganymedea* Barneby & Grimes.

The leaves of *A. piresii* resemble those of the uncommon form of *A. jupunba* with relatively few (2–3) pairs of pinnae, but the moniliform fruit deeply indented from the ventral side into a series of one-seeded lobes, and dehiscent only through the gaping ventral suture, is diagnostic. The species was named in honor of the distinguished botanist João Murça Pires.

10. Abarema auriculata (Bentham) Barneby & Grimes, comb. nov. *Pithecolobium auriculatum* Bentham, London J. Bot. **3:** 217. 1844. — "Borba, on the Río Negro [in reality on lower Río Madeira], in Brazil, [*Riedel*, commun.] *Langsdorff*." — Holotypus, K(herb. bentham.)! = NY Neg, *2010*; isotypi, A!, OXF!. — *Feuilleea auriculata* (Bentham) O. Kuntze, Revis. Gen. Pl. **1:** 187. 1891.

Pithecolobium auriculatum sensu Bentham, 1875: 584; 1876: 438; Ducke, 1949: 37; Spichiger, Boissiera **43:** 344, fig. 172. 1989.

Pithecolobium malacotrichum Harms, Feddes Repert. Spec. Nov. Regni Veg. **17:** 91. 1891. — "Brasilien: Rio de Janeiro (*Glaziou no. 10581*, wohl such no. *9407* . . .)." — Holotypus, †B = F Neg, *1209*!.

Microphyllidious *shrubs* and slender *trees* (2–)3–15 m tall, the young stems and all axes of lvs and inflorescence densely puberulent with forwardly incurved, sordid or golden-brown hairs to 0.1–0.2 mm, the foliage strongly bicolored, the plane, facially glabrous but often minutely ciliolate lfts when dry dark brown and nearly always pallid-spotted above, paler beneath, the capitula of whitish fls single and paired in coeval lf-axils and immersed in foliage. *Stipules* linear, to 0.7–1.8 mm, caducous before expansion of associated lf. *Lf-formula* v–ix/9–12; lf-stk of larger lvs (5–)6–14(–18) mm, the petiole including dilated fuscous pulvinus 5–13(–16) mm, the longer interpinnal segments 8–18 mm; petiolar nectary (rarely 2) immediately below insertion of first pair of pinnae verruci- or mammiform 0.7–1.3 mm diam, a smaller nectary sometimes at tip of lf-stk and yet smaller cupular ones often at furthest 1–2 pairs of lfts; pinnae abruptly decrescent near base of lf-stk, thence subequilong and often not exactly opposite, the longer ones (3.5–)4–7 cm, the longer interfoliolar segments 4–6 mm; pulvinules 0.3–0.5 × 0.5–0.6 mm; lfts (except for obliquely elliptic distal pair) subequiform, in outline obliquely rhombic or rhombic-oblong from inequilaterally cuneate or rectangular base, broadly obtuse or emarginate, 6–13 × 4–6 mm, (1.5–)1.6–2.4 times as long as wide; the diagonal, straight or obscurely sigmoid midrib immersed or shallowly depressed on ventral face of blade, sharply finely prominulous dorsally and there giving rise on anterior side to 4–7 secondary nerves brochidodrome within the almost plane margin, no tertiary venulation visible externally. *Peduncles* 1.5–3.5 cm; capitula 12–24-fld, the subglobose or clavate receptacle 2–4 mm, the fls dimorphic, the terminal one with exserted stamen-tube and free filaments basally thickened; bracts firm, ovate or lanceolate, decrescent upward, the longest ±2–4 mm, persistent into and beyond anthesis; fls normally dimorphic, the perianth of all 5-merous or that of terminal fl 6-merous, all sessile or the lowest of a capitulum elevated on pedicel to 0.7 mm, the perianth of all densely brownish silky-puberulent or the corolla glabrescent below the lobes; PERIPHERAL FLS: calyx 2.3–2.7 × 1.4–2 mm, the tube obtusangulate, the teeth 0.4–0.7 mm; corolla narrowly vase-shaped 4.5–6.2 mm, the lance-ovate lobes 1.2–1.6 mm; androecium 14–18-merous, 17–22 mm, the stemonozone 0.6–1.3 mm, the tube (2.2–)3–5 mm; ovary narrowly ellipsoid, conical at tip, glabrous; stigma scarcely dilated; TERMINAL FL: calyx 3–7.5 mm, corolla 6–7 mm, staminal tube 8–8.5 mm, the free filaments thickened and recurving at base. *Pods* 1–2 per capitulum, subsessile, in outline broad-linear recurved through ⅓–⅔-circle, to 13 cm and 1.2–1.6 cm wide, up to 12(–14)-seeded, the fuscous, smooth or faintly venulose, stiffly papery valves low-convex or hemispherically dilated over each seed, framed by sutures 1.5–2.5 mm wide, orange-red internally; seeds (few seen) plumply lentiform, ±7–8 mm diam, the translucent testa aging dull fawn-brown, pleurogram 0.

In capoeira, at edge of blackwater streams, and in forest-campina ecotone, mostly, perhaps exclusively, on white sands above the annual flood line, below 150 m, scattered along the Amazon River and its immediate tributaries from the Río Caquetá near 72°W in Amazonas, Colombia, and the lower Río Ucayali near 74°W in Loreto (prov. Requena), Peru, downstream to Gurupá near 51°30′W in Pará, Brazil. — Map 12. — Fl. (VII)X–XII. — *Pashaco, pashaquillo* (Peru).

Abarema auriculata differs from sometimes habitally similar *A. jupunba* var. *jupunba* in smaller leaflets (at most 13 mm long) glabrous on both faces, and in specialized campirana ecology. For differential characters of closely related but allopatric *A. ganymedea* see commentary on that species (p. 61).

11. Abarema microcalyx (Bentham) Barneby & Grimes, comb. nov. *Pithecolobium microcalyx* Spruce *ex* Bentham, Trans. Linn. Soc. London **30:** 582. 1875 & *in* Martius, Fl. Bras. **15(2):** 434. 1876. — Typus infra sub var. *microcalyce* indicatur.

Macrophyllidious *trees* 5–25 m with trunk attaining 3.5 dm dbh and grayish annotinous branches, the young stems, lf-stks, and peduncles densely minutely brown-puberulent when young but early glabrescent, only the axillary buds permanently puberulent, the broad chartaceous bicolored lfts facially glabrous, often microscopically ciliolate, lustrous dark brown-olivaceous above, paler brown beneath, the short dense capituliform racemes of exceptionally small, greenish white fls arising solitary and paired from the axil of coeval lvs, immersed in foliage. *Stipules* linear to subulate 1–2 mm, early caducous from base of developing lvs, absent from fruiting specimens. *Lf-formula* i–iii/1–4, lfts 8–32(–36) per lf; lf-stalks of larger lvs 0.5–8.5 cm, the petiole 0.5-2.6 cm, the one or the longer of 2 interpinnal segments (0.7–)1–3.5 cm; petiolar nectary at or shortly below pulvinules of first pinna-pair verruciform, hemispherical or sometimes vertically elongated, 1–2.5(–3) cm diam at base, as or only half as tall, minutely pored, commonly livid and sometimes microscopically puberulent, similar but smaller nectaries sometimes at further nodes of lf-stk and at tip of pinnae, but the last often lacking; pinnae distally accrescent, the rachis of furthest pair 2.7–8 cm, the longer interfoliolar segments 12–32 mm; pulvinules in dorsal view 1.5–2.5 × 0.8–1.7 mm, closely cross-wrinkled; lfts strongly accrescent distally, the blades obliquely obovate, asymmetrically ovate or incipiently rhombic-ovate, somewhat incurved, at base scarcely or moderately inequilateral, broad-cuneate or rounded, obtuse muticous, the distal pair (27–)40–85 × (20–)28–45 mm, 1.3–1.9 times as long as wide; the subcentric midrib gently arched forward, the widely ascending secondary veins brochidodrome well within the loosely revolute margin, all these together with random tertiary venules prominulous on both faces but more sharply so beneath. *Peduncles* 3.7–11.5 cm; racemes ±(20–)30–50-fld, the slenderly clavate receptacle 2–15 mm; bracts not over 1 × 0.5 mm, linear-subulate or subulate, caducous; fls subequiform as to size and shape of 5-merous perianth, but the terminal 1(–2–3) sometimes sessile and with modified androecium; PERIPHERAL FLS: pedicel 0.6–2.5 mm; calyx campanulate 0.7–1.2 mm, the low-deltate teeth 0.1–0.3 mm; corolla greenish 3.2–5 mm, the ovate lobes 1–1.7 mm, microscopically puberulent externally; androecium 10–14-merous, 9–17 mm, the tube including short stemonozone 1.6–2.4 mm; ovary glabrous, conical at apex; style ±2 cm, the stigma scarcely 0.1 mm diam, not dilated; TERMINAL FL: sessile or almost so, the perianth as in peripheral fls, the androecial tube ±2–5 mm. *Pods* mostly solitary, subsessile, in profile broad-linear, evenly incurved through almost to somewhat more than a full circle, in profile ±5.5–14 × 1.1–2.9 cm, (6–)7–12-seeded, the seminiferous suture either evenly recurved or undulately impressed between seeds, the coriaceous or subligneous valves at first plano-compressed, often but not consistently becoming turgid over each seed, fuscous or nigrescent and coarsely but not prominently venulose when mature, smooth and orange internally, the microscopically puberulent sutures dorsally plane and ±2 mm wide to dorsally ridged to 5 mm wide, the valves in cross-section 0.2–2 mm thick, the exocarp and pale crustaceous endocarp always very thin, the pitchlike mesocarp highly variable (as described under each variety); dehiscence of *Abarema*, in thick-walled pods tardy, the valves when of thin texture cracking and irregularly separating, when of thick texture continuous; funicles strap-shaped, distally coiled; seeds not seen.

Among lowland abaremas with few large leaflets *A. microcalyx* is distinguished by the combination of mounded leaf-nectaries and compactly racemose capitula of very small flowers. It is variable in leaf-formula, the leaflets of the largest leaves of a plant varying in number from 8 to 36, and in both width and texture of the pod. Narrower, thinner-textured pods have relatively narrow sutures, the seminiferous one undulately impressed between seeds, and mesocarp <0.5 mm thick over each seed, whereas broader thicker pods have wider sutures not or obscurely constricted between seeds and mesocarp about 1–2 mm thick in cross-section. We have found some variation in morphology of the capitulum, involving length of pedicels and development of a modified central flower, but whereas the characters of leaf-formula and fruit are seemingly well correlated with dispersal and thereby indicate racial differentiation, flowering specimens are as yet too few to establish any significant pattern of variation. Three varieties are here recognized.

Key to the varieties of *A. microcalyx*

1. Pinnae of most larger lvs 2–3 pairs, and the distal pinnae 4–8-foliolate, the whole lf (12–)16–32(–36)-foliolate.
    2. Dispersed on upper rios Negro and Uaupés in NW Amazonas, Brazil, and thence into adjacent Colombia and to the Orinoco divide in T. F. Amazonas, Venezuela; pod relatively narrow and of thin texture, 11–16 mm wide at the seeds, and there in cross-section at most 0.6 mm thick; exterior suture of pod undulately constricted .... 11a. var. *microcalyx*
    2. Dispersed in E-centr. Amazon basin, from Manaus SE to the lower Río Tapajós in W Pará; pod very broad and thick, (11–)15–29 mm wide, and in cross-section ±1–2 mm thick; exterior suture of pod not or obscurely constricted .... 11b. var. *enterolobioides*
1. Pinnae of all or of most lvs 1 pair, in rare lvs 2 pairs but the lfts not over 2 pairs in any lf and the whole lf 8- rarely 12-foliolate; dispersed on the N bank of the lower Amazon in E Pará and Amapá; pod of var. *enterolobioides*......... 11c. var. *parauaquarae*

11a. Abarema microcalyx (Bentham) Barneby & Grimes var. **microcalyx.** *Pithecolobium microcalyx* Spruce *ex* Bentham, ll. cc. supra. 1875 & 1876. — "Habitat prov. do Alto Amazonas ad Rio Negro prope S. Gabriel: Spruce n. 2288; in silvis Capoeiras ad Rio Uaupes [prope Panuré], Spruce n. 2743." — Lectotypus, *Spruce 2288*, K(herb. bentham.) = NY Neg. *2006*; paratypi, 2743 at †B = F Neg. *1211*, F (fragm.)!, K(herb. bentham.)!, P!.

Characters as given in key to varieties; longest *pedicels* ±0.6–1.2 mm; *pods* 11–16 mm wide, the relatively thin valves coarsely venulose externally, the pitchlike mesocarp in cross section (dry) 0.2–0.5 mm thick.

In moist forest on terra firme, commonly on sandy, sometimes clayey soils, below 160 m, apparently not rare on and near the Orinoco–Negro divide in T. F. Amazonas, Venezuela, thence extending S along the Río Negro into NW state of Amazonas, Brazil (S. Gabriel da Cachoeira; lower Uaupés valley) and W just into SE Com. Vaupés, Colombia. — Map 13. — Fl. X–III. — *Maripa, palo de jabón* (Venezuela), the bark furnishes a detergent used in washing.

11b. Abarema microcalyx var. **enterolobioides** Barneby & Grimes, var. nov., a var. *microcalyce*, quoad foliorum formulam et amplitudinem simillima, legumine lato, valvulis obesis (supra semina ±1–2 mm crassis), mesocarpio piceo nigricanti transverse secto saltem 0.9 mm et 2 mm usque lato diversa, a var. *parauaquarae*, quoad legumen simili, foliorum formula altiori abstans. — BRAZIL. Pará: Parque Nacional do Tapajós, km 60 da estrada Itaituba–Jacarecanga; picada para o lago da Bárbara, 21.XI.1978 (fr.), *M. G. Silva & C. Rosário 3861*. — Holotypus, NY; isotypus, US.

Pithecolobium microcalyx sensu Ducke, 1949: 35 [*Ducke 1097, 1412*, NY!].

Characters as given in key to varieties; *pedicels* at anthesis (few seen), 0.5–0.9 mm; *pods* (11)15–29 mm

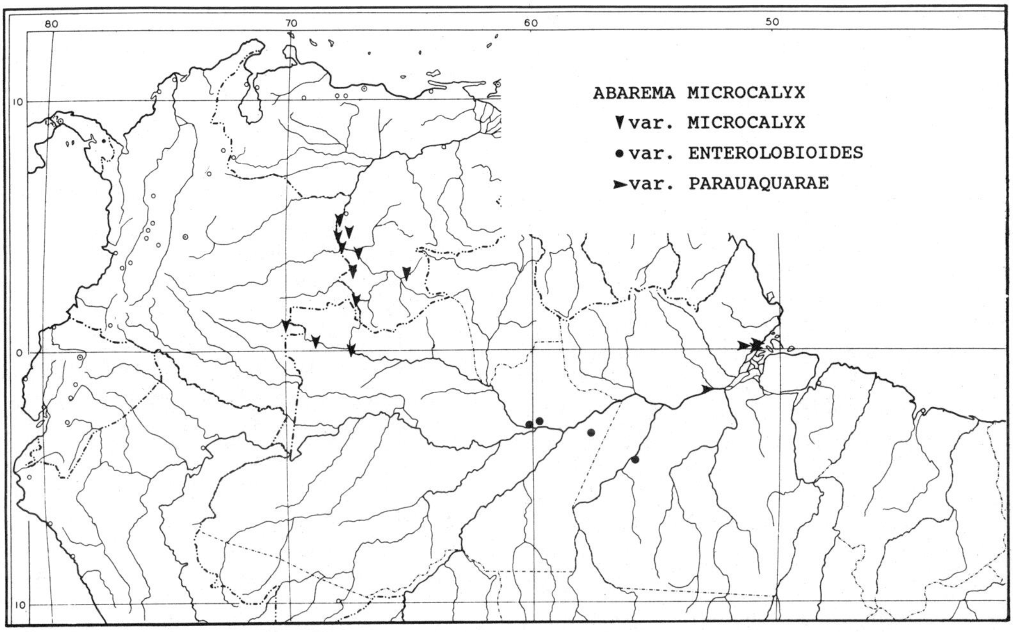

MAP 13. The Orinoco and Amazon basins in Venezuela and Brazil. Range of *Abarema microcalyx* (Bentham) Barneby & Grimes var. *microcalyx*, var. *enterolobioides* Barneby & Grimes, and var. *parauaquarae* (Ducke) Barneby & Grimes.

wide, the grossly thickened valves obscurely venulose, the pitchlike mesocarp 0.9–2 mm thick in section.

In habitats of var. *microcalyx*, sometimes surviving in cut-over and cultivated forest, below 100 m, apparently local on the Amazon River and its affluents from the S in E state of Amazonas (Manaus; Maués) and closely adjacent Pará (Itaituba on Río Tapajós), Brazil. — Map 13. — Fl. dates unknown.

The var. *enterolobioides* combines the leaves of *Pithecellobium microcalyx* sens. str. with the grossly obese pod of *P. parauaquarae* and occupies an intermediate range of dispersal. The three taxa are conveniently treated as geographic varieties of one species. In external aspect the pod forcibly recalls that of *Enterolobium schomburgkii* Bentham, whence the epithet.

11c. Abarema microcalyx var. **parauaquarae** (Ducke) Barneby & Grimes, comb. nov. *Pithecolobium parauaquarae* Ducke, Arch. Jard. Bot. Rio de Janeiro **4:** 28. 1925. — "Habitat loco saxoso aperto in cacumine montis Parauaquara . . . siti inter Prainha et Almeirim ad septentrionem fluvii Amazonum inferioris. *l. A. Ducke 7-10-1919 H[erb]. J[ard]. B[ot]. R[io de Janeiro] n. 10159.*" — Holotypus, RB!; isotypi, P!, U *64419*!, US *1442228*!.

Pithecolobium parauaquarae sensu Ducke, 1949: 35.

Characters as given in key to varieties; longer *pedicels* (few seen) ±1.5–2 mm; *pod* of var. *enterolobioides.*

In thin woodland, on stony hilltops, and in bush islands of campinarana in E Pará (Sas Parauaquara and do Jutaí) and S Amapá (Macapá, Mazagão, Camaipi, campos do Matapí), Brazil. — Map 13. — Fl. X–XI(–?).

This variety seems to differ significantly from var. *enterolobioides* only in simplified leaf-formula, the pinnae mostly only 1 pair (but 2 pairs occur even in one isotype, U!), and the leaflets never more than 2 pairs. More flowering material is required to establish differences, if any exist, in number of flowers per capitulum and length of pedicels.

12. Abarema commutata Barneby & Grimes, sp. nov., cum *A. villifera* (Ducke) Barneby & Grimes hucusque confusa et ei certe affinis, sed ab ea pinnis foliolisque magis numerosis (folii cujusque majoris 24–52, nec 16–20), pinnarum distalium rachi longiori (4–12 nec 2.5–4.5 cm usque longa), capitulorum axi longiori (4–10, nec 2–3 cm longo), necnon flosculis periphericis majoribus (calyce 4.5–6, nec 3–4.2 mm longo), ulterius patria in regione altiori

Guayanae diversum. — GUYANA. Southern Pakaraima Mts: Waipa trail from north Kopinang Savanna, 2750 ft, 4.IX.1961 (fl.), *B. Maguire* (with *C. K. Maguire & G. Wilson-Browne*) *46100*. — Holotypus, NY; isotypus, UC *221602*!, US.

Round-headed macrophyllidious *trees* 8–20 m, the young stems and all axes of lvs and inflorescence densely villous-tomentulose with brown-golden or bronze hairs 0.2–0.8 mm, the bark at length glabrate and blackish mottled, the foliage bicolored, the coriaceous lfts dark-green (when dry brown) glabrous and lustrous above, paler and densely golden-brown-villous beneath, the loose capitula of white fls borne singly and paired in the axil of coevally expanding lvs, early immersed in foliage. *Stipules* early caducous (few seen), elliptic-oblanceolate 3.5–6.5 × 0.8–1.3 mm. *Lf-formula* (i–)ii–iv/3–6; lf-stks stout (1–)2.5–15 cm, the petiole (0.5–)0.8–2.4 cm, the one (or longest) interpinnal segment 0.5–4(–5) cm; petiolar nectary close below first pair of pinnae sessile, low-convex dimpled or almost plane 1.2–2.6 mm diam, smaller ones random on distal segments of lf-stk and on 1–3 furthest segments of pinnae; pinnae much accrescent distally, the rachis of the furthest pair (3–)4–10(–12) cm, the longer interfoliolar segments 10–24 mm; paraphyllidia linear-lanceolate 0.7–2 mm; lft-pulvinules 1–2 × 0.9–1.8 mm; lfts strongly accrescent distally, in outline obovate-elliptic, rhombic-elliptic or obtusely rhombic from inequilaterally cuneate base, broadly obtuse, the distal pair (3–)3.5–6.5 × 2.1–4.5 cm, 1.4–2 times as long as wide; the scarcely displaced, gently incurved midrib giving rise on each side to 8–12 major (and random weak intercalary) secondary nerves brochidodrome within the strongly revolute margin, these simple or rarely branched, only in Surinam proliferating into a mesh of raised veinlets. *Peduncles* 3.5–8 cm; capitula ±12–22-fld, the axis including terminal pedestal 4–9 mm; bracts elliptic-oblanceolate 2.5–4 × 1–1.5 mm, caducous; fls white, dimorphic, the lower peripheral ones raised on a pedicel 0.5–1.2 mm, the terminal fl sessile, the 5-merous perianth of all densely silky externally; PERIPHERAL FLS: calyx turbinate-campanulate 4.5–6 × 3.3–4.5 mm, the ovate-triangular teeth 1.3–2.3 mm; corolla ±9–10 mm, the lobes erect; androecium 26–36-merous, to 25–36 mm, the stemonozone 0.8–1.8 mm, the tube 3–5 mm; ovary subsessile, ellipsoid, densely pubescent; TERMINAL FL: calyx broadly campanulate 5–7 mm; corolla 8–10 mm; stamen-tube 8–12 mm, the stamens to 40. *Pods* (one seen, this perhaps stunted and atypical) sessile, narrowly oblong in profile, recurved through less than ¼-circle, plano-compressed but low-convex over each of ±8 seeds, the valves crustaceous, framed by sutures ±2 mm wide, glabrescent, dull brown outside, reddish overall within; dehiscence through both sutures, the valves coiling; seeds not seen.

In savanna-forest ecotone and in openings in scrub forest, 825–1200 m, localized on the E edge of the Guayana Highland: Gran Sabana in SE state of Bolívar, Venezuela; Pakaraima and Kanuku mts in W-centr. Guyana; and on Tafelberg, Surinam. — Map 14. — Fl. VIII–IX. — *Ercey-euru-yek* (Venezuela; the bark yields saponins for washing).

Since *Maguire 24629* (NY) from Table Mountain in Surinam was identified by Amshoff (in the 1940s) as *Pithecellobium villiferum*, all subsequent collections have been so named in herbaria. The species is obviously close to *A. villifera* in habit, indumentum, and broad obovate or rhombic-obovate leaflets, but differs in dispersal, habitat, average leaf-formula, looser capitula, and larger peripheral flowers, as noted in our diagnosis and key to species. It seems therefore to deserve specific status.

All our material from Venezuela and Guyana has leaflets essentially veinless on the convex upper face and pinnately veined with 8–12 prominent major pairs of almost simple secondary nerves on the dorsal face, in this respect resembling typical *A. villifera* from the upper Río Negro basin. The one collection from Surinam has smaller leaves and secondary nerves prominulous on both faces and moreover branched into a mesh of veinlets, a venation recalling the variant of *A. villifera* described from Río Cueiras in the lower Río Negro basin. Whether these poorly documented variants are individual variants or have a genetic basis remains to be seen.

13. Abarema jupunba (Willdenow) Britton & Killip, Ann. New York Acad. Sci. 35: 126. 1936. *Acacia jupunba* Willdenow, Sp. Pl. **4:** 1067. 1806. — Typus infra sub var. *jupunba* indicatur.

Either macro- or microphyllidious *trees* (3–)5–35 m (in upland Guayana sometimes small bushy trees 1.5–3 m) tall with smooth gray trunk attaining a mature girth of 2–8(10) dm dbh, the young stems, lf-axes, and inflorescence densely golden-brown- or sordid-tomentulose or -puberulent with spreading-incurved or subappressed hairs to ±0.1–0.3 mm, glabrescent in age, the lvs bicolored, the lfts on upper face glossy dark green, often purplish speckled or mottled and (except for sometimes ciliolate costa) glabrous, beneath pallid-papillate dull and either

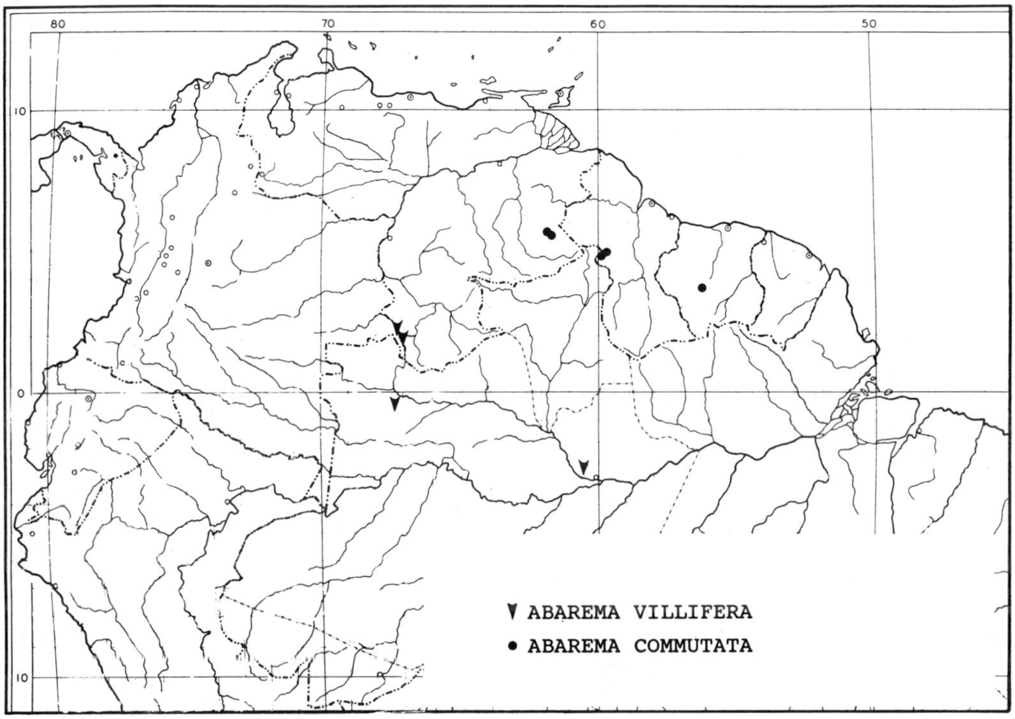

MAP 14. The Guianas and upper Amazon basin of Brazil. Range of *Abarema villifera* (Ducke) Barneby & Grimes and *A. commutata* Barneby & Grimes.

densely or remotely but always minutely strigulose overall, the compact, sometimes subspicate or subracemose, always short capitula of whitish or greenish, white-stamened, commonly fragrant fls axillary to coeval lvs, immersed in foliage. *Stipules* linear or linear-lanceolate or elliptic 1.5–6 × 0.4–0.7(–1) mm, early deciduous from a small pallid scar. *Lf-formula* i–v (in juvenile lf –vii)/3–10(–12); lf-stk of major lvs associated with fls 2.5–11(–18) cm, of var. *trapezifolia* (0.7)1–8(14) cm, of sapling lvs (no further described) up to 18 cm, the petiole (1–)1.3–3.5(–4) cm, the longer interpinnal segments (often lacking in var. *trapezifolia*) 0.8–2.8(–3.5) cm; pinnae (when more than 1 pair) distally accrescent, the rachis of the furthest or penultimate pair longest, 2.5–9 cm, the longer interfoliolar segments 6–13(–14) or in var. *trapezifolia* 12–26(–31) mm; nectary between or close below first pinna-pair sessile or almost so, the stout stipe, if present, concealed by the head, this hemispherical-verruciform or low-convex but dimpled, in vertical view either round or obtusangulately scutiform (0.6–) 0.7–2.1 mm diam, in lateral view not over 0.6 mm tall, similar nectaries near random further pinna-pairs or these lacking, but much smaller, round or vertically elongate cupular ones close below 1–3 furthest pairs of lfts; minute paraphyllidia usually present, sometimes only at posterior side of pinna-pulvini; lft-pulvinules cross-wrinkled 0.6–1.4(–2) × 0.6–1.3 mm; lfts decrescent toward base of pinnae, thence either scarcely or greatly accrescent, in outline obtusely rhombic to rhombic-elliptic or -obovate from inequilaterally cuneate or subrectangular base, obtuse or emarginate, the terminal pair (1.3–)1.6–6.5(–7) × 1–3.6(–4.4) cm, (1.3–)1.5–2.1(–2.3) times as long as wide; the either straight diagonal or subcentric and gently incurved midrib giving rise on each side to ±5–9 major and sometimes as many and almost as strong intercalary secondary nerves brochidodrome within the at least incipiently revolute, always ciliolate margin, these sometimes generating a sinuous tertiary-reticulate venulation, the whole venation variable in strength and prominence, sometimes (especially in older lvs) almost immersed, sometimes sharply prominulous on both faces. *Peduncles* mostly solitary or geminate, seldom ternate (1–)2.5–7.5(–10) cm; capitula or capituliform racemes ±(10–)15–45-fld, either compact or with one or more fls downwardly displaced on peduncle, the receptacle including very short terminal pedestal 2–11 mm; bracts linear-oblanceolate, oblanceolate or elliptic 0.7–1.6 mm, deciduous, or the lowest longer and more persistent; fls ordinarily dimorphic, terminal 1–(2–3) larger and coarser, with modified androecium, but these sometimes modified only as to androecium or abortive and early deciduous, or exceptionally wanting, the perianth of all fls 5-merous, densely brownish silky overall with appressed hairs; PERIPHERAL FLS: pedicel of proximal ones mostly 0.4–1.2 mm, rarely subobsolete, that of distal ones often a little shorter; calyx narrowly or widely turbinate-campanulate 1.7–3 × 1.4–2.1(2.4) mm, the low-deltate teeth 0.3–0.8(–0.9) mm; corolla 4–6.2(–7) mm, the ovate or lance-ovate lobes 1.6–2.8 × 1–1.5 mm; androecium (12–)14–25-merous, 15–23(–31) mm, the stemonozone 0.4–1.1 mm, the tube 2–4(4.5) mm; ovary shortly stipitate, narrowly obovoid-ellipsoid, densely puberulent overall, rather abruptly contracted into the glabrous style but not truncate; style scarcely dilated at tip; TERMINAL FL(S): sessile, the calyx broadly campanulate (1.8–)2.2–3.4 × 2–2.6 mm, the corolla (5.5–)6.5–8.5 mm, the androecial tube 6–11 mm, either scarcely or well exserted, ±1–1.3 mm diam at separation of filaments, these either much or scarcely thickened proximally. *Pods* 1–2 per capitulum, sessile or almost so, in profile broad-linear decurved through ⅔–nearly 2 circles, when well fertilized 6–9.5 × 1–1.6 cm, 8–12-seeded (by random abortion of ovules often shorter), the leathery, fuscous or maroon-red, strigulose but glabrescent, nowhere pulpy valves framed by shallowly (at aborted ovules deeply) undulate ventral and evenly decurved dorsal sutures ±0.9–1.5 mm wide, low-convex over seeds, either distinctly or faintly venulose, internally orange and either smooth or minutely furfuraceous; dehiscence through the length of both sutures, the valves elastically coiling; seeds transverse, basifixed on ribbonlike funicle, plumply lentiform, in broad view as long as or a little longer than wide, 5–8 mm diam, the testa loosely investing the blue embryo, white and opaque near the hilum, thence translucent, pallid or lutescent in age, pleurogram 0.

In upland and non-inundated lowland, primary and secondary forest and forest-savanna ecotone, widespread over tropical South America from NW Colombia to NE Bolivia, W through Venezuela to Trinidad and Tobago and the Guianas, S in Amazonian Brazil to Rondônia and N Mato Grosso, disjunct in coastal forest of Bahia; Lesser Antilles southward from Guadeloupe; range and elevation more precisely stated under the varieties.

Abarema jupunba is distinguished ideally from close relatives by a syndrome of (a) leaflets finely, often only minutely, strigulose beneath, (b) convex and pored petiolar nectaries, (c) small, shortly pedicellate or subsessile, densely silky flowers, and (d) relatively thin-textured pod with valves uniformly cinnabar-red inside and white-and-blue seeds lacking pleurogram. It is exceptionally variable in leaf-formula and in size of leaflets, both between branches of one tree and between trees of different populations. The extreme leaf-forms were treated by Bentham (1875) as *Pithecellobium micradenium* and *P. trapezifolium*, but the numerous intermediates subsequently discovered and a low correlation between pinna number and either size or number of leaflets precludes recognition of more than one, no differences in either flower or fruit being apparent. Our concept of *P. jupunba* is the inclusive one developed by Urban and Britton & Rose. However, analysis of the copious material now accumulated shows that the macrophyllidious variants of the species occur only in the northeastern quarter of its whole range, in fact displacing the microphyllidious ones in all of Venezuela except its Amazonian corner and Maracaibo basin, and extending beyond Venezuela only into immediately adjoining countries and Dutch and French Guianas. It seems appropriate, therefore, to accept two varieties, even though, however they may be defined, there will inevitably be found some specimens ambiguously poised between them.

Because it comprises forms with larger and smaller leaflets and is one of the most widely dispersed members of its genus, *A. jupunba* is a conceptually useful prototype of the group of blue-seeded abaremas with rhombic or rhombic-obovate leaflets and apically conic ovary, and it provides a standard of comparison against which its more local relatives can be evaluated: in the macrophyllidious direction *A. microcalyx*, *longipedunculata*, *villifera* (probably, though pod is unknown), *commutata*, and *adenophora*; in the microphyllidious one *A. auriculata* and polymorphic *barbouriana*. For diagnostic discussion see the accounts of those species.

Key to the varieties of *A. jupunba*

1. Lf-formula iii–v(–vii)/(6–)7–12, and terminal lfts of larger lvs (13)16–35(36) × 10–19(–21) mm; range of the species except most of Venezuela .... 13a. var. *jupunba*
1. Lf-formula i–iii(–iv)/3–6(–7), and terminal lfts (34–)35–65(–70) × 19–36 mm; Venezuela and coterminous countries, with Surinam and French Guiana .... 13b. var. *trapezifolia*

13a. Abarema jupunba (Willdenow) Britton & Killip var. **jupunba.** *Acacia jupunba* Willdenow, 1806, l. c., sens. str. — "*Habitat in provincia Para Brasiliae . . . [Friedrich Wilhelm Sieber 44*, commun.] Com. de Hoffmannsegg." — Holotypus, B-WILLD *19142*, seen in Microform IDC 86. 1388: II.1! — The now defoliate specimens cannot be determined from the negative but were equated with Antillian *Pithecolobium micradenium* by Urban, Symb. Ant. **2:** 258: "Typus Willdenowianus cum plantis antillanis omnino quadrat.," this equation confirmed by the protologue. — *Mimosa jupunba* (Willdenow) Poiret, Encycl., Suppl. **1:** 70. 1810. *Feuilleea jupumba* [*sic*] (Willdenow) O. Kuntze, Revis. Gen. Pl. **1:** 185. 1891. *Pithecolobium jupunba* (Willdenow.) Urban, op. cit. 257. 1900. *Jupunba jupunba* (Willdenow) Britton & Rose, N. Amer. Fl. **23:** 27. 1928, exclus. syn. *Mimosa trapezifolia.*

Pithecolobium micradenium Bentham, London J. Bot **3:** 217. 1844. "Dominica, *Imray.*" — Holotypus, K(herb. hooker.)!; isotypi, GH!, K(herb. hooker.)!. — *Feuilleea micradenia* (Bentham) O. Kuntze, Revis. Gen. Pl. **1:** 188. 1891. — Equated with *P. jupunba* by Urban and by Britton & Rose, 11. cc.

P. brongniartii Duchassaing & Walpers, Flora **36:** 232. 1853. — "Habitat in insulae Guadeloupe sylvis primaevis montanis." — Lectotypus, *Duchassaing s.n.*, P! = NY Neg. *12570*; isotypi, GOET!, P (2 sheets)!. — Equated with *P. micradenium* by Bentham, 1875: 584.

Mimosa vaga sensu Aublet, Hist. Pl. Guiane **2:** 945. 1775; non Linnaeus, 1753. — Spm. authent., BM = NY Neg. *141*!.

Lvs as described in key to varieties.

In virgin and disturbed lowland and upland forest, in forest-savanna ecotone, and on rocky river banks, but not entering seasonally flooded forest, mostly below 400 m but ranging from near sea level on the Atlantic and Pacific coasts of tropical South America to 1200 m in E Peru (San Martín), to 1500 m in Ecuador, and to 600 m on the island of Dominica; interruptedly dispersed around and within the Amazon basin from Ecuador and E Peru E to the delta, S on the Madeira-Beni into NE Bolivia, N through inter-Andean valleys of Colombia to the Pacific coast in Valle, and to the Maracaibo Basin in NW Venezúela; reappearing in lowland forest near the coasts of French Guiana, and between Sergipe and Espírito Santo (in lat. ±10°–20°30′S), and disjunct on Lesser Antilles (Guadeloupe to Grenada, reportedly absent from Martinique). — Map 15. — Fl. in all months of the year, most prolifically IX–III. — *Arepilla* (Venezuela): *dalmaré, savonette* (Grenada); *acacia mâle, tamalin* (French Guiana); *saboeiro* (Pará); *mani* (Bolivia, perhaps missapplied).

13b. Abarema jupunba var. **trapezifolia** (Vahl) Barneby & Grimes, comb. et stat. nov. *Mimosa trapezifolia* Vahl, Eclog. Amer. 3: 36, t. 28. 1807. — "Habitat in insula Trinitatis. Ryan." — Holotypus, C! = NY Neg. *12575*. — *Inga trapeziformis* (Vahl) de Candolle, Prodr. **2:** 441. 1825. *Pithecolobium trapezifolium* (Vahl) Bentham, London J. Bot. **2:** 142. 1843. *Feuilleea jupumba* (Vahl) O. Kuntze, Revis. Gen. Pl. **1:** 185. 1891. *Abarema trapezifolia* (Vahl) Pittier, Trab. Mus. Com. Venezuela **2:** 86. 1927. *Punjuba trapezifolia* (Vahl) Moldenke, Bull. Torrey Bot. Club **59:** 155. 1932.

Mimosa atakta Steudel, Flora, **26:** 758. 1843. — "[*Hostmann & Kappler*] 479 . . . im Innern von Surinam . . . " — Holotypus, P!; isotypi, NY!, P (4 sheets)!. — Equated with the preceding by Bentham, 1875: 583.

Pithecolobium benthamianum Miquel, Linnaea **18:** 592. 1844. — "Crescit prope Paramaribo [Surinam] . . . " — Holotypus, *Focke 812*, U!. — Name overlooked by Bentham (1875) and by Kleinhoonte (1940), but the typus referred by the latter to *A. jupunba* sens. lat.

P. jupunba sensu Urban, 1900: 257, ex parte. *Jupunba jupunba* sensu Britton & Rose, 1928: 27, ex parte.

Lvs as described in key to varieties.

In lowland and upland rain forest, in gallery forest and shrub islands within savanna climax, and occasionally along streams (but not seasonally inundated), supplanting var. *jupunba* in parts of NE South America: in Venezuela in Cord. Costanera, from Carabobo E to the Paria Peninsula; in the Guayana Highland eastward from centr. T. F. Amazonas, attaining on Gran Sabana 1300 m and in Sucre ±700 m, and in the Andean foothills of edo Táchira 1400 m; thence SE to Atlantic lowlands and upland interior of Guyana, Surinam, and NW French Guiana, on Tafelberg attaining 500 m; lowland Trinidad and Tobago; one poorly documented record from Meta, Colombia; apparent intergrades to var. *jupunba* (not mapped) on Ríos Jarí and Pacajá in N and centr. Pará, Brazil. — Map 15. — Fl. X–IV and randomly in all other months. — *Fatura, samán montañero* (Venezuela); *culano* (Guyana); *horoassa, huruassa*, (Arawak in the Guianas); *swampoetamarin* (Surinam).

Three collections from elevations between 500 and 1050 m in Bolívar, Venezuela (*Liesner 19763*, *Prance & Huber 28391*, *Liesner & Holst 20697*; all NY), have pinnae mostly reduced to 1 pair and leaflets glabrous beneath, suggesting passage to *A. longipedunculata.* They have, however, the coiled fruit of var. *trapezifolia* and apparently the small calyx of *A. jupunba*. The populations of var. *trapezifolia* on and around the Gran Sabana are variable in leaf-formula and leaflet-outline and need study in the field.

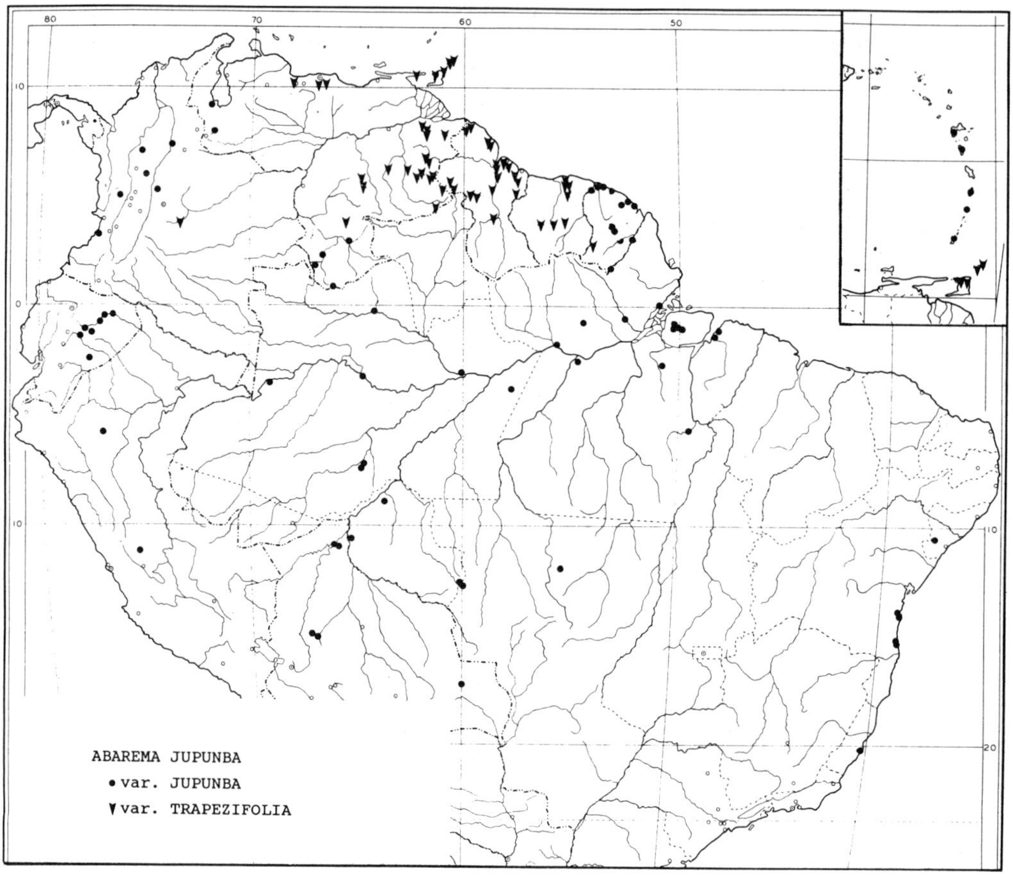

MAP 15. South America and (inset) Lesser Antilles. Range of *Abarema jupunba* (Willdenow) Britton & Killip var. *jupunba* and var. *trapezifolia* (Vahl) Barneby & Grimes.

14. Abarema barbouriana (Standley) Barneby & Grimes, comb. nov. *Pithecolobium barbourianum* Standley, Contr. Arnold Arbor. **5:** 74. 1933. — Typus infra sub var. *barbouriana* indicatur.

Trees attaining 30 m in lowland habitats but often smaller, flowering as bushy treelets at 3 m or in upland habitats remaining bushy at maturity, the trunks attaining 2–3 dm dbh, the young branches and all axes of lvs and inflorescence densely sordid-pilosulous with erect or forwardly incurved, yellowish or reddish brown hairs to 0.2–0.8(–1) mm, the foliage bicolored, the mature lfts firm, on upper face lustrous dark green glabrous (except for sometimes ciliolate midrib), either smooth, or wrinkled, or finely venulose, the concave lower face pallid, pilosulous or softly silky-villosulous with ascending or erect hairs or (locally) glabrous except for strigulose-pilosulous, always cariniform midrib, the capitula of greenish white, white-stamened fls solitary or paired in the axil of coeval lvs, immersed in foliage. *Stipules* subulate, linear-elliptic, or linear 1–3(–4) mm, caducous. *Lf-formula* (ii–)iii–xii/(7–)8–22; lf-stks of larger lvs 2–14.5 cm, the petiole including pulvinus (6–)7–15 (–17) mm, the longer interpinnal segments 7–16(–18) mm; petiolar nectaries plane or shallowly cupulate, round or elliptic, either sessile against lf-stk or stipitate and in profile to 1(–1.6) mm tall, the first one near or well above mid-petiole, sometimes close to first pair of pinnae, (0.4–)0.6–1.7(–2) mm diam, progressively smaller but otherwise similar nectaries below some distal pinna-pairs, and yet smaller, more slenderly stipitate ones below 2–7 pairs of lfts; pinnae strongly decrescent proximally, sometimes a little so distally, the rachis of those near and above mid-lf 3–7 cm, the longer interfoliolar segments 2.5–10 mm;

paraphyllidia to ±0.5 mm nearly always present; pulvinules 0.2–0.65 × 0.4–0.75 mm; lfts decrescent proximally but thence subequilong except for broader and sometimes longer furthest pair, the blades obtusangulately rhombic, rhombic-oblong or -elliptic around a diagonal, almost straight or distally incurved midrib, inequilaterally flabellate-cuneate at base, either obtuse or obtuse-apiculate, the largest (4.5–)5–21 × (2–)2.6–10 mm, 1.9–2.7 times as long as wide; the subcentric midrib giving rise on each side to (2–)3–9 major (but never strong) secondary nerves weakly anastomosing close within the revolute or almost plane margin, these often raised only on hypophyllum but sometimes also, with tertiary venulation, on upper face. *Peduncles* 1.5–8 cm; capitula ±12–20-fld, the clavate receptacle including short terminal pedestal 2–4 mm; bracts ovate-triangular, elliptic or narrowly oblanceolate, 0.75–2.0 × 0.5–0.75 mm, deciduous; fls dimorphic, all sessile or the lowest peripheral ones borne on obscure pedicel not over 0.5(–0.7) mm long (or in Ecuador to 2.5 mm; see discussion of var. *barbouriana*, below), the perianth of all 5-merous (random irregularities), densely silky appressed-pilosulous externally overall; PERIPHERAL FLS: calyx deeply campanulate or turbinate-campanulate (2.5–)3–6 × 1.7–2.4 mm, the triangular, often unequal teeth 0.6–1.3 mm; corolla (4.5–)5–8 mm, the lobes (1–)1.3–3 mm; androecium (12–)14–28(randomly 36, 48)-merous, 21–30(–35) mm, the stemonozone 0.6–1.5 mm, the tube 1.4–6.5 mm; ovary substipitate, conical at apex, densely pubescent; TERMINAL FL: calyx 3.5–6 × 1.6–2.4 mm; corolla 7–12 mm; filament tube 6–16 mm. *Pods* usually solitary, sessile, in profile undulately broad-linear and (when well fertilized) evenly incurved through 1–2.5 circles 3.7–4.4 cm diam, (5–)6–12 × (1–)1.1–1.6 cm, 6–12(–14)-seeded, the stiffly papery, brown or fuscous, finely venulose valves framed by sutures 1.4–2 mm wide, low-convex over each seed, either strigulose or pilosulous overall (but especially along sutures) with sordid or yellowish hairs to 0.3–1.2 mm, the endocarp orange-red either overall or only in the seed-cavities; dehiscence downward through both sutures, the valves coiling; seeds (few seen) plump, ±5–7 mm diam, the translucent testa revealing the bluish embryo, the pleurogram either 0 or complete, ±1.5 mm diam, faintly engraved.

Our concept of *Abarema barbouriana*, which has been reported hitherto only from Panama and northern Colombia, is expanded herein to include *Pithecellobium arenarium* of Venezuelan Guayana and northern Amazonian Brazil, *P. fanshawei* of Guyana and French Guiana, and related forms, never satisfactorily identified, from mountainous northern Venezuela. It now accommodates a range of leaf-formulas comparable to and only slightly higher than those long admitted to the closely related *A. jupunba*, and considerable variation in pubescence of leaflets and pod, in venulation of leaflets, and in size of floral parts. Particular characters appear to be linked to dispersal and are dominant locally but feebly correlated; they suggest incipient racial differentiation, but are incalcitrant to precise definition. As in *A. jupunba*, we recognize only two subdivisions of the species: a relatively microphyllidious var. *barbouriana*, most abundant west of the eastern cordillera in Colombia and in east-central Panama, and thence interruptedly scattered eastward to the Guianas; and a relatively macrophyllidious var. *arenaria*, most frequent on the Gran Sabana and on tepuis of Venezuelan Guayana but recurring in lowland white sand savanna habitats on the sources of Río Negro in Venezuela and in central Amazonian Brazil. Each of these varieties encompasses variant populations that are analyzed below but is not further dismembered taxonomically. It should be emphasized that the fruit, except for the length and orientation of its indumentum, remains extremely constant.

As here redefined, *A. barbouriana* differs from *A. jupunba* primarily in the larger flower, and not quite consistently in the loose silky pubescence of the hypophyllum. Where this indumentum is lacking or sparse, the flower is much larger than that of *A. jupunba* or the leaflets are smaller and more numerous. The fruits are essentially alike.

Key to the varieties of *A. barbouriana*

1. Lf-formula vi–xii/12–19; larger lfts (excluding terminal pair) on fertile branches 4.5–11.5 × 2–5 mm; secondary veins of hypophyllum 2–6 on each side of midrib; Panama and N Colombia, interruptedly E to the Guianas ...... 14a. var. *barbouriana*
1. Lf-formula ii–v(–vi)/(7–)8–11(–13); larger lfts (excluding terminal pair) on fertile branches (9–)10–18 × (3–)5–10 mm; secondary veins of hypophyllum (5–)6–9 on each side of midrib; Venezuelan Guayana and the Río Negro basin in Amazonas, Brazil .... 14b. var. *arenaria*

14a. Abarema barbouriana (Standley) Barneby & Grimes var. **barbouriana.** *Pithecolobium barbourianum* Standley, 1933, l. c., pl. XI. — "Panama: Zetek Trail, Barro Colorado Island, Canal Zone, October 27. 1931, Otis Shattuck 237" — Holotypus, F!; isotypus, MO!. — Erroneously interpreted in the protologue as related to genus *Cojoba* Britton & Rose.

Albizzia (?) *dubia* Britton & Killip, Ann. New York Acad. Sci. **32:** 132. 1936. — "Colombia, 1760–1808, *Mutis 3531*" — Holotypus, US *1560970*!; clastotypus (fragm.), NY!; isotypus, MA!.
Pithecellobium fanshawei Sandwith, Kew Bull. **1948:** 314. 1948. — "BRITISH GUIANA. Bartica–Potaro road, 107 miles . . . Nov. 11th, 1943, *D. B. Fanshawe* in *Forest Dept.* no. 4181 . . . *Ibid.*, 108 m, at Mahida Creek, Potaro River, June 29th, 1942, *Fanshawe* in *Forest Dept.* no. 3490." — Holotypus, FD *4181*, K! = NY Neg. *2037*; isotypus, NY!; paratypi, FD *3490*, K! = NY Neg. *2038*, NY!, U!.

Lf-formula vi–xii/12–19; lfts almost always finely and often densely silky-villosulous dorsally, sometimes only barbellate in anterior basal angle of midrib and along the midrib, the largest ones at most 11.5 × 5 mm (to 15 × 6 mm on juvenile branches), the venation (as given in key) relatively simple; *pods* either strigulose or pilosulous.

In non-inundated forest, from near sea level in Panama and W Colombia to 950 m in Panama (C. Jefe), 1850 m in Antioquia, Colombia, and 1080 m in Aragua, Venezuela, locally plentiful from centr. Panama (Colón, Panamá, canal zone) to NW Colombia (Chocó, Antioquia, Valle de Cauca); in mountainous N Venezuela (Aragua, Miranda); and far disjunct in lowland Guyana (Mazaruni and Potaro valleys) and interior French Guiana. — Map 16. — Fl. VIII–II — *Dormilón, palo paloma* (Colombia).

The var. *barbouriana* is known in Panama and Colombia both from near sea level and from upland forest; there is much variation in form and size of the leaf-nectaries, but we find no correlation between these and elevation of habitat. The pod varies in this same region from strigulose to golden-pilosulous; when pilosulous, it is indistinguishable from that of the remotely allopatric *Pithecellobium fanshawei*, which we here reduce to synonymy. A few individual specimens here referred to var. *barbouriana* deserve brief mention:

a. *Albert de Escobar 5054*, *7708* and *Daly 5959* (all NY) from Guatapé, Antioquia, Colombia, at the upper known limit of the species at 1850 m, have exceptionally small lfts, up to 4.5 mm, rather than the usual range of 7–11 mm.
b. *F. Tamayo 2038* (US) from cloud forest in state of Miranda, Venezuela, at unrecorded elevation, has dorsally glabrate lfts coinciding with unusually numerous filaments (48 in one dissected flower), a number observed once also in var. *arenaria*, but not elsewhere in var. *barbouriana*. Here again the lfts are small in the context of the species. The specimen was annotated by Pittier as a new species of *Calliandra*.
c. *Weitzman & Boom 110* (NY), from Parque Nacional Pittier, the only other collection of var. *barbouriana* known to us from northern Venezuela, has lfts nearly as small as those of the last mentioned, but they are pubescent dorsally as usual in *A. barbouriana*.
d. The specimens from Guyana on which *Pithecellobium fanshawei* was based have the aspect and lf-formula of var. *barbouriana*, but the lfts tend toward a rhombic-oblong rather than elliptic outline and in this respect resemble var. *arenaria*. In the protologue of *P. fanshawei*, Sandwith likened it to the then poorly known *P. arenaria* Ducke. At that time genuine *P. barbourianum* was not represented at Kew. The very similar populations in French Guiana have almost plane lfts reminiscent of *A. jupunba* but slenderly stipitate lf-nectaries and loose yellowish indumentum of *A. barbouriana*.
e. The only collection known from Ecuador (*K. Thomsen 58828*, NY) differs from all others by its peripheral fls on a pedicel to 2.5 mm long; otherwise the plant fits neatly within *barbouriana* var. *barbouriana*. Should the fruit, when it becomes known, present variation yet undocumented in *barbouriana*, the plants may represent a new variety.

14b. Abarema barbouriana var. **arenaria** (Ducke) Barneby & Grimes, comb. et stat. nov. *Pithecolobium arenarium* Ducke, Arq. Inst. Biol. Veg. **2:** 37. 1935. — "Habitat prope Manáos . . . circa Ponte do Mindú . . . 14-3-1932 leg. A. Ducke, H[erb.] J[ard.] B[ot.] R[io de Janeiro] n° 23.233." — Holotypus, RB!; isotypi, P!, US!.

Lf-formula (ii–)iii–v/(7–)8–11; lfts either pilosulous overall on dorsal face, glabrate except for pilosulous midrib, or (locally) glabrous, the largest ones 9–21 × 5–10 (locally only 3–7) mm, the venation (as given in key) commonly better developed than in var. *barbouriana*; *pods* loosely strigulose with hairs up to 0.5 mm.

In savanna and forest-margin habitats at low and montane elevations in Venezuelan Guayana, W-centr. Guyana, and the Río Negro basin in T. F. Amazonas, Venezuela, and state of Amazonas, Brazil: N slope of Mt. Roraima, 800 m, Guyana and on Gran Sabana

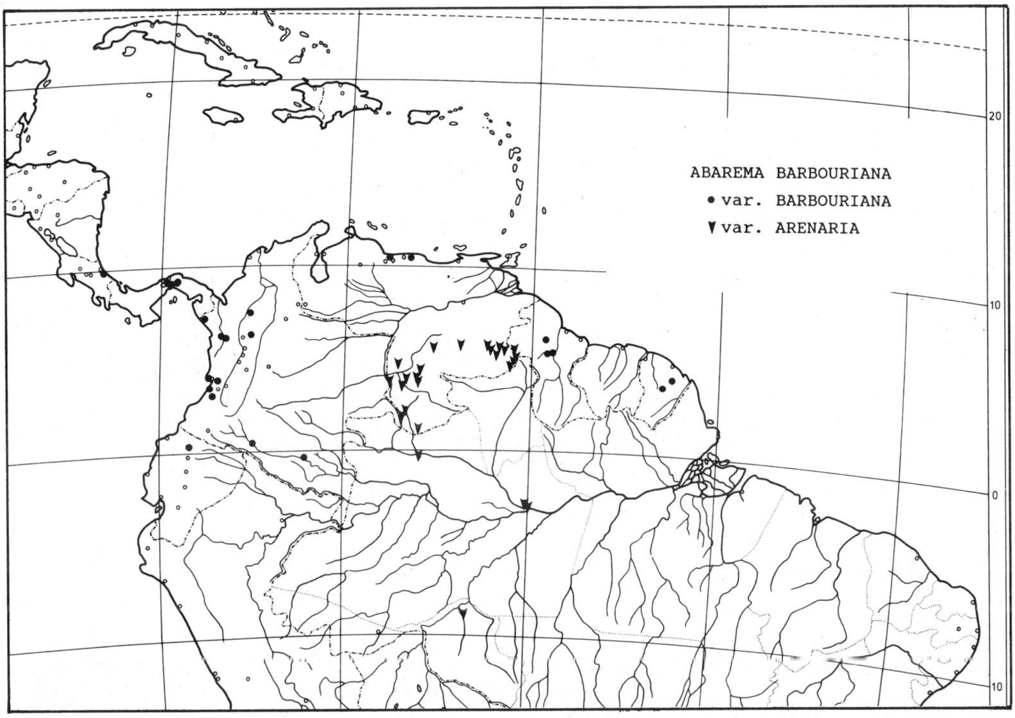

Map 16. Panama and northern South America. Range of *Abarema barbouriana* (Standley) Barneby & Grimes var. *barbouriana* and var. *arenaria* (Ducke) Barneby & Grimes.

and emergent tepuis, at 1250–1900 m, in Bolívar, Venezuela; on cliffs and rocks at 1500–2000 m of tepuis of T. F. Amazonas, Venezuela; in sandy, sometimes inundated savanna and along streams at 120–150 m on the Negro–Orinoco divide in T. F. Amazonas, Venezuela; and collected frequently in moist sandy, non-inundated, pristine and disturbed woodland in vicinity of Manaus, state of Amazonas, Brazil; and disjunct in sandy campo cerrado at ±200 m near Pôrto Velho in N Rondônia, Brazil, and (see discussion below) at 200 m on Río Caquetá, 75°W, in Amazonian Colombia. — Map 16. — Fl. in upland X–II, in lowland nearly throughout the year. — *Guamito* (T. F. Amazonas); *mese-yek* (Bolívar; the bark yields saponins for washing); *angico bravo*, *faveira* (Manaus).

The var. *arenaria* differs from var. *barbouriana* principally in lower leaf-formula, and, for the most part, var. *arenaria* has larger leaflets with more numerous secondary nerves, and locally it has larger flowers. It is racially complex, however, and the number of pinnae, at most 4 or near Manaus occasionally 5 (not 5–12) pairs, is the only dependable criterion. Six groups of populations are feebly distinguished by combinations of habitat, dispersal, and minor features of the indumentum and flower-size, two of them found in lowland white-savannas of the Río Negro basin, two at montane or upland elevations in Venezuelan Guayana, one in Amazonian Colombia, and one in campo of middle Río Madeira basin in N Rondônia:

a. Around Manaus, at 30–40 m: lf-formula (ii–)iii–v/8–11(–13), the largest lfts 11–18 × 5–8 mm; lfts densely pilosulous beneath; peripheral fls small, the calyx 3.2–4 mm, the corolla 5.3–6.5 mm.

b. Headwaters of Río Negro in Venezuela and adjacent Brazil, not over 150 m: lf-formula iii–iv(–v)/10–14, the largest lfts only 9–13 × 3–7 mm, pilosulous overall beneath; peripheral fls of the preceding. In its small, ventrally convex lfts this simulates var. *barbouriana*.

c. In sandy campo near Pôrto Velho, Rondônia: lf-formula iii–iv/7–8, the lfts to 20–24 × 6–7 mm, glabrous dorsally except for pilosulous midrib; fls of the preceding.

d. Tepuis of T. F. Amazonas, Venezuela, 1500–2000 m: lf-formula ii–iv/8–11, the largest lfts 12–16 ×

5–7 mm; lfts glabrous beneath or almost so; peripheral fls of the two preceding.

e. Gran Sabana and tepuis of state of Bolívar, Venezuela, eastward from Río Paragua, 1250–1900 m: lf-formula ii–iv/8–11(–12), the largest lfts (10–)14–21 × (5–)7–10, usually pilosulous beneath but rarely glabrous except for ciliolate midrib; peripheral fls larger, the calyx 4.2–6 mm, the corolla 8–9.5 mm.

f. Solano, on Río Caquetá, 200 m: lf-formula ±iv/8–10, the lfts to 18 × 9 mm; peripheral fls small, the calyx 2.5–4 mm, the corolla 4.5–5 mm.

15. Abarema adenophora (Ducke) Barneby & Grimes, comb. nov. *Pithecolobium adenophorum* Ducke, Arq. Inst. Biol. Veg. **4:** 5. 1938. — "[Brazil, Amazonas:] . . . sat frequens circa Manaos . . . loco Estrada do Aleixo 15-6-1932 leg. A. Ducke (H[erb.] J[ardim] B[otânico do] R[io de Janeiro] 23.238." — Holotypus, RB (2 sheets, fl. & fr.)!; isotypi, NY!, P!, U (2)!, US!.

Pithecolobium adenophorum sensu Ducke, 1949: 36.

Macrophyllidious *trees* 12–30 m with trunk 2–5 dm dbh, the fuscous annotinous branches with epidermis exfoliating in flakes, the mature foliage appearing glabrous but the young branches, petioles and inflorescence densely minutely furfuraceous-puberulent with gold-brown or sordid hairs 0.1 mm or shorter, the bicolored lfts sublustrously brown-olivaceous above, paler beneath, the inflorescence a short terminal panicle of pedunculate, dense and short racemes of greenish white fls arising either singly or geminate from the axil of coevally expanding or quickly hysteranthous lvs, the fruits immersed in foliage. *Stipules* caducous from emergent lvs, linear 1.5–5 mm, absent from mature spms. *Lf-formula* ii–iii(–iv)/2–4(–5), the lfts 16–40(–48) per lf; lf-stks 4.5–33 cm, the petiole proper 2.5–10 cm, the longer (or only) interpinnal segment 2–11.5 cm; a stipitate or sessile, funnel-shaped nectary to 8–13 × 6–12 mm, yellowish when fresh but when dry coriaceous, brown and vertically striate, situated at insertion of first pinna-pair, and often but not always similar but smaller or squatly tube-shaped ones at tip of some pinna-rachises; pinnae of unequal length, the proximal pair shortest, either the distal or the penultimate pair longest, the rachis of these (4–)5–12 cm, its one or its longest interfoliolar segment 1.8–5 cm; paraphyllidia present on nascent pinnae and to 0.7–1 mm, but early caducous from a minute conic protuberance at base of mature ones; lft-pulvinules viewed from dorsal side 2–5 × 1–2.2 mm, closely cross-wrinkled; lfts subequiform but strongly accrescent distally, asymmetrically obovate (or some obscurely rhombic-obovate) from inequilateral, broad-cuneate or rounded base, obtuse muticous or shallowly emarginate, the furthest pair ±5–11 × 3.5–7.5 cm, 1.4–1.8 times as long as wide; the subcentric or moderately displaced midrib evenly arched forward, the close rank of secondary veins widely ascending, brochidodrome well within the incipiently revolute margin, all these and the reticulum of minor veinlets prominulous on both faces but more sharply so beneath. *Peduncles* strongly compressed 2.5–6 cm; racemes (10–)15–25-fld, the slenderly clavate receptacle 2–9 mm; bracts evanescent; fls strongly dimorphic, the lower peripheral ones slenderly pedicellate and greenish, the terminal one substantially stouter, sessile on a short pedestal, and white, the perianth of all 5-merous and brownish puberulent externally; PERIPHERAL FLS: pedicels 4.5–6 × 0.5–0.7 mm; calyx 1.6–3 × 1.5–2.2 mm, the ovate-deltate teeth 0.5–0.9 mm; corolla 3.7–5.5 mm, the lance-ovate, apically cucullate lobes 1.2–2.2 mm; androecium 15–17-merous, ±14–15 mm, the tube (including stemonozone) 2–2.9 mm; TERMINAL FL: calyx 3–5 × 2–2.5 mm; corolla 7–9 × ±2 mm; androecial tube nearly as long as the corolla, the filaments at point of separation dilated and recurved, as in *Abarema leucophylla*; ovary truncate, pubescent on distal half. *Pods* in profile undulately linear but coiled into a ring ±3–3.5 cm in external and ±1.5 cm in internal diam, 6–9-seeded, cuneately contracted at base into a short stipe, obscurely apiculate, the stiffly coriaceous valves at broadest point 9–14 mm wide, low-convex over seeds and there ±0.3 (the crustaceous endocarp ±0.05) mm thick, externally fuscous glabrous venulose, within smooth atrocastaneous or tan, but the seed-cavities then slightly darker and lustrous; dehiscence of *Abarema*; funicle ribbonlike; seeds (not seen fully mature) pallid, the transparent unwrinkled testa loosely investing the forming embryo, pleurogram 0.

In non-inundated equatorial forest at 40–220 m, of highly discontinuous (known) dispersal between 6°N and 6°S latitude in South America: centr. Amazonia in states of Amazonas (Manaus; lower Madeira basin) and W Pará (Trombetas basin); Huallaga Valley in Alto Amazonas of dept. Loreto, Peru; on Río Atrato in depts. Chocó and Valle, Colombia; disjunct in the lowlands of NE Nicaragua (Zelaya). — Map 17. — Fl. V–VI(–?).

In overall aspect and leaf-formula *A. adenophora* resembles *A. microcalyx*, but differs not only in the

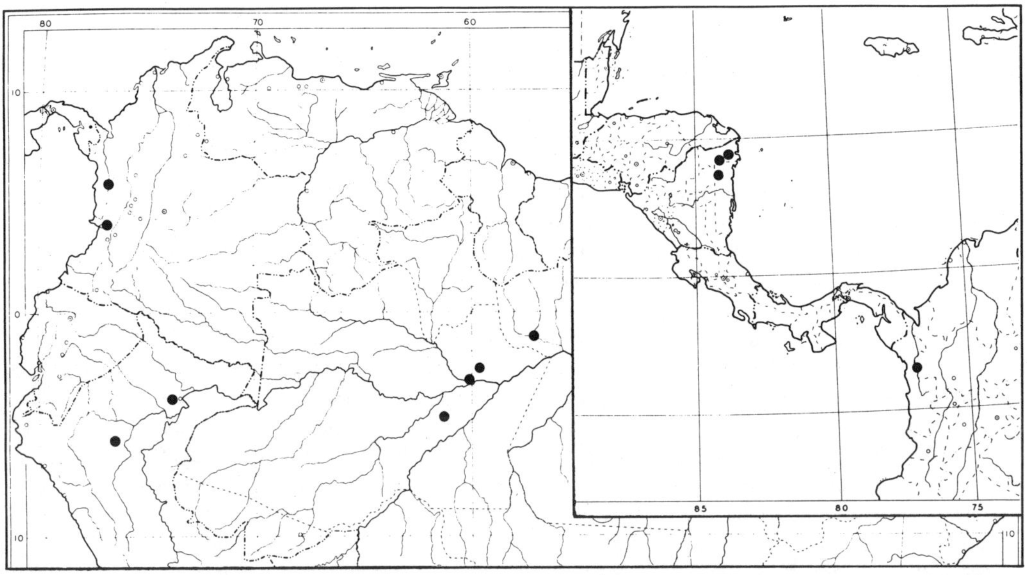

MAP 17. South America and (inset) Central America. Range of *Abarema adenophora* (Ducke) Barneby & Grimes.

hypertrophied petiolar-nectaries but in the much larger central flower of the capitula. The fragmented dispersal of *A. adenophora* might suggest that more than one taxon lies concealed among the disjunct populations, disguised by the prominently modified nectaries, but we have found nothing in their gross morphology to substantiate this possibility. The one other abarema with similarly enlarged nectaries, *A. macradenia*, differs greatly in leaf-formula and ligneous fruit.

A leaf from a stump-sprout collected near Manaus (*Grimes 3118*, NY) has these dimensions: leaf-stalk 4.6 dm; rachis of furthest pinnae 1.9 dm; largest leaflets 12.5–7 cm. An anomalous fruiting collection from the Ducke Reservation at Manaus has poorly developed leaf-nectaries on the primary leaf-stalk but typical cupular ones on the pinna rachises.

A gigantic leaf collected in Zelaya, Nicaragua (*Little 25315*, US), under the same vernacular name, *grano del oro*, as Little's fertile no. *25427* (US) is another example of vigorous sapling growth: lf-formula vi/6; lf-stalk ±4.5 dm, rachis of longer pinnae to 1.6 dm, terminal lfts to 7 × 4 cm. Yet another, (*Poveda 1038*, F, from Costa Rica–Nicaragua frontier) perhaps conspecific, has lf-formula v/6, lf-stalk 7 dm, with rachis of larger pinnae 2.3 dm. These measurements are not introduced into the description of the species.

16. Abarema zollerana (Standley & Steyermark) Barneby & Grimes, comb. nov. *Pithecolobium zollerianum* Standley & Steyermark, Publ. Field Mus. Nat. Hist., Bot. Ser. **22:** 343. 1940. — "Guatemala: Dept. San Marcos: Above Finca El Porvenir . . . lower south-facing slopes of Volcán de Tajumulco, alt. 1,300–1,500 meters, March 9, 1940, *Julian A. Steyermark* 37440 . . . " — Holotypus, F!; isotypus, A!; paratypus, *Steyermark 33295*, F!.

Amply microphyllidious *trees* 5–18 m with smooth trunk, the young branchlets and all axes of lvs and inflorescence densely brown-, bronze- or sordid-puberulent or -tomentulose, the thin-textured lvs strongly bicolored, the lfts above deep green and glabrous except for thinly ciliolate midrib, beneath pale bright green and remotely minutely strigulose overall, the solitary or paired capituliform racemes axillary to coeval or immediately hysteranthous lvs, sometimes at first forming a small efoliate panicle but the fruits immersed in foliage. *Stipules* linear-lanceolate 2–4 mm, very early caducous. *Lf-formula* (iii–)iv–x(–"xiii"; protologue)/11–16(–18); lf-stks (5–)6–20 cm, the stout petiole 2–4.5 cm, the longer interpinnal segments 1.2–2.5 cm; petiolar nectaries sessile, round or elliptic, shallowly cupular with thick wrinkled rim, the first situated shortly below insertion of first pair of pinnae 0.9–3 mm diam, others at some distal pinna- and lft-pairs similar but progressively smaller upward; pinnae accrescent distally but the penultimate pair often longest, their rachis 6–13 cm, the longer interfoliolar segments 5–10(–11) mm; lft-pulvinules in dorsal view 0.5–1 × 0.4–0.8 mm; lfts decrescent proximally, those near and above midrachis, except for the obovate furthest pair (no further mentioned) subequiform, the blades obtusely rhombic-oblong around a diagonal, straight or slightly retroarcuate midrib, at base inequilaterally broad-cuneate, very obtuse or emarginate at broadly rounded apex, the larger ones 12–20 × 5–9 mm, 2–3 times as long as wide; the midrib giving rise on each side to ±6–8 pairs of major (and few weaker intercalary) secondary nerves brochidodrome well within the weakly revolute margin, the tertiary venules few, weak and random, all venation finely prominulous on both faces. *Peduncles* 3–12 cm; racemes ±26–45-fld, the obovoid or scarcely thickened linear axis becoming 3–15 mm; bracts (few seen) very early caducous, linear-spatulate ±1 mm; fls dimorphic, the peripheral ones distinctly pedicellate (but pedicels shorter upward), the terminal one or two subsessile, broader, and with enlarged androecium, the perianth of all 5-merous, finely minutely silky-puberulent either overall or sometimes glabrate proximally; PERIPHERAL FLS: pedicel 0.9–4 mm; calyx turbinate-campanulate 2–2.3 mm, the depressed-deltate teeth 0.3–0.5 mm; corolla 6–6.4 mm, the ovate lobes 2–2.8 × 1.3–1.7 mm; androecium 22–26-merous, 2.3–2.6 cm long, the stemonozone ±1.5 mm, the tube 2.5–3.5 mm; ovary ellipsoid, conical at apex, glabrous; stigma poriform ±0.1 mm diam; TERMINAL FL(S): pedicel 0.2–0.5 × 1 mm; calyx broadly campanulate 2.5–3 × 1.6–3 mm; corolla 6.5–8 mm, the lobes 2–3.5 mm; androecial tube about as long as corolla, dilated at separation of the 23–45, 2–3-seriate whitish filaments. *Pods* 1 per capitulum, sessile, in profile broad-linear, evenly recurved through ¼–¾-circle, 5.5–12.5 × 1.3–1.6 cm, constricted only where ovules abort, the glabrous valves at first plane, framed by thickened sutures, becoming stiffly coriaceous, fuscous, rugulose but externally veinless, the crustaceous mesocarp ±0.4 mm thick, the fully ripe endocarp dull-maroon (in age castaneous) overall, smooth lustrous in the seed-cavities and papillate between them; dehiscence downward through both sutures, the valves recurved but not or little twisted; seeds plumply obovoid or lentiform, at least a trifle compressed, in broad view 7.5–8.5 × 6–7.5 mm, the translucent white testa ±0.2 mm thick, when fresh transmitting the aniline blue color of the horny embryo but when dry dull whitish nearly overall, bearing on each face a complete or hippocrepiform, delicately incised pleurogram 3.5–4 × 2.5–4 mm.

In pine-oak woodland, 1100–1800 m, apparently local in Sa Madre del Sur of S Mexico and adjacent Guatemala, between 91° and 98°W latitude: in district Juquila of SW Oaxaca; and in depts. San Marcos and Quetzaltenango, Guatemala. — Map 18. — Fl. in Mexico V–VII, fr. X–II. — *Coralillo* (Mexico); *plumillo* (Guatemala).

In the context of its dispersal *A. zollerana* is likely to be mistaken only for *A. idiopoda*, which differs, however, in more numerous (to 18–40, not 11–18 pairs in longer pinnae) and narrower (to 2.5–3.5, not 5–9 mm wide) leaflets, in longer peripheral pedicels (the lowest mostly more, not less than 4 mm long) and in the technical character of apically conic, not truncate ovary. The perhaps closely related, though distantly allopatric, *A. jupunba* var. *jupunba* has leaflets emphatically accrescent upward along the pinna rachis, with terminal pair to 1–2 cm (not 5–9 mm) wide, and a more densely silky perianth. The separation between them is weak, and the specific status of *A. zollerana* somewhat precarious.

The epithet preserves the memory of Erich Zoller, proprietor of Finca El Porvenir, who assisted Steyermark in his exploration of Cerro Tajumulco.

17. Abarema leucophylla (Bentham) Barneby & Grimes, comb. nov. *Pithecolobium leucophyllum* Spruce *ex* Bentham, Trans. Linn. Soc. London **30:** 581. 1875 & *in* Martius, Fl. Bras. **15(2):** 433. 1876. — Typus infra sub var. *leucophylla* indicatur.

Macrophyllidious arborescent *shrubs*, treelets, or *trees* 1.5–21 m, attaining a trunk 1.5–4 dm dbh, the developing stems, axillary buds, young lf-stks and inflorescences minutely sordid-puberulent but all but the buds early glabrescent, the broad coriaceous lustrous bicolored lfts glabrous or minutely ciliolate, revolute-margined, the relatively few-fld capitula borne either solitary or geminate in distal lf-axils or in subterminal efoliate pseudoracemes surpassed by foliage. *Stipules* caducous, perhaps lacking in var. *vaupesensis*. *Lf-formula* i–ii(–iii)/1–2(–3); the lfts of larger lvs 4–16 per lf; lf-stks (0.6–)1–11 cm, the petiole 0.6–4 cm, the one (or longer) interpinnal segment, if present, 1–7 cm; lf-nectaries situated immediately below first (or only) pinna-pair sessile, either scutiform, cupular or patelliform, in outline elliptic or subcircular 1.5–3.5 mm in long diameter, similar but smaller nectaries at tip of almost all pinnae; rachis of pinnae 0.8–7 cm, the interfoliolar segment(s) 1.5–3.5 cm; a small conic paraphyllidium sometimes present; lft-pulvinules in dorsal view (2.2–)2.5–6 × 1.3–2.1 mm, wrinkled; lfts 4–11 × (2–)2.6–6.2 cm; venation reticulate. *Inflorescence* axillary or pseudoracemose,

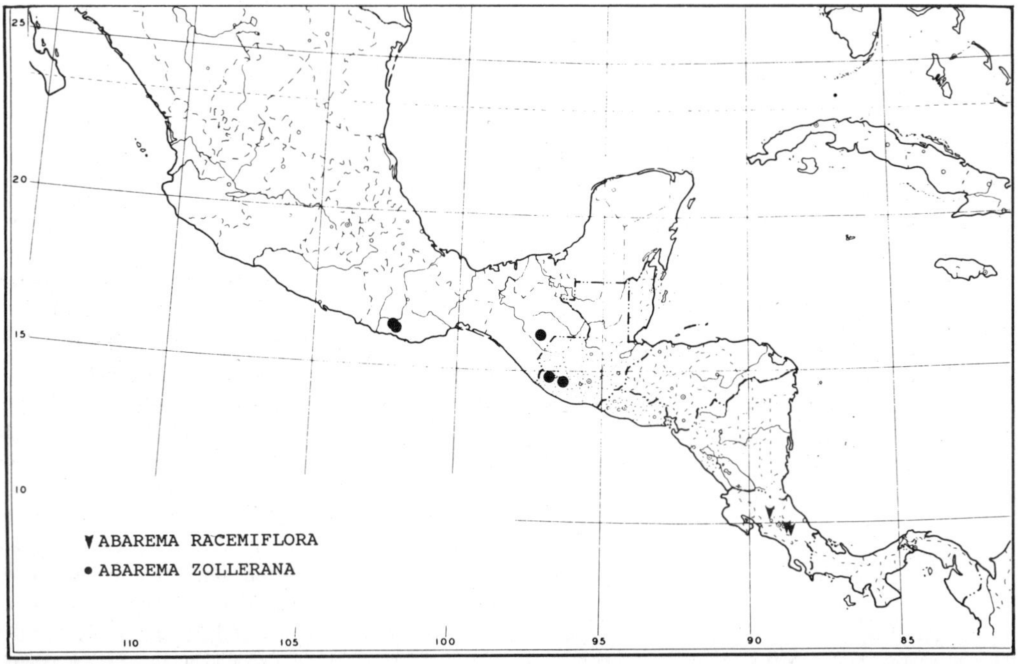

Map 18. Mexico and Central America. Range of *Abarema racemiflora* (Donnell Smith) Barneby & Grimes and *A. zollerana* (Standley & Steyermark) Barneby & Grimes.

the peduncles solitary or 2–3 together, subtended by an expanded leaf or distally by a rudimentary lf-stk; capitula 6–15-fld, the fls dimorphic, the peripheral ones pedicellate, the terminal one sessile on a short pediment and stouter, the receptacle 1.5–5 mm; bracts caducous, linear, ovate or elliptic-obovate 1–2 mm; PERIPHERAL FLS: pedicel 0.2–1.7 mm; perianth 5-merous; calyx campanulate from broad-cuneate or obconic base 2–3.6 mm, the ovate- or deltate-triangular teeth 0.4–0.9 mm; corolla 3–6.5 mm, the lobes ovate, lance-ovate or lanceolate 0.8–1.5 mm; androecium 14–24-merous, the tube 1.5–4 mm, the filaments either pink or reddish distally; ovary truncate, glabrous at anthesis; style as long as or longer than androecium, the stigma scarcely 0.1 mm diam; TERMINAL FL: calyx sessile, broadly campanulate, 2–5 × 1.8–3.5 mm; corolla 3.3–8.5 mm; androecium ±3–7.5 mm, the filaments at point of insertion much thickened and recurved. *Pods* 1–2 per capitulum, ascending, sessile or cuneately narrowed basally into a short neck, in profile narrow- or linear-oblong, scarcely or falcately recurved (but never coiled) 27–50 × 9–14 mm, 6–8-seeded, laterally compressed but strongly turgid, girdled by strong sutures, the valves at first yellow-green but nigrescent when dried, at maturity black or fuscous and coarsely rugulose externally, within dull crimson overall, smooth and lustrous at the seed-cavity but velvety-papillate between them; dehiscence elastic, downward through both sutures, the valves either plane or recurved into a ring (as in *Calliandra*) but not or little twisted; funicles strap-shaped, coiled at apex; seeds obovoid biconvex, marginally compressed, in broad view 6.5–8 × 5–6.5 mm, the very thin translucent testa subhyaline, loosely investing and not concealing the smooth tan or pearl-gray horny embryo, crumpled when dry, pleurogram 0.

Key to the varieties of *A. leucophylla*

1. Trees (4–)5–21 m; peripheral fls with calyx 3–3.6 mm, and corolla 4.5–6.5 mm; terminal fls with calyx 3–5 mm and corolla (5.5–)6–8.5 mm; E Vaupés, Colombia, SW in T. F. Amazonas, Venezuela and NW state of Amazonas, Brazil .... 17a. var. *leucophylla*
1. Shrubs or treelets 1.5–5 m; peripheral fls with calyx 2–2.6 mm and corolla 3–4 mm; terminal fls with calyx 2–2.5 mm and corolla 3.3–3.5 mm; on the upper Río Vaupés and Río Apaporis in SE Com. Vaupés, Colombia................ 17b..var. *vaupesensis*

17a. Abarema leucophylla (Bentham) Barneby & Grimes var. **leucophylla.** *Pithecolobium leucophyllum* Spruce *ex* Bentham, ll. cc. supra. — ". . . *in* *sylvis siccis ad flumen Uaupes prov. do Alto Amazonas* [Brazil], *Spruce 2786*." — Holotypus, K (herb. bentham.)! = NY Neg. *2004*; isotypi, †B = F Neg. *1206*, BM!, F *1540696*!, GH!, GOET!, K(herb. bentham.)!, (herb. hooker.)!, P (2 sheets)!, RB (2 sheets)!. — *Feuilleea leucophylla* (Bentham) O. Kuntze, Revis. Gen. Pl. **1:** 188. 1891.

Inga crassifolia Klotzsch *ex* Bentham, 1875: 582, pro syn. *Pithecolobii filamentosi*, nom. nud. — "Venezuela, *Otto*," K (3 sheets)!.

Characters as given in key to varieties.

In scrub savanna or matorral on white sand, sometimes an element of the savanna-forest ecotone, ±75–180 m, locally plentiful between 4°N and 0°30′S on the upper Orinoco and sources of the Negro in E Vaupés, Colombia, SW T. F. Amazonas, Venezuela, and NW state of Amazonas, Brazil. — Map 19. — Fl. II–VII, X–XI, perhaps irregularly throughout the year, the old fruits often persistent into anthesis of the next season. — *Marepillo* (Venezuela).

17b. Abarema leucophylla var. **vaupesensis** Barneby & Grimes var. nov., a var. *leucophylla* foliolorum formula et flosculis minoribus diversa: foliolorum majorum pinnae 1–2-, foliola plerumque 2-juga; florum periphericorum et terminalis calyx 2–2.5 mm et corolla 3–4 (nec 4.5 et ultra) mm longi; floris terminalis stamina brevius in tubum ±3 (nec 4.5–7.5) cm longum coadunata. Frutices et arbores parvi 1.2–5 m alti. — COLOMBIA. Vaupés: Raudal de Jirijirimo on Río Apaporis, ±0°07′N, 70°36′W, 27.XI.1951 (fl.), *Richard Evans Schultes & Isidoro Cabrera 14643*. — Holotypus, GH; isotypus, NY.

Characters as given in key to varieties.

In open savanna below 300 m, local at the SW edge of the range of var. *leucophylla*, on the upper ríos Vaupés and Apaporis in SE Com. Vaupés, Colombia. — Map 19. — Fl. IX–XI.

The early collections of *A. leucophylla* var. *vaupesensis* were misidentified as *Pithecellobium microcalyx*, from which they are clearly different in plane or shallowly convex, not mounded and small-pored leaf-nectaries, and by larger calyces, more coriaceous leaflets, and basally dilated filaments of the capitulum's terminal flower.

18. Abarema mataybifolia (Sandwith) Barneby & Grimes, comb. nov. *Pithecellobium mataybifolium* Sandwith, Kew Bull. **1948:** 313. 1948. — "BRITISH GUIANA: Bartica-Potaro road, 107 miles, Nov. 15th, 1943, *D. B. Fanshawe* in *Forest Dept.* no. 4232." — Holotypus, K! = NY Neg. *2039*!; isotypi, K!, NY!.

Macrophyllidious *trees* 8–20 m, with trunk 8–30 cm dbh, glabrous except for densely brown-sordid-puberulent axillary buds and inflorescence, the ample chartaceous lfts olivaceous sublustrous above, paler beneath, the inflorescence consisting of capitula borne on solitary and 2–3-nate peduncles either axillary to distal lvs or forming a short terminal pseudo-raceme. *Stipules* not seen, perhaps very small caducous, but possibly lacking (scar obscure). *Lf-formula* i–ii/2–3(4), the lfts 8–16(24) per lf; lf-stk 1–8 cm, the true petiole including wrinkled pulvinus 1–5 cm, the interpinnal segment, when present, to 4.5 cm; petiolar nectaries at insertion of first or only pair of pinnae amorphous, immersed in groove of petiole or shallowly sunken and rugulose, sometimes with a minute cone and pore near middle, lacking from tip of pinna-rachises; rachis of pinnae 4.5–12 cm, the one or the longer of two interfoliolar segments (2–)2.5–6 cm; lf-pulvinules in dorsal view (4.5–)5–9 × 1–2 mm, densely cross-wrinkled; lfts slightly accrescent distally, subequilaterally ovate or elliptic-ovate from broad-cuneate or, especially on anterior side, rounded base, shortly obtusely acuminate, the blade of distal pair 8–14 × 3–7.5 cm, 1.8–2.3(–2.5) times as long as wide; venation densely pinnate, the widely ascending secondary nerves brochidodrome well within the loosely revolute margin, all these and an elaborate reticulum of lesser veinlets prominulous on both faces. *Peduncles* 2–5 cm; capitula ±15–20-fld, the subglobose or shortly clavate receptacle ±3 mm; bracts somewhat dimorphic, the lower ones triangular to lanceolate, 2.5–2.75 × ±1 mm, the upper becoming elliptic, elliptic-spatulate, or spatulate, and smaller, all very quickly deciduous; perianth 5-merous, rusty-pilosulous externally; fls dimorphic, the peripheral ones pedicellate, the terminal one (few seen) sessile and a trifle coarser; PERIPHERAL FLS: calyx campanulate 2–3.6 mm, cuneately narrowed at base into a pedicel 2–3.6 mm, the triangular-ovate teeth 0.5–0.8 mm; corolla narrowly trumpet-shaped 6.3–7.4 mm, the apically cucullate lobes 1.6–2 mm; stamens 18–28, the tube, including the stemonozone, 3–3.5 mm; TERMINAL FL: corolla hardly longer than the rest, but the staminal column exserted to 2 mm; corolla of all fls greenish; the filaments white, brownish in fading; ovary glabrous, at apex a little swollen and obliquely truncate; ovules 10–14. *Pods* 1–2 per capitulum, in

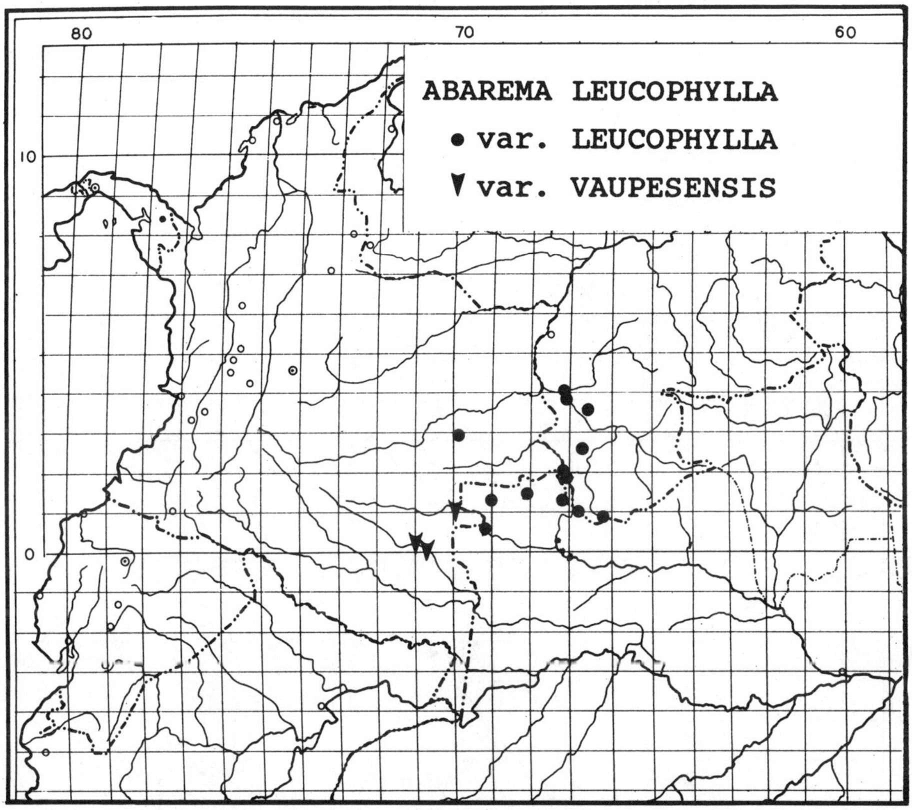

MAP 19. Ríos Orinoco, Negro, and Vaupés in Venezuela, Colombia, and Brazil. Range of *Abarema leucophylla* (Bentham) Barneby & Grimes var. *leucophylla* and var. *vaupesensis* Barneby & Grimes.

profile broad-linear, evenly recurved through half or nearly a complete circle, ±10–20 × (1.3–)1.8–2.6 cm, 9–14-seeded, at first plano-compressed and framed by thickened, very shallowly undulate sutures, the brown or fuscous, glabrous valves becoming low-convex over each seed, coarsely transverse- and reticulate-venose externally, within red-crimson overall; dehiscence of *Abarema;* funicle ribbonlike, contorted distally; seeds plumply lentiform-globose, in broader view 8–9 × 7–8 mm, described when fresh as gray or deep blue, but brunnescent when dried, the membranous translucent testa scarcely 0.05 mm thick in section, closely investing the hard embryo but sometimes a little wrinkled, pleurogram 0; cotyledons plano-convex, closely appressed. Eopylls opposite, bipinnate, subsequent lvs alternate, all of formula i/2.

In virgin and second-growth, non-inundated forest below 170 m, scattered in E equatorial South America in lat. 8°N–5°S, from the Barima River in Guyana SE to Paragominas in NE Pará, Brazil, and west to vicinity of Manaus. — Map 20. — Fl. II, IX–XI. — *Huruassa* (Guyana).

Abarema mataybifolia is notable among macrophyllidious abaremas for the long pulvinule of the leaflets.

Perhaps related to *A. mataybifolia* is *B. Hoffman 3251* (NY), collected in swamp scrub between 1100 and 1500 m on the plateau of Mt. Ayanganna in the Pakaraima Range of Guyana. The material is immature, but the flowers seem compatible with *A. mataybifolia.* The leaf-formula, however, is only i/1–2 and the leaflets are subsessile, not raised on a petiolule 5–9 mm long.

19. Abarema laeta (Bentham) Barneby & Grimes, comb. nov. *Pithecolobium laetum* Bentham, London J. Bot. **3:** 203. 1844. — "Maynas in Brazil, Poeppig." — *Inga laeta* (Bentham) Poeppig *ex* Poeppig & Endlicher, Nov. Gen. Sp. Pl. **3:** 80. 1845.

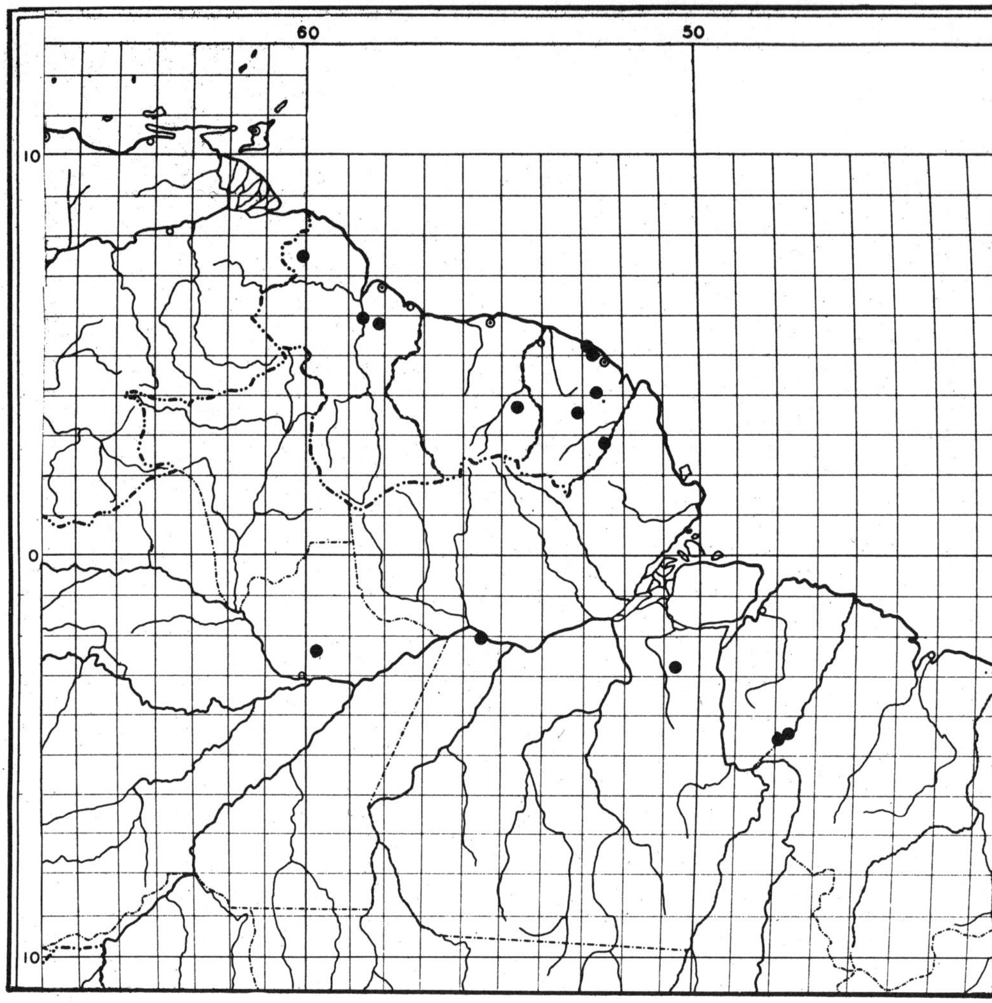

MAP 20. The Guianas and the lower Amazon and Gurupi rivers in Brazil. Range of *Abarema mataybifolia* (Sandwith) Barneby & Grimes.

— "Crescit cum praecedente [in sylvis provinciae Maynas circum Yurimaguas]." — Holotypus, *Pöppig D.2367*, W(herb. poeppig.)! = NY Neg. *12572*; isotypi, †B = F Neg. *1204*, F (fragm.)!, OXF!, W(herb. endlicher.)!. — *Feuilleea laeta* (Bentham) O. Kuntze, Revis. Gen. Pl. **1:** 188. 1891. *Klugiodendron laetum* (Bentham) Britton & Killip, Ann. New York Acad. Sci. **35:** 125. 1936.

?*Pithecolobium polycarpum* Poeppig & Endlicher, Nov. Gen. Sp. Pl. **3:** 81. 1845. — "Crescit in sylvis maynensibus ad Yurimaguas [Alto Amazonas, Loreto, Peru]. Februario [1831] cum fructu maturo lectum." — Equated with *P. laetum* by Bentham, 1875: 582, but no acceptable typus found at W. A candidate for typus, *Poeppig 2245*, W(herb. endlicher.)! disagrees with the protologue in lacking "stipulae . . . spinescentes minutae," in pod valves much less than "sesquipedem longis" [unless *sesquipedem* is a mistake for *semipedem*], and the seeds are well over 6 in number. No *Abarema* is known to have pods 1½ feet (i.e., ±45 cm) long. Perhaps based on mixed materials.

Klugiodendron umbrianum Britton & Killip, Ann. New York Acad. Sci. **35:** 126. 1936. — "Comisario del Putumayo, Umbria, Colombia, in forest at 325 m altitude, December, 1930, G. Klug 1865." — Holotypus, NY!; isotypi, A!, GH!, IAN (fragm.)!, US *1456622*!.

Pithecolobium laetum sensu Bentham, 1875: 582; 1876; 434; Ducke, 1949: 37.

Klugiodendron laetum sensu Nielsen, 1981: 190, fig. 5(4a–4c).

Macrophyllidious arborescent *shrubs*, fertile at (1.5–)2 m but becoming 6(–?) m tall with trunks to

2(–?) cm diam and yellow wood, the hornotinous branchlets, lf-stks, and peduncles pilosulous or subappressed- or incurved-puberulent with rufous or sordid-yellowish trichomes to 0.2–0.7(–0.8) mm, the permanently pale green and subconcolorous, thinly papery lfts lustrous above, dull beneath, either glabrous overall, or minutely ciliolate on margin and on midrib beneath, or (locally in Peru) pilosulous dorsally overall, the loosely capitate or subracemose capitula of dimorphic fls borne singly or in pairs in a few distal lf-axils, or rarely, in default of these lvs, forming a short terminal pseudoraceme. *Stipules* herbaceous, narrowly lanceolate 1.5–5(–7) mm, 1–3-nerved dorsally, thickened at base and early deciduous from a small, sometimes elevated whitish scar. *Lf-formula* i–ii/1, the lfts either exactly 4 or exactly 8 per lf; lf-stks (1.5–)2–14 cm, narrowly shallowly grooved ventrally, the petiole proper (1.5–)2–10(–12) cm, the one interpinnal segment, when present, (2.5–)3–7 cm, ordinarily a little longer than the petiole, sometimes as long or a trifle shorter; first petiolar nectary situated usually somewhat below midpetiole, less often at or above it, rarely near the lf-pulvinus, always sessile, low-convex, in outline narrowly elliptic, 1.5–5 mm in long diameter, without obvious aperture, similar but smaller nectaries immediately below each pair of lfts; pinna-rachises (4–)6–20(–31) mm, widely shallowly grooved ventrally; pulvinule of lfts, seen from dorsal side, (1–)1.5–3 mm; lft-blades of distal (or only) pair of pinnae subsymmetrically elliptic or (ob)ovate-elliptic from cuneate base, caudately acuminate, the larger ones (including acumen) 10–18(–20) × 3–7(–7.7) cm, (2–)2.1–2.5(–3.9) times as long as wide, the anterior one of each pair a little longer than its fellow, all cuneate and usually revolute at very base, the acumen 1–2.5(–3) cm; midrib either straight or very gently antrorsely curved, prominulous on both faces but above less so and sunk into a groove, the secondary nerves from costa numerous, ascending at wide angle, brochidodrome well within the plane or subundulate margin and giving rise to a dorsally prominulous reticulum of veinlets. *Peduncles* 3–7 cm, naked or bearing under the capitulum one or more empty bracts, occasionally with a flower distant from the rest; capitula 12–28-fld, the subglobose or clavate receptacle 2–7 mm long; floral bracts thinly herbaceous, linear-lanceolate 2.5–8 × 0.4–1.5 mm (rarely abbreviated, ovate and only 1.5 mm), tardily deciduous from a small spur; fls sessile or almost so, dimorphic, one or rarely two terminal ones a little larger and with long-exserted stamen-tube, all greenish; perianth (4)5-merous, externally either rufescent-pilosulous overall or glabrate except for minutely barbellate or puberulent calyx- and corolla-lobes; PERIPHERAL FLS: calyx turbinate 2–4 mm, the tube 1.5–2.5 mm, the triangular or depressed-deltate lobes 0.5–1 mm; corolla (4.2–)4.7–7 mm, the ascending, ovate lobes 1.6–2.7 × 1.2–1.9 mm, finely 1-nerved; androecium white, fading orange, 20–34-merous, the filaments united into a tube 1.7–4(–4.5) mm, this shortly coherent to base of corolla, the longer filaments (13–)18–34 mm; TERMINAL FL(S): slightly wider and coarser than the peripheral ones, scarcely longer (to 6–8 mm), but the staminal sheath 7–16 mm long, at least 1 mm longer than corolla-lobes; ovary of all fls glabrous, truncate; ovules 8–14. *Pods* usually one per capitulum, subsessile but abruptly narrowed into a short neck at base, in broad profile oblong or broad-linear (5–)5.5–13(–16) × 1.4–2.2 cm, 2.5–7(–10) times as long as wide, when short nearly straight but when longer evenly recurved through to ¾ circle, at first plano-compressed and framed by the scarcely undulate sutures, becoming umbonate over developing seeds, the stiffly papery valves green turning tan-brown, densely cross-venulose, the venules subcontiguous; dehiscence through both sutures, the valves elastically recurved and coiling, on inner face crimson and densely microscopically papillate (appearing velvety); funicles ribbonlike, coiled but not at all dilated; seeds plumply ellipsoid 10–11 × 6–8 × 5 mm, the translucent testa either colorless (hyaline) or yellowish, pleurogram 0; the horny, plano-convex cotyledons blue or blackish indigo, visible through the testa, their cordate base concealing the plumule.

In the understory of wet, virgin and disturbed equatorial forest, on terra firme, mostly between 100 and 600 m but on Sa Macarena in Colombia attaining 1100 m, in French Guiana 780 m, fairly frequent in W-centr. Amazonia, in SE Colombia, E Peru, extreme SW Venezuela, and Brazil (lat. 3°N–11°S), and outside this range known from remote stations in W Venezuela (sources of Río Apure), in SE Surinam and adjacent French Guiana, and on Sa do Cachimbo in SW Pará, Brazil. — Map 21. — Found in fl. in all months, collected in fr. II, VII–IX. — *Pacay shimbillo, shimbillo*, (Maynas, Peru); *tu ku* (also Maynas; furnishes soap).

Permanently bright green foliage, widely divergent and crowded secondary venation of the obligately geminate, acuminate leaflets, and orange coloration of the dried (or faded) androecia characterize *A. laeta* at anthesis. The densely cross-venulose valves of the pod, green turning light brown externally and

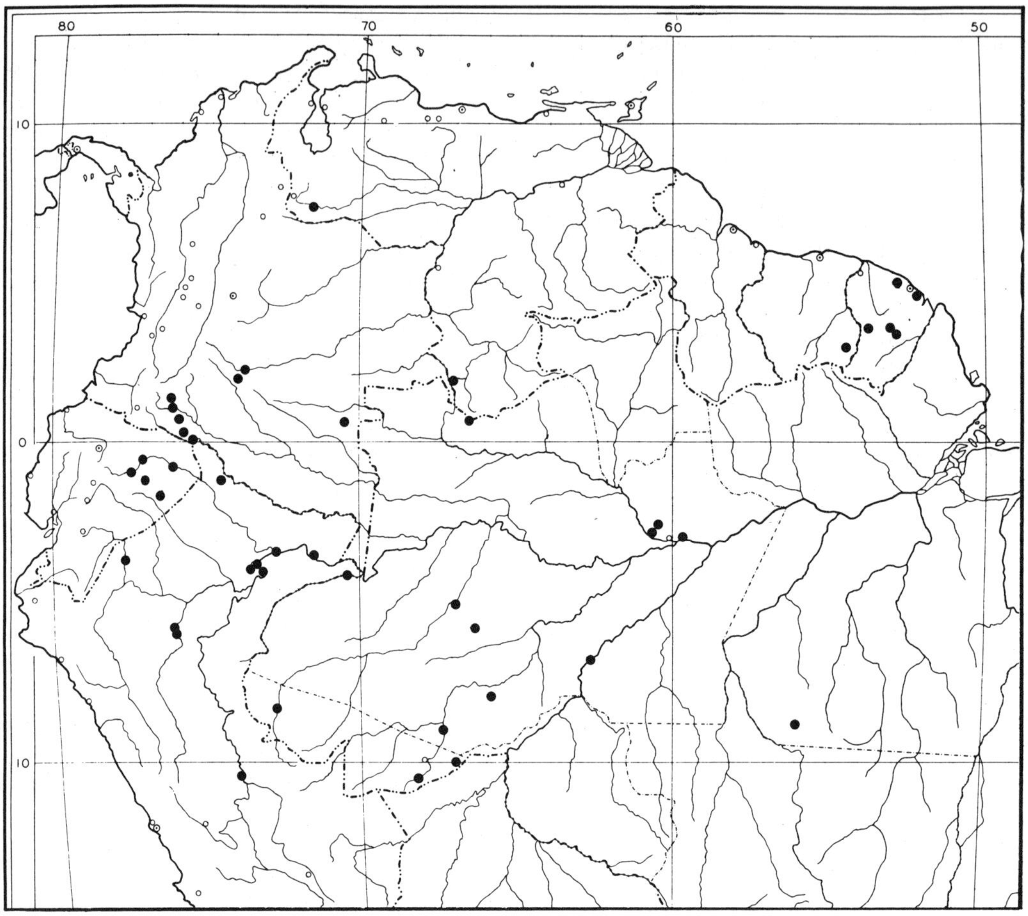

MAP 21. Northern South America. Range of *Abarema laeta* (Bentham) Barneby & Grimes.

velvety-crimson inside, are quintessentially abaremoid, as are the blue exareolate seeds.

20. Abarema campestris (Bentham) Barneby & Grimes, comb. nov. *Pithecolobium campestre* Spruce *ex* Bentham, Trans. Linn. Soc. London **30:** 583. 1875 & *in* Martius, Fl. Bras. **15(2):** 436. 1876. — "Habitat in campis altis Santarem, provincia Para: *Spruce n. 771*." — Holotypus, K(herb. bentham.)! = NY Neg. *2007*; isotypi (some lacking collector's serial number), †B = F Neg. *1183*!, BM!, G!, GH!, NY!, P (3 sheets)!.

Pithecolobium campestre sensu Ducke, 1949: 36.

Macrophyllidious *shrubs* and *treelets* 1–4.5 m with reddish gray scaley bark, the stems, lvs, and peduncles when very young densely puberulent with appressed-ascending or spreading-incurved, pallid or rufous hairs to ±0.1–0.25 mm but all the peduncles early glabrate, the stiffly papery lfts glabrous when fully expanded, dull olivaceous above, paler beneath, the dense capitula of red white-stamened fls solitary or 2–3 together in a few distal lf-axils, immersed in foliage. *Stipules* linear-lanceolate 1.5–4.5 mm, caducous. *Lf-formula* (i–)ii–iii/3–5; lf-stks (3–)7–23 cm, the petiole including dilated lurid pulvinus 2.5–7.5 cm, at middle 1.3–2.8 mm diam, the longer interpinnal segments (2–)2.5–8.5 cm; first nectary contiguous to or close to pulvinus, randomly differentiated, either immersed or mounded, often longer than wide, 1.25–4.3 mm in greatest diameter, similar but smaller nectaries often at insertion of pinnae (but sometimes lacking) and yet smaller ones at tip of some pinna-rachises; pinnae strongly accrescent distally, the rachis of furthest pair ±4–19 cm, the longer interfoliolar segments (1.2–)1.5–5 cm; pulvinules 2–5 × 0.7–1.2 mm, livid and wrinkled; lfts accrescent distally, often much so, ovate, lanceolate, elliptic-lanceolate or rarely elliptic from subequilateral, rounded or subcordate base, shortly obtusely acuminate, the distal ones 3.4–9(–12) × (2–)2.3–4.2(–5) cm, 1.8–2.6 times as long as wide; the straight, almost centric midrib giving rise on each side to 9–14 major (and numerous intercalary minor), widely divergent secondary nerves brochidodrome well within the plane margin, these randomly few-branched to form an open reticulum, the whole venation immersed on upper face, prominent beneath. *Peduncles* 2–6.5 cm; capitula 15–35-fld, the receptacle becoming 2–6 mm; lower bracts ovate-triangular, the upper spatulate, none over 2 mm, all deciduous but the lowest less readily than the rest; fls all sessile or some lower ones contracted at base into a pedicel to 1(–1.5) mm, ordinarily dimorphic, with one or several differentiated fls at apex of spike, but sometimes almost uniform; perianth 5(–6)-merous, the obtusely 5-angulate calyx and the corolla-lobes densely brown-puberulent, but the corolla-tube more thinly so or glabrate; PERIPHERAL FLS: calyx conic-campanulate (2.5–)3–4.5 mm, the low-deltate obtuse teeth 0.3–0.7 mm; corolla tubular-campanulate (6.5–)7–8.5 mm, the lance-ovate lobes 1.7–2.7 mm; androecium (28–)36–46-merous; ovary symmetrically conical at tip, glabrous; stigma poriform; TERMINAL FL(S): calyx 3–4.5 mm; corolla 9–10 mm; stemonozone sometimes to 4 mm, the filament tube 10–13 mm. *Pods* mostly solitary, in profile broad-linear from attenuate base, when well fertilized 10–18 × 1.5–2.4 cm, incurved through a full circle or further, 8–12-seeded, the stiffly papery glabrous, coarsely venulose valves low-convex over each seed, dull fuscous externally, red-orange within, framed by scarcely constricted plane sutures 1.6–2.5 mm wide; dehiscence elastic through the length of both sutures, the valves recurved and coiling; seeds plumply ellipsoid ±7.5–9 × 5.5–7 mm in broad view, bicolored, the testa in lower half firm and white, translucent distally and transmitting the blue color of the embryo, the pleurogram finely engraved on both faces.

In campo thickets on terra firme and in brush-woodland on hills, thriving in capoeira, mostly on sandy soils at ±100–150 m, local in Amazonian Brazil on S affluents of the great river; in E Amazonas (Maués) and W Pará (Ríos Arapiuns and lower Tapajóz), extending S in the Madeira and Tapajóz basins to ±6°30′S on Río Maicoré and 8°S on Río Cururú. — Map 22. — Fl. IX–II.

Abarema campestris is sometimes mistaken for the related and partly sympatric *A. cochleata*, which is similar in equilateral outline of the leaflets. From this it differs decisively in the longer primary leaf-stalks and disproportionately short secondary leaf-axes, which give the whole leaf a characteristic outline, and in the pale green, nearly concolorous leaflets.

21. Abarema cochleata (Willdenow) Barneby & Grimes, comb. nov. *Inga cochleata* Willdenow, Sp. Pl. **4:** 1020. 1806. Typus infra sub var. *cochleata* indicatur.

Macrophyllidious *trees* attaining 20(–25) m with trunk to 5 dm diam, but often flowering as bush or treelet (2–)3 m upward, the annotinous branchlets pallid-gray, the very young stems, axillary buds, and inflorescence minutely tomentellous with rusty or pallid hairs <0.2 mm, the lvs strongly bicolored, the lfts lustrous dark green (when dry dark brown) above, pale olivaceous (when dry tan and dull) beneath, the capitula of greenish white, white-stamened fls solitary and 2–3-nate in distal lf-axils, immersed in foliage. *Stipules* narrowly ovate 0.5–1 mm, caducous, absent from mature spms. *Lf-formula* i–ii(–iii)/3–4; lf-stks 0.6–5(–8) cm, the petiole including livid wrinkled pulvinus 0.6–2.6 cm, at middle 0.8–1.5 mm diam, the one (or rarely the longer of 2) interpinnal segments 8–27(–35) mm; a shallowly cupular, often crumpled nectary 0.7–1.3 mm diam either sessile or depressed into lf-stk immediately below first pair and often below further pairs of pinnae, and similar smaller ones at most nodes of pinna-rachises; pinnae when more than 1 pair accrescent distally, the rachis of further or only pair 2–8(–11) cm, the longer interfoliolar segments 8–27(35) mm; pulvinules (1.5–)2–3.5 × (0.4–)0.5–1.1 mm, wrinkled; paraphyllidia not seen; lfts distally accrescent, narrowly to broadly ovate or ovate-elliptic from either broad-cuneate, or rounded, or shallowly cordate base, shortly bluntly acuminate, the distal pair 4–8(–11.2) × (2.1–)2.3–4.3(–5) cm, 1.7–2.3(–2.7) times as long as wide; the straight centric midrib giving rise on each side to 7–13 major (and random lesser intercalary), widely spreading-incurved secondary veins brochidodrome well within the plane or incipiently revolute margin, these giving rise to a random, open 3–4-nary reticulum, the venation all immersed on upper face, sharply prominulous beneath. *Peduncles* (1–)1.5–4.5(–5.5) cm; capitula (15–)20–27-fld, the receptacle 2–4 mm; bracts dimorphic, 0.6–1.2 mm, the lowest broadly

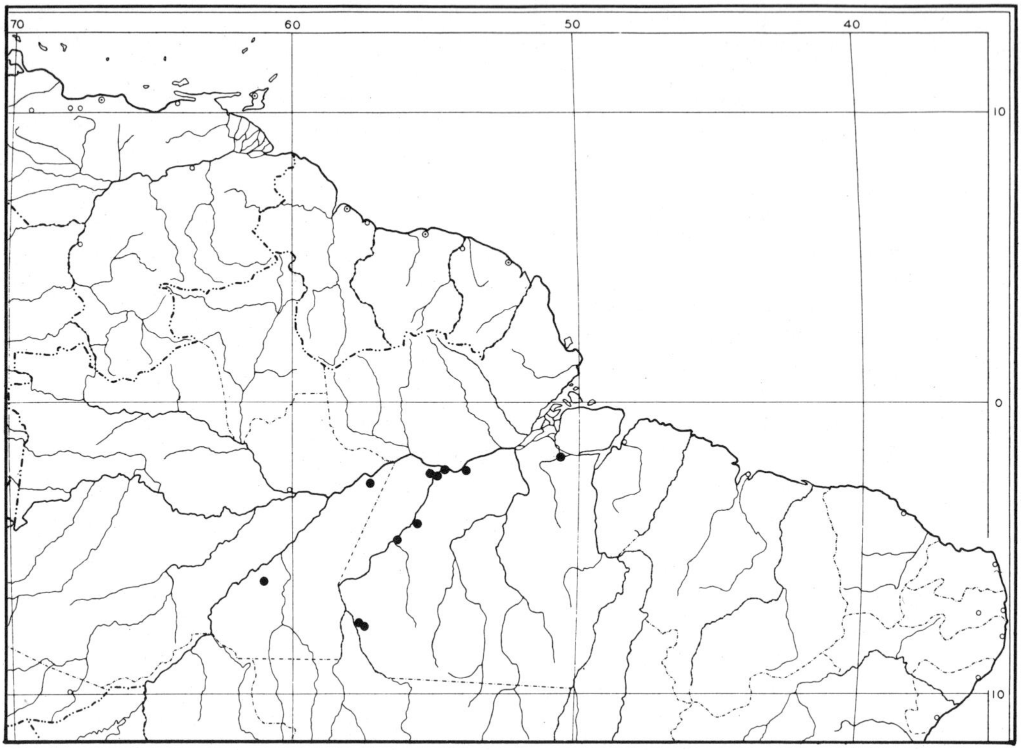

MAP 22. The lower Amazon basin in Brazil. Range of *Abarema campestris* (Bentham) Barneby & Grimes.

ovate, persistent, the further ones spatulate, deciduous; fls dimorphic, the peripheral ones shortly pedicellate, the pedicel (0.8–)1–2.8 mm, the terminal fl sessile or nearly so, with modified androecium; perianth of all fls 5-merous, the calyx and the corolla lobes finely puberulent, the corolla-tube commonly glabrescent; PERIPHERAL FLS: calyx turbinate-campanulate, obtusely 5-angulate 1.8–2.4(–2.8) × 1.1–1.5(–1.7) mm, the triangular or deltate obtuse, dorsally convex teeth 0.25–0.5 mm; corolla 4.5–5.4 (–6.3) mm, the ovate-lanceolate lobes (1.2–)1.4–1.9 mm; androecium 12–22-merous, 14–23 mm, the stemonozone 0.5–0.9 mm, the tube 2–3.2 mm; TERMINAL FL: calyx 2.25–3 mm; corolla 5–7 mm; stemonozone 1.7–2 mm; staminal tube 7–12 mm, at least shortly exserted; ovary of all fls narrowly ellipsoid, symmetrically conical at apex, glabrous. *Pods* 1–3 per capitulum, either sessile or commonly short-stipitate, in profile undulately broad-linear, evenly incurved through ¾-1½ circles, 8.5–15 × (1.2–)1.3–1.7 cm, 7–12(–14)-seeded, the glabrous, at first green then orange or red, when ripe fuscous, stiffly leathery or ligneous valves framed by sutures ±1.5–2 mm wide, convex or domed over each seed, internally red-orange-scurfy; dehiscence through both sutures, the valves ultimately both recurved and coiled; seeds plumply lentiform, in broad view either round or elliptic, 5.5–11 × 5.5–8 mm, the translucent testa when fresh gray or whitish in lower half, appearing blue-gray distally, becoming drab when old, loosely investing the embryo and therefore wrinkled when dry, pleurogram 0.

Abarema cochleata differs from other paucifoliolate abaremas native to middle and eastern Amazonia by leaflets symmetrically ovate and short-acuminate that assume a characteristic brown and tan coloring when dried. Within its range of dispersal *A. microcalyx* var. *parauaquarae* differs in inequilaterally obovate leaflets combined with mounded, small-pored leaf-nectaries; *A. villifera* in asymmetrically obovate leaflets softly pubescent beneath; whereas *A. mataybifolia* has leaflets nearly twice as large, borne on longer pulvinules (5–8, not 1.5–3.5 mm), peripheral flowers of the capitulum a little longer, and a broader pod. The differential characters of *A. campestris* are noted under that species. *Abarema jupunba* var.

trapezifolia, sometimes similar in leaf-formula, has rhombic leaflets puberulent dorsally.

Pithecolobium moniliforme, described from a single tree and collected subsequently only once, differs from typical *A. cochleata* in texture of the pod's valves, but apparently in no other feature, and we consider it here as a variety, not specifically separable. While the exocarp of the pod of *A. cochleata* is ordinarily <1 mm thick in cross-section, that of *P. moniliforme* is thickened over each seed into a woody dome nearly 5 mm thick, and the venation, ordinarily superficial and sharply defined, is differentiated neither superficially nor internally. The pods appear as though galled, but the seeds are well formed. In English commentary Ducke emphasized the peculiar fruit of *P. moniliforme*, but in the Latin diagnosis he stated that it differed from *P. cochleatum* in foliage, though he did not specify in what respect. The leaf-formula is i/3, but this occurs in some plants of the Amazon delta that have a pod typical of *A. cochleata*. In size, outline, texture, color, and venation of the leaflets *A. cochleata* and *P. moniliforme* are identical, as are the capitula and the individual flowers.

Key to the varieties of *A. cochleata*

1. Valves of ripe pod leathery, <2 mm thick in section, differentiated into glabrous, coarsely cross-venulose and reticulate exocarp, thin pitchlike mesocarp, and pallid crustaceous endocarp ±0.2–0.3 mm thick, low-convex over each seed; lower Amazonian Brazil in Pará and Maranhão ....21a. var. *cochleata*
1. Valves of ripe pod woody, not clearly differentiated, strongly domed over each seed and there 5–8 mm thick in section, externally tumulose but not reticulate; central Amazonian Brazil in Amazonas (mun. Manaus) ....21b. var. *moniliformis*

21a. Abarema cochleata (Willdenow) Barneby & Grimes var. **cochleata**. *Inga cochleata* Willdenow, sensu str. — "Habitat in provincia Para Brasiliae, [*Friedrich W. Sieber s.n.*, commun.] com. de Hoffmannsegg." — Holotypus, B-WILLD *19006*, seen in Microform 86: fiche 1380!. *Mimosa cochleata* (Willdenow) Poiret, Encycl., Suppl. **1:** 45. 1810. *Pithecellobium cochleatum* (Willdenow) Martius, Herb. Fl. Bras. 115. 1837. *Feuilleea cochleata* O. Kuntze, Revis. Gen. Pl. **1:** 187. 1891. *Pithecolobium cochleatum* sensu Bentham, 1875: 582; 1876: 434; Ducke, 1949: 36.

As described for the species, modified by key to varieties.

In virgin forest and vigorously surviving in capoeira, in tree-islands and along margin of savannas, occasional in campinarana or in swampy palm-groves, almost exclusively on terra firme below 200 m, common and locally abundant in lower Amazonian Brazil from the lower Ríos Trombetas and Tapajóz downstream to the delta and thence E along the coastal and inland lowlands to central Maranhão. — Map 23. — Fl. almost throughout the year, most profusely VIII–I. — *Inga de lagarta, inga de rosea*; *ingarana* (applied also to *Zygia* and *Macrosamanea* spp.).

21b. Abarema cochleata var. **moniliformis** (Ducke) Barneby & Grimes, comb. et stat. nov. *Pithecolobium moniliforme* Ducke, Bol. Técn. Inst. Agron. N. **2:** 6. 1944. — "[BRAZIL. Amazonas:] Manáos, in silva terris argillosis altis ultra Flores, 27-II-1943 fructiferum, 6-XI-1943 floriferum, Ducke 1428." — Syntypi, ex eadem arbore et sub eodem numero distributi, A!, IAN!, NY!, US *1905904*!. — *Abarema moniliformis* (Ducke) Barneby & Grimes, Acta Amaz. **14** (1/2, Supl.): 95. 1984 [1986].

P. moniliforme sensu Ducke, 1949: 35.

Differing from var. *cochleata* in the woody fruit, as described in key to varieties.

In terra firme forest on yellowish clay soils, ±100 m, apparently very local in mun. Manaus (Flores; distr. Agropecuário da SUFRAMA), centr. Amazonas, Brazil. — Not mapped. — Fl. XI–XII.

22. Abarema callejasii Barneby & Grimes, sp. nov., inter species inflorescentia longe spicato-racemosa praeditas recentius generi *Punjubae* relatis juxta *A. lehmanii* et *A. killipii* ponenda, illius foliolis mediocribus et dispersione antioquiana propior, sed foliis dorso glabris epapillatis (nec crebre papillatis et saltem secus venas majores pilosulis), racemis 25–45 (nec spicis 70–120)-floris, pedicellis 1–2.5 mm longis (nec subnullis), calyce 2.5–2.8 (nec 1.6–2) mm longo, filamentis longioribus ±15 (nec 9.5–11) mm usque longis, necnon legumine falcato 8-spermo (nec spiraliter retorto 10–14-spermo) seminumque embryone fusca (nec coerulea) valde diversa; ulterius ab *A. killipii* (e prov Caldas et Cauca tantum nota) foliolis distalibus 6.5–9 (nec 15) cm usque longis abstat. — COLOMBIA. Antioquia, mun. Briceno: en bosque pluvial montano, 1800–2000 m, desvio a Vereda Manzanares, 4 km O de la Troncal del Caribe sobre la via Ventanas–Briceno (75°30′W, 7°15′N), 21.III.1988 (fl., fr.), *R. Callejas* (with *F. J. Roldan & A. L. Arbaláez*) *6103*. —

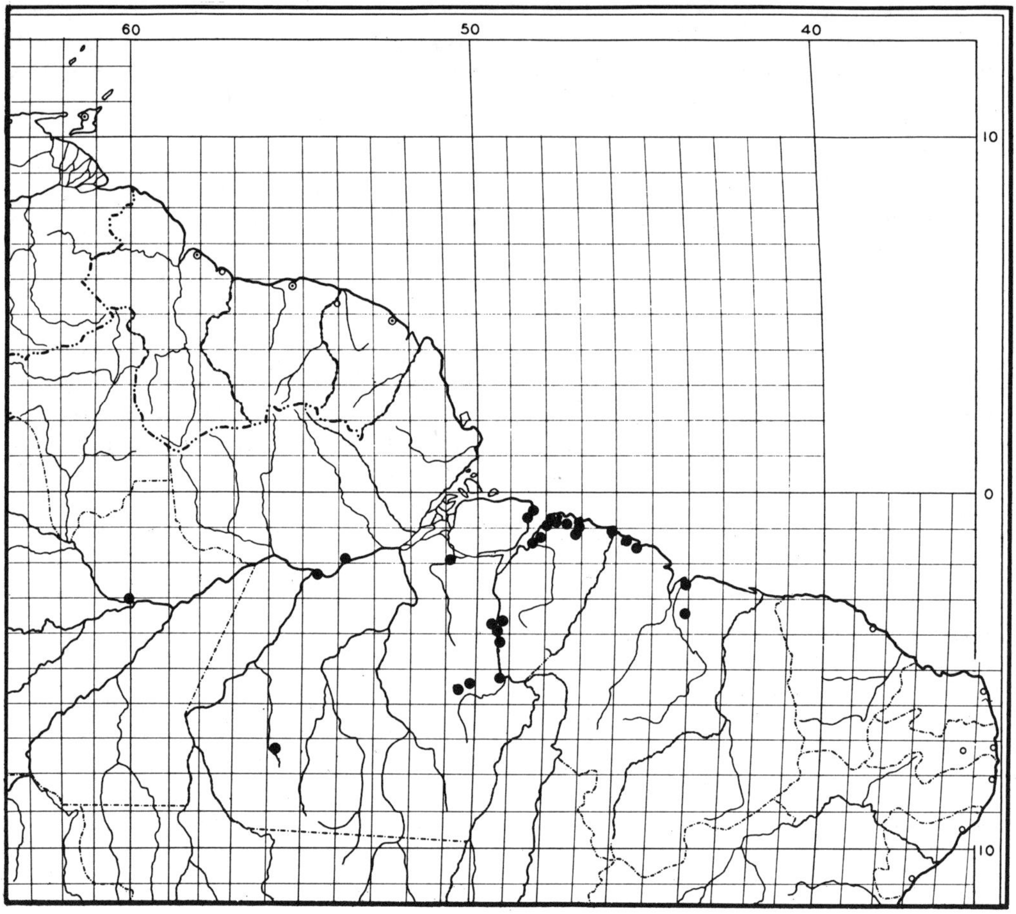

MAP 23. The lower Amazon basin in Brazil. Range of *Abarema cochleata* (Willdenow) Barneby & Grimes var. *cochleata*.

Holotypus, NY (2 sheets); isotypi, F, HUA (not seen), K, MEXU, MO, US. Fig. 7.

Macrophyllidious *trees* 7-14 m with fuscous elenticellate annotinous branchlets, the young parts densely minutely appressed-puberulent with brownish hairs <0.1 mm, early glabrate, the ample papery lfts lustrous olivaceous glabrous, a trifle paler beneath, the few, lax racemes of white-stamened fls arising singly from the axil of coevally expanding lvs toward the apex of new growth. *Stipules* 0. *Lf-formula* i–ii/2–3 (lfts 8–24 per lf); lf-stks (1–)2–5.5 cm, the one interpetiolar segment 1.5–3.5 cm; petiolar nectaries between pulvinules of each pair of pinnae and of each pair of lfts, crateriform with thickened margin, suborbicular 0.6–1.2 mm diam; rachis of one pair or of distal longer pair of pinnae (2.5–)5–7 cm, the longer interfoliolar segment 12–24 mm; lf-pulvinules livid, not cross-wrinkled, 2.5–5 × 0.8–1.2 mm; lfts distally accrescent, in outline subsymmetrically ovate-short-acuminate from broadly cuneate base, obtuse at very apex, the longer ones 6.5–11(–13.5) × 2.2–4 (–6.4) cm, 2.2–2.9 times as long as wide; the straight, centric or very slightly excentric midrib giving rise on each side to 4–7 major secondary veins strongly incurved-ascending to anastomosis shortly within the almost plane margin and to many weaker intercalary ones, these all connected by an elaborate reticulum of finer veinlets, the whole venation prominulous on both faces of lft. *Peduncles* ascending 3–7 cm; racemes 25–45-fld, the fls subhorizontal or widely ascending, the axis becoming 3–6 cm; bracts ovate ±0.4 mm, very early caducous; pedicels 1–2.5 × 0.3–0.4 mm, becoming a little shorter distally along axis; fls 5-merous homomorphic, the perianth minutely sordid-strigulose externally; calyx campanulate

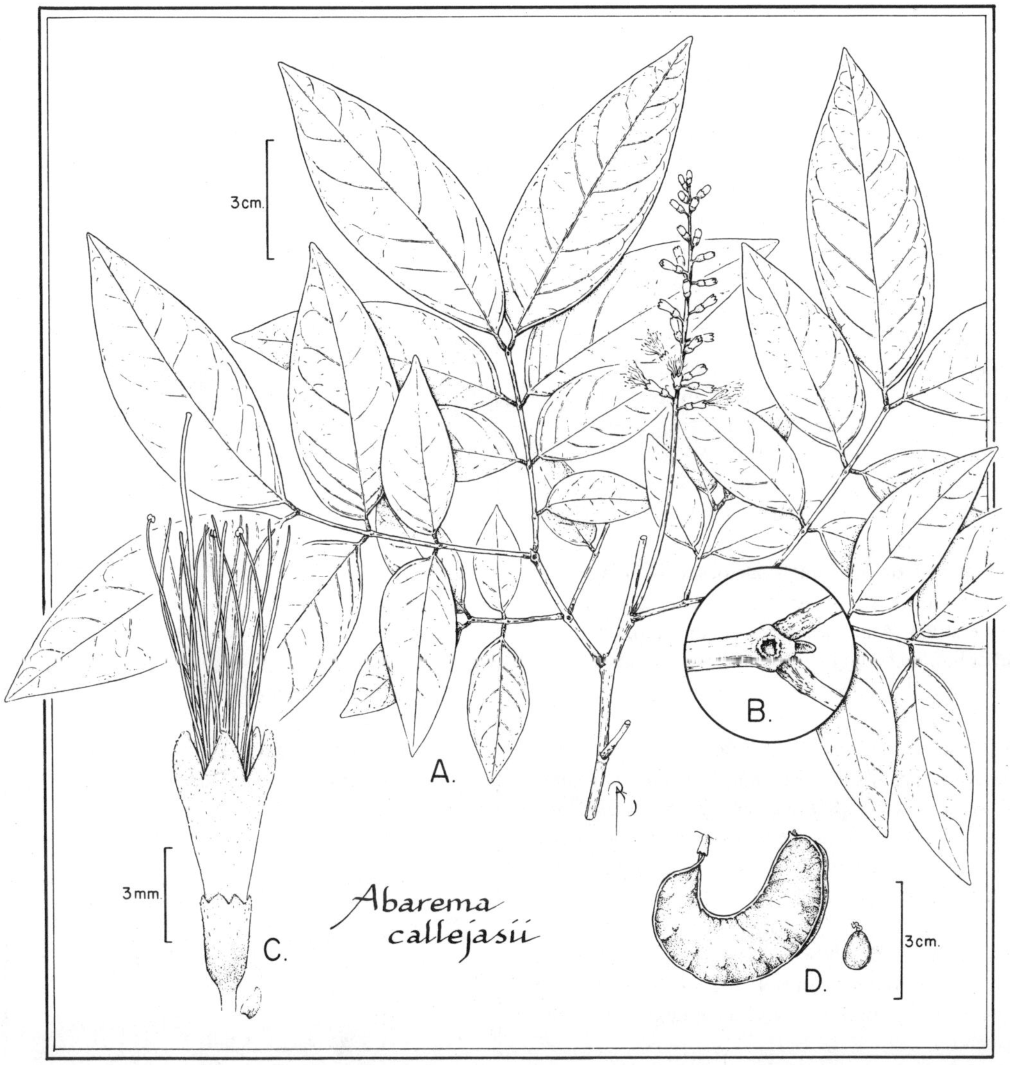

FIG. 7. *Abarema callejasii* Barneby & Grimes. **A.** Habit. **B.** Leaf-nectary between furthest pinna-pair. **C.** Flower. **D.** Fruit and seed. (All drawn from *Callejas 6103* (NY).)

or turbinate-campanulate 2.5–2.8 × 1.6–1.8 mm, the tube firm brownish, the pallid-margined, low-deltate, often unequal teeth 0.2–0.5 mm; corolla very narrowly funnel-shaped 6.3–7.4 mm, the erect lance-ovate lobes 2–2.2 × 1–1.1 mm; androecium 32–42-merous, the stemonozone 1–1.4 mm, the tube 3.5–5 mm, the longest filaments ±15 mm; ovary subsymmetrically fusiform, glabrous; style about 5 mm longer than stamens, the stigma scarcely 0.2 mm diam. *Pods* 1–2 per raceme, in profile falcately broad-linear and evenly recurved through ±¾-circle, cuneately attenuate at base, broadly obtuse and acentrically mucronulate at apex, plano-compressed and framed by scarcely undulate sutures ±1 mm wide in dorsal view, only minutely colliculate over each of 8 seeds, the stiffly papery valves externally dull-fuscous, glabrous and densely ascending-venulose and reticulate overall, internally orange-castaneous smooth; dehiscence primarily through the seminiferous suture, not seen fully realized; funicles narrowly ligulate, coiled at apex; seeds compressed-obovoid-ellipsoid 9–9.5 × 6.5–7 × 2 mm, the pale greenish translucent testa <0.1 mm thick in section, girdled by a slender nerve, loosely investing the hard fuscous embryo and wrinkled when dry, pleurogram 0; endosperm 0.

In montane rain forest at 1600–2000 m, known only from the E slope of Cord. Central in mun. Briceno, Amaltí, and S. Luis, between 6° and 7° 15′N in state of Antioquia, Colombia. — Map 24B. — Fl. III–V.

Abarema callejasii is the only known Colombian species with spicate inflorescence and slenderly pedicellate flowers, and differs further from its neighbor *A. lehmannii* in dorsally glabrous leaflets, in relatively few-flowered racemes, in longer calyces and androecia, and in falcate rather than spirally coiled fruits. The species was first collected in May 1944 on the Colombia Cinchona Mission by Earle E. Core (*635*, NY, US). The specimens, in young flower, were identified as *Pithecellobium* by R. S. Cowan, but have since lain among indeterminate Mimosaceae. In 1988, Ricardo Callejas (HUA) at our request revisited Core's locality and obtained the fine flowering and fruiting material on which our description is principally based. It is a pleasure to dedicate this remarkably distinct, Antioquian tree to an admired friend and obliging colleague.

23. Abarema lehmannii (Britton & Killip) Barneby & Grimes, comb. nov. *Punjuba lehmannii* Britton & Rose *ex* Britton & Killip, Ann. New York Acad. Sci. **35:** 127. 1936. — "Nopaya, Colombia, 1000–2000 m altitude, *Lehmann B. T. 364*." — Holotypus, NY (fl.)!; isotypi, GH (fl., 2 sheets)!, K (fr.) = photo *s.n.*, NY!, US *1242750* (fragm.)!. — *Pithecellobium popayanense* C. Barbosa, Caldasia **15:** 192. 1986; non *P. lehmannii* Harms, 1921. — *Pithecellobium lehmannii* C. Barbosa *ex* Gray Index, April 1988 (sphalm.).

Macrophyllidious *trees* 7–12 m, the young branches, lf-axes and inflorescence at first densely, later more thinly silky-pilosulous with either erect or narrowly ascending, weak, pale yellowish or rusty hairs, the foliage bicolored, the chartaceous lfts rich green, brunnescent when dry and glabrescent or glabrate above, beneath densely pallid-papillate overall and pilosulous at least along veins, sometimes overall, the small bronze-gold-strigulose fls borne in moderately dense spikes arising singly or geminate from the axil of coeval lvs and surpassed by them. *Stipules* 0 (none seen in vernation). *Lf-formula* i–ii/3–5(–6), the lfts (12–)16–36(–44) per lf; lf-stks stout 1.2–7.5(–10) cm, commonly consisting of petiole only, this 1.2–5.5 cm, the one interpinnal segment of few lvs ±2.0–4.5

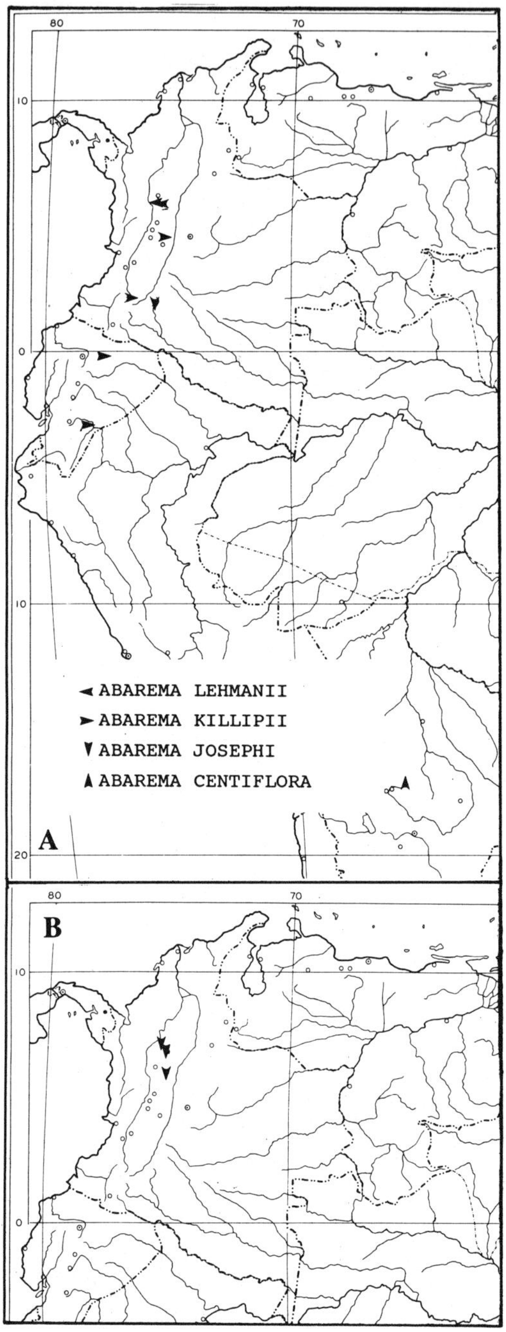

Map 24. The Andean Cordillera. **A.** Range of *Abarema lehmannii* (Britton & Killip) Barneby & Grimes, *A. killipii* (Britton & Killip) Barneby & Grimes, *A. josephi* Barneby & Grimes, and *A. centiflora* Barneby & Grimes. **B.** Range of *Abarema callejasii* Barneby & Grimes.

cm; petiolar nectaries of irregular development, either sessile or immersed in epidermis, flat or shallowly concave, round or transversely dilate 0.5–2 mm diam, present at tip of lf-stk and at insertion of many lft-pairs, sometimes at insertion of minute paraphyllidia; rachis of pinnae 4.5–17(–21) cm, the longer interfoliolar segments (1.2–)1.5–2.7 cm; lft-pulvinules in dorsal view ±2–3.5 × 1.2–1.7 mm; lfts accrescent distally, the blade symmetrically ovate or elliptic from rounded or broad-cuneate base, shortly obtusely acuminate, the distal and penultimate pairs (4.5–)5–9(–11.5) × 3–6.5 cm, 1.4–3 times as long as wide; the centric straight or slightly displaced and forwardly curved midrib giving rise to ±7–11 pairs of incurved-ascending secondary nerves brochidodrome well within the incipiently revolute margin and to random intercalary minor ones, all these and the slender tertiary and reticular venules immersed but discolored on upper face, sharply elevated beneath. *Peduncles* 1.5–3.5 cm; spikes ±70–130-fld, the axis becoming 5–13 cm; bracts ovate 0.3–0.7 mm, early dry caducous; perianth 5-merous, homomorphic in all fls; pedicels subobsolete, wider than long, at most 0.3 mm; calyx membranous campanulate 1.3–3(–3.4) × 1.2–1.5 mm, the depressed-deltate or triangular lobes 0.2–0.8 mm; corolla subtubular, only slightly expanded distally, (4–)4.5–7 mm, the suberect ovate lobes 0.8–1.5 × 0.55–1.1 mm; androecium 40–45-merous, 9.5–11 mm, the stemonozone ±1 mm, the tube 4–5.5 mm (about as long as or a little shorter than corolla); ovary slenderly ellipsoid, at anthesis glabrous, puberulent after fertilization; style about as long as longer stamens, the stigma scarcely dilated, 0.1–0.15 mm diam. *Pods* 1–4 per spike, in profile undulately broad-linear, recurved through 1–1.5 circles, ±12–16 (–"20") × 1–1.5 cm, 10–14-seeded, the fuscous glabrescent valves at first plano-compressed and framed by thickened, more densely golden-strigulose sutures, becoming elaborately reticulate and low-convex over seeds, the crustaceous endocarp in section ±0.3 mm thick, internally brown-red overall, smooth and lustrous in seed-cavities, somewhat furfuraceous between them, resinous mesocarp 0; funicles ribbonlike, coiled at apex; seeds plumply obovoid, in broader view ±7–8 × 6–6.5 mm, the translucent [grapeskin] testa ±0.1 mm thick in section, loosely investing the horny aniline-blue, wrinkled embryo, pleurogram 0.

At margin of humid montane woodland, known precisely only from 2200 to 2440 m in mun. Caldas and La Ceja of Antioquia, near 6°–6°05′N in the Cordillera Central of Colombia, the type-locality "Nopaya" not identified. — Map 24A. — Fl. IV–V; fr. I–IV(–?).

Abarema lehmannii is most closely related to allopatric *A. josephi*, which differs in higher leaf-formula, and *A. killipii*. The one other *Abarema* with long-spicate inflorescences known from cordilleran Colombia, *A. callejasii*, has slenderly pedicellate, not sessile flowers, and a falcate, not coiled, pod.

We are indebted to Linda Albert de Escobar for fresh and abundant flowers and fruits of *A. lehmannii* from mun. Caldas near Medellín, and for color photographs, now filed with the specimens at NY, showing the blue seeds against the background of red-brown endocarp that are characteristic of *Abarema*.

The epithet *popayanense* substituted by Barbosa for *lehmannii*, which is preoccupied in *Pithecellobium* but available in *Abarema*, was suggested (according to annotation) by *Lehmann 8623* (GH) from Popayán, herein referred to var. *killipii*.

24. Abarema killipii (Britton & Killip) Barneby & Grimes, comb. nov. *Punjuba killipii* Britton & Rose *ex* Britton & Killip, Ann. New York Acad. Sci. **35:** 127. 1936. — "Open hills, Salento, Caldas [W foothills of Nevado de Tolima at ±4°40′N], Colombia, 1700–1900 m altitude, July 25–31, 1922, [*E. P.*] *Killip 9043*." — Holotypus, NY!; isotypi, F!, GH!, US!. — *Pithecellobium killipii* (Britton & Killip) C. Barbosa, Caldasia **15:** 191. 1986.

Macrophyllous *trees* 7–10 m with trunk 1.6–2 dm dbh, the young branches, lf-stks and inflorescence densely minutely silky-puberulent with sordid-yellowish hairs to 0.25 mm, the broad thin-textured lfts glabrate except along principal nerves or fully glabrous in age, the relatively dense amentiform spikes of homomorphic, greenish white fls borne singly and geminate in the axil of coeval lvs. *Stipules* 0. *Lf-formula* i/4–6, the lfts 16–24 per lf; lf-stks 2.5–4.5 cm, at middle 1.5–2.5 mm diam, the ovate terminal appendage thickened, caducous; a sessile, shallowly cupular thick-rimmed, crumpled and transversely dilated nectary ±1.2 mm diam at tip of lf-stk and similar, scarcely smaller ones between all but the first lft-pair; rachis of pinnae 8–16 cm, one of the pair often a little shorter than the other, the interfoliolar segments ±2–4 cm; lft-pulvinules 2.5–3.5 × 1.4 mm, not wrinkled; lfts a little distally accrescent, the blade ovate from broad-cuneate base, shortly bluntly acuminate, the largest (either furthest or penultimate) pair ±9–15 × 4–4.7 cm, 2.1–2.4 times as long as wide; venation obtusely prominulous on either face of lft, the centric, almost straight midrib giving rise to 6–7 major (but slender), incurved-ascending secondary nerves weakly brochidodrome well within the almost plane margin, the tertiary and reticular venulation fine and sinuous. *Peduncle* and raceme-axis together 10–13 cm; spikes ±120–165-fld, the sessile, horizontally spreading fls crowded but at full anthesis not contiguous; bracts subulate 0.5 mm or less, caducous; perianth finely yellowish silky-strigulose externally; calyx deeply campanulate 2.8–3 × 1.1–1.3 mm, the broad-triangular or subulate teeth unequal, 0.1–0.35 mm; corolla narrowly trumpet-shaped 7.5–8 mm, the erect, unqual, narrow-ovate obtuse lobes 0.7–2 × 0.6–0.7 mm; androecium 34-merous, 12–16 mm, the stemonozone 0.7–0.9 mm, the tube 6–6.5 mm, the tassel greenish white; no intrastaminal disc. *Pods* (of type) sessile, undulately linear and recurved through more than one full circle, at each of ±10 seeds 10–12 mm wide in profile, the stiffly papery valves stramineous externally when ripe, internally pinkish orange at the seed-cavities; dehiscence elastic, the valves separating the full length and individually coiling; seeds unknown.

In moist montane and premontane forest, surviving deforestation in meadows and on open hillsides, known by one collection from foothills of Nevado de Tolima, near 4°40′N in prov. Caldas, Colombia, and from two stations on the E slope of the Ecuadorean Andes on the headwaers of ríos Napo and Santiago, within lat. 0°–3°S in prov. Napo and Santiago-Zamora. — Map 24A. — Fl. VII–XI.

25. Abarema racemiflora (Donnell Smith) Barneby & Grimes, comb. nov. *Pithecolobium racemiflorum* Donnell Smith, Bot. Gaz. **56:** 57. 1913. — "In silvis collinis ad Las Vueltas, Tucurrique, Comarea [comarca] de Puntarenas [9°51′N, 83°42′W, in prov. Cartago], Costa Rica, alt. 600–750 m, Febr. 1899, A. Tonduz n. 13060." — Holotypus, US *942404*!; isotypi, F!, G (11 sheets)!, GH (2 sheets)!, NY (3 sheets)!, LL!, P!, US *942405*!. — *Punjuba racemiflora* (Donnell Smith) Britton & Rose, N. Amer. Fl. **23:** 28. 1928.

Macrophyllidious *trees* flowering when 6–7 m but attaining 27 m with trunk 4 dm dbh, the innovating branchlets and foliage densely brown-puberulent-tomentulose but adult stems and lvs glabrous, the ample chartaceous bright green lfts subconcolorous, the inflorescence composed of very long, laxly many-fld racemes arising in the axil of coeval lvs, paired with but distal to an axillary bud. *Stipules* 0. *Lf-formula* i–ii/2–3, the lfts 12–20 per lf; lf-stks (1–)2–5.5 cm, the petiole (1–)2–3.7 cm, the one interpetiolar segment, when present, 1.5–2 cm; petiolar nectaries poorly developed, the first one forming a shallow cup <1.5 mm diam at tip of lf-stk or between proximal pair of pinnae, sometimes obsolete or immersed, the interfoliolar nectaries similar but smaller, often obscure; rachis of pinnae 4–10.5 cm, the longer interfoliolar segments (1.5–)2–3.5 cm; lft-pulvinules 2.5–5 × 1–1.5 mm; lfts strongly accrescent distally, elliptic-acuminate from equilateral or slightly asymmetrical, cuneate base, the acumen obtuse muticous, the blade of furthest pair (7–)10–15.5 × (3–)3.5–5.5 cm, 2.3–2.8 times as long as wide, the centric, straight or very slightly forwardly incurved midrib immersed above, cariniform beneath, giving rise to ±5–8 pairs of major secondary veins incurved-ascending to anastomosis close within the plane margin and many minor, more widely divergent secondary ones, all these and an intricate reticulum of veinlets prominulous on both faces. *Peduncle* and raceme-axis together 2–3.5 dm, much surpassing the foliage; the raceme remotely ±55–95-fld, the fls homomorphic, widely ascending or subhorizontal; bracts subulate 0.4–0.7 mm, dry and caducous long before anthesis; pedicels straight 3–6.5 × 0.3–0.4 mm, dilated at base; perianth 5-merous, thinly puberulent externally, a little more densely so on lobes; calyx deeply narrowly campanulate 3.2–4.5 × 1.2–1.5 mm, weakly 5-nerved, the ovate-mucronulate, unequally divided teeth 0.3–0.8 mm; corolla whitish, tubular, 7 mm, the almost erect, narrowly lanceolate lobes 2–2.7 × 0.7–1 mm; androecium 32–36-merous, 14–15 mm, the stemonozone ±1.5–2 mm, the tube 6–6.5 mm; ovary glabrous, obliquely conical at apex; style about as long as filaments, at apex dilated into a funnel-shaped stigma 0.15 mm diam. *Pods* few per spike, subsessile, in profile undulately broad-linear evenly recurved through at least one circle, 15 × 1.4 (!, described as 15–19 × 0.7–1.3 in.) cm, 5–8-seeded, the sutures scarcely thicker than the valves, these coriaceous, low-convex over seeds, externally pale brown, faintly deeper-mottled, dull and papillate, within smooth, dull-ochraceous between seeds and lustrous castaneous in the seed-cavities, the exocarp very thin, resinous mesocarp 0, crustaceous endocarp 0.4–0.5 mm thick in section; dehiscence of *Abarema*; funicle ribbonlike, coiled; seeds obovoid, moderately compressed, in broad view ±10 × 6.5 mm, the hard, when old fuscous (fresh *could* be blue), coarsely wrinkled embryo loosely invested by a grapeskin-like translucent testa, pleurogram 0; no endosperm.

On wooded hills near 700–900 m, known only from the vicinity of Turrialba in the province of Cartago (E of Turrialba, Tucurrique) and at 10°15′N, 84°22′W in prov. Alajuela (Villa Quesada), Costa Rica. — Map 18. — *Chaperno.*

Abarema racemiflora is the one Central American species of its genus with loosely racemose flowers and is remarkable in its immediate group, otherwise all Andean, for length and slenderness of the individual pedicels.

26. Abarema obovata (Bentham) Barneby & Grimes, comb. nov. *Calliandra (?) obovata* Bentham, London J. Bot. **3:** 101. 1844. — "Brazil, *Lobb.*" — Holotypus, *Lobb 83*, from southern Brazil, the locality unknown, K(herb. hooker.)! = NY Neg. *2008*. — Erroneously equated by Bentham, 1875: 583 with *Pithecolobium lusorium*.

Trees 4–15 m, closely resembling and related to *A. filamentosa*, but the lfts never fully glabrous, either pilosulous beneath or merely ciliolate, and sometimes with a secondary nerve from the pulvinule, in fruit resembling both *A. filamentosa* and *A. brachystachya*. *Stipules* 0.5–3 mm. *Lf-formula* (i–)ii–iii(–iv)/3–4(–5, exceptionally –7), the lfts of ampler lvs to 24–40(–60 mm); lf-stks (1.5–)2.5–8 cm, the petiole 1–3 cm, the longer interpinnal segments 8–35 mm; nectary between first pair of pinnae either shallowly cupulate or distinctly pored, ±0.8–1.2 mm diam; rachis of longer pinnae 4–8 cm, the interfoliolar segments 12–22 mm; lft-pulvinules 0.8–2 × 1–1.2 mm; lfts obovate or rhombic-obovate, obtuse or subemarginate, the larger ones 3–6 × 2–3.5 cm. *Inflorescences* and fruit essentially as in *A. brachystachya*, the calyx of peripheral fls ±3 mm and corolla ±5.5–6.5 mm.

On wooded hillsides, sometimes in mata do cipó, 400-1030 m, known with certainty from E- and N-centr Minas Gerais, on both slopes of Sa. do Espinhaço around Belo Horizonte and Viçosa, N to Grão Mogol, in lat. 16°30′–21°S. — Map 25. — Fl. I–II; fr. ripe VI–X.

For commentary on kinship to *A. brachystachya* and *A. filamentosa* see the following discussions of those species.

All but one collection of *A. obovata* known to us have at most 4 pairs of leaflets in distal pinnae, several not more than 3 pairs, but the blades vary from glabrous or brownish pilosulous dorsally. The exceptional *Saint-Hilaire B[1]/1310* (P), from an unknown locality in Minas Gerais, has up to 6, even 7 pairs, densely pilosulous on the back and ciliate. Provisionally we include this in *A. obovata*, with which it is otherwise in full agreement.

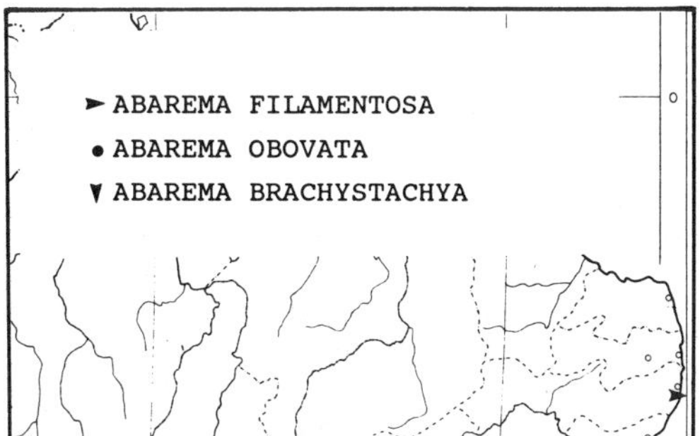

MAP 25. Coastal Brazil. Range of *Abarema filamentosa* (Bentham) Pittier, *A. brachystachya* (de Candolle) Barneby & Grimes, and *A. obovata* (Bentham) Barneby & Grimes.

27. Abarema brachystachya (de Candolle) Barneby & Grimes, comb. nov. *Inga brachystachya* de Candolle, Prodr. **2:** 440. 1825. — ". . . in Brasilia." — Holotypus, G-DC! = F Neg. 6973.

Mimosa lusoria Vellozo, Fl. Flumin. **11:** t. 14. 1829 & Arq. Mus. Nac. Rio de Janeiro **5:** 433. 1881. — "*Habitat loco supra citato* ["maritimis silvis," presumably restinga near the city of Rio de Janeiro]." — Holotypus, the plate cited. — *Pithecolobium lusorium* (Vahl) Bentham, London J. Bot. **3:** 204. 1844.

Pithecolobium rhombeum Bentham, London J. Bot. **3:** 217. 1844 — "Brazil, *Sello.*" — Holotypus, not found at K in

1990; presumed isotypi, †B = F Neg. *1217*; *Sello 890*, BM!. — *Feuilleea rhombea* (Bentham) O. Kuntze, Revis. Gen. Pl. **1:** 189. 1891.

Pithecolobium lusorium sensu Bentham, 1875: 583; 1876 437.

Pithecolobium rhombeum sensu Bentham, 1875: 584; 1876: 437.

Macrophyllidious *shrubs* and *treelets* 1–3 m, the young branches, all lf-stks, and peduncles sordid- or rufous-pilosulous with incurved-ascending or spreading hairs to 0.2–0.4 mm, the chartaceous venulose bicolored lfts glabrous except for minutely ciliolate, loosely revolute margin, bright green or olivaceous and lustrous above, paler beneath, the short dense capituliform racemes solitary or 2–3-nate in the axil of contemporary lvs, or, by reduction of these forming a short terminal pseudoraceme. *Stipules* linear or linear-attenuate (1.5–)2–5 × 0.2–0.4 mm, caducous. *Lf-formula* iii–vi(–vii)/4–7(–10), the lfts of larger lvs (48–)76–164; primary lf-stks (4–)6–20 cm, the petiole proper (1–)2–4.5 cm, the longer interpinnal segments (1–)2–4.5 cm; lf-nectaries sessile, orbicular or variously crumpled and sometimes extended lengthwise, shallowly concave, usually pallid, the first (in random lvs wanting) situated immediately below insertion of proximal pair of pinnae and 0.8–1.5 mm diam, similar but mostly smaller ones at insertion of distal pairs of pinnae and lfts; pinnae decrescent proximally but either the distal or the penultimate pair longest, the rachis of these (4–)4.5–9.5 cm, the longer interpinnal segments 10–21 mm, a pair of minute paraphyllidia often present but caducous; lft-pulvinules in dorsal view 1–1.7 × 0.4–0.9 mm; lfts subequiform but accrescent distally, very obtusely rhombic from inequilateral base, obtuse or obscurely emarginate, the penultimate pair (18–)20–35 × (9–)10–21 mm, the furthest pair to 2.5–4.5 cm; venation of a diagonal midrib slightly curved forward and pinnately branched throughout, the posterior nerve from pulvinule no longer than secondary ones, all these and a reticulum of lesser venules prominulous on both faces. *Peduncles* stout 4-15 cm; capitula ±20–45-fld, at full anthesis more or less hemispherical, the fls ascending and spreading on pedicels a little decrescent upward along receptacle, this becoming narrowly clavate and 5–11 mm; bracts oblanceolate ±1.5–2.5 mm, early dry caducous; fls dimorphic, the peripheral ones pedicellate, the terminal one sessile, the perianth of all 5-merous (random abnormalities), greenish white or ochroleucous, minutely puberulent overall; PERIPHERAL FLS: pedicel of lowest ones (1.5–)2.3–4.7(–5) mm; calyx turbinate-campanulate (2.3–)3.2–4.7(–5) mm, the obtusely triangular teeth 0.5–1.1 mm; corolla 6.4–9.7 mm, the lobes 2.3–3.2 mm; androecium white 32–56-merous, the longer filaments 24–42 mm, the stemonozone 0.9–1.5 mm, the tube 2.5–5 mm; ovary glabrous truncate; ovules 8–13; TERMINAL FL(S): pedicel subobsolete; calyx campanulate, 4–5.5 mm, the teeth 0.7–1 mm; corolla 8.5–12 mm, the lobes 2–2.6 mm; staminal tube as long as or to 2.5 mm longer than corolla. *Pods* of *A. filamentosa*, 14–16 mm wide; seeds of *A. filamentosa*, 7.5–8.5 × 6–6.5 mm, the pleurogram ±6 × 3.5 mm.

In restinga and brushy woodland along the Atlantic coast of S Brazil, below 100 m, locally plentiful in latitude 22°40′–25°30′S between Espírito Santo and Paraná. — Map 25. — Fl. IX–III(–V).

Abarema brachystachya, *A. obovata*, and *A. filamentosa* form a replacement series that runs northward in a narrow ribbon along the Brazilian coast from Paraná to El Salvador, but is interrupted in Espírito Santo, where the line turns inland into submontane woodland. This south-north sequence corresponds more or less exactly with a morphological cline expressed in leaf-formula and in size of leaflets, which are most numerous and smallest in *A. brachystachya*, fewest and largest in *A. filamentosa*, and of intermediate size and number in *A. obovata*. Taking into account only the larger, fully expanded leaves of any plant, we found in *A. brachystachya* 48 leaflets per leaf and upwards, in *A. obovata* 24 to 36 per leaf, and in *A. filamentosa* 24 or less. The smaller, more numerous leaflets tend to be rhombic in outline, the larger ones obliquely obovate. Leaflets of *A. filamentosa* are glabrous throughout, whereas those of *A. obovata* and *A. brachystachya* are at least finely ciliolate, and those of *A. obovata* often pilosulous on one or both faces.

The basionym *Inga brachystachya* was known by Bentham (1875: 583) to be a taxonomic synonym of *Mimosa lusoria* Vellozo, but following the so-called Kew rule, which recognized the first combination in the accepted genus, he treated the species as *Pithecolobium lusorium*, a name unacceptable under the present code.

28. Abarema filamentosa (Bentham) Pittier, Trab. Mus. Com. Venezuela **2:** 86. 1846. *Pithecolobium filamentosum* Bentham, London J. Bot. **5:** 106. 1846. — "Rio Doce in Brasil, *Martius* in herb. Hooker under the name of *Inga filamentosa* [nom nud. in syn.]." — Holotypus, K! = NY Neg. *2005*.

— *Feuilleea filamentosa* (Bentham) O. Kuntze, Revis. Gen. Pl. **1:** 187. 1891.

Pithecolobium filamentosum sensu Bentham, 1875: 581; 1876: 433, exclus. pl. venezol. ab Otto lecta (quae ≡ *Abarema leucophylla*) et syn. *P. turbinato* Bentham (quae ≡ *Abarema turbinata*); *Pithecellobium filamentosum* sensu Lewis, 1987: 179.

Macrophyllidious *trees* with mottled bark, attaining 4–9 m, but sometimes flowering precociously as a shrub at 2–3 m, the twigs, all lf-axes and peduncles densely puberulent when young with brown-gold trichomes at most 0.2 mm, early glabrate, the broad coriaceous bicolored lfts lustrously brown-olivaceous above, paler dull beneath, glabrous overall, the short capituliform racemes arising solitary or geminate from the axil of contemporary lvs. *Stipules* subulate <1 mm, caducous from incipient lvs, none visible on most branchlets at or after anthesis. *Lf-formula* i or i–ii(–iii)/2–3, lfts 8–24 per lf; lf-stks (3–)4–47 mm, the petiole proper (3–)4–14 mm, either flattened or widely sulcate ventrally, the one interpinnal segment (or the longer of two) 16–35 mm; lf-nectaries (often pastured and incomplete) sessile verruciform, either hemispherical or depressed-dome-shaped, or subangulate, crumpled and only obscurely pored, lurid-brown or blackish, the first situated immediately below first (or only) pinna-pair and 1.2–2.5 mm diam, similar but smaller ones at insertion of further pinnae and always between the furthest pair of lfts; rachis of distal or only pair of pinnae (2–)2.5–8 cm, the distal or only interfoliolar segment 1.3–3.1 cm; lfts subequiform but larger upward, the blades symmetrically or a little obliquely obovate to broad-oblanceolate from scarcely or moderately inequilateral, broad-cuneate or subcordate base, broadly rounded and muticous or shallowly emarginate at apex, the anterior one of the furthest pair 4–8.5(–10) × 2.2–5(–5.4) cm, (1.3–)1.5–2.3(–2.4) times as long as wide; venation of subcentric, straight or gently incurved midrib pinnately branched its whole length, the secondary veins brochidodrome well within the loosely revolute margin and giving rise to an elaborate reticulum of venules, the whole venation prominulous on both faces. *Peduncles* (2.5–)3–9 cm; racemes densely 20–45-fld, the narrowly clavate or almost linear receptacular axis becoming 6–19 cm; bracts ovate-triangular 0.15–0.5 mm, densely brown-puberulent dorsally, caducous; fls dimorphic, maturing in rapid succession upward, the proximal ones smallest and contracted at base into a distinct pedicel, the distal ones progressively a little larger and more shortly pedicellate, 1-3 terminal fls largest and appearing subsessile (the short pedicel nearly as stout as the calyx-tube), these fls with exserted filament tube but not precociously expanding; calyx and corolla of all fls greenish when fresh, minutely puberulent especially on the lobes, brunnescent in drying, the filaments white; pedicels 0.3–1.3(–2) mm; perianth of all fls 5-merous; PERIPHERAL FLS: calyx turbinate 1.7–3.2(–3.6) × 1.2–2 mm, the broad-triangular or depressed-deltate teeth 0.2–0.8(–0.9) mm; corolla (4–)4.2–6.2(–7.3) mm, the lanceolate, apically cucullate lobes 1.8–2.2(–2.6) mm; androecium 16–28(–32)-merous, 21–36 mm, the stemonozone ±1 mm, the tube 1.6–3.4 mm, shorter than corolla; ovary glabrous truncate; style scarcely dilated at tip, the stigma 0.1–0.15 mm diam; ovules 8–12; TERMINAL FLS: calyx campanulate 1.8–3.7 mm; corolla 6.5–9 mm; filaments united into a tube exceeding the corolla by 0.5–4 mm. *Pods* 1–3 per capitulum, subsessile, in profile broad-linear recurved through at least half and often through more than one circle, 7.5–13 × (1.3–) 1.5–2 mm, the leathery valves fuscous or fuscous-castaneous, papillate but glabrous, at first plano-compressed and framed by thickened sutures, becoming low-umbonate over seeds, the seminiferous (outer) suture undulately impressed between seeds, the dorsal (inner) one more evenly curved, the ripe endocarp lustrously castaneous in the seed-locules but tan and dull elsewhere; dehiscence elastic, through both sutures, the valves at once recurved and twisted; seeds suspended on a ribbonlike, distally coiled funicle, plumply obovoid, in broader profile 6–7 × 5.5–6 mm, the smooth, white or ivory testa closely investing the indigo-blue embryo, ±0.3 mm thick in section, largely opaque and lustrous but differentiated around the distal pole into a translucent zone which forms a U-shaped gray-blue blot lying between finely engraved elliptic pleurograms 4–4.5 × 3.3 mm.

In wet Atlantic forest and restinga, below 50 m, formerly common and continuously dispersed along coastal Bahia from the latitude of Salvador southward to the mouth of Río Doce in NE Espírito Santo; apparently isolated at 8°20′S in Pernambuco; an old report from Venezuela (*Otto s.n.*, K!) is based on misidentified *A. leucophylla.* — Map 25. — Fl. II–VIII; fr. VII–II.

The cladistic analysis indicates that *A. filamentosa* is most closely related to *A. cochliacarpos*, on the basis of the externally papillate fruit. Characters of the leaf not amenable to cladistic analysis hint at a relationship to *A. brachystachya* and *A. obovata*; for commentary on this relationship, see under *A.*

brachystachya (p. 92). *Abarema filamentosa* is indistinguishable from *A. turbinata* by the characters used in the cladistic analysis, but differs from it also in leaf-formula and size of leaflets, characters not amenable to scoring.

Lewis (1987: 179) suggested that the Amazonian *Pithecolobium microcalyx* and *P. parauaquarae* (here treated as two varieties of *Abarema microcalyx*) are conspecific with *Abarema filamentosa*, but he must have overlooked substantial differences in shape of petiolar nectaries, organization of the pinnae, differences in morphology of the terminal flower, ovary, and internal coloration of the valves.

29. Abarema turbinata (Bentham) Barneby & Grimes, comb. nov. *Pithecolobium turbinatum* Bentham, London J. Bot. 5: 106. 1846. — "Bahia, *Luschnath*." — Holotypus, K!. — Erroneously equated by Bentham, 1875: 582, with *Pithecolobium filamentosum.*

Pithecellobium filamentosum sensu Lewis, 1987: 179, ex parte ("large-leafleted form"); non Bentham, s. str.

Closely resembling *A. filamentosa* in habit, stature, indumentum, in verruciform lf-nectaries, inflorescence, and facially glabrous, eciliolate lfts, differing in lf-formula and ample lfts. *Lf-formula* i/1–2, the lfts 4–8 per lf; lf-stks 6–24(–55) mm; rachis of the geminate pinnae 1.5–7.5 cm, the one interfoliolar segment, when present, 1.8–5 cm; lft-pulvinules 2–3.5 × 1–1.4 mm; lfts of distal (or only) pair broadly elliptic-obovate 8–14 × 3.5–7 cm, 1.6–2.5 times as long as wide, in texture and venation like those of *A. filamentosa*, the anterior lfts of the smaller proximal pair often lacking. *Peduncles* 7–17 cm; axis of capitulum 1–2 cm; fls dimorphic, the perianth of all greenish white brunnescent, minutely puberulent distally; PERIPHERAL FLS: pedicels scarcely 1.5 mm; calyx turbinate 3.5–4.5 × 1.3–1.6 mm, the teeth 0.7–1.2 mm; corolla ±6.5 mm; androecium and gynoecium of *A. filamentosa*; TERMINAL FL(S): 1–3, sessile, slightly larger than the peripheral ones. *Pods* (few seen) of *A. filamentosa* but sometimes wider, ±15–19 × 1.4–2.4 cm, to 15-ovulate.

In restinga and Atlantic coastal forest, below 30 m, within the range of *A. filamentosa* but localized in coastal Bahia between 14° and 16°S (Maraú, Olivença, Una, Belmonte). — Fl. (V–)VII–IX.

Bentham gave no reason for equating his *Pithecellobium turbinatum* with *P. filamentosum*. Lewis (1987: 179) seemingly construed the substance, but not the name, of *P. turbinatum* as a form of *P. filamentosum* with large leaflets, but this needs demonstration.

30. Abarema cochliacarpos (Gomes) Barneby & Grimes, comb. nov. *Mimosa cochliacarpos* B. A. Gomes, Obs. Bot.-Med. Bras. pl. 34, t. 4 ("*cochleacarpos*"), fig. 3 [a leaf]. 1803 [seen in facsimile: Falcão, Brasiliensia Documenta vol. 5. 1972]. — ". . . *habitat in montibus, frequens, ut traditur, in Brasiliae provinciis* S. Paulo *et* Minas Geraes; *occurrit raro in Riojaneria*." — Typus [?; Urban wrote that Gomes's plants were in herb. Lisbon Polytechnic School]. — *Inga cochlocarpos* (Gomes) Martius *in* Spix & Martius, Reise Bras. **1:** 549. 1823, pro syn. *Feuilleea cochlocarpa* (Gomes) O. Kuntze, Revis. Gen. Pl. **1:** 185. 1891. *Pithecolobium cochliocarpum* [*sic*] (Gomes) Macbride, Contrib. Gray Herb., n. ser. **59:** 3. 1919. — Doubtfully equated by Bentham, 1875: 583, with *Pithecolobium auaremotemo* Martius.

Inga nandinaefolia de Candolle, Prodr. **2:** 440. 1825. — ". . . in Brasilia. Raddi . . . (. . . in hb. Moricand.)." — Holotypus, G-DC! = F Neg. 6974; isotypus, GH (herb. moricand e fl)! — Equated by Bentham 1875: 583, with *Pithecolobium auaremotemo* Martius.
Mimosa vaga Vellozo, Fl. Flumin. 11, t. 13. 1829 & Arq. Mus. Nac. Rio de Janeiro **5:** 433. 1880. — "*Habitat maritimis silvis* [near Rio de Janeiro]." — Holotypus, the cited plate!. — *Pithecolobium auaremotemo* Martius, Flora **20(2,** Beibl. **8** [= Herb. Fl. Bras.]): 115. 1837, the epithet derived from the next following, but technically a legitimate substitute for *M. vaga* Vellozo, *non* Linnaeus, 1753, and only provisionally equated with the prior *M. cochliacarpos* Gomes. — Spm. authent.: "Habitat in Sebastianopoli," *Martius*, M! = IPA Neg. *1316*. — Treated by Bentham, 1875: 583, as *P. auaremotemo.*
Avaramo-temó Piso, Hist. Nat. Bras. 168, ic. 1648.
Pithecellobium cochliocarpum sensu Lewis, 1987: 178.

Macrophyllidious *trees* and arborescent shrubs, potentially attaining 30 m with trunk to 4.5 cm dbh, but sometimes flowering as a shrub or small bushy tree only 2–5 m, the young branchlets and most young lf-axes puberulent with fine, forwardly curved, either pallid, sordid or rusty hairs at most 0.15 mm mixed with minute thickened or claviform, dark reddish trichomes, glabrescent in age, the glabrous chartaceous lfts bicolored, dark olive-green and lustrous above, paler beneath, when dry brunnescent or fuscescent, the dense capitula of greenish white fls borne solitary or 2–3 together in the axil of coevally developing lvs, the fruits long persistent on annotinous branches. *Stipules* very small and obscure, firm triangular, less than 1.2 mm, more or less adnate to stem. *Lf-formula* ii–iv(–v)/3–4(–5); lf-stks (2–)3–11(–15) cm, the petiole (5–)8–30(–36) mm, the longer interpinnal segments 1.2–3(–4.5) cm; petiolar nectaries sessile or immersed in epidermis, shallowly cupular or plane, the first one situated between proximal pinna-pair 0.5–1.1 mm diam, those between distal pinnae and between most pairs of lfts similar but a little smaller; pinnae strongly accrescent distally but either the furthest or the penultimate pair longest, the rachis of these 3–7.5(–8.5) cm, the longer interfoliolar segments 10–22(–28) mm; lft-pulvinules in dorsal view 0.8–1.8 × 0.5–1 mm; lfts strongly accrescent distally, variable in outline from obovate or obovate-elliptic to elliptic and shortly acuminate, at base subequilaterally cuneate, at apex very obtuse or emarginate, the blades of the penultimate pair 2.5–6 × 1.4–2.6(–3) cm, 1.6–2.6(–2.8) times as long as wide; the almost centric and almost straight midrib giving rise to ±6–9 pairs of major secondary nerves ascending at ±45° to anastomosis well within the incipiently revolute margin, the tertiary venulation ordinarily weak and random, the whole venation either weakly prominulous or impressed on upper face, a little more sharply prominulous beneath. *Peduncles* (3–)4–11.5 cm; capitula ±25–45-fld, the subspherical or shortly clavate receptacle 2.5–4 mm diam (rarely one or more random flowers remote on peduncle); bracts ovate or spatulate 0.4–0.8 mm, early caducous; fls homomorphic, all sessile or nearly so, glabrous except for minutely ciliolate calyx-teeth and pallid-papillate or -ciliolate tip or distal margin of corolla-lobes, the perianth 5-merous, greenish white, the filaments white (brown when dry); calyx membranous campanulate 1.6–2.2(–2.8) × 1.1–1.6(–1.8) mm, the depressed-deltate obtuse teeth 0.2–0.5 × 0.5–0.7 mm; corolla 5.5–6.5 mm, the nearly erect, distally involute lobes 1.5–2.2 × 0.7–1 mm; androecium 20–25-merous, 13–16 mm, the stemonozone 1.1–1.6 mm, the tube (3.5–)4–7 mm, usually about as long as, rarely a little longer or shorter than the corolla; ovary obliquely truncate, glabrous; style either longer or shorter than longest stamens, the little-dilated stigma 0.1–0.2 mm diam. *Pods* 1 or rarely 2 per capitulum, sessile at cuneate base, in profile undulately linear and evenly recurved through (¾–)1–3 complete turns into a ring or flat spiral ±3.5–4 cm in exterior diam, at fertile locules (9–)10–15(–17) mm wide, deeply constricted only where ovules abort, 8–12-seeded, the valves at first plano-compressed and framed by stout sutures, becoming convex over each seed, the glabrous, dull fuscous or almost black exocarp rugulose and densely papillate overall, the endocarp orange-brown, smooth in the seed cavities but furfuraceous between them; dehiscence of *Abarema*; funicle ligulate, coiled at apex; seeds broadly obovoid, compressed but plump, in broad view ±7 × 6 mm, the testa bicolored, in lower ⅓ ivory-white and faintly areolate (pleurogram ±3 × 4 mm), translucent distally and there blackish when fresh but brown when dry, the color derived from fuscous bony embryo within, the crustaceous brittle outermost shell hardly 0.4 mm thick in section lined with gelatinous tissue ±0.2 mm thick.

In coastal woodland, sometimes on the slope of morros, and inland in disturbed mata do cipó and cerrado or campo rupestre, discontinuously dispersed below 200 m along the Atlantic lowlands of Brazil from Paraiba S to Rio de Janeiro, in Bahia locally inland and on Chapada Diamantina ascending to 1100 m. — Map 26. — Fl. IX–III. — *Barbatimõ* (NE Brazil, a name shared with *Stryphnodendron adstringens*, from the common property of the bark).

Lewis (1987: 178) distinguished the coastal phase of *A. cochliacarpos* from the inland one, which tends to be smaller and to have leaflets obovate rather than elliptic and shortly bluntly acuminate; but some material from Rio de Janeiro has similarly shaped leaflets and is said to flower as a bushy treelet. The species differs from *Abarema filamentosa* in (inter alia) dense capitula of sessile or subsessile flowers and dull blackish, densely papillate pod-valves. The pod of populations in Paraiba and Pernambuco tends to be wider (to 14–17 mm) than in those in Bahia southward (10–15 mm).

The epithet *cochliacarpos* was so spelled in the original description, but a variant *cochleacarpos* appeared in the title of the accompanying illustration. We do not here adopt the spelling preferred by Macbride when he transferred Gomes's epithet to *Pithecellobium*, but instead choose the stem vowel *-a* and the final *-os*, which seems to indicate a noun in apposition rather than a declinable adjective.

31. Abarema langsdorfii (Bentham) Barneby & Grimes, comb. nov. *Pithecolobium langsdorfii* Bentham, London J. Bot. **3:** 205. 1844. — "High mountains in Minas Geraes, *Langsdorff*; Brazil, *Sello*." — Lectotypus, *Riedel s.n., comm. Langsdorff*, K(herb. hooker.)! = NY Neg. *2011*; isotypi, F (fragm. e P)!, K(herb. bentham.)!, LE!. — Syntypi, *Sello s.n.*, K!; in herb. gaudichaud. *208* e Sta Catarina, G! = F Neg *28266*. — Burkart, 1979, inappropriately cited Sello's collection as the type, though did not expressly typify the name thereby. We have preferred

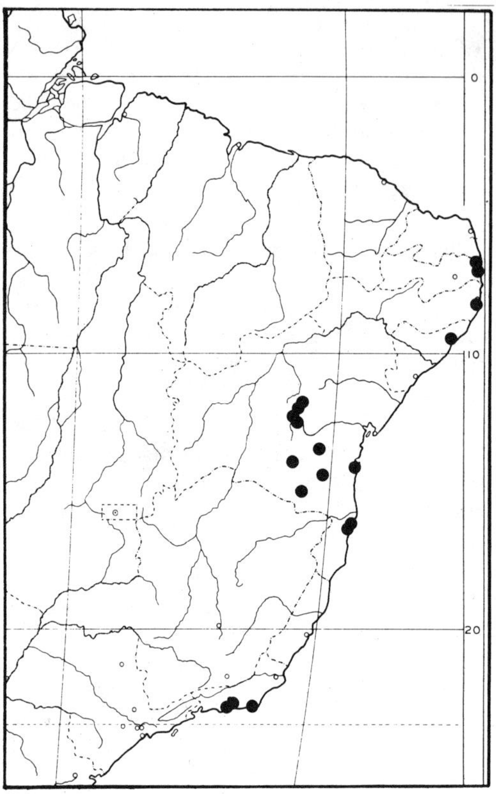

MAP 26. Northeastern Brazil. Range of *Abarema cochliacarpos* (Gomes) Barneby & Grimes.

the original spelling of the epithet to that adopted in *Flora Brasiliensis* and subsequently. — *Feuilleea langsdorfii* (Bentham) O. Kuntze, Revis. Gen. Pl. **1:** 188. 1891.

Pithecolobium langsdorfii sensu Bentham, 1875: 584; 1876: 438 ("*langsdorffii*"). *Pithecellobium langsdorffii* sensu Burkart, 1979: 116, est. 16 + map (Sta Catarina only); Lewis, Leg. Bahia 179. 1987.

Amply microphyllidious *trees* 3–15(–17) m with trunk attaining 2–3.5 dm dbh, the young stems and all lf- and inflorescence-axes densely pilosulous or tomentellous with either straight ascending or forwardly incurved, brown, golden-brown, or sordid hairs to 0.15–0.35 mm, the lvs strongly bicolored, the firm lfts ventrally lustrous and deeply brown-tinged either overall, or toward middle only, or discontinuously brown-spotted, dorsally dull pale olivaceous, commonly glabrous but sometimes remotely minutely ciliolate and less often thinly strigulose, the short racemes of greenish, white-stamened fls solitary or geminate in the axil of coeval lvs, immersed or nearly so in foliage. *Stipules* linear or linear-lanceolate 0.6–3 mm, caducous from early expanding lvs, absent from fruiting spms. *Lf-formula* iii-viii(–xi, in Bahia –xiv)/(14–)16–28(–30); lf-stks (3–)4–14(–17) cm, the petiole including discolored pulvinus 11–32 mm, at middle 0.7–1.8 mm diam, the longer interpinnal segments 8–18(–23) mm; first petiolar nectary at or close below first pinna-pair, either cupular sessile or turbinate and incipiently stout-stipitate, round or elliptic 1–2 mm diam, always thick-rimmed and often corrugate, similar ones at all succeeding pairs or only at tip of lf-stk, or scattered, a much smaller but otherwise similar nectary randomly at tip of pinna-rachises; pinnae a little decrescent distally, the rachis of the penultimate pair (3.5–)4.5–8.5(–11) cm, the longer interfoliolar segments 1.7–3.6(–4.5) mm; pulvinules 0.3–0.6 mm, nearly as wide; lfts subequilong except at very ends of pinnae, narrowly oblong, narrowly lance- or elliptic-oblong from inequilateral, antically obtusangulate, postically cuneate base, at apex either obtuse, or obtuse apiculate, or triangular-acute, those at midrachis 7–12(–14.5) × 2–4 mm, (2.6–)2.9–4 times as long as wide; venation of 1–2(–3) nerves from pulvinule, the nearly centric but proximally incurved, thence straight midrib giving rise on the anterior side or on both to 5–8 weak ascending, mostly simple secondary nerves either weakly brochidodrome or expiring close within the scarcely revolute margin, the posterior primary nerve(s) seldom readily apparent, always very short and weak. *Peduncles* (3–)4–11 cm; racemes (rarely condensed capituliform) 12–27(–30)-fld, the narrowly clavate or linear receptacle (4–)6–20 mm, the fls not or little dimorphic except for pedicels progressively shorter or obsolescent distally, the 1–2 furthest fls sessile but hardly larger than the rest, their stamen-tube sometimes but not invariably a little longer; bracts obsolete; pedicels of lower fls 0.4–2(–2.5) mm; perianth 5-merous (random exceptions), sericeous-strigulose externally, the corolla sometimes more densely so than the calyx; PERIPHERAL FLS: calyx either campanulate or deeply campanulate, even subcylindric, (2.7–)3–5 (–5.6) × 1.7–2.2 mm, the unequal, obtuse teeth 0.5–1.6 mm; corolla (5.7–)6–8.5(–9.5) mm, the often very unequal lobes (1–)1.7–2.8 × 1.1–1.3 mm; androecium (20–)24–38(–40)-merous, to 23–44 mm, the stemonozone 0.7–1.6 mm, the tube 2–5(–6) mm; ovary sessile or obscurely stipitate, truncate at apex, puberulent above middle or only around the top; stigma poriform or scarcely dilated; TERMINAL FL(S): like the peripheral, but the androecial tube a trifle longer, sometimes to 2.5 mm longer than corolla. *Pods* usually 1 per capitulum, in profile undulately linear and evenly recurved through ±1–2 circles into a ring, the body 5.5–12 × 0.7–1 cm, 6–9(–10)-seeded, each segment ±8–15 mm long, the stiffly papery glabrous, when ripe fuscous and coarsely reticulate valves framed by sutures ±1.2 mm wide, low-convex over each seed, not over 0.2 mm thick in section, internally orange-brown at the seed-cavities, tan between them; dehiscence downward through both sutures, the valves elastically recurved and coiling; seeds (few seen) plumply lentiform ±5–5.5 × 4.5 mm, the translucent testa loosely investing the brown embryo, charged on each face with a delicately incised, complete or almost complete pleurogram 2–2.6 × 1.6 mm, this displaced from center of seed-face toward its base.

On rocky summits, open hillsides, and in disturbed valleys of the coast ranges of SE Brazil, descending from 1050 m into restinga forest near sea level, scattered and locally plentiful from the morros of Guanabara SW into E Sta Catarina, and northward, always at submontane elevations, becoming rarer, in seasonally dry woodland habitats, through Sa. da Mantiqueira and E Minas Gerais to the sources of Río Catolé near 14°50′S in SE Bahia. — Map 27. — Fl. nearly throughout the year, most prolifically IX–II. — *Pau gamba*; *raposeira*.

The combination of many narrow, emphatically bicolored leaflets, brownish above and bright green beneath, and short dense racemes of nearly or quite homomorphic flowers distinguishes *A. langsdorfii* from all other abaremas native in Atlantic southern Brazil, and the species is consequently seldom misidentified. The displacement of the pleurogram toward base of the seed is a feature unique within *Abarema*, though encountered rarely elsewhere in Ingeae. The cladogram (Fig. 4) suggests that on the basis of the homomorphic stamens and pubescent ovary, both homoplasious character states, *A. langsdorfii* is more closely related to the West Indian members of the genus than to other Brazilian ones.

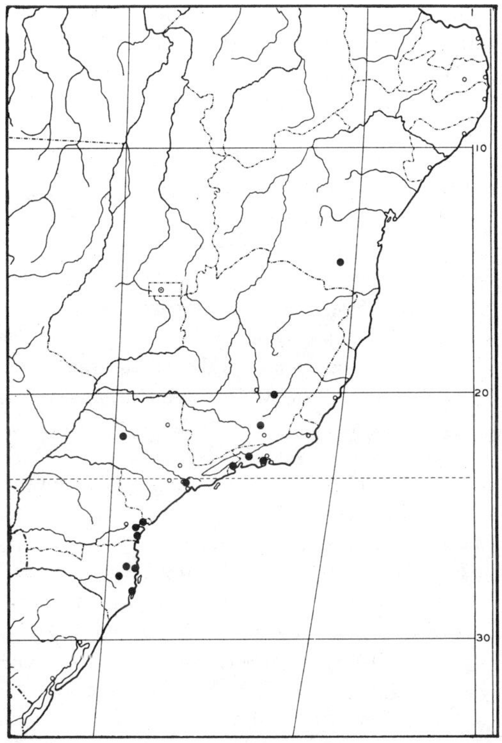

Map 27. Southeastern Brazil. Range of *Abarema langsdorfii* (Bentham) Barneby & Grimes.

32. Abarema asplenifolia (Grisebach) Barneby & Grimes, comb. nov. *Pithecolobium asplenifolium* Grisebach, Cat. Pl. Cub. 83. 1866. — "Cuba occ. (W[right] 2403)." — Holotypus, to be sought at GOET; isotypi, G!, GH!, K!, P! labeled "Pinal, Mayarí Abajo, Cuba orientalis 6 aug" (fl & fr jun), MO!, NY!, US (fragm. *ex* herb. urban.)! — *Feuilleea asplenifolia* (Grisebach) O. Kuntze, Revis. Gen. Pl. **1:** 187. 1891. *Jupunba asplenifolia* (Grisebach) Britton & Rose, N. Amer. Fl. **23:** 27. 1928.

Pithecellobium asplenifolium subsp. *mayarense* Borhidi, Acta Bot. Acad. Sci. Hung. **22:** 298. 1976. — "Cuba; Sierra del Cristal, Nicaro, Mayari, Montes United Fruit. Leg. J. ACUÑA et [F.] ZAYAS 19663 . . . 26–27. Mai. 1955." — Holotypus, SV (not seen); isotypus, NY!.
P. asplenifolium sensu Bisse, 1988: 225.

Arborescent *shrubs* of unknown maximum stature, the young growth densely minutely brown-puberulent, the alternate or subopposite, multifoliolate lvs bicolored, the firmly papery, ventrally convex lfts dark lustrous green (when dry, brown) and glabrous above, beneath densely microscopically pallid-papillate and either glabrous or minutely remotely strigulose (especially along midrib, and sometimes incipiently barbellate in the anterior basal angle), the short racemes of long-pedicellate fls solitary in the axil of few coeval lvs, immersed in foliage. *Stipules* 0 (no scars). *Lf-formula* (ii–)iii–vi/9–12; longer lf-stks 5–7.5 cm, the petiole and the longer interpinnal segments each 1–2 cm; petiolar nectaries sessile, shallowly cupular thick-rimmed 0.7–1.4 mm diam, situated between pinnae of first pair, all pairs, or only the distal pair of pinnae, much smaller ones between few furthest pairs of lfts; pinnae decrescent proximally, the rachis of furthest and penultimate pairs 4.5–7 cm, the longer interfoliolar segments 5–7 mm; lft-pulvinules in dorsal view 0.7–1.2 × 0.6–0.7 mm; lfts gradually accrescent distally but except for proportionately wider distal pair subequiform, in outline obliquely oblong- or oblance-elliptic from inequilaterally cuneate base, broadly rounded muticous at apex, the longer ones (11)12–18 × 5–8 (–"10," Borhidi) mm, 1.7–2.6 times as long as wide; the subcentric midrib gently retroarcuate near base, straight thereafter, giving rise to ±5–8 pairs of secondary nerves brochidodrome within the revolute margin, the tertiary venulation weak and random, the whole venation immersed or impressed on upper face of blade, prominulous on lower. *Peduncles* 4.5–10 cm; racemes ±17–24-fld, the almost homomorphic fls either scattered or subverticellate, the axis becoming 5–11 mm; bracts minute caducous; lower pedicels subhorizontal and distally incurved 5.5–11 × 0.4–0.5 mm, the further ones progressively shorter, that of furthest only slightly larger but not otherwise modified fls 1.5–3 mm; LOWER AND MEDIAN FLS: perianth 5-merous, either sordid-puberulent overall or the calyx-tube glabrate; calyx turbinate-campanulate ±4–5 × 1.2–1.8 mm; corolla turbinate-campanulate ±7.5–9 × 1.5–2.2 mm, the erect or gently arcuate lobes externally brown-puberulent; androecium 40–58-merous, to 2.6–3.0 cm, the stemonozone 1–1.6 mm, the tube 3–6 mm; ovary truncate, densely puberulent. *Pods* (little known) like that of *Abarema glauca*, 8–10-seeded, the valves 7–8 mm wide at seeds, constricted between them; seeds 5.3 × 3.5 mm, pleurogram 2.6 × 1.3 mm.

In montane woodland and coastal scrub thickets, apparently endemic to NE prov. Oriente, Cuba: attaining 800 m on Sa. de Nipe and Sa. del Cristal, and on low hills and cayos of the N coast between Pta Cabanas and Pta Gorda. — Fl. V–VIII(–?).

In foliage characters *A. asplenifolia* is intermediate between the microphyll species *A. oppositifolia*, *A. nipenesis*, and *A. asplenifolia* and the macrophyll ones *A. glauca* and *A. obovalis*; for measurements and numbers of leaflets see our key to West Indian *Abarema* (p. 45). The subspecies *mayarense*, which we confidently reduce to synonomy, was supposed to differ from typical *asplenifolia* in elliptic-obovate leaflets 10–18 × 5–10 mm, lustrous and bullescent on upper face, concave and densely appressed-pilose beneath, and in leaf-rachis 4–8 (not 2.5–5.5) cm long; and it was thought to be geographically disjunct from the west-Cuban var. *asplenifolia*. The supposed differences in leaflets are not borne out by the material that we have seen and are not all expressed in the isotype of subsp. *mayarense* at NY. It is true that Grisebach described *P. asplenifolium* from "Cuba occ[identalis]", but this was a mistake; the isotype (*Wright 2403*) at GH is labeled "Mayari Abajo," and no doubt was collected in Oriente Province, as are all modern specimens, and close to the type-locality of subsp. *mayarense*.

From its dispersal so far as exactly known it is likely that *A. asplenifolia* is confined to serpentine soils, although the substrate is seldom specifically recorded by collectors.

33. Abarema glauca (Urban) Barneby & Grimes, comb. nov. *Pithecolobium glaucum* Urban, Symb. Ant. **7:** 277. 1912. — "Hab. in Sto. Domingo austr. prope Barahona ad Bahoruco 50 m. alt., m April. flor.: Fuertes n. 183, prope Cabral 25 m alt., m. Mart. flor.: idem n. 857." — Syntypi, †B; lectotypus, no. *183*, NY!; isotypi, G!, US!; paratypi, no. *857*, A!, F!, K!, MO!, US!, U!. — *Jupunba glauca* (Urban) Britton & Rose, N. Amer. Fl. **23:** 25. 1928.

Pithecolobium savannarum Britton, Bull. Torrey Bot. Club **41:** 4. 1914. — "Along a watercourse on barren savannas southeast of Holguin, Oriente [Cuba], April 7, 1909 ([*J. A.*] *Shafer 1194*)." — Holotypus, NY!; isotypi F!, NY (2 sheets)!, US *792269*!. — *Jupunba savannarum* (Britton) Britton & Rose, N. Amer. Fl. **23:** 25. 1928. — Equated with *P. glaucum* by Liogier, Fl. Cuba Suppl. 72. 1969.

Pithecolobium discolor Britton, Bull. Torrey Bot. Club **41:** 4. 1914. — "Type from Batabano [prov. La Habana, Cuba], April 10, 1903, ([*J. A.*] *Shafer 161*); apparently the same species at Old Kerr's Point, Abaco, Bahamas ([*L. J. K.*] *Brace 2017*)." — Holotypus, NY!; paratypi, NY!, US!. — *Jupunba discolor* (Britton) Britton & Rose, N. Amer. Fl. **23:** 25. 1928. — Equated with *P. glaucum* by Liogier, Fl. Cuba Suppl. 72. 1969; and by Correll & Correll, Fl. Bahama Archipel. 678. 1982.

Pithecolobium glaucum sensu Urban, Ark. Bot. **24A(4):** 2. 1932, descr. ampliat. ex *Ekman 2698, 4222*; *Pithecellobium glaucum* sensu Bisse, 1988: 224.

P. savannarum sensu Bisse, 1988: 224.

Trees potentially attaining 30 m in height and 9 dm dbh, most often seen only 4–9 m and occasionally flowering precociously as a shrublike sapling, the mature bark nearly smooth, exfoliating in large plates, the innovations densely bronze-puberulent or -furfuraceous, the lf-axes and inflorescence thinly inconspicuously strigulose-pilosulous with straight appressed or ascending-incurved hairs ±0.1–0.2 mm, the lvs almost always strongly bicolored, the thinly chartaceous lfts on upper face lustrously dark green, brown, purplish brown, or brown-mottled on pallid ground and uniformly glabrous, on lower face pallid, either glabrous or minutely remotely strigulose and often with a tuft of sordid hairs in the anterior basal angle of midrib, the short racemes of whitish fls arising singly or geminate from the axil of coeval lvs and immersed in foliage, the fruiting peduncles and dehisced fruit often long persisting below the new foliage. *Stipules* 0. *Lf-formula* (ii–)iii–vi(–vii)/(5–)6–8(–10); lf-stks of vigorous young branches 6–15(–17) cm, of some on late branchlets condensed and only 1.5–5 cm, the petiole (0.6–)1–3.5(–4) cm, the longer interpinnal segments (0.5–)1–3(–3.7) cm; petiolar nectaries randomly variable in development, always sessile or some almost completely immersed, the first one, situated close below insertion of proximal pinna-pair, either mounded and small-pored or shallowly patelliform (then either round or vertically elliptic) and ±1.3–2.7 mm diam, or grossly verruciform, woody, wrinkled and (2–)3–6(–9) mm diam, exceptionally suppressed, smaller pored nectaries between furthest pairs of pinnae and lfts and sometimes between others; pinnae, when 2 pairs, distally accrescent, when 3 or more the proximal pair decrescent, the distal pairs thence subequilong; lft-pulvinules in dorsal view 0.8–1.4 × 0.5–0.9 mm; lfts only a little accrescent distally, in outline obliquely obovate or oblong-obovate from broad-cuneate base, on anterior side incipiently rhombic by dilation below midblade, at apex broadly rounded or emarginate, muticous, the penultimate pair 20–32 × (8–)9–19 mm, (1.4–)1.5–2.2(–2.5) times as long as wide; venation essentially pinnate (occasionally a short anterior primary nerve from pulvinule), the nearly centric and straight midrib giving rise on anterior side to ±5–8 (on posterior side to slightly fewer) major secondary nerves brochidodrome well within the plane or incipiently revolute margin and to random weaker intercalary ones, all these and erratic tertiary venules prominulous on both faces. *Peduncles* 3–8(–10) cm; racemes 20–65-fld, the axis becoming 5–22 mm, linear or linear-fusiform, the fls all homomorphic except for length of pedicel; bracts very quickly deciduous (none seen); pedicels widely ascending or the outermost horizontal, the proximal ones 2.5–7 mm, the distal progressively (but only slightly) shorter; perianth 5-merous, minutely puberulent externally, especially on lobes of corolla; calyx campanulate from either turbinate or rounded base, 2–4.5 × 1.6–2.4 mm, the usually unequal teeth broadly ovate to deltate, 0.5–1.2 mm; corolla (6.2–)6.5–9.5 mm, narrowly trumpet-shaped, the ovate lobes 1.8–3 × 1.1–1.8 mm; androecium 28–42-merous, 16–24.5 mm long, the stemonozone 1.2–2.3 mm, the tube 3.5–6.5 mm; ovary sessile, truncate at apex, silky-strigulose beyond glabrous base; style either longer or shorter than the longer stamens, the stigma scarcely dilated, 0.1–0.15 mm diam. *Pods* 1–2 per spike, subsessile, in profile undulately linear and (well fertilized) evenly coiled into one or two complete circles, (5.5–)8.5–12.5 cm × (6–)7–10(–11) mm, at interseminal isthmi (2–)3–8 mm wide, the leathery valves at first fuscous, plano-compressed, framed by thickened sutures, becoming low-convex over each seed and the exocarp becoming gray-tan and irregularly roughened-papillate, the crustaceous endocarp not over 0.2 mm thick in section, internally tan between seeds and dark chestnut brown in the cavities, these either round or vertically elliptic 7–10 × 5.5 –9 mm; dehiscence elastic, through both sutures; funicle ribbonlike, coiled at apex; seeds plumply pealike, a little compressed, in facial view subcircular and 4–4.5 mm diam, the translucent testa white when fresh, yellowing in age, closely investing the aniline-blue embryo, to 0.2 mm thick proximally but thinner distally where revealing a patch of blue from within, each face finely engraved with a closed pleurogram 2.5–3 × 1.8–2.5 mm.

In coastal thickets, coppice, and along streams in savanna, often on limestone, below 50 m, interruptedly widespread over Cuba from Pinar del Río E to W Oriente, N in the Bahamas to Andros and Great Abaco islands, and E locally into Hispaniola, in Haiti on the NW coast and Île de la Tortue, in Dominican Republic on the NW coast (Pto Plata) and Barahona Peninsula. — Map 28. — Fl. I–VII, perhaps sporadically throughout the year.

The epithet *discolor* is very appropriate to *A. glauca*, for the leaflets, especially when dry, are notably bicolored, the glaucescent-purplish-gray upper face contrasting with the pale, fresh green hypophyllum. This characteristic foliage combined with umbelliform capitula make it one of the more easily recognized species of the genus. Contrary to what might be expected from the synonymy, *A. glauca* is not exceptionally variable; the features stressed by Britton & Rose (1928: 24) as diagnostic for *Jupunba discolor* and *J. savannarum*, that is the shape of the calyx and the outline or coloration of the seeds, are insignificant in the context of the larger sample of specimens now available.

34. Abarema obovalis (A. Richard) Barneby & Grimes, comb. nov. *Inga obovalis* A. Richard, Hist. Phys. Cuba, Pl. Vasc. 472. 1846. — "Crescit in *Vuelta de Abajo*, ubi detexit clar. don José María

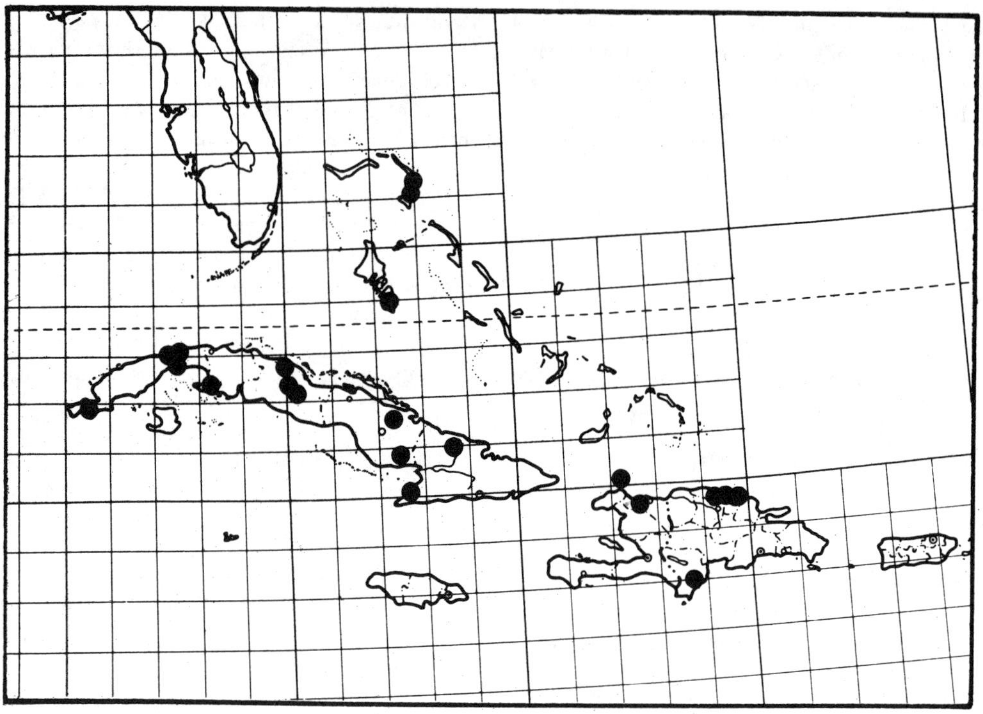

MAP 28. Cuba, Hispaniola, and the Bahamas. Range of *Abarema glauca* (Urban) Barneby & Grimes.

Valenzuela." — Holotypus, P(herb. richard.), 2 sheets!. — *Calliandra revoluta* Grisebach, Cat. Pl. Cub. 83. 1866, substit. illegit. *Pithecolobium obovale* (A. Richard.) C. Wright *in* Sauvalle, Anales Acad. Ci. Méd. Habana **5:** 407. 1868. *Feuilleea obovalis* (A. Richard) O. Kuntze, Revis. Gen. Pl. **1:** 188. 1891. *Jupunba obovalis* (A. Richard) Britton & Rose, N. Amer. Fl. **23:** 26. 1928.

Pithecolobium truncatum Britton, Bull. Torrey Club. **41:** 5. 1914. — "[Cuba.] Southern Oriente. Type from gravelly hills, El Cobre, west of Santiago, March 23, 1912 (*Britton, Cowell & Shafer 12874*)." — Holotypus, NY!; isotypi, MO!, U!, US *1265305* (fragm.)!, *698536*!. — *Jupunba truncata* (Britton) Britton & Rose, N. Amer. Fl. **23:** 26. 1928. — Equated with *P. obovale* by Liogier, Fl. Cuba Suppl. 72. 1969.

Pithecolobium. pinetorum Britton, Bull. Torrey Club **41:** 6. 1914. — "[Cuba.] Mountain pinelands of northern Oriente. Type from southeast of Paso Estancia, May 1–2, 1909 (*Shafer 1725*); fruit and seeds described from *Shafer 3096*, collected in pinelands of Sierra Nipe." — Holotypus, NY!; isotypi, GH!, US *659898*!; paratypi, F!, NY!. — *Jupunba pinetorum* (Britton) Britton & Rose, N. Amer. Fl. **23:** 26. 1928. — Equated with *P. obovale* by Liogier, Fl. Cuba Suppl. 72. 1969.

Pithecellobium. obovale sensu Bisse, 1988: 225.

Pithecellobium. pinetorum sensu Bisse, 1988: 225.

Macrophyllidious *trees* attaining 6–15 m but sometimes fertile at 2 m and upward, the very young branchlets and lf-stks densely puberulent with bronze hairs ±0.1 mm but only the axillary buds remaining so, the broad, stiffly chartaceous lfts conspicuously bicolored, above dark lustrous green, brunnescent when dry except for contrasting pallid venation and fully glabrous, beneath pallid and often thinly minutely appressed-puberulent, the very short or subumbellate racemes of white (when dried golden-strigulose) fls arising singly or paired from the axil of coeval lvs, immersed in foliage. *Stipules* 0, even in lf-vernation. *Lf-formula* i–ii(–iii)/3–5(–6), lfts (12–)16–56 per lf; lf-stks stout (0.7–)1–8 cm, the true petiole 0.7–3.2 cm, the longer (or single) interpinnal segment 1.2–3.5 cm; lf-nectaries variable in development, the first situated immediately below insertion of proximal pinna-pair, sessile, most often grossly verruciform, round or vertically elongated 2–8 × (1.5–)2–4 mm, when dry nigrescent, finely wrinkled and with or without a dimple or small pore near or below middle, but sometimes obscure and sunken in the epidermis, those between further pinna-pairs smaller or 0, those at tip of pinna-rachises often lacking; pinnae usually accrescent distally, the rachis of distal (or only) pair

(4–)5–11 cm, the longer interfoliolar segments 12–25(–32) mm; lft-pulvinules in dorsal view 1.3–2.6 × 0.9–1.8 mm; lfts usually much accrescent distally, the furthest pair subsymmetrically, the lateral ones inequilaterally obovate (broad-oblanceolate), wider on anterior side but not trapeziform-angulate, all at base broad-cuneate or flabellate, at apex broadly obtuse muticous or shallowly emarginate, the largest 4–6.5 × (2.8–)3–5(–5.5) cm, (1.1–)1.2–2.8(–3.2) times as long as wide, all when old incipiently bullate and marginally revolute; the midrib gently curved forward, the secondary veins ascending at ±45° and brochidodrome well within the margin, all these together with random tertiary venules finely prominulous on both faces but more conspicuous by pallid coloration above. *Peduncles* (3–)4–8 cm; racemes 7–20-fld, subcorymbosely hemispherical at full anthesis, the fls all nearly equiform but the pedicels progressively shorter upward, the terminal fl(s) sometimes sessile and a trifle larger but their androecium not differentiated; bracts very quickly caducous, none seen; pedicels of lower fls 3–6 mm; calyx microscopically puberulent or glabrate, at anthesis (3–)3.5–5 mm, in young bud pyriform, the orifice then truncate, minutely 5(–6)-toothed, the whole at full anthesis turbinate-campanulate, the thick-textured tube 2.7–3.3 × 2.3–3 mm, the teeth (often of different lengths) either ovate-deltate or depressed-deltate and subtruncate-apiculate 0.3–0.4 mm, rarely the teeth all adherent and the orifice then truncate and minutely 5-apiculate; corolla 6.5–10 mm, the tube externally glabrate, the lanceolate, externally silky-puberulent, apically cucullate lobes 2.6–5 mm; androecium (33–)36–60-merous, to 2.5–3.4 cm, the stemonozone ±1.5–2.5 mm, the white filaments united into a tube 3–5.5 mm; ovary densely puberulent, truncate at apex; style exserted from longer stamens, at apex 0.1–0.2 mm diam. *Pods* 1–4 per capitulum, subsessile, in profile broad-linear, evenly recurved through ±¾ to nearly a full circle, framed by thickened sutures, 3.5–6.5 × 0.9–1.3 cm, the exterior (seminiferous) suture not or very shallowly undulate-impressed between the 8–10 seeds, the valves stiffly coriaceous, the thin exocarp at first dusky and venulose, becoming light brown and roughened, strongly umbonate over each seed, the pitchlike mesocarp ±0.5 mm and the crustaceous white endocarp ±0.2 mm thick, internally deep-crimson and papillate between the smooth seed-cavities; dehiscence of *Abarema*; funicles ligulate; seeds plump, firm and shiny, in broad view 4.4–5.6 × 3.7–5 mm, the thin translucent testa closely investing the hard indigo-blue embryo, this most visible as a blue patch on each face of seed above its middle, pleurogram delicately incised, complete, 3–3.6 × 2–2.7 mm.

In woods and thickets, sometimes along streams, in Cuba from near sea level on Isla de la Juventud up to 900 (in Camaguey 1100) m, in Hispaniola at 250–720 m, scattered in Cuba from Pinar del Río and Isla de la Juventud (de los Pinos) to upland Oriente, in Hispaniola localized on Massif de la Hotte in Dépt. Sud, Haiti, and in the central highlands of prov. La Vega in Dominican Republic. — Map 29. — Fl. X–XI, II–VIII, perhaps randomly throughout the year. — *Roble blanco* (Isla de la Juventud, Cuba).

In the flora of Cuba and Hispaniola *A. obovalis* is instantly identified by low leaf-formula and proportionately ample leaflets to 4–6.5 cm long. Its calyx is spathiform in vernation, the teeth developing only as the membranous margin is split by the swelling corolla and androecium. The 5, rarely 6 teeth, marked only by minute apiculations in vernation, may become simply low-deltate-ovate, or by vertical split at the sinus depressed-deltate and apiculate, and may be of uniform size or appear united in pairs, due to imperfect separation. The typus of *Pithecolobium truncatum* is a rare instance of the teeth failing to separate. No further example is known to us from Oriente, but we have one from Isla de los Pinos, where the normal form is prevalent. The staminal column varies from 3 to 5.5 mm in length, including the stemonozone. That of *P. pinetorum* is at the upper extreme of the observed range of variation, but the typus of this species, except that it is stressed by pervasive insect damage, is not in any other way remarkable.

35. Abarema alexandri (Urban) Barneby & Grimes, comb. nov. *Pithecolobium jupunba* (Willdenow) Urban var. *alexandri* Urban, Symb. Ant. **2:** 258. 1900. ≡ *Jupunba alexandri* (Urban) Britton & Rose. *Pithecolobium alexandri* (Urban) Urban, Symb. Ant. **5:** 358. 1908. — Typus infra sub var. *alexandri* indicatur.

Trees attaining 12 m, either macro- or microphyllidious, the young branchlets, lf-axes and inflorescence at first villosulous with brown-gold spreading hairs ±0.1–0.25 mm but early glabrescent or finally glabrate, only the axillary buds of mature stems tomentulose, the thin-textured, delicately venulose lfts bicolored, bright green above, paler duller beneath, the loose spiciform racemes of small greenish-yellowish fls arising singly or 2–3 together from the

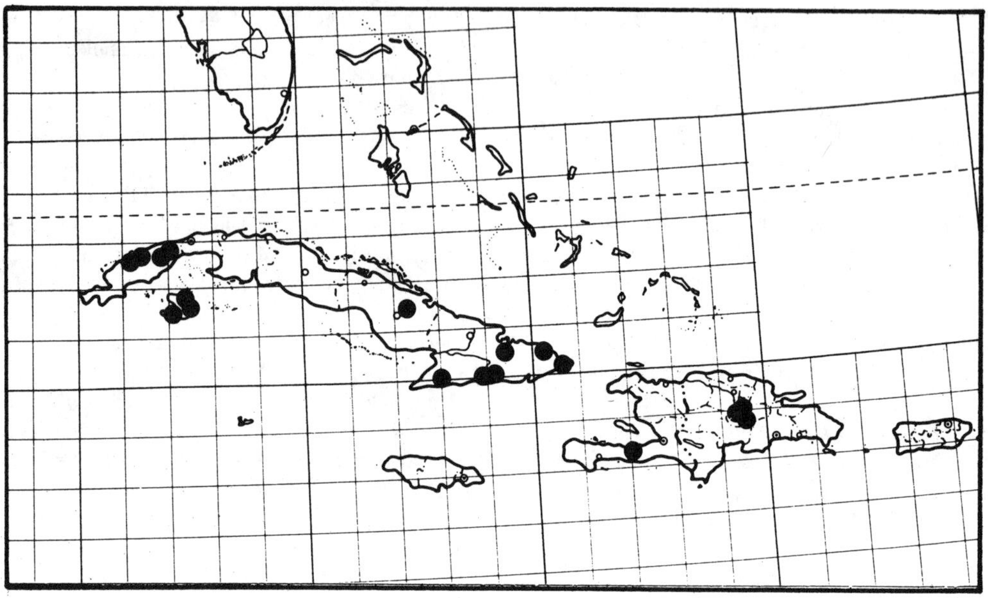

MAP 29. Cuba and Hispaniola. Range of *Abarema obovalis* (A. Richard) Barneby & Grimes.

axil of coeval lvs, the fruits long persistent, immersed in foliage. *Stipules* 0. *Lf-formula* (i–)ii–vii/4–15; lf-stks (2–)3–15 cm, the petiole 1.3-4.5 cm, the longer interpinnal segments 1.2–3 cm; nectary immediately below insertion of first pinna-pair sessile, shallow-cupular or almost plane, in vertical view round or broadly elliptic 0.8-1.7 mm diam, similar but smaller ones at some further nodes of lf-rachis and at all or most nodes of pinna-rachises; pinnae accrescent distally but the penultimate pair often longest, the rachis of these 5–12 cm, the longer interfoliolar segments 6–25 mm; lft-pulvinules in dorsal view 0.3–1.6 × 0.5–0.9 mm; lfts decrescent proximally but otherwise, except for proportionately narrower furthest pair (no further described), subequiform and all obtusely rhombic or rhombic-elliptic around a diagonal midrib, at base inequilaterally broad-cuneate (or postically rounded), obtuse muticous at apex, those at and beyond midrachis 11–60 × 6–41 mm, 1.5–2.4 times as long as wide; the midrib straight or slightly arched forward, the secondary veins ascending at ±45° to anastomosis well within the plane margin, these and random tertiary veinlets subequally prominulous on both faces of blade. *Peduncles* 3.5–8 cm; raceme-axis 2.5–8 cm, the fls horizontal or a little declined; bracts very quickly deciduous (none seen); pedicels of proximal fls 0.5–4 mm, those of distal fls progressively shorter, the furthest fls sessile or almost so; fls seemingly homomorphic, no dimorphic terminal one seen; perianth 5-merous, thinly puberulent externally or glabrate; calyx campanulate 2.2–3.8 × 1.3–2.5 mm, the ovate teeth (often unequal) to 0.5–0.9 mm; corolla narrowly trumpet-shaped 5.5–7.5 mm, the ovate lobes (1–)1.7–3 × 0.8–1.6 mm; androecium 32–40-merous, the longest filaments 11–14.5 mm, the stemonozone 1.5–2.3 mm, the tube 3.5–5 mm; ovary glabrous, truncate at apex; style a little exserted, the stigma poriform or a trifle dilated, 0.1–0.2 mm diam. *Pods* 1–4 per spike, in profile undulately broad-linear recurved through ¾ to 1¼ circle, 7–10.5 × 1–1.3 cm, contracted at base into a very short stipe or cuneate neck, minutely apiculate, the valves at first dark green and venulose, when ripe gray-tan and roughened, framed by thickened sutures shallowly impressed between seeds, low-convex over each of 6–10(–11) seeds, within tan between seeds but lustrous-castaneous in the seed-cavities, the interseminal constrictions 11–16 mm apart; dehiscence elastic, the valves coiling and the seeds dangling; funicles ribbonlike, coiled distally; seeds plumply obovoid, a trifle compressed, in broad view 5.5–7.5 × 4.3–6 mm, when fresh slate-blue or particolored blue and whitish, the more or less translucent testa 0.3–0.6 mm (restored by soaking) thick, in older seeds lustrous cream-colored proximally, brown or brown and indigo in a transverse band around the distal hemisphere outside the finely engraved, continuous pleurogram, this 3.3–3.8 × 2–3 mm; embryo aniline-blue.

In woods or thickets on limestone substrates, 350–830 m, endemic to Jamaica, where represented by two, apparently partly sympatric varieties. — *Tamarind shadbark.*

The openly racemose units of inflorescence of *A. alexandri*, which are found in no other West Indian abarema, are technically indistinguishable from those of the cordilleran species *A. callejasii*, *A. lehmanii*, *A. killipii*, and *A. racemiflora*. However, the cladogram (Fig. 4B) supports Britton & Rose in suggesting that this species is more closely related to Caribbean species of *Abarema*.

The foliage of *A. alexandri* is suprisingly variable. We maintain here two of the three infraspecific taxa traditionally based on leaf-formula.

Key to the varieties of *A. alexandri*

1. Relatively microphyllidious; lf-formula iv–vii/7–15; larger lfts 11–27 × 6–11 mm; Hanover, St. Catherine, and Westmoreland parishes................35a. var. *alexandri*
1. Relatively macrophyllidious; lf-formula (i–)ii–iii/4–6; larger lfts 35–60 × 21–41 mm; St. Andrew, St. Elizabeth, and Trelawny parishes ....35b. var. *troyana*

35a. Abarema alexandri (Urban) Barneby & Grimes var. **alexandri.** *Pithecolobium jupunba* var. *alexandri* Urban, 1900, l. c., sens. str. — "Habitat in Jamaica: Alexander [Prior]." — Holotypus, presumably †B; neotypus (dated 1850, fl.), *Prior s.n.*, GOET!; isotypi, K!, NY!. — *Pithecolobium alexandri* var. α Urban (by implication), Symb. Ant. **5:** 358. 1908. *Jupunba alexandri* (Urban) Britton & Rose, N. Amer. Fl. **23:** 24. 1928.

Pithecolobium alexandri var. *intermedium* Urban, Symb. Ant. **5:** 359. 1908. "Jamaica . . . prope Catadupa in Lapland 500 m, [*Wm*] *Harris* n. 9182." — Holotypus, presumably †B; isotypi, NY!, US!. — Equated with *J. alexandri* by Britton & Rose, 1928, l. c.; overlooked by Adams, 1972: 332.

Lf-formula and lft-size as given in key to varieties; interfoliolar segments of rachis 6–11 mm; lft-pulvinules 0.3–0.6 × 0.5–0.7 mm; longer pedicels 2–4 mm; *calyx* 2–2.5 mm diam; corolla 6.5–7.5 mm; otherwise as given for the species.

Distribution as given in key.

35b. Abarema alexandri var. **troyana** (Urban) Barneby & Grimes, comb. nov. *Pithecolobium alexandri* var. *troyanum* Urban, Symb. Ant. **5:** 359. 1908. — "Jamaica . . . prope Troy 600–830 m alt., m. Majo flor., Nov., Dec, fruct., [Wm] Harris no. 8800, 8960, 9077, 9263, 9921, 9948." — Syntypi presumably all †B; lectotypus, *Wm. Harris 8960*, collected 15 May 1905 (fl.), NY!; isolectotypi, F (2 sheets)!, NY!; paratypi from: "near Troy," *Wm. Harris 8800* 7 Dec 1904 (fr.) NY!, US!, *Wm. Harris 9007*, 21 Nov 1905 (fr.) NY!; "SE slope of Dolphin Head" *Wm. Harris 9263* 24 Dec 1906 (fl.) NY!; "Malvern to Mountain Side" *Wm. Harris 9921* 12 Sep 1907 (fl.) NY!; "Stanmore Hill" *Wm. Harris 9948* 11 Sep 1907 (fr.) NY!. — *Jupunba troyana* (Urban) Britton & Rose, N. Amer. Fl. **23:** 25. 1928.

Lf-formula and lft-size as given in key to varieties; interfoliolar segments of pinna-rachis to 12–25 mm; lft-pulvinules 1–1.6 × 0.5–0.7 mm; longer pedicels 0.5–2 mm; *calyx* 1.3–1.5 mm diam; corolla 5.5–6.5 mm; otherwise as given for the species.

Distribution as given in key.

36. Abarema oppositifolia (Urban) Barneby & Grimes, comb. nov. *Pithecolobium oppositifolium* Urban, Symb. Ant. **2:** 258. 1900. — "Hab. in Haiti prope Port-au-Prince in montibus Furcy, m. April. flor.: Picarda n. 1024." — Holotypus, †B; neotypus, *Leonard 4327*, annotated by Urban, US *1076785*!; isotypi, US *1411959*, *1411960*!. — *Jupunba oppositifolia* (Urban) Britton & Rose, N. Amer. Fl. **23:** 28. 1928.

Pithecolobium trinitense Britton, Bull. Torrey Bot. Club **41:** 7. 1914. — "Hillsides, El Porvenir to Aguacate, Trinidad Mountains, Santa Clara [Las Villas, Cuba], at 700–900 m altitude, March 10, 1910 (*Britton & Wilson 5346* . . .)." — Holotypus, NY!; isotypus, US *1265304* (fragm.)!. — *Jupunba trinitensis* (Britton) Britton & Rose, N. Amer. Fl. 23: **28.** 1928.

P. trinitense sensu Bisse, 1988: 223.

Amply densely microphyllidious *trees* 3–12 m with trunk attaining 6 dm dbh, the leaves and branchlets often appearing opposite but seldom exactly so and not at every node, the young growth, all lf-axes, and the inflorescence rusty-golden-puberulent with incurved hairs <0.25 mm, the lvs strongly bicolored, above dark green (brunnescent or nigrescent when dried), glabrous and lustrous, beneath pale, dull and minutely or sparsely strigulose and often in addition barbellate in anterior basal angle of midrib, the short umbelliform capitula solitary and long-pedunculate in the axil of coeval lvs, immersed in foliage, the dehisced fruits often persisting on annotinous defoliate branchlets. *Stipules* obsolete or perhaps sometimes very early caducous, no scar seen. *Lf-formula* iv–xiii/16–25(–30); lf-stks 4–13 cm, the petiole 6–20 mm, the longer interpinnal segments 5–10 mm; lf-nectaries capriciously variable, always sessile, the first of each lf situated immediately below proximal pinna-pair mostly 1–3 mm diam, either grossly verruciform small-pored, when as wide as or wider than the petiole, or low-convex, or immersed in petiolar tissue, random smaller nectaries between some distal pinna-pairs or lacking, yet smaller ones between many pairs of leaflets or only between few distal ones, or lacking; pinnae decrescent at each end of lf-stk, those near mid-lf 3.5–6 cm, the longer interfoliolar segments 2–4.5 mm; lft-pulvinules subobsolete, in dorsal view not over 0.25 mm; lfts subequal except for some smaller proximal pairs and the slightly wider furthest pair, the blade oblong or oblong-elliptic from inequilateral semicordate base, very obtuse or obscurely emarginate at apex, those at midrachis 5–7.8 × 2.1–3.5 mm, (2–)2.1–2.8 times as long as wide; the midrib recurved immediately above base and thence straight or nearly so, subcentric, giving rise on each side to 3–4(–5) widely ascending or subhorizontal secondary venules weakly brochidodrome within the revolute margin, tertiary venulation 0 (rarely a weak posterior primary nerve from pulvinule). *Peduncles* 3–9 cm; capitula ±15–25-fld, the narrowly clavate or hemispherical receptacle ±3–5 mm; bracts oblanceolate <0.5 mm, very early caducous; fls homomorphic or very slightly longer distally, the pedicel of proximal ones 1–2 mm, that of distal ones progressively shorter or 0; perianth 5-merous, puberulent externally; calyx 3.4–4.2 × 2–2.5 mm, the low-deltate teeth 0.4–0.7 mm; corolla narrowly vase-shaped 6–7.2 mm, the ascending, lance-ovate lobes 2.4–2.9 × 1.2–1.5 mm; androecium 43–52-merous, 12–15 mm long, the stemonozone 1–1.2 mm, the tube 3.5–4.5 mm; ovary truncate glabrous; stigma not or scarcely dilated. *Pods* solitary or 2 per capitulum, sessile but backwardly attenuate below the first seed, in profile undulately broad-linear 6–11 × 0.6–1(–1.2) cm, (6–)7–9(–10)-seeded, either falcately or spirally recurved through ¾ to 1½ circles, the ligneous valves at first plano-compressed becoming low-convex and only a trifle thickened over each seed, externally dull fuscous-brown and sometimes bluntly venulose, either glabrous or minutely remotely puberulent, the endocarp crimson in the seed-cavities and tan between them, the seed-locules 7–13(–15) mm long; dehiscence of *Abarema*; funicles narrow-ligulate, twisted at apex; seeds basipetal, plumply lentiform, in broad profile 4.3–6 × 3.4–5 mm, the smooth translucent testa at first colorless and transmitting the blue coloring of the embryo, later in part or wholly pearly-gray or pale yellowish tan, in cross-section 0.2–0.3 mm thick, closely investing the embryo, charged on each broad face with a finely engraved, continuous pleurogram 2.5–4 × 1.4–3 mm.

In thickets and woodland or mountain pine-forest, 800-1600 m, discontinuously dispersed from centr. Cuba E to centr. Hispaniola: in Cuba on Sa. de Trinidad and Loma de Banao in Las Villas, and on Sa. Maestra in S Oriente; in Haiti on Massif de la Selle in dept. Ouest and its prolongation into prov. Independencia and Barahona in Dominican Republic, thence NE into Cordillera Central of prov. Santiago and La Vega. — Map 30. — Fl. IX–XI, III–VI, perhaps irregularly throughout the year, the fruits long persistent.

Abarema oppositifolia is the most widely dispersed of the Antillian microphyll abaremas and offers a convenient standard by which the more local, perhaps derived species can be evaluated. The leaves tend to be crowded toward the tip of the branchlets but are fortuitously opposite in some specimens only. *Jupunba trinitensis* was supposed by Britton and Rose (1928: 24) to differ in slightly larger, thinner-textured, less lustrous leaflets, but these are ineffective criteria in the material that has accumulated since.

37. Abarema nipensis (Britton) Barneby & Grimes, comb. nov. *Pithecolobium nipense* Britton, Bull. Torrey Bot. Club **41:** 6. 1914. — "Mountains of northern Oriente [Cuba]. Type from near Woodfred, Sierra Nipe, 450–550 m altitude, Dec. 20, 1909 [fr.], ([*J. A.*] *Shafer 3220*)." — Holotypus, NY!; isotypus, US *792379*!. — *Jupunba nipensis* (Britton) Britton & Rose, N. Amer. Fl. **23:** 28. 1928.

P. nipense sensu Bisse, 1988: 223.

Arborescent *shrubs*, flowering first as shrubs as small as 1.2 m, but attaining 9 m, closely related to *A. oppositifolia* and resembling it in indumentum and in subsessile, strongly bicolored lfts, but differing in smaller lvs, in smaller, dorsally 1-nerved lfts, in smaller slender-pedicellate fls, and in many fewer stamens. *Lf-formula* iv–vi/(13–)15–18; lf-stks 2–4 cm, the petiole 2.5–6 mm, the longer interpinnal segments 1–2 mm; first nectary of lf-stk <1 mm diam, sometimes lacking; lf-rachis 2–3.5(–4) cm, and the longer interfoliolar segments 1–2 mm; larger lfts 3–5 × 1.3–1.8 mm. *Pedicels* of proximal fls 4–6 × 0.2–0.3 mm; calyx turbinate-campanulate ±2.4–2.5 × 1.3–1.4 mm, the obtusely deltate teeth ±0.5 mm; corolla ±5.5 mm, the narrowly ovate lobes 2–2.2 × 0.9–1.3 mm;

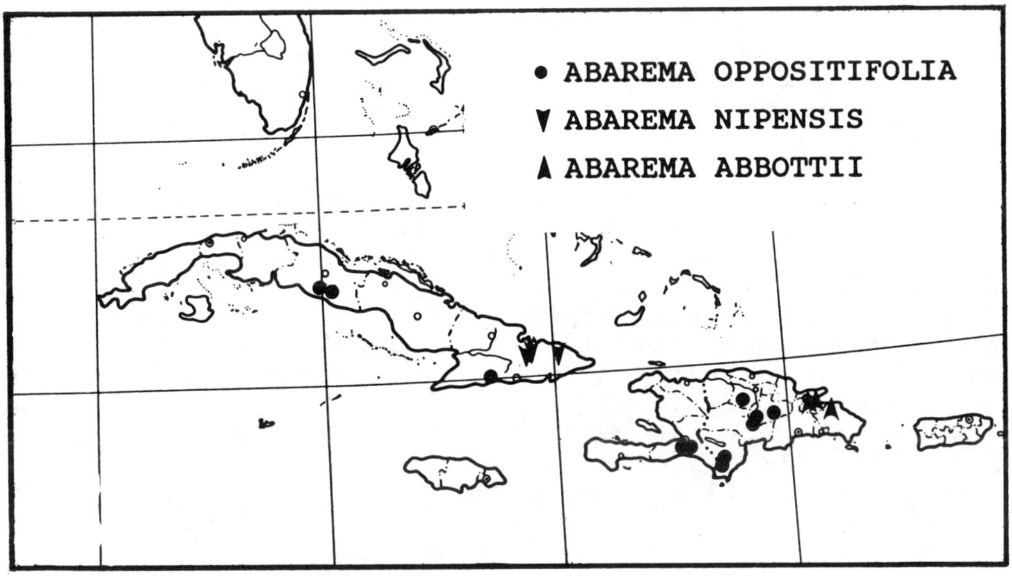

MAP 30. Cuba and Hispaniola. Range of *Abarema oppositifolia* (Urban) Barneby & Grimes, *A. nipensis* (Britton) Barneby & Grimes, and *A. abbottii* (Rose & Leonard) Barneby & Grimes.

androecium 18–24-merous, the stemonozone ±1 mm, the tube 3–3.5 mm, the longest filaments ±16–19 mm. *Pods* and seeds essentially like those of *A. oppositifolia.*

In open pine woods, thickets, and on rocky ridges on stream banks, probably always on serpentine bedrock, ±450–800 m, known only from Sas de Nipe and Moa and vicinity in E prov. Oriente, Cuba. — Map 30. — Fl. V–X (–?).

Abarema nipensis may be derived, like *A. abbottii*, from *A. oppositifolia* or an immediate precursor, but it is adapted to serpentine, not calcareous, bedrock. In terms of the characters used in the cladistic analysis, it is identical to *A. oppositifolia* but has tiny one-nerved leaflets, smaller flowers borne on longer pedicels, and only half as many stamens to the androecium.

38. Abarema abbottii (Rose & Leonard) Barneby & Grimes, comb. nov. *Pithecolobium abbottii* Rose & Leonard, J. Wash. Acad. Sci. **17:** 254. 1927. — "Collected near San Lorenzo Bay, on the south coast of Samaná Bay, Dominican Republic, April 29, 1922, by W. L. Abbott (no. 2258)." — Holotypus, US *1079252*!; isotypi, NY!, US!. — *Jupunba abbottii* (Rose & Leonard) Britton & Rose, N. Amer. Fl. **23:** 27. 1928.

Microphyllidious *trees* 4–12 m with rough gray furrowed bark, the young branches densely, the lf-axes and inflorescence more thinly furfuraceous-puberulent with incurved, brownish or sordid hairs <0.2 mm, the crowded subsessile, commonly subopposite lvs strongly bicolored, the firm, ventrally convex lfts bright green (when dry dark brown) glabrous and lustrous above, beneath pallid dull and glabrous except for a tuft of hairs in anterior basal angle of midrib, the subumbellate capitula of white fls borne solitary in few distal lf-axils, immersed in foliage. *Stipules* 0 (no scars seen). *Lf-formula* of adult fertile branches iv–vi/8–12 (of some sterile branches, no further described, to x–xi/13); lf-stks 2–6 cm, low-convex dorsally, narrowly ridged and grooved ventrally, shallow-sulcate laterally, the true petiole almost 0 (and pulvinus obscurely differentiated), the longer interpinnal segments 7–12(–14) mm; petiolar nectaries sessile or almost so, cupular thick-rimmed 0.3–0.7 mm diam, lacking between first and often between second pair of pinnae but always present distally, smaller nectaries between some distal lft-pairs; pinnae decrescent at each end of lf-stalk but more abruptly so distally, the longer ones 2–4 cm, the longer interfoliolar segments 2.5–5 mm; lft-pulvinules in dorsal view 0.2–0.4 mm; lfts sessile against pinna-rachis, a little decrescent proximally but otherwise subequiform, the blade oblong-obovate or broadly oblong-elliptic from inequilateral broad-cuneate base, broadly rounded at apex, the longer ones 5–8.5(–10) × 3–5 mm, (1.4–)1.5–2.2(–2.5) times as long as wide; the straight subcentric midrib depressed ventrally, prominulous dorsally and there giving rise on each side to 2–3 weak secondary venules faintly brochidodrome within the revolute margin, tertiary venulation fully immersed. *Peduncles* more or less compressed (1–)1.5–4 cm; capitula ±20–30-fld, the terminal fl sometimes sessile and a trifle larger but its androecium scarcely modified, the receptacle narrowly clavate 2–3 mm long; bracts subulate 0.4–0.9 mm, puberulent, caducous; proximal pedicels 2–4 × 0.4 mm, the distal ones either nearly as long or progressively shorter; perianth 5-merous, externally glabrous except for microscopically ciliolate calyx-teeth and thinly puberulent corolla-lobes; calyx turbinate-campanulate 2.2–3 × 1.4–1.8 mm, the depressed-deltate teeth 0.2–0.3 mm; corolla 6–7 mm, the ovate lobes 1.8–2.5 × 1–1.3 mm; androecium 27–36-merous, the longest filaments 16–21 mm, the stemonozone 0.6–1 mm, the tube 2.5–4 mm; ovary glabrous truncate. *Pods* usually solitary, subsessile but narrowed into a basal neck, the body in profile undulately linear, 5–8 × 0.8 cm, evenly recurved through a full circle, to 12-seeded, the prominent sutures <1 mm wide, the thinly leathery, fuscous glabrous, venulose valves low-convex over each seed, the endocarp dull chestnut-red; seeds described as blue (not seen mature), on a coiled funicle, pleurogram complete.

In broadleaf woodland on limestone, ranging from near sea level up to 800 m, localized in NE Dominican Republic, around the S shore of Samaná Bay and W into Sa Quita-espuela, in depts. Samaná and El Seibo. — Map 30. — Fl. IV–VII.

Abarema abbottii is closely related to *A. oppositifolia*, of which it could be interpreted as a subspecies modified in response to a xeric environment. Smaller, more crowded leaves, leaflets fewer by a few pairs, suppression of the petiolar nectary, slightly smaller flowers, and endocarp pigmented overall are the weak morphological criteria that coincide with its allopatric dispersal in northeastern Hispaniola.

The following species (nos. 39–45) are unknown in fruit and were not included in the phylogenetic analysis.

39. Abarema levelii (Cowan) Barneby & Grimes, comb. nov. *Pithecellobium levelii* Cowan, Mem. New York Bot. Gard. **10:** 69, fig. 48. 1961. — "VENEZUELA: Amazonas: Caño Cupaven, opposite mouth of Río Atabapo, alt. 150 m . . . 11 May 1954, *J. Silverio Level 73* . . . along Río Orinoco just about [above] mouth of Río Atabapo, alt. 125 m. . . 1 Jun 1959, *Wurdack & Adderley 42730*." — Holotypus, *Level 73*, US *2281812*!; isotypi, NY (2)!; paratypi, (*Wurdack & Adderley 42730*), NY!, P!, US!, VEN!.

Macrophyllidious unarmed *trees* ±8–10 m with vertically wrinkled branchlets, the young stems and all axes of the terminal, shortly exserted panicle of few- but long-fld capitula densely brownish-puberulent, the amply 4-foliolate lvs glabrous, the lfts lustrously olivaceous and venulose on both faces, a little darker above, the capitula borne singly or 2–3-nate in the axil of a short efoliolate lf-stk charged near apex on ventral side with a gross verruciform nectary. *Stipules* 0 (no scar). *Lf-formula* i/1; lf-stks including livid wrinkled pulvinus 5–23 mm, at middle 1–1.6 mm diam, shallowly openly sulcate ventrally; a sessile or impressed, shallowly concave, elliptic nectary 2–3 mm in long diameter at apex of petiole, and a similar, slightly smaller one at tip of each pinna-rachis; rachis of pinnae 12–35 mm; lft-pulvinules livid wrinkled 4–8 × 1.1–1.7 mm; paraphyllidia 0; lfts ovate from equilaterally rounded or slightly oblique base, bluntly short-acuminate, 7–12 × 3–6 cm, 1.9–2.8 times as long as wide; the straight centric midrib giving rise on each side to 3–5 strong incurved-ascending and many, much finer, widely ascending secondary nerves, the strong ones brochidodrome close within the plane corneous margin, these all engendering a mesh of 3–4-nary venules, the whole venation prominulous on both faces, strongly so beneath. *Primary axis* of inflorescence ±4–10 cm, its branches about half as long; peduncles ±20–33 mm; capitula densely 6–14-fld, the fls ascending, the receptacle including terminal pedestal 1.5–3 mm; bracts deltate-ovate ±0.5–0.75 mm, caducous; fls dimorphic, all sessile, the perianth of all 5-merous, densely appressed-strigulose with minute sordid-bronze hairs; PERIPHERAL FLS: calyx campanulate 3–3.5 × 1.4–1.8 mm, the ovate, often unequal teeth to 0.5 mm; corolla tubular-infundibuliform 11.5–15.5 mm, the involutely lanceolate lobes ±4 × 1.3 mm; androecium in the type collection ±30-merous, nearly 4.5 cm, the filaments white, the stemonozone ±1.7 mm, the tube ±14 mm, in Brazil (scarcely known) shorter, the tube ±8 mm; ovary very slenderly fusiform, minutely puberulent; ovules 9; TERMINAL FL: calyx 3–4 × 2.5–3 mm, corolla ±13 mm; androecium ±45-merous, the tube 11.5 × 2 mm. *Pods* unknown.

In blackwater igapó woodland, 120–150 m, known from only two localities: at mouth of Río Atabapo in

T. F. Amazonas, Venezuela (near 4°N); and on Río Abacaxis in SE state of Amazonas, Brazil (near 4°22′S). — Map 31. — Fl. V–VII.

In the protologue of *Abarema levelii*, Cowan found no close relative in *Pithecellobium* sens. lat., but remarked that the species differed from all congeners in leaflet-number and inflorescence. In reality it is not essentially different in either of those features from *A. leucophylla*, the reduction of distal leaves to a rudimentary leaf-stalk with nectary being the same in both, and a leaf-formula of i/1, apparently stablized in all leaves of *A. levelii*, being frequent in distal leaves and sometimes uniform in all leaves of *A. leucophylla*. Both species have the dimorphic flowers of *Abarema*. A shadow of doubt will linger over the attribution of *Pithecellobium levelii* to *Abarema* until the fruit and seeds are procured, as flowering specimens are a little unusual in *Abarema* for the length of the peripheral corollas. However, the species is otherwise fully compatible with our concept and definition of *Abarema*, and it cannot be accommodated in any other described genus of Ingeae.

The one Brazilian collection of *A. levelii* (*S. Hill 12937*, GH, US), from the middle Amazon basin at a point 1400 km distant from the type locality, seems to differ from the Venezuelan plants in slightly shorter corolla and androecium, but it is a good match for the type in other respects. Obviously much remains to be learned about the morphology and dispersal of *A. levelii*.

Because of a technical difference in leaf-apex, *A. levelii* is contrasted in our key to species with *A. laeta*, but not with its seemingly closer relative *A. leucophylla* var. *leucophylla*, which may be separated as follows:

1. Lfts bicolored, lustrous above, pallid dull beneath, broadly rounded at apex; corolla of peripheral fls of the capitulum 3–6.5 mm, not more than twice as long as calyx .... *A. leucophylla* var. *leucophylla*
1. Lfts concolorous, lustrous on both faces, shortly bluntly acuminate; corolla of peripheral fls 11.5–15.5 mm, 3–4.5 times as long as calyx .... *A. levelii*

40. Abarema oxyphyllidia Barneby & Grimes, sp. nov., *Ab. barbourianam* foliolorum numero et magnitudine simulans, sed ab ea imprimis foliolis anguste ovatis acutis, capituli floribus periphericis pedicello 2.5 mm usque elevatis, necnon ovario glabro distincta, ulterius a var. *barbouriana* (costaricensi et panamensi) corolla fere 10 (nec 5–7) mm longa distans. — HONDURAS. La Paz: Cordillera Guajiquiro, 5 km a Sabanetas, 21.V.1964 (fl.), *A.*

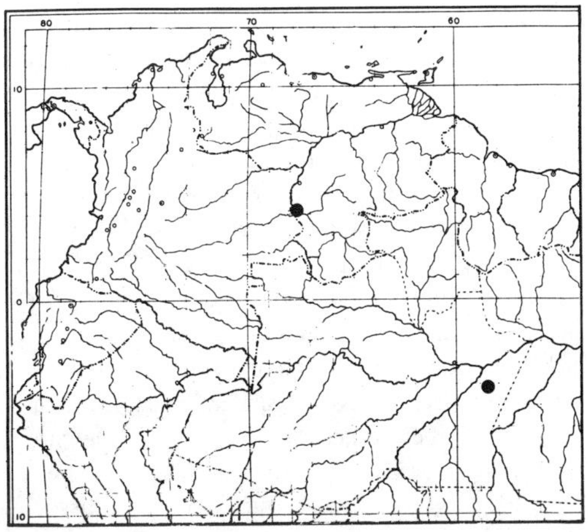

Map 31. The upper Orinoco, Venezuela, and central Amazonia, Brazil. Range of *Abarema levelii* (Cowan) Barneby & Grimes.

Molina R. & Albertina R. Molina 13921. — Holotypus, NY; isotypus, F *1638723*!.

Microphyllidious *trees* 5–20 m with trunk to 5 dm dbh, resembling *A. barbouriana* in habit, lf-formula, and brown indumentum of young stems, lf-axes and inflorescence, but the bicolored lfts acute and dorsally barbellate below middle on anterior side of midrib only, otherwise facially glabrous, remotely ciliolate, the peripheral fls pedicellate, and the ovary glabrous. *Stipules* fugacious, not seen. *Lf-formula* viii–ix/17–21; lf-stks 8–11 cm, the petiole and longer interpinnal segments ±1 cm; nectaries at or close below each pair of pinnae, sessile cupular, the first ±1 mm diam, the rest hardly smaller, a very small nectary at tip of few pinnae but this commonly lacking; pinnae a little decrescent at each end of lf-stk, the rachis of longer ones 5–6 cm, the longer interfoliolar segments ±3 mm; pulvinules 0.1 × 0.3–0.4 mm, the lfts sessile against pinnae-rachis; lfts slightly decrescent at each end of rachis, in outline obliquely lance-ovate from inequilaterally flabellate base, sharply triangular-acute, the larger ones ±6.5–7.5 × 2.3–2.5 mm; the subcentric, obscurely sigmoid midrib shallowly depressed ventrally, cariniform dorsally, giving rise on each side to 3–5 faint secondary veins almost imperceptibly anastomosing close within the almost plane margin. *Peduncles* solitary and paired in coeval lf-axils, 3–4.5 cm; capitula ±15–20-fld, the narrowly clavate receptacle attaining 5 mm; bracts not seen; the peripheral fls pedicellate, the terminal one (not seen expanded, apparently not greatly differentiated) sessile, the

5-merous perianth of all densely brown-strigulose-pilosulous; PERIPHERAL FLS: calyx deeply campanulate 5 × 2.3 mm, the lance-subulate teeth ±1.5 mm; corolla ±9.7 mm, the lobes 3 × 1.2 mm; androecium white, 40-merous, the stemonozone 1.5 mm, the tube 5–7.5 mm, the longest filaments 19 mm; ovary subsessile, conic at apex, glabrous. *Pods* unknown.

In mixed forest near 2100 m, known only from Cordillera (Sierra) Guajiquiro, ±14°10′N, 87°50′W, in SW Honduras. — Fl. V–VI.

While it would obviously be desirable to have more complete material, especially the fruit, of *A. oxyphyllidia* before describing it as new, the sum of its known characters clearly sets it apart from any ingeoid species in Central America. In general facies, due to great number of pinnae and leaflets, it resembles *A. barbouriana*, but differs in narrowly ovate, sharply acute, (not oblong-elliptic obtuse) leaflets, subumbellate (not sessile) flowers, longer peripheral corollas, and a perfectly glabrous ovary. *Abarema idiopoda* is more easily distinguished by the small campanulate calyx and long slender pedicels of the peripheral flowers combined with obtuse leaflets and mostly lowland habitat.

41. Abarema villifera (Ducke) Barneby & Grimes, comb. nov. *Pithecolobium villiferum* Ducke, Arq. Inst. Biol. **4:** 5. 1938.–"Habitat ad ripas inundabiles fluminis Curicuriary (Rio Negro affluentis) super cataractas, 26-2-1936 leg. *A. Ducke*, H[erb]. J[ard]. B[ot]. R[io de Janeiro] 35534." — Holotypus, RB!; isotypi, P!, U!, US!.

Macrophyllidious *trees* 4–20 m, the new growth, except for lustrous glabrous upper face of strongly bicolored, at maturity subcoriaceous lfts, villosulous throughout with erect or curved, mostly golden-brown or sordid-whitish hairs to 0.4–0.7 mm, the capitula either solitary or paired in the axil of coevally expanding lvs, early immersed in foliage. *Stipules* linear or linear-lanceolate 2–3 × 0.3 mm, caducous. *Lf-formula* i–ii/2–3; lf-stks, including a scarcely differentiated pulvinus, 1–3 cm, the one interpinnal segment, when present, 7–23 mm, about as long as petiole; petiolar nectaries sessile or almost so, the first at or close below first pair of pinnae, plane or low-convex 1–2.4 mm diam, a similar one often near tip of lf-stk and yet smaller, shallow-concave ones on 1–2 furthest segments of pinna-rachis; pinnae when 2 pairs strongly accrescent, the rachis of the one or of the further one 25–45 mm, the longer interfoliolar segment 9–18 mm; subulate paraphyllidia at base of each pinna-rachis; lft-pulvinules 1–2.2 × 0.8–1 mm; lfts strongly accrescent distally, obliquely obovate-elliptic, or the proximal ones rhombic-ovate, from asymmetrically broad-cuneate or semicordate base, broadly rounded or shallowly emarginate, conspicuously revolute, the distal pair 28–70 × 13–36 mm, 1.4–2.1 times as long as wide; the gently incurved midrib little displaced, the secondary venation variable, described in key to variants. *Peduncles* 3–6 cm; capitula 9–15-fld, the receptacle including terminal pediment 2–3 mm; bracts linear or narrowly oblance-elliptic 1.2–3 × 0.3–0.6 mm, deciduous; fls dimorphic but all sessile or almost so, the white or greenish perianth of all 5-merous and densely subappressed-silky externally; PERIPHERAL FLS: calyx turbinate-campanulate 3–4.2 × 1.8–2.4 mm, the triangular teeth ±1 mm; corolla ±5–6.5 mm, the lanceolate lobes 1.8–2.3 mm; androecium 20–24-merous, 21–31 mm, the stemonozone <1 mm, the tube 1.7–3 mm; TERMINAL FL: usually solitary, sometimes paired, the calyx broadly campanulate 4–5.5 × 2.5–3 mm, the corolla 7–8 mm, the staminal tube 8–10 mm; ovary symmetrically ellipsoid, conical apically, densely silky-pubescent overall. *Pods* unknown.

In riparian forest subject to flooding, below 175 m, local on the upper forks of Río Negro in NW Amazonas, Brazil, and adjoining T. F. Amazonas, Venezuela, between 3°N and 1°20′S latitude; and (variant, see commentary below) in terra firme forest of lower Río Negro basin (Río Cueiras) in ±2°45′S. — Map 14. — Fl. II, IV, VII, the whole cycle unknown.

Abarema villifera is readily recognized by few ample leaflets softly silky-villous beneath and by few-flowered capitula. In foliage it closely resembles the upland and allopatric *A. commutata*, which see for comment. As *A. villifera* is still unknown in fruit, it is not included in the cladistic analysis, and any relationship implied by its striking similarity to *A. commutata* remains unconfirmed.

On the upper Río Negro, the leaflets of *A. villifera* are pinnately veined, all minor venation being immersed in the epidermis. One collection from Río Cueiras, a tributary of the lower Río Negro shortly upstream from Manaus, differs strikingly in elaborately venulose leaflets and when better known will probably emerge as taxonomically distinct. The observed differences are as follows:

1. Venation of lfts simply pinnate, the secondary nerves from midrib 6–7 pairs, unbranched; upper Río Negro in SW Venezuela and adjacent Brazil ....variant a, typical

1. Venation of lfts reticulate, the secondary nerves from midrib 12 pairs or more, giving rise to a mesh of tertiary and quaternary venules prominulous on both faces; lower Río Negro in central Amazonas, Brazil .... variant b

42. Abarema josephi Barneby & Grimes, sp. nov., *A. lehmannii* arcte affinis, imprimis foliorum formula diversa: pinnae 2–3-, nec 1–2-jugis; foliola 5–7-, nec 3–5-juga, ergo cujusque folii 44–68 (nec 12–36). — COLOMBIA. Comisaría del Caquetá: Cordillera Oriental, vertiente orental, Quebrada del Río Hacha, bosques en Ruidosa, 2000 m, 26 Mar 1940 (fl jun), *J. Cuatrecasas 8707.* — Holotypus, US *1796398*; isotypus, *1299131* = photo (*s.n.*), NY.

Macrophyllidious *trees* 6 m, the young stems densely brownish golden-velutinous, the stiffly papery bicolored lfts brown-olivaceous and strigulose along major veins above, beneath pallid and thinly strigulose overall, the flower-spikes arising singly from the axil of coeval lvs and surpassed by them. *Stipules* 0. *Lf-formula* ii–iii/5–7, the lfts 44–68 per lf; lf-stks 4.5–8.5 cm, the petiole 30–40 × 1.8–2.6 mm, the longer interpinnal segments 2–2.5 cm; petiolar nectaries between each pair of pinnae sessile orbicular, shallowly cupular 0.8–1.4 mm diam, and similar ones between 1–3 furthest pairs of lfts; pinnae distally accrescent, the rachis of furthest pair 10–12 cm, the interfoliolar segments ±14–19 mm; pulvinules 1.5–2 × 1–1.2 mm; lfts strongly accrescent distally, subsymmetrically ovate or broad-elliptic from inequilaterally cordate or broadly cuneate base, very shortly or obscurely deltate-acuminate at apex, the distal pair 4.5–6.5 × 2.6–3.2 cm, 1.7–2.1 times as long as wide; the midrib almost straight, the 6–7 pairs of major secondary veins brochidodrome well within the revolute margin, these and a reticulum of venules all immersed or engraved on upper face, sharply raised beneath. *Spikes* (few seen) 70–95-fld, the peduncle and rachis together ±9 cm; bracts 0.4–0.6 mm, deciduous; fls sessile, subhorizontal, the 5-merous perianth brown-velutinous overall; calyx campanulate ±2.2 × 1.6 mm, the depressed-deltate teeth 0.3–0.5 mm; corolla tubular slightly dilated distally 6.5 mm, the ovate unequal lobes 1–1.7 × 1.2 mm; androecium 35-merous ±11 mm, the tube a little shorter than corolla, the filaments presumably white when fresh, brownish when dried; ovary narrowly ellipsoid, conical at apex, glabrous at anthesis. *Pods* unknown.

In montane woodland at 2000 m, known only from the type locality near the W boundary of Comisaría del Caquetá at ±1°52′N, 75°40′W in Colombia. — Map 24A. — Fl. III–IV(–?).

Abarema josephi is distinguished from the nearly related *A. lehmannii* by higher leaf-formula and consequently by more numerous, smaller leaflets. Additional collections, and mature fruits, are required to confirm its specific status. The species is dedicated to Dr. José Cuatrecasas, whose name, like that of Celestino Mutis, is imperishably connected with the Colombian flora.

43. Abarema centiflora Barneby & Grimes, sp. nov., foliorum pinnis 1-jugis foliolisque 5-jugis ovato-acuminatis 6.5–8 cm longis usque, necnon racemis multifloris densis ±7 cm usque longis insignis, ab affini *A. lehmannii* floribus racemosis (nec spicatis) absimilis. — BOLIVIA. Depto Cochambamba, prov. Chapare: El Limbo, 200 m, 25 Nov 1966 (fl), *Roy F. Steinbach 538.* — Holotypus, NY; isotypi, F *1645245*, GH, MICH, US *2533435*.

Macrophyllidious *trees* 8 m, with longitudinally furrowed branchlets, except for lf-faces minutely sordid-puberulent throughout and the indumentum extending upward along the primary lf-veins, the foliage bright green subconcolorous, the relatively dense spiciform racemes of small greenish, white-stamened fls arising solitary in the axil of few coeval lvs, paired with and distal to an inhibited branch-bud. *Stipules* apparently 0 (no scars). *Lf-formula* i/(2–)4–5; lf-stks 2–6.5 cm; lf-nectaries at tip of lf-stk and at insertion of lfts plane round ±1–1.5 mm diam, flush with or almost embedded in epidermis; rachis of pinnae to 13–15 cm, the longer interpinnal segments 2.5–3 cm; lft-pulvinules in dorsal view 3.5–4 × 1–1.5 mm; lfts moderately accrescent distally, the blades ovate-short-acuminate from cuneate-flabellate base, the larger ones ±6.5–9 × 3–3.5 cm; the subcentric midrib either straight or gently curved forward toward lft-apex, giving rise to 5–7 pairs of major, incurved-ascending secondary veins brochidodrome well within the nearly plane margin and to many minor intercalary, widely divergent ones, all these together with connecting tertiary and reticular venules prominulous on both faces. *Peduncles* 1.5–4 cm; racemes ±80–100-fld, the axis ±7–9 cm; bracts minute, shed long before anthesis; pedicels subhorizontal, the lowest ±1 × 0.6 mm, the distal ones progressively a little shorter; perianth 5-merous, finely minutely puberulent externally; calyx deeply campanulate ±2.6 × 1.6 m, the low-deltate obtuse teeth 0.3–0.5 mm;

corolla 4.8 mm, the lance-ovate, nearly erect lobes 1.6 × 1 mm; androecium 32-merous, 11 mm long, the stemonozone scarcely 1 mm long, the tube 4.5 mm (nearly as long as corolla); ovary glabrous, obliquely conical at apex; style a little shorter than longest stamens, the stigma scarcely dilated, a little more than 0.1 mm diam. *Pods* unknown.

In humid montane woodland near 2200 m, known only from the E slope of the Bolivian Andes near 17°S in prov. Chapare of dept. Cochabamba. — Map 24A. — Fl. XI–XII.

Abarema centiflora, the southernmost known species with long-spicate inflorescences, appears fully distinct from its Colombian and Ecuadoran relatives either in leaf-formula or in racemose (not spiciform) units of inflorescence, or in both, but it is known as yet only from the flowering type collection and will require reassessment when the fruit is secured.

Species non satis cognitae

44. Abarema ricoae Barneby & Grimes, nom. provis.

NICARAGUA. Zelaya: mun. Rama, camino desde "Santa Julia" hasta "La Palmera." 12°08′–09′N, 84°14′–15′W, 60–95 m, 24 mayo 1984 (fr.), *Walter Robleto 678* (BM).

Macrophyllidious unarmed *trees* 17 m tall, 1.5 dm dbh, the lf-axes and lf-buds densely puberulent with incurved sordid hairs to ±0.1 mm, the lvs moderately bicolored, the lfts lustrous above, paler beneath, glabrous except for microciliolate midrib and margin, the fls (as adduced from fr specimens) borne in dense corymbiform capitula on long axillary peduncles. *Stipules* early deciduous from small pallid scar, the blade not seen. *Lf-formula* (from 2 lvs) iv–v/6; lf-stk ±18 cm, the petiole ±6 cm, the interpinnal segments scarcely shorter; a sessile round, almost plane nectary 2–2.5 mm diam near midpetiole and one between each pair of pinnae, smaller elliptic, more deeply concave ones on pinna-rachises close below furthest 2–3 pairs of lfts; pinnae distally accrescent, the rachis of longer ones 11–12 cm, the longer interfoliolar segments 18–23 mm; lft-pulvinules 2–2.5 × 1.2 mm; lfts accrescent distally, rhombic-ovate around an incurved subcentric diagonal midrib, at inequilateral base cuneate postically, subtruncate antically, at apex bluntly deltate, the penultimate pairs ±3.5–4.5 × 1.7–2.3 mm and about twice as long as wide; the dorsally prominent midrib giving rise to ±10–14 pairs of ascending major primary nerves brochicocrome close within the subrevolute margin and to a reticulum of sinuous venules finely prominulous on both faces. *Peduncles* (1 seen) in fruit ±11 cm; fruiting receptacle together with small terminal pedestal ±5 × 4 mm, the pedicel-scars densely crowded; bracts deciduous, not seen; fls unknown, but the pedestal indicating a terminal fl, and the peripheral ones certainly pedicellate. *Pods* solitary, undulately broad-linear, evenly recurved through more than 1 circle, in profile ±12 × 1.6–1.8 cm and ±12-seeded, the stiffly leathery valves nearly black externally, faintly venulose, framed by sutures ±1.5 mm wide, glabrous, biconvex over each seed, the endocarp smooth and lustrous red-brown in seed locule but this girdled with a band of fuscous papilli, the interseminal segments tan; dehiscence of *Abarema*, the valves elastically separating, coiling and twisting; funicle compressed, sigmoid; seeds not seen.

In disturbed forest at 60–90 m, known only from the one collection, bearing dehisced fruit in May.

In foliage the described specimen closely resembles *Balizia acreana*, and was so identified in 1988 by L. Rico (BM). The fruit, however, borne on a stout peduncle 11 cm long, a woody pedicel nearly 12 mm long, and a densely flowered claviform receptacle, seems incompatible with Amazonian *B. acreana*, the inflorescence of which is a terminal effuse, leafless corymbiform panicle composed of very numerous little corymbiform flower-spikes, with individual peduncle at most 2.5 cm and peripheral pedicels less than 4 mm long. Since the peduncles and pedicels of other abaremas do not elongate appreciably after anthesis, the contrasts in length between those of *Balizia acreana* and "*Abarema ricoae*" are most probably valid, even though made between plants at very different stages of maturity. The position of the peduncle of "*A. ricoae*" cannot certainly be made out from this unique specimen, but it seems to have been borne in the axil of a coeval leaf or immediately below the foliage of the current year. It could not possibly arise as part of a terminal panicle.

The incomplete material precludes full description, but this striking plant cannot go unnoticed. For the present the name "*Abarema ricoae*" is only provisionally proposed, in honor of Lourdes Rico Arce.

45. Abarema agropecuaria Barneby & Grimes, nom. provis.

BRAZIL, Amazonas, Distrito Agropecuario de SUFRAMA north of Manaus, 2°19′S, 60°05′W, *A. V. V. Silva* in INPA/WWF *2206.1860.2* (NY).

Trees of unknown stature, the young branchlets, lf-axes, and units of inflorescence densely brownish silky-velutinous overall, only the ventral face of lfts glabrous. *Stipules* narrowly elliptic, ±3 mm, ephemeral. *Lf-formula* ii/4–6; lf-stks 5–7.6 cm, the petiole 3–4 cm, the one interpinnal segment 2–3.5 cm; a nectary between each pair of pinnae mounded-hemispherical but crumpled when dry, ±1.2–2 mm diam, nigrescent; rachis of distal pinnae 3.5–7 cm, charged at base with a pair of subulate paraphyllidia, the longer interfoliolar segments 11–14 mm; lft-pulvinules 1–1.4 mm; lfts subrhombic-oblong from inequilateral base, attaining ±20 × 9 mm but perhaps longer when fully mature, all faintly pinnate-veined. *Peduncles* solitary or 2–3-nate in axil of coevally expanding lvs, 3–5 cm; capitula incipiently umbelliform, the peripheral fls shortly pedicellate, the terminal fl sessile; perianth in bud resembling that of *A. floribunda*; ovary truncate. *Pods* unknown.

The specimen cited has flush foliage and flowers ready to expand; it is not referable to any known Amazonian species of *Abarema*. The flowers are reminiscent of those of sympatric *A. floribunda* at the same stage of immaturity, but the leaf-formula is significantly lower and the young leaflets are densely brown-velvety dorsally. The specimen is from tree no. *1860* in Reserva of Projeto Dinâmica Biológica de Fragmentos Forestais at ±2°19′S, 60°05′W.

The *Samanea*-alliance

The *Samanea*-alliance was a second group that formed a fully resolved clade in the exemplar analysis (Fig. 3B, no. 2). The outgroup is *Parachidendron pruinosum* Koorders. This clade is distinguished by auxotelic meristems and delayed development of the leaves; that is, the leaves form contemporaneously with the inflorescences, but they complete development more slowly. Most of these species have been variously referred to *Samanea*, *Pseudosamanea* and *Albizia*, and they include *Albizia occidentalis* T. Brandegee, *Pseudosamanea guachapele* (Kunth) Harms, *Albizia cubana* Britton & Rose, *Samanea saman* (Jacquin) Merrill, *Albizia inopinata* (Harms) G. P. Lewis, and *Calliandra tubulosa* Bentham. The six species were included in a phylogenetic analysis. HENNIG86 was again used, but, because of the small size of the data set, ie* (an exhaustive search option) was used. The data set is shown in Table 2; the characters used are as follows (again numbered according to the conventions of HENNIG86):

Table 2

Data set used in the analysis of the *Samanea*-alliance

	0	5	1 0
PARARCHIDENDRON†	0*000	00000	??0
OCCIDENTALE	00100	00010	001
GUACHAPELE	00110	00010	012
CUBANA	00110	-0010	012
SAMAN	01111	11011	111
TUBULOSA	11111	11111	111
INOPINATA	11111	11111	111

Note: † = outgroup; * = taxon polymorphic for this character; ? = state not observed in this terminal; - = state inapplicable or of uncertain homology.

0. Nectaries: 0 = near tip of leaf-stalk; 1 = no nectaries near tip of leaf-stalk.
1. Leaflets: 0 = broadest at or above midblade; 1 = broadest below mid-blade.
2. Leaflets: 0 = alternate; 1 = opposite.
3. Flowers: 0 = homomorphic; 1 = dimorphic.
4. Sutures: 0 = plane; 1 = differentiated into broad keels.
5. Pod: 0 = not septate; 1 = septate between seeds.
6. Mesocarp: 0 = thin and dry; 1 = thick and pitch-like.
7. Exocarp: 0 = smooth and continuous; 1 = transversely fibrous and wrinkled.
8. Valves: 0 = internally red; 1 =internally tan.
9. Endocarp: 0 = forming a continuous chamber; 1 = forming "bubbles" around seeds. State 1 is indistinguishable from the "bubbles" surrounding the seeds in *Enterolobium*. *Samanea saman*, *Albizia inopinata*, and *Calliandra tubulosa* all have seeds surrounded in endocarp packets. The endocarp of one valve wall does not form a continuous layer and the packet appears as a complete balloon, rather than as two layers adhering together.
10. Seeds: 0 = laterally compressed; 1 = more or less terete or quadrangular in cross-section.
11. Seeds in profile: 0 = round- or oblong-elliptic; 1 = narrowly elliptic.
12. Testa: 0 = brown(-yellow); 1 = buff or ivory-white.

The analysis gave one tree of 14 steps and a consistency and a retention index both of 1.00. The tree is shown in Figure 8. *Albizia occidentalis* comes off immediately a sister-group to the rest of the alliance, distinguished solely on the basis of homoplasious

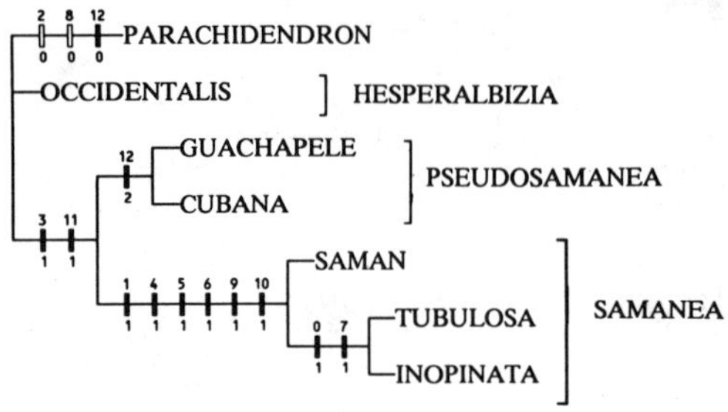

FIG. 8. The tree resulting from an analysis of the *Samanea*-alliance.

characters by which all members of the *Samanea*-alliance differ from the outgroup: the opposite leaflets (no. 2), internally tan valves (no. 8), and brown seed-testa. *Albizia occidentalis* has several unique features in the context of the taxa included in this analysis, including sessile flowers and a complete pleurogram. There are no synapomorphies that place *A. occidentalis* in a clade with any other species. Likewise, other than the developmentally delayed leaves, which are probably homoplastic in the context of the tribe, there are no synapomorphies uniting all six species. As there are significant differences in morphology of fruits and inflorescences, we feel we have no option but to recognize the monotypic genus *Hesperalbizia* Barneby & Grimes (no. IV) for *A. occidentalis.*

The remainder of the taxa form a clade united by dimorphic flowers (no. 3) and seeds narrowly elliptic in profile (no. 11). *Albizia cubana* and *Pseudosamanea guachapele* form a clade with buff or ivory-white seed-coats. They are treated below in genus *Pseudosamanea* (no. V). It should be noted that the fruits of these two species are quite different: those of *Ps. guachapele* are inertly dehiscent through both sutures and aseptate inside, whereas those of *Ps. cubana* are (sub)indehiscent and sometimes septate inside. The septa of *cubana* represent, however, an autapomorphy distinct from the septa found in genus *Samanea* (see below). In *Samanea*, the septa apparently derive from the mesocarp, which pushes through and around and totally surrounds the "bubbles" containing the seeds. In *Ps. cubana* the septa are weakly formed, and it appears that they are derived from endocarp, not mesocarp. The development of the fruits of both groups need study.

Samanea saman (Jacquin) Merrill, *S. inopinata* (Harms) G. P. Lewis, and *Calliandra tubulosa* Bentham form a monophyletic clade united by five synapomorphies: leaflets broadest below midblade (no. 1), sutures differentiated into broad keels (no. 4), a pitchlike mesocarp (no. 6); an endocarp forming bubbles around the seeds (no. 9), and seeds more or less terete or quadrangular in cross-secton (no. 10). These taxa are included in genus *Samanea* (no. VI).

IV. HESPERALBIZIA
BARNEBY & GRIMES

Hesperalbizia Barneby & Grimes, gen. nov. monotypicum, signis sequentibus inter affines praestans: stipulae parvae membranaceae basi cordatae; foliola pinnatim venulosa; cujusque capituli flores parvi homomorphi sessiles: ulterius a *Pseudosamanea* legumine longistipitato pleurumque 8–12 (nec 12–30)-spermo seminumque areola clausa absimile. — Sp. unica: *H. occidentalis* (T. S. Brandegee) Barneby & Grimes - *Albizzia occidentalis* T. S. Brandegee. — Etymology: Gr. *hesperos*, the evening star, by extension, the West + the generic name *Albizia.*

Xerophilous but macrophyllidious, drought-deciduous *trees*, the *inflorescence* precocious, preceding or coeval with flush of new foliage, the terminal meristem continous beyond it. *Stipules* small, membranous (deciduous), cordate at base. *Lf-formula* iii–viii/5–10; lft-venation pinnate. *Fls* capitulate, sessile, homomorphic. *Pods* long-stipitate, broad-linear plano-compressed, 8–12(–13)-seeded, dehiscence inert, through both sutures; pleurogram of seed-faces complete. — Sp. 1, of SW Mexico.

Although the plano-compressed, stiffly papery pod of *H. occidentalis* is of a type conventional in *Albizia* sens. lat., the species is morphologically isolated by the syndrome of characters listed in the diagnosis. Especially noteworthy is the absence of a modified central flower, which we interpret, in the context of the albizioid Ingeae, as a derived state. In general aspect, *H. occidentalis* somewhat resembles the paleotropical *Albizia lebbeck*, but this differs in linear-oblong stipules, palmate-pinnate (not simply pinnate) venation of leaflets, and pediceled peripheral flowers.

1. Hesperalbizia occidentalis (T. S. Brandegee) Barneby & Grimes, comb. nov. *Albizia occidentalis*, Proc. Calif. Acad. Sci., ser. 2, **3:** 222. 1892. — " . . . in the Cape Region of Baja California . . . along the base of the mountains from Todos Santos on the west to San Jose on the south and Miraflores on the east . . . first collected by Xantus . . . . Señor Cypriano Dodero . . . sent me flowers from the trees growing along the plaza of San Jose." — Lectotypus, "San José del Cabo, from Cipriano Dodero" *UC 80581*; paratypi, "San Jose del Cabo, Sept. 18. 1890, leg. T. S. Brandegee," *UC 80580*, and "Sierra de la Laguna, Jan. 27. 1890, leg. T. S. Brandegee," *UC 80579*.

(?)*Leucaena plurijuga* Standley, Contr. U.S. Natl. Herb. **20:** 189. 1919. — ". . . collected at Monte León, Michoacán, Mexico, November 12, 1892 by C. G. Pringle (n. 5352)." — Holotypus, US *246386*!. — *Albizzia plurijuga* (Standley) Britton & Rose, N. Amer. Fl. **23:** 48. 1928.

Albizzia (?) *obliqua* Britton & Rose, N. Amer. Fl. **23:** 47. 1928. — "Ymala, Sinaloa, September or October, 1891, *E. Palmer 1723*." — Holotypus (fr.), NY!; isotypus, US *305335*!. — Equated with *A. occidentalis* by McVaugh, Fl. Novo-Galiciana **5:** 144, fig. 19. 1987.

Albizzia occidentalis sensu Standley, 1922: 390; Wiggins, 1964: 591.

Macrophyllidious, drought-deciduous *trees* attaining 20(–30) m but flowering when much smaller, the trunk smooth gray, the branchlets brown pallid-lenticellate, the foliage bicolored, the papery-membranous lfts light green and glabrous above, beneath paler and thinly puberulent along veins to finely subappressed-pilosulous, the inflorescence of small green and whitish capitula arising either from the axil of hysteranthous lvs or from efoliate nodes of new branchlets, the papery pods long persistent below mature foliage and on defoliate branches. *Stipules* membranous cordate ±1–1.5 mm, weakly 5–7-nerved, deciduous as associated lf expands, absent from mature spms. *Lf-formula* iii-v/5–8 [in Chiapas vii-viii/10]; lf-stks (6–)8–23.5 cm, the petiole (2.5–) 3–8 cm, at middle 0.8–2.4 mm diam, the longer interpinnal segments ±1.5–4.5 cm, the narrow ventral groove continuous between pinna-pairs; nectary below mid-petiole, sometimes close to the lf-pulvinus, sessile, round or elliptic, shallowly concave, 1.6–3.3 mm diam, in profile 0.2–0.7 mm tall, and commonly a smaller one near tip of lf-stk and a yet smaller one near tip of pinna-rachises; pinnae proximally decrescent, the rachis of further pairs 4–12(–15) cm, the longer interfoliolar segments ±1–2 cm; lft-pulvinules 1–3 × 0.4–0.8 mm; lfts accrescent distally, ovate-apiculate to subrhombic-ovate or -elliptic or oblance-elliptic and abruptly acuminulate, all from moderately inequilateral base, the penultimate pair 2.5–5.3 × (0.9–)1–2.4 cm; venation pinnate, the nerve arising from pulvinule on each side of midrib shorter and weaker than the first pair of secondaries, the gently incurved, centric or slightly forwardly displaced midrib giving rise on each side to (4–)5–7 secondary nerves incurved-ascending to anastomosis shortly within the plane or incipiently revolute margin, the whole venation pallid and sharply finely prominulous on both faces. *Peduncles* solitary or 2–5-nate, (2–)3–6.5 cm; capitula ±10–20(–?)-fld, the fls sessile homomorphic; bracts very small and early caducous; perianth pale green, 5-merous, either glabrous or microscopically puberulent externally; calyx campanulate 2.2–3.3 × 1.1–2.5 mm, the ovate or low-deltate acute teeth 0.5–0.6 mm; corolla narrowly vase-shaped 8.5–10 mm, the lobes 1–2.8 mm; androecium 52–76-merous, 20–24 mm, the stemonozone 0.8–2.3 mm, the tube 5–11 mm, either included or well exserted, free filaments white capillary; stipe of linear-elliptic glabrous ovary ±1–1.5 mm; style shortly exserted from longer stamens, very slightly dilated at tip. *Pods* solitary or paired, broad-linear or linear-elliptic plano-compressed, the body 14–25(–27) × 2.2–4.2 cm, cuneately contracted at base into a stipe 1–3 [in Chiapas –4] cm, irregularly attenuate and often sterile toward apex, 8–12(–13)-seeded, the stiffly papery, stramineous or purplish-castaneous, densely cross-venulose valves framed by essentially straight, dorsally plane sutures 1.2–1.7 mm wide, scarcely convex [but in Chiapas bullate over seeds], internally pallid smooth, the cavity continuous; seeds transverse at middle of pod on filiform funicle, plumply discoid or oblong-ellipsoid 8–9 × 6–8 × 2.5–3 mm, the crustaceous testa brown sublustrous (sometimes cracking spontaneously when dried), the complete pleurogram 2–4 mm diam.

In deciduous woodland, from near sea level to 1500 m, locally plentiful in tropical and marginally subtropical W Mexico: Cape Region of Baja California Sur, southward from La Paz, and sometimes planted; coastal plain and foothills of Pacific slope from centr. Sinaloa S and E to W Jalisco, Colima, and Oaxaca; a probably allied form at middle elevations in Chiapas; cultivated in Honduras (Comayagua). — Map 32. — Fl. after rains, the fr. long retained on leafless branches. — *Palo fierro* (Sinaloa).

The populations of *H. occidentalis* in Chiapas, as noted parenthetically in our description of the species, seem to differ somewhat from those farther north, but we have seen only incomplete material and must leave their taxonomic status an open question.

V. PSEUDOSAMANEA Harms

Pseudosamanea Harms, Notizbl. Bot. Gart. Berlin-Dahlem **11:** 54. 1930. — Sp. typica (unica): *Ps. guachapele* (Kunth) Harms ≡ *Albizzia guachapele* (Kunth) Dugand.

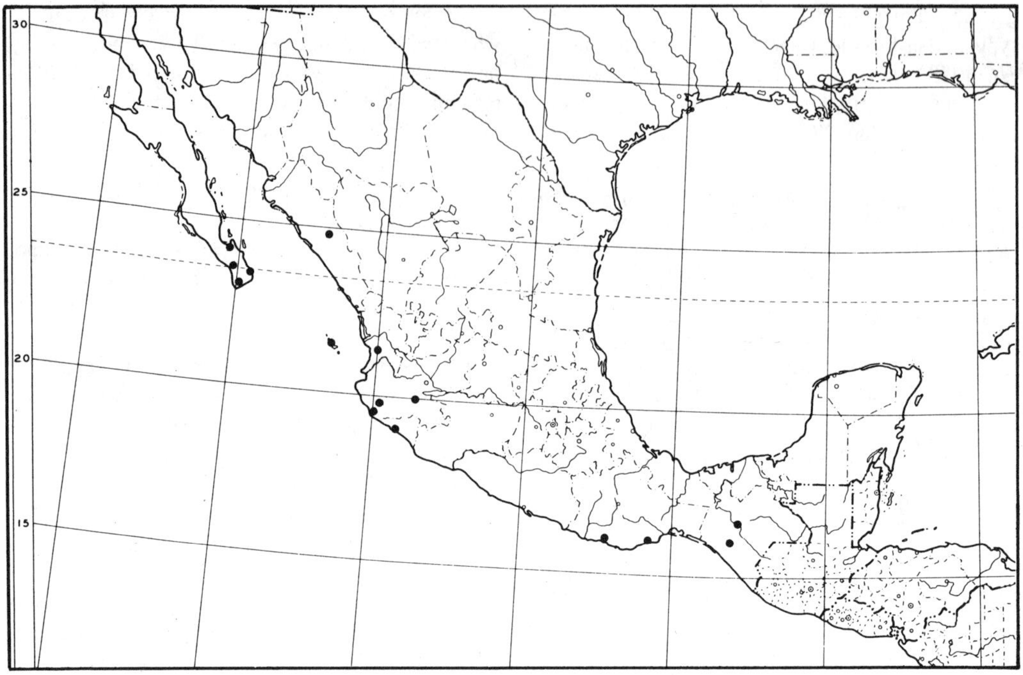

MAP 32. Mexico. Range of *Hesperalbizia occidentalis* (T. S. Brandegee) Barneby & Grimes.

Pseudosamanea sensu Britton & Killip, 1936; 130.
Albizia sensu Nielsen, 1981; 180, minima ex parte, fig. 1(10) sub titulo *Pseudosamanea*.

Macrophyllidious deciduous *trees*, the inflorescence composed of umbelliform capitula axillary to coeval lvs, the meristem continuous beyond them. *Stipules* lanceolate caducous. *Lf-formula* iii–vi/5–8, the longer lfts 2.3–5 cm; lft-venation pinnate. *Fls* dimorphic, the peripheral ones slenderly pedicellate, the terminal one sessile, much larger. *Pods* sessile or shortly stipitate, in profile broad-linear plano-compressed, nearly straight, ±12–22 × 2–4.5 cm, either dehiscent through ventral suture or subindehiscent; seeds transverse on subfiliform funicle, the crustaceous testa buff or ivory-white, the pleurogram U-shaped or complete. — Spp. 2, one continental, widespread from SE Mexico to NE Peru, centr. Venezuela, and in cultivation, the other endemic to Cuba.

The two species of *Pseudosamanea* resemble some paleotropical members of *Albizia* that are planted or naturalized in the Americas, but are distinguished by essentially pinnate, not palmate-pinnate leaflet-venation, and by organization of the inflorescence, the units of which arise directly from lf-axils and do not form a terminal panicle of pseudoracemose capitula. At anthesis the pseudosamaneas may easily be mistaken, and *Ps. guachapele* often has been, with *Samanea saman*. Where sympatric, this pair may be separated, in absence of characteristic fruits or where the inflorescence architecture is not readily apparent, by length of peripheral pedicels, 11–22 mm long in *Ps. guachapele* and <5 mm long in *S. saman*. Genuine albizias that occur in the range of *Ps. guachapele* have either many more and smaller leaflets, or sessile peripheral flowers, or an inflorescence of efoliate pseudoracemose panicles and smaller flowers, or incipiently segmented pods, or some combination of those features.

Cassens and Miller (1981) stated that the wood of *Pseudosamanea* can be told from other woods with nonseptate rays and nonconfluent parenchyma on the basis of its mostly uniseriate rays and the size of the intervascular pits.

Conspectus of genus *Pseudosamanea*

¶ Of continental dispersal, widespread.
  1. *Ps. guachapele* (Kunth) Harms
¶¶ Local on Cuba only.
  2. *Ps. cubana* (Britton & Rose) Barneby & Grimes

Key to the species of genus *Pseudosamanea*

1. Pedicel of outer peripheral fls 11–22 mm; pods sessile 10–20-seeded; SE Mexico to Venezuela and NE Peru . . . . 1. *Ps. guachapele* (p. 115)
1. Pedicel of outer peripheral fls 4–6.5 mm; pod stipitate 24–30-seeded; Cuba . . . . 2. *Ps. cubana* (p. 116)

1. Pseudosamanea guachapele (Kunth) Harms, Notizbl. Bot. Gart. Berlin-Dahlem **11:** 54. 1930. *Acacia guachapele* Kunth *in* Humboldt, Bonpland & Kunth, Nov. Gen. Sp. **6(qu):** 281. 1824. — "Crescit in sylvis, prope Guayaquil Quitensium." — Holotypus, *Humboldt & Bonpland 3805* P(HBK)! = photo, NY!. — *Lysiloma guachapele* (Kunth) Bentham, Trans. Linn. Soc. London **30:** 533. 1875. *Pithecolobium guachapele* (Kunth) Macbride, Publ. Field Mus. Nat. Hist., Bot. Ser. **13(3):** 54. 1943. *Pithecellobium guachapele* (Kunth) Cowan, Mem. New York Bot. Gard. **10:** 144. 1958, comb. superfl. *Albizia guachapele* (Kunth) Dugand, Phytologia **13:** 389. 1966.

Pithecollobium longepedatum Pittier, Contr. U.S. Natl. Herb. **20:** 464. 1922. — "Type . . . [US] 860140, collected at El Coyolar, near Orotina, Costa Rica . . . 100 to 200 meters, by Carlos Werckle." — Holotypus, US *860140*! isotypus, NY!. — *Albizia longipedata* [*sic*] (Pittier) Britton & Rose *ex* Record, Trop. Woods **11:** 14. 1927; Britton & Rose, N. Amer. Fl. **23:** 46. 1928. — Equated with *Lysiloma guachapele* by Standley, Contr. U.S. Natl. Herb. **27:** 195. 1928; Harms, Notizbl. Bot. Gart. Berlin-Dahlem **11:** 52–55. 1930.

Samanea samanigua Pittier, Bol. Ci. Técn. Mus. Com. Venez. **1:** 54. 1926. — "[VENEZUELA.] Alrededores El Sombrero, 120 m., Guárico . . . fl. & fr. Febrero 1924 (Pittier 11442, tipo)." — Holotypus, VEN!; isotypus, US *1193251*!. — Two paratypi, *Pittier 10165, 11430*; topotype, "from type tree," collected 10.IX.1927 (fl.), *Pittier 12467*, NY!. — Equated with *A. guachapele* by Harms, 1930, l. c.

Albizzia guachapele sensu Witsberger, 1982: 31, fig. 31 (lf.).

Macrophyllidious, drought-deciduous *trees* 8–25 (–30) m with roughened trunk 2–8 dm dbh (often precociously fertile), the bark exfoliating in plates, the thick annotinous branchlets angulate by prominent 3-lobed lf-scars, in youth together with lvs and inflorescence silky-pilosulous or -tomentulose with incurved sordid or yellowish hairs to 0.15–0.35 mm, the foliage bicolored, the papery lfts above darker green sublustrous, when mature only thinly puberulent, beneath paler and more densely pubescent (especially along veins), the umbelliform capitula arising single and geminate (rarely ternate) from the axil of coevally expanding lvs at early nodes of the current season's branchlets, the flat brown pubescent papery fruits long-persistent below adult foliage and often after lf-fall, the terminal meristem produced beyond the flowers. *Stipules* lanceolate 2.5–4.5 mm, caducous long before expansion of associated lf, silky dorsally. *Lf-formula* (iii–)iv–vi/5–8; lf-stks (9–)10–22(–25) cm, the petiole 3–9 cm, at middle 1.2–3 mm diam, the longer interpinnal segments (1.7–)2–4(–5) cm, the shallow ventral groove continuous; nectary at or a little above or below midpetiole, sessile, elliptic or almost round, shallowly concave, thin-margined, 1–2.5 mm diam, in profile 0.2–0.4 mm tall, sometimes a second smaller nectary near it and another at or close to apex of lf-stk, yet smaller ones on 2–3 furthest segments of pinna-rachis; pinnae strongly decrescent proximally, the rachis of the furthest and the antepenultimate ones (5–)6–10.5(–12) cm, their longest interfoliolar segments 11–21(–25) mm; paraphyllidia usually 0, rarely 1(–2), and exceptionally the posterior one developed into a small blade; lfts accrescent distally, the discolored pulvinules (0.8–) 1.2–2 × 0.25–0.55(–0.7) mm, the blades (disregarding broad terminal pair) subrhombic-obovate from asymmetric, postically rounded and antically cuneate base, broadly rounded or shallowly emarginate at apex, the larger ones (2.3–)2.5–4.7(–5.2) × 1.5–2.4 (–2.8) cm, 1.5–2.2(–2.4) times as long as wide; venation pinnate, the straight or gently incurved midrib subcentric, giving rise on each side to 7–10 major secondary veins brochidodrome shortly within the incipiently revolute margin and these in turn to an open irregular reticulum of veinlets, the whole venation finely prominulous on both faces, more sharply so beneath. *Peduncles* (2.5–)3–7.5(–8.5) cm; capitula (16–)20–34-fld, the fls strongly heteromorphic, the peripheral subumbellately ascending on slender pedicels, the terminal one much larger and sessile, the clavate or orbicular receptacle at anthesis 2–5 × 2–3.5 mm; bracts linear-oblanceolate 3–5.5 mm, silky dorsally, glabrous within, early dry caducous; perianth of all fls 5-merous or randomly 6–7-merous, greenish or tinged with reddish brown, densely silky-tomentulose externally, the filaments white or greenish-white fading yellowish or finally brown; PERIPHERAL FLS: pedicel 11–22 × 0.3–0.5(–0.65) mm; calyx narrowly campanulate or vase-shaped from turbinate base, (4–) 4.5–7 mm, the hypanthium ±1 mm deep, solid except for stipe of ovary, the tube (including hypanthium)

3.5–5.8 × 1.6–2.6(–3.4) mm, the ovate, often unequal teeth to (0.6–)0.8–1.9 mm; corolla 9.5–11 mm, the often unequal lobes ±1–3 mm; androecium (16–)18–31-merous, (24–)30–42 mm, the stemonozone 0.6–1.7 mm, the tube (4.5–)5–8 mm; ovary stipitate, the stipe as long as hypanthium, the body linear-ellipsoid 3–5.5 mm, densely puberulent, tapering into glabrous style a little longer or shorter than longest filament, at tip not or scarcely dilated, the stigmatic orifice 0.1–0.3 mm diam; TERMINAL FL: calyx cylindro-campanulate 9–15.5 × 3.5–5.5 mm, sometimes contracted at base into an obscure pedicel not longer than wide, the teeth 1.2–2.5 mm; corolla 18–26 mm; androecium 54–110-merous, the longest filaments 41–45 mm, the tube a little longer or shorter than the corolla, ±23–25 mm; ovary nonfunctional or rudimentary. *Pods* solitary or commonly 2–6 per capitulum, sessile, in profile broad-linear, straight or nearly so, at base broadly cuneate or asymmetrically rounded, at apex subtruncate and thence slenderly rostellate, (10–)12–20.5 × 2–3.5 (in Peru –4.6) cm, (10–)12–20-seeded, framed by straight or on adaxial side shallowly undulate, dorsally plane sutures 1.2–2 mm wide, the thin green, finally papery, rufous-brown or discolored valves low-convex over each seed, densely softly puberulent overall, pallid and satiny within, the cavity continuous; dehiscence (on the tree or after falling) through the ventral suture, the valves gently recurving to emit the seeds; funicles subfiliform, once or twice sigmoid at apex; seeds horizontal at middle of pod, attached subsymmetrically at base, narrowly oblong-ellipsoid but plump, moderately compressed, in broad view ±7–10 × 3.7–4.6 mm, the testa when ripe hard, pale bluff or ivory-white, the narrowly U-shaped pleurogram 4.5–6 × 0.9–1.7 mm.

In drought-deciduous woodland and at the drier edge of gallery forest, surviving as a shade tree in pastures and along hedgerows, sometimes cultivated, native from near sea level to 550 and in Colombia to 1000 m, interruptedly widespread from S Mexico (Tabasco, Campeche, Chiapas) SE through Central America, mostly on the Pacific slope, to Panama, thence S through inter-Andean valleys of Colombia to Pacific lowland Ecuador and immediately adjacent Peru (Tumbes), and once recorded (? introduced) from Amazonian Ecuador (near Coca); from N Colombia eastward through N Venezuela, from Zulia through the coastal cordillera and the N slope of the Orinoco basin to Anzoátegui, S to the mouth of Río Meta in Apure and to the lower Caroní Valley in Bolívar; cultivated in Bahia and Sao Paulo, Brazil, in Cameroon, and probably elsewhere. — Map 33. — Fl. at end of dry season, with or just prior to renewal of foliage, in North America mostly XII–III; in South America I–IV, VII–IX. — *Cadeno* (Guatemala); *carreto* (El Salvador; wood used by wheelwrights); *guaramillo, samán* (Colombia); *lara blanca, masaguaro, samanigua, tabaca* [*taboca*] (Venezuela); *guachapelí* [*guachapelé, huachapelí*] (Ecuador).

2. Pseudosamanea cubana (Britton & Rose) Barneby & Grimes, comb. nov. *Albizia cubana* Britton & Wilson *ex* Britton & Rose, N. Amer. Fl. **23:** 46. 1928 (11 feb). — "Southern Oriente, Cuba. Type from United States Naval Station, Guantánamo Bay, March 1909, *Britton 2050*." — Holotypus, NY!; isotypus, US *847726*!.

Pithecolobium (?) *bacona* Urban, Symb. Ant. **9:** 434. 1928 (15 mar). — "[Cuba.] Prov. Habana prope Batabanó ad La Mora in palm-savannis . . . m. Dec. Fruct.: [Ekman] n. 12929 (typus) . . . [paratypi, steriles *Ekman 8247, 8415* et fructifer, no. *9473*, omnes e prov. Oriente citantur]." — Holotypus + paratypi, †B; paratypus, *Ekman 8415*, NY!. — Equated with *A. cubana* by Britton & Rose, 1928: 193.
Albizzia cubana sensu Bisse, 1988: 226.

Macrophyllidious *trees* 6–12(–?) m with trunk to 8(–?) dm dbh and flaking bark, the branchlets in vernation densely pilosulous but at maturity glabrescent, the dull olivaceous lvs bicolored, the papery lfts glabrous and venulose above, beneath paler and at least along some veins, often overall, thinly strigulose-pilosulous with fine pallid, mostly procumbent or some subappressed hairs to 0.4–0.9 mm, the inflorescence thinly tomentulose, the umbelliform capitula borne on a stout peduncle solitary in the axil of few distal lvs, immersed in foliage. *Stipules* (few seen) triangular-lanceolate 2–4 mm, very early caducous, lacking from fertile spms. *Lf-formula* iv–v(–vi)/5–8; lf-stks of well-developed lvs (others not described) 8–20 cm, the petiole (2.5–)3.5–7.5 cm, at middle 1.2–2 mm diam, the ventral groove continuous between pinnae, the longer interpinnal segments 2–4 cm; petiolar nectary commonly below, less often near or shortly above midpetiole, sessile cupular 0.7–1.2 mm diam or subimmersed in ventral sulcus, rarely obsolescent, and random smaller nectaries at insertion of some distal pinnae, with yet smaller ones near insertion of most lft-pairs; pinnae unequal, the first and last shorter, the rachis of median ones 8–12.5 cm, the longer interfoliolar segments 10–21 mm; pulvinules 1.4–2.5 × 0.4–0.6 mm, scarcely wrinkled; lfts a little or scarcely accrescent upward, the first pair

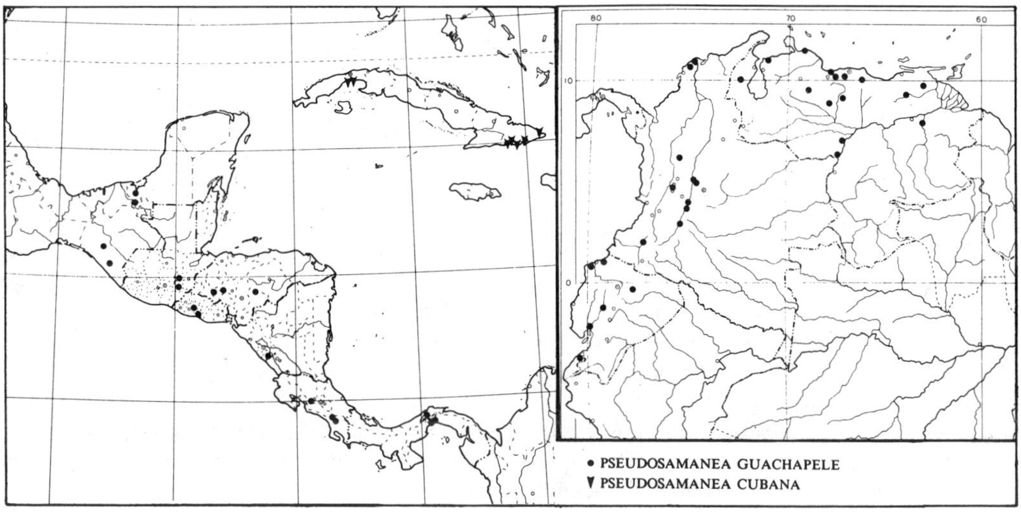

MAP 33. Southern Mexico, Central America, Cuba, and (inset) northern South America. Range of *Pseudosamanea guachapele* (Kunth) Harms and *Ps. cubana* (Britton & Rose) Barneby & Grimes.

represented by minute paraphyllidia, the blades of the rest subinequilaterally obovate or broadly oblanceolate from broad-cuneate base, broadly rounded at apex, the longer ones 23–36 × 13–24 mm, 1.4–2.2 times as long as wide; venation of essentially pinnate type (rarely one very short primary nerve either anterior or posterior to midrib), the subcentric, straight or slightly incurved midrib giving rise to 5–7 pairs of major secondary nerves brochidodrome well within the incipiently revolute margin, these in turn generating an open reticulum of veinlets, the whole venation finely prominulous on both faces, a little more sharply defined beneath. *Peduncles* 5–9.5 cm; capitula umbelliform ±20–25-fld, the narrowly clavate receptacle ±4–5 mm; bracts caducous (none seen); fls strongly heteromorphic, the peripheral ones elevated on a slender pedicel (4–)5–6.5 × 0.3–0.6 mm, 5-merous, the terminal one sessile, the perianth 6–7-merous, all silky-pilosulous externally; PERIPHERAL FL: calyx slenderly campanulate 5–7.5 × 1.6–2.5 mm, the unequal, triangular-lanceolate obtuse teeth 0.5–2 mm; corolla 11–13 mm, the unequal, lanceolate lobes 2–4.5 mm; androecium 20–28-merous, ±2.5–2.7 cm, the stemonozone ±1 mm, the tube 8 mm, the tassel white, fading yellow; TERMINAL FL: calyx 7–10 × 3–4 mm; corolla 12–19 mm; stamens to 44, ±3 cm, the stemonozone to 3 mm, the tube 16 mm; ovary of all fls slenderly fusiform, symmetrical at tip; style about as long as longest stamens, turbinately dilated at apex. *Pods* solitary or to 6 per capitulum, at once stipitate (the stout stipe 2–4 mm) and narrowed at base into a pseudo-stipe to ±1 cm, the body broad-linear, plano-compressed, straight or almost so, 15–22 × 2.2–3.2 cm, 24–30-seeded, at apex rounded and apiculate, the brown, stiffly papery, low-corrugate, densely tomentulose, internally pallid but not satiny valves framed by almost straight sutures 1–1.6 mm wide; dehiscence essentially none, the fruit falling tardily and disintegrating on the ground, the sutures not separating, the valves cracking transversely only under pressure; seeds transverse at middle of pod on filiform funicle, ±6–7 mm apart, compressed but plump, in broad outline narrowly oblong-elliptic 6.5–7.4 × 3.4–4.3 mm, the crustaceous testa smooth, ocher-buff, the complete pleurogram 3.5–4 × 1.7–2.3 mm.

In palm-savanna and along watercourses below 50 m, endemic to the S coast of Cuba, best known from prov. Oriente, where extending from Ensenada de Mora E to the S Baracoa region (74°20′–77°20′W), and apparently isolated near Batabanó (82°18′W) in S prov. La Habana. — Map 33. — Fl. I–III, VII; fr. mostly III–V, but long persistent on the tree. — *aimiquí, bacona.*

VI. SAMANEA MERRILL

Samanea Merrill, J. Wash. Acad. Sci. **6:** 46. 1916. — Sp. typica: *Samanea saman* (Jacquin) Merrill ≡ *Mimosa saman* Jacquin ≡ *Pithecolobium saman* (Jacquin) Bentham ≡ *Albizzia saman* (Jacquin) F. v.

Mueller. *Calliandra* sect. *Samanea* Grisebach, Fl. Brit. W. I. 225. 1864.

Pithecolobium sect. *Samanea* ser. *Carnosae* [*sic*] Bentham, Trans. Linn. Soc. London **30:** 587. 1875, ex parte, omnibus praeter lectotypum, *P. saman*, exclusis.
Samanea sensu Britton & Rose, 1928: 34. quoad sp. typ., caeteris expulsis; Mohlenbrock, 1963: 435, parva ex parte, seriebus *Coriaceis* et *Parvifloris* omissis.
Albizia sensu Nielsen, 1981: 180, fig. 1(2), minori ex parte.

Macrophyllidious *trees*, some attaining great age and size, the branching sympodial, new growth arising yearly from buds below the prior season's determinate inflorescence, the lfts folding forward slowly at night or after shock. *Stipules* herbaceous, lanceolate, early deciduous. *Lf-formula* iii–vi(–vii)/3–8(–9); lft-venation pinnate. *Capitula* axillary to coeval or quickly hysteranthous lvs, umbelliform, the fls dimorphic, the peripheral ones at least shortly pedicellate, the terminal one stouter, sessile; calyx of peripheral fls vase-shaped 4.5–11 mm, the corolla 7–14.5 mm; androecium (16–)20–36-merous, the tube included, the tassel pink or reddish; ovary (sub)sessile, narrowly ellipsoid. *Pods* subsessile, in profile broad-linear, straight or nearly so, either compressed but plump or biconvex, the valves composed of thin, glabrous or densely puberulent exocarp, thick pulpy, when ripe pitchlike mesocarp, and crustaceous-lignescent endocarp either adherent or narrowly septiferous between seeds; dehiscence 0, the seeds released by weathering, or by predators, or following excretion by cattle; seeds transverse on slender compressed funicle, oblong-ellipsoid, the testa hard opaque, closely investing the embryo, areolate. Germination in *S. tubulosa* (*M. Nee 39105*, NY) is epigeal, the first two eophylls subopposite, the proximal one simply pinnate, 4–6-foliolate, the distal one bipinnate, each pinna 4-foliolate. — Spp. 3, endemic to tropical continental North and South America, native from El Salvador southeastward to E Brazil, Paraguay, and NE Bolivia, but one species (*S. saman*) planted and subspontaneous northward into Mexico and long established in the West Indies and in parks, gardens, and agroforestry plantations of the Old World.

The genus *Samanea*, internally uniform in characters of substance, belongs unequivocally among the albizioid Ingeae, of which it has the dimorphic terminal flower and the hard areolate seeds. Indehiscence of the fruit, on which emphasis has been placed in former classifications, is a mechanical consequence of an incrassate pericarp. It has evolved independently in various neotropical Ingeae (e.g., *Enterolobium*, *Balizia*, and *Albizia* ex parte, as well as in *Inga* itself) and also in some Madagascan *Albizia* (Capuron, 1970: 22 passim). Indehiscence becomes taxonomically significant only within particular geographic and coincident morphological contexts. In Madagascar differentiation of *Albizia* must have proceeded pari passu with speciation, for the endemic species, while extraordinarily diverse in form, texture, and dehiscence of the fruits, have preserved evidence of a common lineage in the peculiar displacement of the areole to the base of the seed, a feature encountered nowhere else in the genus. *Samanea* possesses no known feature unique in tribe Ingeae. However, though the fruits of *Samanea* are technically indistinguishable from those of *Enterolobium*, they are never confused. Evolution of its fruit, by acquisition of sweet nutritive pulp in the mesocarp, loss of dehiscence, and hard seeds that pass undigested through the gut of herbivores, appears to be a single (if plurifaceted) biological specialization that arose prior to the differentiation of the three modern species. *Samanea* is certainly monophyletic, and on the basis of cladistic analysis sectional status within a more comprehensive *Albizia* is indefensible.

Samanea is most closely related to *Pseudosamanea*, and at anthesis *S. saman* and *Ps. guachapele* are often mistaken for one another. *Samanea* differs from *Enterolobium*, which has a similar dispersal mechanism, in organization of the inflorescence, and in the straight fruit. If *Samanea* were subordinated to *Albizia* nothing but tradition would prevent subordination of *Enterolobium* to equivalent rank.

When Merrill proposed a genus *Samanea* to accommodate the Rain Tree (and perhaps other species, obliquely alluded to), he traced in detail its checkered taxonomic history, which involved successive transfers from genus to genus; it is superfluous to repeat the story, implicit in the synonymy of *S. saman* set out below. Britton & Rose (1928: 1, in clave, 34) expanded the definition of *Samanea* so as to include along with the type species *Balizia leucocalyx*, *Abarema macradenia*, and *Cojoba filipes*. Mohlenbrock (1963: 435) expanded *Samanea* yet further to embrace all of Bentham's series *Carnosa*, *Coriacea*, and *Parviflora* of *Pithecolobium*. He seems to have been unaware that the fruits of all *Coriacea* (= our genus *Macrosamanea*) as well as those of some *Parviflora* are inertly dehiscent, as is that typical of *Albizia*. Capuron (1970) recognized *Samanea* in Merrill's sense, whereas Cárdenas (1974: 125) reverted to an orthodox *Pithecellobium* ser. *Carnosa* Bentham. Finally Nielsen (1981: 180) returned *Samanea* to a comprehensive circumtropical *Albizia*.

Cassens and Miller (1981) differentiated the wood of *Samanea* from that of all other New World Ingeae on the basis of the following syndrome of character states: septate fibers and confluent parenchyma both lacking, intervascular pits 7–10 μm, and mostly biseriate rays.

Conspectus of genus *Samanea*

¶ Petiolar nectaries between first and furthest pairs of pinnae; lfts of longer pinnae 6–9 pairs; interfloral bracts spatulate; seeds 12–20.
1. *Samanea saman* (Jacquin) Merrill

¶¶ First petiolar nectary at or close to base of lf-stk, none between pinna-pairs; lfts of longer pinnae 3–5(–6) pairs; interfloral bracts linear or linear-lanceolate; seeds 20–30.
2. *Samanea tubulosa* (Bentham) Barneby & Grimes
3. *Samanea inopinata* (Harms) Barneby & Grimes

Key to the species of genus *Samanea*

1. Lfts of longer pinnae 6–8(–9) pairs, the blade glabrous lustrous on upper face; petiolar nectaries always between first and furthest pinna-pairs and often between all (none at base of lf-stk); interfloral bracts spatulate, with linear claw and rhombic blade 1–2.3 mm wide, in early praefloration clasping the young fl-buds; ovary grooved along each side, at anthesis glabrous in lower ⅔, distally papillate or minutely puberulent (but becoming pilosulous overall following fertilization); corolla of terminal fl 9.5–14.5 mm; pod (14–)17–23 mm wide, 6–10 mm thick, the face of each valve depressed and nearly plane, but shallowly corrugate lengthwise, the exocarp continuous, neither transversely fibrous nor wrinkled; seeds 12–20, in broad view 8–11.5 × 5–7.5 mm; native mostly N of the Equator (W Ecuador, Colombia, Venezuela, and northward, now circum-Caribbean) but widely introduced elsewhere . . . . 1. *S. saman* (p. 119)

1. Lfts of longer pinnae 3–5(–6) pairs, the blade usually dull and minutely or remotely puberulent on upper face; one petiolar nectary at or close to base of lf-stk, none between pinna-pairs; interfloral bracts linear or linear-oblanceolate 0.25–1 mm wide, never concealing the fl-buds; ovary at anthesis subterete, densely silky-strigulose from base upward; corolla of terminal fl 14–27 mm; pod variable in diameter, but the valves strongly convex, the exocarp not as above; seeds 20–30, in broad view 6.5–8.5 × 3.5–8 mm; native of South America southward from the Equator to E Bolivia, Paraguay, and Bahia, sparingly N into N Brazil, E Venezuelan Guayana, S Guyana, and W to Pacific Ecuador, seldom planted.
   2. Ripe pod in section 12–18 × 9–13 mm, the fuscous exocarp continuous, smooth except where transversely fibrous near the sutures, when dry becoming ridged lengthwise or coarsely wrinkled; SW periphery of Amazon basin in Peru, Brazil, and thence S to SW Brazil and Paraguay, disjunct in Pacific Ecuador and in lower Amazonian Brazil . . . . 2. *S. tubulosa* (p. 121)
   2. Ripe pod in section either like the last or to 35 × 20 mm, the velutinously pilosulous exocarp of the valves horizontally wrinkled and deeply fissured across its whole width, from suture to suture, not wrinkled lengthwise; of bicentric dispersal, E Guayana Highland and periphery in N Brazil, SE Venezuela, and S Guyana; SE Brazil, from Maranhão to Pernambuco and Bahia. . . . . 3. *S. inopinata* (p. 123)

1. Samanea saman (Jacquin) Merrill, J. Wash. Acad. Sci. **6:** 47. 1916. *Mimosa saman* Jacquin, Fragm. Bot. 15, t. 9. 1801. — Described from plants grown at Vienna from seeds "ex patria Caracas [Venezuela]." — Typus, W, not seen, but the illustrated protologue decisive. — *Inga saman* (Jacquin) Willdenow, Sp. Pl. **4:** 1024. 1806. *Pithecolobium saman* (Jacquin) Bentham, London J. Bot. **3:** 216. 1844. *Calliandra saman* (Jacquin) Grisebach, Fl. Brit. W. I. 225. 1864. *Albizzia saman* (Jacquin) F. v. Mueller, Select [Extratropical] Pl., ed. 8: 27. 1891. *Feuilleea saman* (Jacquin) O. Kuntze, Revis. Gen. Pl. **1:** 189. 1891. *Enterolobium saman* (Jacquin) Prain *ex* King, J. Asiat. Soc. Bengal, pt. 2, Nat. Hist. **66:** 352. 1897. *Zygia saman* (Jacquin) A. Lyons, Pl. Nam., ed. 2: 503. 1907.

Inga cinerea Humboldt & Bonpland *ex* Willdenow, Sp. Pl. **4:** 1024. 1806. — "*Habitat ad* Caracas." — Holotypus, *Humboldt 603*, B-WILLD *19052*, seen in Microform 86. 1382: III. 1!. — *Pithecolobium cinereum* Bentham, London J. Bot. **3:** 216. 1844. — Equated with *Pithecolobium saman* by Bentham, 1875: 587.

Inga salutaris Kunth *in* Humboldt, Bonpland & Kunth, Nov. Gen. Sp. **6:** 304. 1824. — "Prope Turbaco et Carthagena, alt. 30–180 hex. (Regno Novo-Granatensi.)" — Holotypus, P-HBK!; isotypus, *Bonpland 1382,* P! = F. Neg. *39892.* — Equated with *Pithecolobium saman* by Bentham, 1875: 587.

Acacia propinqua A. Richard, Hist. Phys. Cuba, Pl. Vasc. 466. 1846. — "In insula Cuba." — Holotypus, *R de la Sagra* s. num., P(herb. richard.)!.

Pithecolobium saman sensu Cook & Collins, Contr. U.S. Natl. Herb. **8(2):** 220–222. 1903; sensu Witsberger, 1982: 40, fig. 40.

Albizia saman sensu A. C. Smith, Fl. Viti. Nova **3:** 79. 1985.

Samanea saman sensu Bisse, 1988: 226.

Macrophyllidious *trees* of potentially great age and size, flowering as treelets but becoming 25(–30), exceptionally 40–50 m tall with short stout trunk to

2(–3) m dbh and broad low crown nearly twice as wide as high, the efoliate annotinous branchlets fuscous, either lenticellate or not, the hornotinous branchlets, all lf-axes and peduncles densely pilosulous with fine whitish, either straight spreading or shorter incumbent hairs to 0.2–0.7 mm, the lvs bicolored, the lfts dark-olivaceous, glabrous and lustrous on upper face, beneath pallid dull and pilosulous, the young inflorescences yellowish tomentulose, the umbelliform capitula fasciculate by 2–5 in the axil of coevally expanding or quickly hysteranthous lvs and early immersed in foliage, the fruits persistent for nearly a year on the tree, falling as the new growth starts, the new branchlets arising from lateral buds below the determinate inflorescence of the new year. *Stipules* herbaceous, lanceolate or oblanceolate 2.5–7 mm, densely pilosulous, caducous with expansion of the subtended lf. *Lf-formula* iii–vi(–vii)/6–8(–9); lf-stks of fully developed lvs (4.5–)6–25 cm, the petiole 3–7.5(–8.5) cm, at middle 1.4–3 mm diam, the ventral groove open and shallow, the first interpinnal segment (1.2–)1.7–4.7 cm, the further ones progressively shorter; petiolar nectaries close below first, furthest, and often all pinna-pairs, sessile, shallowly cupular thick-rimmed, in ventral view round or bluntly triangular (distorted, didymous), the first one 0.8–1.4 mm diam, similar but smaller nectaries on pinna-rachises close below each lft-pair; pinnae strongly accrescent distally, the rachis of furthest and penultimate pairs 6–14.5 cm, the longer interfoliolar segments 10–22(–25) mm; minute membranous paraphyllidia caducous from small scars at apex of corrugate, ventrally sulcate pinna-pulvini; lft-pulvinules 1–1.7(–2) × 0.5–1(–1.2) mm; lfts conspicuously accrescent distally, all but the rhombic-obovate and largest terminal pair obtusely rhombic-oblong or -elliptic from inequilateral base, obtuse but sometimes minutely mucronulate, the blades of penultimate pair (20–)24–62 × (8–)10–25(–27) mm, (1.9–) 2.1–2.8 times as long as wide; venation pinnate, the subcentric, either straight diagonal or more often gently incurved midrib giving rise on each side to (7–)8–12(–13) widely incurved-ascending secondary veins brochidodrome within the narrowly revolute margin, these in turn generating a tertiary and reticulate venulation, the costa, secondary and tertiary venules of fully mature lfts prominulous dorsally, on upper face more finely so and the areoles of the reticulum smaller. *Peduncles* (3.5–)4–8.5 cm, woody in fruit; capitula hemispherical (12–)15–22(–25)-fld; bracts 3–7 × 1–2.3 mm, dimorphic, the lowest lance- or oblance-elliptic, the interfloral ones spatulate, the linear claw abruptly expanded into a rhombic blade, all deciduous at anthesis, when young folded over and concealing the young fl-buds; fls dimorphic, the peripheral ones at least shortly pedicellate, the terminal one stout, sessile, the perianth of all 5–6-, or that of terminal fl 5–9-merous, densely yellowish pilosulous-tomentulose overall (the corolla tube sometimes glabrescent); PERIPHERAL FLS: pedicel compressed, to 1–3.5(–5.5) × 0.3–0.6(–0.7) mm; calyx slenderly vase-shaped (4.4–)4.8–7(–7.8) × 1.3–2 mm, the deltate-ovate obtuse, often unequal teeth 0.6–1.3 mm; corolla 9–13(–13.5) mm, slenderly trumpet-shaped, the lobes (often unequal) up to 1.4–3.7 × 1–1.7 mm; androecium (18–)21–30(–32)-merous, (26–)28–37(–42) mm long, the filaments white proximally, pink or red distally, the stemonozone 1.2–2.1 mm, the tube 5–7.5(–8) mm; ovary obscurely stipitate, linear-ellipsoid, on each face shallowly grooved lengthwise, prior to fertilization minutely papillate or puberulent in upper third, becoming when fertile densely pallid-tomentulose; style a trifle longer than longest filament, the stigma commonly porose, rarely dilated to 0.25 mm diam; TERMINAL FL: sessile, the calyx (5.5–)7–10.2 × 2.5–4.3 mm, the teeth 1–1.6 mm; corolla 9.5–14.5 mm, the lobes ± 3–3.5 mm; androecium ±50–80-merous, the tube (7.5–)9–13 × 1.8–2.6 mm, a little shorter or longer than corolla, the free part of filaments proximally dilated and recurved. *Pods* 1(–2) per capitulum, broad-linear abruptly contracted at base into a short neck and as abruptly cuspidate at apex, compressed but plumply fleshy, when well fertilized straight or almost so and 10–22 × (1.4–)1.7–2.3 × 0.6–1 cm, (10–)12–20-seeded, coarsely bicarinate by plane or shallowly sulcate, woody sutures ±2.5–3.5 mm wide, the low-convex valves consisting of (a) thin, continuous, when dry papery, livid-castaneous or blackish, somewhat lustrous, irregularly crumpled, glabrate or minutely puberulent exocarp; (b) thin crustaceous endocarp, adherent between seeds to form discreet seed-cavities but not septiferous; and (c) thick alveolate mesocarp filled with black, when dry pitchlike (sweet, nutritious) pulp; dehiscence 0, the seeds released only by weathering or predators, and passing undigested through the gut of cattle and other herbivores; seeds transverse, obliquely basifixed on slender funicle, plumply oblong-ellipsoid, 8–11.5 × 5–7.5 mm, the smooth castaneous (or, especially within the pleurogram, fuscous) testa leathery-crustaceous, brittle, opaque, closely investing the ivory-colored embryo, the elongately U-shaped pleurogram 5–8.3 × 2–4.4 mm.

A stately tree, native in and at margins of seasonally dry, deciduous and semideciduous as well as moister evergreen woodland and savanna (llano), below 1450 and mostly below 450 m, in the Orinoco Valley and on the Caribbean slope in Venezuela, in N Colombia, and in Central America northwestward perhaps to El Salvador, but extensively cultivated both within and outside its presumed natural range, this consequently no longer demonstrable; planted for shade and ornament on streets, in parks, and on roadsides throughout much of lowland continental Latin America and West Indies, locally used (at least in the past) as coffee-shade, and often preserved in pastures for cattle-shade and for the nutritious pods, thence locally naturalized; planted in Hawaii and locally in Old World Tropics. — Map 34. — Fl. most abundantly toward the end of the dry season, but sporadically through the year. — *Campano, saman* (*zaman*), (Venezuela, Colombia); *rain tree*, whence *árbol de lluvia* (Spanish) and cognate *chorona* (Portuguese); *carreto, cenicero, dormilón, zarza* (Central America), *carabelí, coují, lara, urero* (Venezuela); *French tamarind* (Guyana); *algarrobo* (Cuba), *guango* (Jamaica); *guannegoule* (Haiti); *coco tamarind, cow tamarind* (Grenada).

Once disentangled from *S. tubulosa*, from which it is infallibly distinguished at anthesis by position of the petiolar nectary and by form of the floral bracts, *S. saman* is sharply defined and in most features a stable species. The fruit varies in length and girth, but scarcely in other respects. The leaves are sensitive to light, folding at dusk and at the approach of storms, whence the vernacular name rain tree.

2. Samanea tubulosa (Bentham) Barneby & Grimes, comb. nov. *Calliandra tubulosa* Bentham, London J. Bot. **3:** 101. 1844. — "Brazil: banks of the Rio San Francisco, near Villa Nova [NE Bahia], Gardner, n. 1280; Cuyaba [Mato Grosso, *Riedel*, commun.] Langsdorff." — Lectotypus, labeled by Bentham "Sierra Chapada [= Chapada dos Guimarães]. Riedel" accompanied by a ticket reading "1830, [no.] 22, Cuyaba. Hb. Mus. Petr[op.]," K(herb. bentham.)! = photo s.n., NY!; presumed isotypi, *Riedel s.n.* e Cuiabá, A!, NY! — Paratypus, *Gardner 1280*, K! = NY Neg. *2016* = probably *S. inopinata* (Harms) Barneby & Grimes.

Pithecolobium saman var. (β) *acutifolium* Bentham *in* Martius, Fl. Bras. **15(2):** 442. 1876 "acutifolia". — "Habitat in prov. Para prope Santarem: Spruce n. 671; in Peruvia prope Tarapoto: Spruce n. 4405." — Syntypi, K: *671* (fl.), herb. bentham.! = NY Neg. *2071*!, *671* (fr), herb. bentham.! = NY Neg. *2018*!, and *4405* (fr), herb. bentham.! K(herb. hooker.)!, NY!, no. *4450*, NY!, P!.

Pithecolobium venosum Rusby, Mem. Torrey Bot. Club **6:** 29. 1896. — "[Bolivia.] Between Guanai and Tipuani, Apr.–June, 1892 ([*H. H. Rusby*] 1392)." — Holotypus, NY!; isotypi, DS-CAS!, K!, MINN!, UC *946443*!.

Samanea saman sensu Bernardi, 1984: 192.

Macrophyllidious *trees* commonly 8–25 m but in relatively dry forest conditions only (2–)3–8 m tall, with trunk attaining (1–)2–10 dm dbh, fissured, light-colored, thick, spongy or often corky bark, and angulate branchlets, the young stems, all lf-axes and peduncles puberulent or pilosulous with pallid, either straight and spreading-ascending or shorter incumbent hairs to 0.15–0.5 mm, the foliage bicolored, the lfts dull olivaceous and minutely or remotely puberulent on upper face, on lower face paler and more densely pilosulous (especially along veins), the umbelliform capitula of greenish, sometimes pink-tinged flowers with long, distally pink filaments arising singly or 2–3-nate in the axil of coeval or quickly hysteranthous lvs, sometimes in early anthesis appearing terminally paniculate but quickly surpassed by adult foliage, the innovations of each year arising laterally from below the annotinous inflorescence. *Stipules* narrowly lanceolate 3–9 × 0.7–1.3 mm, pubescent like adjacent stem, deciduous with early expansion of the associated lf. *Lf-formula* iii–vi (–vii)/3–5(–6); lf-stks (7–)8–28 cm, the petiole (2–)2.5–8(–11) cm, at middle (1–)1.2–2.7 mm diam, the ventral groove continuous between pinna-pairs, the longer interpinnal segments (2–)2.5–7 cm; nectary near base (well below middle) of lf-stk and often almost adjacent to lf-pulvinus sessile, in vertical view either round or (commonly) elliptic (1.2–)1.4–4 mm diam, either shallowly patelliform, or plane, or rarely low-convex, usually thick-rimmed, in profile ±0.2–0.4 mm tall; no nectary between any pinna-pair but smaller, shallow-cupular ones on pinna-rachises close below the furthest 1–2 or rarely between all lft-pairs; pinnae greatly accrescent distally, the rachis of furthest pair (4–)5–11 cm, the longer interfoliolar segments (1.1–)1.3–3(–3.6) cm; paraphyllidia at apex of pinna-pulvini linear caducous; lft-pulvinules (1–)1.2–2.2(–2.5) × 0.5–1 mm, not wrinkled; lfts strongly accrescent distally, obliquely oblong- or rhombic-obovate from inequilateral base, either broadly obtuse, or obtuse-apiculate, or contracted into a deltate-apiculate tip, the blades of penultimate pair (2.5–) 2.8–6 × (1.3–)1.4–3.8 cm, (1.4–)1.5–2.2 times as long as wide, the furthest pair commonly yet larger and often proportionately broader; venation pinnate,

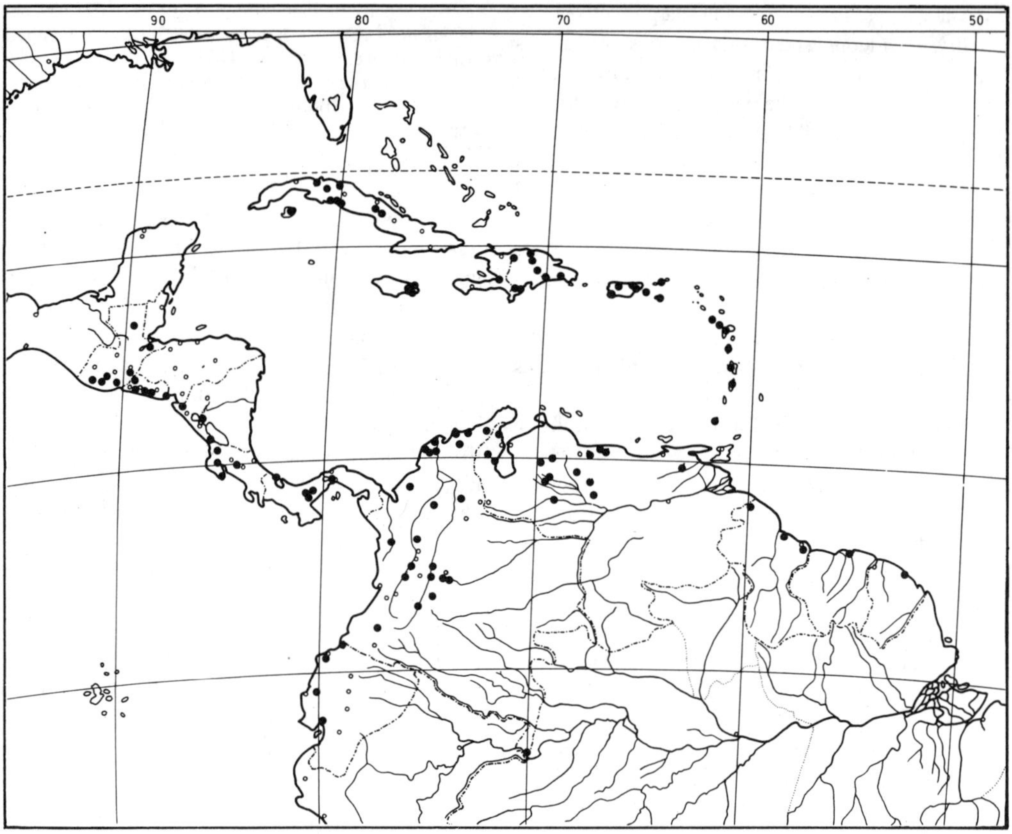

MAP 34. Central America, the West Indies, and northern South America. Range of *Samanea saman* (Jacquin) Merrill.

the straight diagonal or more often gently incurved and slightly excentric midrib giving rise on each side to (5–)6–9(–10) major (and sometimes few random minor) incurved-ascending secondary nerves brochidodrome shortly within the plane margin, and these in turn to a reticulum of venules, the whole venation pallidly discolorous and prominulous on both faces of lft, more sharply so beneath. *Peduncles* 3–10.5(–15) cm; capitula ±12–20-fld, the axis including pedestal of terminal fl usually 1.5–4 mm but the pedestal itself sometimes elongate to 5 mm and an occasional fl downwardly displaced onto peduncle; interfloral bracts linear-oblanceolate (1.7–)2–7 × 0.25–1 mm; fls dimorphic, the peripheral ones pedicellate, the terminal one sessile, the perianth of all 5-merous (rare abnormalities) and densely silky-pilosulous externally, the vesture of the corolla usually paler and more lustrous than that of calyx; PERIPHERAL FLS: proximal pedicels (1.4–)1.7–5.5(–7) × 0.35–0.55 mm; calyx including hypanthium narrowly infundibuliform 5-ribbed (4–)4.6–7.5(–8.5) × 1.5–2.7(–3) mm, the often unequal, ovate or obtusely deltate teeth 0.6–1.8(–2) mm; corolla 9.5–14.5 mm, the ovate or narrowly oblong, apically cucullate and fimbriolate lobes (2.6–)3–5.5(–6.5) mm; androecium (16–)22–30 (–36)-merous, 34–46 mm, the stemonozone 1.5–2.3 mm, the tube 5–9(–10) mm, the free filaments white or pallid proximally, deep rose, pink, or reddish purple distally; ovary sessile, linear-elliptic tapering into the style, densely silky-strigulose at anthesis, densely pilosulous following fertilization; TERMINAL FL: calyx 7–12 × (1.8–)2–3.4 mm; corolla (13.5–)14–22 mm. *Pods* 1(–2) per capitulum, sessile, in profile broad-linear, straight or randomly bent (but not recurved or coiled) girdled by stout plane sutures 2.5–4 mm wide, when well fertilized (7–)10–18.5 × 1.2–1.8 × 0.9–1.3 cm, 20–31-seeded, the finely or minutely but often at least at first densely puberulent valves convex, consisting of thin fuscous exocarp, a broad layer of alveolar pulp (pitchlike when dry), and a dry crustaceous endocarp inwardly either impressed or produced to divide the cavity into 1-seeded locules, the fleshy valves becoming bluntly ridged lengthwise but not or only faintly depressed between seeds, when ripe transversely cracking (but not sulcate) toward sutures; seeds transverse on slender but compressed funicle, basifixed, plumply compressed-oblong-ellipsoid, in broad view 6.5–8.5 × 3.5–4.7 mm, the smooth hard testa fuscous-castaneous, the narrowly U-shaped pleurogram ±5–6.5 × 2–2.4 mm.

In seasonally dry, semideciduous or drought-deciduous forest and in savanna enclaves within wet evergreen forest, mostly below 500 m, but ascending in inter-Andean valleys of Peru to 600–950 m, and in the Bolivian Yungas to 1300–1800 m, discontinuously widespread around the W and SW periphery of the Amazon basin and in the upper Paraguai basin, in Peru, Brazil westward from the Tocantins, Bolivia, and Paraguay, and reappearing locally along the lower Amazon and its immediate tributaries in Pará, Brazil; one isolated station in Pacific Ecuador (estado Manabí, there sympatric with *S. saman*, perhaps introduced), cultivated in NE Argentina (Corrientes). — Map 35. — Fl. (IX–)X–II(–III), immediately following flush of new lvs, the fruit shed simultaneously or a little earlier. — *Lluicho vaina, lluicho vainilla* (Peru); *burdão de velho* (Acre); *ichizozo, penoco* (Bolivia); *niño azote* (Paraguay).

Samanea tubulosa, which as a native tree replaces *S. saman* in South America southward from the equator, is readily distinguished from it, when in flower, by the nectary at base of each leafstalk and by (sub)linear floral bracts. The leaflets of longer pinnae are usually fewer by one or two pairs, and all are puberulent rather than glabrous on the upper face. Differences between the fruits collected by Spruce at Tarapoto in Peru and those of the genuine samán from Caracas were noted by Bentham in his revision of Mimoseae (1875: 587). In *Flora Brasiliensis* he voiced the suspicion that more than one species lay concealed in his concept of *Pithecolobium saman*, but for lack of complete material he could go no further than describe Spruce's plant as a var. *acutifolia* [*sic*], supposedly different in characters of foliage rather than fruit. He was unaware that he had earlier described the same taxon as *Calliandra tubulosa*. The protologue of *C. tubulosa* was derived from two flowering collections, almost certainly not conspecific: *Riedel s.n.* from Chapada dos Guimarães near Cuiabá in Mato Grosso; and *Gardner 1280* from near the mouth of Río São Francisco in northern Bahia. The identity of Riedel's plant is now firmly established by fruiting topotypes, but that of Gardner's plant will remain ambiguous until we have confirmatory pods from the region of the type locality. The known facts of geographical dispersal (cf. Maps 35 and 36) suggest that Gardner's plant is in reality *S. inopinata*. Our lectotypification of *C. tubulosa* makes the epithet available and unavoidable for the widespread species described here and bypasses the enigmatic identity of *Gardner 1280*.

Samanea tubulosa and *S. saman* are allopatric, so far as known (cf. Maps 34 and 35), but were collected close together on the same day near Chone in Manabí, Ecuador, by L. Holm-Nielsen (nos. *27947* and *27918* respectively, both NY). *Samanea saman* is a well-known and presumably native element of seasonally dry forest in Pacific Ecuador, but *S. tubulosa* is, except for this one record, entirely trans-Andean. Its status in Ecuador requires observation.

3. Samanea inopinata (Harms) Barneby & Grimes, comb. nov. *Serianthes inopinata* Harms, Notizbl. Bot. Gart. Berlin-Dahlem **11:** 55. 1930. — "Brasilien: Kultiviert im Bot. Garten Rio de Janeiro unter dem Namen *Pithecolobium saman* (Oktober 1928 — A. Ducke n. 15248)." — Holotypus, †B, isotypi, K!, RB!, UB!. — *Pithecolobium inopinatum* (Harms) Ducke, Mem. Inst. Oswaldo Cruz **51:** 426. 1953. *Albizia inopinata* (Harms) Lewis, Leg. Bahia 182. 1987.

Pithecellobium nuriensis [sic] Irwin, Mem. New York Bot. Gard. **15:** 107, fig. 4. 1966. — "Venezuela, Edo. Bolívar, between Rancho Alegre and base of cerro, on trail to Quebrada Cabeza Burro, 5 km east of Chicharras. *J. A. Steyermark 89317*." — Holotypus, NY (2 sheets)!; isotypus, VEN *58840*!.

Calliandra tubulosa Bentham, London J. Bot. **3:** 101. 1844, ex parte, quoad *Gardner 1280*, exclus. lectotypo.

Albizia sp. Lewis & Owen, Leg. Ilha de Maracá 43. 1989. — Spm. authent.: *Lewis 1422* (K!, NY!), *Ratter 5723* (K!, NY!).

Macrophyllidious *trees* potentially attaining 40 m with trunk to 1.5 m diam, but commonly smaller, sometimes flowering as treelets 2.5 m, the annotinous branchlets becoming corky and fissured; in almost all aspects other than the distinctive fruits resembling *S. tubulosa*, the pubescence similar in quality and distribution, either pallid or yellowish, the dimensions of lf-parts and fls sometimes but not consistently greater. *Stipules* ±4–6.5 × 1–1.3 mm. *Lf-formula* iii–iv/3–5; lf-stk of lvs on flowering branchlets 12–20 cm (of lvs from sterile or juvenile stems, no further mentioned, to 45–55 cm), the petiole 4–8 cm, at middle 1.7–2.8 mm diam, the longer interpinnal segments 3–5.5(–6.5) cm; lf-nectaries of *S. tubulosa*, the petiolar one at or shortly above base of petiole 1.2–3.5 mm

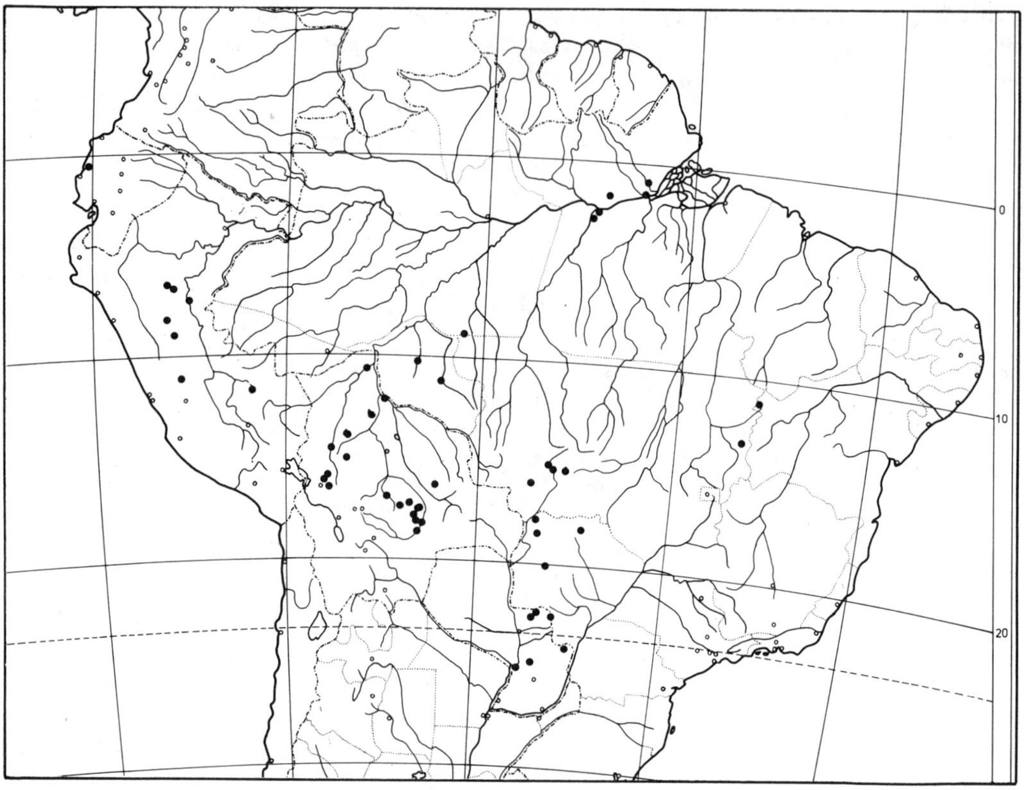

MAP 35. Tropical and subtropical South America. Range of *Samanea tubulosa* (Bentham) Barneby & Grimes.

diam; rachis of distal pinnae 4.5–11 cm, the longer interfoliolar segments 1.7–3 cm; lfts at apex either broadly obtuse muticous, or obtuse mucronate, or low-deltate apiculate, the blades of the penultimate pair ±4–6 × 2.1–2.5 cm, the venation as in *S. tubulosa. Peduncles* 6.5–15 cm; capitula 12–20-fld, the receptacle including terminal pedestal 1.5–4 mm, interfloral bracts linear-oblanceolate 3–10.5 × 0.45–1 mm; PERIPHERAL FLS: pedicel of lower ones 4–9.5 × 0.4–1 mm; calyx 7-11 × 2.7–4 mm, the teeth up to 0.9–2.6 mm; corolla 14–18.5 mm, the lobes 3–6 mm; androecium 24–35-merous, 42–54 mm long, the stemonozone 2–3.5 mm, the tube 8–12 mm; ovary sessile, linear in profile, tapering into the style, densely silky-strigulose overall; TERMINAL FL: differentiated apparently as that of *S. saman* but not studied in detail. *Pods* 1–2 per peduncle, sessile or almost so, in profile broad-linear 11–25 × 1.5–3.5 cm, only a trifle compressed and 1–2 cm thick, straight or randomly bent sidewise but not arcuately recurved, girdled by prominent, dorsally either plane or shallowly sulcate sutures 4–6 mm wide, the densely puberulent, often subvelutinous valves closely and deeply transversely wrinkled from suture to suture, internally composed of crustaceous or ligneous endocarp 0.2–1 mm thick in section and alveolate pulpy (pitchlike) mesocarp 2.5–6 mm thick, the cavity divided by septa into 1-seeded locules 5–6 mm long; dehiscence 0; seeds resembling those of *S. tubulosa*.

In forest, or southward in caatinga-forest, 100–710 m, discontinuously dispersed in E South America both N and S of the Amazonian Hylaea: in E Venezuelan Guayana (lower Río Paragua and Altiplanicie de Nuria in estado Bolívar) and adjacent Guyana (Kanuku Mts) and Brazil (Ilha de Maracá in N Roraima); and in E Brazil from centr. Maranhão to Pernambuco (Tapera) S to ±13°30′ in E-centr. Bahia (Maracás, Jequié, Jaguaquara); described from a tree, of unknown provenance, cultivated at Rio de Janeiro. — Map 36. — Fl. N of the equator III–V, in E Brazil X–III. — *Samán* (Venezuela); *coronha* (Roraima); *casqueiro* (Bahia).

Samanea inopinata is a cryptic species, not certainly distinguishable at anthesis from the closely

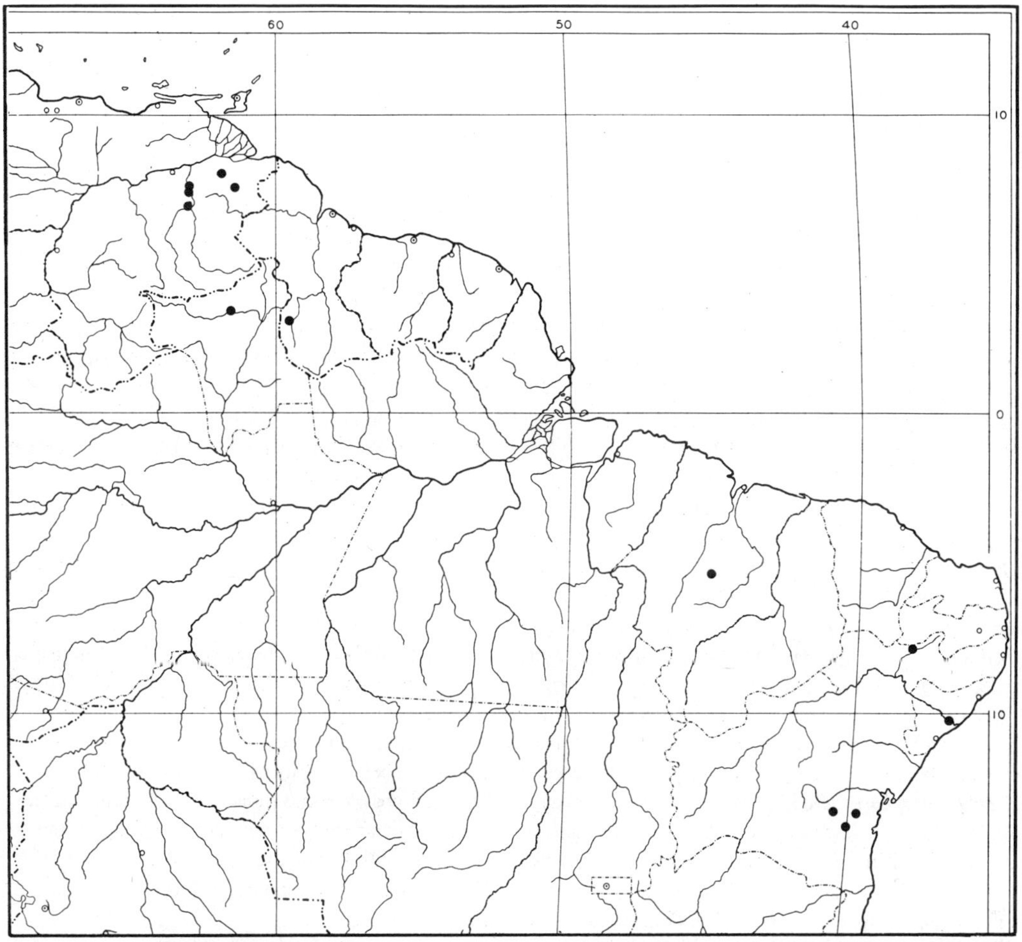

MAP 36. Northeastern and northern South America. Range of *Samanea inopinata* (Harms) Barneby & Grimes.

related *S. tubulosa.* The peripheral flowers of the capitula are potentially larger and borne on potentially longer pedicels, and the range in leaf-formula is narrower, but we have found overlap in all vegetative and floral characters that we have measured. The fruit, however, is instantly recognized, for the valves lack the continuous livid exocarp common to *S. tubulosa* and *S. saman* and instead are deeply and closely wrinkled horizontally from one suture to the other. Fruits of this sort seem to coincide with dispersal shown in Maps 35 and 36. In our assessment of allopatry we initially ignored flowering specimens, which are inherently ambiguous.

In its two foci of dispersal, as described above, the flower is equally variable in size, but the pod appears to have diverged into two forms. In eastern Brazil (*Mori 12191*, NY; the cultivated type collection of *Serianthes inopinata*), the pod is about 3.5 cm wide and 2 cm thick, with pulpy mesocarp about 6 mm and ligneous endocarp nearly 1 mm thick in section. By contrast, the populations in and around the Guayana Highland (type collection of *Pithecellobium nuriense*; *Lewis 1422*, NY) have pods only 1.5–1.8 cm wide and 1–1.3 cm thick, with mesocarp 2.5–4 mm and thinner, crustaceous endocarpic walls about 0.2 mm thick in section. Lewis and Owen (1989: 43) discussed the problem, but postponed a taxonomic conclusion until the relationship between these plants and the true *saman*, which herein is considered irrelevant, might be clarified. If the apparent difference in fruit-size is confirmed by new collections from the northern and southern lobes of the range of *S. inopinata*, we would admit *P. nuriense* as varietally distinct. The disjunction in range is less compelling insofar as precedents of similar bifocal dispersal without morphological differentiation are well established: for examples in Caesalpiniaceae see *Senna velutina* (Vogel) Irwin & Barneby and *Chamaecrista roraimae* (Bentham) Irwin & Barneby, described by Irwin and Barneby (1982: 232, 669).

The *Chloroleucon*-alliance

The exemplar analysis indentified another clade (Fig. 3A, no. 3) including *Enterolobium schomburgkii* Bentham (ENTSCHOMB), *Pithecellobium incuriale* (Vellozo) Bentham (INCURIAL), and *Chloroleucon mangense* (Jacquin) Macbride (CHLOROLE). This clade is supported by the following characters: the presence of perulate buds and leaves suppressed at the earliest nodes. Added to these taxa is *Pithecellobium blanchetii* (Bentham) Bentham (PBLANCHE), a species that in many previous analyses was sister-group to (INCURIAL + CHLOROLE) and one that also has perulate bud-scales and leaves suppressed at the earliest floriferous nodes. In the exemplar analyses both *E. schomburgkii* and *P. incuriale* were often sister-group to other taxa, and when *E. schomburgkii* alone was used as outgroup in this analysis, the cladograms were poorly resolved. As the possibility existed that, in the context of a species-level analysis, *E. schomburgkii* is actually a member of the ingroup, *Parkia* was added as outgroup. Also, *P. blanchetii* is also included just to be sure that in a species-level analysis it is not nested in the taxa comprising CHLOROLE and INCURIAL. The data set is shown in Table 3; the characters used in the analysis follow.

0. Vegetative shoots: 0 = monomorphic; 1 = dimorphic. Some species have ephemeral shoots that erupt from brachyblast-like buds. These shoots are both vegetative and reproductive and apparently do not persist more than one season.
1. Stems: 0 = unarmed; 1 = armed with sterile lignescent thorns derived from peduncles.
2. Perulate bud-scales: 0 = absent; 1 = present.
3. Stipules: 0 = quickly deciduous; 1 -lacking.
4. Leaves: 0 = late suppressed (suppressed at distalmost, younger nodes); 1 = coeval with inflorescences; 2 = early suppressed (suppressed at proximalmost, oldest nodes).
5. Pinnae: 0 = decrescent at both ends; 1 = proximally decrescent, thence subequilong; 2 = distally accrescent; 3 = one pair (rarely 2, then distally accrescent).
6. Venation of leaflets: 0 = pinnate; 1 = palmate; 2 = palmate-pinnate.
7. Veins from pulvinule (when venation palmate): 0 = reduced to midrib; 1 = always 3; 2 = 3 to 5.
8. Flowers: 0 = andromonoecious; 1 = all hermaphrodite.
9. Flowers: 0 = monomorphic; 1 = dimorphic.
10. Staminal tube: 0 = lacking; 1 = included in corolla; 2 = exserted from corolla.
11. Ovary: 0 = conic; 1 = truncate at apex.
12. Pod: 0 = straight to retrofalcate; 1 = coiled.
13. Valves: 0 = not twisting at dehiscence; 1 = twisting at dehiscence.
14. Sutures: 0 = both straight or shallowly undulate; 1 = ventral one undulate, the dorsal one straight.
15. Valves: 0 = not cracking between seeds; 1 = cracking between seeds.
16. Valves: 0 = not separating from sutures; 1 = separating from sutures.
17. Valves: 0 = not gaping to release seeds; 1 = gaping to release seeds.
18. Endocarp: 0 = not mealy; 1 = mealy or pithy and expanding. Note that *Enterolobium schomburgkii* has a mealy mesocarp, not endocarp.
19. Seeds: 0 = transverse in pod; 1 = with basipetal orientation; 2 = obliquely or subvertically descending.
20. Seeds: 0 = ellipsoid; 1 = round or lenticellar.
21. Seed-coat: 0 = hard; 1 = grapeskin-like.
22. Seed-coat: 0 = not translucent; 1 = translucent.
23. Seed-coat: 0 = not winged; 1 = winged and girdled with a submarginal nerve.
24. Pleurogram: 0 = hippocrepiform; 1 = lacking; 2 = complete.

The analysis yielded three trees of 66 steps, with a consistency index of 0.50 and a retention index of 0.63. One tree is shown in Figure 9, the alternative topologies, which will be discussed below, are shown in Figure 10. *Parkia* and *Enterolobium schomburgkii* are outgroups to the remainder of the species, which form a clade united by early-suppressed leaves (no. 4, with three homoplasious changes above this node), pinnae proximally decrescent thence subequilong (no. 5, four homoplasious changes above this node), leaflets with 3–5 veins from pulvinule (no. 7, two homoplasious changes above this node), and round or lenticellar seeds (no. 20). There are two major clades above the two outgroups. The first includes species distinguished by a grapeskin-like (no. 21), translucent (no. 22) seed-coat. *Pithecellobium blanchetii* (Bentham) Bentham is sister-group to the remainder of the species in this first clade. This species was arbitrarily included in the analysis (see above) as a precaution to be sure it is not nested among the other species in-

Table 3

Data set used in the analysis of the *Chloroleucon*-alliance

			1	1	2
	0	5	0	5	0
PARKIA†	0000-	0110-	00000	00100	00000
SCHOMBURGKII†	00100	0-011	111-1	00000	00001
INCURIALE	00112	12-10	10000	00100	11111
LIMAE	00112	22-10	10000	01000	11111
MINARUM	?0111	12-10	10000	00100	11111
FOEDERALE	?0111	30-10	10000	00100	11111
BLANCHETII	00100	11211	20000	00000	11101
FOLIOLOSUM	11111	11111	21001	10010	10000
MANGENSE	1*111	11*11	21000	1012	10000
EURYCYCLUM	1111?	1101?	??111	01000	10002
ACACIOIDES	11112	11011	11110	00002	10000
EXTORTUM	1111?	1111?	??010	00101	10000
GUANTANAMENSE	1111?	31210	10100	11000	10002
DUMOSUM	11112	21210	10011	1101?	10002
TORTUM	11111	11211	20111	00111	10000
TENUIFLORUM	11112	21210	10011	00011	10002
CHACOENSE	11112	31210	10000	00000	1000?

Note: † = outgroup; * = taxon polymorphic for this character; ? = state not observed in this terminal; - = state inapplicable or of uncertain homology.

cluded in this analysis. *Pithecellobium blanchetii* has annual, axillary, strictly reproductive branch systems, like those in species of *Albizia* and *Enterolobium*, and in previous analysis it often was sister-group to these taxa. In our opinion, given the weak resolution of the exemplar analysis, its relationships are still unresolved. Given the number of homoplasious characters not shared with the remainder of the clade, we are placing it in the monotypic genus *Blanchetiodendron* Barneby & Grimes (no. VII). The remaining species in this clade share the following synapomorphies: venation palmate-pinnate (no. 6), but changes to pinnate in FOEDERALE) and seed-coat winged and girdled with a submarginal nerve (no. 23). These taxa are placed in genus *Leucochloron* (no. VIII).

The remaining species share the following synapomorphies: dimorphic vegetative shoots (no. 0), stems armed with lignescent peduncles (no. 1), and a complete pleurogram (no. 24), and are put in *Chloroleucon* (no. IX). The first sister-group relationship is between *guantanamense* and *chacoënse*, both of which have leaves reduced to one or rarely two pairs of pinnae, and the remaining species, all of which have pods that twist at dehiscence (no. 13). Above this, one clade includes *eurycyclum* and *acacioides*, both with leaf-venation reduced to a midrib. The relationships of the other six species are less well resolved, and account for the three trees. The differences concern one species, *extortum*, and two characters: the condition of the endocarp (no. 18, mealy or not), and the position of the seeds (no. 19, transverse, basipetal or obliquely descending). Figure 10A shows the topology when a mealy endocarp (no. 18, state 1) and basipetally oriented seeds (no. 19, state 1) are synapomorphic. *Extortum* is then sister to *foliolosum* + *mangense*, and there is a reversal in character 18 from state 1 to state 0 in *extortum* and two homoplasious changes in character 19 above *extortum*. In Figure 10A character 19 state 1 is synapomorphic for all six species, and *extortum* is sister to the remaining five, which have character 18, state 1 as a synapomorphy. In Figure 10B character 18 state 1 is synapomorphic for all six species, and character 19 has three homoplasious changes above this node. *Extortum* then is sister taxon to *tortum*, and has a reversal in character 18. When the flowers of *extortum* become known, its relationships to the remaining species may well be resolved. Three characters concerning the flowers (no. 9, 10 and 11) all show changes among this group of species.

VII. BLANCHETIODENDRON
Barneby & Grimes

Blanchetiodendron Barneby & Grimes, gen. nov. monotypicum pro *Enterolobium* seu *Pithecolobium blanchetii* Bentham conditum, inter *Leucochloron* et *Albiziam* sect. *Arthrosamaneam* quasi intermedium, illius forsan propius: *Leucochlori* habet

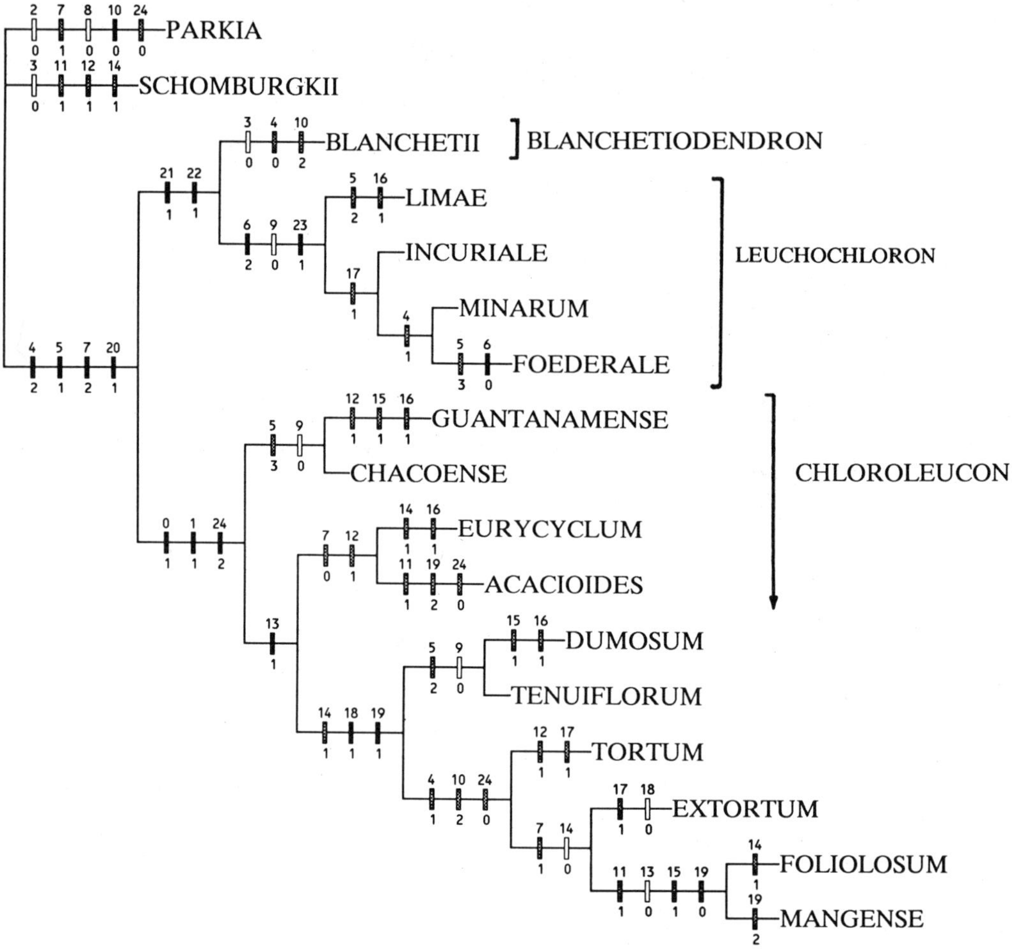

FIG. 9. The tree resulting from an analysis of the *Chloroleucon*-alliance.

gemmas perulatas striatas, legumen plano-compressum necnon semina disciformia tenuitestata pleurogrammate orbata, *Albiziae* autem inflorescentiam pseudoracemosum axillarem efoliatam vel terminalem paniculatam et flosculum terminalem heteromorphum, ideo ab utroque genere sejungendum. — Sp. unica: *B. blanchetii* (Bentham) Barneby & Grimes ≡ *Enterolobium blanchetii* Bentham.

Description of this monotypic genus is implicit in that of its one species given below. Its diagnostic features, shared as it were between *Leucochloron* and *Albizia* sect. *Arthrosamanea* are: striate axillary perules; an efoliate inflorescence consisting of axillary pseudoracemes (simple or sometimes remotely branched) and often a small terminal panicle; flowers dimorphic, the peripheral ones of each unit pedicellate, the terminal ones sessile; pod broad-linear plano-compressed, the valves papery, dehiscent inertly through both sutures; seeds thinly disciform, the lustrous testa translucent, narrowly winged by a prominent peripheral nerve, exareolate. The perules, pod, and seed are characteristic of *Leucochloron*, but the organization of the inflorescence and the umbelliform capitula including an enlarged sessile terminal flower would be anomalous in that setting. On the other hand, the inflorescence and pod, except for the seed, are compatible with *Albizia* sect. *Arthrosamanea*, but the perulate resting-buds and the thin exareolate seed-coat would be inappropriate exceptions in that genus. In effect, *Blanchetiodendron blanchetii* is clearly related on the one hand to *Leucochloron* and *Chloro-*

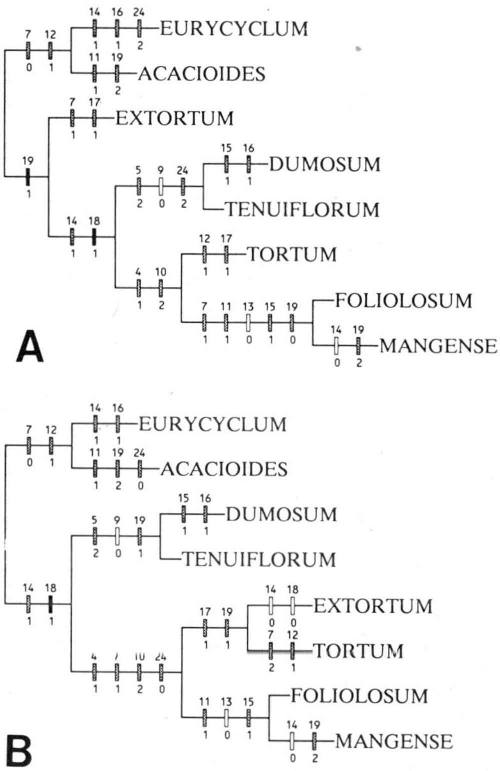

FIG. 10. Alternative topologies in an analysis of the *Chloroleucon*-alliance. **A.** Character 19, state 1, synapomorphic for six species; character 18, state 1, for five. **B.** Character 18, state 1, synapomorphic for six species; character 19 with three homoplasious changes.

leucon, and on the other to *Albizia* sect. *Arthrosamanea*, but morphologically isolated between these taxa. This ambiguous character is reflected in the taxonomic history. The species was first described, from flowering material, as an *Enterolobium*, and it was transferred to *Pithecellobium* when the fruit became known. Clearly not conspecific with *Pithecellobium* sens. str., it has recently been accommodated in *Albizia* sens. lat., but there also appears to us peripheral.

1. **Blanchetiodendron blanchetii** (Bentham) Barneby & Grimes, comb. nov. *Enterolobium* (?) *blanchetii* Bentham, London J. Bot. **3:** 224. 1844. — "Serra Acurua, in Bahia, *Blanchet*, n. 2776." — Holotypus K(herb. bentham.)! = NY Neg. *2015*; isotypi, †B = F Neg. *1182*, BM ("Sertão de S. Francisco")!, F!, NY(herb. meisner., the locality given as "Fazenda de Utinga, in Certão [sertão] de S. Francisco")!, P (3 sheets)!. — *Pithecolobium blanchetii* (Bentham) Bentham, Trans. Linn. Soc. London **30:** 587. 1875 & *in* Mart., Fl. Bras. **15(2):** 441. 1876. *Feuilleea incerta* O. Kuntze, Revis. Gen. Pl. **1:** 185. 1891. *Albizia blanchetii* (Bentham) Lewis, Leg. Bahia 162. 1987.

Macrophyllidious *trees* 5–12 m with trunk 1.5–2.5 cm dbh and lenticellate branchlets, appearing glabrous but the young branchlets and all lf- and inflorescence-axes thinly puberulent with incurved pallid hairs to 0.1–0.3 mm, the olivaceous lvs moderately bicolored, the lfts dull on both faces, a little paler beneath, commonly glabrous facially and minutely ciliolate but rarely (in Minas Gerais) rather densely puberulent dorsally overall, the inflorescence a series of short efoliate simple pseudoracemes or narrow panicles of few-fld umbelliform capitula, the lower pseudoracemes arising from the axil of coeval lvs and shorter than them, the upper ones forming a shortly exserted efoliate panicle; perulate buds axillary to many mature lf-stks, ovoid, 1–5 mm, the castaneous scales imbricate, dry, striately nerved, glabrous dorsally, sometimes minutely ciliolate. *Bud-scales* ovate to broadly triangular 0.3–2.5 mm long, proximal ones shorter than distal ones, striate, green or brown, thrown off as the new shoot begins to elongate, each leaving a linear transverse scar at base of branch. *Stipules* 0. *Lf-formula* (ii–)iii(–iv)/(8–)9–15; lf-stks 4–10.5 cm, the petiole 2.5–5.5 cm, the longer interpinnal segment 1.1–2.4 cm, the ventral groove continuous between pinna-pairs; petiolar nectary near or above midpetiole, sessile, cupular or shallow-cupular 1–1.6 mm diam, in profile 0.3–0.6 mm tall, often imperfect or eaten and in random lvs apparently immersed or even obsolete, minute nectaries at tip of some pinna-rachises, none between pinna-pairs; first pair of pinnae somewhat shorter than further ones, the rachis of these 5.5–10 cm, its longer interfoliolar segments (5–)6–10 mm; pulvinules in dorsal view 0.6–1.1 × 0.4–0.6 mm; lfts proximally decrescent, the anterior one of the first pair often greatly reduced or absent, the rest subequilong, the blades narrow-oblong from inequilaterally cuneate or wide-flabellate base, sharply deltate-apiculate, the larger ones 12–20 × 4–6 mm, 2.7–3.5 times as long as wide; venation of 4–5 primary nerves from pulvinule, the dorsally prominulous pallid midrib a trifle forwardly displaced, either straight or gently incurved, the inner posterior nerve expiring below midblade, the outer posterior and one anterior primary nerves very short, the midrib giving rise on each side to 5–8 ascending, very slender secondary nerves mostly expiring or only obscurely camptodrome within the plane margin, the whole venation of upper face immersed or finely engraved, prominulous only beneath. *Primary axis* of each pseudoracemose or paniculate element of the inflorescence 2–6 cm, the secondary axes, when present, not over 1.5 cm, some of them arising from same node as pedicels; peduncles 3–6 per node of inflorescence, the longest 8–13 mm, the capitula umbelliform 5–11-fld, the fls dimorphic, the peripheral ones slenderly pedicellate, the terminal one sessile on a short pedestal, the receptacular axis 1–2(–2.5) mm; PERIPHERAL FLS; pedicel 1.5–6 × 0.2–0.5 mm; perianth 5-merous; calyx turbinate-campanulate or somewhat urn-shaped 2–3 mm, externally either glabrous or minutely puberulent, the depressed-deltate ciliolate teeth 0.1–0.25 mm; corolla 4–5 mm, silky-strigulose externally, the ovate lobes 1.4–2.2 × 1–1.2 mm; androecium 26–38-merous 11–16 mm, the stemonozone 0.3–0.45 mm, the tube 1.5–2.2 mm, included in corolla; ovary glabrous, symmetrically conical at apex; style about as long as the white filaments, the stigma poriform; TERMINAL FL: calyx 2.2–3 × 2 mm; corolla 4.5–6 mm; androecium 30–46-merous, the tube 3.5–6.5 mm, the free portion of some filaments dilated at base. *Pods* solitary pendulous, the plano-compressed body in profile broad-linear straight 12–22 × 2.7–4.3 cm, abruptly contracted at base into a stipe 4–8 mm, rounded and mucronate at apex, when well fertilized 8–10-seeded, framed by almost straight, 2-ribbed and centrally sulcate sutures ±1.3 mm wide, the papery, brown or livid-brown valves glabrous but often minutely granular along the sutures, externally finely and sharply transverse-venulose, whitish and smooth within, quite free from each other between seeds; dehiscence through both sutures, inert; funicles filiform, sigmoidally contorted at apex; seeds transverse at middle of pod, disciform 10.5–12 × 8–10 × 0.8–2 mm, a trifle longer than wide, the pallid grapeskin testa translucent, produced around the seed's periphery as a wing near 1 mm wide, pleurogram 0; the pale green cotyledons concealing the radicle; endosperm 0.

In semideciduous forest, mata de cipó, and cerrado, sometimes on granitic outcrops, between 400 and 1000 m, not uncommon along the crest and valleys of Chapada Diamantina between 10° and 14°S in Bahia, Brazil, extending W to Sa. do Açuruá and E down the Paraguaçú to Cruz de Almas; somewhat isolated in middle valley of Río Jequitinhonha in NE Minas Gerais at latitude 16–17°S. — Map 37. — Fl. X–XII; fr. II–VIII. — *Canela de burro*; *canzil*.

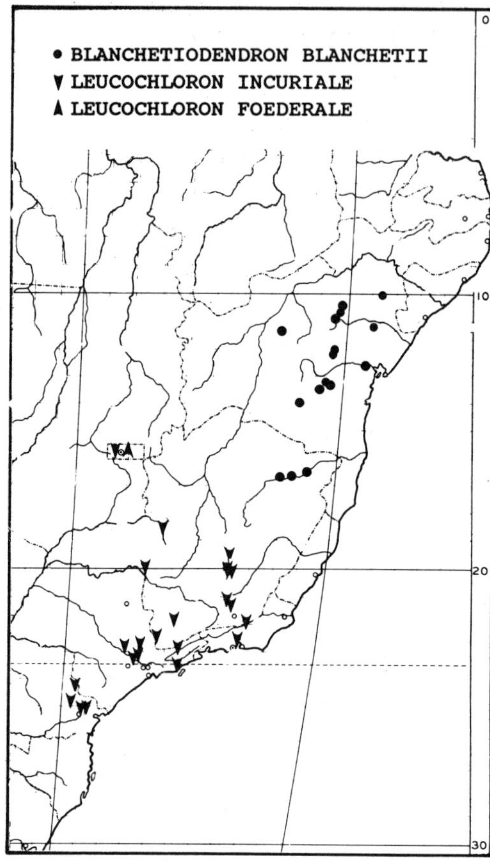

MAP 37. Eastern upland Brazil. Range of *Blanchetiodendron blanchetii* (Bentham) Barneby & Grimes, *Leucochloron incuriale* (Vellozo) Barneby & Grimes, and *L. foederale* (Barneby & Grimes) Barneby & Grimes.

VIII. LEUCOCHLORON
BARNEBY & GRIMES

Leucochloron Barneby & Grimes, gen. nov., hinc *Chloroleuco*, illinc *Blanchetiodendro* affine et utroque gemmis perulatis simile, sed a priori spinis axillaribus (pedunculis sterilibus induratis) nullis, legumine lato-lineari plano-conpresso necnon seminum disciformium testa membranacea exareolata, a *Blanchetiodendro*, quoad legumen et semina congruo, capitulis axillaribus (nec pseudoracemosis paniculatisve) flosculisque capitatis homomorphis (nec umbellatis heteromorphis) distans. — Sp. typica: *L. incuriale* (Vellozo) Barneby & Grimes ≡ *Mimosa incurialis* Vellozo. — Etymology: Gr. *leucos*, white + *chloros*, greenish yellow; anagram of *Chloroleucon*.

Unarmed, multi- or plurifoliolate *trees* with monopodial branches, the young growth puberulent or pilosulous with pallid, yellowish or rusty hairs; a perulate bud of often striate scales in the axil of many mature lvs and at apex of branchlets. *Stipules* caducous. *Lf-formula* (i–)ii–ix/4–27; petiolar nectaries sessile, cupular or almost plane, the first one borne usually between or close below proximal pinna-pair, but sometimes below midpetiole; lft-pulvinules mostly <1 mm; venation of lfts palmate-pinnate, or in *L. foederale* primarily pinnate. *Flowers* capitulate homomorphic, the capitula either solitary or fasciculate, borne on hornotinous branchlets either immediately below or axillary to coeval lvs; perianth 5-merous, the calyx deeply campanulate, the corolla tubular-infundibuliform; androecium 20–40-merous, the tube about as long as corolla or often well exserted, the filaments white; no intrastaminal disc; ovary sessile or almost so, in profile elliptic; stigma scarcely or not dilated. *Pods* (unknown in *L. minarum*) sessile or shortly stipitate, in profile broad-linear, either straight or gently decurved, the valves stiffly papery or coriaceous; dehiscence tardy, inert, through one or both sutures, the valves narrowly gaping to emit the seeds; funicle filiform; seeds transverse, discoid (<1 mm thick), either discrete in one rank or imbricate, the thin, lustrous and translucent testa narrowly winged peripherally and girdled with a submarginal nerve, lacking pleurogram; endosperm 0. — Spp. 4, of Atlantic and planatine Brazil, from NE Bahia S to Paraná, inland to montane Minas Gerais, and to Distrito Federal; a partially known mimosoid tree found in the middle Orinoco Valley in state of Guárico, Venezuela, is suggestive of a fifth, distantly disjunct species of the genus.

The generitype of *Leucochloron* was misplaced by Bentham (1875: 591) in *Pithecolobium* ser. *Coriacea* (= *Macrosamanea* of this revision), from which it differs in perulate resting-buds, in inflorescence architecture, in lack of nectary on lower bracts of the capitulum, and in thin shiny disciform seeds. Nielsen (1981: 190) treated it as of unknown affinity, characterized by floral traits of *Albizia* but by seeds reminiscent of *Zygia*. We have not confirmed the presence of a hexamerous terminal flower with two ovaries, described by Nielsen, and suppose he must have happened on a random anomalous flower such as occurs rather frequently in pithecellobioid Ingeae. *Leucochloron incuriale* forms an apparently natural group with *L. minarum* and *L. foederale*, which collectively resemble *Chloroleucon* in striate perules and in inflorescence-type, but differ in the lack of thorny axillary branchlets, in the broad, stiffly papery pod, and in the discoid exareolate seeds. Its inflorescence is essentially that of *Enterolobium*, which differs most obviously in the obese indehiscent fruit.

A potential addition to the four species described below is the *couvi*, mentioned by Lewis (1987: 180) as related to *L. incuriale* but differing in the pod, which is contricted between seeds as in *Pseudopiptadenia*. This puzzle is known from the São Francisco valley near Barreiras in western Bahia; we have seen no specimens. A fragmentary branchlet bearing two leaves and immature pods, collected in the state of Guárico, Venezuela (Río Tamanaco, *H. M. Curran 703*, NY), has the perules of *Leucochloron* and a broad, plano-compressed fruit compatible with this genus (seeds not seen). We cannot assimilate this specimen into any mimosoid described or recorded from Venezuela and hazard the guess that it may represent another instance of amphi-Amazonian disjunction.

As the species of *Leucochloron* differ primarily in number of pinna-pairs or leaflets, or by autapomorphic characters, their relationships are not resolved (Fig. 10). Furthermore, the fruit of *L. minarum* is still unknown.

Conspectus of genus *Leucochloron*

¶ Lf-formula iii–ix/8–27; largest lfts 5–12 mm; stipules of each pair alike; Bahia, Minas Gerais, and S to Paraná.
1. *L. limae* Barneby & Grimes
2. *L. incuriale* (Vellozo) Barneby & Grimes
3. *L. minarum* (Harms) Barneby & Grimes

¶¶ Lf-formula (i–)ii(–iii)/2–6; largest lfts 16–38 mm; stipules of each pair dissimilar; Distrito Federal.
4. *L. foederale* (Barneby & Grimes) Barneby & Grimes

Key to the described species of genus *Leucochloron*

1. Lf-formula iii–ix/8–27; stipules homomorphic; venation of lfts palmate-pinnate; Atlantic slope of Brazil from Bahia to Paraná, inland only in Minas Gerais.
    2. Stipules 1–3 mm; capitula globose; corolla silky-puberulent.
        3. Trees (3–)5–12 m; petiolar nectary between or close below first pinna-pair; lfts of longer pinnae (11–)14–27-jug., at apex sharply triangular or deltate-apiculate, (2.6–)2.9–4 times as long as wide; pod 10–19 cm, the 12–18 seeds 19–23 mm diam, either discrete in one file or narrowly irregularly imbricate; Minas Gerais southward. . . . . 2. *L. incuriale* (p. 134)

3. Shrubs 1–3 m; petiolar nectary close to or well below midpetiole; lfts of longer pinnae 8–14-jug., at apex broadly obtuse; pod ±7 cm, the ±8 seeds 14 mm diam, strongly imbricate; E centr. Bahia . . . . 1. *L. limae* (p. 132)

2. Stipules at lower nodes of new shoots, beyond the perular scars, lance-elliptic, to 5–7 mm; capitula hemispherical; corolla glabrous; Minas Gerais, the precise dispersal and pod unknown . . . 3. *L. minarum* (p. 135)

1. Lf-formula (i–)ii(–iii)/2–6; each pair of stipules dimorphic; venation of lfts pinnate; Distrito Federal, to be sought in adjacent Goiás . . . . 4. *L. foederale* (p. 135)

1. Leucochloron limae Barneby & Grimes, sp. nov., *L. incuriali* arcte affinis, sed frutex (nec arborescens), foliorum nectario infra vel juxta medium petiolum sito, pinnarum majorum foliolis minus numerosis 8–14 (nec 14–27)-jugis apice late obtusis, necnon leguminis minoris seminibus ±8 (nec 12–18) ±14 (nec 19–23) mm longis; ulterius patria distanti diversa. — BRAZIL. Bahia: Serra Preta, Rod. BA-052, 35 km antes de Ipirá, 7.IX.1990 (fr.), *H. C. de Lima & al. 3877.* — Holotypus, RB. — Ibid., eodem die (fl.), *H. C. de Lima 3875*; paratypus, NY. Fig. 11.

Shrubs 1–3.5 m with lenticellate glabrate annotinous and older branches, the young stems and all axes of lvs and inflorescence pilosulous with gray, often rufescent hairs to 0.2–0.35 mm, the lvs bicolored, when dry brownish, olivaceous and either glabrous or transiently puberulent on upper face, paler and often barbellate dorsally in anterior basal angle of midrib, ciliolate, the globose capitula of whitish fls arising either solitary or geminate from the lowest, efoliate nodes of new branchlets; resting buds perulate, ovoid ±1–1.7 mm, pubescent, the scales not perceptibly nerved externally. *Stipules* triangular 2 mm, caducous. *Lf-formula* iv–vii/8–14; lf-stks 3–7 cm, the petiole 8–18 mm, at middle 0.7–1.1 mm diam, the longer interpinnal segments 4–15 mm; a round or elliptic, cupular thick-rimmed nectary 0.6–1.5 mm diam near or well below midpetiole, and similar ones between 1–2 furthest pinna-pairs, a much smaller one at tip of most pinna-rachises; pinnae usually a little accrescent distally, the rachis of distal ones 2.5–6 cm, the longer interfoliolar segments 2–4.5 mm; lft-pulvinules discolored, transversely elliptic ±0.35–0.5 × 0.6–0.8 mm; lfts a little decrescent proximally and the first anterior one usually reduced to a paraphyllidium, all except the obliquely obovate terminal pair oblong from obliquely truncate base, broadly obtuse (some very obscurely apiculate), the longer ones 5–11.5 × 2.5–5.5 mm, 2–2.2 times as long as wide; venation at first palmate, the nearly straight midrib displaced to divide blade ±1:1.5–2, the inner posterior primary nerve incurved to anastomosis near or short of midblade, the outer one(s) very short, the midrib giving rise on each side to 3–4 sinuous secondary and random tertiary nerves, the venation immersed or almost so above, bluntly prominulous beneath. *Peduncles* 2–2.5 cm; capitula ±25–35-flowered, without filaments 9.5–10 mm diam, the sessile homomorphic fls finely yellowish-puberulent overall, the receptacle ±1.5 mm; bracts spatulate-oblanceolate ±1–1.5 mm, deciduous; calyx narrowly campanulate from rounded or cuneate-obconic base, 3.4–3.7 × 1.2–1.5 mm, the unequal teeth 0.2–0.7 mm; corolla tubular 5.7 mm, scarcely wider upward, the 5 lobes nearly 1 mm; androecium 36-merous, 12 mm, the stemonozone 0.6 mm, the tube 5.5 mm, a little thickened at base but lacking a disc; ovary sessile, narrowly ellipsoid, glabrous. *Pods* 1 per capitulum, apparently ascending, sessile, in profile linear-oblong ±7 × 1.7–1.8 cm, obtuse at each end, straight or almost so, plano-compressed, the leathery valves framed by plane sutures ±1.2 mm wide, externally dull brown, obscurely venulose, densely minutely puberulent overall, internally dull stramineous, the cavity continuous; seeds ±8, transverse on filiform funicle, imbricate lengthwise, disciform ±14 mm diam, the thin lustrous translucent testa produced as a peripheral wing ±0.8 mm wide, the plano-compressed greenish embryo not quite filling the seed-cavity, its suborbicular-obcordate cotyledons ±12 mm diam, their basal auricles not concealing the radicle.

In mata de cipó and caatinga, at unrecorded elevations, locally plentiful in the lower Paraguaçú basin (muns. Ipirá, Itaberaba) near 40°W, 12°–12°30′S, in NE Bahia, Brazil. — Fl. IX–X.

Leucochloron limae was first known to us through flowering specimens from Itaberaba (*Hatschbach 48194*, NY), which we had laid aside as representing a probably undescribed congener of *L. incuriale*, the known range of which extends no farther north than the São Francisco-Paraná divide near 18°S. Young flowering and ripe fruiting material collected in the same region by Dr. Haraldo de Lima in 1990 confirmed our first impression of its distinct nature, and we have pleasure in dedicating the species to this generous colleague and noted student of South American Leguminosae. *Leucochloron limae* resembles *L. incuriale* in essential characters, but differs in lesser

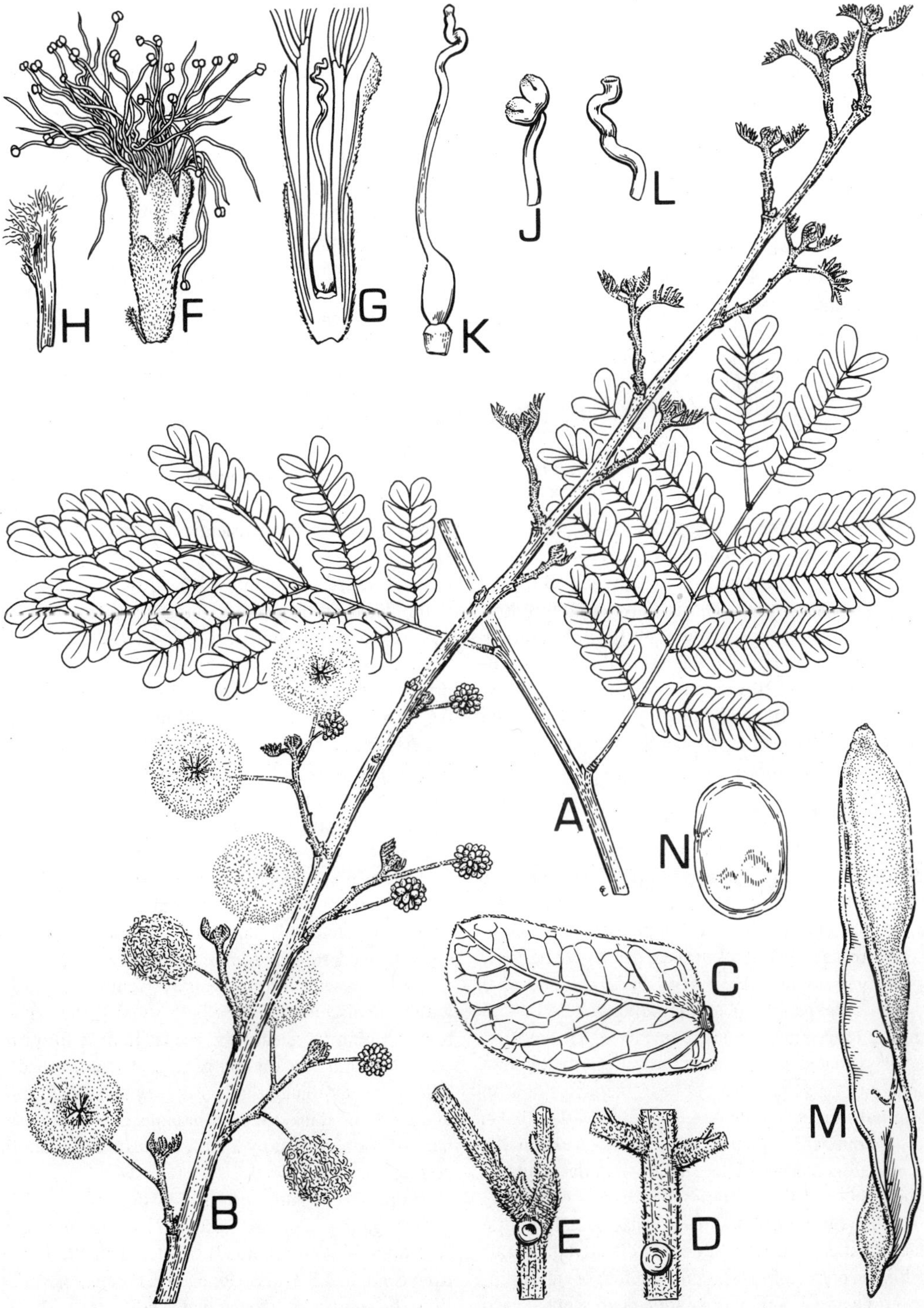

FIG. 11. *Leucochloron limae* Barneby & Grimes. **A.** Mature leaves. **B.** Flowering branch showing precocious inflorescences. **C.** Leaflet, lower surface. **D.** Petiolar gland. **E.** Gland between terminal pinnae-pair. **F.** Flower. **G.** Longitudinal section of flower. **H.** Bract. **J.** Stamen. **K.** Gynoecium. **L.** Stigma and style. **M.** Dehisced fruit. **N.** Seed. (All from *Lewis* in *Carvalho 5339.* Drawn by Eleanor Catherine (K); contributed by G. P. Lewis.

stature, in position of the petiolar nectary, in fewer, proportionately broader, more obtuse leaflets, and in the small, relatively few-seeded fruit.

2. Leucochloron incuriale (Vellozo) Barneby & Grimes, comb. nov. *Mimosa incurialis* Vellozo, Fl. Flum. Icon. **11:** t. 24. 1829. — No locality given, Vellozo's *Observationes* on this species inadvertently omitted from the text in Arch. Mus. Nac. **5:** 435 [*Mimosa* no. 22]. 1881. — Holotypus, the cited plate, as interpreted by Bentham (1844). — *Pithecolobium incuriale* (Vellozo) Bentham, London J. Bot. **3:** 212. 1844. *Feuilleea incurialis* (Vellozo) O. Kuntze, Revis. Gen. Pl. **1:** 188. 1891.

Pithecolobium martianum Bentham, London J. Bot. 3: 218. 1844. — "Brazil, *Martius, Herb. Fl. Bras.* n. 1101." — Holotypus, K(herb. bentham.)! = NY Neg. *2021*; isotypus, NY!. — Equated with the preceding by Bentham, 1875: 581.

Amply microphyllidious, unarmed *trees*, fertile when (3–)5–12 m, the young branches and all lf-axes and peduncles densely sordid- or rusty-pilosulous with more or less sinuous hairs to 0.4–0.8 mm, the lvs bicolored, the facially glabrous but minutely ciliolate lfts brown-olivaceous above, paler green beneath, the globose capitula of silky fls arising singly or 2–4 together on hornotinous branchlets, either from the earliest efoliate nodes or from more distal foliate ones, or from both; the perulate buds terminal or 2–4(–5) in axils, depressed- to elongate-ovoid, 1.5–5 × 1.25–2.25 mm, the scales short-persistent after rupture of bud. *Stipules* caducous (lacking from many herbarium specimens), ovate from broad clasping base, 1–3 mm, externally nerveless or striately 3–several-nerved, densely pubescent. *Lf-formula* v–viii(–ix)/(11–)14–25(–27); lf-stks (5–) 6–13 cm, the petiole including pulvinus (1–)1.5–3.5 cm, the longer interpinnal segments 7–15 mm; first nectary situated on petiole proper shortly below first pinna-pair or sometimes between them, variable in form but usually sessile or subsessile 0.3–1 × (0.4–) 0.7–1.8 mm, either glabrous or pilosulous externally, the secretory surface either shallow-cupular or plane, similar but smaller nectaries between one or more distal pairs of pinnae and lfts; pinnae usually a little decrescent proximally, scarcely so distally, the rachis of longer ones (3.5–)4–10 cm, the longer interfoliolar segments ±2–4 mm; lfts a little decrescent at each end of rachis or subequilong, the first pair either arising contiguous to the pinna-pulvinus or represented there, either on both sides or on anterior side only, by linear paraphyllidia 0.5–1.4 mm, the blades subfalcately lanceolate or lance-oblong from an obliquely truncate base, obtusely cordate on proximal side, at apex sharply triangular or acutely deltate-apiculate, the larger ones (5–)6–10(–11) × 1.4–3.4(–3.7) mm, (2.6–)2.9–4 times as long as wide; venation of 3(–4) primary nerves from pulvinule, the midrib a little forwardly displaced from midblade, 3–4-branched on each side beyond midblade, the secondary nerves brochidodrome well within the nearly plane margin, the inner posterior primary nerve incurved to or beyond midblade, the outer one(s) very short, a few weak tertiary venules commonly developed and all venation finely prominulous dorsally, only the stronger veins raised on upper face. *Peduncles* 10–22 mm, ebracteolate: capitula globose, densely many-fld, prior to anthesis moriform, without filaments 9–12 mm diam, the subglobose, nonalveolate receptacle 1.2–2.4 mm diam; bracts linear-spatulate 1–1.4 mm, pilosulous distally, sometimes persistent after fall of fls; fls homomorphic, sessile, silky-pubescent externally, greenish white when fresh, the filaments then also whitish; calyx narrowly campanulate, scarcely dilated distally, 2.3–3.2 × 1.6–2 mm, the 5 ovate teeth 0.3–0.8 mm, subincurved; corolla cylindric, slightly dilated upward, 4–6 mm, the ovate teeth 0.6–1 mm; androecium 24–27-merous, 9–13 mm, the staminal tube 4–7 mm, as long as or to 1.5 mm longer than corolla, the stemonozone 1 mm, the free filaments 4.5–6 mm; style longer than the stamens, the very slightly dilated stigma ±0.15 mm diam; ovules 14–18. *Pods* usually 1 per capitulum, sessile, in profile broad-linear, either straight or gently decurved, 10–19 × 3.2 cm, plano-compressed, the straight or shallowly undulate, 3-costate or incipiently winged sutures framing the flat valves, these leathery brown dull, obscurely venulose, pallid within; dehiscence through both sutures, these narrowly gaping to release the seeds; funicle filiform; seeds horizontal at middle of pod, discoid, broadly elliptic or suborbicular in broad view and 19–23 mm diam, <1 mm thick, the membranous, lustrously castaneous testa girdled by a dark nerve, pleurogram 0; no endosperm.

In open woodland, bush-islands in campo, and in cerrado-gallery ecotone, mostly ±710–1500 m (exact elevations seldom recorded), scattered in SE Brazil from centr. and S Minas Gerais to E Paraná, in Minas W to the sources of Río Paranaíba at Patos de Minas. — Map 37. — Fl. (VIII–)IX–XII. — *Angico rajado, Chico Pires, cortiça, corticeira, páo de cortiça, sucupira do campo* (in Monte Alegre, S. P.).

3. Leucochloron minarum (Harms) Barneby & Grimes, comb. nov. *Pithecolobium minarum* Glaziou [Bull. Soc. Bot. France, Mém. **3B:** 190. 1906, nom. nud.] *ex* Harms, Feddes Repert. Spec. Nov. Regni Veg. **17:** 92. 1921. — "Brasilien: Minas Geraes, São José do Lagedo, im Campo (Glaziou no. 13789, 13790)." — Syntypi, †B, one (no. *13789*) surviving as fragm. and as F Neg. 1212!, this here designated as lectotypus; isotypi, K!, P!; paratypi, *Glaziou 13790*, K!, P (3 sheets)!.

Microphyllidious *trees* ("grand arbre") of unrecorded stature, with rough grayish lenticellate annotinous branchlets, the new ones, like all lf-axes and peduncles, pilosulous with loose, forwardly incurved hairs to ±0.15–0.5 mm, the foliage strongly bicolored, the lfts when dry dark brown above, pale fawn beneath, glabrous facially but microscopically ciliolate, the small, loosely flowered capitula arising singly in the first 1–2 lf-axils of expanding branchlets; perulate buds 3–7 mm, the striate castaneous scales minutely puberulent dorsally. *Stipules* at nodes distal to the perules linear-elliptic, to 5–7 mm, caducous. *Lf-formula* iii–v/12–16(–"17" ex char.); lf-stks 2–4.5 cm, the slender petiole 8–15 mm, the longer interpinnal segments 6–10 mm; a sessile cupular nectary 0.6–1.4 mm diam situated above midpetiole, sometimes close to first pair of pinnae, smaller ones at tip of lf-stk and of pinna-rachises; pinnae decrescent proximally, the axis of further ones 3.5–5 cm, the longer interfoliolar segments 2–2.6 mm; lfts in outline and venation like those of *L. incuriale*, but more acutely auriculate at base and porrect at apex, the largest 5.5–7.5 × 2–3 mm. *Peduncles* 17–32 mm; capitula hemispherical, laxly ±20–28-fld, the receptacle obovoid 1.5–2 mm, the homomorphic fls at anthesis all ascending; bracts filiform 1.5–2 mm, puberulent, caducous; perianth 5-merous; calyx campanulate, a little dilated upward, 2–2.5 × 0.6–1.2 mm, thinly hispidulous externally, the deltate or subulate teeth 0.3–0.9 mm; corolla tubular, slightly dilated upward, 3.8–4.4 mm, glabrous, the tube weakly nerved, the ovate lobes 0.6–0.8 mm; androecium ±32–40-merous, 10–11.5 mm long, the stemonozone 0.4–1 mm, the tube 5–6.5 mm, well exserted from corolla; ovary sessile linear-ellipsoid glabrous; stigma ±0.1 mm diam. *Pods* unknown.

In campo, at unrecorded elevation, known only from the type locality at São José do Lagedo, this not presently located, and from "Olho d'Agua, Minas Gerais" (*St. Hilaire s.n.*, P), possibly the place of that name near 22°S latitude in NE São Paulo, visited by St. Hilaire in mid-October 1819.

In addition to the type and syntype from São José do Lagedo we have seen of *L. minarum* only one other collection, obtained also in Minas Gerais by Auguste de St. Hilaire in the early nineteenth century. Unfortunately, neither locality has been identified with assurance, only flowering material was collected, and our picture of the species is thus lamentably incomplete. Until the fruit is discovered a shadow of doubt remains as to the affinity of the species. The resemblance to *L. incuriale* is, however, so close, that we feel justified in proposing the transfer from *Pithecellobium* to *Leucochloron*. Relatively large stipules at nodes immediately succeeding the perulate scales, relatively low leaf-formula and short leaves, looser hemispherical rather than dense globose capitula, and glabrous corolla are the features on which we depend at present to distinguish *L. minarum*, but each and all are in need of confirmation.

4. Leucochloron foederale (Barneby & Grimes) Barneby & Grimes, comb. nov., *Pithecellobium foederale* Barneby & Grimes, Brittonia **36:** 238, fig 2. 1984. — "BRAZIL. Distrito Federal: Côrrego Landim ±20 km NE of Brasília, 7 May 1966 (fr vix maturo); *H. S. Irwin, Grear, Sousa & Reis dos Santos 15672*." — Holotypus, UB!; isotypi, C!, G!, GH!, K!, LE!, NY!, P!, R!, S!, US!.

Macrophyllidious *trees* (2–)5–15 m with trunk attaining 2 dm diam, the slender annotinous branches fissured gray, the reddish hornotinous ones with all lf-axes and peduncles villosulous with mostly incurved yellowish hairs 0.2–0.5 mm, the bicolored lfts olivaceous and except for puberulent midrib glabrous above, beneath paler dull and either glabrous or pilosulous, the margin ciliolate, the subglobose capitula axillary, solitary, immersed in foliage; ovoid perulate buds in most lf-axils. *Stipules* dimorphic, one of each pair lance-ovate, ±4–8 × 1.5–2 mm, papery, castaneous, striate, the other much narrower, all ciliolate distally, caducous. *Lf-formula* (i–)ii(–iii)/2–4, or 5–6, the inner lft of first pair often suppressed; lf-stks including livid pulvinus 12–20 mm; petiole 3–8 mm, the 1(–2) interpinnal segments 7–12 mm; cupuliform nectaries between or immediately below each pair of pinnae 0.5–1.2 mm diam, 0.2–0.4 mm tall; pinnae accrescent upward, the rachis of distal pair 3–4.5 cm; lfts upwardly accrescent, the blades all asymmetrically trapezoid, semicordate at base, obtuse mucronulate at apex, those of distal pair 16–38 × 5–15 mm; venation pinnate, the straight oblique midrib giving rise on each side to several widely ascending secondary venules and a reticulum prominulous on both faces of young lfts but later immersed. *Peduncles* (1.5–)2–3 cm; capitula 30–40-fld, the fls homomorphic, all sessile, the conic receptacle ± 1.5 mm; bracts linear-filiform 2–5 mm; calyx cylindro-campanulate, 2–2.8 mm, a little dilated in upper third, sordid-puberulous, the tube 5-nerved, the depressed-deltate teeth ±0.2 mm; corolla 4.5–6 mm, the tube 2.5–4 mm, glabrous, the campanulate limb ± 2 mm, its ovate lobes ±0.5 mm, externally puberulous; androecium 20–25-merous, the greenish-white filaments united into a tube 7–8.5 mm, free through 6(–7) mm; anthers ±0.3 mm diam; ovary elliptic, substipitate, puberulent distally; style glabrous, 10–12.5 mm. *Pods* indehiscent while on the tree, in profile broad-linear, plano-compressed 11–18 × 1.9–2.5 cm, rounded mucronulate at apex, tapering at base into a stipe 1–5 mm, (3–)7–8(–9)-seeded, the slightly raised sutural rim shallowly undulate and crispate, the stiffly papery, glabrous or toward base puberulent valves becoming brown and finely venulose, scarcely elevated over each seed; funicle filiform; seeds plano-compressed, in broad profile obovate 9 × 7 mm, subacutely marginate, the brown testa minutely punctate overall, pleurogram 0.

In cerrado and gallery woodland, ±900–1000 m, known only from Chapada do Contagem (Côrrego Landim; Fercal) in Distrito Federal, Brazil. — Map 37. — Fl. IX–X, the fruit slowly maturing in V–VIII.

This is a well-defined taxon, differing from *L. incuriale* and *L. minarum* in dimorphic stipules and relatively few, pinnately venulose leaflets, and from the former in solitary, not fasciculate capitula. In the protologue it was misplaced, following Bentham's disposition of *P. incuriale*, in *Pithecellobium* ser. *Coriacea*, herein equated with the genus *Macrosamanea*.

IX. CHLOROLEUCON (BENTHAM) BRITTON & ROSE

Chloroleucon (Bentham) Britton & Rose, N. Amer. Fl. **23:** 36. 1928. *Pithecolobium* sect. *Chloroleucon* Bentham, London J. Bot. **3:** 197, 221, 1844. — Lectogeneritypus (Britton & Rose, 1928, l. c.): *Ch. vincentis* (Bentham) Britton & Rose ≡ *Pithecolobium vincentis* Bentham ≡ *Chloroleucon mangense* var. *vincentis* (Bentham) Barneby & Grimes.

Chloroleucum (Bentham) Britton & Rose *ex* Record, Trop. Woods **10:** 24. 1927, sine descriptione generica. — Following *Index Nominum Genericorum* this antedates *Chloroleucon* Britton & Rose and is typified by *Ch. guatamalense*, a then newly described taxon obviously long posterior to the basionymic section *Chloroleucon* Bentham. Record merely provided a description for a plant that had been identified by Britton & Rose as a new species of the then undescribed genus *Chloroleucon* and misspelled the name.

Cathormion sensu Pittier, Arb. Legum. **1:** 54–56. 1927; non Hasskarl, 1855.

Trees and *shrubs* adapted to an annual dry season, the lvs then partly or wholly deciduous, the annual growth differentiated into long- and short-shoots, long-shoots floriferous only their first year, the annotinous branches stiffly flexuous, lenticellate, randomly armed with solitary or paired, axillary thorns derived from lignescent sterile peduncles (but some branches, or some rare individual plants, unarmed), the flowering branchlets rapidly emergent from a cone of imbricate, striately nerved, caducous perules, the flowers at anthesis seldom coeval with mature lvs; indumentum of plain, forwardly curved hairs, sometimes lacking, never glandular. *Stipules* most commonly lacking, when present subulate, filiform, linear, or oblanceolate-elliptic, quickly deciduous. *Lf-formula* i–ix/1–46, the subsessile lfts linear to obovate, the size and outline reciprocally adjusted to number; petiolar nectary concave, sessile or almost so, usually near or below midpetiole, rarely between first (or only) pair of pinnae, smaller nectaries at tip of lf-stk and pinna-rachises. *Peduncles* either solitary or fasciculate at 1–4 lower, foliate or efoliate nodes of brachyblasts, surpassed by mature lvs, ebracteolate; fls mostly ascending, either capitate or densely shortly racemose-spicate, the receptacle usually obovate, sometimes linear-fusiform, or obovate and produced into terminal pedestal of a heteromorphic fl; bracts very small, usually deciduous before anthesis; flowers of each unit of inflorescence either homomorphic or commonly dimorphic, the always sessile terminal one then a little stouter and with modified androecium; perianth normally 5-merous (random exceptions); calyx narrowly campanulate, 5-nerved, short-toothed; corolla narrowly tubular with dilated limb, the tube striately nerved when dry; androecium 10–30-, in *Ch. chacoënse* to 52-merous, the tube weakly united at base to corolla, in peripheral fls included to distinctly exserted, in terminal fls always exserted, strongly dilated at orifice and there randomly fimbriate by auxiliary sterile filaments; ovary sessile, either conic or truncate at apex; style a little longer or shorter than filaments, the stigma minutely dilated. *Pods* sessile but sometimes basally attenuate, linear or narrow-oblong either straight, or backwardly falcate, or coiled into a compressed or open helix, or erratically decurved and twisted, the sutures either evenly decurved or undulately constricted between seeds, the valves when ripe either ligneous or coriaceous (thinly so in *Ch. chacoënse*), coarsely venulose; dehiscence tardy and inert, through one or both sutures, but these seldom gaping widely enough to release the seeds, the valves in age sometimes cracking horizontally between seeds; endocarp not colored, coherent between successive seeds but seldom produced there as incipient septa; funicle filiform, sinuous; seeds either transverse (in wider pods) or basipetal (in narrower ones), compressed-lenticular, with hard pallid testa and finely engraved, often incomplete pleurogram; endosperm 0. — Spp. 10 (1 pluriracial), of lowland and less often submontane, xeromorphic brush-woodlands, seasonally dry forests, savannas, and deserts in warm-temperate and tropical America, from NW Mexico (27°N) to the Antilles, Argentina (28°S) and SE Brazil, in equatorial latitudes found mostly in savanna enclaves within forest climax. Fl. at onset of rainy season(s), in desert following erratic rains.

The genus *Chloroleucon* was first described by Bentham as a section of *Pithecolobium*, and, even with later additions that have expanded its boundaries, it remains firmly characterized by axillary spines, striate resting-buds, and proteranthous capitula. The spines are modified sterile peduncles, such as occur in no other Ingeae. They have been carelessly interpreted (Mohlenbrock, 1963: 439; Nielsen, 1981: 182) as spinescent stipules, such as occur in *Pithecellobium* sens. str. and *Havardia*, but are identified by axillary origin and by occasional rudimentary flower-buds at tip. The flowers of each capitulum are nearly always heteromorphic (either one or several distal ones larger than rest, with modified androecium), which we take to indicate relationship to *Albizia*, as already suggested by Nielsen (1981). Buds with striate perules that protect the primordia of floriferous branchlets or long-shoots are found also in the seemingly related genus *Leucochloron*, but this differs in absence of spines and in a characteristic membranous seed-coat lacking pleurogram. The fruits of *Chloroleucon* are variable in texture and extremely diverse in curvature, passing from almost straight to falcate or turned into a flattened spiral. The fruit is sometimes necessary for certain identification to species.

It must be borne in mind that some herbarium specimens do not show the armature of spines, and M. Nee informs us that in Bolivia some whole trees of *Ch. mangense* var. *mathewsii* are spineless. Unarmed flowering specimens of *Ch. guantanamense* from Hispaniola (*García 4837*, NY) came from trees armed on the trunk with thorny efoliate branchlets. Also some capitula seem to lack a heteromorphic flower, but we have seen few of these, and some specimens in young leaf do not retain the deciduous perules. These cannot be referred by formula (or by our key) to their genus, and will readily be recognized as *Chloroleucon* only by close students of Ingeae.

Cassens and Miller (1981) distinguished the wood of *Chloroleucon* from other taxa that lacked septate fibers and confluent parenchyma on the basis of the color of the heartwood and the specific gravity. *Chloroleucon dumosum* is unlike other members of the genus they examined in having biseriate, not uniseriate rays.

Conspectus of genus *Chloroleucon*

1. *Ch. guantanamense* (Britton) Britton & Rose
2. *Ch. chacoënse* (Burkart) Barneby & Grimes
3. *Ch. eurycyclum* Barneby & Grimes
4. *Ch. acacioides* (Ducke) Barneby & Grimes
5. *Ch. extortum* Barneby & Grimes
6. *Ch. dumosum* (Bentham) G. P. Lewis
7. *Ch. tenuiflorum* (Bentham) Barneby & Grimes
8. *Ch. tortum* (Martius) Pittier *ex* Barneby & Grimes
9. *Ch. foliolosum* (Bentham) G. P. Lewis
10. *Ch. mangense* (Jacquin) Britton & Rose
    a. var. *mangense*
    b. var. *mathewsii* (Bentham) Barneby & Grimes
    c. var. *vincentis* (Bentham) Barneby & Grimes
    d. var. *leucospermum* (T. Brandegee) Barneby & Grimes
    e. var. *lentiscifolium* (A. Richard) Barneby & Grimes
    f. var. *tetrazyx* Barneby & Grimes

The most useful and definitive specific characters of individual species in the genus are found in the mature leaves and fully formed fruits. The plants are conspicuous and often collected when in flower only, and specimens of this age, because differences in the inflorescences are few and subtle and the foliage is not yet fully expanded, will always be difficult to name. When the geographic origin is known, however, the choice of possibilities is much reduced. The most practical approach to a key to species of *Chloroleucon* is a preliminary sorting by allopatric or vicariant range of dispersal. Measurements of leaflets recorded herein are taken from mature, fully expanded leaves, such as almost never coincide with flowers at full anthesis.

Key to the species of genus *Chloroleucon*

1. Lf-formula i(–ii)/1–2; in chaco forest about the common boundaries of Bolivia, Argentina, and Paraguay . . . . 2. *Ch. chacoënse* (p. 140)
1. Either pinnae or lfts more numerous, the latter not less than 4 pairs in longer pinnae; range of the genus.
    2. Localized on the S coast of prov. Oriente, Cuba, and in SW Dominican Republic (Barahona); lf-formula i or i–iii/9–21, the lfts linear or linear-oblong and not over 2.5 mm wide; pod coiled through 2–3 circles into a vertically compressed spiral ±2.5 cm diam . . . . 1. *Ch. guantanamense* (p. 139)
    2. Allopatric, and in Cuba only westward from prov. Las Villas, thence to some other Antilles and to continental North and South America; lf-formula i–x/5–46, in Cuba the lfts of longer pinnae only 5–8 pairs and the larger ones 3–7.5 mm wide; pod either straight, or falcate, or spirally coiled, but in Cuba and Hispaniola recurved through <1 circle.
        3. Distribution N of the Equator, from NE Brazil and Venezuela to Mexico and the Antilles.
            4. Lfts of longer pinnae at once many and narrow, in longer pinnae at least 14 pairs and the largest 2 mm wide or less.
                5. Lfts of longer pinnae 25–46 pairs, if <30 pairs in longest pinnae then the larger ones 1–2 mm wide; pod coiled through 1.5–4 circles into a vertically compressed spiral.
                    6. Valves of pod 6–7 mm wide, coiled into a spiral 1.6–2 cm diam; larger lfts 3–6.5 × 0.5–0.9 mm; NE Brazil, French Guiana . . . . 4. *Ch. acacioides* (p. 141)
                    6. Valves of pod 11–13 mm wide, coiled into a spiral ±3 cm diam; larger lfts 7.5–9 × 1.6–2 mm; N-centr. state of Bolívar in Venezuela . . . . 3. *Ch. eurycyclum* (p. 140)
                5. Lfts of longer pinnae 14–25(–29) pairs, the largest 1.0–1.8(2) mm wide; pod almost straight or gently falcate; N Colombia and NW Venezuela, W sporadically to S Mexico (Chiapas); var. *mangense* of. . . . . 10. *Ch. mangense* (p. 149)
            4. Lfts of longer pinnae 4–13 pairs, rarely to 16 pairs but then the larger ones well over 2 mm wide; other varieties of. . . . . 10. *Ch. mangense* (p. 149)
        3. Distribution S of the equator, in tropical E and SW Brazil, sub-Andean Peru, Bolivia, Argentina, and Paraguay.
            7. Lfts of longer pinnae at once many and narrow, at least 17 pairs and <1.5 mm wide.
                8. Lfts of longer pinnae 30–66 pairs, the longest 3–6.5 mm long and 5–7 times as long as wide; pod coiled through 1.5–4 circles into a compressed spiral 1.6–2 cm diam, the valves thinly coriaceous and ±6–7 mm wide; NE Brazil, from the lower Amazon valley in Pará S to extreme NE Mato Grosso and adjacent Goiás, E to E Pernambuco, cultivated elsewhere . . . . 4. *Ch. acacioides* (p. 141)
                8. Lfts of longer pinnae 17–37 pairs, the longest 2.7–4.5 mm long and 3.2–5 times as long as wide; pod either much broader than 6–7 mm and simply falcate, or compressed-moniliform and randomly twisted but not coiled; E Brazil Sward from Pernambuco and S Ceará, and disjunct in W Mato Grosso do Sul, far E Bolivia, and extreme NW Argentina.
                    9. Pod simply falcate and its valves coriaceous 13–18 mm wide at the seeds, the ventral suture alone impressed between seeds, these horizontal to the pod's long axis; widespread over the range just defined . . . . 9. *Ch. foliolosum* (p. 148)
                    9. Pod sinuously linear-moniliform and its valves stiffly papery ± 7 mm wide at seeds, both sutures constricted, the seeds basipetal; local in N-centr. Bahia. . . 5. *Ch. extortum* (p. 142)
            7. Lfts of longer pinnae either fewer than 13 pairs or the larger ones over 2 mm wide, or both.
                10. Localized on the middle río Huallaga in San Martín, Peru, and on the lower Beni and adjacent Madeira rivers in NE Bolivia and adjacent Rondônia, Brazil; lfts porrect at tip; pods almost straight; bark of trunk thin and smooth; fls of each capitulum abruptly heteromorphic, the androecium of the terminal ones strongly differentiated; var. *mathewsii* of. . . . . 10. *Ch. mangense* (p. 149)
                10. Allopatric; where approaching the last in Bolivia the lfts symmetrical at tip, the pod coiled, and the bark corky and furrowed; fls of each capitulum subhomomorphic or only slightly modified upward along the receptacle.
                    11. Bark corky, deeply fissured; pod coiled through 4–8 circles into a loose spiral or equivalently contorted; widespread in Argentina northward from Tucumán and N Corrientes, in Paraguay, and in E Bolivia, disjunct eastward in Brazil on upper Río São. Francisco in Minas Gerais and on the São Francisco-Tocantins divide in E-centr. Goiás. . . . . 7. *Ch. tenuiflorum* (p. 145)
                    11. Bark thin and smooth; pods various; E Brazil, from S Ceará and R. G. Norte southward, in Minas Gerais only E of Sa do Espinhaço.
                        12. Lfts of longer pinnae (13–)14–26 pairs; pod strongly falcate but not coiled; caatinga and comparable brush-woodland inland from the coast, from SE Minas Gerais N to S Ceará and Río Grande do Norte . . . . 6. *Ch. dumosum* (p. 144)
                        12. Lfts of longer pinnae 6–9 pairs; pod coiled through several circles into a springlike spiral; coastal restinga and brush–woodland of the coastal plain in states of Rio de Janeiro and Guanabara . . . . 8. *Ch. tortum* (p. 146)

1. Chloroleucon guantanamense (Britton) Britton & Rose, N. Amer. Fl. **23:** 38. 1928. *Pithecolobium* (?) *guantanamense* Britton, Bull. Torrey Bot. Club **41:** 7. 1914. — "United States Naval Station, Guantanamo Bay, Oriente [Cuba], March, 1909 [fr.] (*Britton 2051*)." — Holotypus, NY!; isotypus, US *791190*!.

Chloroleucon (?) *baracoënse* Britton & Rose, N. Amer. Fl. **23:** 38. 1928. — "Caletico, southern Oriente, Cuba, July–August, 1924 [fl.], *Brother León 12379*." — Holotypus, NY!; isotypus (fragm.), US *1301948*!.
Chloroleucon guantanamense sensu Mejía & al., Moscosoa **8:** 36, figs. 2, 3. 1994, nom. nud.

Stiffly branched, microphyllidious, *arborescent shrubs* 4–10 m with habit and inflorescence of the genus, the trunk sometimes armed with sterile thorny branchlets, the fertile branches either randomly armed with supra-axillary thorns to ±3–8 mm or unarmed, glabrous except for minutely thinly ciliolate lf-axes and lfts, the foliage olivaceous subconcolorous; perulate buds (scarcely known) apparently <1 mm. *Stipules* lacking (perhaps small caducous, but not present on specimens seen). *Lf-formula* i iii/9 21; lf-stks 4–34 mm, the petiole 4–20 mm; a sessile cupular nectary 0.5–0.8 mm diam and 0.2 mm tall near midpetiole in lvs with 2–3-jugate pinnae and a similar one at tip of lf-stk, this alone present when pinnae 1 pair, and smaller nectaries also at tip of pinna-rachises; rachis of longer pinnae 2.5–5.5 cm, the longer interfoliolar segments 1.6–4.5 mm; lfts widely ascending from rachis, decrescent at each end of it, in outline linear, narrowly oblong or narrowly oblong-obovate from inequilaterally obtusanagulate base, asymmetrically deltate-acute, those near midrachis 6.5–9 × 1.4–2.5 mm, 3.4–4.9 times as long as wide; venation of 4–5 primary nerves from pulvinule, the midrib displaced to divide blade ±1:2, on each side 1–2-branched above middle, the anterior primary nerve faint or almost obsolete, the inner posterior one produced to anastomosis well beyond midblade, the outer posterior ones short and weak, the whole venation immersed on upper face of blade, pallid and prominulous on lower. *Peduncles* solitary or 2–4 together, 6–17 mm; units of inflorescence either subcapitate or compactly racemiform ±15–20-fld, the fls homomorphic, the receptacle either obovoid ±1 mm, lacking a terminal pedestal, or slenderly fusiform and to 6 mm; pedicels 0.2–1 mm; fls externally glabrous, greenish or whitish, the perianth 5-merous; calyx ±2.6–3 × 1 mm, the deltate teeth 0.2–0.5 mm; corolla 6.4–7 mm, the erect lobes 1.4–1.5 × 0.7 mm; androecium ±14–24(–?)-merous, up to 15–16 mm long, the tube ±6–8 mm; ovary glabrous, conical at apex; stigma a little dilated, 0.15 mm diam. *Pods* solitary, broad-linear from cuneate base, recurved through ±2 circles into a tight, vertically compressed spiral 2–2.5 cm diam or equivalently distorted, the sutures 0.7–1.0 mm wide, the exterior one undulately constricted between seeds and the interior one describing a circle 0.3–0.5 cm diam, the thinly leathery, when ripe brittle, venulose and glabrous valves bullately dilated (alternately on each side of pod) over each of 14–18 seeds, at broadest point 8–13 mm wide; dehiscence inert, through both sutures, the valves also cracking transversely between seeds, and the segments tardily separating from each other and from the valves (incipiently craspedial); seeds transverse either at middle of valves or displaced toward their perimeter, lenticular, in broad outline obovate or suborbicular 3.5–4.4 × 3.3–3.5 mm, the testa smooth, greenish fawn-colored, faintly fuscescent within the pleurogram, this 2.5–3.1 × 1.5–2.4 mm, complete.

In scrub thickets on limestone, sometimes along watercourses, below 450 m, known only from E Cuba and SW Hispaniola: in Cuba on the S coast of Oriente province between Guantánamo Bay and Punta Caleto (long. 74°15′–75°10′W); in Hispaniola on Sa de Bahoruco in Barahona (71°25′W). — Fl. VI–VIII. — *Aroma*.

Britton described *Pithecellobium guantanamense* from fruiting material and *Ch. baracoënse* from a specimen in flower. The typus of *P. guantanamense* has leaves of long-shoots with two pairs of pinnae and those of short-shoots with only one pair; that of *Ch. baracoënse* has most leaves with one pair of pinnae but a developing long-shoot, still immature, shows a leaf with two pairs. The fruit of *Ch. baracoënse* was unknown to Britton, but a fruiting collection from close by the type-locality (between Playa Blanca and Jauco, *León 13300*, NY) has the small coiled pod of *P. guantanamense*. It seems certain that they represent the same species. The floral receptacle in Cuba is contracted, and the pedicels are almost obsolete, whereas in Hispaniola the receptacle is fusiform and the pedicels are up to 1 mm long.

In the two known flowering specimens we could find no heteromorphic central flower in the capitula, but this may occur in others. In general aspect and especially in form of the pod *Ch. guantanamense* seems very closely related to *Ch. acacioides*, from which it differs little except in the reduced number of pinnae. From the other Cuban chloroleucon, *Ch. mangense* var. *lentiscifolium*, it differs greatly in leaf-formula, size of leaflets, and spiral pod.

2. Chloroleucon chacoënse (Burkart) Barneby & Grimes, comb. nov. *Pithecellobium chacoënse* Burkart, Darwiniana **7:** 513, fig. 2. 1947. — "ARGENTINA. Prov. de Salta: Jaquiasmé, F. Devoto y F. Rial Alberti, 22–XI–1938 [deflor., fr.]; Herb. Dir. Forestal Min. Agricultura n° 1661 . . . PARAGUAY. Chaco central . . . 22° lat. S y 60°35′O, T. Rojas 8397, X.1938." — Holotypus, Hb. D.F.M.A. *1661*, SI! (seen in photo, NY, kindly supplied by Profesora N. Bacigalupo, V. 1989).

Pithecellobium chacoënse sensu Bernardi, 1984: 188, fig. 8.

Macrophyllidious deciduous bushy *treelets* to 2–3.5 m, with stiffly flexuous fuscous, sparingly lenticellate annotinous branchlets, either unarmed or armed at random nodes with stout supra-axillary thorns to ±1(–?) cm, glabrous throughout or the new branchlets and dorsal face of lfts puberulent, the foliage pale green subglaucescent concolorous, the capitula of creamy-white fls borne singly on short-shoots lateral to annotinous branchlets or exceptionally also in lf-axils of long-shoots: perulate buds (not evident on some young branchlets) ovoid or hemispherical 0.5–1 mm, castaneous glabrous. *Stipules* mostly obsolete, rarely subulate and to 2–4 mm, the blades thin-membranous caducous. *Lf-formula* i–ii/1–2(–3), commonly i/1–2; lf-stks 8–16(–24) mm, the petiole 8–20 mm, the interpinnal segment, if present, to 12 mm; a sessile elliptic nectary near middle, below middle, or rarely at apex of petiole 0.6–1.4 mm diam, similar smaller nectaries at insertion of lfts; rachis of pinnae 4–20 mm, the interfoliolar segment, if present, 4–10 mm; lfts when more than one pair accrescent distally, the further or only pair obliquely obovate from shallowly semicordate base, broadly obtuse, 15–29 × 12–22 mm, 1.2–1.6(–1.8) times as long as wide; venation of 4–5 primary nerves from the pulvinule, the almost straight midrib displaced to divide blade ±1:2, all primary nerves forked several times and giving rise to an open reticulum of venules prominulous on both faces of blade. *Peduncles* characteristically solitary at first node of brachyblasts, randomly also in lf-axils of long-shoots, 13–25 mm; capitula ±16–25-fld, without filaments 10–12 mm diam, the monomorphic fls sessile, glabrous except for minutely papillate-ciliolate corolla-lobes, the clavate receptacle not over 1.5 mm diam; bracts membranous minute, caducous long before anthesis of associated fl; perianth 5-, randomly 6-merous; calyx campanulate 2–2.4 × 1–1.1 mm, the depressed-deltate teeth 0.15–0.25 mm; corollas 5–6 mm, the tube striate-nerved, the ovate lobes variable in length, at most 1.7 mm, the blade concave ventrally; androecium 28–52-merous, ±12–13 mm long, the filaments united into a tube 5.5–6 mm, about as long as or distinctly exserted from corolla; ovary glabrous, conical at apex. *Pods* usually 1 per capitulum, narrowly oblong or oblong-elliptic from cuneate-attenuate or even short-stipitate base, 5.5–10.5 × 1.4–2(–2.3) cm, abruptly apiculate at apex, falcately decurved through scarcely ¼ circle, 5–9-seeded, the sutures 0.4–0.6 mm wide, the ventral one sometimes shallowly undulate but both often evenly curved, the stiffly papery, elaborately venulose glabrous valves fuscous dull; dehiscence apparently inert, through both sutures; seeds transverse on filiform funicle, not seen ripe.

In open thorn forest, 350–900 m, apparently localized astride the Bolivia-Argentina-Paraguay boundaries in lat. 20°–23°S, 60°–64°W. — Map 38. — Fl. IX–XI; fr. III–IV(–?). — *Palo barroso.*

Chloroleucon chacoënse stands alone in its genus in greatly reduced leaf-formula and proportionately ample leaflets. Comparable leaf-reduction is seen in *Zygia pithecolobioides* (Hassler) Barneby & Grimes, ined. and *Microlobius foetidus* (Jacquin) Sousa.

3. Chloroleucon eurycyclum Barneby & Grimes, sp. nov., foliis microphyllidiis et legumine spiraliter torto *Ch. acacioidem* simulans, sed foliolis majusculis dissitioribus 7.5–9 × 1.6–2 (nec 3–6.5 × 0.5–0.9) mm usque et praecipue leguminis valvulis sublignosis 11–13 (nec 6–7) mm latis in helicem ±3 (nec 1.6–2) cm diam circinatis, ulterius patria aliena diversa. — VENEZUELA. Bolívar: cercanías del campamento florestal "Potreritos" en la Reserva Forestal Paragua, II.1970 (fr.), *Carlos Blanco 655.* — Holotypus, MER *01700*; isotypi, NY, P, U, US.

Trees to 24 m tall with trunk to 4.5 dm diam, resembling *Ch. acacioides* except for coarser foliage and larger, woodier pod. *Stipules* not seen. *Lf-formula* v–vi/25–31; lf-stks to 9.5 cm, the petiole to 3 cm; the first nectary near midpetiole sessile, cupular 1.2–1.8 diam, ±0.5 mm tall; rachis of longer pinnae 4–7.5 cm, the interfoliolar segments to 2–2.5 mm; larger lfts linear 7.5–9 × 1.6–2 mm, 4.5–5.2 times as long as wide. *Inflorescence* unknown. *Pods* 11–13 mm wide, evenly coiled into a compressed spiral ±3 cm diam, the sutures 1–1.4 mm wide, the perimetric one sinuously constricted between seeds, the valves becoming woody, coarsely venulose, castaneous glabrous; seeds 5.6–6.5 × 4.5–5 mm, the complete pleurogram 4–4.5 × 2.7–2.9 mm.

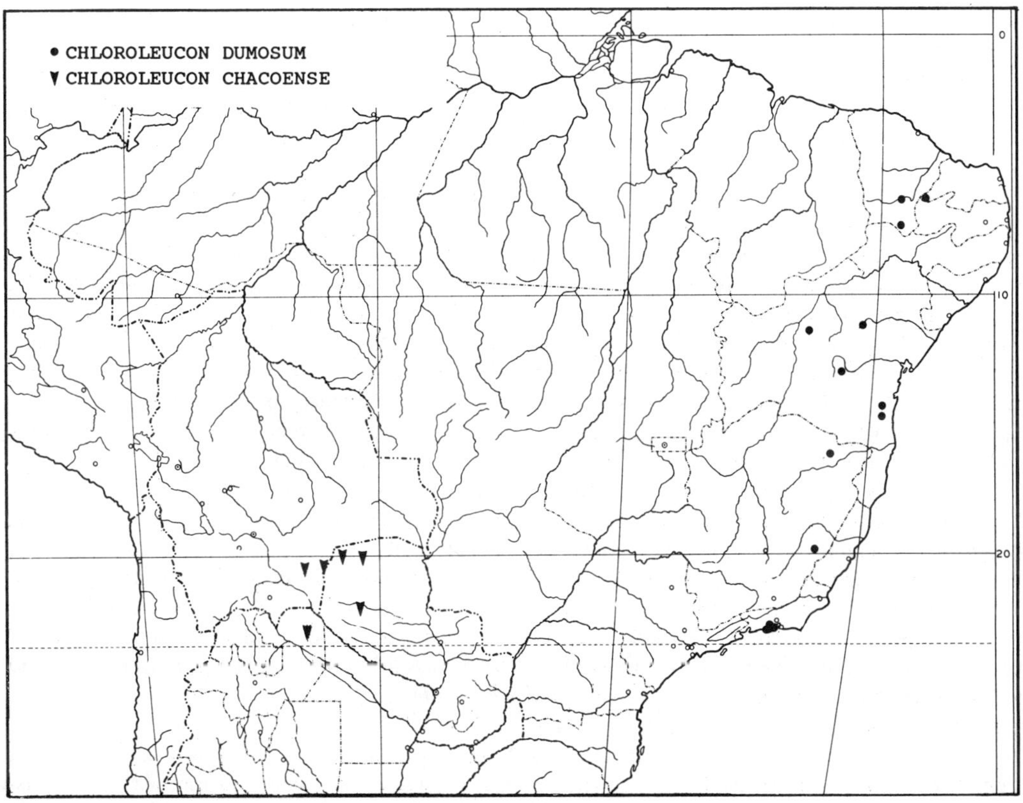

MAP 38. Central South America. Range of *Chloroleucon dumosum* (Bentham) G. P. Lewis and *C. chacoënse* (Burkart) Barneby & Grimes.

In unrecorded habitat, but to be expected in seasonally dry, semideciduous tropical woodland below 250 m, known only from the type locality on lower Río Paragua in state of Bolívar, Venezuela. — Fr. ripe in II. — *Montañero, quiebrahacho.*

This poorly known species has the multifoliolate leaves of *Ch. acacioides* combined with the spirally coiled woody pod almost of *Ch. tenuiflorum* and is in these respects obviously distinct from all other members of the genus found in northern South America. The only chloroleucon known to be nearly sympatric with *Ch. eurycyclum* is *Ch. mangense* var. *tetrazyx*, grossly different in leaf-formula and in gently falcate pod.

4. Chloroleucon acacioides (Ducke) Barneby & Grimes, comb. nov. *Pithecellobium acacioides* Ducke, Arch. Jard. Bot. Rio de Janeiro **3:** 69. 1922. — Ten syntypes are listed, from which we here designate as lectotypus: MG *10207*, labeled "Santarem (Pará), 19.VII.1916 [fr]. *A. Ducke 16358*"!; isotypi, G!, US *1040876*!.

Pithecolobium foliolosum Bentham, London J. Bot. **3:** 223. 1844, ex parte, quoad pl. pernambucensem Gardneranam, lectotypo excluso. = *Gardner s.n.* in May, 1838, e Río Capibara, BM!.
Pithecolobium parvifolium sensu Bentham *in* Martius, Fl. Bras. **15(2):** 454. 1876, ex parte, quoad *Spruce 882* (K!, NY!) cum nom. nud. in synonomia *Calliandra spruceana* Bentham; non *P. foliolosum* (Swartz) Bentham, quae = *Ch. mangense* var. *mangense*.
Pithecolobium acacioides sensu Ducke, 1949: 46.

Broad-crowned drought-deciduous microphyllidious *trees* 5–30 m with smooth trunk attaining 2–5 dm diam, either randomly armed at nodes with stout tapering, solitary or paired, ascending thorns 4–27 mm or more often unarmed, the stiffly flexuous, densely lenticellate branchlets early glabrate, the lf-axes and peduncles pilosulous with fine, forwardly incurved pallid hairs to 0.15–0.35 mm, the lfts bicolored when mature, paler beneath, at least thinly ciliolate but glabrous facially, the inflorescence of the genus; perulate buds axillary to mature lvs ovoid (1.5–)2–4 mm, the scales dorsally castaneous glabrous, striately nerved, microscopically ciliolate. *Stipules* often obsolete, when developed either linear or linear-oblanceolate and attaining 12 mm, the submembranous blade finely several-nerved, early caducous. *Lf-formula* (vi–)vii–x/30–46; lf-stks 4.5–8.5(–12) cm, the petiole proper 8–24 mm, the longer interpinnal segments 4–8(–11) mm; nectary at or shortly below midpetiole sessile, shallowly cupular (0.5–)0.6–1.3(–1.6) mm diam, in profile 0.2–0.5 mm tall, a smaller nectary at tip of lf-stk and yet smaller ones between 1–3 furthest pairs of lfts; rachis of longer pinnae 2.5–4(–6) cm, the longer interfoliolar segments 0.5–0.8(–1.2) mm; lfts subequilong except at very extremities of rachis, at first subvertically imbricate but spreading at maturity, the blades linear from obliquely obtusangulate base, straight or slightly porrect at deltate or triangular acute apex, the larger ones 3–6.5 × 0.5–0.9 mm, 5–7 times as long as wide, the almost centric midrib 1(–2)-branched above midblade, flanked on either side by an intramarginal primary nerve produced almost to blade's apex, the venation immersed on upper face, finely prominulous and often pallid beneath. *Peduncles* solitary or geminate 5–16 mm; capitula hemispherical 16–27-fld, the fls dimorphic, the axis of receptacle, including the pedestal of the sessile heteromorphic central fl, 2–3 mm; bracts minute or obsolete; PERIPHERAL FLS: sessile or the lowest raised on pedicel to 0.45 mm; perianth 5-merous, the calyx and corolla greenish, the anthers white at early anthesis, yellowing in age; calyx campanulate 5-nerved 1.3–2 × 0.5–0.7 mm, either glabrous or thinly pilosulous externally, the deltate teeth 0.1–0.25 mm; corolla narrowly trumpet-shaped 3.6–5.4 mm, the lance-ovate lobes 1.2–2 × 0.6–0.8 mm, microscopically papillate on margin; androecium 10–12-merous, 11–14.5 mm long, the filaments united into a tube 1.6–2.4 mm surrounding the glabrous truncate ovary; TERMINAL FL: sessile, the calyx 1.5–1.7 × 1–1.7 mm, the corolla 5.5–6.5 × ±1 mm; androecium of same length as that of peripheral fls but the tube 6–8.5 mm, distally 0.8–1.5 mm diam, fimbriately dissected at orifice, but fertile stamens no more numerous; stigma slightly enlarged, ±0.1 mm diam. *Pods* usually solitary, subsessile, undulately linear 6–7 mm wide, spirally decurved and coiled through 1.5–4 circles into a compressed helix 1.6–2 cm diam, the sutures ±0.5–0.8 mm wide, the exterior (seminiferous) one sinuously constricted between seeds, the thinly leathery green glabrous venulose valves becoming crustaceous fuscous, bullately extended over each seed, these lying transverse in discrete chambers but not separated by raised septa; dehiscence through both sutures but tardy and irregular, the seeds mostly released by weathering; seeds plumply lentiform, in broad view 3.8–4.5 × 3.6 mm, the testa smooth, pale tan, fuscous over the pleurogram, this nearly complete 2.5–3 × 1.5–1.7 mm.

In seasonally dry tropical woodlands and at edge of campo or on river terraces, in Pará going out to sands of the coast, below 200 m, discontinuously widespread in NE Brazil and French Guiana: in Brazil, from the lower Amazon valley in Amapá and Pará upstream to the mouth of Río Trombetas, S along the Tocantins-Araguaia to near 11°S in the NE corner of Mato Grosso and adjoining Goiás, E through Maranhão and N Ceará to E Pernambuco, in French Guiana along and near the coast from Cayenne W to the mouth of the Marowijne, to be sought in adjacent Surinam; occasionally cultivated elsewhere (Medellín, Colombia, *Duque 1834*, US; Rio de Janeiro, Brazil, *H. M. Curran 357*, US). — Map 39. — Fl. VI, IX–XII, simultaneously with flush of new foliage. — *Jurema, jurema branca, pereira.*

Ducke described *Ch. acacioides* as a spiny tree of low or median stature with broad umbrella-shaped crown, resembling some African acacias and like them leafless in the dry season. The small, tightly spiral pod recalls that of *Enterolobium schomburgkii* in form, but is of thinner texture.

5. Chloroleucon extortum Barneby & Grimes, sp. nov. *Ch. acacioidi* et *Ch. folioloso* affinis, foliolis angustis numerosis cum utraque fere congrua, ab ambabus legumine valde diverso abiens: legumen moniliforme compressum undulato-lineare ±17 × 0.7 cm, hinc inde sinuatim contortum sed haud spirale, valvulis chartaceis puberulis, maturis a sese separatis. — BRAZIL. Bahia: Pindobacu, 3.IX.1981 (fr.), *J. D. C. Arouck Ferreira 82.* — Holotypus, K; isotypus, NY.

Trees of unknown dimensions with fuscous unarmed branches, glabrous except for minutely pilosulous lf-axes, lft-margins, and fruit, the foliage bicolored, the small narrow lfts dull olivaceous above, pallid beneath, the inflorescence (presumably of the genus) unknown; perulate buds plumply ovoid 1–3 mm, the striate greenish castaneous scales microscopically ciliolate. *Stipules* lacking. *Lf-formula* iv–v/22–25; lf-stk 19–25 mm, the petiole 5–8 mm, the

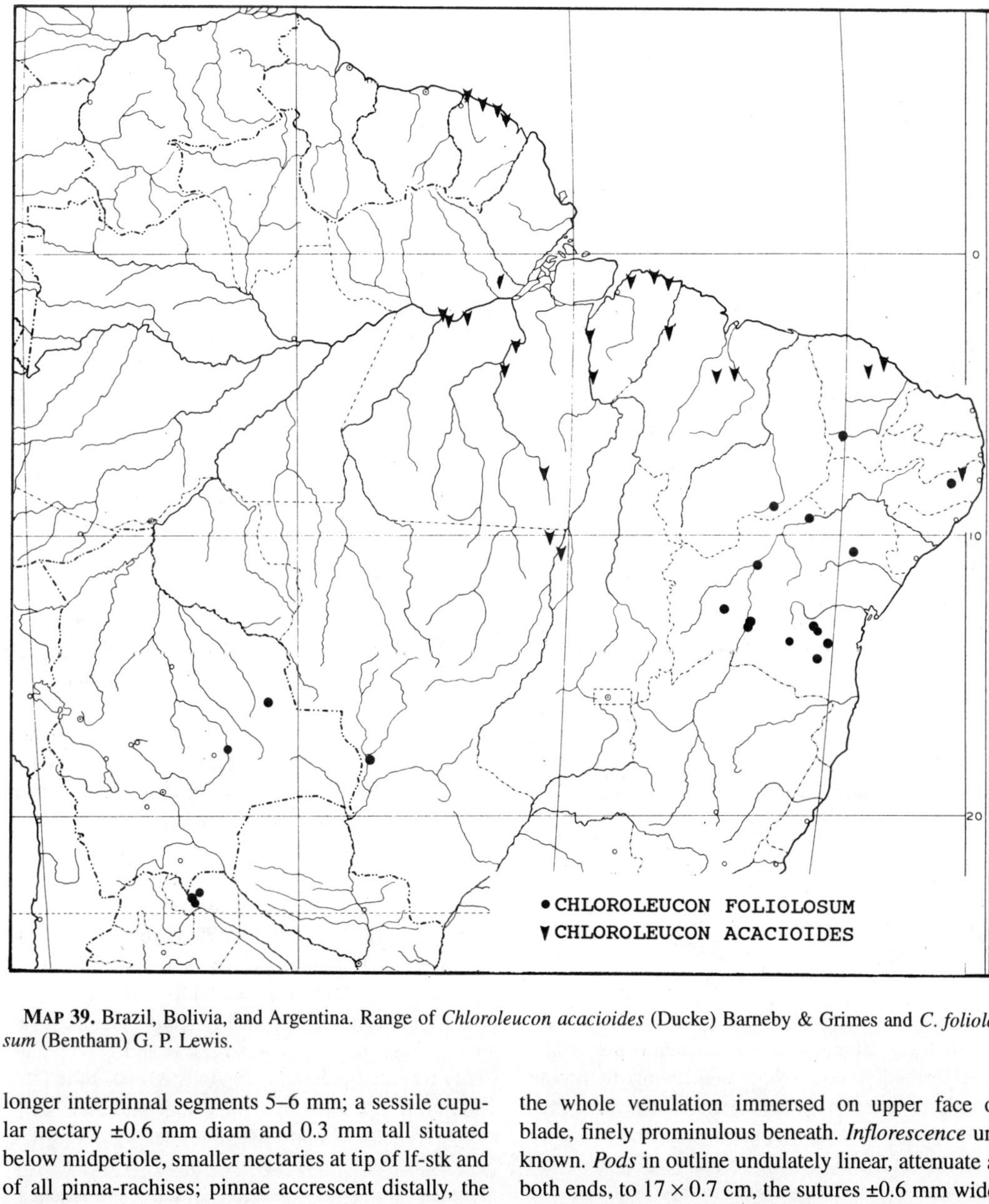

MAP 39. Brazil, Bolivia, and Argentina. Range of *Chloroleucon acacioides* (Ducke) Barneby & Grimes and *C. foliolosum* (Bentham) G. P. Lewis.

longer interpinnal segments 5–6 mm; a sessile cupular nectary ±0.6 mm diam and 0.3 mm tall situated below midpetiole, smaller nectaries at tip of lf-stk and of all pinna-rachises; pinnae accrescent distally, the longest 2–3 cm, the longer interfoliolar segments 0.8–1 mm; lfts decrescent only at extremities of rachis, otherwise subequilong, linear from obliquely rectangulate base, triangular-acute, slightly arched forward, those near midrachis 4.5–6 × 0.9–1.1 mm, 5–5.5 times as long as wide; venation of 3 nerves from pulvinule, the subcentric midrib porrect at tip, giving rise on each side to 1–2 weak secondary branches, the anterior primary nerve tenuous, the posterior one produced to anastomosis beyond midblade, the whole venulation immersed on upper face of blade, finely prominulous beneath. *Inflorescence* unknown. *Pods* in outline undulately linear, attenuate at both ends, to 17 × 0.7 cm, the sutures ±0.6 mm wide, each shallowly widely constricted between the 13–14 seeds, the interseminal isthmi plane, mostly ±3–5 mm wide, the stiffly papery, fuscous-brown, venulose, sparsely puberulent valves low-convex over each seed, when ripe randomly decurved and twisted (ribbonlike) and fully dehiscent along both sutures, the valves separating to release the seeds; seeds basipetal on a filiform contorted funicle, in broad outline ±5 × 3.7 mm, the smooth testa pale fawn-color, the U-shaped pleurogram ±2.5 × 1.4 mm.

In semideciduous woodland (probably caatinga) near 500 m, known only from the type locality at 10°57′S, 40°27′W in N Bahia. — Ripe fr. in IX.

Although *Ch. extortum* is known to us through only one collection, we have no hesitation in describing it. Its slender, sinuously linear pod, constricted between seeds and after dehiscence randomly twisted like that of *Pithecellobium unguis-cati* (but not red inside) is unlike that of any previously known *Chloroleucon.*

6. Chloroleucon dumosum (Bentham) G. P. Lewis, Leg. Bahia 165. 1987. *Pithecolobium dumosum* Bentham, London J. Bot. **3:** 223. 1844. — "Near Barra do Jardim, in Piauhy [properly in far S Ceará, near 39°W], Gardner, n. 1746 [properly *1946*]." — Holotypus, K(herb. bentham.)! = NY Neg. *2032*; isotypi, NY (2 sheets)!. — *Feuilleea dumosa* (Bentham) O. Kuntze, Revis. Gen. Pl. **1:** 187. 1891.

Pithecellobium glazioui Bentham, Trans. Linn. Soc. London **30:** 597. 1875 & *in* Martius, Fl. Bras. **15(2):** 454. 1876 "*glaziovii*". — ". . . in monte Corcovado provinciae Rio de Janeiro: Glaziou n. 1565." — Holotypus, K (annotated by Sandwith as type specimen, but lacking collector's serial number and locality-data) = NY Neg. *2033*!; isotypi, C(herb. warming.) = F Neg. *21904*!, P (dated 17 sept. 1867)!, US *453717*!. — *Feuilleea glazioui* (Bentham) O. Kuntze, Revis. Gen. Pl. **1:** 188. 1891. *Chloroleucon glazioui* (Bentham) Lewis, Leg. Bahia 166. 1987.

P. vinhatico Record *in* Record & Mell, Timbers Trop. Amer. 211. 1924. — "Yale Nos. 1806 and 4698 collected by [H.M.] Curran with botanical material, Bahia, Brazil" — Authentic material, labeled "*vinhatica de espino*, Rio Gongogy, 100–500 m, State of Bahia": *Curran 28*, NY (fragm. + photo)!, US!, and *59*, GH!, NY (fragm. + photo)!, US *704860*!. — *Chloroleucon vinhatico* (Record) Killip *ex* Record, Trop. Woods **63:** 6. 1940.

Trees flowering when (2.5–)3–30 m tall, with smooth trunk attaining 2–5.5 dm diam, potentially armed at random nodes with stout thorns to 1.5 cm (but often shorter, or on most branchlets lacking), the young stems, lf-axes, lft-margin, and often one or both faces of lfts pilosulous with spreading-incurved pallid hairs to 0.1–0.25 mm, the foliage bicolored, the brown-olivaceous plane lfts paler beneath, the hemispherical capitula or short ovoid racemes of fragrant white fls borne usually 2–4 together (rarely solitary) at the first 2–4, either foliate or efoliate nodes of new branchlets; perulate buds 0.5–3 mm, the scales either puberulent or glabrous dorsally. *Stipules* usually 0, when present either filiform 1-nerved or elliptic several-nerved, to 2.5–4 mm, caducous long before maturity of associated lf. *Lf-formula* (ii–)iii–v(–vi)/13–26; lf-stks 2.5–8 cm, the petiole (8–)10–33 mm, the longer interpinnal segments 7–18 mm; a sessile cupular nectary either near, or above, or below midpetiole (0.5–)0.6–1.2 mm diam, in profile 0.15–0.4 mm tall, similar but smaller nectaries at apex of lf-stk and of pinna-rachises; pinnae accrescent distally, the rachis of longer ones 3–9 mm, the longer interfoliolar segments 2–4(–4.5) mm; lfts oblong or linear-oblong from obtusangulate base, obtuse or deltate subacute, the larger ones 7–15 × (1.7–)2–4.5 mm, 2.7–4.7(–5) times as long as wide; venation of 3–5, commonly 4 primary nerves from pulvinule, the midrib displaced to divide blade a little less than 1:2, either straight or gently porrect distally, on one or both sides 3–4-branched, the anterior primary nerve faint, narrowly intramarginal, the inner posterior one produced well beyond midblade, the outer ones short and weak, all these and random connecting venules sharply prominulous, less so beneath but distinctly so above. *Peduncles* 7–30(–40) mm; capitula 25–50-fld, the fls either all sessile or the lower ones elevated on slender pedicel to 1.1 mm, this becoming shorter distally, all fls either almost homomorphic or the terminal one or two differing in slightly longer androecial tube, but not sharply distinguished from the rest; receptacle obovoid or narrowly clavate 2.5–12 mm; bracts linear-subulate to 1 mm, caducous before anthesis; perianth 5-merous, the calyx hispidulous overall or only toward the minutely ciliolate orifice, the corolla either glabrous except for papillate margin of lobes or thinly puberulent distally; calyx (1–)1.2–2.3(–2.5) × 0.5–0.8 mm, the deltate or subulate teeth 0.1–0.25 mm; corolla (4.5–)5–8 mm, the tube striate when dry, expanded into the ascending limb; androecium 16–20-merous 15.5–17.5 mm, the tube 6.5–9 mm, exserted either in all or only in some distal fls; ovary glabrous, conical at apex. *Pods* usually solitary, sessile at cuneate base, in profile broad-linear and at once falcately recurved and randomly twisted, 10–14 × 1.1–1.4 cm, the stout sutures 1.2–2 mm wide, the convexly arched seminiferous one deeply undulate-constricted between seeds, the opposed one evenly recurved, the stiffly leathery, finally subligneous, venulose, glabrous valves low-convex over each seed, when weathered liberated from the stringlike sutures but not breaking up into free-falling segments, the mesocarp becoming somewhat mealy; dehiscence tardy, inert, the narrowly gaping sutures not readily releasing the seeds; funicle filiform, sinuous; seeds (few seen completely ripe) transverse, plumply obovoid-ellipsoid, in broad profile ±6 × 4.5 mm, the hard testa ±0.13 mm, the obscure hippocrepiform pleurogram ±3 × 2.5 mm.

In caatinga thickets and disturbed coastal forest, in Bahia often in cocais, below 500 m, interruptedly widespread from S Ceará and Río Grande do Norte S through Bahia and E Minas Gerais to Guanabara. — Map 38. — Fl. erratically through most of the year, in the NE following rains. — *Arapiraca* (shared with *Ch. foliolosum*), *jurema branca* (shared with *Ch. acacioides*), *pau rósea*, *vinhatico de espino*.

Chloroleucon dumosum, as here defined, is quite variable in width of leaflets, in length of floral receptacle, and in size of flowers, but we have not discovered any convincing discontinuity in any of these features or any clear correlations between them. There may, however, be some clinal variation along a north-south axis from smaller denser to longer looser inflorescence, while the broadest leaflets are yet known from Bahia only. We have not been able to confirm the great difference in flower-size between *Ch. dumosum* and *Ch. glazioui* that was described by Lewis (1987: 165, in clave) and can discern no other criteria. *Pithecellobium vinhatico* corresponds with the broad-leaflet type just mentioned.

In the common range, *Ch. dumosum* differs from *Ch. foliolosum* and *Ch. acacioides* in much wider leaflets, but more nearly resembles the first in the form of the fruit. It lacks the abrupt heteromorphism of the flowers, which is a constant character of *Ch. foliolosum* and *acacioides*. Leaf-formula decisively distinguishes the pubescent form of *Ch. tenuiflorum*, which in Minas Gerais replaces *Ch. dumosum* to the west of Sa do Espinhaço, as does the spirally coiled fruit.

7. Chloroleucon tenuiflorum (Bentham) Barneby & Grimes, comb. nov. *Pithecolobium tenuiflorum* Bentham, London J. Bot. **3:** 222. 1844. — "Brazil, *Pohl*; near Jaquara, *Langsdorff*.", the data expanded in 1876 (l. c. infra) to: "in provincia Goyaz ad Arrayas: Pohl; ad Taquara [= Jaquaré, centr. Minas Gerais], *Riedel*." — Lectotypus, *Pohl 2343*, collected in Sept. 1819 (fl.) at Eng. Sarfedas near Arraias, Goiás (±13°S, 47°W), K(herb. bentham.)!; isotypi NY!, W!, paratypi *Riedel 757*, collected in Oct. 1824 (fl.) at Jacquaré near Belo Horizonte, Minas Gerais, K(herb. bentham., herb. hooker.)!, NY!. — *Pithecolobium tortum* var. *pubescens* Bentham *in* Martius, Fl. bras. **15(2):** 453. 1876.

Pithecolobium scalare Grisebach, Symb. Fl. Argent. 123. 1879. — "[*Lorentz & Hieronymus*] 709 . . . T[ucumán]: in sylvis subtropicis pr. La Cruz. J[ujuy], O[rán]: in sylvis virgineis Tabaccal; Gr[an] Chaco, Laguna del Palmar. (Paraguay: Bal[ansa] 1427. 1428)." — Lectotypus, *Lorentz & Hieronymus 1014*, Jujuy, IV.1873 (fr.), GOET!; paratypi, all GOET!: *Lorentz & Heironymus 334*, La Cruz, IV.1872 (sterile); *510*, Tabaccal pr. Oran, 17.VII.1873 (fr.); *567*, Gran Chaco, VIII.1873. (fr.); *Balansa 1427*, L'Assomption, Paraguay, IX.1875 (fl.); *1428* (also at P!), Cerro de Paraguarí, VII.1875 (fr)! spm. authent., *Lorentz s.n., misit Grisebach*, K!. — *Feuilleea scalaris* (Grisebach) O. Kuntze, Revis. Gen. Pl. **1:** 189. 1891. *Pithecolobium tortum* fo. *scalare* (Grisebach) Hassler, Feddes Repert. Spec. Nov. Regni Veg. **16:** 155. 1919.

Pithecolobium scalare fo. *parvifolium* Chodat & Hassler, Bull. Herb. Boissier, ser. 2, **4:** 482. 1904, "*parvifolia*". — "[Paraguay:] . . . in campis salsis pr. flumen Salado, Oct., n. 3400." — Holotypus, G!. — Reduced to the species by Hoc, 1992a: 91.

Pithecolobium scalare var. *hirsutum* Chodat & Hassler, Bull. Herb. Boissier, ser. 2, **4:** 482. 1904, "*hirsuta*". — "[Paraguay:] . . . pr. Concepcion, Sept., n. 7468." — Holotypus, *Hassler 7468*, G!; isotypus, NY!. — Reduced to the species by Hoc, 1992a: 91.

Calliandra brachyandra Spegazzini, Revista Argent. Bot. **1:** 185. 1926. — "[Argentina.] In dumetis collinis loco Cuesta de Trancas vocato, Jan. 1892, et secus Rio San Francisco, Jujuy, Mrt. 1906." — Holotypus, LP, not seen; examined by Hoc (1992a: 91), and identified as *Pithecellobium scalare* (as foreseen by Burkart 1952: 110).

Pithecellobium anajuliae Rizzini, Arch. Jard. Bot. Rio de Janeiro **26:** 28. 1983. — "Arbor culta in Horto Universitatis Brasiliae, D.F., e seminibus reportatis e Januaria, Minas Gerais, ab. E. P. Heringer; legit Heringer 16.237 (20–X–76)." — Holotypus, RB *193689* (2 sheets) = NY Neg. 12576!

Pithecellobium scalare sensu Lindman, Bih. Kongl. Svenska Vetensk.-Akad. Handl. 24, Afd **III(7):** 52, fig. 13. 1898; Burkart, 1952: 115, fig. 14; Hoc, 1981: 542, figs. 6, 10 (seed); Bernardi, 1984: 190; Hoc, 1992a: 90.

Broad-crowned drought-deciduous *trees* to 5–18 m with deeply furrowed corky bark, either unarmed or armed at random nodes with stout, widely ascending thorns 3–15 mm, the stiffly flexuous annotinous branchlets brown, pallid-lenticellate, the new growth either glabrous throughout or the young stems and lf-axes ventrally ciliolate with fine recurved hairs, the firm plane lfts light green and a trifle paler beneath, commonly glabrous or glabrous ciliolate but sometimes (in the principal range only rarely, eastward in Goiás and Minas Gerais consistently) thinly puberulent overall, the dense capitula of small whitish, homomorphic fls borne singly or more often in fascicles of 2–5 at nodes of either efoliate or depauperately foliate, lateral brachyblasts, the whole inflorescence becoming immersed in foliage, the coiled woody pods long persistent on annotinous defoliate branches; perulate buds ovoid 0.7–1.4 mm. *Stipules* not seen, possibly sometimes developed but if so precociously deciduous. *Lf-formula* ii–iv(–v)/7–10 (–11,13); lf-stks 2–6(–7) cm, the petiole (1–)1.2–2.5 cm, the one or the longer interpinnal segment 8–20 (–26) mm; petiolar nectaries sessile or almost so, shallow-cupular or squatly tub-shaped, the first, situated near or well below midpetiole, 0.5–0.9(–1.2) mm diam, 0.2–0.45(–0.5) mm tall, a small nectary at tip of most lf-stks and yet smaller ones at tip of pinnae; pinnae subequilong or the distal somewhat longer, the rachis of the furthest one (2.5–)3–6.5 (–7.5) cm, its longer interfoliolar segments (3–)4–7(–8) mm; lft-pulvinules in dorsal view 0.5–0.8 × 0.4–0.6 mm; lfts decrescent proximally or subequilong and uniform except for slightly broader distal pair, in outline narrowly oblong or broad-linear from inequilaterally obtusangulate base, straight or almost so, either deltately subacute or obtuse at apex, the larger ones (10–)11–20 × (2.7–)3–5.5 mm, 2.7–4.8 times as long as wide; venation primarily palmate, of 4–5 nerves from pulvinule, the straight or gently incurved midrib forwardly displaced to divide blade ±1:2 and giving rise on each side to 2–4 strong and few weak intercalary secondary veins, the one anterior primary vein very short, the inner posterior one strong and incurved to anastomosis well beyond midblade, the outer posterior primaries progressively shorter, the tertiary venules random, the whole venation prominulous on both faces of mature lfts but more sharply so beneath. *Peduncles* 12–26 mm, compressed and often a little dilated distally; capitula ±(20–)25–50-fld, the fls continuous, sessile or almost so (pedicels at most 0.2 mm), homomorphic, the globular or lance-ovoid receptacle (1.5–)2–4.5 mm; bracts linear-filiform ±1–1.5 mm, <0.2 mm wide, caducous before anthesis; fls usually glabrous except for minutely ciliolate or marginally papillate calyx- and corolla-lobes, the perianth 5-merous; calyx membranous campanulate 1.7–3.3(–3.5) × 0.7–1.2 mm, the triangular, either obtuse or acute, often unequal teeth 0.2–0.7 mm; corolla subtubular or narrowly trumpet-shaped (4.3–)4.8–6.5 mm, the tube (when dry) strongly 5-nerved, the lance-ovate lobes 0.7–1.3 mm; androecium (18–)23–31-merous, free from corolla, 9–14.5 mm long, the tube subcylindric or commonly dilated distally, (4.7–)5.5–9.5 mm, at least as long as and usually a little longer than corolla; ovary linear glabrous; stigma scarcely dilated, ±0.1–0.15 mm diam. *Pods* sessile, abruptly cuneate or attenuate at base, broad-linear coiled through 4–8 complete turns (or equivalently contorted) into a loose corkscrew spiral 2–2.8 cm diam, the strong prominent, tan or yellowish sutures 1.2–1.8 mm wide, the interior one evenly recurved, the exterior one shallowly undulate, the stiffly leathery or ligneous, glabrous, fuscous or brown, coarsely venulose valves (10–)11–15 mm wide at seed-locules and there low-convex, consisting largely of pitchlike mesocarp to ±0.5 mm thick, the inner face pallid (not red); dehiscence tardy and inelastic, primarily through the seminiferous suture but this scarcely gaping, the seeds often released only after weathering of the pod on the ground; funicle filiform, sinuous; seeds 12–24, basipetal, compressed-obovoid-ellipsoid, in broad profile 5.5–6.5 × 3.8–5 mm, hard (unchanged when soaked), the testa 0.3–0.35 mm thick in section, externally smooth but scarcely lustrous, light brown except for a pallid halo bordering the pleurogram, this elliptic complete 3.5–4 × 1.2–1.9 mm; cotyledons plano-convex, closely appressed.

In semideciduous woodland, in chaparral, on sandy plains, sometimes in seasonally flooded grassland, mostly between 200 and 700 m but ascending in Bolivia to 1400–2000 m, rather frequent along the Andean foothills and adjacent plains between 11° and 28°S in Bolivia and NW Argentina, thence interruptedly E in Chaco Forest to the Paraná basin in Paraguay and NE Argentina (Misiones, N Corrientes); disjunct in cerrado on upper Río São Francisco in N and centr. Minas Gerais and on the São Francisco-Tocantins divide in E-centr. Goiás. — Map 40. — Fl. VIII–XII(II). — *Juno, kellu taku, paycasillo, tuno,* (Bolivia); *espinillo, palo cascarudo, tatané, tataré, tutaré,* (Argentina, Paraguay).

Chloroleucon tenuiflorum is the species long known as *Pithecellobium scalare*; the change of name follows reassessment of the prior *Pithecolobium tenuiflorum* Bentham. The species is distinguished by corky fissured bark and a spiral pod similar to that of *Ch. tortum*, a very close relative allopatric along and near the coast of Rio de Janeiro. As described above, *Ch. tenuiflorum* varies much in pubescence, the planaltine Brazilian populations having pilosulous or at least puberulent young stems and leaflets, whereas those of the Paraná basin southward and westward are, with exceptions such as the typus of *P. scalare* var. *hirsutum* Chodat & Hassler, nearly or quite glabrous. These variants might deserve some intraspecific rank, but we can find nothing to separate them at the specific level.

According to Hoc (1981: 542), the trees of *P. tenuiflorum* increase by sprouting from adventitious buds on the roots.

8. Chloroleucon tortum (Martius) Pittier [Estud. Prod. Flor. Venez. 16. 1927, nomen] *ex* Barneby & Grimes, comb. nov. *Pithecellobium tortum* Martius, Flora **20(2**, Beibl. **8** [= Herb. Fl. Bras.]): 114. 1837.

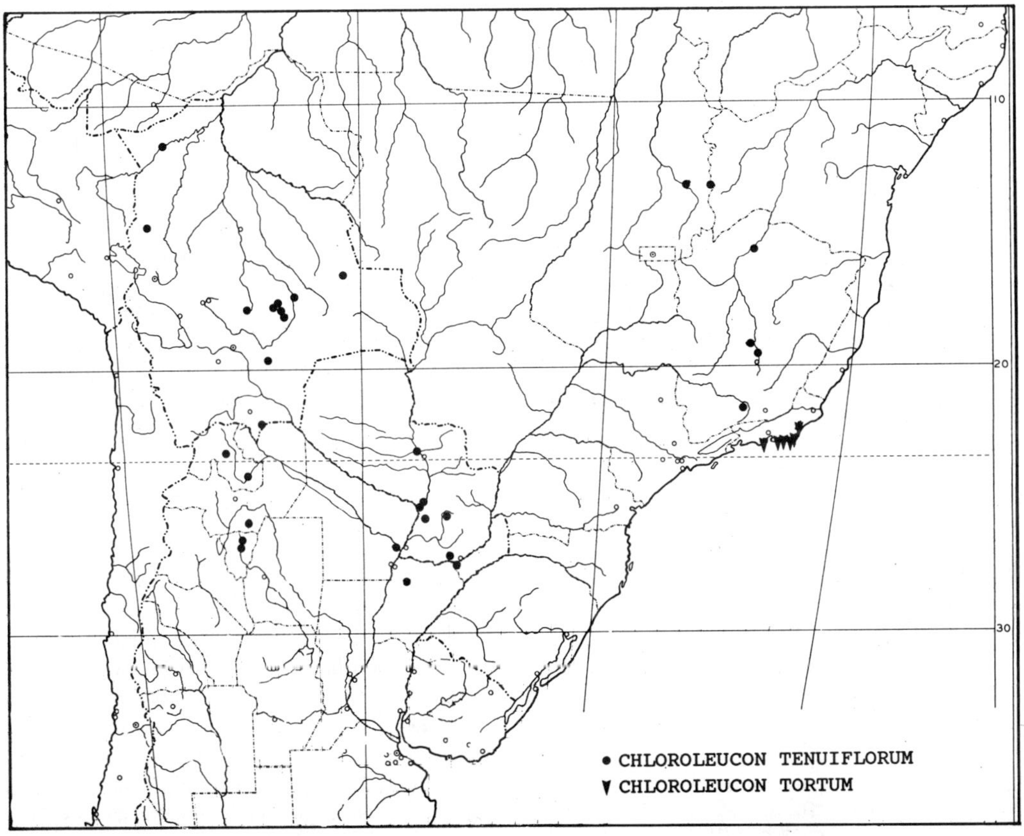

MAP 40. Tropical and subtropical South America. Range of *Chloroleucon tenuiflorum* (Bentham) Barneby & Grimes and *C. tortum* (Martius) Pittier *ex* Barneby & Grimes.

— "In sylvis siccioribus depressis, ad sepes, locis calidis, apricis e.g. ad Lagoa Sacoareme [Saquarema] in prov. Sebastianopolitana [= Rio de Janeiro] prope Villam de campos, in deserto Bahiensi rel.: fl. Sept." — Typus not indicated; spm. authent.: "in silvis praemitivis", *Martius 1827*, K!. — *Feuilleea torta* (Martius) O. Kuntze, Revis. Gen. Pl. **1:** 188. 1891. *Cathormion tortum* (Martius) Pittier, Arb. Legum. **1:** 55 (without citation of protologue). 1927.

Diffuse *shrubs* and bushy *trees* 1.5–10 m, closely resembling *Ch. tenuiflorum* but more commonly armed than not, the thorns 0.5–2 cm, and the foliage glabrous, and the fls heteromorphic; perules 1.5–4.5 mm. *Stipules* linear 2–2.5 × 0.25–0.33 mm, quickly deciduous. *Lf-formula* (ii–)iii–vi/6–8(–9); lf-stks 2.5–5.5 cm, the longer interpinnal segments 7–16 mm; rachis of longer pinnae 2.5–4.5 cm, the longer interfoliolar segments 4–7 mm; petiolar nectaries obconic or tub-shaped, the first situated well below midpetiole, sometimes close to lf-pulvinus, 0.4–0.7 mm diam, 0.3–0.6 mm tall; lfts of *Ch. tenuiflorum*, the larger ones 8–14 × 3.2–5.5 mm, 2.2–3 times as long as wide. *Peduncles* 8–16 mm; fls heteromorphic, the peripheral ones either sessile or the outermost raised on a pedicel to 0.7 mm, the terminal one sessile on a receptacular pedestal to 1 mm; perianth of *Ch. tenuiflorum*, the calyx 1.8–2.3 mm, the corolla 5.5–6.5 mm; filaments ±18, to 14–17 mm long, the tube in peripheral fls ±3–4.5 mm, shorter than the corolla, that of the terminal fl well exserted. *Pods* of *Ch. tenuiflorum*, the sutures 1–1.5 mm wide, the stiffly leathery or finally ligneous valves 11–15 mm wide, coiled into a compressed or more open and springlike helix 2.5–3 mm diam; seeds of *Ch. tenuiflorum* but the pleurogram incomplete.

In restinga along the ocean front and in thickets on sandy soils of the coastal plain below 70 m, apparently endemic to the coast of Rio de Janeiro state and Guanabara, from the city and bay eastward to cabos

Frio and dos Buzios. — Map 40. — Fl. IX–I. — *Yacarí* (Glaziou).

Chloroleucon tortum resembles *Ch. tenuiflorum* in the helicoid pod but has the smooth bark and branches of *Ch. glazioui*, vicariant inland, of which it may be an ecologically specialized derivative adapted to littoral habitats.

9. Chloroleucon foliolosum (Bentham) G. P. Lewis, Leg. Bahia 166, quoad nomen solum. 1987. *Pithecolobium foliolosum* Bentham, London J. Bot. **3:** 223. 1844. — "Banks of the Capibamba [probably = Capiberibe] in Pernambuco, *Gardner*; Villa do Barra [on Río São Francisco near 11°S in NW Bahia], *Blanchet*, no. 3136." — Lectotypus, *Blanchet 3136*, K(herb. bentham.)!; isotypi, †B = F Neg. *1196*, NY!. — The syntype, *Gardner, s.n.*, represents *Ch. acacioides*. — *Feuilleea bahiensis* O. Kuntze, Revis. Gen. Pl. **1:** 185. 1891.

Pithecolobium grisebachianum Harms, Feddes Repert. Spec. Nov. Regni Veg. **17:** 92. 1921. "Argentina: In der Umgebung von Oran (Lorentz et Hieronymus no. 283 et 468, 1873." — Syntypi, †B, one (*468*) = F Neg. *1197*!. — *Calliandra grisebachiana* (Bentham) Spegazzini, Revista Argent. Bot. **1:** 79. 1925. — *C. portoricensis* sensu Grisebach, Symb. Fl. Argent. 122. 1879, non Bentham.

Pithecellobium myriophyllum Malme, Ark. Bot. **23A(13):** 44, fig. 7. 1931. — "[Brazil, Matto Grosso do Sul:] Corumbá . . . pr. oppidum, 19¾03 ([Regnellsch. Reise] II: 3030, fructibus submaturis ornata)." — Presumed holotypus, dated 6 apr. 1903, S! = NY Neg. *12582*.

Pithecellobium oligandrum Rizzini, Leandra **3–4:** 10. 1974. — "Ad Casa Nova, Bahia [on Río São Francisco upstream from Joàzeiro], colecta a F. B. Ramalho 176 (29–III–1973)" — Holotypus, RB *168526*!. — Equated with *Ch. mangense* by Lewis, 1987: 166.

Calliandra aristulata Rizzini, Rodriguésia **41:** 166. 1976. — "Habitat in caatinga ad São Raimundo Nonato (Piauí), collegit D. P. Lima 13.235 (24–IX–1973)." — Holotypus, RB *190615*!. — Equated with *Ch. foliolosum* by Lewis, 1987; l. c.

Chloroleucon mangense sensu Lewis, Leg. Bahia 166, fig 11F. 1987, quoad pl. brasil., non *Mimosa mangensis* Jacquin.

Pithecellobium grisebachianum sensu S. Martínez, Darwiniana **19:** 468–471, figs. 4, 5 (bud morphology). 1975; Hoc, 1981: 546, figs. 7, 10 (seed).

Drought-deciduous, microphyllidious *arborescent shrubs* 3–8(–10) m with smooth trunks attaining 3 dm diam and stiffly flexuous, fuscous or castaneous, lenticellate annotinous branchlets, either unarmed or randomly armed at nodes with at first setiform but early lignescent and vulnerant, solitary or paired thorns to ±1 cm (mostly shorter), the lf-axes, lft-margin and often one or both faces of the thin-textured, subconcolorous lfts puberulent-pilosulous with incurved-ascending hairs to 0.15–0.25 mm but sometimes glabrous overall, the small hemispherical capitula of white or greenish white, often fragrant fls arising singly or rarely geminate from the axil of coevally expanding lvs on short-shoots; perulate buds narrowly ovoid compressed 2–4 mm, the castaneous striate scales glabrous dorsally, microscopically ciliolate. *Stipules* usually obsolete, exceptionally developed, then membranous oblanceolate to 5(–?) mm, caducous long before maturity of associated lf. *Lf-formula* v–vii/(15–)17–27; lf-stk of longer lvs at maturity 3.5–6 cm (often shorter at anthesis), the longer interpinnal segments 5–9(–10) mm; nectary situated near or well below midpetiole, sessile or almost so, shallowly cupular 0.2–0.5(–0.6) mm diam, in profile 0.2–0.5 mm tall, a similar but smaller one at tip of lf-stk and yet smaller ones at tip of pinnae; pinnae decrescent toward base of lf-stk, thence subequilong, the rachis of longer ones 15–34 mm, the longer interfoliolar segments 0.5–1.1 mm; lfts decrescent only at extremities of pinnae, the rest subequilong, at first forwardly imbricate but spreading in age, in outline linear-oblong or very narrowly elliptic-oblong from auriculate base, at apex either obtuse or subacute, the larger ones 2.7–4.5(5) × (0.6)0.7–1.3 mm, 3.2–5 times as long as wide; venation of 3 (and sometimes a very short posterior fourth) primary nerves from pulvinule, the subcentric midrib flanked on each side by a weak intramarginal nerve produced almost to apex of blade and connected to them by 1 or 2 transverse secondary venules, the whole venation immersed above, pallid and finely prominulous beneath. *Peduncles* 8–19 mm; capitula 12–21-fld, the receptacle, including pedestal of terminal fl, 2–4 mm, the fls heteromorphic, the peripheral ones either sessile or very shortly pedicellate, the terminal one always sessile, a little stouter but hardly longer, its androecium much modified; bracts minute, membranous, very early caducous; PERIPHERAL FLS: perianth 5-merous, either glabrous or the calyx thinly puberulent externally, the corolla-lobes papillate-ciliolate; calyx membranous campanulate 0.9–1.7 × 0.6–0.8 mm, the depressed-deltate teeth 0.1–0.2 mm; corolla 3.2–5.2 mm, the lance-ovate lobes often unequal, 0.7–2 × 0.4–1 mm; androecium commonly 10–12-, rarely 20-merous, 10–14.5 mm, the tube 1.5–3.5 mm, always a trifle shorter than corolla; TERMINAL FL: calyx 1–2 × 0.8–1.3 mm, the teeth to 0.2 mm; corolla 4–7 mm; androecium 5.5–8.5 mm, the tube exserted from corolla, 5.5–8.5 mm, dilated and fimbriate at apex, the fertile stamens as in the peripheral flrs; ovary truncate glabrous; style not dilated at apex, the stigma ±0.1 mm diam. *Pods* usually solitary, in profile broad-linear from narrowly cuneate base, ±8.5–14 × 1.3–2.4 cm, averaging wider (1.8–2.4 cm) in Argentina than in Bahia (mostly 1.3–1.8 cm) but not otherwise different, evenly recurved through ¼ to a full circle (but not spirally contorted), sometimes also randomly horizontally twisted, 14–19-seeded, the sutures 2–5 mm wide framing the at first plane, either green, brown or fuscous, stiffly leathery or eventually fibrous-woody, transversely venulose valves, these becoming low-convex over seeds and shallowly cross-sulcate between them, the mesocarp mealy, somewhat thickened; dehiscence inert, through both sutures, and the valves tending to crack between seeds (but not dissolved into free-falling segments), the seeds transverse in separated locules; ventral suture often undulate, but deeply constricted only where seeds abort, the dorsal (concavely arched) suture more evenly decurved; seeds (few seen) in broad view oblong-elliptic ±7–8.5 × 4.5–6 mm, the testa smooth, tan, somewhat fuscous-shadowed beneath a hippocrepiform pleurogram ±4.5–5 × 2–3 mm.

In caatinga woodland, spiny shrub thickets, and short-tree forest of "chaco" type, mostly at 400–700 m but lower in the Pantanal of SW Brazil, of bicentric range in extra-Amazonian tropical South America: best known from E Brazil, between 6° and 15°S from S Ceará and W Pernambuco to S-centr. Bahia; and scattered within 15°–23°S on the upper Paraná in Mato Grosso do Sul, Brazil, on the Paraná-Beni divide in E Bolivia, and on the headwaters of Río Bermejo in Salta, Argentina. — Map 39. — Fl. X–II. — *Guaiacán blanco, palo barroso* (Argentina, both used for other species); *arapiraca* (Bahia); *triadim.*

In South America south of the Equator there are two species of *Chloroleucon* deceptively similar in finely divided microphyllidious foliage: *Ch. foliosum* and *Ch. acacioides.* The former has a relatively broad, toughly coriaceous pod 13–24 mm wide, either falcate or recurved into an open ring, the latter a thinly leathery pod only 6–7 mm wide, tightly coiled into a compressed spiral. As shown on Map 39, their ranges are vicariant, but on the whole well separated. One fruiting specimen from the Paraguaçú basin in Bahia (Machado Portela, *J. N. Rose 19964* [NY, US], which we have referred to *Ch. foliolosum* because of the falcate lignescent fruit, has atypical foliage, with leaf-formula iii–v/17–20 and leaflets to 7 × 1.8 mm, suggesting passage to *Ch. dumosum.* Since *Ch. foliolosum* and *Ch. dumosum* are sympatric in this region the possibility of hybridization cannot be dismissed.

We have found no significant difference between Bahian *Ch. foliolosum* and the populations distantly disjunct in Mato Grosso and northern Argentina that have been described as *Pithecellobium myriophyllum* and *P. grisebachianum.* A similar discontinuous dispersal is known in *Mimosa hexandra* M. Micheli (Barneby, 1991: map 16).

10. Chloroleucon mangense (Jacquin) Britton & Rose, N. Amer. Fl. **23:** 38. 1928. — Typus infra sub var. *mangensi* indicatur.

Wide-crowned, drought-deciduous *trees* and *arborescent shrubs,* at maturity (2)3–8 m, with smooth bark peeling in flakes and densely lenticellate annotinous branchlets, randomly armed with solitary or rarely geminate, divergent, tapering and vulnerant thorns to ±0.6–2.5 mm, very variable in lf-formula and in pubescence, either glabrous throughout or the lfts pilosulous with fine white, loosely spreading-ascending hairs to 0.15–0.4 mm, the hemispherical capitula of narrow whitish or greenish white fls expanding coevally with new foliage at lowest nodes of new branchlets, the lvs at anthesis commonly not fully expanded (membranous, smaller than at maturity), usually a little paler beneath than above; perulate buds axillary to older lvs ovoid 1.5–5 mm, their scales glabrous dorsally, striate-nerved, often microscopically ciliolate, each deciduous from a linear, transversely semiamplexicaul scar. *Stipules* at foliate and floriferous nodes commonly obsolete, when present submembranous, linear or narrowly oblanceolate-elliptic, 5–13 × 0.5–4 mm, the blade finely nerved, quickly deciduous. *Lf-formula* ii–ix/4–25(–29), the numbers more exactly given under each variety; lf-stk of mature lvs (0.5–)2–9.5 cm (of lvs coeval with fls, no further described, often shorter), the petiole proper (8–)9–24 mm, the longer interpinnal segments 6–20 mm; a convex nectary variable in form either near midpetiole or well below it, either sessile or shortly stipitate, 0.2–0.8 mm tall and 0.3–1 mm diam, similar but smaller nectaries on some distal segments of lf-stk and yet smaller ones at tip of each pinna-rachis; pinnae usually subdecrescent proximally, the rachis of distal ones 2.5–7 cm, the longer interfoliolar segments 0.8–17(–22) mm; lfts linear to broadly obovate or oblong-elliptic, the largest of a plant (always disregarding the proportionately broader furthest pair) 3.5–22 × 1–14 mm, 1.4–5 times as long as wide; venation of (2)3–5 primary nerves from very short pulvinule, the midrib subcentric to displaced to divide blade to ±1:2, either straight or less often distally curved forward, on each side 2–5-branched, the inner posterior primary nerve brochidodrome usually well above midblade, the outer ones very short, a tertiary venulation or reticulum developed in broader lfts, all venation sharply prominulous beneath, less so or immersed above. *Peduncles* solitary or paired with a thorn in the first 1–3-foliate axils of each new branchlet, (7–)10–30 mm, very slender at anthesis, much thickened in fruit; capitula 15–30-fld, the fls erect-ascending, the receptacle becoming 1–3 mm; bracts commonly rudimentary or absent, exceptionally membranous, spatulate-oblanceolate to 1.6 mm; fls dimorphic, 1–3 central ones not or scarcely longer than the rest but their androecium much modified; perianth of all fls glabrous except for always densely pallid-papillate margin of corolla-lobes, or the calyx, rarely the corolla also, thinly pilosulous; PERIPHERAL FLS: perianth 5-merous; calyx narrowly campanulate 1.7–2.8 × 0.8–1.2 mm, the ovate-deltate teeth 0.2–0.5 mm; corolla 4.4–7.3 mm, the tube 0.8–1 mm diam, the ovate lobes 0.8–1.5 mm, their slender midrib once or twice branched; androecium 10–18-merous, 12–18 mm, the tube 2.3–3.2 mm, free from and included within the corolla; TERMINAL FL(S): calyx and corolla not or scarcely longer but a little broader than in peripheral fls; antheriferous stamens no more numerous than in peripheral fls but the filaments united into trumpet-shaped tube 7–10 mm, its flaring orifice fimbriate with short sterile filaments; ovary of all fls truncate, glabrous; stigma slightly dilated, ±0.2 mm diam. *Pods* often solitary but sometimes to 5 per capitulum, sessile but basally attenuate, in profile broad-linear, nearly straight or falcately recurved (or equivalently contorted) through to ¾ circle (but not spirally coiled), when well fertilized 8–26 × 0.6–1.5 cm, at first flat and thin-walled, becoming thicker and lignescent at maturity, the strong sutures either straight or undulately constricted between the seeds, the thin, reddish brown or greenish, always glabrous exocarp coarsely venulose, the crustaceous endocarp to 0.3 mm thick, internally mealy between seeds, these lying in discrete cavities; funicle filiform, coiled distally; dehiscence tardy, the sutures narrowly gaping but the seeds apparently released only by weathering on the ground, the valves also developing transverse fracture lines between seeds as though incipiently lomentiform; seeds either obliquely or subvertically descending, compressed-oblong-ellipsoid, in broad view 5.5–7.5 × 3–5.1 mm, smooth and pale fawn- or putty-colored, fuscous over the pleurogram, this elongately U-shaped 3.2–5 × 1.7–3 mm; seed-testa ±0.4 mm thick in section, composed of a thin translucent coat gelatinous when wetted and a hard impervious inner coat; no endosperm.

In deciduous and semideciduous, more or less xeromorphic woodland or chaparral at elevations mostly below 600 m, exceptionally to 1200 m, widespread in locally differentiated forms from the S margin of the Sonoran Desert in NW Mexico to Colombia, Venezuela and the Antilles, thence S along the E foothills and intermontane valleys of the Andes into NE Bolivia. Flowering with the flush of new foliage following rains, becoming leafless or nearly so in the dry season; filaments white on expansion, becoming ochroleucous or yellowish when faded.

Excepting only the localized Cuban *Ch. guantanamense* we here include within *Ch. mangense* all segregates of *Chloroleucon* native to Mexico and Central America, Colombia and Venezuela, and the Antilles; and we further extend it to embrace *Ch. mathewsii*, found around the western edge of the Amazon basin in Peru, northeastern Bolivia, and adjoining Brazil. In continental South America north of the Amazon River it is the only species of *Chloroleucon*, other than *Ch. foliolosum* and *Ch. eurycyclum*, readily distinguished either by very numerous small leaflets or by tightly coiled fruit, or by both.

Within this extensive range of dispersal, the inflorescence of *Ch. mangense* remains monotonously uniform within a quite narrow size range, but there is conspicuous variation in pubescence of the leaves, in number of pinnae and leaflets per leaf, and in both length and width of fruit, which have consequences in width and length of interseminal septa and in orientation of the seeds. The variation in pubescence appears almost everywhere random and independent of other factors; despite the importance attached to it in the past, it has no taxonomic significance. Within each subunit of the species, such as occupies a phytogeographically natural range and is in other respects uniform, there occur perfectly glabrous and more or less densely pubescent populations (or individuals). These are far from evenly distributed within the species as a whole: Cuban var. *lentiscifolia*, for example, is prevailingly glabrous but occasionally pilosulous, whereas the reverse is true of var. *vincentis* and var. *leucospermum*, and all known examples of var. *mangense* and var. *mathewsii* are pubescent to some degree. Length and width of the fully mature pod are, within narrow limits, correlated with dispersal. However, the length of the fruit is not determined by number of ovules or seeds. In the long narrow pod of var. *leucospermum* the seeds are no more numerous than in the shorter and proportionately broader one of var. *vincentis*, but they stand further apart in narrower longer locules separated by distinct constrictions of the sutures. Because of vertical extension and horizontal constriction of the valves, the seeds are turned mechanically into a nearly vertical orientation. Number and reciprocally adjusted size of leaflets are the most obvious differences between geographic races of *Ch. mangense*, but care must be taken in interpreting the size. Stems collected when the flowers are in full anthesis normally bear only immature leaves, with small tender leaflets not yet fully expanded and materially different in size and texture from those of mature leaves coeval with the fruit. The smallest leaflets are linear, whereas larger ones vary imperceptibly from narrowly oblong to oblong-elliptic or obovate, and the venation increases in complexity with increase in size. Marked discontinuities in leaflet-shape between geographically vicariant varieties of *Ch. mangense* are generally lacking, but in conjunction with fruit characters the average leaf-formula is distinctive.

Geographically central to the species is var. *mangense*, characterized by high leaf-formula, narrow leaflets, and relatively broad pod with straight sutures. We hypothesize that in context of this species a multifoliolate, microphyll leaf and a relatively wide pod with straight sutures are primitive features. Passing outward in any direction from the geographic center of the species one encounters progressively fewer leaflets, or narrower pods, or both, and the leaf-modification appears to have followed parallel courses in areas now distantly disjunct. On account of the parallelism, the identity of a given specimen of unknown origin may be unrecognizable by characters of the foliage, but easily known by the pod. The varieties of *Ch. mangense* with relatively low leaf-formula and broad leaflets are decisively distinguished from South American *Ch. tortum* and *Ch. tenuiflorum* only by straight or falcate but never spirally coiled pod.

Correlation between morphology and dispersal is so nearly perfect in *Ch. mangense* that we are led to interpret occasional discordant records as due to dispersal by humans. The single records of var. *leucospermum* in Haiti and at the port of Kingston in Jamaica are examples. There are some genuinely puzzling intermediate forms, the identity of which we may have misinterpreted. Some plants from northern Falcón, Venezuela, and from Yucatán Peninsula seem intermediate between Cuban var. *lentiscifolium* and continental vars. *vincentis* and *leucospermum* respectively. An imperfect flowering specimen from Ecuador (Portovelo, perhaps = Portoviejo, but even the province unknown, *J. N. Rose 23452*, NY) is ambiguous between vars. *mangense* and *mathewsii*.

Key to the varieties of *Ch. mangense*

1. Lfts at once numerous and narrow, the lf-formula iv–ix/14–25(–28), the longer lfts (disregarding the terminal pair) 3.5–8 × 1–1.8(–2) mm, not incurved at apex; pod relatively short and broad, 9–13 × (1–) 1.1–1.5 cm; Caribbean N Colombia, centr. Panama, and NW Venezuela (Maracaibo basin), disjunct on Pacific coast of Honduras (Bahía de S. Lorenzo), Chiapas, Mexico (Arriaga), and on E Jamaica (vicinity of Kingston) . . . . 10a. var. *mangense*
1. Lfts fewer, except S of the equator and occasionally in Central America, and always broader, the pinnae not more than 5 pairs and the larger lfts 2–14 mm wide; pod various; almost wholly allopatric.
    2. Native of Amazonian Peru and NE Bolivia; lfts relatively many, in longer pinnae 13–20 pairs, narrowly oblong, truncate at base, the deltate apex forwardly incurved . . . . 10b. var. *mathewsii*
    2. Native N of the equator, from upper Magdalena valley in Colombia to NW Mexico, E to NE Venezuela and the Antilles; lfts of longer pinnae mostly 4–13, in Central America sometimes to 16 pairs, but then symmetrical at apex.
        3. Lfts of longer pinnae at least 5 pairs, and the longest <8 mm wide; range as given, but in E Venezuela only N of the Orinoco.
            4. Pod relatively short and broad, 9–15 × 1.1–1.5 cm; Venezuela N of the Orinoco and W of Maracaibo basin, Lesser Antilles, Virgin Islands (St. Croix), and disjunct in upper Magdalena Valley, Colombia (Cundinamarca, Huila) . . . . 10c. var. *vincentis*
            4. Pod narrower, 0.7–10.5(–11) mm wide, but variable in length; Mexico and Central America, Cuba, and single records from E Jamaica, S Haiti, and Virgin Islands, where perhaps introduced.
                5. Fully fertilized pod 14–26 cm long, the sutures undulately constricted beween the relatively distant seeds, these subvertically descending; lfts of longer pinnae 7–13(–16) pairs; Mexico (S Sonora and Baja California Sur to S Tamaulipas, Veracruz,Yucatán, and Chiapas) to Nicaragua, one record each from SW Haiti and SE Jamaica, where perhaps introduced . . . . 10d. var. *leucospermum*
                5. Fully fertilized pod 8–15 cm long, the sutures straight or almost so, not constricted between seeds, these obliquely descending across the cavity; lfts of longer pinnae

5–8 pairs; W half of Cuba, Belize, N Honduras (Olanchito) forms intermediate to the last in Yucatán. . . . . 10e. var. *lentiscifolium*

3. Lfts of longer pinnae 4 pairs, the largest 11–14 mm wide; lower Caroní basin in Bolívar, Venezuela . . . . 10f. var. *tetrazyx*

10a. Chloroleucon mangense (Jacquin) Britton & Rose var. **mangense.** *Mimosa mangensis* Jacquin, Enum. Syst. Pl. 34. 1760. — "In insulis Caribaeis vicinaque Americes continenti.", elaborated in Select. Stirp. Amer. Hist. **2(qu):** 267. 1763 & ibid. (fol) t. 262, fig. 70. 1781. — "Habitat Carthagenae, frequens in insula Mango [= Isla Manga, today a suburb of the city of Cartagena, Colombia]." — Holotypus, not seen, to be sought at W; the protologue and associated plate are unambiguous. — *Mimosa antillarum* Lamarck *ex* Poiret, Encycl. Suppl. **1:** 80. 1810, nom. illegit. *Feuilleea mangensis* (Jacquin) O. Kuntze, Revis. Gen. Pl. **1:** 186. 1891. *Pithecolobium mangense* (Jacquin) Macbride, Contr. Gray Herb. **59:** 3. 1919, based on the description of 1763. *Enterolobium mangense* (Jacquin) Fawcett & Rendle, Fl. Jamaica **4:** 151 (excl. syn. *Acacia micrantha* Desv.). 1920. *Cathormion mangensis* [sic] (Jacquin) Dugand, Trop. Woods **40:** 11. 1934, nom. nud.

Mimosa parvifolia Swartz, Flora Ind. Occid. **2(2):** 948. 1800. —". . . in arenosis apricis non procul a littoribus maris Antillas affluentis." — Holotypus, S(herb. swartz., 2 sheets)! one = NY Neg. *12581*. — *Acacia parvifolia* (Swartz) Willdenow, Sp. Pl. **4:** 1086. 1806. *Pithecolobium parvifolium* (Swartz) Bentham, London J. Bot. **3:** 223. 1844.

Inga marthae K. Sprengel *ex* de Candolle, Prodr. **2:** 441. 1825. — "ad Sanctam-Martham. Bertero [in herb. balbis.]." — Holotypus "misit Balbis, 1822", fl., G-DC! = F Neg. *6971*! — Equated with *Pithecolobium parvifolium* by Bentham, 1844: 223.

Albizzia marthae Britton & Killip, Ann. New York Acad. Sci. **35:** 133. 1936. — "Santa Marta, Magdalena, Colombia, about 80 m elevation, *H. H. Smith 296*." — Holotypus, US!; isotypi A!, GH!, NY (one sheet labeled: "Near Bonda, below 500 ft," one without locality-data)!, U!, UC *1219189*!. - *Pithecellobium marthae* (Britton & Killip) Niezgoda & Nevling, Phytologia **44:** 379. 1979.

Lf-formula iv–ix/14–25(29); lfts linear from obtusangulate base, deltately acute, the larger ones (3.5)5–8 × 1–1.8 mm, (3.1)3.5–5 times as long as wide. *Pods* almost straight to falcately recurved, when well fertilized 9–13 × (1–)1.1–1.5 cm, the sutures not or scarcely constricted, the valves coherent between seeds but not produced inwardly as septa.

In thickets and disturbed matorral, sometimes surviving in hedges, below 150 m, locally frequent around Lago Maracaibo in NW Venezuela (Zulia), in the Caribbean lowlands of N Colombia (Magdalena, Atlántico, Bolívar), and in central Panama, and known from single collections in Pacific Honduras, from Chiapas, Mexico (Arriaga), and in coastal SW Ecuador; disjunct in SE Jamaica, where perhaps an introduction of long standing. — Map 41. — Fl. (X–XI) II–V. — *Caudero, quebrahacha* (Venezuela); *carbonero, hoyo de zorra, vainillo* (Colombia).

In the protologue, Jacquin ascribed to *Mimosa mangensis* a leaf-formula of v/16, but in 1791 illustrated it by a leaf showing only 12 pairs of leaflets. Swartz at first identified his Jamaican plant with Jacquin's, but was dissuaded by an apparent disparity in leaflet-number, redescribing the species as *M. parvifolia.* Modern material from the western Caribbean and from Jamaica shows almost the same range of variation in leaf-formula and confirms Bentham's conclusion that these two names apply to one taxon.

The continental dispersal of *Ch. mangense* follows a long-recognized pattern of dispersal that coincides with a belt of drought-deciduous woodland. The stations in Jamaica are seemingly anomalous, and the species may not be aboriginally native there. It is not recorded where on Jamaica Swartz collected *M. parvifolia*, but all subsequent collections are from Kingston or the hill country of parishes immediately adjoining the city. It was unknown to (or not recorded by) Patrick Browne, and James Macfadyen, writing in 1837, had it only from Kingston and Linguanea, now part of the city itself.

Inga marthae was recognized as a synonym of *Pithecellobium parvifolium* by Bentham as early as 1844. *Albizia marthae* was transferred by Niezgoda and Nevling to *Pithecellobium* on examination of the pollen, which is discordant with that of typical *Albizia*; they made no comment on the validity of the species within *Pithecellobium.*

10b. Chloroleucon mangense var. **mathewsii** (Bentham) Barneby & Grimes, comb. et stat nov. *Pithecolobium mathewsii* Bentham, London J. Bot. **3:** 222. 1844 "*Mathewsi*". — "Mozabamba in Peru [= Moyobamba, in N San Martín, 6°S, 77°W], *Mathews*." — Holotypus, collected in 1836, K(herb. hooker.)! = NY Neg. *2045*; isotypus, *Mathews 4185*, P!. — *Feuilleea mathewsii* (Bentham) O. Kuntze, Revis. Gen. Pl. **1:** 188. 1891.

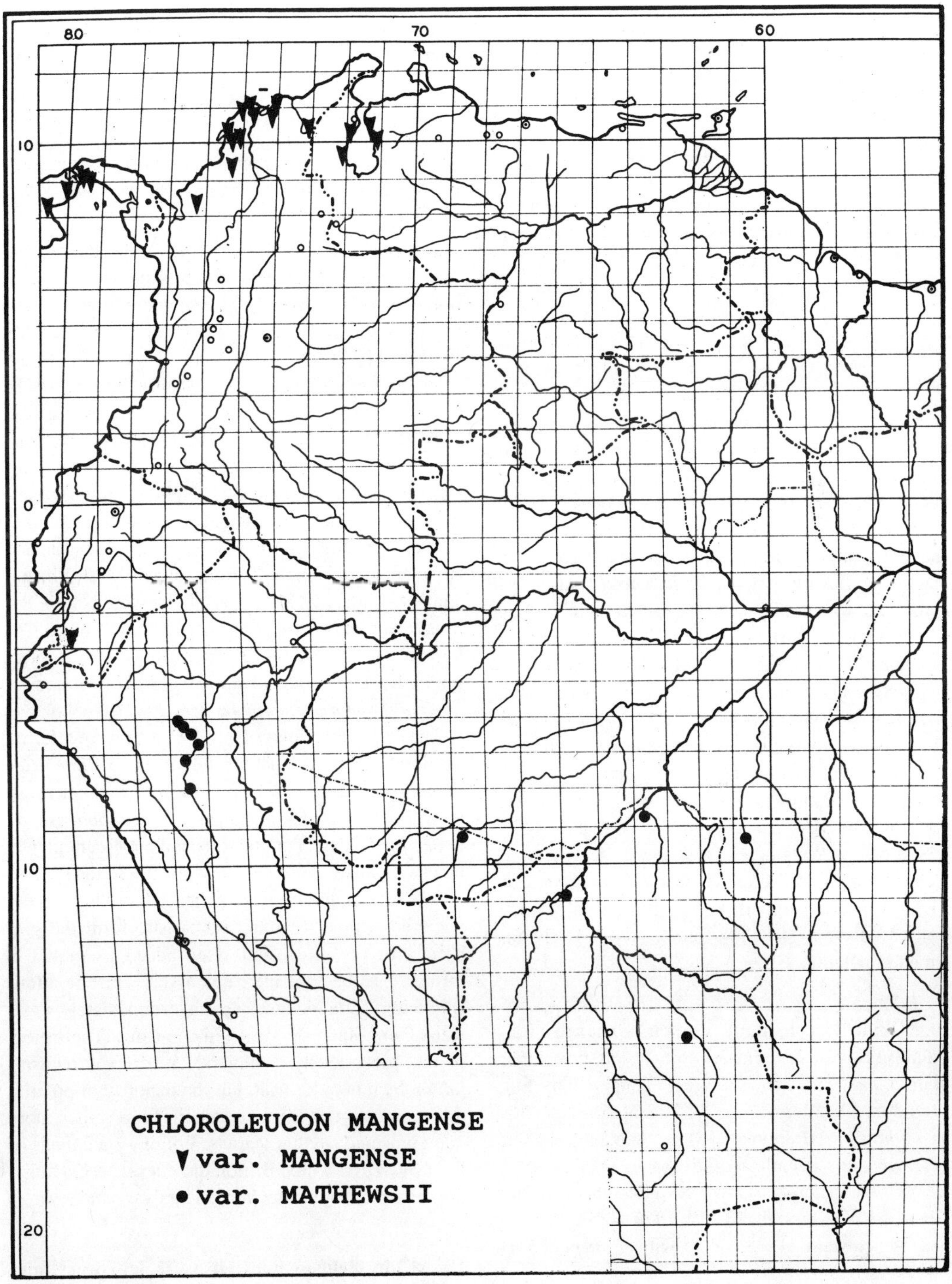

MAP 41. Tropical South America. Range of *Chloroleucon mangense* (Jacquin) Britton & Rose var. *mangense* and var. *mathewsii* (Bentham) Barneby & Grimes.

Lf-formula v–vi/13–19(20); lf-axes and at least the lower face of lfts pilosulous; lfts obtusely rhombic-oblong from truncate base, porrect at apex, the larger ones (4.5–)5–9.5 × (1.5–)2–3.4 mm, 2.6–3.5 times as long as wide, ±5-nerved from pulvinule, the midrib displaced to divide blade ±1:2, incurved distally. *Pods* (little known) linear, nearly straight, to 23.5 × 0.85–1.5 cm, to 28-seeded.

In disturbed dry forest, sometimes on river banks, below 400 m, locally bicentric around the W edge of the Amazon basin in lat. 6°–15°S: middle Huallaga valley in dept. San Martín (prov. Moyobamba, San Martín and Mariscal Cáceres), Peru; and in the upper Purús and Madeira-Beni basins in Acre, N Rondônia, and NW Mato Grosso, Brazil and Beni, Bolivia. — Map 41. — Fl. VIII, IX, XII, II, the full season unknown. — *Palo de estribo* (Bolivia).

Comparison of Bentham's (1875) diagnoses of *Pithecolobium parvifolium* (our var. *mangense*) and *P. mathewsii* reveals objective differences between them only in flower size, in width of leaflets, and in length and width of pods. The floral differences have turned out to be illusory, but the leaflets of *P. mathewsii* really are wider and have a characteristic outline, oblong verging on rhomboid, with distinctly porrect, asymmetric tip unlike those of other forms of *Ch. mangense*. The fruit of *P. mathewsii* is still known from few examples, but seems to be longer and a trifle narrower than that of var. *mangense* and has scarcely a hint of pinched interseminal isthmi. The multiple similarities to *Ch. mangense* indicate in our view an essentially conspecific relationship.

Some specimens of var. *mathewsii* lack the spinescent thorns that ordinarily characterize the genus. At the variety's southern known limit at 14°44′S in Bolivia Michael Nee has found a completely unarmed population.

10c. Chloroleucon mangense var. **vincentis** (Bentham) Barneby & Grimes, comb. et stat. nov. *Pithecolobium vincentis* Bentham, Trans Linn. Soc. London **30:** 597. 1875. — "Isle of St. Vincent's, *Guilding*." — Holotypus, K(herb. hooker)!. — *Acacia vincentis* (Bentham) Grisebach, Fl. Brit. W. I. 222. 1860. *Chloroleucon vincentis* (Bentham) Britton & Rose, N. Amer. Fl. **23:** 37. 1928. — Equated by R. A. Howard, 1988: 372, with *Pithecellobium tortum* Martius.

Pithecellobium caraboboense Harms, Notizbl. Bot. Gart. Berlin-Dahlem **8:** 51. 1921. — "Venezuela: Staat Carabobo, zwisch. Las Trincheras und La Entrada, bei P[uerto] Cabello (PITTIER n. 8820 — Mai 1920 . . .)." — Holotypus, presumably †B; isotypi, NY!, US *1065245*!. — *Cathormium caraboboense* (Harms) Pittier, Arb. Legum. **1:** 55. 1927. *Chloroleucon carabobense* [sic] (Harms) Pittier, Man. Pl. Usual. Venez., Supl. 35. 1939, without direct citation of protologue.

Chloroleucon bogotense Britton & Killip, Ann. New York Acad. Sci. **35:** 129. 1936. — "Anapoima, near Bogotá, Cundinamarca, Colombia, 500 m altitude, *Triana 4481* . . ." — Holotypus, NY!; isotypi, US *939711*!, W!.

Pithecolobium tortum sensu Pittier, Man. Pl. Usual. Venez. 198. 1926, non Martius.

Lvs glabrous, or more commonly the lf-axes and often one or both faces of lfts pilosulous; *lf-formula* ii–iv/(5–)6–11(–12), the lfts oblong or oblong-elliptic to -obovate, the larger ones becoming at maturity (6.5–)7–14(–20) × 2.5–7.5(–9) mm, (1.7)1.9–2.8 times as long as wide. *Pods* usually falcate, 10–14 mm wide, the sutures not constricted between seeds.

In deciduous scrub-woodland, surviving disturbance in fences, in Venezuela and Lesser Antilles from near sea level up to ±400 m, in Colombia attaining 800 m, along the coast and on hillsides in the interior of Venezuela from Falcón E to Monagas and Anzoátegui, disjunctly N to the islands of St. Vincent and Martinique, and S in xerophytic woodland of the Magdalena valley in Cundinamarca and Huila, Colombia. — Map 42. — Fl. abundantly IV–VI, intermittently later. — *Angarillo, raspayuco* (Colombia); *cogicillo, quebrado, quiebra-hacha* (Venezuela).

The early collections of var. *vincentis* were referred by Bentham, in his mature judgment, to *Pithecellobium tortum*, and his example was followed until very recently. Harms distinguished *P. caraboboense* by pubescent foliage, but glabrous and pubescent plants, otherwise alike, collected subsequently in the state of Carabobo are clearly conspecific. Before the falcately recurved fruit takes its characteristic form the glabrous state of var. *vincentis* does indeed resemble *Ch. tortum*; but the fruit indicates a very close relationship with var. *mangense*, vicariant immediately westward, from Maracaibo basin into Panama. The variety is only precariously distinguished from var. *lentiscifolium* by a broader fruit, but the majority of populations, outside the state of Falcón, have rather more numerous and smaller leaflets. Fruiting specimens of *Ch. bogotense*, which Britton did not see, seem indistinguishable from some Venezuelan var. *vincentis*.

10d. Chloroleucon mangense var. **leucospermum** (T. Brandegee) Barneby & Grimes, comb. et stat. nov. *Pithecolobium leucospermum* T. Brandegee, Univ. Calif. Publ. Bot. **10:** 182. 1922. — ". . . in a barranca near Cameron [Camarón], Vera Cruz. [C.

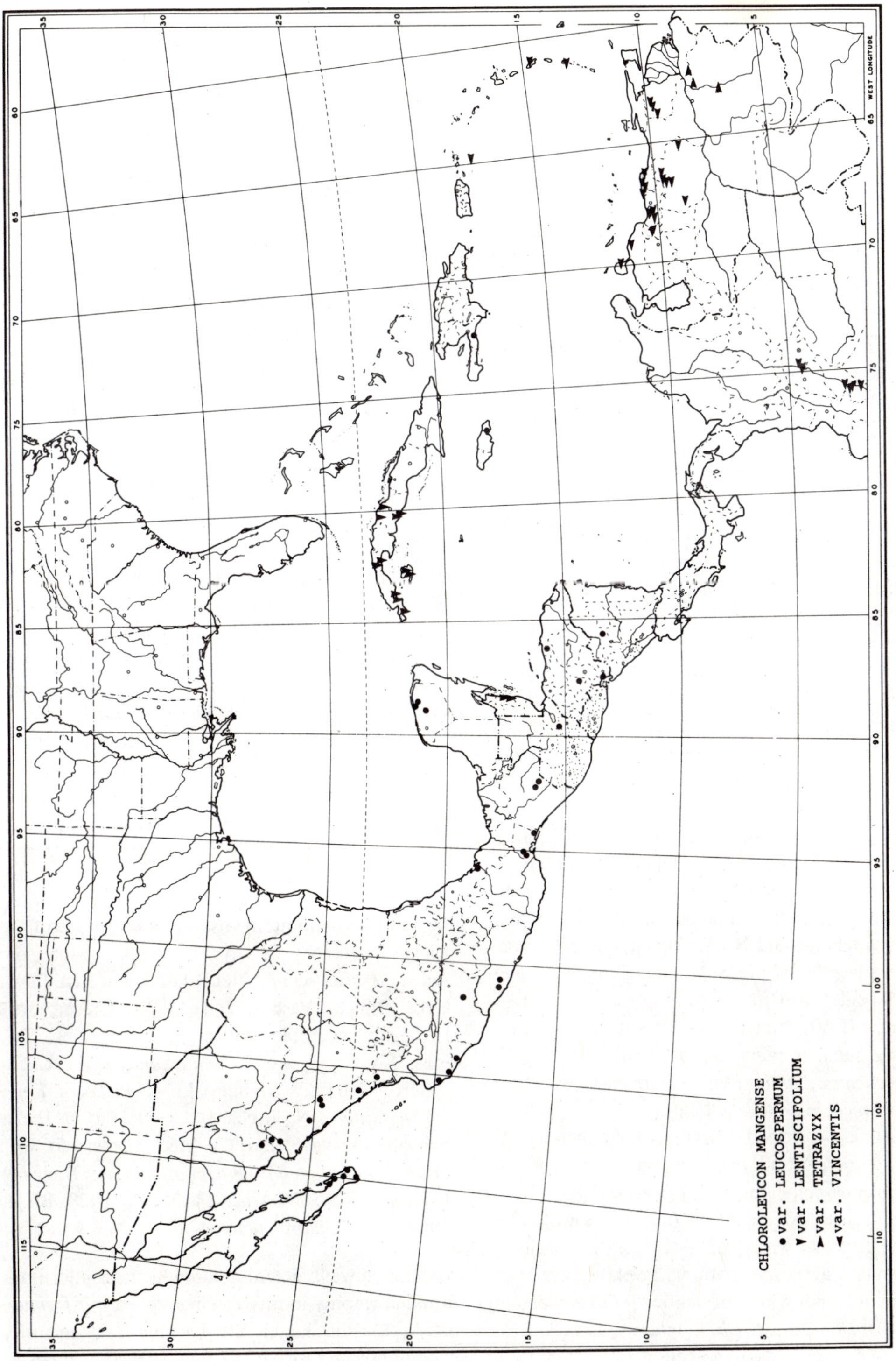

MAP 42. Mexico, Central America, the West Indies, and northwestern South America. Range of *Chloroleucon mangense* var. *leucospermum* (T. Brandegee) Barneby & Grimes, var. *lentiscifolium* (A. Richard) Barneby & Grimes, var. *tetrazyx* Barneby & Grimes, and var. *vincentis* (Bentham) Barneby & Grimes.

A. Purpus] No. 8717." — Holotypus, collected in IV.1922 (fl., fr.), UC *206765*!; isotypi, NY! = NY Neg. *10374*, US *1169660*!.

Chloroleucum [*sic*] *guatemalense* Britton & Rose *ex* Record, Trop. Woods **10:** 24. 1927. — "Collected by Record & Kuylen near Olanchito [Honduras]. (No. 59; Yale No. 10,005)." — Holotypus, dated 14.II.1927 (fr.), NY!; isotypus, US *1315413*!. — *Pithecolobium guatemalense* (Record) Standley, Publ. Field Mus. Nat. Hist., Bot. Ser. **4:** 308. 1929.

Chloroleucon undulatum Britton & Rose, N. Amer. Fl. **23:** 37. 1928. — "Type from Sierra de Alamos, Sonora, March 18, 1910, *Rose, Standley & Russell 13041*." — Holotypus (fr.), NY!; isotypi, F!, US *635861*!. — *Pithecolobium undulatum* (Britton & Rose) H. S. Gentry, Publ. Carnegie Inst. Wash. **527:** 121. 1942.

Chloroleucon langlassei Britton & Rose, N. Amer. Fl. 23: 37. 1928. — "Petatlan [on the Pacific coast of Guerrero near 101°15′W], Michoacan or Guerrero, November 23, 1898, *E. Langlassé 687a*." — Holotypus, US *385639*!; isotypi, NY (fragm. + photo)!, P (no. 687)!. — *Pithecolobium langlassei* (Britton & Rose) Standley, Publ. Field Mus. Nat. Hist., Bot. Ser. **4:** 309. 1929.

Foliage, or at least the lf-axes, commonly pilosulous, but in W and S Mexico randomly glabrous; *lf-formula* ii–v/(6–)7–16, the larger lfts at maturity 7–16(–17) × 2–6(–10) mm. *Pods* 14–26 × (0.5–)0.6–1(–1.5) cm, straight or gently decurved, the sutures shallowly constricted between the subvertically descending seeds, the interseminal isthmi flat.

In short-tree forest, desert chaparral, and semideciduous tropical forest on the coastal plain, ascending on dry hills and along rocky arroyos in the foothills or hilly interior to 600, in Chiapas to 1200 m, widely scattered over the Pacific lowlands of Mexico from S Sonora (Sa. de Alamos) and S Baja California Sur to Chiapas, thence E and SE in scattered stations to centr. Veracruz, Yucatán Peninsula, upland Chiapas, interior Honduras, and N Nicaragua; single records, perhaps adventive, from S Haiti, SE Jamaica, and Virgin Islands (St. Croix). — Map 42. — Fl. most abundantly II–VI, sometimes again VII–X, especially following summer rains. — *Palo ebán* (Baja California), *cucharo*, *ebano blanco* (Sinaloa); *borcelano* (Guerrero); *cachi de toro* (Chiapas).

Except that we extend the synonymy to include *Ch. guatemalense* and *Ch. langlassei*, our concept of var. *leucospermum* coincides with *Pithecellobium leucospermum* sensu McVaugh (1987: 236). As would be expected from its wide dispersal, there is considerable variation at the populational level, especially in indumentum and width of leaflets. *Pithecolobium leucospermum* sens. str., *Ch. langlassei*, and *Ch. guatemalense* answer to the pubescent phase that prevails over the variety's whole range, and *Ch. undulatum* to the glabrous one, which occurs randomly through most of the Mexican range of the variety, from Sonora to Oaxaca. Among these segregates *Ch. langlassei* has relatively narrow leaflets, suggestive of var. *mangense*, but they are less numerous, as are the pinnae. A long narrow pod, undulately constricted between seeds at the flattened isthmi, combined with the low leaf-formula given above, effectively characterizes the taxon. In Pacific Oaxaca some populations with relatively many and small leaflets might be thought of as intergradient with var. *mangense*, but have, so far as known to us, the pod of var. *leucospermum*. Plants of Yucután Peninsula with relatively few and ample leaflets are either intermediate to or perhaps as well referred to var. *lentiscifolia*.

There is no strong evidence one way or the other bearing on the status of var. *leucospermum* on Jamaica, Hispaniola, or St. Croix, but, as it is known today by only one record from each island, a hypothesis of casual introduction from the mainland is plausible. The locality in Jamaica (*Maxon & Killip 312*, NY) is at the port of Kingston where, if native, it could hardly have escaped notice of James Macfadyen, William Harris, or modern collectors; that on Hispaniola (*Ekman 5964*, NY) is on the south coast of Haiti near the port of Jacmel. Both of these plants represent the less common glabrous variant mentioned above. The record from St. Croix (*Liogier 34208*, (NY, in bud) requires fruit for confirmation of variety.

The seeds of var. *leucospermum* are pallid but not really white. Ordinarily the testa is pale fawn-color with a fuscous patch on each face within the pleurogram, but it is sometimes putty-colored overall.

10e. Chloroleucon mangense var. **lentiscifolium** (A. Richard) Barneby & Grimes, comb. et stat. nov. *Acacia lentiscifolia* A. Richard *in* Sagra, Hist. Phys. Cuba, Bot. Pl. Vasc. 469. 1846. — "Crescit circa Havanam, mensibus novembre et decembre florens." — Holotypus, *R. de la Sagra s.n.* e Cuba, P(herb. richard., 2 sheets)!. — *Pithecolobium lentiscifolium* (A. Richard) C. Wright *in* F. A. Sauvalle, Anales Acad. Ci. Méd. Habana **5:** 407. 1869. *Chloroleucon lentiscifolium* (A. Richard) Britton & Rose, N. Amer. Fl. **23:** 37. 1928; Bisse, 1982: 221 ("*Chloroleucum*").

Almost always glabrous throughout, a pilosulous form known only locally in Pinar del Río; *lf-formula* iii/5–7(–8), the larger lfts becoming at maturity oblong-obovate 8–15 × 3–7 mm, (1.9–)2–3 times as long as wide. *Pods* 9–14 × 0.8–1.1 cm, nearly straight to strongly falcate, the sutures not constricted between seeds, these crowded along the pod, obliquely descending.

In dry thickets, often on limestone, sometimes in savanna, along watercourses, or on dunes, below 250 m, widespread and common over the W half of Cuba, from Pinar de Río and Isla de la Juventud to Las Villas; Belize (Corozal distr.). — Map 42. — Fl. III–V, X, perhaps at other times. — *Guayabillo* (Cuba).

The var. *lentiscifolium* is feebly distinguished by the combination of few, relatively large leaflets and a pod at once short, narrow, and not constricted between seeds. Specimens intermediate to var. *leucospermum* in Yucután and to var. *vincentis* on Paraguaná peninsula in northwest Venezuela connect var. *lentiscifolium* with the more populous and widespread varieties of *Ch. mangense*.

10f. Chloroleuson mangense var. **tetrazyx** Barneby & Grimes, var. nov., ab aliis speciei formis foliolis tam paucis quam amplissimis cujusque pinnae distalis (3–)4-jugis, 15–22 × 11–14 mm usque diversa. — VENEZUELA. Bolívar, depto Heres: tramo inferior del río Paragua, entre río Chiguaó y Caño el Gato, 31.I.1980 (fr.) *L. Marcano Berti* in MER *43–980*. — Holotypus, MER *33690*.

Trees ±2 m, glabrous; perules 2–2.5 mm; *lf-formula* ii–iii/(3–)4, the larger lfts 15–22 × 11–14 mm. *Pods* ±12 × 0.75 cm.

In xeromorphic scrub woodland at 200–300 m along Río Caroní and affluent streams in the districts of Heres and Piar, E to foothills of Altiplanície de Nuria, state of Bolívar, Venezuela. — Map 42.

The var. *tetrazyx* is clearly different from var. *vincentis*, the form of *Ch. mangense* prevalent in Venezuela north of the Orinoco, in the few, very large, glabrous leaflets and in the relatively narrow pod. It has more in common with the distantly allopatric var. *lentiscifolium*, but even fewer and ampler leaflets.

The *Pithecellobium*-alliance

The exemplar analysis gave resolved relationships for another group of species included in this monograph (Fig. 3A, no. 4). These species are included in a larger clade characterized by stipular spines and short-shoot systems that are either strictly vegetative or vegetative and reproductive. This larger clade includes species referred to *Acacia* and *Pithecellobium*. The species for which we provide treatments were split into several genera by Britton and Rose (1928), including *Havardia*, *Painteria*, *Ebenopsis*, and *Pithecellobium* sensu stricto. The exemplar analysis showed that *Acacia farnesiana* (Linnaeus) Willdenow and *A. nilotica* (Linnaeus) Willdenow are the outgroups. The data set is shown in Table 4; the characters used in the analysis are as follows.

0. Vegetative branches: 0 = arising proleptically; 1 = arising sylleptically and proleptically.
1. Leaf-nectary: 0 = sessile; 1 = shortly stipitate.
2. Leaf-nectary: 0 = on leaf-stalk; 1 = between pinna-pairs.
3. Pinnae: 0 = proximally decrescent thence subequilong; 1 = decrescent at each end; 2 distally accrescent; 3 = one pair.
4. Venation: 0 = pinnate; 1 = palmate.
5. Inflorescences: 0 = on short-shoots; 1 = on long shoots.
6. Inflorescences: 0 = with pseudoterminal flower; 1 = without pseudoterminal flower.
7. Receptacle of inflorescences with pseudoterminal flowers: 0 = spicate; 1 = capitulate.
8. Flower-buds: 0 = ovoid-pyriform; 1 = flask-shaped.
9. Flowers: 0 = not white vespertine; 1 = white vespertine.
10. Corolla lobes: 0 = erect-ascending; 1 = recurved.
11. Corolla-lobes: 0 = not apically hooded; 1 = apically hooded.
12. Disc: 0 = lacking; 1 = as callosities; 2 = as a well-formed disc.
13. Anthers: 0 = round; 1 = elliptic.
14. Ovary: 0 = sessile; 1 = stipitate.
15. Pod: 0 = laterally flattened; 1 = subterete.
16. Pod: 0 = stiffly papery; 1 = leathery; 2 = ligneous.
17. Pod: 0 = not inertly dehiscent through both sutures; 1 = inertly dehiscent through both sutures; 2 = follicular.
18. Valves: 0 = not framed by enlarged sutures; 1 = framed by enlarged sutures.
19. Pod: 0 = aseptate; 1 = septate.
20. Pod: 0 = recurved; 1 = straight.
21. Valves: 0 = not reflexed with age; 1 = elastically reflexed with age.
22. Valves: 0 = not coiling at dehiscence; 1 = coiling at dehiscence.
23. Funicle: 0 = straight or sinuous; 1 = distally sigmoid. A sinuous funicle is more or less uniformly sinuous throughout and is continuously variable to a straight funicle. The distally sigmoid funicles are straight at the suture, proceed to the seed, and at the seed become strongly sigmoid.
24. Aril: 0 = lacking; 1 = attached at the hilum only; 2 = attached along the sarcotesta.

Table 4

Data set used in the analysis of the *Pithecellobium*-alliance

			1	1	2	2
	0	5	0	5	0	5
NILOTICA†	?000?	?0000	00?00	00001	10000	01?00
FARNESIANA†	0001?	00000	00000	10000	00000	00?10
UNGUISCATI	0?130	10000	00011	01000	00111	10111
LANCEOLATUM	00130	11-00	00010	02100	00112	12?11
FURCATUM	??130	?1-00	00010	02100	00112	12?11
MACRANDRIUM	00110	11-01	00211	02000	00112	12?11
EBANO	01121	00100	0011?	12001	10000	01011
CONFINIS	01121	00100	00210	12001	10000	01011
CAESALPINIOIDES	01121	00000	00?11	12001	10000	01?11
PALLENS	10011	10000	10111	00000	10010	01011
ALBICANS	10011	10000	10010	00000	10010	00011
MEXICANA	10011	10100	10010	00000	10010	01?11
SONORAE	10021	10100	00210	00000	10010	01111
CAMPYLACANTHUS	10021	?0100	00110	00000	10010	00011
LEPTOPHYLLA	0?111	00000	10111	01000	01000	01011
ELACHISTOPHYLLA	0?131	00000	00111	01000	01000	01011
REVOLUTA	0?131	?0000	?0111	01000	01000	01?11
PLATYLOBA	00020	?0011	01211	00010	10010	01111
ACATLENSIS	0002?	00011	01211	00010	10010	01111
PREHENSILIS	00?30	00011	01210	00010	10010	01011

Note: † = outgroup; ? = state not observed in this terminal; - = state inapplicable or of uncertain homology.

25. Seed: 0 = deciduous from funicle; 1 = persistent on funicle.
26. Pleurogram: 0 = incomplete; 1 = complete; 2 = lacking.
27. Mucilage on seed-coat: 0 = lacking; 1 = present.
28. Pollen: 0 = no supplementary sets of pores; 1 = supplementary sets of pores faced by fours. This character and the following are taken from Guinet (1981, and pers. comm.).
29. Nonequatorial pollen apertures: 0 = colpori; 1 = pori.

The analysis yielded two equally parsimonious trees of 65 steps with a consistency index of 0.55 and a retention index of 0.75. The two trees were generated by an unresolved trichotomy involving *Painteria elachistophylla*, *P. leptophylla*, and *P. revoluta*. One tree is shown in Figure 12. The ingroup of the *Pithecellobium*-alliance forms a clade sister-group to *Acacia farnesiana* and *A. nilotica* on the basis of pinna-characters (no. 3, state 2; five homoplasious changes in this character above this node); a disc of callosities (no. 12, six homoplasious changes above this node); elliptic anthers (no. 13); and non-equatorial apertures of the pollen as pori (no. 29). Above this are groups of clades. The first, distinct on the basis of a homoplas (no. 23), includes two monophyletic groups. The first is distiguished by three synapomorphies: ovoid-pyriform buds (no. 8); apically hooded corolla-lobes (no. 11) and valves framed by sutures (no. 18). This clade includes three species placed in *Havardia* by Britton and Rose. On the basis of the three characters listed, which are morphologlical characters associated presumably with moth pollination, we are separating these three species into a new genus *Sphinga* (no. X). Sister group to *Sphinga* are the remainder of *Havardia* sensu Britton and Rose, distinguished by branches arising proleptically (no. 0). These are treated in *Havardia* (no. XI) below.

The second major group arising above the outgroups are distinguished by leaf-nectaries shortly stipitate (no. 1). One of the clades above this node has a synapomorphy of shortly stipitate leaf-nectaries (no. 1), and three homoplasious changes. The three species in this clade are placed in *Ebenopsis* (no. XII). Sister-group to *Ebenopsis* are those species united by a synapomorphy of stiffly leathery pods (no. 16, with a homoplasious change above) and three homoplasious changes. Three of these species have pod valves elastically reflexed with age (no. 21) and are included in *Painteria* following (no. XIII). The relationships are unresolved among these species because the inflorescence of *P. revoluta* is still unknown. The remaining species are referred to *Pithecellobium* sensu stricto, which awaits treatments of the Mexican and Central American species, and will appear in Part II of our treatment.

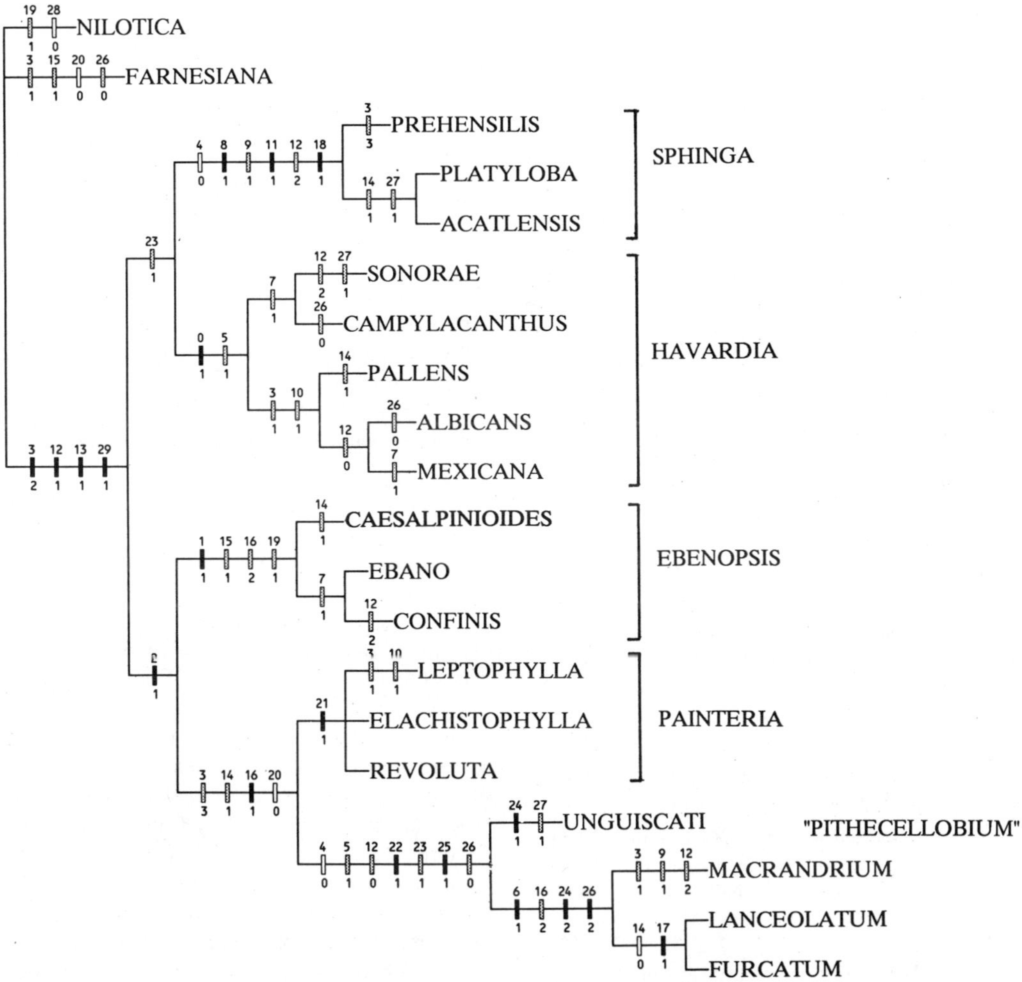

FIG. 12. One of two trees resulting from an analysis of the *Pithecellobium*-alliance. The second tree is generated by a different topology at the trichotomy involving *Painteria elachistophylla, P. revoluta,* and *P. leptophylla.*

Sphinga, Havardia, Ebenopsis, and *Painteria* differ from *Albizia* sens. lat., which they often superficially resemble, in the syndrome of spinescent stipules and lack of a dimorphic flower terminal to each unit of inflorescence. They form a series of progressively xeromorphic groups of species, three of which were described as genera by Britton and Rose. *Havardia* has nectaries on the leaf-stalk below the first pair of pinnae, basically pinnate venulation, a relatively unspecialized, facultatively pseudoracemose-paniculate inflorescence, and short flowers with poorly developed intrastaminal disc. Its straight, papery, plano-compressed and unilocular pod with compressed-lentiform seeds on dilated funicle is indistinguishable from that of some genuine *Albiziae*. The five species of *Havardia* are distributed in sesonally dry, deciduous or semideciduous woodland at low elevations on both the Gulf and the Pacific slopes in Mexico, one extending to Nicaragua. Hardly different, except in the firmer-walled, decurved pod, is *Painteria*, but here the capitula are confined to brachyblasts, and the leaflet-venation is basically palmate. The species of *Painteria* are generally smaller (<2 m), stiffer shrubs endemic to upland interior Mexico. The genus *Ebenopsis*, containing one arborescent species distributed around the Gulf lowlands from southern Texas to Yucatán and two smaller distributed in desert around the Gulf of California, is distinguished from the foregoing by a straight, ligneous, subcylindric pod, obese seeds separated by adventitious septa, and at least shortly stipitate leaf-nectaries. The genus *Sphinga*, hitherto included in *Havardia*, with which it agrees in form and texture of the fruit, is distinguished from the rest by a greatly elongate, proportionately narrow perianth, a far-extruded androecium, a strongly developed intrastaminal disc, flower-buds flask-shaped rather than obovoid in praefloration, and vespertine anthesis. Although diverse in detail, the four genera have much in common, both morphologically and ecogeographically, and appear to form a natural alliance within the albizioid Ingeae. This indeed was already perceived by Bentham, who, although he had fruits only of the *Havardia-Sphinga* type, established a section *Ortholobium* in *Pithecolobium* to accommodate them.

Since Britton and Rose divided sect. *Ortholobium* into three genera, Mohlenbrock (1963) recommended the maintenance of *Havardia* and *Ebenopsis* as genera independent of *Pithecellobium*, but suggested that *Painteria* could best be reduced to *Zygia*, with which it has little in common. Nielsen (1981; 184) adopted a broadly defined genus *Havardia* essentially equivalent to our four (and to *Pithecolobium* sect. *Ortholobium*) so far as the New World is concerned but included three Asiatic elements: *Pithecellobium tenue* Craib (transferred by Kostermans, first to *Acacia*, and later to a monotypic genus *Thailandentadopsis*); *Painteria nitida* (Vahl) Kostermans; and *Pithecellobium vietnamense* Nielsen. We have seen only scanty and incomplete material of these species, but provisionally exclude them from *Havardia* as probably parallel, paraphyletic offshoots of the Mimosaceae Ingeae.

X. SPHINGA BARNEBY & GRIMES

Sphinga Barneby & Grimes, gen. nov., legumine recto planocompresso crasse papyraceo cum *Havardia* congrua, sed floribus elongatis, calyce profunde campanulato ante corollam emersam fusiformi acuto, corolla ipsa 11.5–27 (nec 3–5.5) mm longa, necnon androecio longissimo 4–9 (nec 0.9–1.5) cm longo diversa. — Sp. typica: *S. platyloba* (de Candolle) Barneby & Grimes ≡ *Acacia platyloba* de Candolle. — Etymology: confected from *Sphinx* (Sphingidae), the putative pollinator + Tupi *ingá*, vernacular for several Mimosaceae.

Pithecolobium sect. *Ortholobium* sensu Bentham, 1875; 593, ex parte (spp. 84–86), lectotypo exclus.

Shrubs and *trees*, one sometimes sarmentose. *Stipules* of primary leaves subulate, spinescent. *Lf-formula* i–v/1–14; lf-nectary below midpetiole; the lfts opposite, the venation pinnate or (*E. acatlensis*) subpalmate. *Capitula* all or almost all arising from brachyblasts, the fls sessile or nearly so; fl-buds before exsertion of corolla flask-shaped, acute-acuminate; calyx cylindro-campanulate 5.5–12.5 mm; corolla narrowly trumpet-shaped 11.5–27 mm; androecium 34–176-merous, 4–9 cm, the tube usually far exserted; intrastaminal disc clasping stipe of ovary 0.5–1.3 mm tall. *Pods* broad-linear plano-compressed 6–10-seeded, the stiffly papery valves framed by sutures, the cavity continuous, dehiscence inert, through both sutures; seeds transverse on dilated, contorted or sigmoid funicle. — Spp. 3, two of arid or seasonally dry, tropical lowland and submontane (interior) Mexico, adjacent Guatemala, and NW South America (NE Colombia, NW Venezuela, Dutch W. I.), one endemic to Cuba.

The three species of *Sphinga* were all known to Bentham and grouped by him into an informal division of his *Pithecolobium* sect. *Ortholobium*, from which they were transferred by Britton and Rose to *Havardia*. The relationship to *Havardia* is not disputed, but these species differ collectively from the core of *Havardia* in the greatly elongated perianth (which recalls that of *Macrosamanea* in the context of *Zygia*), with long, silky corolla expanding at nightfall.

For notes on wood anatomy see discussion in *Pithecellobium*.

Conspectus of genus *Sphinga*

¶ Pinnae of all lvs 1-jug.; capitula 1–3-fld; Cuba.
1. *S. prehensilis* (C. Wright) Barneby & Grimes

¶¶ Pinnae of larger lvs 2–5-jug.; capitula 4–15-fld; continental North and NW South America.
2. *S. acatlensis* (Bentham) Barneby & Grimes
3. *S. platyloba* (de Candolle) Barneby & Grimes

Key to the species of genus *Sphinga*

1. Lf-formula (i–)ii–v/2–14, if lfts as few as 2 pairs per pinna then the longer ones 11–30 mm; capitula 4–15-fld; Mexico to N South America.
    2. Lfts 2–5 pairs per pinna, the longest 11–30 mm; androecium 34–82-merous; valves of pod glabrous; range as given above . . . . . 3. *S. platyloba* (p. 164)
    2. Lfts of longer pinnae 7–14 pairs, the longest 4–9.5 mm; androecium (86–)140–176-merous; valves of pod densely pilosulous overall; Mexico only . . . . . 2. *S. acatlensis* (p. 161)
1. Lf-formula i/1–2, the larger lfts 2.5–7 mm; capitula 1–3-fld; pod glabrous; Cuba. . . . . . 1. *S. prehensilis* (p. 161)

1. Sphinga prehensilis (C. Wright) Barneby & Grimes, comb. nov. *Calliandra prehensilis* C. Wright in Sauvalle, Anales. Acad. Ci. Méd. Habana **5:** 406. 1869. — "[Cuba.] Potrero Manatí, Trinidad. . ." — Type collection, *C. Wright 3543*, some elements distributed as "*Calliandra apoda* Wright [nom. nud.]"; lectotypus, GH!; isotypi, K!, NY (2 sheets)!. — *Pithecolobium prehensile* (C. Wright) Bentham, Trans. Linn. Soc. London **30:** 593. 1875. *Feuilleea prehensilis* (C. Wright) O. Kuntze, Revis. Gen. Pl. **1:** 188. 1891. *Havardia prehensilis* (C. Wright) Britton & Rose, N. Amer. Fl. **23:** 42. 1928. Fig. 13.

Effusely branched, microphyllous prickly *shrubs* ±1–1.5 m, either glabrous up to the thinly silky fls or minutely puberulent on young branchlets, the stiffly flexuous long-shoots armed at each node with a pair of ascending-recurved, rigidly lignified, smooth fuscous stipules, the minute foliage drought-deciduous, the capitula of few long narrow fls subsessile either in the axil of primary lvs or fasciculate with lvs on axillary short-shoots. *Stipules* of primary lvs 2.5–5 mm, those of diminished lvs of short-shoots much shorter, erect 1-nerved, not or scarcely spinescent. *Lf-formula* i/1–2; lf-stks 1–7 mm, at middle ±0.2–0.3 mm diam; nectary commonly near midpetiole sessile round, rarely (*Acuña 18679*, NY) at tip of petiole and squatly turbinate, in either case 0.15–0.3 mm diam, dimpled or shallowly cupular; rachis of pinnae 1–3 mm, the one interfoliolar segment, when present, to 1.5 mm; lft-pulvinules 0.15–0.4 × 0.15–0.3 mm; lfts obovate or oblong-obovate from inequilateral, often shallowly semicordate base, broadly obtuse or emarginate and sometimes obscurely apiculate, 2.5–7 × (1–)1.4–5 mm; venation pinnate, the straight centric or subcentric midrib giving rise on each side to 2–4 simple secondary nerves brochidodrome well within the plane margin, the whole venation bluntly prominulous on lower face, more faintly so or immersed on the upper. *Peduncles* solitary, 1–4 mm; capitula 1–3-fld, the homomorphic fls erect, the receptacle not over 1.5 mm; bracts either ovate or lanceolate 1–3 mm, 1-nerved, subpersistent; pedicels ±0.5–0.8 mm, nearly as thick; perianth essentially 5-merous, but some corolla-lobes often fused, the greenish calyx minutely puberulent, the white corolla silky-pilosulous distally, either overall or along middle of lobes; calyx subcylindric a little dilated beyond middle, striately nerved, ±7–9 × 2.3–3 mm, the triangular acute teeth 1.2–3 mm; corolla narrowly trumpet-shaped 19–27 mm, the erect lanceolate, apically hooded lobes 5–9 mm; androecium ±55–75-merous, 54–90 mm long, the stemonozone ±1 mm, the tube 31–54 × 1.6–2 mm; disc embracing base of ovary 0.5–0.7 mm tall; ovary glabrous, tapering into the style, shallowly sulcate facially; stigma poriform. *Pods* solitary, in profile broad-linear 5.5–8.5 × 1.2–2 cm, plano-compressed (6–)7–8-seeded, at base cuneately contracted into a pseudostipe about as long as calyx and at apex into triangular-lanceolate beak, the papery fuscous-brown valves framed by plane or low-carinate sutures 1.2–1.5 mm wide, low-convex over each seed, delicately cross-venulose glabrous, internally pallid dull, the cavity continuous; seeds transverse near middle of pod, basifixed on compressed, sigmoid funicle, strongly compressed lentiform, in broad view round or broadly elliptic, 7–10 × 5.8–7.5 × 1.4–1.8 mm, the firm crustaceous testa castaneous lustrous, smooth or minutely pitted, in section ±0.1 mm thick, tightly investing the pallid embryo, the periphery only a trifle thickened, the U-shaped pleurogram 2–5 × 1.2–4 mm.

Forming thickets in savanna, on streambanks, and on dry plains, reportedly tolerating saline soils, below 200 m, local in centr. and E Cuba (Las Villas; Oriente). — Map 43. — Fl. X–XI(–?).

2. Sphinga acatlensis (Bentham) Barneby & Grimes, comb. nov. *Pithecolobium acatlense* Bentham, Trans. Linn. Soc. London **30:** 593, 1875. — "Mexico, Acatlan, in the Pueblo district [estado de Puebla], *Andrieux*, n. 395." — Holotypus, K!; isotypus, P!. — *Feuilleea acatlensis* (Bentham) O. Kuntze, Revis. Gen. Pl. **1:** 187. 1891. *Havardia acatlensis* (Bentham) Britton & Rose, N. Amer. Fl. **23:** 42. 1928.

Pithecolobium acatlense sensu Rose, Contr. U.S. Natl. Herb. **5:** 195. 1899; M. Micheli, Mém. Soc. Phys. Genève **34:** 284, t. 27. 1903 (= *Langlassé 1031*, GH! sub nom. ined. *Calliandra eriopetala* M. Micheli); Standley, 1922: 395; McVaugh, 1987: 233.

Xeromorphic, drought-deciduous, microphyllous *shrubs* and *trees* 2–7 m with stout stiff virgate, fuscous pallid-lenticellate, early glabrate branches armed at each node of long-shoots with a pair of ascending, straight or commonly gently recurved, glabrous atropurpureous spinescent stipules, the axes of new lvs softly pilosulous with fine spreading white hairs to 0.3–0.5 mm, the pallidly olivaceous, scarcely bicolored, firm-textured lfts glabrous or facially glabrous but remotely ciliolate to thinly pilosulous on both faces, the few-fld capitula of long, densely pilosulous whitish fls arising singly from short-shoots

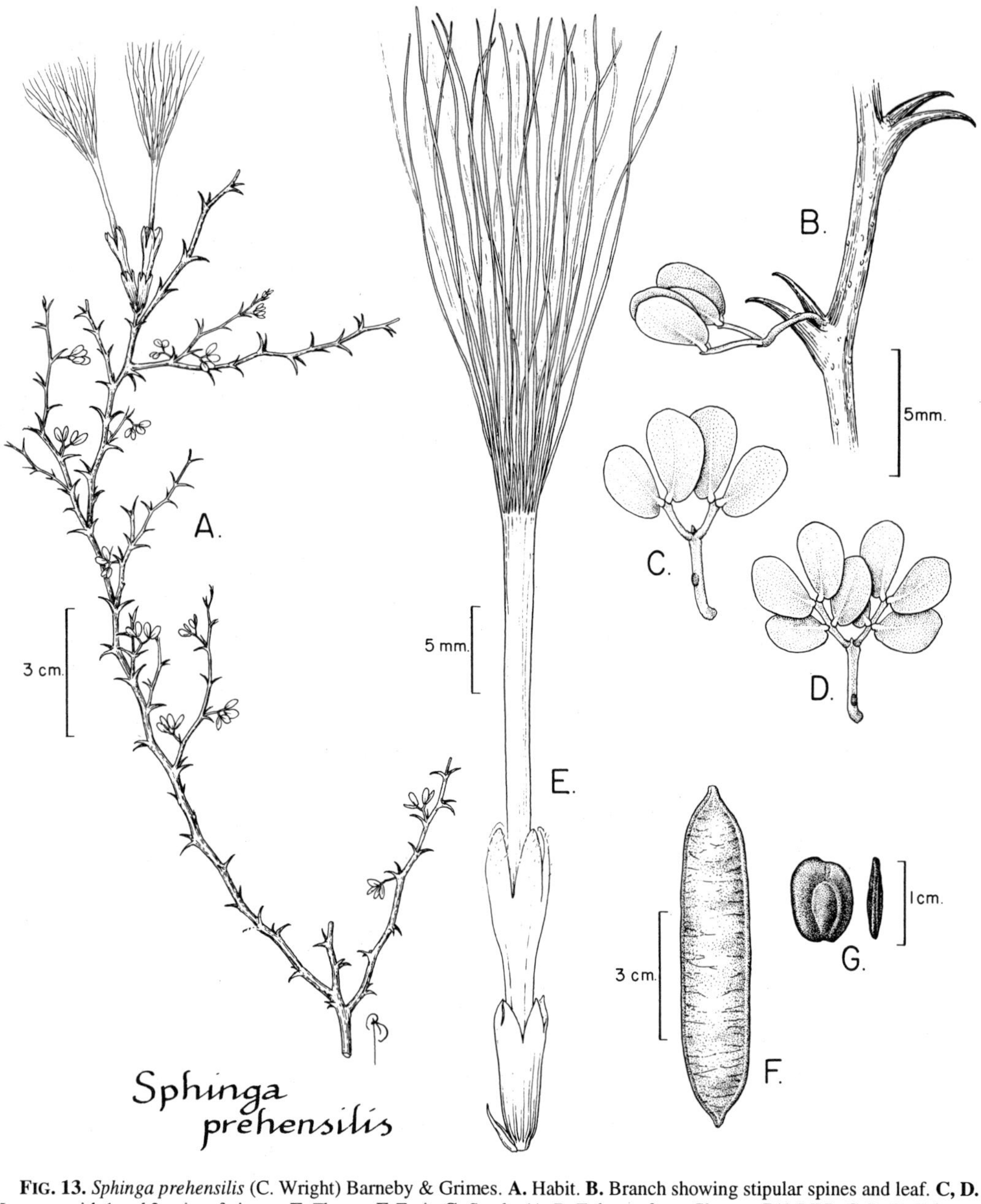

FIG. 13. *Sphinga prehensilis* (C. Wright) Barneby & Grimes. **A.** Habit. **B.** Branch showing stipular spines and leaf. **C, D.** Leaves, with 1 and 2 pairs of pinnae. **E.** Flower. **F.** Fruit. **G.** Seeds. (A, B, E drawn from *Clement 741* (NY); C, D, F, G drawn from *Britton 5572* (NY).)

either coevally with or well before a fascicle of hornotinous lvs, the long, densely tomentulose pods pendulous. *Stipules* of primary lvs on long-shoots 3.5–9 mm, at base ±1–1.5 mm diam and obscurely trigonous, tapering and vulnerant distally, long-persistent after fall of associated lf, the stipules of fasciculate lvs much shorter and more slender. *Lf-formula* ii–v/7–14(–"16"); lf-stks slender, those of major lvs (1–)1.5–4 cm, (of random fasciculate or drought-inhibited lvs, no further described, shorter), the petiole (5–)6–26 mm, at middle 0.35–0.6 mm diam, the longer interpinnal segments (1.5–)2.5–8 mm, the narrow ventral groove continuous; nectary near or well below midpetiole sessile, shallowly cupular, round or

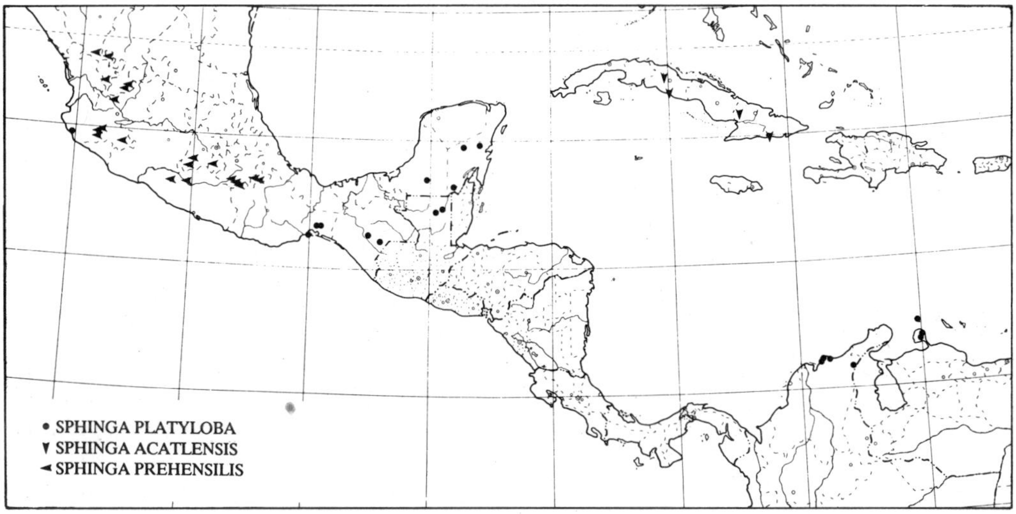

Map 43. Mexico, Central America, the West Indies, and northwestern South America. Range of *Sphinga platyloba* (de Candolle) Barneby & Grimes, *S. acatlensis* (Bentham) Barneby & Grimes, and *S. prehensilis* (C. Wright) Barneby & Grimes.

elliptic 0.4–1.6 mm diam, thick-rimmed, in profile not over 0.3 mm tall, no nectaries on upper lf-stk, rare minute ones at tip of some pinna-rachises; pinnae little graduated in length, the rachis of longer ones (6–)9–36 mm, the longer interfoliolar segments (0.9–)1.2–3 mm; lft-pulvinules 0.3–0.5 × 0.25–0.3 mm; lfts a little decrescent at each end of rachis, elsewhere subequilong, narrowly ovate or lance-oblong from asymmetrically shallow-cordate or semicordate base, contracted into a sharply acuminulate tip 0.3–0.5 mm, a little forwardly curved, those near mid rachis (3.7–)4–9.5 × 1.4–3.4 mm, 2.4–3.6(–4.1) times as long as wide; venation of 2–3 nerves from pulvinule, the gently incurved midrib displaced to divide blade ±1.2–2 and giving rise on posterior side to (2–)3–4 secondary nerves brochidodrome close within the plane margin, the one (or inner) posterior nerve produced to anastomosis well short of mid-blade, the outermost when present much shorter, the whole venation immersed on upper face, bluntly, sometimes faintly prominulous beneath. *Peduncles* 1–8 mm, rigid and woody in fruit; capitula (3–)4–8-fld, the fls ascending homomorphic, the receptacle scarcely dilated 1–1.5 mm; bracts 2–4 mm, either linear, or oblanceolate, or narrowly ovate, becoming papery and deciduous at anthesis; perianth basically 5-merous, but corolla-lobes often fused and the sinuses between calyx-teeth often very unequal, the calyx yellowish striate, thinly pilosulous, the corolla densely white-silky-pilose; calyx in early bud narrowly flask-shaped acute or acuminate, at anthesis cylindro-campanulate 9.5–12.5 × 3–4.5 mm, contracted at base into a drum-shaped stipe not over 1 mm, the lanceolate acute teeth to 3–4.5 mm; corolla 14.5–18.5 mm, the hooded lobes separated by sinuses ±1–6 mm deep, some united to tip; androecium (86–)140–176-merous, (43–)48–57 mm, the stemonozone 1.5–2.5 mm, the tube 10–19 × 1.8–3 mm, ±1 mm shorter to 3.5 mm longer than corolla; disc (0.6–)0.8–1.3 mm, sometimes charged internally with a few weak hairs; ovary shortly stipitate, at anthesis glabrous, tapering at each end, shallowly sulcate laterally; stigma a little dilated, ±0.2–0.25 mm diam. *Pods* 1 per capitulum, technically sessile but attenuate at base into a laterally grooved pseudostipe 5–20 mm, in profile broad-linear straight (10–)12–18 × 1.8–2.4(–2.9) cm, (7–)8–10-seeded, abruptly contracted at apex into a fragile (often broken) erect cusp 3.5–9(–15) mm, the plano-compressed, stiffly papery or leathery valves framed by obtusely 3-ribbed sutures ±3–4 mm wide, low-convex over each seed, densely gray- or brownish-pilosulous overall, internally pallid dull, the cavity continuous; seeds transverse, basifixed on dilate sigmoid funicle, plumply discoid, in broad view round or bluntly oblong, 9–13 × 8.5–12 × 2–2.5 mm, the leathery or crustaceous, castaneous and highly lustrous testa 0.2–0.3 mm thick in section, closely investing the embryo, a little thickened at the periphery, the pleurogram irregularly U-shaped ±5–8 × 3.5–6.5 mm.

On open hillsides and plains in mesquite-grassland, matorral and thorn-scrub, or tropical deciduous woodland, 500–1650 m, scattered in Mexico from the S lobe of Durango and W Jalisco E through the Balsas basin to the Tehuacán desert and headwaters of Río Papaloapan in Puebla. — Map 43. — Fl. III–VII, with or preceding new foliage. — *Asinchete, cinchete* (México, Guerrero).

3. Sphinga platyloba (de Candolle) Barneby & Grimes, comb. nov. *Acacia platyloba* Bertero *ex* de Candolle, Prodr. **2:** 472. 1825. — "ad Sanctam-Martham (Bertero)." — Holotypus, G-DC! = F Neg. *28268*. — *Feuilleea platyloba* (de Candolle) O. Kuntze, Revis. Gen. Pl. **1:** 186. 1891. *Pithecolobium platylobum* (de Candolle) Urban, Symb. ant. **5:** 360. 1908. *Havardia platyloba* (de Candolle) Britton & Rose, N. Amer. Fl. **23:** 43. 1928.

Pithecolobium sericiflorum Bentham, London J. Bot. **5:** 105. 1846. — "Venezuela, *Funcke*, n. 536." — Holotypus, K(herb. hooker)! = NY Neg. *2043*; isotypus, OXF!.

Acacia platyloba K. Sprengel, Syst. Veg. **3:** 144. 1826, hom. illeg. — "Ad fl. Magdalen. Bertero." — Holotypus not known to survive, presumably an isotypus of *A. platyloba* Bertero. — *Havardia platyloba* (K. Sprengel) Britton & Rose, N. Amer. Fl. **23:** 43. 1928, hom. illeg. — Equated with *A. platyloba* de Candolle by Bentham, 1875: 593, and synonymized with the generitypus.

Pithecollobium leiocalyx Standley, Contr. U.S. Natl. Herb. **20:** 189. 1919. — "... US ... 567316, collected at Salina Cruz, Oaxaca, Mexico, April 28, 1910, by C. R. Orcutt (no. 3288)." — Holotypus, US *567316*!; clastotypus (fragm.), NY!. — *Havardia leiocalyx* (Standley) Britton & Rose, N. Amer. Fl. **23:** 42. 1928.

Pithecolobium falconiense Pittier, Trab. Mus. Com. Venez. **5:** 5. 1927. — "[Venezuela.] Falcón: Cerca de Pecaya, distrito Bolívar; flores Mayo 13 de 1927. (Dr. P. Christ n. 24, tipo)." — Holotypus, US *1344349*!; probable clastotypus (fragm., numbered *Christ 26*), NY!.

Stiffly branched *arborescent shrubs* of open, semi-deciduous scrub-woodland, at anthesis (1–)2–7 m, and in closed woodland becoming sarmentose or lianoid, in either case armed at nodes of long-shoots with stout, usually recurved, lignified and vulnerant stipules, the annotinous stems fuscous, pallid-lenticellate, the young stems and lvs glabrous to pilosulous overall with fine white hairs to 0.15–0.3 mm, the lvs bicolored, the firm plane lfts dull olivaceous (sometimes pruinose) above, pallid beneath, the loose capitula of long white vespertine fls arising mostly from short-shoots axillary to an annotinous (or fallen) lf, rarely solitary in random coeval lf-axils, the corolla notably white-silky throughout or distally. *Stipules* triangular-subulate 0.5–5 mm, at first ascending, those subtending primary lvs of long-shoots mostly >1.5 mm, early lignified and slightly to strongly recurved, persistent on defoliate stems, those of short-shoots mostly <1.5 mm, remaining erect, not or scarcely indurate. *Lf-formula* (i–)ii–iii(–iv)/(1–)2–5; lf-stk of primary lvs (1.5–)2–7.5(–9) cm (of some fasciculate lvs, no further described, shorter) the petiole (1–)1.3–4.3(–6) cm, at middle (0.35–)0.4–1.1 mm diam, the first (or only) interpinnal segment 5–14(–18) mm; nectary below midpetiole elliptic patelliform thick-rimmed, in long diameter 0.9–3.4 mm, in profile 0.15–0.45 mm tall, no nectary between pinnae but a small one on furthest segment of pinna-rachises; pinnae scarcely or decidedly accrescent distally, the rachis of longer ones (0.6–)1.2–5 cm, the one or the longer interfoliolar segment 4–12(–15) mm; lft-pulvinules 0.4–1 (–1.1) × 0.35–0.6 mm; lfts often much, sometimes scarcely accrescent distally, in outline oblong-elliptic or -oblanceolate to broadly obovate from inequilaterally rounded to broad-cuneate, seldom shallowly semicordate base, broadly obtuse but often minutely apiculate, the penultimate pair 11–30 × (4–)5–15 mm, (1.5–)1.6–2.4 times as long as wide, the furthest pair either as big or bigger and proportionately wider; venation pinnate, the straight or scarcely incurved midrib either centric or displaced to divide blade 1:1.5, giving rise on each side to 4–6 slender secondary nerves brochidodrome well within the sometimes cartilaginous margin, a weak random tertiary venulation sometimes developed, the whole venation immersed (often pallidly discolored) on upper face, delicately prominulous beneath. *Peduncles* solitary and 2–3-nate 2–4(–4.4) cm, lignescent in fruit; capitula 7–15-fld, the subsessile fls ascending, the receptacle 1.5–5(–6) mm; bracts linear-lanceolate or narrowly ovate 0.6–3 mm, rather tardily deciduous; fls homomorphic; pedicels when present stout, not over 1 mm, as long as or longer than broad; fl-buds narrowly flask-shaped acute; perianth basically 5-merous but some lobes either of calyx or corolla often fused, then appearing (2–)3–4-merous, or the corolla randomly 5-merous, the calyx glabrous or remotely pilosulous distally, the exserted part of corolla or at very least its lobes densely white-silky-pilose; calyx cylindric, striately nerved (5.4–)6–10 × (1.2–)1.5–2 mm, the triangular or lanceolate acute teeth unequal, the sinuses 0.8–4.5 mm deep, one of them often much deeper than the rest, the calyx then appearing spathiform; corolla narrowly trumpet-shaped (11.5–)13–17.5 mm, the ovate or lanceolate, apically hooded lobes 2.5–6 × 1.2–3.2 mm; androecium 34–82-merous, 4–6.7 cm, the far-exserted tube 20–28 × 0.9–1.7 mm, the stemonozone 1.2–2.7 mm; disc 0.6–1.5 × 0.6–1.3 mm, clasping the short stipe of the ovary, this glabrous at anthesis, shallowly sulcate laterally, gradually tapering into the exserted style; stigma poriform. *Pods* solitary, broad-linear, straight or almost so, cuneately contracted at base into a pseudostipe 3–16 mm and at apex broadly rounded and cuspidate, when well fertilized 11–18(–20) × 2.1–3.1(–3.6) cm, 6–9(–10)-seeded, the stiffly papery, brown or fuscous, glabrous valves framed by dorsally plane or low-carinate (but not winged) sutures ±1.2–2.4 mm wide, low-convex over each seed, finely transverse-venulose, internally pallid, smooth but not lustrous, the cavity continuous; seeds transverse, basifixed on contorted, strongly compressed funicle, lentiform, in broad view either round or broad-elliptic 9.5–14 mm diam, strongly compressed and 1.5–3.5 mm thick, the leathery, lustrously castaneous testa loosely investing the pallid embryo, girdled around the periphery with a thickened rim ±0.5–0.7 mm wide, the finely engraved pleurogram incomplete, (6–)6.5–10 mm long.

In drought-deciduous or semideciduous, primary and disturbed forest or scrub-woodland, mostly between sea level and 300 m, but locally in Chiapas, Mexico, at 700–1100 m, discontinuously dispersed in S Mexico and NE Guatemala, and in far NW South America: known in Mexico from coastal hill-country in S Jalisco, from the Pacific lowlands of Isthmus of Tehuantepec in Oaxaca, from the upper Mexcalapa basin in Chiapas, and from the Yucatán Peninsula, whence S shortly into Petén, Guatemala; in South America localized on the lower foothills of Sa de Sta Marta in Magdalena, Colombia, the Paraguaná peninsula in Falcón, Venezuela, and the adjacent island of Aruba in Dutch West Indies. — Map 43. — Fl. in Mexico and Guatemala III–VII, IX, in South America II–IV, VIII–X. — *Dabaruida, uña de gaber* (Aruba).

Sphinga platyloba is variable in four respects: habit of growth, leaf-formula, pubescence, and number of stamens. These variable features are, however, poorly correlated and only partially coincide with discontinuities in dispersal. In Caribbean South America, whence came the typus, the plants are free-standing, arborescent shrubs, nearly glabrous, with few and relatively ample leaflets (not more than 3 pairs, the largest at least 18 mm long), and stamens about 55–80 in number. In lowland Yucatán the habit and vesture are essentially as in Venezuela and Colombia, but the stamens are fewer, about 35–45 per androecium. In Oaxaca the plant is reportedly either shrubby or sarmentose, the leaflets are smaller and more numerous (3–7 pairs in distal pinnae and the largest seldom attaining 2 cm in length) and vary moreover from glabrous except for a basal tuft of hairs dorsally to randomly pilosulous on both faces, and the androecium is about 60–77-merous. The seemingly isolated populations in coastal Jalisco hardly differ from those in Oaxaca. In upland Chiapas, *S. platyloba* is described in field notes by Breedlove as either a tree, a vine, or a sprawling shrub. Here the leaflets are relatively small and not less than four pairs in distal pinnae, and commonly are pubescent to some degree, although sometimes glabrate, while random samples of androecia ranged from 40 to 67. Although differential evolution seems to be proceeding in these segments of the species' range, there is not yet the sort of disparity between them that justifies taxonomic segregation.

XI. HAVARDIA SMALL

Havardia Small, Bull. New York Bot. Gard. **2:** 91. 1901. sens. str. — Generitypus: *H. brevifolia* (Bentham) Small ≡ *Pithecellobium brevifolium* Bentham = *H. pallens* (Bentham) Britton & Rose.

Pithecolobium sect. *Ortholobium* Bentham, Trans. Linn. Soc. London **30:** 592, ex parte (spp. 82, 83). 1875. — Sectionis lectotypus: *P. albicans* (Kunth) Bentham ≡ *Havardia albicans* (Kunth) Britton & Rose.
Havardia sensu Britton & Rose, N. Amer. Fl. **23:** 40, ex parte (spp. 1–5). 1928; Nielsen, 1981: 184, ex parte, exclus. syn. plur.

Trees (2–)3–11 m. *Stipules* associated with primary lvs subulate or conical, spinescent. *Lf-formula* i–ix(–xiii)/7–36, the lfts either opposite or (*H. albicans*) alternate along pinna-rachis, the largest ±2.5–10(12) mm; lf-nectaries sessile, the first situated below proximal pinna-pair; venation of lfts pinnate or simple, exceptionally obscurely palmate-pinnate. *Inflorescence* various, the capitula or (*H. mexicana*) capituliform racemes arising either from lf-axils of long-shoots, coevally with or preceding the lf, or from brachyblasts, or from both; calyx either shallowly or deeply campanulate 1–3.4 mm; corolla 3–5.5(–6.5) mm, the lobes at full anthesis either erect or recurved; androecium (28–)30–52-merous, 9–14.5 mm, the basal callosities within the tube often obscure or 0, rarely forming a pronounced nectarial disc; ovary subsessile or (*H. pallens*) shortly stipitate. *Pods* broad-linear straight, plano-compressed, 8–13-seeded, the chartaceous or thinly coriaceous, stiffly papery valves low-convex over each seed on one face or (*H.

campylacanthus) on both, the cavity continuous; dehiscence tardy, inert, through both sutures; seeds transverse on dilated, distally sigmoid funicle. —Spp. 5, of warm-temperate, and seasonally dry tropical, Mexico, adjacent Texas, and Central America.

Except for spinescent stipules and lack of a dimorphic flower in the capitula, the species of *Havardia* resemble *Albizia* sect. *Arthrosamanea*, their young inflorescence recalling the panicle of pseudoracemes prevalent in *Albizia*. The inflorescence of the section shows forms transitional between the panicle of pseudoracemes prevalent in *Albizia* and that of the xerophytically modified genera *Sphinga*, *Painteria*, and *Ebenopsis*: at one extreme *H. albicans* and *H. pallens*, where most capitula are borne on new growth, axillary to hysteranthous leaves, at the other *H. campylacanthus*, which forms a passage to *Sphinga*. In all but *H. campylacanthus*, in which the valves of the ripe pod are biconvex over each seed, the pods of *Havardia* are undulately corrugate lengthwise, as in the genus (Mimoseae) *Parapiptadenia* Brenan. Specialized features of individual species of *Havardia* are the pronounced ovary-stipe of *H. pallens*, the alternate leaflets of *H. albicans*, and the pedicellate flowers of *H. mexicana*.

Conspectus of genus *Havardia*

¶ Pinnae of larger lvs (2–)3–7 pairs; lfts opposite along pinna-rachis, in longer pinnae 12 pairs or more; calyx shallowly campanulate; corolla-lobes recurved at full anthesis; ovary-stipe 0.9–1.8 mm; valves of pod alternately convex and concave over successive seeds.
3. *H. pallens* (Bentham) Britton & Rose

¶¶ As ¶, but pinnae of larger lvs 6–12 pairs, lfts alternate along rachis, and ovary-stipe 0.3 mm or less.
4. *H. albicans* (Kunth) Britton & Rose

¶¶¶ As ¶, but lfts of longer pinnae only 7–11 pairs, fls slenderly pedicellate, and ovary-stipe 0.3 mm or less.
5. *H. mexicana* (Rose) Britton & Rose

¶¶¶¶ As ¶, but pinnae of larger lvs 1–4 pairs, calyx deeply campanulate, corolla-lobes erect at full anthesis, and ovary-stipe <0.3 mm.
1. *H. sonorae* (S. Watson) Britton & Rose

¶¶¶¶¶ As ¶, but pinnae of larger lvs 5–13 pairs, lfts of longer pinnae 18–31 pairs, calyx deeply campanulate, corolla-lobes erect at full anthesis, ovary-stipe <0.25 mm, and valves of pod low-convex over each seed on both faces.
2. *H. campylacanthus* (Rico & Sousa) Barneby & Grimes

Key to the species of genus *Havardia*

1. Pinnae of larger lvs (1–)2–6(7) pairs; S Texas to S Mexico.
   2. Lfts of longer pinnae (10–)12–36 pairs; fls sessile or almost so; if of NW continental Mexico the calyx deeply campanulate and the corolla-lobes erect.
      3. Lf-stk of longer lvs (2.5–)3–7.5 cm and rachis of longer pinnae 2–5.5(–8) cm; largest lfts (5–)5.5–10.5(–12) × (1.6–)1.8–3.3 mm; calyx shallowly campanulate 1–1.7(–1.9) × 1–1.5 mm; corolla-lobes at full anthesis recurved; stipe of ovary 0.9–1.8 mm; S Texas S through E Mexico to Oaxaca and Chiapas . . . . 3. *H. pallens* (p. 169)
      3. Lf-stk of longer lvs (0.2–)0.4–1.5 cm and rachis of longer pinnae 1.2–2.4 cm; largest lfts 2.2–5.5 × 0.8–1.3 mm; calyx deeply campanulate 2.8–3 × 1.4–1.6 mm; corolla-lobes at full anthesis erect; stipe of ovary 0.2–0.25 mm; NW Mexico (Sonora, Sinaloa) . . . . 1. *H. sonorae* (p. 166)
   2. Lfts of longer pinnae 7–11 pairs; fls elevated on slender pedicel 1–3 mm; calyx shallowly campanulate (less than half as long as corolla) and corolla-lobes recurved at full anthesis; NW Mexico (Sonora, Sinaloa, Baja California Sur) . . . . 5. *H. mexicana* (p. 172)
1. Pinnae of larger lvs 6–12(–13) pairs; S Mexico to Nicaragua.
   4. Lfts alternate along pinna-rachis; corolla-lobes recurved at full anthesis; SE Mexico (Yucatán Peninsula) and adjacent Guatemala . . . . 4. *H. albicans* (p. 171)
   4. Lfts opposite along pinna-rachis; corolla-lobes erect at full anthesis; Pacific S Mexico (Guerrero, Oaxaca), Honduras, Nicaragua . . . . 2. *H. campylacanthus* (p. 167)

1. Havardia sonorae (S. Watson) Britton & Rose, N. Amer. Fl. **23:** 42. 1928. *Pithecolobium sonorae* S. Watson, Proc. Amer. Acad. Arts **24:** 49. 1889. — "Common at Guaymas [Sonora, Mexico] and at Loreto and Mulege [Baja California Sur]. ([*E. Palmer*] 58 [in 1887].)" — Holotypus (from Guaymas), GH!; isotypi, NY (2 sheets)!, UC *84451*!, US *46935*!.

Pithecellobium sonorae sensu Wiggins, 1964: 174, 590.

Microphyllidious drought-deciduous *trees* 2–7 m with narrow crown and gray, vertically striped lenticellate, stiffly flexuous branches, commonly but not always armed at each node of long-shoots with a pair of either erect subconical or longer recurved spinescent cat's-claw stipules, the new growth pilosulous, sometimes only thinly or residually, with fine erect whitish hairs to 0.1–0.3 mm, the foliage dull-olivaceous to bright green and the plane, moderately bicolored lfts from pilosulous overall to facially glabrous ciliolate, the small capitula of whitish fls borne solitary or 2–3-nate in primary lf-axils either near base of each year's long-shoots or of lateral, hysteranthously foliate branchlets, and sometimes most lvs fasciculate on axillary brachyblasts. *Stipules* at most primary nodes 1.5–4 mm, either tapering and recurved from swollen base or conical erect, persistent on defoliate branches, but at random nodes merely stiffly subulate and <1 mm, often densely puberulent when young, becoming fuscous glabrate. *Lf-formula* i–iv/(10–)13–21, the pinnae of primary lvs often geminate, of fasciculate lvs mostly 2–4-jug.; lf-stks of major lvs (2–)4–15 mm, the petiole 2–7.5 × 0.3–0.45 mm, the ventral groove obscure, the one or the longest interpinnal segment 1–2.5 mm; petiolar nectary either below, at, or above midpetiole, sessile, shallow-cupular thick-rimmed, commonly 0.3–0.6 mm diam and 0.1–0.15 mm tall, in random lvs minute or lacking, sometimes a similar gland at tip of lf-stk, but none on pinna-rachises; pinnae when not geminate subaccrescent distally, the rachis of distal pair 12–24 mm, the longer interfoliolar segments 0.6–1.3(–1.6) mm; lft-pulvinules ±0.15–0.2 mm diam; lfts abruptly decrescent at each end of rachis, otherwise subequilong, the blade narrow-oblong or -oblong-elliptic from shallowly semicordate base, obtuse but often minutely mucronulate, those near midrachis 2.2–5.5 × 0.8–1.3 mm, 2.7–4 times as long as wide; venation weakly pinnate, elevated and visible only dorsally, the moderately displaced, straight or slightly incurved midrib either simple or giving rise to 2–3 pairs of faint secondary nerves feebly brochidodrome within the plane margin. *Peduncles* 8–20 mm; capitula ±10–17-fld, the receptacle <2 mm; bracts ovate 0.6–0.9 mm, persistent into anthesis, then dry deciduous; fls sessile, homomorphic, 5-merous (or the calyx sometimes 6–7-merous), strigose overall externally, the calyx brownish gray, the corolla white-silky; calyx deeply campanulate 2.8–3 × 1.4–1.6 mm, the deltate teeth ±0.3–0.6 mm; corolla 4.8–5.2 mm, the erect lobes 1–1.5 × 0.6–0.7 mm; androecium 46–52-merous, 12–13 mm, the stemonozone 0.4–0.5 mm, the tube 5 mm; the 5-lobed nectarial disc 0.25–0.3 mm tall; ovary glabrous ellipsoid, conic at each end, the stipe 0.2–0.25 mm. *Pods* 1(–2) per capitulum, in profile broad-linear contracted at base into a combined stipe and pseudostipe 5–9 mm and at apex into an erect cusp 2–8 mm, the straight plano-compressed body 6.5–10.5 × 1.2–2 cm, the stiffly papery, fuscous- or reddish-brown valves framed by dorsally plane sutures 0.9–1.4 mm wide, low-convex (alternately on one and the other face) over each seed, both minutely puberulent and granular with dark red clavate trichomes, smooth pallid (but not satiny) within, the cavity continuous; seeds transverse on compressed funicle, disciform, in broad view bluntly oblong or subcircular (5.3–)7–9.8 × (5–)5.6–7 mm, in section 1.2–2 mm thick, when ripe castaneous lustrous, the seed-coat consisting of a translucent outer membrance and a crustaceous testa ±0.1–0.15 mm thick at middle of seed, thickened at the margins to form an obtuse peripheral keel, the U-shaped pleurogram 2.5–4.2 × 2–2.8 mm.

On plains and foothills below 400 m, in deciduous brush-woodland and thornscrub, and along washes in mesquite grassland, forming small colonies or thickets, locally plentiful but of restricted range between 24° and 28°N along the coastal plain of S Sonora and N Sinaloa, Mexico, and on the coast of Gulf of California near 25°N (Bahia S. Carlos). — Map 44. — Fl. VI–VII(–?) — *Guló, palo jocono, uña-de-gato* (applied to various mimosoid legumes armed with cat's-claw spines or prickles).

2. Havardia campylacanthus (L. Rico & M. Sousa) Barneby & Grimes, comb. nov. *Pithecellobium campylacanthus* L. Rico & M. Sousa, Ann. Missouri Bot. Gard. **73:** 722. 1986. — "7 km al O-NO de Tehuantepec, distrito de Tehuantepec, Oaxaca, Mexico, 17 marzo de 1981, *M. Sousa 11938*." — Holotypus, MEXU (not seen); isotypi, BM!, CAS!. — The epithet as published appears to be a noun in apposition and is so retained.

Microphyllidious arborescent *shrubs* and *trees* 2–14 m with gray lenticellate branches, of sympodial growth, armed on trunk and at each node of straight or weakly flexuous long-shoots with stout recurved lignescent stipules, the young growth variably puberulent, or strigulose, or pilosulous with either forwardly incumbent, or straight subappressed, or straight erect pallid hairs to 0.1–0.45 mm, the lfts either glabrous ciliolate or facially pilosulous, either strongly or scarcely bicolored and either matte or lustrous on upper face, the capitula of white-stamened fls either fasciculate on brachyblasts or axillary to lower lvs of lateral branchlets. *Stipules* of primary lvs mostly widely spreading or gently recurved, tapering from dilated base into a vulnerant point, 3–10 mm, those on brachyblasts triangular or subulate scarcely 1 mm. *Lf-formula* (v–)vi–xi(–xiii)/(17–)19–31; lf-stks (2.5–)3–7.5 cm, the petiole 8–24 mm, at middle 0.4–0.9 mm diam, the longer interpinnal segments 4–9.5 mm; nectary near or below midpetiole sessile,

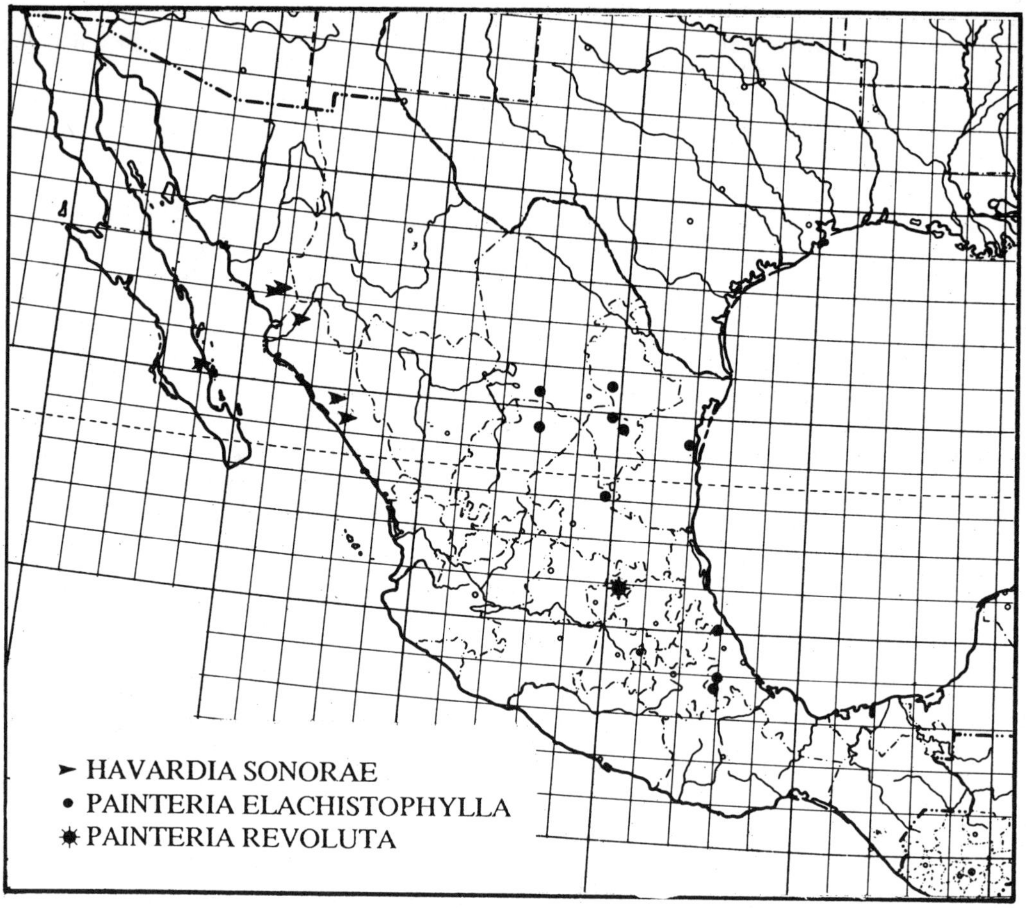

MAP 44. Mexico. Range of *Havardia sonorae* (S. Watson) Britton & Rose, *Painteria elachistophylla* (S. Watson) Britton & Rose, and *P. revoluta* (Rose) Britton & Rose.

shallow-cupular thick-rimmed or plane and dimpled, round or elliptic (0.5–)0.6–1.2(–1.5) mm diam, in profile 0.15–0.4 mm tall, in some lvs similar but smaller nectaries between 1–2(–3) furthest pinna-pairs, and a minute one at tip of some pinna-rachises; pinnae accrescent distally, the rachis of longest ones 2.7–4 cm, the longer interfoliolar segments 0.9–1.4 mm; lfts opposite, decrescent proximally, otherwise subequilong, the blade linear or linear-oblong from deeply semicordate-auriculate base, obtuse or obscurely deltate-apiculate, either straight or incipiently porrect distally, the longer ones 3.3–6.5 × 0.8–2 mm, 3–4.4(–4.8) times as long as wide; venation pinnate, the slightly forwardly displaced midrib giving rise on each side to 3–5 weak secondary venules expiring or faintly brochidodrome within the plane margin, the whole venation prominulous only beneath. *Peduncles* (1.3–)2–4 cm; capitula ±12–37-fld, the clavate receptacle 1.5–2.5 mm diam; bracts minute, <1 mm; fls sessile, homomorphic, the perianth 5-merous, either glabrous except for minutely ciliolate calyx-teeth and sometimes distally puberulent corolla-lobes, or the latter white-silky dorsally; calyx deeply campanulate ±2.5–3.4 × 1.2–1.6 mm, the deltate-ovate teeth 0.4–0.8 mm; corolla subcylindric ±5 mm, the erect lobes 0.8–1 mm; androecium 40–50-merous, 9–11.5 mm, the stemonozone 0.7–0.8 mm, the tube 3.6–4 mm, the callosities at base 0.2–0.35 mm; ovary subsessile, slenderly ellipsoid, the stipe 0.1–0.2 mm; style about as long as stamens, the stigma poriform. *Pods* solitary or to 3 per capitulum, in profile oblong or (commonly) broad-linear, contracted at base either gradually or abruptly into a pseudostipe ±7–14 mm and abruptly at apex into an erect cusp 1.5–8 mm, the straight or almost straight fertile body 7–13 × (1.3–) 1.4–2.4(–2.6) cm, 8–13-seeded, plano-compressed, the stiff, somewhat brittle, brownish green or fuscous, either glabrous red-granular or both granular and puberulent valves framed by bluntly 3-angulate sutures ±1.5–2 mm wide, externally veinless, the cavity continuous; funicles dilated, sigmoid under the seed; seeds in broad profile orbicular or oblong-elliptic 9–12 × 7–10 mm, plumply disciform and strongly compressed around the periphery, the U-shaped pleurogram ±5.5–6.5 × 3–3.5 mm.

In tropical deciduous brush-woodland, especially along intermittent streams, mostly between 1 and 200 m but ascending to 700 m in Honduras, discontinuously scattered along the Pacific lowlands from Guerrero and Oaxaca in Mexico SE to the Atlantic drainage in Honduras (Comayagua) and the Managua basin in Nicaragua (dept. Boaco). — Map 45. — Fl. II–V. — *Uña-de-gato*; *guachumil* (Oaxaca); *chicharón* (Nicaragua).

3. Havardia pallens (Bentham) Britton & Rose, N. Amer. Fl. **23:** 42. 1928. *Calliandra pallens* Bentham, London J. Bot. 5: 102. 1846. — "[Mexico. Hidalgo:] Zimapan, *Coulter*." — Holotypus, n.v.; isotypus, TCD, seen in photo at NY!. — *Pithecellobium pallens* (Bentham) Standley, Trop. Woods **34:** 39. 1933. — Erroneously referred by Bentham, 1875: 592 to *P. albicans*.

Pithecolobium brevifolium Bentham *in* A. Gray, Pl. Wright. **1:** 67. 1852. — "[Mexico. Nuevo León:] between Cerralvo and Monterey . . . *Wislizenus* (No. 362, 355). East of Rinconada and Papagallo; also between Cerralvo and Maria, *Gregg*." — Lectotypus, *Wislizenus, s.n.*, a packeted fruit accompanied by Bentham's holographic description, GH!. — *Feuilleea brevifolia* (Bentham) O. Kuntze, Revis. Gen. Pl. **1:** 187. 1891. *Zygia brevifolia* (Bentham) Sudworth, U.S.D.A. Div. Forest. Bull. **14:** 248. 1897. *Havardia brevifolia* (Bentham) Small, Bull. New York Bot. Gard. **2:** 92. 1901.

Acacia nueciana Buckley, Proc. Acad. Nat. Sci. Philadelphia **1861:** 453. 1862. — "[Texas.] Near the Nueces River." — Holotypus, not found. — Equated by Bentham, 1875: 592, on authority of A. Gray, with *P. brevifolium*.

Havardia nelsonii Britton & Rose, N. Amer. Fl. **23:** 41. 1928. — "[Mexico.] Between San Geronimo and La Venta, Oaxaca, July 13, 1898, *E. W. Nelson 2776*." — Holotypus, US!; clastotypus (fragm.), NY!; isotypus, NY! = NY Neg. *9330*. — *Pithecollobium nelsonii* (Britton & Rose) Standley, Trop. Woods **34:** 40. 1933.

Pithecolobium brevifolium sensu Sargent, Silva **3:** 135, t. 146. 1892.

Pithecellobium pallens sensu B. L. Turner, Leg. Texas 28. 1959; Isely, 1973: 114; Estrada & Marroquín de la Fuente, 1992: 72.

Microphyllidious *trees* or bushy *treelets* 2–9(–12) m, erratically armed at nodes with spinulose or ligneous stipules (on new branchlets often 1 mm or less and strictly ascending, on older branches woody, either ascending or spreading, to 15 mm), the young branchlets and lf- and inflorescence-axes either loosely strigulose or pilosulous with incumbent or spreading, extremely fine whitish hairs to 0.1–0.2 (–0.25) mm, the pallidly matte-olivaceous lvs moderately bicolored, the lfts a little paler beneath than above, either glabrous, or glabrous ciliolate, or finely puberulent on lower or on both faces and often with a tuft of hairs in anterior basal angle of midrib dorsally, the rather small capitula of whitish fls fasciculate by 2–5(–7), the fascicles axillary either to coevally expanded or to hysteranthous lvs along hornotinous branchlets, together forming at first efoliate but later leafy pseudoracemes, the axis of most determinate, the branching therefore sympodial, no lateral branchlets reduced to brachybasts. *Stipules* 0.5–15 mm, when shorter than 2 mm erect, often with stiff burnished tip, when longer spreading widely or subhorizontal and to 3 mm wide near laterally compressed base, the longer ones lignescent and rigid, all straight or almost so, commonly puberulent like the stems but glabrescent distally, the stouter ones long-persistent. *Lf-formula* (ii–)iii–vi(–vii)/(10–)12–21; lf-stks of larger lvs (2.5–)3–7.5(–8.5) cm, the petiole (13–)15–32 mm, at middle 0.5–0.7(–0.85) mm diam, the longer interpinnal segments 5–13 mm; petiolar nectary sessile, shallowly cupular thick-rimmed, round or elliptic, 0.5–1.4(–1.9) mm diam, in profile 0.15–0.4 mm tall, variably situated but most commonly below midpetiole, rarely at very base, sometimes a little above midpetiole, sometimes rudimentary, a second similar nectary near first pinna-pair, often very small ones at tip of pinna-rachises, or between 2–5 furthest pairs of lfts; pinnae little graduated, the rachis of penultimate pair 2–5.5(–8) cm, the longer interfoliolar segments 1.2–3.3(–4) mm; lft-pulvinules 0.3–0.7 × 0.2–0.4 mm, not wrinkled; lfts opposite, subequilong except at very ends of rachis, the first pair either very small or commonly represented by a linear paraphyllidium, the blades narrow-oblong or linear from semicordate-auriculate base, very shortly or obscurely depressed-deltate-acuminate, those near midrachis (5–)5.5–10.5(–12) × (1.6–)1.8–3.3(–3.5) mm, (2.5–)2.7–4.1(–4.4) times as long as wide; venation pinnate, the slightly displaced, straight or distally porrect midrib giving rise on each side to 3–5(–6) secondary nerves weakly brochidodrome well within the plane margin, the whole venation immersed on

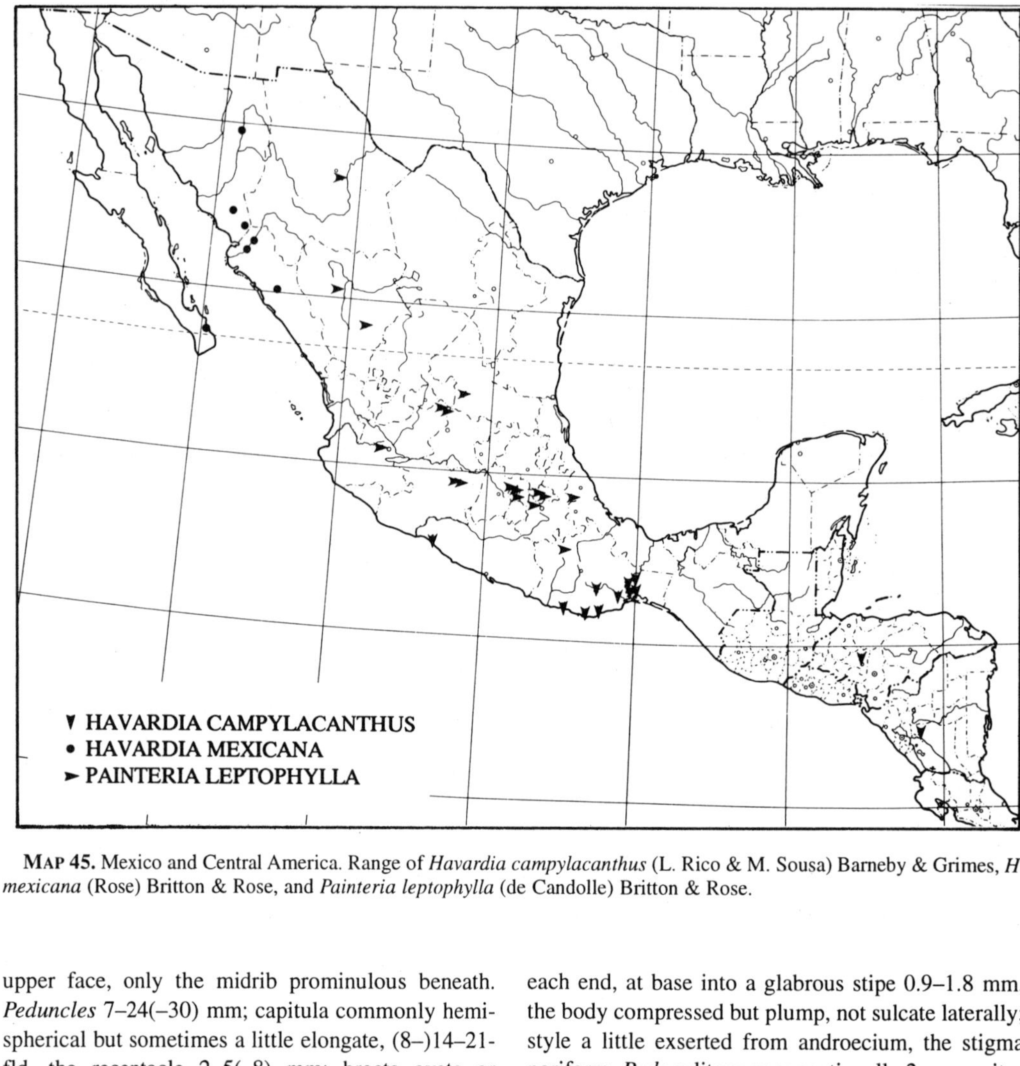

MAP 45. Mexico and Central America. Range of *Havardia campylacanthus* (L. Rico & M. Sousa) Barneby & Grimes, *H. mexicana* (Rose) Britton & Rose, and *Painteria leptophylla* (de Candolle) Britton & Rose.

upper face, only the midrib prominulous beneath. *Peduncles* 7–24(–30) mm; capitula commonly hemispherical but sometimes a little elongate, (8–)14–21-fld, the receptacle 2–5(–8) mm; bracts ovate or oblanceolate 0.6–1.2 mm, persistent into anthesis, then dry deciduous; fls sessile, homomorphic, the perianth 5(randomly 6)-merous, the calyx densely silvery-puberulent or -pilosulous externally, the greenish white corolla glabrous up to the lobes, these puberulent overall or commonly only toward apex; calyx shallowly campanulate or hemispherical 1–1.7 (–1.9) × 1–1.5 mm, the ovate or deltate teeth 0.2–0.5 mm; corolla narrowly vase-shaped (3.5–)4–5.3(–6.3) mm, the spreading-recurved lobes (1–)1.2–2 × 0.6–1.1 mm; androecium (28–)30–52-merous, (9–)10.5–14.5 mm, the stemonozone 0.4–0.6 mm, the tube (1.5–)1.7–2.7(–3.6) mm, thickened internally at base into callosities 0.2–0.4 mm; ovary either glabrous or densely minutely puberulent, cuneately contracted at each end, at base into a glabrous stipe 0.9–1.8 mm, the body compressed but plump, not sulcate laterally; style a little exserted from androecium, the stigma poriform. *Pods* solitary or exceptionally 2 per capitulum, both stipitate and pseudostipitate, the stipe and pseudostipe together 6–16 mm, the body broad-linear straight, abruptly contracted into an erect cusp 2–7 mm, when well fertilized 7–12 × 1.2–1.8(–2) cm, 8–13-seeded, the stiffly papery, when ripe brown, glabrate, closely coarsely cross-venulose valves framed by bluntly 3-carinate (not winged), straight or scarcely undulate sutures 1.2–1.8 mm wide, low-convex (on alternate faces of pod) over each seed, pallid smooth (but not satiny) within, the cavity continuous; seeds transverse at middle of pod on compressed, distally sigmoid funicle, compressed-lentiform, in broad view oblong-elliptic or nearly round 5.3–7 × 4–6 mm, the smooth or finely pitted, scarcely lustrous, dark brown or fuscous testa ±0.1–0.15 mm thick, closely investing the pallid embryo, the pleurogram not quite complete.

In tropical deciduous woodland and thorn-scrub mostly below 450 but attaining 1100–1200 m in desert mountains of the Chihuahua Desert and 1170 m in Chiapas, widespread and common on the coastal plain and foothills of the Gulf of Mexico from S Texas to N Veracruz, inland to NE Chihuahua, E Coahuila, E-centr. Durango and E Nuevo León, apparently disjunct in Pacific lowland Oaxaca, and at the head of the Río Grande basin near 16°N, 92°W in Chiapas (mun. La Trinitaria), Mexico. — Map 46. — Fl. V–VIII. — *guajillo* (*huajillo*, *-a*), *tenaza*.

4. Havardia albicans (Kunth) Britton & Rose, N. Amer. Fl. **23:** 41. 1928. *Acacia albicans* Kunth, Mimoses, t. 27. 1820 & p. 87. 1821. — "... in litore Maris Antillarum, prope Campeche Mexicanorum." — Holotypus, P-HBK!. — *Pithecolobium albicans* (Kunth) Bentham, Trans. Linn. Soc. London **30:** 592. 1875. *Feuilleea albicans* (Kunth) O. Kuntze, Revis. Gen. Pl. **1:** 187. 1891.

Albizzia rubiginosa Standley, Publ. Carnegie Inst. Wash. **461:** 58. 1935, nom. illeg., non Miquel, Fl. Ned. Ind. 1855. — "Campeche: Tuxpeña, in "bajo", October 26, 1931, *C. L. Lundell* 866 . . ." — Holotypus, F *655193*! = F. Neg. *57054*; isotypi, DS-CAS!, NY!, UC *486954*!, US *1494719*!. — *A. lundellii* Standley, Trop. Woods **52:** 26. 1937.

Microphyllidious *trees* 4.5–12 m, with trunk attaining 2–3 dm dbh, erratically armed at nodes with stiffly ascending, either subulate, or slenderly conical, or acicular, or in age lignescent and vulnerant stipules, but the young stems often unarmed; vesture of fine short pallid hairs, the alternate lfts strongly bicolored, dark matte-green above, finely silky-strigulose beneath and often with tuft of hairs in antrorse basal angle of midrib, the small, few-fld capitula fasciculate, mostly in the axils of hysteranthous lvs, on branches of a terminal panicle, but some axillary to developed lvs, the branching pattern of each season sympodial. *Stipules* ascending, straight, linear-subulate or acicular and terete 0.8–3 mm to woody and sometimes compressed and up to 7.5 mm, these long-persistent. *Lf-formula* vi–xi(–xii)/23–36; lf-stks of major lvs 5–9 cm, the petiole 16–28 mm, the longer interpinnal segments ±5–10 mm; nectary near midpetiole sessile, hemispherical or half-ellipsoid, obtuse, dimpled at apex, 1.1–2.2 mm diam, in profile 0.4–0.8 mm tall, a similar scarcely smaller nectary at tip of some lf-stks and a yet smaller, more cupular one at tip of random pinna-rachises; pinnae a little decrescent at each end of lf-stk, the rachis of the penultimate pair 3–5.5 cm, the longer interfoliolate segments 1–2.2 mm; a pair of minute ascending paraphyllidia at apex of each pinna-pulvinus; lft-pulvinules 0.2–0.5 mm; lfts subequilong except at very ends of rachis, linear or linear-oblong from deeply semicordate-auriculate base, deltately subacute, straight or slightly porrect, the larger ones (3.5–)4–8 × (0.8–) 1.1–2 mm, ±3.5–5 times as long as wide; venation simple, of usually porrectly arcuate midrib strongly excentric at base becoming nearly centric distally, this prominulous beneath only, sometimes giving rise to 2–3 faint pairs of secondary nerves, the whole venation then incipiently pinnate. *Peduncles* very slender, fasciculate by 2–8, the longer ones of a fascicle 7–12 mm; capitula densely 6–10(–11)-fld, the receptacle not over 1 mm, the sessile fls homomorphic, the 5-merous perianth finely gray-silky-strigulose overall; bracts ovate or oblanceolate, convex dorsally, 0.7–1.3 mm, becoming dry but persistent into anthesis, then deciduous; calyx subhemispherical 1.7–2 mm, nearly as wide, the depressed deltate obtuse teeth 0.25–0.4 mm; corolla campanulate 3–3.8 mm, the recurving lobes 1.2–2 × 0.9–1.3 mm; androecium 34–42-merous, 10–12.5 mm, the stemonozone obscure or to 0.45 mm, the tube 1.5–2 mm, sometimes thickened internally at very base, but nectariferous disc not developed, the exserted filaments white; ovary shortly stipitate, glabrous, not sulcate laterally, tapering into the style. *Pods* mostly solitary, occasionally 2–3 per capitulum, in profile broad-linear straight 10–15 × 1.8–2 cm, cuneately contracted at base into a pseudostipe 5–9 mm and more abruptly so at apex into an erect, narrowly subulate (fragile, easily broken) beak 3–7 mm, 8–12-seeded, the stiff crustaceous valves framed by straight, bluntly tricarinate sutures ±1.5–2 mm wide, low-convex (alternately depressed and elevated lengthwise) over each seed, densely pubescent with variously proportioned short plain hairs and clavate or dilated and coralloid, red or rufous trichomes, appearing fuscous-gray-tomentulose or rufous-tomentulose overall, within pallid but not satiny, the cavity continuous; dehiscence tardy, inert, through both sutures; funicle dilated, sigmoid under the seed, this not seen fully ripe but seemingly discoid, the pleurogram complete.

In drought-deciduous woodland or thorn-scrub, and sometimes in swamp-forest, below 150 m, SE Mexico (Campeche, Yucatán, Quintana Roo, N Chiapas), NE Guatemala (Petén), and N Belize (Corozal distr.). — Map 46. — Fl. VI–VIII. — *Chucum*, *salom*.

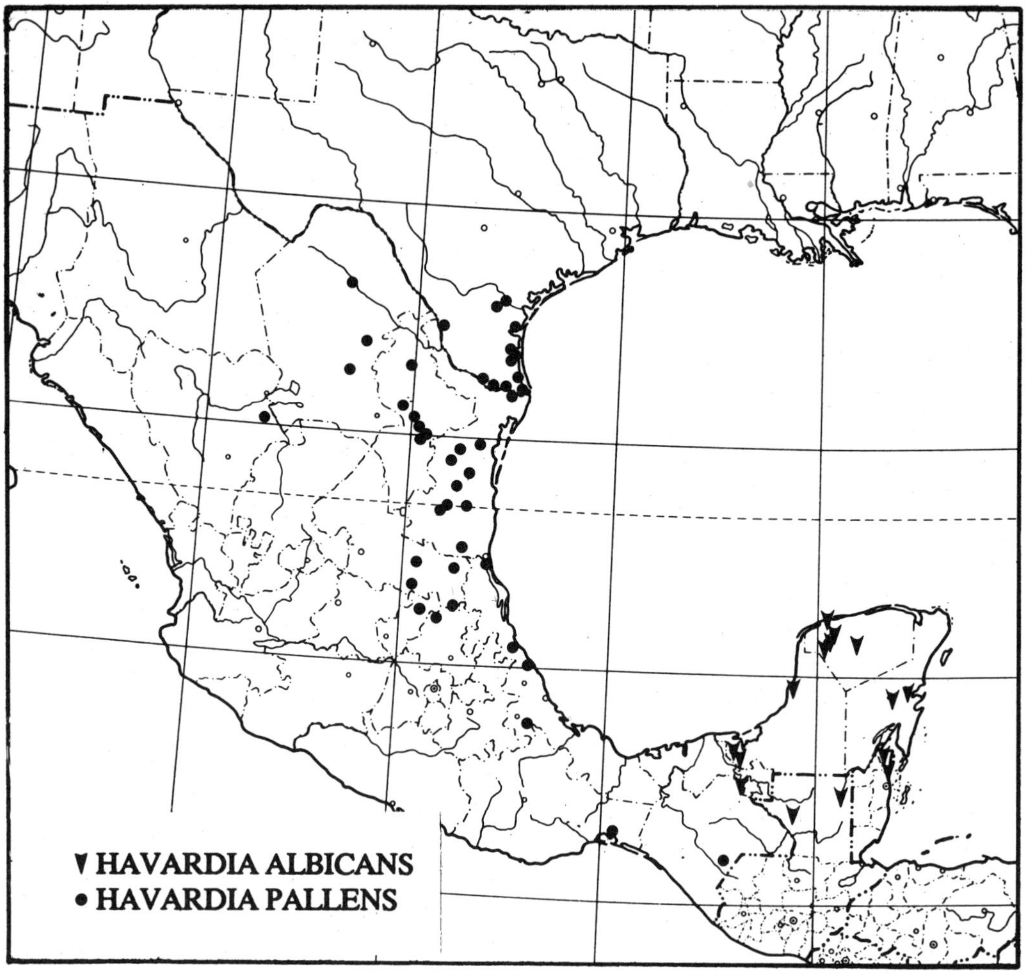

MAP 46. Southern Texas, Mexico, and northern Central America. Range of *Havardia albicans* (Kunth) Britton & Rose and *H. pallens* (Bentham) Britton & Rose.

5. Havardia mexicana (Rose) Britton & Rose, N. Amer. Fl. **23:** 41. 1928. *Pithecolobium mexicanum* Rose, Contr. U.S. Natl. Herb. **1:** 100. 1891. — "[Mexico. Sonora:] In the Alamos Mountain. March 26 to April 8 [1890, *E. Palmer*] No. 297." — Holotypus, US *46937*!; isotypus, NY!, UC *84419*!.

Pithecollobium mexicanum sensu Standley, 1922: 397; *Pithecellobium mexicanum* sensu Wiggins, 1964: 590.

Microphyllidious drought-deciduous *trees* attaining 10 m, with smooth, transversely lenticellate bark, erratically armed at nodes of new branchlets with a pair of small, stiffly ascending, subulate or acicular stipules and at those of annotinous branches with lignescent, sometimes longer, either ascending or spreading-decurved ones, the young growth finely pilosulous with more or less erect white hairs to 0.25–0.45 mm, the foliage pallidly gray-olivaceous, only a trifle paler beneath, the lfts commonly pubescent beneath only, but sometimes puberulent also above, and sometimes glabrous facially but ciliolate, the shortly racemose capituliform units of inflorescence arising 1–3 together at nodes of either efoliate, partly foliate, or hysteranthously foliate pseudoracemes lateral to the main axis of long-shoots. *Stipules* of young growth 0.5–2 mm, of older branches 1–4 mm, long-persistent. *Lf-formula* ii–v/7–11; lf-stks of primary lvs 1–4 cm, the petiole 4–16 mm, at middle 0.35–0.5 mm diam, the longer interpinnal segments 3–8 mm; a nectary, rarely 2, randomly situated near base of petiole, near midpetiole, or close below first pinna-pair, sessile, shallow-cupular thick-rimmed, 0.3–0.5 mm diam, in profile scarcely 0.2 mm tall, none at tip of pinna-rachises; pinnae a little decrescent proximally, the rachis of penultimate pair 9–19 mm, the longer interfoliolar segments 0.9–2 mm; lft-pulvinules 0.2–0.4 × 0.2–0.25 mm, not wrinkled; lfts decrescent proximally, less so distally, the blade linear-oblong to narrowly oblong-obovate from semicordate-auriculate base, obtuse or obtusely apiculate, those near midrachis 3.3–6.5 × 1.3–2.2 mm, (2–)2.4–3.9 times as long as wide; venation pinnate (sometimes a relatively strong anterior primary nerve almost as long as the first secondary one), the straight midrib only a trifle forwardly displaced from center, giving rise on each side to 2–4(–5) weak secondary nerves brochidodrome close within the plane margin, the whole venation immersed on upper face, finely prominulous beneath. *Peduncles* 4–16 mm; capitula (technically short racemes) subhemispherical 7–15-fld., the linear receptacle (1–)1.5–4.5 mm; bracts linear-oblanceolate, early deciduous; pedicels 1–3.1 mm, at middle 0.25–0.4 mm diam; fls homomorphic, 5-merous, the perianth pilosulous overall or the corolla only so beyond middle; calyx shallowly campanulate 1.3–1.8 × 1.4–1.8(–2) mm, faintly 5-nerved, the broadly 3-angular teeth 0.4–0.6 mm; corolla 3.2–4.4 mm, the recurving lobes 1.5–2.3 × 0.8–1.1 mm; androecium 37–48-merous, 9–11 mm, the stemonozone 0.4–0.6 mm, the tube 0.7–1.8 mm, lacking callosities at base within; ovary glabrous at anthesis (but puberulent after fertilization), the stipe ±0.3 mm, the body ±1.3 mm, conical at each end, not sulcate laterally; stigma scarcely dilated. *Pods* shortly stipitate, the body in profile broad-linear, straight, 6–10 × 1.7–2.5 cm, 7–8-seeded, the papery valves dull brown glabrous externally, pallid within; seeds 6–10 × 4.5–6.5 mm, the pleurogram 4–6 × 2.5–3 mm.

In thorn forest below 200 m, apparently localized on the Pacific lowlands of Mexico in S Sonora and N Sinaloa, and on the Gulf slope of Baja California Sur, in lat. 23°30′–30°N. — Map 45. — Fl. II–IV. — *Chino, palo chino.*

XII. EBENOPSIS BRITTON & ROSE

Ebenopsis Britton & Rose, N. Amer. Fl. **23:** 33. 1928, a legitimate substitute for *Siderocarpos* Small, Bull. New York Bot. Gard. **2:** 91. 1901; non *Siderocarpus* Pierre, 1890. — Sp. typica: *Siderocarpos flexicaulis* (Bentham) Small ≡ *Acacia flexicaulis* Bentham ≡ *Ebenopsis flexicaulis* (Bentham) Britton & Rose = *Ebenopsis ebano* (Berlandier) Barneby & Grimes.

Ebenopsis sensu Mohlenbrock, 1963: 439.
Havardia sensu Nielsen in Polhill & Raven, 1981: 184, parva ex parte.
Chloroleucon sensu Rico, 1991: 519, parva ex parte.

Xeromorphic *trees* and *shrubs* with stiffly flexuous branches. *Stipules* associated with primary lvs subulate from swollen base, early lignescent, vulnerant, persistent. *Lf-formula* i–iv/2–7; lf-nectaries interpinnal, cupular short-stipitate; lfts opposite, of relatively broad outline, the primary venation palmate, the midrib often expiring short of blade-tip. *Inflorescence* of spikes or capitula arising from brachyblasts, the fls sessile; calyx 1–1.5 mm; corolla ±3–4.5 mm, the lobes erect; androecium 32–66-merous, the tube included or barely exserted, charged internally at base with small callosities sometimes cohering into a continuous lobed disc. *Pods* massive, compressed-sausagelike (reminiscent of *Tamarindus*), the woody valves produced inwardly as pithy interseminal septa, the exocarp in age breaking into polygons; dehiscence tardy, inert, the sutures at first separating at each end of pod but not gaping, ultimately separating through their whole length; seeds transverse on straight, subterete funicle, plumply obese, resembling chickpeas (but red-castaneous in color). — Spp. 3, of warm temperate and dry tropical Mexico and S Texas.

Exomorphic features shared by *Ebenopsis* with some or all species of *Havardia* and kindred genera are: dimorphic branches, the units of inflorescence arising from axillary brachyblasts along annotinous long-shoots; spinescent stipules; low leaf-formula; a nectarial disc or callosities around stipe of ovary. But the massive, ligneous, internally septate pod and the obese reddish seeds are unique. Although the pod in many specimens seems indehiscent, the valves do tardily separate, at first at each end of the fruit but finally throughout its length, into two canoe-shaped halves. The status of the group depends on subjective evaluation of carpological characters.

The transfer to *Chloroleucon* of *Ebenopsis ebano*, recently proposed by Rico (1991: 519), is summarily rejected. Rico speaks (l. c.) of *Chloroleucon* as one of three American pithecellobioid genera with spinescent stipules. This mistake, perhaps copied thoughtlessly from Nielsen (1981: 182), could have been avoided by paying attention, if not to specimens, to Bentham's protologue of *Pithecolobium* sect. *Chloroleucon*, where the armament was correctly described as *spinae . . . axillares*. These axillary thorns are barren and indurated inflorescence axes, in no way homologous to the genuine indurated stipules of *Ebenopsis* or *Havardia*. The real stipules of *Chloroleucon* are submembranous, precocious, and evanescent, but when these are lacking the true nature of the thorns can be inferred from the supra-axillary insertion. The same misinterpretation of armament was made by Nielsen et al. (1983a: 323) in the case of *Cathormion* Hasskarl.

Cassens and Miller (1981) noted that the wood of *Ebenopsis* is distinguished by the color of the heartwood and the high specific gravity.

Conspectus of genus *Ebenopsis*

¶ Fl-spikes relatively loose, the axis with peduncle 2–5 cm; pinnae of most lvs 2 pairs; around Gulf of Mexico, S Texas to Yucatán.
2. *E. ebano* (Berlandier) Barneby & Grimes

¶¶ Fls compactly spicate or capitate, the receptacle and peduncle together <1.5 cm; pinnae of most primary lvs either exactly 1 pair or 3–4 pairs; around Gulf of California, in S Baja California, Baja California Sur, and Sinaloa.
3. *E. confinis* (Standley) Barneby & Grimes
1. *E. caesalpinioides* (Standley) Britton & Rose

Key to the species of genus *Ebenopsis*

1. Trees 2–8(–20) m of S Texas, NE Mexico, and Yucatán Peninsula; pinnae 1–3, but of most lvs exactly 2 pairs; largest lfts 7–16.5 mm, ±2–3 times as long as wide; peduncle and axis of narrow fl-spike together 2–4.5(–5) cm; calyx 0.8–1.2 mm; seeds compressed (conventionally) parallel to the plane of the cotyledons, the pleurogram consequently on the two broader faces . . . . 2. *E. ebano* (p. 175)
1. Shrubs and treelets 0.8–3(–4) m of S Baja California, Baja California Sur, and adjacent Sinaloa; pinnae either 1 pair in almost all lvs or <2 pairs in larger lvs; larger lfts 5–12 mm, ±1–2 times as long as wide; peduncle and axis of condensed and short spike or capitulum together <1.5 cm; calyx 1.2–1.6 mm; seeds (unknown in rare *E. caesalpinioides*) compressed contrary to the plane of the cotyledons, the pleurogram consequently on the two narrower faces.
   2. Lf-formula i(–ii)/2–5; midrib of lfts mostly expiring short of blade-tip; fls capitulate, the axis of capitulum shorter than peduncle; pod in profile 4–10 × 2–3 cm; peninsular Baja California . . . . . 3. *E. confinis* (p. 177)
   2. Lf-formula iii–iv/6–7; midrib of lfts attaining blade-tip; fls shortly densely spicate, the peduncle shorter than floral axis; pod in profile ±18 × 4–4.5 cm; S Sinaloa . . . . 1. *E. caesalpinioides* (p. 174)

1. Ebenopsis caesalpinioides (Standley) Britton & Rose, N. Amer. Fl. **23:** 33. 1928. *Pithecollobium caesalpinioides* Standley, Proc. Biol. Soc. Wash. **37:** 45. 1924. — "Type . . . [US] no. 1,111,319, collected at El Zapote, Municipalidad [*sic*] de Mazatlan, Sinaloa, Mexico, in 1923 [fl.] by Sr. J. G. Ortega (no. 5128)." — Holotypus, US *1111319*!; clastotypus (fragm.), NY!; syntypus (*Ortega 5218*, fr.), US *1111327*!.

Stiffly branched, microphyllidious *trees* or *shrubs* of unrecorded stature with stout, abruptly flexuous, densely lenticellate long-shoots armed at each node with a pair of ascending spinescent stipules, the young growth puberulous with pallid flexuous hairs to ±0.25 mm, the leaves ephemeral, the pale green, subconcolorous lfts glabrous facially, minutely ciliolate, the short dense fl-spikes arising from thickened brachyblasts axillary to primary lvs. Primary *stipules* tapering from thickened base, 5–10 mm, those of brachyblasts subulate 1–1.5 mm, closely imbricate. *Lf-formula* iii–iv/6–7; lf-stks of primary lvs (few seen) 4–5.5 cm, the petiole 12–18 × 0.5 mm, the interpinnal segments 8–12 mm; nectaries at insertion of each pinna-pair peg-shaped 0.4–0.8 × 0.2–0.5 mm, the head cupular thick-rimmed; pinnae accrescent distally, the rachises of distal pair 15–32 mm, the longer interfoliolar segments 3–5 mm; lft-pulvinules ±0.2–0.3 × 0.5 mm, the lfts appearing sessile against rachis; lfts a little accrescent distally, in outline oblong- or obovate-elliptic from postically semicordate, antically broad-cuneate base, obtuse or broadly deltate-subacute, the penultimate pair 6–12 × 2.2–4 mm, ±2–3 times as long as wide; venation palmate, the subcentric midrib produced to lft-apex, 1–2(–3) branched on each side, the posterior primary nerves usually 2, the inner one produced to or beyond mid-blade, the outer one shorter, the whole venation bluntly prominulous dorsally, immersed on upper face. *Axis of fl-spikes* including short peduncle ±10–12 mm, the fls subcontiguous, sessile, homomorphic; bracts spatulate or rhombic-spatulate 1–1.4 mm, persistent into anthesis; perianth 5-merous, the calyx densely puberulent overall, the corolla glabrous or nearly so below middle, the lobes puberulent like calyx, the whole stramineous when dried; calyx campanulate, obscurely angulate ±1.4 × 1.2 mm, the obtuse teeth 0.2 mm; corolla narrowly vase-shaped 4 mm, the erect obtusely ovate lobes ±0.8 mm; androecium ±37-merous, 11 mm, the tube 4.5 mm, shortly exserted from corolla; ovary stipitate, at anthesis glabrous, the stipe 1 mm, the slenderly compressed-ellipsoid body

1.4 mm, 12-ovulate; style about as long as androecium, the stigma poriform. *Pods* (one seen complete, dehisced, probably a year old) sessile, broad-linear in profile, slightly decurved, ±18 × 4–4.5 cm, compressed but strongly biconvex, horizontally shallow-sulcate between seeds but the immersed sutures not undulate, the thin fuscous epidermis finally breaking into small polygons, the woody valves at maturity ±3 mm thick, apparently narrowly septiferous, the seed-cavities smooth, not colored, ±1.3–1.5 cm long, 2.5–4 cm wide; dehiscence through the length of both sutures; funicle and seeds unknown.

In unrecorded habitat, but to be expected in thorn forest below 100 m, known only from the coastal plain of Sinaloa, Mexico, near Matzatlán. The type was collected at "El Zapote"; a second collection at "Los Zapotes." — Map 47. — Fl. presumably following rains. — *Guaypinole*; *tempisque*. One collector (*Ortega 5218*) noted that the roasted seeds are eaten.

Higher leaf-formula, somewhat longer floral axis of the capitulum, and a much larger pod distinguish the two available collections of *H. caesalpinioides* from closely related *H. confinis*. The seeds are unknown, but the relatively narrow cavities in the pod suggest that they are obcompressed like those of *H. confinis*, with areole on the narrower faces.

2. Ebenopsis ebano (Berlandier) Barneby & Grimes, comb. nov. *Mimosa ebano* Berlandier, Mosaico Mexicano **4:** 418. 1840 fide C. H. Muller, Phytologia **41(6):** 386. 1979. — Original text not seen; lectotypus (Muller, l. c.): s. loc., *Berlandier 2262* (PH, herb. short.), the locality given on isotypus at F: "[El Encinal] de San Fernando à Santander [Tamaulipas, ±25°N] . . . 8bre 1830. Mexique." — Isotypi, G!, K! (see next entry), P!. — *Chloroleucon ebano* (Berlandier) L. Rico, Kew Bull. **46:** 519. 1991.

Acacia flexicaulis Bentham, London J. Bot **1:** 505. 1842. — "Texas, *Berlandier*. 2262." — Holotypus, K! = isotypus of the preceding. — *Pithecolobium flexicaule* (Bentham) Coulter, Contr. U.S. Natl. Herb. **2:** 101. 1891. *Zygia flexicaulis* (Bentham) Sudworth, U.S.D.A. Div. Forest. Bull. **14:** 248. 1897. *Siderocarpos flexicaulis* (Bentham) Small, Bull. New York Bot. Gard. **2:** 91. 1901, nom. illegit. *Samanea flexicaulis* (Bentham) Macbride, Contr. Gray Herb., N. Ser. **59:** 2. 1919. *Ebenopsis flexicaulis* (Bentham) Britton & Rose, N. Amer. Fl. **23:** 33. 1928.

Hoopesia arborea Buckley, Proc. Acad. Nat. Sci. Philadelphia **1861:** 453. 1862, ex parte, exclus. flor. caesalpiniaceo. — Equated by Bentham, 1875: 514, with *Acacia flexicaulis*, but based on a mixture, the essential character referable mostly to *Cercidium*.

Pithecolobium texense Coulter, Contr. U.S. Natl. Herb. **1:** 27. 1890. — "[*Nealley*] 153 . . . Near Rome (Starr Co. [Texas])." — Holotypus, US *15159*!. — Equated with *P. flexicaule* by Coulter, 1891, l. c.

Calliandra geniculata Wendland *ex* Bentham, Proc. Linn. Soc. London **30:** 514. 1875, pro syn. (cult.).

Pithecolobium flexicaule sensu Sargent, Silva **3:** 137, t. 147. 1892; Man. trees 538, Fig. 445. 1905 & ed. 2, p. 598, fig. 540. 1922; Standley, 1922: 396.

Pithecellobium flexicaule sensu B. L. Turner, 1959: 28, map 2 (Texas); Pennington & Sarukhán 1968: 184 & fig.; Isely, 1973: 114, map 38 (Texas); Vines, Trees of E. Texas 255 & fig. 1977.

Microphyllous, drought-deciduous *trees* commonly 2–8 m, with 1–several trunks 1.5–4 dm dbh but in undisturbed valley woodland attaining a height of 20 m and girth of 1 m, the bark fissured and flaking, the coarse stiffly flexuous branches armed at most nodes with a pair of straight ascending, ligneous, distally purplish-black or -castaneous and lustrous stipules subtending the primary lvs of long-shoots, most lvs fasciculate on scaley brachyblasts, the foliage appearing glabrous but the lf-axes and exceptionally also the plane, thinly papery olivaceous lfts thinly ciliolate with pallid hairs <0.25 mm, the amentiform spikes of small greenish ochroleucous, fragrant fls fasciculate on brachyblasts, mostly shorter than associated lvs. *Stipules* of primary lvs 2–10 mm, narrowed upward from ultimately swollen base into a vulnerant tip, those of verticillate lvs much shorter, imbricate. *Lf-formula* (i–)ii(–iii)/4–6; lf-stks 7–25(–28) mm, the petiole 3–14(–17) mm, at middle 0.4–0.7 mm diam, the longer interpinnal segments 4–11 mm; nectary between each pair of pinnae stoutly stipitate, in profile 0.45–1 mm tall, the shallow-cupular head 0.3–0.65 mm diam, no nectary at tip of pinna-rachises; pinnae when more than 1 pair accrescent distally, the rachis of furthest pair 11–27(–33) mm, the longer interfoliolar segments (2.2–)3–7 mm; lft-pulvinules 0.25–0.4 × 0.3–0.5 mm, finely wrinkled; lfts opposite, accrescent distally, the blade oblong or bluntly rhombic-oblong from semi-cordate-auriculate or asymmetrically shallow-cordate base, broadly obtuse or widely emarginate, the penultimate pair 7–16.5 × 3–6(–6.4) mm, (1.8–)2.1–2.9 times as long as wide; venation palmate, of (3–)4–5(–6) nerves from pulvinule, the open, often somewhat sinuous midrib a little forwardly displaced from center, V-forked well beyond midblade and usually not quite reaching the blade-apex, the inner posterior primary nerve produced to anastomosis at or well beyond midblade, the outer ones progressively shorter, the stronger primary nerves weakly branched and connected by random feeble tertiaries. *Fl-spikes*

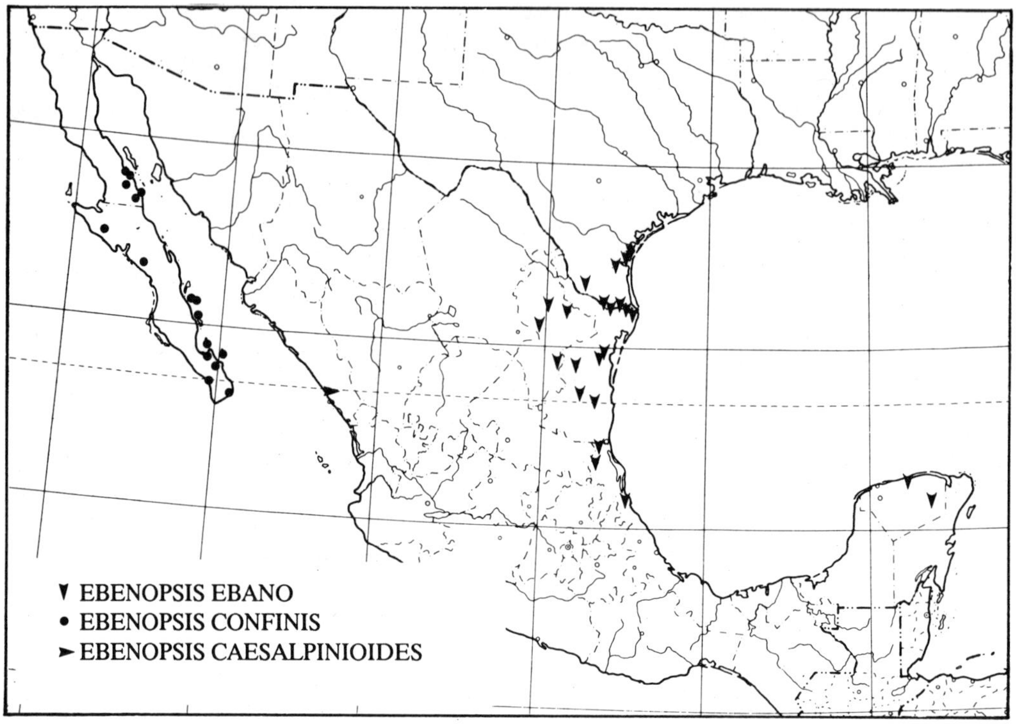

MAP 47. Mexico and southern Texas. Range of *Ebenopsis ebano* (Berlandier) Barneby & Grimes, *E. confinis* (Standley) Barneby & Grimes, and *E. caesalpinioides* (Standley) Britton & Rose.

including peduncle 2–4.5(–5) cm, densely yellowish- or sordid-puberulent overall, the rachis compressed, the fls either subcontiguous or well-spaced; bracts ovate 0.4–0.8 mm, tardily deciduous; perianth 5-merous or the calyx randomly 4–7-merous, the latter puberulent overall, the corolla sometimes glabrate proximally but always puberulent on lobes; calyx hemispherical 0.8–1.2 × 1.1–1.5 mm, bluntly ribbed, the depressed-deltate teeth 0.2–0.4 mm, one sinus sometimes recessed almost to base; corolla funnel-shaped 3.3–4.6 mm, the ascending lobes ±1–1.5 × 0.7–1.2 mm; androecium (34–)42–66-merous, (6.5–)7–19 mm, the stemonozone 0.4–0.6 mm, the tube 3–4 mm, the callosities at base within 0.2–0.25 mm tall, sometimes poorly developed; ovary 1.1–1.4 mm, cuneately contracted at each end, at base into a stipe ±0.3 mm, at anthesis either glabrous or microscopically papillate, puberulent after fertilization. *Pods* almost always solitary, of slow maturation and hanging on the tree for nearly a full year, in aspect recalling the fruit of tamarind, sessile or almost so, compressed-sausagelike, straight or gently decurved, abruptly apiculate, when well fertilized 9–19 × (1.6–)1.8–3 cm, 9–17 mm thick, the broad sutures immersed, the woody valves ±1.5–2.5 mm thick, produced between seeds as pithy septa, rough and densely minutely brown-puberulent externally, appearing fuscous (or blackish)-felted, when weathered breaking into a mosaic of small tesserae; dehiscence very tardy, after falling, the valves at first separating at each end of pod, not or scarcely gaping, finally separating their whole length; seeds obliquely descending on filiform, straight funicles, plumply oblate-spherical or obtusely rhomboid 10–15 × 9.5–10 × 6–8 mm, the red-brown smooth testa 0.6–0.9 mm thick, closely investing the lentil-like pallid embryo, consisting of a thin outer coat becoming translucent when soaked, a thin crustaceous undercoat, and a much thickened fuscous or blackish endotesta, the not quite complete pleurogram 8–11.5 × 6.5–9.5 mm.

In thickets and subtropical deciduous woodland, on caliche and gritty clay soils of upland hillsides and sometimes on richer soils of river bottoms, mostly below 500 m, locally plentiful in NE Mexico (Tamaulipas and adjoining states) and SE Texas (N to the Nueces valley); disjunct and local in Yucatán; cultivated in S Texas, Florida, and elsewhere. — Map 47. — Fl. V–VII(XII). — *Ebano* (Mexico); *Texas ebony* (U.S.A.), *acte* (S. L. Potosí, Huasteco).

Stiffly flexuous branches, spinescent stipules, characterisitc leaflet-venation (see description), and spicate units of inflorescence arising from scaly brachyblasts identify *E. ebano* at anthesis, and the woody pod is unmistakable. The fruit ripens slowly and has been described as indehiscent or almost so, but after falling the valves in fact separate their whole length into a pair of boat-shaped vessels that spill the seed on the ground. Texas ebony is most commonly seen as a shrubby treelet, dark green when foliate, forming thickets in thorn-scrub and chaparral on impermeable caliche substrates, but Isely (1973) reported that in rich bottomland it can attain a stature of 20 m. The hard wood is used for fenceposts and in Mexico provides excellent charcoal.

3. Ebenopsis confinis (Standley) Barneby & Grimes, comb. nov. *Pithecollobium confine* Standley, Contr. U.S. Natl. Herb. **20:** 191. 1919. — "Type [US] no. 638936, collected at Cape San Lucas, Baja California, Mexico, March 23, 1911, by J. N. Rose (no. 16339)." — Holotypus, US *638936*!; isotypus, NY!. — *Havardia confinis* (Standley) Britton & Rose, N. Amer. Fl. **23:** 33. 1928. Fig. 14.

Pithecollobium confine sensu I. Johnston, Proc. Calif. Acad. Sci., ser. **12:** 1034. 1924; Wiggins, 1964: 589 (sub *Pithecellobio*).

Stiffly branched microphyllous, drought-deciduous *shrubs* and *treelets* of rounded, or broad and depressed, flat-topped outline, 0.8–3(4) m, the coarse gray flexuous long-shoots armed at each node with a pair of straight divergent fuscous-castaneous woody stipules tapering from a swollen base into a vulnerant point, the new growth, except for often glabrous faces of the pallidly olivaceous lfts, densely minutely puberulent overall with fine spreading-ascending, gray or sordid hairs 0.1–0.2 mm, the primary lvs of long-shoots quickly deciduous (absent from most specimens), those associated with fls and pods fasciculate on hemispherical or short-cylindric brachyblasts, the small greenish white fls crowded into shortly pedunculate capitula or capituliform spikes, the massive woody fruits long-persistent and slowly maturing on the branches. *Stipules* of primary lvs 2–8 mm, long-persistent, those of fasciculate lvs much smaller, closely imbricate. *Lf-formula* of primary and fasciculate lvs alike i–ii/2–5; lf-stks 2–20 mm; petiole 2–10 × 0.2–0.5 mm, the interpinnal segment when present 5–12 mm; nectary between each pair of pinnae commonly stipitate, shallow-cupular 0.2–0.5 mm diam, in profile 0.2–0.7 mm tall, or in some lvs sessile or obsolescent; rachis of one or furthest pinna-pair 2.5–19(–24) mm, the longest interfoliolar segment 1.5–4.5 mm; lft-pulvinules 0.2–0.5 × 0.3–0.5 mm, finely wrinkled; lfts either equilong or subaccrescent distally, erratic in outline, oblong-obovate or -elliptic to suborbicular, semicordate at base, broadly obtuse or widely shallowly emarginate, the larger ones (4–)5.5–11 × (2.7–)3.2–6.5 mm, 1–1.9(–2.2) times as long as wide; primary venation of 3–5 nerves from pulvinule, the moderately displaced, straight or sinuous midrib commonly V-forked beyond middle and expiring or attenuate short of the plane margin and giving rise on each side to 1–3 randomly brochidodrome secondary nerves, the inner posterior primary produced more often to well beyond midblade, the outer ones progressively shorter, tertiary venules few and random, the whole venation immersed or weakly raised on upper face, bluntly prominulous beneath. *Peduncles* 1–3 per short-shoot, 3–8 mm; capitula ±17–35-fld, the axis 2–10 mm; bracts obovate-cuneate or spatulate 0.6–1 mm, persisting into anthesis; perianth 5- or the calyx 5–7-merous, the calyx minutely gray-puberulent at and above middle, the corolla so on lobes only; calyx campanulate, bluntly ribbed, 1.2–1.6 × 1.1–1.3 mm, the depressed-deltate teeth 0.2–0.25 mm; corolla 3–4.4 mm, the ovate erect lobes 0.9–1.6 × 0.7–0.9 mm; androecium 32–52-merous, 6.5–8.5 mm, the stemonozone 0.4–0.7 mm, the tube 2.1–4 mm, the discolored disc 0.2–0.3 mm tall; ovary at anthesis glabrous 1–1.3 mm, becoming papillate-puberulent after fertilization, the stipe 0.3–0.6 mm; style shorter than longest stamens, the stigma poriform. *Pods* solitary, sessile, in profile oblong or broad-linear, either straight or (most commonly) gently decurved, essentially like that of *E. ebano* in woody texture and fuscous or blackish coloration, (4–)6–10 × 2–3 cm, 8–10-seeded, 1.5–2 cm thick; dehiscence of *E. ebano*; seeds massive, 11–16 × 10–13 × 4–5 mm, like those of *E. ebano* in funicle, testa, and coloration, but crowded into narrower loculi and compressed contrary to the plane of the cotyledons, the pleurogram consequently on the two narrower (not the broader) faces; testa either smooth or pitted.

On desert hillsides, along washes, and in thin chaparral below 250 m, locally plentiful in Baja California Sur, especially but not exclusively on the Gulf slope and adjacent islands, and SE Baja California, from Bahia de los Angeles to the Cape (lat. 23°–29°N). —

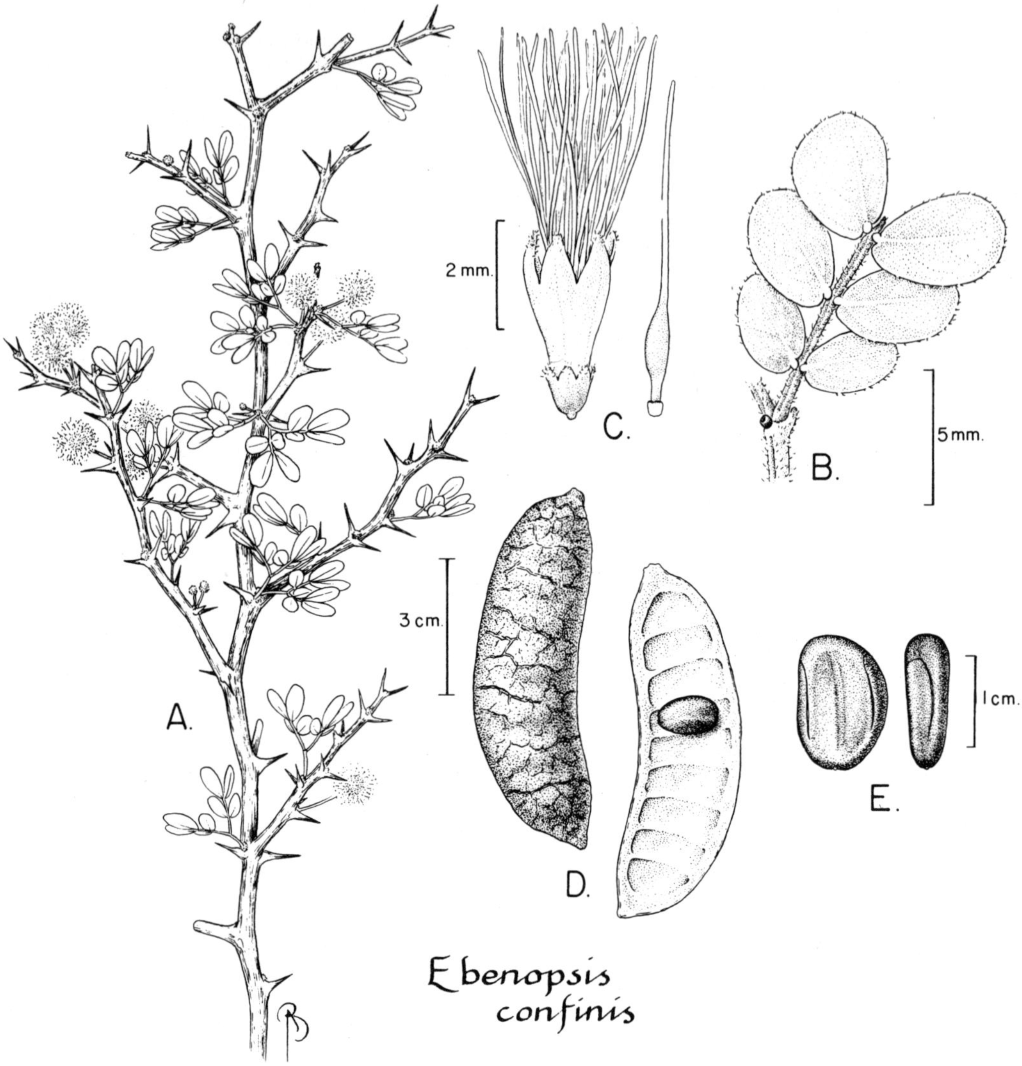

FIG. 14. *Ebenopsis confinis* (Standley) Barneby & Grimes. **A.** Habit. **B.** Leaf. **C.** Flower and ovary. **D.** Fruit. **E.** Seed. (A, B, and C drawn from *Wiggins 5445* (NY); D drawn from *Johnston 3498* (NY); and E drawn from *Rose 16896* (NY).)

Map 47. — Fl. I–IV, the fruit persisting into XII. — *Palo fierro*.

Ebenopsis confinis is readily distinguished over the greater part of its range by geminate pinnae of all leaves. The type collection, from Cabo San Lucas, is exceptional in having some leaves with two pairs of pinnae. The simplified foliage together with condensed capitula rather than amentiform spikes of flowers and the allopatric dispersal in desert environment amply differentiate it from the related *E. ebano*. The apparent displacement of the pleurogram from the broad face to the edge of the seed is due to deformation of the seed itself by confinement in a relatively narrow locule.

XIII. PAINTERIA BRITTON & ROSE

Painteria Britton & Rose, N. Amer. Fl. **23:** 35. 1928. — Sp. typica: *P. revoluta* (Rose) Britton & Rose ≡ *Pithecolobium revolutum* Rose.

Zygia ex parte sensu Mohlenbrock, 1963: 434.
Havardia sensu Nielsen, 1981: 184, fig. 2(4b), ex parte, quoad spp. amer.

Microphyllous xeromorphic *shrubs* ±3–15 dm. *Stipules* of long-shoots lignescent and vulnerant, of short-shoots only acicular or subulate, rather closely imbricate. *Lf-formula* i–vi(–vii)/3–20; lf-nectary between proximal (or only) pinna-pair, rarely between 2 pairs, not on petiole proper; venation of lfts either palmate or almost simple. *Inflorescence* of (shortly spiciform) capitula arising from brachyblasts, the fls (unknown in *P. revoluta*) sessile or almost so; calyx campanulate or hemispherical ±1–3 mm; corolla ±3.5–6 mm, the lobes either ascending or recurving-spreading; androecium 28–76-merous, 6–12 mm, the tube either included or shortly exserted, bearing internally at base 5 small, sometimes obscure (or perhaps wanting) callosities; stipe of ovary 1–2.4 mm. *Pods* compressed but turgid, retrofalcate or retrocircinnate, the leathery valves biconvex over each seed, the cavity continuous; dehiscence tardy through both sutures; funicles straight or sinuous, not sigmoid; seeds plumply lentiform. — Spp. 3, of the Mexican Plateau, one of them disjunct in lowland Tamaulipas.

Painteria is feebly distinguished from *Havardia* by tougher-walled, falcately or further recurved pods, and by straight or sinuous but not distally sigmoid seed-funicles. The painterias are smaller treelets or shrubs that have the pronounced dimorphism of growth into long- and short-shoots and the deciduous microphyllidious foliage appropriate to (and presumably induced by) the arid climate of the Mexican Plateau.

Conspectus of genus *Painteria*

¶ Lf-formula ii–vii/10–20; lfts plane; calyx shortly pedicellate, 1.5–3.2 mm.
1. *P. leptophylla* (de Candolle) Britton & Rose

¶¶ Lf-formula i(–ii)/3–10(–12); lfts plane; calyx sessile, 0.9–1.5 mm.
2. *P. elachistophylla* (S. Watson) Britton & Rose

¶¶¶ Lf-formula i/3–5; lfts strongly revolute; fls not known.
3. *P. revoluta* (Rose) Britton & Rose

Key to the species of genus *Painteria*

1. Lf-formula iii–vii/10–20 . . . . 1. *P. leptophylla* (p. 179)
1. Lf-formula i(–ii)/3–10(–12).
   2. Lfts suborbicular or broadly oblong, plane, palmately 2–3-nerved, the longer ones ±2–6 × 1.3–3.3 mm; widespread in the range of the genus . . . . 2. *P. elachistophylla* (p. 180)
   2. Lfts elliptic but revolute into a cylinder 2–3 × 0.5–0.8 mm, the venation immersed; local in Querétaro . . . . 3. *P. revoluta* (p. 181)

1. Painteria leptophylla (de Candolle) Britton & Rose, N. Amer. Fl. **23:** 36. 1928. *Acacia leptophylla* de Candolle, Cat. Pl. Horti Monsp. 74. 1813. — ". . . verosimiliter in Hispanorum territorio Americano. Ex horto Madritensi semina habuit cl. Broussonet sub *Mimosae leptophyllae* nomine." — Described from sterile plants in the garden at Montpellier. — *Mimosa leptophilla* [*sic*] Cavanilles, Elench. Pl. Horti Matr. 24. 1803, nom. nud. *Pithecolobium leptophyllum* (de Candolle) Daveau, Bull. Soc. Bot. France **59:** 635, t. XVI (photos of *Acacia leptophylla* and *A. mauroceana* cultivated at Montpellier), t. XVII (photos of *Mimosa leptophylla* cultivated in the Madrid Botanic Garden). 1912.

Acacia mauroceana de Candolle, Cat. Pl. Horti Monsp. 74. 1813. — "Habitat in regno Maroccano unde semina retulit clar. Aug. Broussonet." — Described from plants cultivated at Montpellier, erroneously thought to be of Maroccan provenance, but as shown by Daveau, 1912, l. c., of the same stock as the preceding, ultimately derived from seeds brought to Madrid by L. Née (see next entry). — *Mimosa mauroceana* Desfontaines, Tabl. École Bot. 181. 1804, nom. nud. — Equated by Daveau, Bull. Soc. Bot. France **59:** 629–636. 1912, with *Pithecolobium leptophyllum* (de Candolle) Daveau; confirmed by Mathez, Willdenowia **19:** 29. 1989.

Inga leptophylla Lagasca, Gen. Sp. Pl. 16. 1816. — "Hab. in America. Semina missit cel. L. Nee." — As shown by Daveau, 1912, based on the same cultivated stock as *Acacia leptophylla* de Candolle, but cultivated at Madrid.

Pithecolobium palmeri Hemsley, Diagn. Pl. Nov. Mexic. 50. 1878. — "Mexico: in regione San Luis Potosi, alt. 6000–8000 ped., *Parry et Palmer* 220." — Holotypus, K!; isotypus, GH!. — Equated with *Pithecolobium leptophyllum* by Harms *ex* Daveau, Bull. Soc. Bot. France **59:** 635. 1912; Standley, 1922: 396; Britton & Rose, 1928: 36. — Erroneously referred by Hemsley to sect. *Chloroleucon*.

Pithecolobium palmeri var. *recurvum* S. Watson, Proc. Amer. Acad. Arts **23:** 272. 1888. — ". . . in the Mapula Mountains, Chihuahua, by C. G. Pringle, April 1887." — Holotypus, *Pringle 1461*, GH!; isotypus, NY!. — Equated with *Pithecollobium leptophyllum* by Standley, 1922: 396 (sphalm. "*recurvatum*"); Britton & Rose, 1928: 36.

Pithecolobium leptophyllum sensu Espinosa *in* Rzedowski & Rzedowski, Fl. Fanerog. Valle de México **1:** 286. 1979; McVaugh, 1987: 235, the epithet there derived from *Inga leptophylla* Lagasca (1816), not *Acacia leptophylla* de Candolle (1813).

Low xeromorphic, stiffly branched microphyllous *shrubs* 4–15 dm, armed at each node of flexuous long-shoots with a pair of ascending or spreading, either straight or gently recurved lignescent stipules, the young growth, except for often facially glabrous lfts, densely puberulent with procumbent or spreading, whitish hairs to 0.1–0.25 mm, the old stems gray glabrate, the lvs olivaceous subconcolorous, the subglobose capitula of white-stamened fls arising single or 2–3-nate from brachyblasts, either accompanied by or antecedent to fasciculate lvs. *Stipules* of primary lvs tapering from dilate, often decurrent base to a vulnerant, pallid or glossy-castaneous point, (3–) 4–9 mm, those of brachyblasts stiffly subulate, not more than 1.5 mm. *Lf-formula* of primary lvs iii–vi(–vii)/(10–)12–20, the pinnae of lvs on brachyblasts often 2–3 pairs; lf-stks 0.5–4(–5) cm, the petiole 2.5–12(–18) mm, at middle 0.5–0.8 mm diam, the narrow ventral sulcus continuous between all but first pair of pinnae, the longer interpinnal segments 2.5–8 mm; a dimpled, shortly stout-stipitate or subsessile nectary between proximal pinna-pair 0.25–0.5 mm diam and about as tall, none on pinna-rachises; pinnae distally accrescent, the rachis of furthest pair 9–27(–36) mm, the longer interfoliolar segments 0.8–2.2 mm; lft-pulvinules 0.2–0.3 mm diam, not wrinkled; lfts proximally decrescent and also sometimes less so at very end of rachis, the plane blades narrowly oblong, linear-oblong or -lanceolate from shallowly semicordate base, obtuse or obscurely apiculate, those near midrachis 2.4–4.5 × 0.8–1.2 mm, (2.7–)3–4.5 times as long as wide; venation weakly developed superficially, the straight, simple or 1-branched midrib prominulous only dorsally, subcentric or displaced to divide blade to 1:2, in some relatively broad lfts one faint posterior primary nerve ascending to or beyond midblade, the ventral face of blade essentially nerveless. *Peduncles* (1–)3–13(–18) mm; capitula 16–35-fld, the clavate receptacle 1.5–2.5 mm, sometimes produced at apex into a filiform sterile appendage; bracts linear-oblanceolate or -spatulate 0.8–2.1 mm, persistent into anthesis; fls externally homomorphic but some functionally staminate; perianth 5-merous (rare abnormalities), the calyx minutely puberulent either overall or only on teeth, the reddish or greenish corolla glabrous up to the often minutely silky and always densely fimbriolate lobes; calyx campanulate, contracted at base into a stout pedicel 0.2–0.6 mm long and about as wide, (1.3–) 1.5–3.2 × 1.4–1.8 mm, the ovate or deltate teeth 0.2–0.7(–0.9) mm at early anthesis, the membranous sinuses sometimes splitting in age; corolla (3.5–)3.8–5 mm, the ovate recurving lobes 1.2–1.9 × 0.9–1.2 mm; androecium 40–76-merous, 6–10.5 mm, the well-defined stemonozone (0.5–)0.6–1 mm, the tube 2–4 mm, the callosities at base 0.25–0.4 mm tall or sometimes (especially in staminate fls) subobsolete; ovary slenderly ellipsoid compressed glabrous, the stipe 1–1.4 mm; style of bisexual fls about as long as androecium, the stigma punctiform. *Pods* 1–2(–4) per capitulum, in profile falcately or subcircinnately broad-linear, attenuate into an erect cusp 2–6 mm, the fertile body 7–11.5 × 1.1–1.9 cm, 8–10-seeded, the stiffly leathery valves at first plano-compressed, becoming turgid and low-convex (on both faces of pod) over each seed, in section 0.3–0.5 mm thick, composed of thin black mesocarp and thin crustaceous endocarp, at first or permanently densely gray-puberulent overall or finally glabrescent and fuscous-castaneous, indistinctly venulose, pallid but not satiny within, the cavity continuous; dehiscence inert, through both sutures, tardy; seeds obliquely descending on scarcely dilated, straight or sinuous (not sigmoid) funicle, compressed but plump, 7.5–11 × 7–9 × 3–4 mm, the smooth hard, moderately lustrous testa fuscous-castaneous, the pleurogram 5–8 × 3–7 mm, incomplete.

In grassland, matorral, and at the lower edge of the pine-oak belt, on both andesitic and calcareous substrates, sometimes in badlands, 1850–2750 m, scattered over the Mexican Meseta Central from centr. Chihuahua to W San Luis Potosí, the Santiago valley in centr. Jalisco, and the N slope of Cord. Volcánica from Michoacán to Tlaxcala and Puebla; one record from N Oaxaca (distr. Teposcolula). — Map 45. — Fl. III–VI(–VIII). — *Huisache*, a name applied more commonly to *Acacia farnesiana* Linn. and relatives.

2. Painteria elachistophylla (S. Watson) Britton & Rose, N. Amer. Fl. **23:** 36. 1928 "*elachisthophylla*". 1928. *Pithecolobium elachistophyllum* A. Gray *ex* S. Watson, Proc. Amer. Acad. Arts **17:** 352. 1882. — "At Monterey, Nuevo Leon, ([*E. Palmer*] 289 [in Feb. 1880])." — Holotypus, GH!; isotypus, NY!.

Pithecolobium compactum Rose, Contr. U.S. Natl. Herb. **8:** 33. 1903. — "Collected by J. N. Rose on the flat plain west of Tehuacan, Puebla [Mexico], August 1, 1901 (no. 5840)." — Holotypus, US!; clastotypus (fragm.), NY!. — *Painteria compacta* (Rose) Britton & Rose, N. Amer. Fl. **23:** 36. 1928.

Pithecolobium purpusii T. S. Brandegee, Univ. Calif. Publ. Bot. 4: 85. 1910. — "[Mexico. Puebla:] . . . near Tlacuiloltepec [7000–8000 ft, Jun–Jul. 1907, *C. A. Purpus*] No. 3869." — Holotypus, UC *137969*!; isotypus, NY!. — Equated with *Painteria compacta* by Britton & Rose, 1928: 36.

Pithecollobium elastichophyllum [sic] sensu Standley, 1922: 395; Estrada & Marroquín de la Fuente, 1992: 71, fig. 20.

Pithecollobium compactum sensu Standley, 1922: 395.

Stiffly branched microphyllous *shrubs* ±3–12(25) dm with pallid or fuscous, flexuous branches erratically armed at nodes with pairs of straight ascending, straight spreading, or rarely subrecurved spinescent stipules subtending primary lvs of long-shoots, most lvs fasciculate on hemispherical or short-cylindric scaley brachyblasts, the annotinous stems glabrate, the new branchlets and lf-stks puberulent or pilosulous with fine whitish hairs 0.1–0.3 mm, the firm olivaceous, scarcely bicolored lfts either thinly pilosulous on both faces, on lower face only, or glabrous except for minute cilia, the capitula or short spikes of dull reddish, white-stamened fls solitary or fasciculate on brachyblasts, either subsessile or very shortly pedunculate. *Stipules* of most primary lvs 2–6 mm, tapering to a fuscous vulnerant point, of some lvs conic and only 1–2 mm, those of brachyblasts acicular or stiffly subulate <1.5 mm. *Lf-formula* i(–ii)/3–10(–12), the pinnae bijugate only in rare primary lvs; lf-stks 1.5–5(–9) mm, the petiole 1.5–3(–5) mm; a subsessile or shortly stipitate nectary at tip of petiole round, dimpled, 0.15–0.3 mm diam, in profile narrowly drum-shaped or obovoid 0.3–0.5 mm tall, no nectary at tip of pinnae; rachis of pinnae mostly (5–) 6–15(–16) mm, in one example (*Fryxell 3610,* NY) to 30 mm, the longer interfoliolar segments 1.4–2.4(–3) mm; lft-pulvinules 0.15–0.3 × 0.2–0.3 mm, faintly wrinkled; lfts opposite, either subequilong except for often longer furthest pair, or a little decrescent proximally, the blade suborbicular or broadly oblong from shallowly semicordate base, broadly obtuse, the penultimate pair 1.7–6 × 1.3–3.3 mm, 1.1–1.8 times as long as wide; venation palmate, of 2–3(4) nerves from pulvinule, the straight or subsinuous midrib only a trifle forwardly displaced, produced to blade-apex or V-forked close below it, the inner posterior primary nerve incurved-ascending to anastomosis approximately at or a little beyond midblade, the whole venation usually immersed on upper face, bluntly prominulous beneath. *Peduncles* 0–3(7) mm; capitula ±14–26-fld, the receptacle 2–5(11) mm; bracts ovate acute 0.4–0.6 mm, persistent into anthesis; fls sessile, homomorphic, the calyx 5–7-, the corolla 4–5-merous, the calyx minutely gray-puberulent in distal ⅓, the corolla only at tip of lobes; calyx campanulate or hemispherical 0.9–1.5 × 1–1.3 mm, the deltate, often unequal, more or less incurved teeth 0.1–0.35 mm; corolla narrowly vase-shaped 3.6–5.8 mm, the ascending, ovate lobes 1–1.8 × 0.7–1.2 mm; androecium 28–44- or in one example 68–70-merous, 9–12 mm, the stemonozone ±0.4 mm, the tube 3.4–5.6 mm, either shortly or not exserted, the callosities at base of tube at most 0.2 mm tall, often obscure or wanting; ovary in profile narrowly ellipsoid, either glabrous or minutely puberulent at anthesis, becoming densely puberulent after fertilization, the stipe 1.1–2.4 mm; style about as long as androecium, slightly dilated at stigma. *Pods* solitary, in profile undulately broad-linear, long-attenuate proximally into pseudostipe and stipe, abruptly cuspidate, (5.5–)7–11 × 1–1.4 cm, (4–)5–8(–10)-seeded, evenly recurved through ¼–½ or even one full circle, the sutures ±2–2.4 mm wide, not prominent, shallowly constricted between most seeds, the stiffly leathery nigrescent valves low-convex over each seed, densely minutely puberulent overall, scarcely venulose, in age (after dehiscence) randomly fissured; dehiscence inert, through both sutures; seeds (few seen) descending on straight funicle, plumply lentiform 9–10 mm diam in broad view, nearly half as thick, the pleurogram incomplete.

On plains and hillsides in desert brush communities, mostly between 1400 and 2360 m but descending in Tamaulipas to unknown elevation in the Gulf lowlands, locally plentiful but of discontinuous range in NE Mexico: Chihuahuan Desert and W slope of Sa Madre Oriental in S Coahuila, N Zacatecas, Nuevo León and San Luis Potosí; E-centr. Tamaulipas (Rancho Loreto, 24°20′N); and dry interior valleys of states of Puebla (Tlacuilotepec; Tehuacán) and Oaxaca (distr. Teposcolula). — Map 44. — Fl. II–V(VIII).

Painteria elachistophylla is known from disjunct, desert or desertic enclaves, and the leaflet-number and pubescence show some differentiation between populations. In the Chihuahuan Desert the leaflets are mostly 3–5 pairs per pinna and thinly pilosulous, at least on the lower face. In San Luis Potosí and Puebla the leaflets are 5–7, rarely 8 or randomly up to 12 pairs per pinna and glabrous except for cilia. These states correspond with *P. elachistophylla* and *P. compacta* as defined by Britton and Rose. In Nuevo León we find instances (e.g., *Mueller 961*, NY) of 6–7 pairs of pubescent leaflets and in Tamaulipas one (*M. C. Johnston 10469*, NY) of 3–4 pairs of glabrous leaflets. Britton and Rose assigned to their two species peduncles essentially none as opposed to very short, but we have not found this a useful differential character. An exceptionally robust plant from municipio Rayón, San Luis Potosí (*Fryxell 3610*, NY) has shortly but distinctly spiciform capitula.

3. Painteria revoluta (Rose) Britton & Rose, N. Amer. Fl. **23:** 35. 1928. *Pithecolobium revolutum* Rose, Contr. U.S. Natl. Herb. **10:** 96, XXVII. 1906. — "[US] no. 453274, collected by Rose and Painter south of Higuerillas, Queretaro, August 23, 1905 (no. 9784)." — Holotypus, US *453274*!; clastotypus (fragm.), NY!.

Small, stiffly branched xeromorphic microphyllous *shrub* with habit and armament of *P. elachistophylla*, the lvs pilosulous with erect hairs to 0.2–0.4 mm, the capitula subsessile on brachyblasts. *Stipules* of primary lvs to 2–3 mm, widely ascending or gently recurved, lignescent and vulnerant, those of brachyblasts 1 mm or less. *Lf-formula* i/3–5; petiole 1.5–3.5 × 0.3–0.6 mm; a substipitate cupular nectary at tip of lf-stk 0.2–0.25 mm diam and about as tall, the head orange, no nectary on pinna-rachises; rachis of pinnae 2–4.5 mm, the longer interfoliolar segments 0.4–1 mm; lfts opposite sessile, elliptic but so completely revolute as to become cylindric, convex ventrally, sulcate dorsally, the longest 2–3 × 0.5–0.8 mm, venation immersed. *Fls* unknown. *Pods* in profile broad-linear or linear-oblanceolate, falcately recurved through ±⅓–½ circle, when well fertilized 6–7.5 × 1–1.2 cm, 6–7-seeded, the rigid lignescent valves low-convex (on each face equally) over each seed, shortly densely gray-pilosulous externally, the cavity continuous; funicles sinuous but not sigmoid; seeds plumply lentiform ±8.5–9 × 7–7.5 × 3–3.5 mm, the hard black or atrocastaneous testa either smooth or pitted, the U-shaped pleurogram ±5–6 × 3–3.5 mm.

In desert grassland near 1900 m, known only from the type locality on S Meseta Central in Querétaro, Mexico. — Map 44.

This diminutive shrublet seems most closely related to *P. elachistophylla* but is instantly recognized by the tiny leaflets rolled backward into a tube narrowly sulcate dorsally. According to Rose, it was locally common at the type locality.

XIV. MACROSAMANEA BRITTON & KILLIP

The exemplar analysis gave a large clade that has inflorescence meristems that are persistent to some degree (Fig. 3B). That is, the plants are ramiflorous or cauliflorous. Within this clade occur species of *Macrosamanea* Britton & Killip (Fig. 3B, no 5). *Calliandra* Bentham and *Zapoteca* H. Hernández are the outgroups. The data set is shown in Table 5; the characters used in the analysis are as follows.

0. Resting buds: 0 = covered by stipules; 1 = not covered by stipules.
1. Stipules: 0 = deciduous; 1 = persistent.
2. Petiolar nectaries: 0 = lacking; 1 = present.
3. Pinnae: 0 = proximally decrescent thence subequilong; 1 = distally accrescent; 2 = reduced to 1 pair.
4. Paraphyllidia: 0 = present; 1 = absent.
5. Leaflet apex: 0 = obtuse to mucronulate; 1 = sigmoid-porrect.
6. Leaflets: 0 = distally accrescent; 1 = proximally decrescent thence subequilong.
7. Venation: 0 = brochidodrome; 1 = camptodrome.
8. Veins: 0 = 2 from pulvinule; 1 = 1 from pulvinule; 2 = 5 or more from pulvinule.
9. Leaves: 0 = coeval; 1 = suppressed at distalmost, youngest nodes of the inflorescence.
10. Inflorescence meristem: 0 = annual; 1 = persistent for some seasons.
11. Floral bracts: 0 = monomorphic; 1 = dimorphic, the lower nectariferous, the upper not.
12. Dimorphic bracts: 0 = lacking; 1 = deciduous; 2 = persistent.
13. Flowers: 0 = pedicellate; 1 = sessile.
14. Polyad number: 0 = 8; 1 = 16; 2 = 24 or more.
15. Ovary: 0 = single; 1 = reduplicated.
16. Valves: 0 = elastically reflexed at dehiscence; 1 = inertly dehiscent through both sutures.
17. Exotesta: 0 = crustaceous; 1 = papery.
18. Exotesta: 0 = with a pleurogram; 1 = lacking pleurogram.
19. Seed-coat; 0 = lacking a peripheral nerve; 1 = with a peripheral nerve.

Table 5

Data set used in the analysis of *Macrosamanea*

	0	5	1 0	1 5
CHOUSTONIANA†	00000	01?11	00000	00000
ZTETRAGONA†	01010	01010	00001	00000
PUBIRAMEA	10101	01001	0110?	01111
DISCOLOR	10111	01011	11102	01111
DUCKEI	10111	01011	11112	01111
AMPLISSIMA	1?101	01000	11?02	01111
CONSANGUINEA	11110	11011	11102	11111
FROESII	10100	11111	?1112	11111
SPRUCEANA	10100	11011	01102	01111
KEGELII	11100	01101	10002	0????
MACROCALYX	10110	00020	?0002	11111
SIMABIFOLIA	10121	0?100	01212	01111
PRANCEI	10121	00110	11202	01111

Note: † = outgroup; ? = state not observed in this terminal.

The analysis gave 1 tree of 42 steps with a consistency index of 0.57 and a retention index of 0.62. The tree is shown in Figure 15. The topology in Figure 15

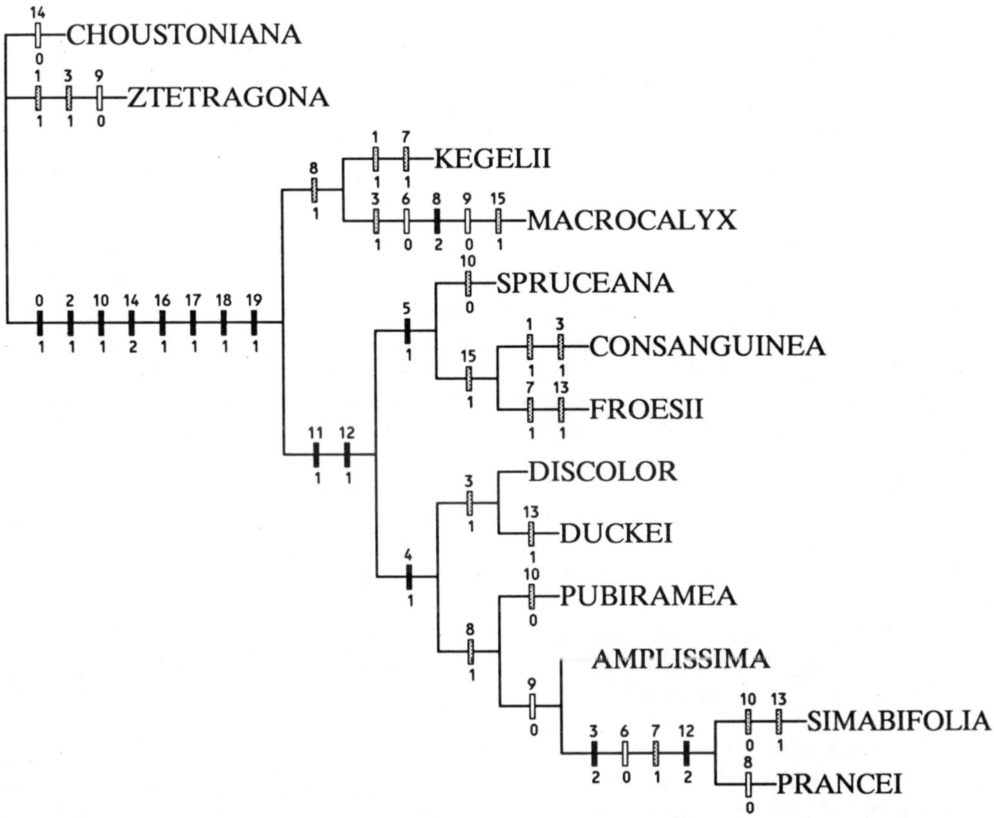

FIG. 15. One of three trees from a phylogenetic analysis of *Macrosamanea* Britton & Rose. The fruit of *Macrosamanea kegelii* is unknown. This topology results if the fruit of *M. kegelii* is assumed to be similar to that of all other species of *Macrosamanea* and if the inflorescence meristems of *M. kegelii* and *M. macrocalyx* are persistent (unknown owing to scarcity of material).

results if the fruit of *M. kegelii* should be like the fruit of other *Macrosamaneae* and if *M. macrocalyx* has persistent inflorescence meristems. In this tree, monophyly of *Macrosamanea* is supported by eight synapomorphies: naked buds (no. 0); presence of petiolar nectaries (no. 2); persistent inflorescence meristems (no. 10); polyads of 24 grains or more (no. 14); fruit inertly dehiscent through both sutures (no. 16); papery (no. 17) testa lacking pleurogram (no. 18), and seed-coat with peripheral nerve (no. 19). All these states are homoplastic in the context of the whole tribe. *Macrosamanea macrocalyx* and *M. kegelii* form a pair of species sister-group to the remainder of the genus. Both these species are known from very few collections: *Macrosamanea kegelii* is still unknown in fruit, and it is not known whether *M. macrocalyx* has persistent inflorescence meristems.

The remaining species of *Macrosamanea* differ from *M. macrocalyx* and *M. kegelii* by the dimorphic bracts (nos. 11, 12) and form two clades. The first, including *M. spruceana*, *M. consanguinea*, and *M. froesii* contains those species with porrect leaflets (no. 5), the second, those species lacking paraphyllidia (no. 4). In both these clades, relationships are resolved by homoplasious characters, save the sister-group *simabifolia-prancei*, which share pinnae reduced to one pair (no. 3) and persistent dimorphic bracts (no. 12).

Macrosamanea Britton & Rose *ex* Britton & Killip, Ann. New York Acad. Sci. **35:** 131. 1936. — Sp. typica: *M. discolor* (Willdenow) Britton & Killip ≡ *Inga discolor* Humboldt & Bonpland *ex* Willdenow ≡ *Inga adiantifolia* Kunth, nom. illegit. — *Mimosaceae* trib. *Ingeae* gen. D., Nielsen *in* Polhill & Raven, Advances Legume Syst. **1:** 190. 1981. — Etymology: Gr. *macros*, great + *Samanea*, from the aboriginal samán (rain-tree).

Pithecolobium sect. *Samanea* ser. *Coriaceae* [sic] Bentham, Trans. Linn. Soc. London **30:** 589, sp. ult. exclusa. 1875 & *in* Martius, Fl. Bras. **15(2):** 443. 1876 "*Coriacea*". — Sp. lectotypica: *P. adiantifolium* (Kunth) Bentham ≡ *Macrosamanea discolor* (Willdenow) Britton & Killip.

Samanea sensu Pittier, 1927: 61, minori ex parte (spp. no. 6, 7).

Pithecolobium ser. *Coriaceae* sensu Ducke, 1949: 39–41.

Unarmed, slender *trees*, *arborescent shrubs* either free-standing or sarmentose when crowded, or (in savanna and campirana) slender erect *shrubs,* either macro- or microphyllidious, the lfts mostly of firm texture, the capitula of relatively long and narrow, homomorphic, either sessile or pedicellate fls borne solitary or fasciculate by 2–5 in the axil of coeval or lately fallen lvs, or by suppression of the subtending lvs forming small, axillary or pseudoterminal (in *M. kegelii* cauliflorous), efoliate pseudoracemes of capitula, the branching essentially monopodial, but the terminal shoot often abruptly inhibited beyond the last node of the year's growth; indumentum of short, gray or sordid hairs, sometimes scanty, or wanting below the fls. *Stipules* firm, deltate, triangular or linear-lanceolate, deciduous or exceptionally persistent. *Lf-formula* i–ix/2–26(–31); petiolar nectaries sessile, scutiform or patelliform or shallowly cupular, the first situated either next to lf-pulvinus, or between first pinna-pair, similar nectaries often between further pinna-pairs and near tip of pinna-rachises; lft-pulvinules usually broader than long, wrinkled; venation palmate and pinnate. *Receptacle* of capitula either subglobose, or pyriform, or stoutly linear-claviform, alveolate; bracts at base of capitulum either subtending the lowest 2–5 fls or immediately below them, commonly reflexed and charged ventrally with a nectary, but in *M. amplissima* the lowest bracts lacking nectary and deciduous but bracts of further fls bearing a nectary and persistent, and in *M. kegelii* and *M. macrocalyx* all bracts lacking nectary; perianth normally 5-merous but the teeth and lobes unequally cleft, sometimes confluent; calyx cylindric or cylindro-campanulate (4.5–)5–20(–26) mm, or in *M. macrocalyx* tumid and 20–29 mm; corolla narrowly trumpet-shaped, usually white and externally silky-pubescent, 11–63 mm; androecium 45–248-merous, the tube adnate at base to corolla, nearly as long as it or far exserted, the tassel either white or distally pink-tinged; ovary usually solitary, exceptionally 2 or 3 in *M. froesii* and *M. consanguinea* and so far as known 2–3 in *M. macrocalyx*, symmetrically linear-ellipsoid, tapering into style, surrounded at base by a variably developed but always perceptible disc; style shortly exserted from longer stamens, the stigma either porose or a little dilated; ovules 6–26. *Pods* sessile or almost so, in profile narrowly oblong or broad-linear, straight or gently decurved but never annular or coiled, either plano-compressed or becoming biconvex where elevated over obese seeds, the sutures not greatly thickened, the valves leathery; dehiscence inert, through the seminiferous suture and ultimately both sutures, the valves narrowly gaping to discharge the seeds; funicles subfiliform; seeds ellipsoid, usually strongly compressed and at least narrowly imbricate along the continuous seed-cavity, where overlapping and deformed by mutual pressure, the papery-membranous castaneous testa girdled by a prominent peripheral nerve, becoming fragile, wrinkled over faces of seed but pleurogram 0, loosely investing the horny fuscous embryo; endosperm 0. — A South American genus of 11 species, the majority of riparian, seasonally flooded forest, two of seasonally wet sandy savanna or campirana, most diverse and numerous in the Amazon basin, extending N into the Orinoco valley and the Guianas, in an equatorial belt between 10°N and 10°S latitude.

The genus *Macrosamanea* has the homomorphic flowers, the compressed, inertly dehiscent pod, and the papery, exareolate seed-coat of *Zygia*, but differs in the long narrow perianth and with two exceptions (*M. kegelii, M. macrocalyx*) in characteristic nectaries on the inner face of bracts that subtend either the capitula or their individual flowers. The inflorescence of *Zygia* is with few exceptions (these unlike *Macrosamanea* in short calyx and slender corolla) cauliflorous, whereas the capitula of *Macrosamanea* are initially borne (except in *M. kegelii*) either directly pedunculate or in short pseudoracemes from coeval leaf-axils, the meristems of which may continue to function for some short time after abscission of the leaf. Leaves reduced to one pair of pinnae are stabilized in *Zygia* sect. *Zygia* whereas in *Macrosamanea* the pinnae are actually or potentially two or more pairs except in *M. simabifolia.* Nectaries on floral bracts that appear to be homologous with those of *Macrosamanea* occur in *Archidendron* ser. *Stipulatae* (Mohlenbrock) Nielsen (Opera Bot. **76:** 9, fig 56(4). 1984), though in *Archidendron* when nectaries are present they are present on all bracts. This feature is unique to *Macrosamanea* among neotropical Ingeae, and by itself identifies to genus the great majority of specimens. The two species that lack this signature of *Macrosamanea* are nevertheless referred here because they share synapomorphies of the genus. The cauliflorous inflorescence and pale green coloration of the foliage of *M. kegelii* suggest a remote connection with *Zygia* sect. *Marmaroxylon*, where it would be anomalous in the form of the perianth. *Macrosamanea macrocalyx* is by any criterion an isolated species, notably distinct from the rest in the tumid calyx about two-thirds as long as the corolla, two or three ovaries per flower, and few ample, basally semicordate leaflets. This was described by Ducke as a member of *Pithecolobium* ser. *Coriacea*, and if not accommodated therein must be referred to a monotypic genus.

We exclude from *Macrosamanea* the anomalous *Pithecellobium inundabile* Ducke, first described by Ducke, on account of its simply pinnate leaves, as an *Inga* of sect. *Leptinga*, but later found to have dehiscent fruit similar to that of *Macrosamanea* and consequently referred to *Pithecellobium*. This species will be transferred to *Zygia* in Part II of this revision.

Macrosamanea was explicitly typified from the outset by *M. discolor* (Willdenow) Britton & Killip, which (under the invalid title of *Pithecolobium adiantifolium*) was a characteristic member of *Pithecolobium* ser. *Coriacea* Bentham. Kleinhoonte (1940: 329) extended the concept of *Macrosamanea* so far as to admit *Balizia pedicellaris*, a species of albizioid, not zygioid affinity, hence alien to the original core of *Macrosamanea*. It is in the sense of *M. pedicellaris* that Nielsen (1981: 182, fig. 1/4a) subordinated *Macrosamanea* to *Albizia*, for he simultaneously recognized ser. *Coriacea* as a distinct but undescribed "genus D." We here adopt, with minor additions and adjustments, Nielsen's evaluation of the genus, but adopt the already available name *Macrosamanea*.

Cassens and Miller (1981) stated that the wood of *Macrosamanea* is unique in the Ingeae in having both septate fibers and confluent parenchyma.

Conspectus of genus *Macrosamanea*

¶ Lfts of longer pinnae 6–30 pairs; blade of lfts either straight or obscurely sigmoid, and if sigmoid then obtuse or obtuse mucronulate; fls capitulate; a nectary on bracts at base of capitulum, but not on bracts subtending further fls; corolla at least twice as long as calyx, this never tumid; ovary solitary.

- 5. *M. discolor* (Willdenow) Britton & Killip
  - a. var. *discolor*
  - b. var. *arenicola* (Cowan) Barneby & Grimes
- 6. *M. duckei* (Huber) Barneby & Grimes
- 7. *M. pubiramea* (Steudel) Barneby & Grimes
  - a. var. *pubiramea*
  - b. var. *lindsaeifolia* (Bentham) Barneby & Grimes

¶¶ Lfts of longer pinnae 9–14 pairs; blades of lfts as in ¶; fls densely racemose; a nectary both on sterile bracts at base of raceme and on some bracts subtending further fls; corolla and ovary of ¶.

- 8. *M. amplissima* (Ducke) Barneby & Grimes

¶¶¶ Lfts of longer pinnae (8–)11–21 pairs; blade of lfts sigmoid, at apex turned forward and sharply acute; bracts, capitula, nectaries, and corolla as ¶.

- 2. *M. spruceana* (Bentham) Record
- 3. *M. consanguinea* (Cowan) Barneby & Grimes
- 4. *M. froesii* Barneby & Grimes

¶¶¶¶ Lfts of ¶¶¶; bracts lacking nectary; capitula pseudoracemose from defoliate nodes below contemporary foliage; corolla of ¶.

- 11. *M. kegelii* (Meisner) Kleinhoonte

¶¶¶¶¶ Lfts of longer pinnae 2–4 pairs, the blade broadly semicordate at base; inflorescence of ¶ but bracts lacking nectary; calyx tumid, more than half as long as corolla; ovaries 2(–3).

- 1. *M. macrocalyx* (Ducke) Barneby & Grimes

¶¶¶¶¶¶ Lfts of longer pinnae 2–4(–5) pairs, the blade obovate or oblanceolate, not semicordate at base; inflorescence and nectaries of ¶.

- 9. *M. simabifolia* (Bentham) Pittier
- 10. *M. prancei* (Barneby) Barneby & Grimes

Key to the species of genus *Macrosamanea*

1. A nectary on bracts at base of each unit of inflorescence.
   2. Lfts of longer pinnae 6 pairs or more.
      3. Lfts straight or obscurely sigmoid, if at all porrect at apex then either obtuse or merely mucronulate.
         4. Fls 12–50 per unit of inflorescence; pinnae of larger lvs mostly 3 pairs or more.
            5. Fls capitulate, the receptacle 2.5–6.5 mm long; bracts subtending fls all similar and caducous, lacking a nectary; pedicels 0–1.5 mm; calyx 4.5–12 mm and corolla (16–)17–34 mm; pod 2.5–5.5 cm wide.
               6. Fls relatively large, the calyx 6.5–12 mm and corolla 18–34 mm; androecium 80–106-merous . . . . 7. *M. pubiramea* (p. 194)
               6. Fls smaller, the calyx 4.5–7.5(–8) mm and corolla 11–16 mm; androecium 45–50-merous . . . . 6. *M. duckei* (p. 193)
            5. Fls densely shortly racemose, the receptacle 8–25 mm long; bracts subtending distal fls bearing a nectary like that of sterile bracts at base of raceme; pedicels 3.5–6.5 mm; calyx 13–18 mm and corolla 40–53 mm; pod 5.5–7 cm wide. . . . . . 8. *M. amplissima* (p. 198)
         4. Fls fewer than 12 per capitulum; pinnae of larger lvs 1 or 1–2 pairs. . . . . 5. *M. discolor* (p. 192)
      3. Lfts sigmoidally arcuate, beyond midblade turned forward and at apex sharply acute.
         7. Fls relatively short, the calyx 8–12 mm and corolla 22–30(–40) mm; pinnae of larger lvs mostly <6 pairs.

8. Lfts of longer pinnae (6–)7–12 pairs and the longer ones 20–35 × 9–20 mm; pod 2.7–3.3 cm wide; SE T. F. Amazonas in Venezuela and in adjacent Guainía and Vaupés in Colombia . . . . 3. *M. consanguinea* (p. 189)
8. Lfts of longer pinnae (12–)14–18 pairs and the longer ones 12–21 × 3.5–9 mm; pod 5–6.5 cm wide; Río Urubú in NE Amazonas, Brazil and on upper Río Negro in T. F. Amazonas, Venezuela . . . . 4. *M. froesii* (p. 191)
7. Fls extremely long and narrow, the calyx (14–)15–27 mm and corolla 36–63 mm; pinnae of larger lvs 6–9(–10) pairs and largest lfts 14–26 × 4.5–9(–10) mm; pod (4.3–)5–7 cm wide . . . . 2. *M. spruceana* (p. 188)
2. Lfts of longer pinnae not more than 4 pairs.
9. Pedicels obsolete; calyx 5–6.5 × 1.5–2 mm; corolla 13–22 mm; SW T. F. Amazonas in Venezuela and upper Río Negro in Brazil. . . . . 9. *M. simabifolia* (p. 199)
9. Pedicels 2–3 mm; calyx 8–10 × 3–5 mm; corolla 26–30 mm; N. state of Amazonas, Brazil. . . . . 10. *M. prancei* (p. 200)
1. No nectary on bracts.
10. Lf-formula v–viii/12–16; lfts rhombic from broad-cuneate base, the larger ones 2–3 cm; inflorescence a depauperate pseudoraceme of capitula arising from defoliate nodes below current foliage; calyx subcylindric, 3.5–6 mm diam; ovary solitary; local in Surinam and French Guiana . . . . 11. *M. kegelii* (p. 202)
10. Lf-formula i–iii/2–4; lfts ovate-elliptic from semi-cordate base, the larger ones 5–8 cm; capitula axillary to coeval lvs; calyx tumid, ±7–10 mm diam; ovaries 2–3; lower rio Xingú in Pará, Brazil. . . . . 1. *M. macrocalyx* (p. 186)

1. Macrosamanea macrocalyx (Ducke) Barneby & Grimes, comb. nov., *Pithecolobium macrocalyx* Ducke, Arch. Jard. Bot. Rio de Janeiro **3**: 64. 1922. — "Habitat in regione fluminis Xingu medii, ad ripas paludosas rivuli Ambe infra villam Altamira [±3°10′S, 52°10′W], l[egit] A. Ducke 16–12–1916, n. 16.622." — Holotypus, RB!; isotypi, BM!, MG! = F Neg. *s.n.* in F *601752* + fragm.!, G! = F Neg. *28267* + fragm.!, P!, U *64412*!, US *104947*!.
Fig. 16.

Macrophyllidious, potentially sarmentose *shrubs* and *treelets* with slender glabrate lenticellate branches, either glabrous throughout or the lf-axes thinly minutely sordid-puberulent, the ample, reticulately venulose lfts truly glabrous, olivaceous above, scarcely paler beneath, the loose capitula of large whitish glabrous fls subsessile or shortly pedunculate, solitary or paired either in a primary lf-axil or in the axil of the last lf of the branchlets, appearing terminal. *Stipules* triangular-subulate to ±2 mm, deciduous. *Lf-formula* i–ii(–iii)/2–4, of most lvs ii/2–3; lf-stks 1.5–6.5 cm, the petiole 4–15 mm, the longer interpinnal segments to 25–40 mm; nectaries sessile thick-rimmed cupular 0.6–2 mm diam between each pair of pinnae and smaller ones between at least the furthest lft-pair; pinnae accrescent distally, the rachis of furthest pair ±3–6.5 cm, the longer interfoliolar segments 1.8–2.8 cm; pulvinules transversely elliptic 1–2 × 1.8–2.7 mm; lfts accrescent distally, ovate or broadly lance-elliptic from broadly semicordate base, the distal ones ±5–8 × 2.3–3.2 cm, 1.8–2.5 times as long as wide; venation of 5–7 nerves from pulvinule, the costa a little forwardly displaced from midblade and either straight or gently evenly incurved, the other primary nerves all posterior, the innermost incurved-ascending almost to or shortly beyond midblade, the midrib giving rise to ±6–8 pairs of major and indefinite intercalary secondary nerves brochidodrome well within the plane margin, these is turn generating a 3- and 4-nary reticulum, the whole venation sharply raised on both faces. *Capitula* 9–18-fld, the fls ascending, homomorphic, the receptacle ±4–8 mm; bracts papery, linear-oblanceolate 7–10 × 1.8–3 mm, several-nerved, without nectary; perianth 5-merous, glabrous; calyx tumidly cylindro-campanulate, 20–29 mm long, contracted at base into a turbinate pedicel 1–2 mm, laxly investing the corolla, marcescent around the base of pod, when pressed 7–10 mm wide, papery-membranous pale green, the faint nerves irregularly branched, the lobes connivent at tip 3–6 × 4–5 mm, sharply acute at apex; corolla white, 4–11 mm longer than calyx, the lance-ovate acute lobes ±4.5–9 × 3–4 mm; androecium ±160–248-merous 4.8–6.8 cm, the stemonozone 2.5–3 mm, the tube about as long as the corolla or to 6 mm longer, the exserted filaments white; callosities surrounding base of ovaries ±1 mm tall; ovaries 2 or 3, glabrous, slenderly fusiform; style ±5 mm longer than longer stamens, the stigma punctiform, not dilated. *Pods* (1 seen RB, not fully ripe) sessile, in profile oblong, broadly obtuse at both ends, slightly decurved, ±6 × 2.2 cm, the sutures ±1–1.5 mm wide, the stiffly leathery, plano-compressed brownish valves

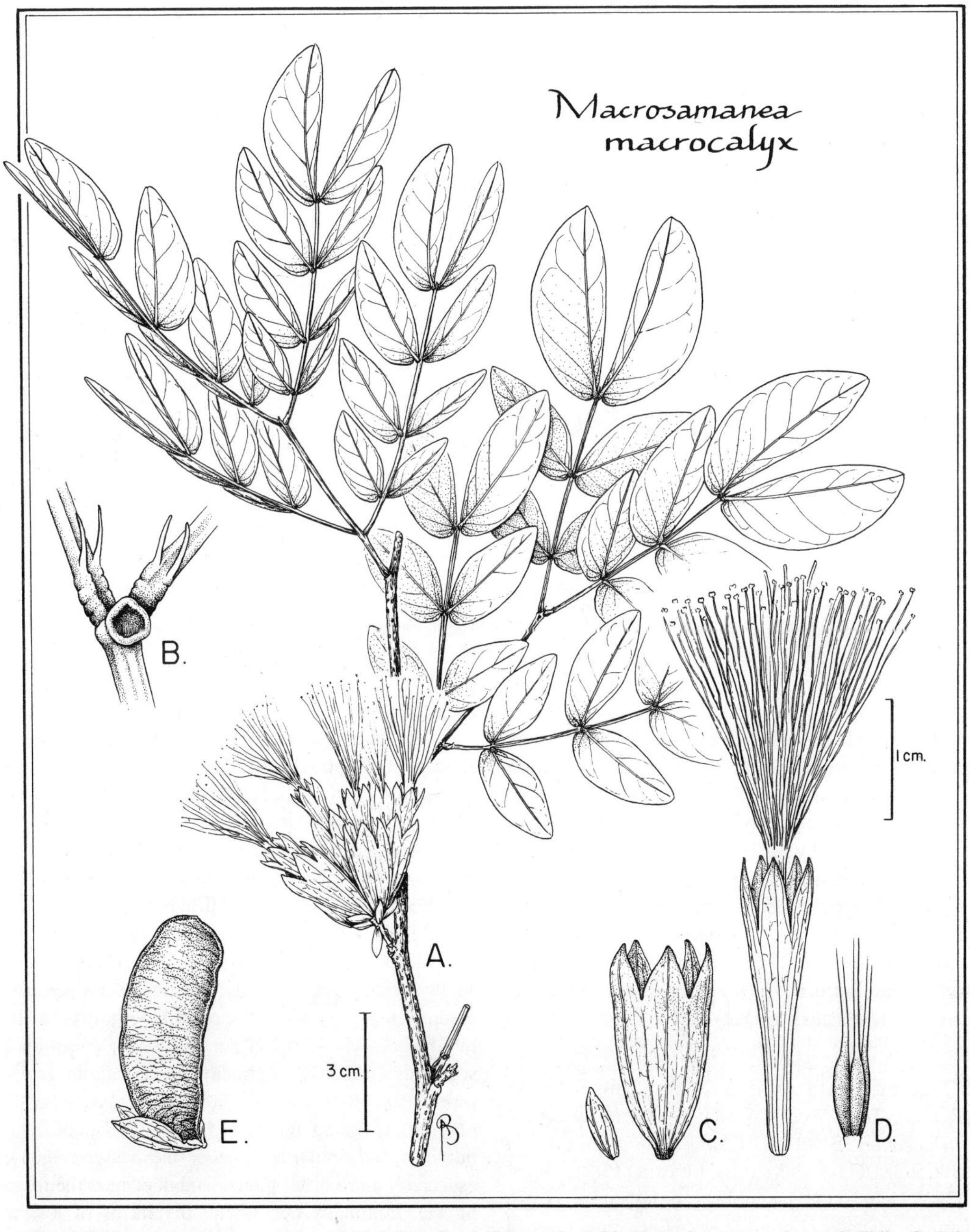

FIG. 16. *Macrosamanea macrocalyx* (Ducke) Barneby & Grimes. **A.** Habit. **B.** Leaf-nectary and paraphyllidia. **C.** Calyx and flower. **D.** Ovaries. **E.** Fruit. (Drawn from *Ducke 16622* (BM) and *Kuhlmann 2035* (RB).)

glabrous, coarsely cross-venulose; dehiscence and seeds unknown.

At edge of flooded forest, known only from the banks and tributary creeks of Río Xingú in the vicinity of Altamira, Pará, Brazil. — Map 48. — Fl. XII, IV.

Macrosamanea macrocalyx has the habit and long whitish, externally silky perianth of its genus, but differs from the original nucleus of antecedent *Pithecellobium* ser. *Coriacea* in a tumid calyx (resembling that of *Inga inflata* Ducke) long in proportion to the

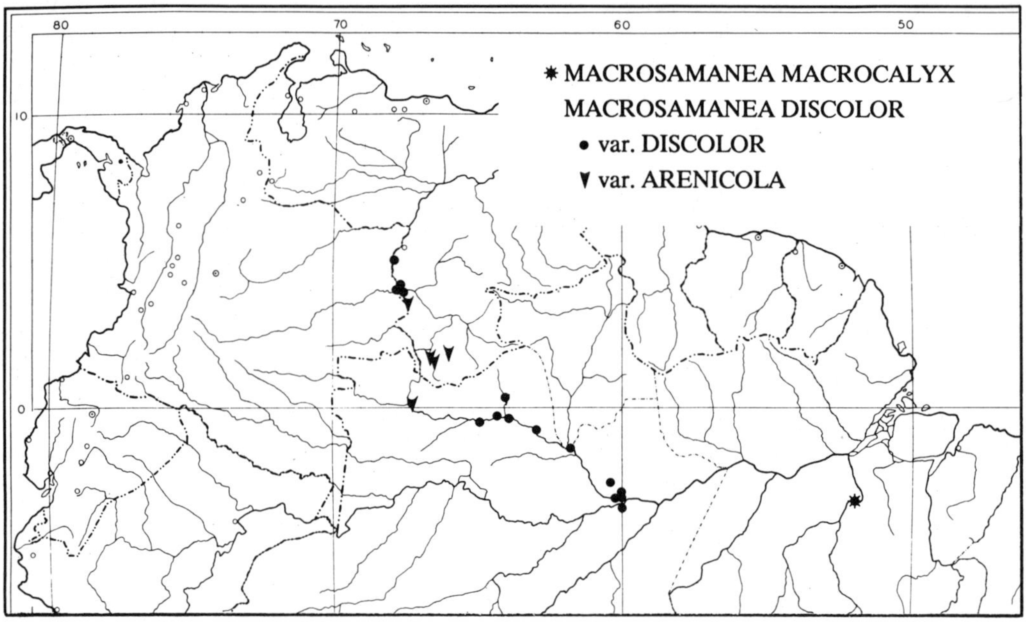

MAP 48. Ríos Orinoco, Negro and Xingú. Range of *Macrosamanea macrocalyx* (Ducke) Barneby & Grimes and of *M. discolor* (Willdenow) Britton & Killip var. *discolor* and var. *arenicola* (Cowan) Barneby & Grimes.

corolla, in lack of nectary on bracts of the capitula, and in having two or three gynoecia (like *Affonsea* Saint Hilaire) to a flower. Of these features only the inflated calyx is unique to *M. macrocalyx*; the naked bracts are shared by *M. kegelii*, and random flowers of *M. consanguinea* and *M. froesii* have twinned pistils. The varzea habitat is consistent with *Macrosamanea*. Ducke, who discovered and named the species, referred it to ser. *Coriacea*, from which it could be excluded only by overemphasizing the significance of the modified calyx.

2. Macrosamanea spruceana (Bentham) Killip *ex* Record, Trop. Woods **63:** 6. 1940. *Pithecolobium spruceanum* Bentham, Trans. Linn. Soc. London **30:** 590. 1875 & *in* Martius, Fl. Bras. **15(2):** 443, t. CXV. 1876. — "... in rupibus ad flumen Rio Negro prope Manaos: Spruce." — Holotypus, *Spruce 1810*, collected in IX.1851 (fl.), K(herb. bentham.)! = NY Neg. *2020*; isotypi, NY!, OXF!, P!. — *Feuilleea spruceana* (Bentham) O. Kuntze, Revis. Gen. Pl. **1:** 1891. 1891.

Amply leafy sarmentose *shrubs* and slender *treelets* flowering when 2–8 or even 16 m tall, the hornotinous branches subterete, smooth or verruculose, like all lf-axes and whole inflorescence densely shortly sordid-pilosulous, the foliage drying brown-olivaceous, the lfts sublustrous above, a trifle paler beneath, the umbelliform capitula of long, ascending fls either axillary to and much surpassed by upper lvs or forming a condensed efoliate terminal pseudoraceme immersed in foliage. *Stipules* firm but early caducous, erect, linear-lanceolate 2.5–4 mm. *Lf-formula* vii–ix/(13–)15–21; lf-stks 13–31 cm, terete or obscurely shallowly grooved ventrally, the petiole proper (including stout pulvinus) on Río Solimoes mostly 8–20 mm but elsewhere to 3–5 cm, the longer interpinnal segments 2–5.5 cm; a vertically dilated elliptic, scutiform nectary 3–5 mm in long diameter at base of petiole, usually contiguous to and never far distant from pulvinus, and similar but smaller ones between many, especially more distal pinnae and at or near their tip; pinnae proximally decrescent, the rachis of longer distal ones 8–15 cm, the first pair of pinnae distant (0–)1–5.5 mm from pulvinule (no paraphyllidia), the interfoliolar segments as long as or a little longer than the width of the lfts, the rachis narrowly marginate and grooved ventrally; pulvinules of lfts transversely elliptic 0.4–0.7 × 1–1.5 mm, horizontally striate-rugulose; lft-blades, except for smaller proximal pairs and the notably longer and narrower distal pair, subequiform, in outline rhombic around a diagonal midrib, the anterior basal side lying parallel to the pinna-rachis, the base bluntly short-auriculate on posterior side, the apex sharply triangular and turned forward through nearly 90°, the larger blades 14–24 × 4.5–9 mm, 2.6–3.1 times as long as wide, the blades of distal pair obliquely elliptic acuminate, nearly straight on anterior side, obtusangulate on posterior one, all blades appearing glabrous but often microscopically ciliolate and sometimes remotely minutely puberulent; venation of 2 primary nerves from pulvinule, the gently sigmoid midrib giving rise on each side to 5–9 widely ascending secondary venules and a weaker, narrowly intramarginal posterior nerve, tertiary venulation weak and uncertain, all veins immersed above, prominulous beneath. *Peduncles* solitary and 2–3 together, stout, 1–6 cm, the lower axillary ones longer than those of the terminal panicle, the subglobose receptacle ±4 mm diam; bracts dimorphic, the lowest 2–4 persistent, deltate 2–3.5 mm, charged on inner face with a depressed horny nectary about as wide itself, the upper ones caducous, oblanceolate; fls ascending, ±10–15 per capitulum; pedicels at anthesis stout, 1.5–4 × 1.2–1.7 mm; calyx tubular or a little expanded distally, (14–)15–20(–26) mm, at widest ±4–6 m, the narrowly ovate or lanceolate (normally 5) teeth 4–8 mm; corolla narrowly trumpet-shaped 36–63 mm, externally densely pallid-sordid-puberulent overall, the narrowly lanceolate lobes 6–11.5 mm; androecium ±70–85-merous, 5.5–9 cm long, the tube ±3.5–6 cm, either much or not exserted, toward base lightly adherent to the corolla, the white filaments free for ±2–3 cm; style well exserted beyond longest filaments, at apex either linear or dilated, the stigma 0.2–0.5 mm diam; ovules to 26. *Pods* subsessile, in profile broad-linear almost straight or a trifle decurved, attaining 17–31 × 4.2–7 cm, plano-compressed except where dilated by mature seeds, framed by thickened sutures, the leathery brown valves when young densely puberulent overall with short brownish or pallid and sometimes with minute translucent red trichomes, glabrate and in age prominently cross-reticulate, inertly dehiscent primarily through the seminiferous suture; seeds transverse at middle of pod and imbricate lengthwise, ellipsoid strongly compressed, in broad view 24–33 × 17–25 mm, the testa thinly papery castaneous fragile, with strong peripheral nerve; cotyledons horny fuscous.

In seasonally flooded forest, where occurring either as a sarmentose shrub of the understory or as a free-standing tree in open places, of apparently bicentric dispersal: at ±120–250 m along the Solimões river and its affluents (Apaporís, Japurá) upstream from the mouth of Río Negro in Brazil to Iquitos in Peru and far SE Colombia, twice recorded from the lower Amazon near Santarém and Gurupá; and disjunct below 150 m in and around the Orinoco delta in T. F. Delta-Amacuro, Venezuela. — Map 49. — Fl. (IV) VII–XII. — *Pashaquilla, yaku-pashuca* (in Maynas, Peru).

Ample multifoliolate leaves with sharply acute sigmoid leaflets coinciding with extremely long narrow flowers and broad pod make *M. spruceana* virtually unmistakable at all seasons. It is the commonest and most often collected *Macrosamanea* of western Amazonia, from where it may likely have been brought down to Pará by floodwater. Its remotely disjunct occurrence in the Orinoco delta is less easily explained. These Venezuelan populations are represented in herbaria by indifferent material, the identity of which needs confirmation.

3. Macrosamanea consanguinea (Cowan) Barneby & Grimes, comb. nov. *Pithecellobium consanguineum* Cowan, Mem. New York Bot. Gard. **10:** 68, fig. 47. 1961. — "VENEZUELA: Amazonas: Río Guainía, alt. 120–140 m, clearing at Limoncito, left back of Cano San Miguel 15 km above mouth; 14 Oct 1957, *Bassett Maguire, John J. Wurdack & William M. Keith 41878*." — Holotypus, US *2267420*!; isotypi, GH!, NY!, VEN!.

Macrophyllidious *trees* 4–7 m, the young branchlets and lf-axes minutely sordid-pubescent, the ample lvs bicolored, the firm glabrous lfts when dry dark-brown-olivaceous above, pale dull beneath, the long narrow, densely strigulose fls borne in short terminal or sometimes lateral (? cauliflorous) efoliate pseudoracemes of few-fld capitula surpassed by distal lvs. *Stipules* firm, either triangular or lanceolate 2–3 mm, either persistent or at length deciduous. *Lf-formula* iii–vii/8–14; lf-stks 8–33 cm, the petiole 2.5–4 cm, at middle 2–4 mm diam, the longer interpinnal segments 2.5–5 cm; nectary near base of petiole sessile patelliform, elliptic 2.5–5 mm in long diam, coarsely callous-marginate, smaller subcircular nectaries at insertion of several distal pinnae and yet smaller ones on rachis between several distal lft-pairs; pinnae accrescent distally, the rachis of penultimate pair 10–18 cm, the longer interfoliolar segments 7–17 mm; lft-pulvinules in dorsal view transversely elliptic 0.5–0.8 × 1.3–2 mm, cross-wrinkled; lfts decrescent toward base of rachis and the terminal pair at once longest of all and semi-ovate or -elliptic, the rest subequiform,

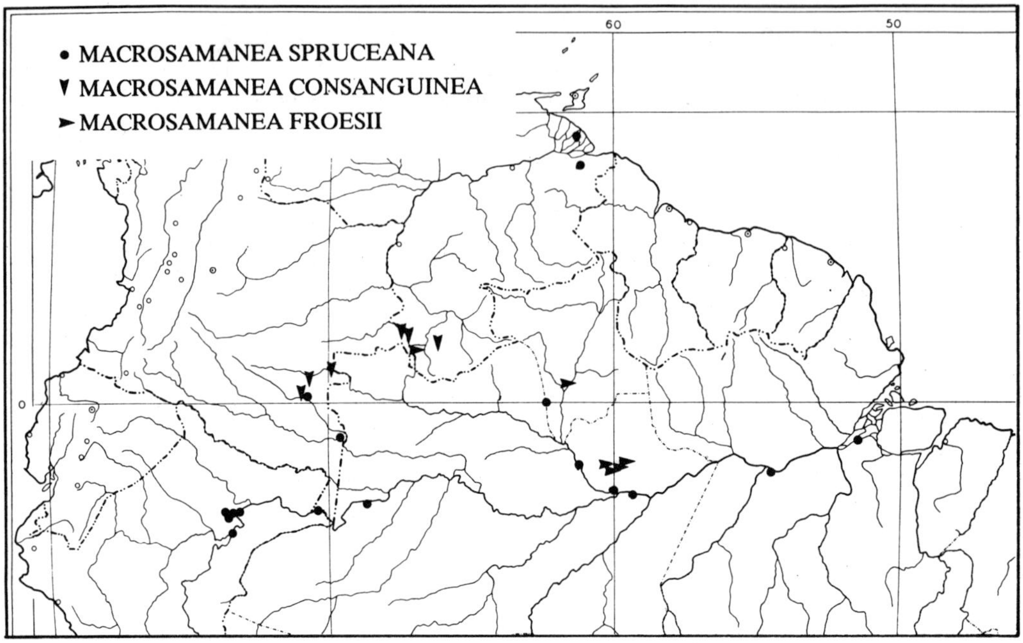

MAP 49. Northern South America. Range of *Macrosamanea spruceana* (Bentham) Killip, *M. consanguinea* (Cowan) Barneby & Grimes, and *M. froesii* Barneby & Grimes.

oblong-trapezoid around a diagonal midrib, obtusangulate except at the porrect, sharply triangular-acute apex, those at and above midrachis 20–40 × 9–18 mm, 2.1–2.7 times as long as wide; venation of 2(–3) primary nerves from pulvinule, the shallowly sigmoid, faintly pinnate midrib and a weaker posterior nerve produced well above midblade, the secondary venules brochidodrome well within the plane margin, a faint tertiary venulation sometimes externally visible, the whole venation immersed or almost so on upper side, feebly prominulous beneath. *Peduncles* solitary or geminate, stout, 1.5–4 cm, some subtended by a rudimentary nectariferous lf-stk; capitula 3–15-fld, the pitted subglobose receptacle 1.8–4.5 mm diam; bracts dimorphic, one or two outer ones deltate deflexed and charged ventrally with low-convex horny fuscous nectary, persistent into fruit, the rest oblanceolate 1.5–2.5 mm caducous; pedicels 0–1.5 mm, to 1.5 mm diam; perianth 5-merous; calyx firm, brown when dry and only obscurely nerved, deeply campanulate 8–12 × 3–4.5 mm, minutely sordid-puberulent overall, the teeth unequal, the sinuses between them 1–3.5 mm deep; corolla tubular, slightly dilated upward, (21–)24–30 mm, when fresh pale yellowish green, densely strigulose externally, the almost erect, lanceolate or nearly ovate lobes 4–6 mm; androecium white, 160–214-merous, 48–55 mm, the stemonozone 3–4 mm, the tube 22–28 mm; style to 2–3 mm longer than longer stamens, the stigma not or scarcely dilated 0.2–0.3 mm diam. *Pods* (few seen) broad-linear nearly straight, 18–20 × 3–3.4 cm, contracted at base into a stout stipe 7–9 mm, abruptly obtuse but mucronulate at very apex, ±18-seeded, the dull-brown coriaceous valves framed by sutures ±2 mm wide, weakly venulose, plane except where low-colliculate over maturing ovules, not seen fully ripe; the dehiscence and seeds unknown.

In igapó woodland below 250 m, known at present from upper Río Vaupés in long. 70°00′–70°45′W, Com. Vaupés, Colombia; from a creek affluent to Río Guainía near Limoncito (±2°40′N, 67° 30′W) near San Carlos, and at mouth of Río Casiquiare in SW T. F. Amazonas, Venezuela; and from Río Preto, affluent of Río Negro at ±64°W in Amazonas, Brazil. — Map 49. — Fl. VI–IX(–?). — *Arepillo, guamo del río.*

Macrosamanea consanguinea was described from specimens in bud, and its status, other than a close relationship to *M. spruceana*, has remained questionable. Four collections from Mitú, on the upper Vaupés in Colombia, in foliage closely comparable with the type but bearing mature flowers and almost ripe fruits, confirm the tentative diagnosis by Cowan. The species now appears to differ from *M. spruceana* in proportionately wider leaflets, materially shorter calyx and corolla, and an androecium of twice as many filaments. Fruits of *Macrosamanea* are variable in both length and width, and distinctions at this level are precarious; nevertheless one example of fruiting *M. consanguinea* (*Zarucchi 2188*, MO, US) has a pod much narrower than anything yet seen in *M. spruceana*, and this may provide a further diagnostic character. The difference in color of the flowers' indumentum mentioned in the protologue is seemingly due to immaturity rather than to any inherent quality of the hairs.

In one flower bud of the type collection there are three ovaries and in a flower of *Aymard 9078* (NY) there are two.

4. Macrosamanea froesii Barneby & Grimes, sp. nov., habitu toto et praesertim foliolis sigmoideo-rhombicis apice porrecto acutissimis *M. spruceanae* et *M. consanguineae* manifeste affinis, calyce corollaque mediocribus (illo 8–11 mm, hac 22–30(–31) mm longis) propior, sed pinnarum majorum foliolis (12–)14–18 (nec 7–12)-jugis et 12–21 × 3.5–9 (nec 22–30 × 9–12) mm usque diversa. — Brazil. Amazonas: margem inundada do Río Urubú, a montante do igarapé dos Patos [±2°04′S, 59°50′W], 13.XII.1956, *W. Rodrigues 350.* — Holotypus, INPA *4460*; isotypi, MG *23,540*, US *2368346* = NY Neg. 12703.

Slender, amply multifoliolate sarmentose *shrubs* of unrecorded stature, with smooth gray annontinous branches, the young stems, lf-axes and inflorescence densely pallid- or brown-puberulent, the bicolored lfts facially glabrous, minutely ciliolate, when dry dark brown above, paler beneath, the inflorescence a short pseudoterminal efoliate pseudoraceme of few-fld capitula immersed in foliage. *Stipules* narrowly lanceolate 2–4 mm, early deciduous. *Lf-formula* iv–v(–vi)/(12–)14–18; lf-stks 6–10.5 cm, the petiole including the swollen but not wrinkled pulvinus 12–20 mm, at middle 1.3–1.6 mm diam, the longer interpinnal segments 14–20 mm; a sessile elliptic, plane thick-marginate nectary 1.7–3 mm in long diameter borne near midpetiole, smaller rounder ones between 2–3 distal pinna-pairs, and yet smaller ones between few distal pairs of lfts; pinnae decrescent proximally but the rest subequilong, the rachis of distal pairs 6–9 cm, the longer interfoliolar segments 3.5–4.5 mm; pulvinules transversely elliptic ±0.5 × 1–1.3 mm, nigrescent; lfts sessile against rachis, decrescent at very base of pinna but otherwise (except for longer and obliquely lanceolate furthest pair, no further described) subequiform, rhombic-oblong around the diagonal, shallowly sigmoid costa, at apex sharply acute and porrect, those at midrachis 12–21 × 3.5–9 mm; venation of 2 nerves from pulvinule, a weak posterior primary one produced more or less to mid-blade and a pinnately branched midrib, the 5–7 pairs of secondary nerves faintly camptodrome within the plane margin, the whole venation immersed on ventral face of lft, only weakly finely prominulous beneath. Primary *axis of pseudoraceme* ±2–4 cm; peduncles 2–5 at most nodes, 6–11 mm; capitula 2–6-fld, the axis <2 mm; bracts deltate ±1 mm, the first charged ventrally with a horny nectary; fls sessile, the perianth 5-merous, externally brown-puberulent; calyx campanulate 8–11 × 3.5–4.5 mm, the triangular teeth to 1.5–3.5 mm; corolla 22–30(–31) mm; androecium ±112-merous and 4–4.5 cm, the stemonozone nearly 2 mm, the tube a trifle longer than the corolla; ovaries sometimes 2; the style ±5 cm, the stigma punctiform. *Pods* (*Damião 2640*, INPA) broad-linear-oblanceolate, plano-compressed, in profile ±30 × 6.5 cm (reminiscent of *Campsiandra*), abruptly contracted at each end, the stiffly coriaceous valves puberulent along the sutures, facially glabrous, transversely venulose; dehiscence and seeds unknown.

On river banks in igapó woodland below 120 m, apparently local in NE Amazonas and adjacent Roraima, Brazil; one record from San Carlos do Río Negro in T. F. Amazonas, Venezuela. — Map 49. — Fl. X–XII. — *Sucurijuzinho* (employed as nerve tonic in baths).

Macrosamanea froesii, first collected by R. Lemos Fróes at igarapé Jatuarana on lower Rio Urubú in 1949, has been confused with *Pithecellobium lindsaeifolum* (*M. pubiramea* var. of this revision) and *M. spruceana.* It has the porrectly acute leaflets of the latter, but calices only half as long and depauperate capitula of only two to six flowers; the capitula and leaflet-shape are incompatible with *M. pubiramea.* The closest relative is *M. consanguinea*, allopatric in southwestern Venezuela and adjoining Colombia, which has the porrect leaflet-tip of *M. froesii* and can have on occasion comparably few-flowered capitula, but is different in fewer and wider leaflets, as described in the foregoing diagnosis. As material accumulates, these two species may prove collectively to constitute a small-flowered variety or subspecies of *M. spruceana.*

One flower from the type specimen was found on dissection to contain two perfect ovaries, but we suppose this to be an anomaly.

5. Macrosamanea discolor (Willdenow) Britton & Rose *ex* Britton & Killip, Ann. New York Acad. Sci. **35:** 131. 1936. — Typus infra sub var. *discolori* indicatur.

Awkwardly branched *arborescent shrubs* mostly 2–5 m but sometimes (in savanna) flowering as erect shrubs 0.3–1.5 m, the older trunks and branches gray, lenticellate and glabrous, the young branchlets and all axes of lvs and inflorescence sordidly pilosulous with short spreading-ascending hairs, the stiffly papery or in age coriaceous lfts usually glabrous or almost so but rarely microhirtellous on both faces, the capitula of long whitish fls borne, either singly or 2–3 together, on long peduncles in the axil of coeval lvs or more shortly pedunculate in small functionally terminal and subterminal pseudoracemes. *Stipules* erect firm, narrowly lanceolate, 1.5–4(–5.5) mm, caducous. *Lf-formula* i or i–ii/14–22(–29); lf-stks of lvs with 1 pair of pinnae 5–16(–17) mm, of those with a second pair 2–5 cm, the interpinnal segment then 1.5–3.5 cm; a depressed elliptic, grossly corneous-margined nectary 1.6–4 mm in long diameter inserted immediately below first (or only) pair of pinnae, and a smaller suborbicular one below second pair (when present), further, much smaller ones between distal or between all but a few proximal pairs of lfts; rachis of the one pair or of the longer, distal pair of pinnae 7–15.5 cm, the longer interfoliolar segments 3–7.5 mm; paraphyllidia either linear-subulate 0.3–1 mm, or reduced to minute conic points, or absent, the posterior one exceptionally represented by a small lft; lft-pulvinules transversely elliptic 0.5–1 × 0.8–1.7 mm; lfts subequilong except for a few smaller proximal pairs and the longer narrower distal pair, the blades lanceolate from asymmetrical base where dilated and obtuse on anterior side and cuneate on posterior side, the longest ones (15–)18–31 × (3.8–)4–6(7) mm, (3.9–)4–5.5(–5.7) times as long as wide, all obtuse but mucronulate at apex, prominently 2-nerved from pulvinule, the nearly straight or outwardly curved (but not sigmoid) midrib either subcentric or forwardly displaced to divide blade ±1:2, pinnately branched and the secondary nerves brochidodrome well within the plane margin, the posterior basal nerve produced to or beyond midblade, the tertiary connecting venules few and weak, the whole venation prominulous either on both faces or only beneath. *Peduncles* 4–45(–55) cm, charged immediately below the first fls with 1–3 deltate bracts bearing on inner face a nectary like those of lf-stks; capitula 3–12-fld, the globose or shortly clavate receptacle 2–5 mm diam, the ascending fls either subsessile or elevated on stout pedicel up to ±1 mm; floral bracts caducous (few seen), narrowly elliptic or linear-oblanceolate 2.5–5.5 mm; calyx either cylindric or tumid and cylindro-campanulate (6.5–)7.5–15 × (3–)3.5–5 mm, densely minutely sordid-puberulent externally, many-nerved but the nerves not very prominent, the ovate or lanceolate teeth quite variable even at times in one calyx, 2.5–6 mm; corolla narrowly trumpet-shaped, pubescent like the calyx but in early anthesis a little more densely so (thus appearing paler), (15)17–32 mm, the lanceolate, in age subinvolute lobes (3)4–8 mm; androecium 30–53 mm, the stamen-tube a little shorter or a little longer than the corolla, the exserted filaments white or white tinged with pink distally; style up to 1 cm longer than longest filaments, a trifle dilated at apex; ovules ±14–20, often many aborting. *Pods* solitary or up to 3–4 per capitulum, subsessile, in profile narrowly oblong or broad-linear, straight or gently recurved, extremely variable in absolute width and in width relative to length, (6.5)7.5–21 × 1.5–3.2 cm, (2.6)3–14 times as long as wide, plano-compressed, the coriaceous brown valves plane except where low-umbonate over maturing seeds, framed by thickened sutures, reticulate-venulose and minutely sordid-puberulent overall; dehiscence and ripe seeds unknown.

Macrosamanea discolor is closely related to *M. pubiramea*, of which it has nearly the individual flower, but is distinguished, at least in typical form, by reduction of pinnae to one or two pairs per leaf, and by depauperate capitula of at most a dozen flowers. In width proportionate to length, its leaflets resemble those of *M. pubiramea* var. *lindsaeifolia*, with which it is rather widely sympatric and from which it may be derived. With some hesitation we recognize two varieties of *M. discolor*, seemingly differentiated by habitat but not always cleanly separable in the herbarium. The var. *arenicola* appears to occupy an enclave within the range of var. *discolor* and to be found in wet savanna rather than on flooded river banks, but its ecology requires study in the field.

The description of *Macrosamanea aquatica* and *Pithecellobium subaquaticum* from parts of one collection was an accidental duplication. The types are unquestionably conspecific with the original *Inga discolor*, discovered by Humboldt on the Orinoco a little more than 100 km upstream from Sanariapo. Bentham had lost sight of the genuine *M. discolor*, which lay buried in his concept of *Pithecellobium adiantifolium*, and when rediscovered was unidentifiable in Bentham's monograph.

Key to the varieties of *M. discolor*

1. Lfts, even at advanced maturity, venulose on both faces, and their midrib only a trifle forwardly displaced; peduncles all or almost all axillary to fully developed lvs, 2–4.5(–5) cm long; fls relatively large, the calyx (9–)10–15 mm, the corolla 21–32 mm; of bicentric dispersal: on the upper Río Orinoco in Venezuela, and on the middle and lower Río Negro in Brazil, the range interrupted in SW Venezuela by that of var. *arenicola*; apparently confined to seasonally flooded river banks and igapó forest....................................5a. var. *discolor*
1. Lfts venulose beneath, but smooth and veinless above, their midrib centric or forwardly displaced to divide blade ±1:2; peduncles all or almost all disposed in short terminal or subterminal pseudoracemes and only 4–15 mm long; fls on the average smaller, the calyx (6.5–)7–13 mm, the corolla 15–24 mm; local on both sides of the Orinoco-Negro divide in SW T. F. Amazonas, Venezuela, and in Amazonian caatinga near S. Gabriel do Cachoeira on upper Río Negro in Amazonas, Brazil, apparently confined to white sand savanna and savanna-forest ecotone.......................................5b. var. *arenicola*

5a. Macrosamanea discolor (Willdenow) Britton & Killip var. **discolor.** *Inga discolor* Humboldt & Bonpland *ex* Willdenow, Sp. Pl. **4:** 1023. 1806. — "*. . .in America meridionali ad flumen* Atabapo." — Holotypus, *Humboldt & Bonpland 995*, B-WILLD *19049*, seen in Microform!; isotypi, P (2 sheets)!. — *Mimosa discolor* (Willdenow) Poiret, Encycl., Suppl. **1:** 46. 1810. *Inga adiantifolia* Kunth, Mimoses 66, t. 21. 1820, nom. substitutum illegit. *Pithecellobium adiantifolium* (Kunth) Bentham *in* Hooker, London J. Bot. **3:** 218. 1844. *Feuilleea discolor* (Willdenow) O. Kuntze, Revis. Gen. Pl. **1:** 185. 1891. *Samanea adiantifolia* (Kunth) Pittier, Arb. Arbust. Venez. **5:** 55, in clave. 1925. *Pithecellobium discolor* (Willdenow) Macbride, Publ. Field Mus. Nat. Hist., Bot. Ser. **13(3)** [= Fl. Peru]: 63. 1943.

Macrosamanea aquatica Pittier, Bol. Soc. Venez. Ci. Nat. **11:** 15. 1947. — "[Venezuela.] *Alto Orinoco*: Boca de Sanariapo, 120 m, entre Pto. Ayacucho y San Fernando . . . ([Llewellyn] Williams 15946, tipo; 13243)." — Holotypus, VEN *25073*!; isotypi, F!, NY (2 sheets)!, RB!, paratypus, (no. *13243*), VEN!. — *Pithecellobium aquaticum* (Pittier) Cowan, Mem. New York Bot. Gard. **10:** 67. 1961.

Pithecellobium subaquaticum Schery, Fieldiana, Bot. **28:** 259. 1952. ". . . at mouth of Sanariapo river, Territorio Federal Amazonas . . . *Llewellyn Williams 15946*." — Holotypus = isotypus of *M. aquatica*, F!; isotypi as given under the preceding.

Characters as given in key to varieties.

In seasonally inundated riparian forest and on river banks below high water, below 160 m, of seemingly bicentric range in SW Venezuela, immediately adjacent Vaupés, Colombia, and N Amazonian Brazil: in Venezuela and Colombia on banks of upper Orinoco River from Sanariapo upstream to San Fernando de Atabapo; in Brazil along the Río Negro between Ilha Grande and Manaus (long. 60°–65°W). — Map 48. — Fl. XI–VI. — *Cabibi* (Río Negro); *guamita orillera.*

5b. Macrosamanea discolor var. **arenicola** (Cowan) Barneby & Grimes, comb. et stat. nov. *Pithecellobium arenicola* Cowan, Mem. New York Bot. Gard. **10:** 67, fig. 46. 1961. — "VENEZUELA: Amazonas: . . . on right bank of Río Pacimoni 50 km above mouth, alt. 100–140 m . . . 7 Feb 1954, *Bassett Maguire, John J. Wurdack & George S. Bunting 37568 . . .*" — Holotypus US *2281823, 4*!; isotypus, NY!; paratypus, *Maguire et al. 37264*, NY!, US!.

Characters as given in key to varieties.

In savanna on white sand, below 150 m, vicariant north and southward with var. *discolor* on the Orinoco-Negro divide in SW T. F. Amazonas, Venezuela, on the middle Atabapo, Siapo, and Pacimoni rivers. — Map 48. — Fl. II–V.

6. Macrosamanea duckei (Huber) Barneby & Grimes, comb. nov. *Pithecellobium duckei* Huber, Bol. Mus. Paraense Hist. Nat. **5:** 377. 1909. — "Hab. Lago de Faro [Pará, long. ±56°30′W], ad ripam, 15 VIII 07 leg. A. Ducke (8333)." — Holotypus, MG! = F *Neg. 8333* + fragm.!; isotypi, BM!, G!, US *1442408*!.

Pithecellobium scandens Ducke, Arch. Jard. Bot. Rio de Janeiro **4:** 29. 1925. — ". . . in ripis inundatis fluminis Tapajoz medii infra locum Quataquara [Coatáquará, ±5°S], 12–8–1923 (H[erb.] J[ard.]. B[ot.] R[io de Janeiro] n. 16.775); prope ipsius fluminis cursum inferiorem circa Itaituba [±4°10′S] in paludibus, 20–10–1922 (H. J. B. R. n. 16.774); 1. A. Ducke." — Lectotypus, RB *16775*!, isotypi, K!, U!, US *1442283*!, paratypi, (*16774*), RB!, U!.
Pithecellobium duckei sensu Ducke, 1949: 40.
Pithecellobium scandens sensu Ducke, 1949: 40.

Amply coarsely leafy, potentially arborescent or sarmentose *shrubs* (2–)2.5–6(–8) m with terete, at first brown but finally grayish and densely lenticellate branchlets, the young stems with all lf-axes and peduncles sordidly pilosulous with erect and forwardly incurved hairs <0.3 mm, the broad, stiffly chartaceous lfts scarcely or moderately bicolored, lustrous above, duller beneath, glabrous or only microscopically ciliolate, the inflorescence composed of pedunculate capitula either solitary or 2–3 together in coeval lf-axils, or pseudoterminally pseudoracemose. *Stipules* erect appressed, firmly herbaceous, lanceolate or ovate-triangular 1–4 × 0.6–2 mm, deciduous. *Lf-formula* i–iv/6–11(–13); lf-stks (0.6–)1–12(–18) cm, the true petiole 0.6–2.5 cm, the longer interpinnal segments (when present) 2.5–6(–7) cm, the first one always much longer than the true petiole; first petiolar nectary sessile, elliptic, scutellate or patelliform, 2.5–4 mm in long diameter, variably situated on lf-stk either immediately to shortly below first pair of pinnae, or contiguous to lf-pulvinus, in conjugately pinnate lvs often extending from pulvinus to the pinna-pulvini, smaller, suborbicular but otherwise similar nectaries between distal pairs of pinnae and lfts; rachis of longer pinnae 9–18(–21) cm, the longer interfoliolar segments (8–)10–22 mm; paraphyllidia at base of pinna-rachis either 1–2, subulate and <1 mm, or reduced to minute papillae, or obsolete; pulvinules transversely elliptic cross-wrinkled 0.6–1.5 × 1.7–2.5 mm; lfts except for always longer and proportionately narrower, more oblique terminal pair subequiform but gently decrescent proximally, the blade rhombic around a diagonal, straight or very slightly sigmoid midrib, at base asymmetrically cuneate, at apex either obtuse or emarginate but in either case sharply minutely apiculate, the larger ones (24–)26–50 × 11–23 mm, 2.1–2.5 times as long as wide; primary venation of 2 major veins from pulvinule, the costa and one narrowly intramarginal posterior nerve produced beyond midblade, the costa pinnately branched and the ascending secondary nerves brochidodrome well within the recurved margin and giving rise to random connecting tertiary venules, the whole venation finely prominulous on both faces, more sharply so beneath. *Peduncles* 1.5–6.5(–7.5) cm, bearing at apex 2–several reflexed bracteoles charged on adaxial face with a depressed horny nectary; capitula densely 14–30-fld, prior to anthesis globose and the fls at anthesis radiating from a globose or ellipsoid, alveolate receptacle 2.5–5 mm diam; floral bracts narrowly elliptic or linear-spatulate 2–4 mm, caducous; calyx sessile, sordid-puberulent overall, either cylindric or cylindro-campanulate and dilated in distal ⅓, ±4.5–7.5(–8) mm long and at middle (1–)1.2–1.8(–2) mm diam, weakly 10-nerved, the 5 ovate-deltate, subinvolute lobes 1–2 mm; corolla slenderly trumpet-shaped, pallidly sordid-silky overall, 11–16(–17) mm, the lance-ovate teeth 2–3.5 mm; androecium 45–50-merous, 2.5–4 cm, white, the filament tube as long as or to 3 mm longer than the corolla; ovary at anthesis glabrous, subsessile; style longer than filaments, only a trifle dilated at the stigma. *Pods* (few seen, and only one of them fully ripe) subsessile, in profile oblong, straight or slightly decurved, 5–11.5 × 2.8–4 cm, broadly rounded at each end, abruptly apiculate, the plane leathery brown valves framed by thickened sutures, closely transverse-venulose, densely minutely sordid-puberulent overall; dehiscence through the ventral suture, the valves scarcely opening to shed the seeds; seeds compressed-ellipsoid ±24 × 16 × 4 mm, the thin-papery testa becoming wrinkled, castaneous.

On river banks and lake shores and in seasonally flooded (varzea) or permanently inundated (igapó) forest below 200 m, frequent along rivers of centr. Amazonian Brazil: along the Amazon itself and immediate tributaries between the mouth of Río Purús and that of Río Trombetas in Amazonas and adjacent Pará, thence S along the Purús, Madeira, and Tapajoz, on the Madeira reaching 10°S in state of Rondônia. — Map 50. — Fl. IV–X.

Macrosamanea duckei is in foliage somewhat intermediate between *M. amplissima* and *M. pubiramea* var. *lindsaeifolia*, having relatively broad rhombic leaflets that recall *M. amplissima* but a leaf-formula more nearly compatible with that of var. *lindsaeifolia*. It differs decisively from both in the small flowers, and further from *M. amplissima* in lack of pedicels. We have found no reliable specific differences between typical *M. duckei* and *Pithecellobium scandens*, which Ducke himself (1949: 40) admitted were similar. When first described, they appeared to differ in leaf-formula, in *P. duckei* i–ii/6–9, in *P. scandens* ii–iv/7–13, but the small gap in numbers is now spanned by intermediates. Following the pattern prevalent in the Mimosaceae, the size of leaflets increases as the number delines, and vice versa. The collective range of the two forms is a very natural one.

7. Macrosamanea pubiramea (Steudel) Barneby & Grimes, comb. nov. *Inga pubiramea* Steudel, Flora **26:** 759. 1843. — Typus infra sub var. *pubiramea* indicatur.

Amply foliate, potentially arborescent and when crowded sarmentose or vinelike *shrubs* (1.5–)2–6(–10) m tall, the old branches and trunks gray glabrate, the hornotinous ones together with axes of lvs and inflorescence sordidly pilosulous with short, incurved-ascending hairs, the brown-olivaceous lfts glabrous or microscopically ciliolate along midrib above, when mature stiffly chartaceous, the low-convex upper face smooth lustrous, veinless or almost so, the lower face a little paler, finely venulose, the

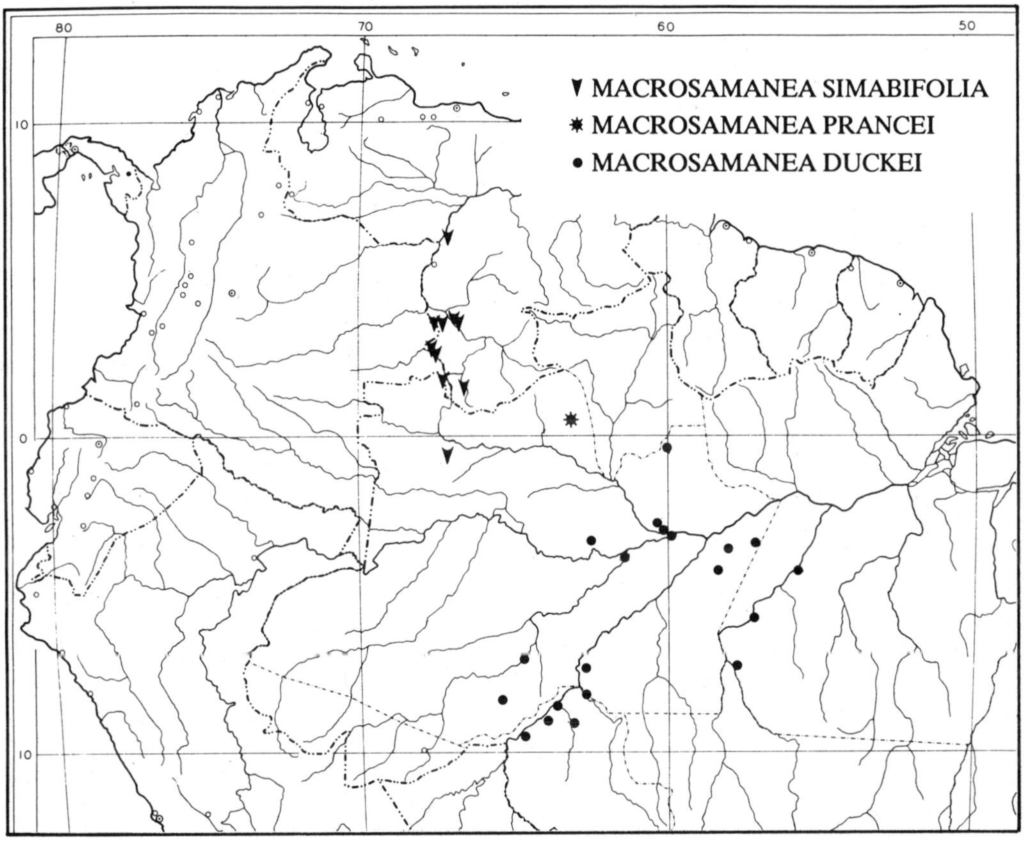

MAP 50. Northern South America. Range of *Macrosamanea simabifolia* (Bentham) Pittier, *M. prancei* (Barneby) Barneby & Grimes, and *M. duckei* (Huber) Barneby & Grimes.

well-furnished capitula of ascending, long and narrow, whitish or pinkish fls either axillary to coeval lvs or forming a very short terminal pseudoraceme immersed in foliage. *Stipules* erect-appressed, firm, lance-triangular 0.8–2.8 mm, at most faintly nerved dorsally, caducous. *Lf-formula* (ii–)iii–ix(–xiv)/(9–)10–26(–31); lf-stks including coarsely wrinkled pulvinus (3–)5–21 cm, the true petiole commonly no longer than the first nectary, exceptionally to 6–15 mm, but always shorter than the interpinnal segments, these to 12–45(–55) mm; a sessile elliptic, shallowly concave, grossly callous-marginate nectary (2–)2.5–5 mm in long diameter situated immediately below first pair of pinnae, often impinging on or contiguous to lf-pulvinus, rarely distant from it, and similar but progressively smaller, mostly suborbicular ones between some distal pairs of pinnae and lfts; pinnae strongly decrescent proximally, less or not so distally, the short first pair ordinarily reflexed, the rachis of longer pairs (6–)7–15(–18) cm, their longer interfoliolar segments (2.5–)3–12(–17) mm; lfts except for few, gently decrescent proximal pairs and for the longer narrower distal pair all subequilong, the first pair either contiguous to the pinna-pulvinus or represented by minute subulate paraphyllidia, the blades linear-lanceolate to oblong or rhombic-oblong, mostly obtuse or obtuse and mucronulate, deltately acute when relatively narrow, at base cuneate on proximal and obtusangulate on distal side, the largest (10–)12–30(–38) × (2–)2.5–17 mm, 2.1–5.2(–6) times as long as wide, the midrib of proportionately narrow lfts subcentric, straight or obscurely sigmoid, that of proportionately broad ones diagonal, one weak (and sometimes an outermost very short), narrowly intramarginal, posterior primary nerve produced to or beyond midblade, the costa pinnate, the secondary veins giving rise to weak sinuous tertiary venules, all venation usually immersed on upper face, always prominulous on the lower. *Peduncles* either solitary or 2–3 together, (1–)1.5–5.5(–9) cm, charged immediately below the capitulum with 1–2(–3) cycles of reflexed deltate bracts, each of these bearing on its adaxial (exposed) face a sessile, reddish or brownish, plane or crumpled, corneously margined nectary as wide as itself or wider; capitula (12–)15–25(–40)-fld, the ellipsoid or globose, alveolate receptacle 3–6.5 mm in greatest diameter; floral bracts fugacious, dry and brown before anthesis or possibly sometimes lacking, when present linear or linear-spatulate 0.7–2.5 mm; calyx and corolla faintly many-nerved externally, sordid- or pallid-puberulent overall, the corolla often a little more densely so, but sometimes both glabrescent in age; calyx either sessile or contracted at the often slightly bulbous base into a stout pedicel to 1.5 mm, overall 6.5–12 mm long, in lower ⅔ cylindric but expanded distally, the tube at middle 1.2–2(–2.4) mm diam, the ovate-apiculate or lanceolate teeth (1–)1.3–2.3 mm; corolla (16–)18–32(–34) mm, the slender tube very gradually dilated upward, the lance-ovate teeth 2–5(–6) mm; androecium ±80–106-merous (some interior filaments sterile), the staminal tube (at maturity of fl) usually exserted 2–10(–13) mm from corolla, but sometimes no longer or even a trifle shorter than it, the filaments white throughout or less often distally tinged with pink or carmine; ovary at anthesis glabrous; style exserted from the tassel of stamens, at tip either poriform or a little dilated and funnel-shaped, 0.2–0.3 mm diam; ovules 18–24. *Pods* 1–4 per capitulum, sessile, in profile oblong or broad-linear, usually a little decurved, (6–)8–14 × (2.2–)2.5–5.6 cm, broadly rounded at each end, apiculate, plano-compressed but early umbonate and finally tumescent over developing seeds, the brown, stiffly chartaceous valves framed by thickened sutures, thinly puberulent overall, transversely venulose; dehiscence through the ventral suture, the valves narrowly gaping to release seeds; seeds transverse on filiform funicle, imbricate by their margins and there becoming deformed by mutual pressure, plumply discoid and in broad profile elliptic-obovate 16–19 × 10–14 mm, the thinly papery, lustrously castaneous, fragile testa rugulose at middle of each face, carinate or incipiently winged around the periphery; cotyledons horny, cordate at base, the basal sinus either closed over or exposing the tip of the radicle.

Macrosamanea pubiramea is the most widely dispersed species of the genus and also the most variable in number of pinnae and in number and size of leaflets. As defined by the foregoing description it includes Bentham's *Pithecolobium adiantifolium* var. *multipinnum*, *P. longiflorum*, and *P. lindsaeifolium*; modern collectors have not confirmed the habital differences that Bentham (1875: 590) thought might distinguish *P. adiantifolium* from the rest. Over much of the range of the species (Map 51) there is no difficulty in separating a var. *pubiramea*, with relatively narrow leaflets, from a var. *lindsaeifolia*, with broader ones. The var. *pubiramea* is the only form encountered in the Guianas or in Amazonian Brazil downstream from Manaus, whereas var. *lindsaeifolia* almost completely replaces it northwestward from the mouth of the Río Negro. We have from the banks of the lower Río Negro, however, examples of both varieties, and also of the related *M. discolor*, and some ambiguous forms that may represent introgression among the three. The riparian habitat is the same for all of them, so the opportunity for hybridization may easily arise. A few random records of one variety deep within the range of the other are perhaps based on individual variation too literally interpreted.

The name *P. adiantifolium* is illegitimate (see under *M. discolor*), and *P. longiflorum* is junior by three years to *Inga pubiramea*, which provides the unfamiliar, but unavoidable epithet in *Macrosamanea*.

Key to the varieties of *M. pubiramea*

1. Larger lfts of any plant either narrowly lanceolate or narrowly rhombic-oblong and (2–)2.5–6 mm wide, mostly 3–5(–6) times as long as wide; longer interfoliolar segments of pinna-rachis likewise 3–6 mm long; Guyana to lower Amazon valley in SW T. F. Amazonas (Venezuela), E Amazonas and W Pará, Brazil, thence SW through Sa do Cachimbo to middle Madeira valley in S state of Amazonas, in and near Manaus (Brazil), intergradient with the next..7a. var. *pubiramea*
1. Larger lfts of any plant rhombic-oblong or trapezoidal, 6.5–17 mm wide, 2–3.1(–4) times as long as wide; longer interfoliolar segments of pinna-rachis likewise 6–12(17) mm long; middle and upper Amazon valley in Brazil, and thence NW into Vaupés, Colombia, and N through the Río Negro-Orinoco divide to far W Bolívar and E Apure in Venezuela......................7b.var. *lindsaeifolia*

7a. Macrosamanea pubiramea (Steudel) Barneby & Grimes var. **pubiramea.** *Inga pubiramea* Steudel, 1843, l. c., sens. str. — "Hostmann und Kappler in Innern von Surinam gesammelt haben . . . nr 171." — Holotypus, P(herb. steudel.)!; isotypi, MO (herb. bernhardi.)!, P (2 sheets, one mixed with *Calliandra*)!. — Tentatively interpreted by Bentham,

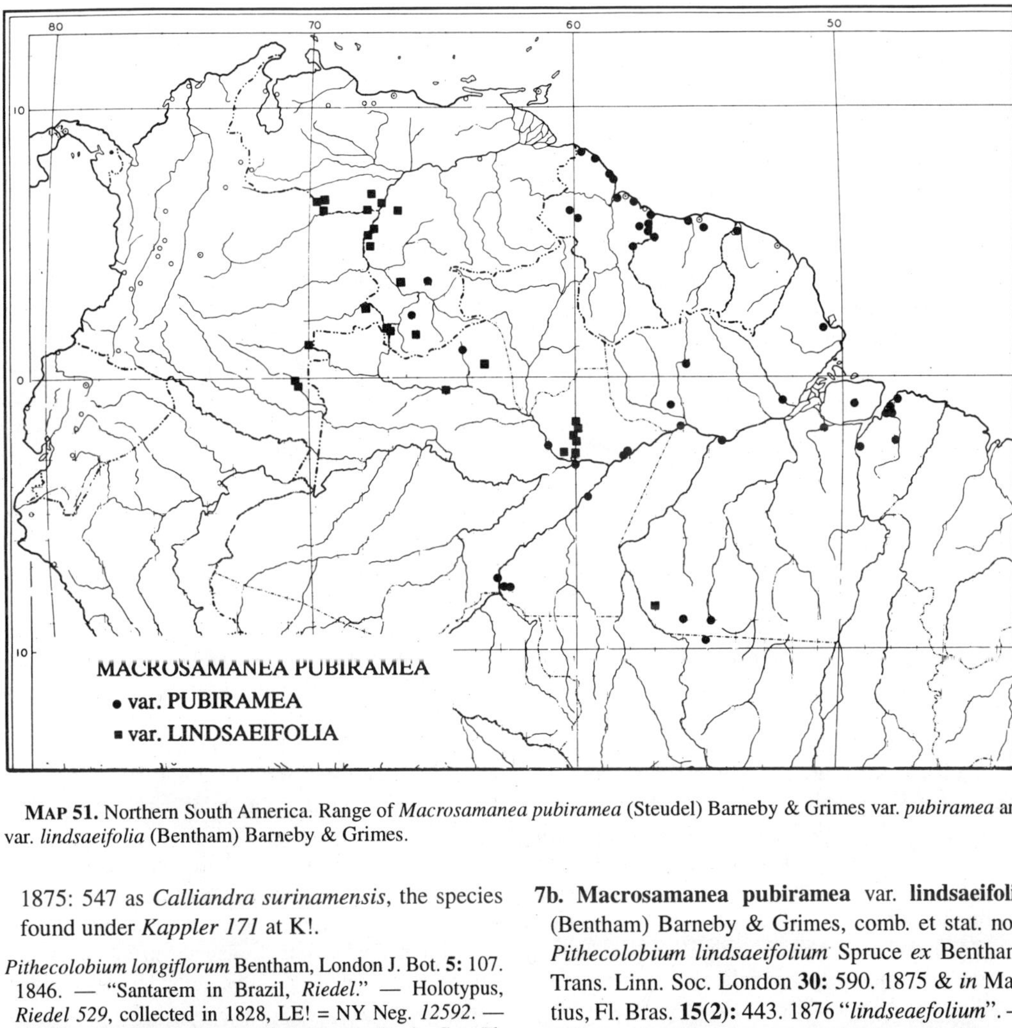

MAP 51. Northern South America. Range of *Macrosamanea pubiramea* (Steudel) Barneby & Grimes var. *pubiramea* and var. *lindsaeifolia* (Bentham) Barneby & Grimes.

1875: 547 as *Calliandra surinamensis*, the species found under *Kappler 171* at K!.

Pithecolobium longiflorum Bentham, London J. Bot. **5:** 107. 1846. — "Santarem in Brazil, *Riedel*." — Holotypus, *Riedel 529*, collected in 1828, LE! = NY Neg. *12592*. — *Feuilleea longiflora* (Bentham) O. Kuntze, Revis. Gen. Pl. **1:** 188. 1891. *Macrosamanea longiflora* (Bentham) Pittier, 3rd Conf. Interam. Agric. Caracas 358. 1945.
Pithecolobium miquelianum Meisner, Linnaea **21:** 250. 1848. — "[Surinam:] In silvis humidis prope Joden-Savanna, m. Nov. 1845 legit Kegel n. 1162." — Holotypus, NY (herb. meisner.)! isotypus, GOET!. — Equated provisionally by Bentham, 1875: 590, with *P. adiantifolum*, and by Kleinhoonte, 1940: 331, with *M. discolor*.
Pithecolobium adiantifolium var. *multipinnum* Bentham *in* Martius, Fl. Bras. **15(2):** 445. 1876. — "In Guiana frequens . . ." — Lectotypus, *Hostmann & Kappler 1277*, K (2 sheets)!; isotypi, BM!, G-DC!, NY!, U!; paratypi, *Schomburgk 387*, BM!, G!, NY!, US!.
Pithecellobium adiantifolium sensu Lindeman & Mennega, Bomenboek Surinam 189. 1963.

Lf-formula (iv–)v–ix(–xiv)/(12–)15–26(–31); interpinnal segments of lf-stk up to 12–26(–31) mm; lfts as given in key to varieties.

Range as given in key. — Map 51.

7b. Macrosamanea pubiramea var. **lindsaeifolia** (Bentham) Barneby & Grimes, comb. et stat. nov.
Pithecolobium lindsaeifolium Spruce *ex* Bentham, Trans. Linn. Soc. London **30:** 590. 1875 & *in* Martius, Fl. Bras. **15(2):** 443. 1876 "*lindseaefolium*". — "Habitat in ripis fluminis Uaupés prope Panuré: Spruce n. 2925, et (specimen unicum) in silvis ripariis ad Manáos prov. do Alto Amazonas: Spruce; ad Ega: Poeppig." — Lectotypus, *Spruce 2925*, K(herb. bentham.) = NY Neg. 2019!; isotypi, K(herb. hooker.)!, W!; paratypi, Spruce *s.n.* in 1855, Barra do Rio Negro, †B = F Neg. *1207*, NY!, *Poeppig s.n. in 1831/2*, NY!, *Poeppig 2899*, GOET!. — *Feuilleea lindseaefolia* (Bentham) O. Kuntze, Revis. Gen. Pl. **1:** 188. 1891.

Lf-formula iii–v(–vi)/(9–)10–17(–20); interpinnal segments of lf-stk (12–)15–45(–55) mm; lfts as given in key to varieties.

Range as given in key. — Map 51. — *Cancha de velho* (Tefé).

A collection (*Domínguez 313*, VEN) from the Guiania valley near 2°30′N, 68°15′W in Colombia has a leaf-formula (iv/6) compatible with *M. duckei* but flowers of var. *lindsaeifolia*, of which it may be a form with few leaflets. It possibly represents a distinct taxon.

8. Macrosamanea amplissima (Ducke) Barneby & Grimes, comb. nov. *Pithecolobium amplissimum* Ducke, Arq. Inst. Biol. Veg. **4:** 6. 1938.– ". . . in silva a Rio Negro periodice inundabile prope Cucuhy (civ. Amazonas [±1°N, 67°W]), 21–9–1935 leg. A. Ducke H[erb]. J[ard]. B[ot]. R[io de Janeiro] 35527." — Holotypus, RB (2 sheets)!; isotypus, US *1740710*!.

Pithecolobium amplissimum sensu Ducke, 1949: 39.

Coarsely macrophyllidious weak *trees* and *bush-ropes* attaining 5–15 m, with obtusely 4-angular woody branches, either glabrous below the inflorescences or the young branchlets and lf-axes minutely thinly puberulent, the ample subcoriaceous lfts bicolored, lustrously olivaceous above, paler (when dry brunnescent) beneath, the dense capituliform racemes of ascending, very long and narrow fls either solitary or paired in the axil of a few distal, coeval or lately fallen lvs, the pedicels, calyces, and corollas all densely minutely furfuraceous-puberulent with sordid or gray hairs. *Stipules* (few seen) thick-textured, triangular 1–1.5 mm, either persistent or deciduous. *Lf-formula* (iv–)v–vii/(9–)10–14; lf-stks 2.5–6 cm, subterete, the petiole including coarsely wrinkled pulvinus 3–10 cm, at middle 4–6 mm diam, the longer interpinnal segments 7–10 cm; first nectary situated next to lf-pulvinus, sessile, patelliform, or button-shaped 4–6 mm diam, similar but progressively smaller ones between each pair of pinnae and between or immediately below each pair of lfts, the latter sometimes incipiently stalked; pinnae a little decrescent at each end of lf-stk, the rachis of longest ones 1.5–2.6 cm, the longer interfoliolar segments (13–)17–24 mm; pulvinules transversely linear-elliptic 0.5–0.7 × 2–3 mm; lfts a little decrescent proximally, the first pair arising 3–7 mm distant from minute conic caducous paraphyllidia (one of these exceptionally dilated into a diminutive lft), the blades, except for the always larger, obliquely rhombic-elliptic distal pair, subequiform, exactly or subfalcately rhombic from inequilaterally cuneate base, obtusangulate but often minutely mucronulate at apex, those near and above midrachis 30–65 × 13–25 mm, 2.2–2.6 times as long as wide; costa diagonal, gently or obscurely sigmoid, giving rise on each side to ±8–12 major secondary nerves brochidodrome shortly within the revolute margin, these and some weak and irregular tertiary venules finely prominulous on both faces but more sharply so beneath. *Peduncles* stout 1.5–5.5 cm, bearing at apex ±3–4 deltate reflexed bracts charged ventrally with a horny patelliform nectary; fls ±36–60 per capitulum, crowded on a deeply pitted and wrinkled receptacle 8–25 × 4–7 mm; floral bracts dimorphic, the lower ones deltate-ovate or oblance-spatulate, early dry caducous, the upper ones resembling bracts of peduncle though smaller, reflexed deltate persistent, charged ventrally with a small plane nectary; pedicels 3.5–8 mm; calyx cylindric, distally dilated, 13–18 mm, at middle 2–3.5 mm diam, the ovate deltate lobes 2.5–3 mm; corolla slenderly trumpet-shaped 40–53 mm, the lanceolate lobes 3.5–7 mm; androecium ±110–150-merous but many shorter stamens sterile, the longest 6.5–8.5 cm, the tube adnate to corolla through ±5–10 mm, either as long as or to 10 mm longer than it; style exserted beyond longer stamens, the dilated tip 0.2 mm diam. *Pods* usually solitary, subsessile, oblong or broadly linear in profile, rounded at both ends but excentrically apiculate, (12–)14–26 × 5.5–7 cm, the leathery brown valves at first plane, framed by the narrowly elevated sutures, tumescent as the seeds swell, transversely venose, minutely puberulent overall but glabrate in age; dehiscence through the length of both sutures, the valves narrowly gaping to release the seeds; funicle filiform; seeds 10–16 per pod, transverse at middle of the valves, narrowly imbricate lengthwise, plumply discoid, in broad profile broad-elliptic and 29–34 × 21–24 mm, the lustrous castaneous, membranous and brittle testa crumpled facially, girdled by a prominent dark nerve.

A bush-rope in dense forest, becoming independently treelike in more open places, scattered in lowland terra firme forest and in igapó at ±90–250 m within the NW quarter of the Amazon basin: in E Vaupés, Colombia (ríos Vaupés and Apoporís); in T. F. Amazonas, Venezuela (ríos Guainía-Negro); and in state of Amazonas, Brazil (Río Negro and immediate affluents downstream to Río Cueiras; middle Río Madeira). — Map 52. — Fl. VII–IX(–?); fr. V, IX–X, apparently maturing slowly over several months.

The aptly named *Macrosamanea amplissima* is distinguished in its genus by coarse leaves, ample coriaceous leaflets, broad tumid pod, pedicellate flowers crowded along a clavate receptacle and thus technically racemose, and by the nectary on the bract subtending some flowers as well as on the sterile bracts at base of the capitulum. The foliage of *M.*

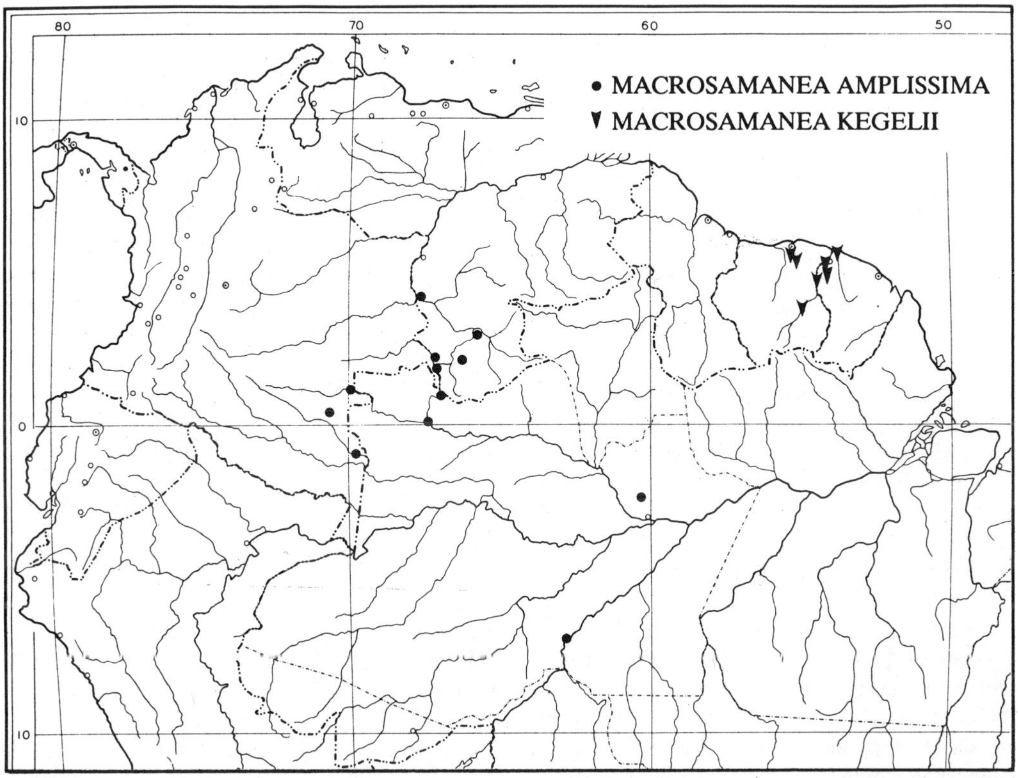

MAP 52. Northern South America. Range of *Macrosamanea amplissima* (Ducke) Barneby & Grimes, and *M. kegelii* (Meisner) Kleinhoonte.

duckei is sometimes similar, but its flowers are both sessile and much shorter.

9. Macrosamanea simabifolia (Bentham) Pittier, 3rd Conf. Interam. Agric. Caracas 358. 1945 "*simabaefolia*". *Pithecolobium simabifolium* Spruce *ex* Bentham, Trans. Linn. Soc. London **30:** 589. 1875 & *in* Martius, Fl. Bras. **15(2):** 443. 1876. — "Venezuela, near the Brazilian borders on the river Pacimoni, *Spruce*." — Holotypus, *Spruce 3329*, K(herb. bentham.)!; isotypi, †B = F. Neg. *1219*!, NY!, P!. — *Feuilleea simabaefolia* (Bentham) O. Kuntze, Revis. Gen. Pl. **1:** 189. 1891. *Samanea simabaefolia* (Bentham) Pittier, Bol. Ci. Técn. Mus. Com. Venez. **1:** 55. 1926.

Slender *treelets* and erect *shrubs* 2–6(–8) m with dark purplish brown terete, in age defoliate, smooth or sparsely lenticellate, virgate or few-branched hornotinous branchlets, the coriaceous or firmly chartaceous, when dry brown-olivaceous lvs relatively few and amply paucifoliolate, appearing glabrous but the ventral face of lf-axes and the axillary peduncles minutely puberulent or villosulous and the lfts sometimes microscopically ciliolate, the few-fld capitula solitary or paired (rarely to 5 together) in the axil of coeval lvs on peduncle shorter than subtending lf. *Stipules* appressed to stem, triangular-deltate 0.6–1.5 mm, deciduous. *Lf-formula* i/2–4(–5) very commonly and often exclusively i/3; petiole including coarsely rugulose pulvinus 4–9(–10) × 1–2 mm, openly shallowly sulcate ventrally, charged just below insertion of the one pair of pinnae with a depressed broad-elliptic, wide-pored nectary 1.2–2.4 mm in long diameter, evidently wider than tall; rachis of pinnae (2.5–)3–6(–7) cm, charged between pairs of lfts with a nectary like that of lf-stk but smaller, the longer (distal) interfoliolar segments 12–26 mm, a pair of erect subulate paraphyllidia 0.1–0.5 mm at base of first segment; lfts subequilong or somewhat accrescent distally, in outline elliptic-oblanceolate to broadly obovate or incipiently rhombic-obovate, the furthest pair 27–55(–65) × (6–)8–26(–37) mm, (1.7–)2–3.7(4) times as long as wide, all at base inequilaterally cuneate into a deeply rugulose pulvinule (0.5–)0.7–2 × 1.3–2.5 mm, at apex shallowly emarginate and mucronulate, the margins either plane or weakly revolute; venation of lfts immersed or almost so on upper face, becoming prominulous on lower face, the midrib subcentric and almost straight, giving rise to ±5–8 pairs of major secondary nerves camptodrome shortly within the margin and to many lesser intercalary ones, the tertiary venules sinuous. *Peduncles* (2–)2.5–5 cm; capitula ±9–15-fld, the clavate or globose, pitted receptacle becoming 2–3.5 mm diam; bracts ovate or the upper ones oblanceolate, 1–2.5 mm, the always small lower ones charged ventrally with a depressed horny nectary. Fls sessile, widely ascending, the greenish, when dry papery brown calyx minutely puberulent, the pinkish white or brownish pink corolla a little more densely appressed-puberulent externally; calyx cylindric or slightly dilated above middle, 5–6.5 × 1.5–2 mm, weakly nerved, the tube ±4–5 mm, the triangular teeth 0.8–1.3 mm; corolla narrowly trumpet-shaped 13–22 mm, the triangular or lanceolate lobes 2–4 mm; filaments ±50–55 but only ±30 antheriferous, the tube 14–26 mm, shortly fused at base with the corolla, either slightly shorter or to 8 mm longer than corolla; ovary glabrous; ovules 6–12(–13); style filiform, the stigma poriform. *Pods* 1–2 per capitulum, subsessile, in profile narrowly oblong or broad-linear (6.5–)7.5–14(–18) × (1.4–)1.8–3(–3.7) cm, straight or gently decurved, obtuse at both ends or cuneate at base, mucronulate at apex, plano-compressed, the leathery, when ripe brown and externally low-venulose valves framed by narrowly dilated sutures; dehiscence through the seminiferous suture, the valves separating but only narrowly gaping to release the seeds, the endocarp pallid; seeds transverse near middle of the continuous cavity, contiguous or shortly imbricate (and then mutually distorted by pressure), in broad profile oblong or oblong-obovate 20–24 × 10–14 mm, compressed but plump, the papery-membranous testa with 1 strong peripheral nerve, the cotyledons bony.

Along streams in shrub savanna and in low forest on quartzite outcrops, locally plentiful at 100–200 m on and near the Orinoco-Negro divide in SW T. F. Amazonas, Venezuela, from the middle Atabapo and C. Yapacana S to Río Pacimoni (within 1°–4°N, 66°30′–67°30′W), and in immediately adjoining Colombia (Guainía) and Brazil (Río Curicuriarí, Amazonas, ±0°20′S, 67°W). — Map 50. — Fl. VI; fr. ripe in IV–VI, XI.

Macrosamanea simabifolia and *M. prancei*, next following, differ from all other species of the genus in relatively few leaflets, the distal pinnae mostly having two to four, exceptionally five pairs. The differences between this closely related pair are presented in the key to the species. The leaflets of *M. simabifolia* are variable in outline, most often narrowly obovate, sometimes broadly so (*Wurdack & Adderley 42992*, NY), sometimes narrowly so or oblanceolate (*Davidse 17280*, NY). The species has an idiosyncratic facies and is seldom misidentified.

10. Macrosamanea prancei (Barneby) Barneby & Grimes, comb. nov. *Pithecellobium prancei* Barneby, Acta Amaz. **16/17:** 141. 1986/87. — "Brazil. Amazonas: sandy margin of rio Aracá near Serrinha, 0°25′N, 63°23′W, 24 Jul 1985 (fl.), *G. T. Prance* (with *C. D. Mota & J. A. C. da Silva*) *28744*." — Holotypus, INPA!; isotypi, INPA (5 sheets), K!, MO!, NY!, US!. Fig. 17.

Macrophyllidious unarmed *shrubs* and *treelets* at anthesis 1–4(–"10") m, with coarsely lenticellate branches at first lustrously castaneous but graying in age, glabrous below the minutely strigulose inflorescence, the coriaceous lfts bicolored, lustrous green above, paler duller beneath, the umbelliform capitula of cream-white flowers either solitary or fasciculate by 2–3 in the axil of a few distal contemporary leaves. *Stipules* triangular 1–2 mm, carnosulous, deciduous. *Lf-formula* i/3–4; lf-stks including wrinkled pulvinus and short-subulate terminal appendage 9–14 × 3–5 mm, charged ventrally near apex with a gross depressed-cupular thick-walled nectary 4.5–7 × 3.5–4.5 mm; rachis of pinnae 5–10 cm, openly sulcate ventrally and charged immediately below each pair of lfts with a thick-walled crateriform nectary ±1–2 mm diam, and at very base, next to pulvinus, with a pair of subulate-conic paraphyllidia <1 mm; lfts a little accrescent upward, the blades obliquely rhombic-obovate from inequilateral broad-cuneate base, obtuse-emarginate, the distal pair ±6–7.5 × 2.5–4 cm, all 2-nerved from pulvinule, the scarcely excentric, dorsally prominulous midrib straight, pinnately branched its whole length, the secondary nerves ascending at angles near 45° and camptodrome well within the blade's subrevolute corneous margin, the posterior primary nerve almost simple, ascending to or beyond midblade. *Peduncles* 2.5–6.5 cm, glabrous or minutely puberulent; receptacle clavate 4–5 × 2–3 mm, deeply alveolate at attachment of each fl; bracts dimorphic, the lower ones deltate ±1.5 mm, each

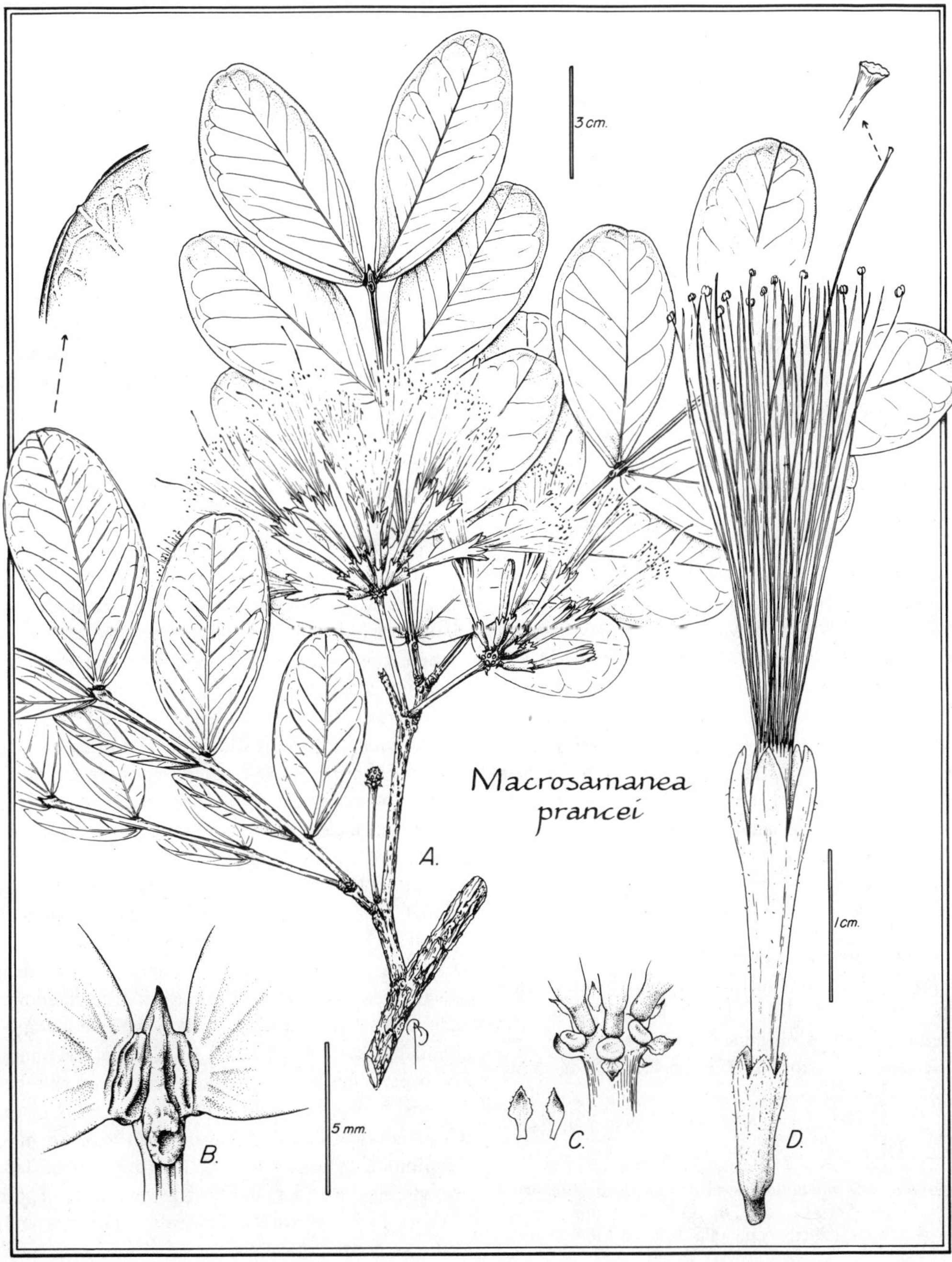

FIG. 17. Illustration of *Macrosamanea prancei* (Barneby) Barneby & Grimes. **A.** Flowering branch. **B.** Tip of leaf-stalk showing nectary. **C.** Floral receptacle, showing nectaries on lower bracts. **D.** Flower. (All drawn from *Prance 28744*.)

charged on inner face with an obese plane or subdepressed nectary as wide as or wider than itself, the upper bracts oblanceolate, all persistent; fls erect and widely ascending, homomorphic, 5-merous, the calyx and corolla externally minutely strigulose; pedicels stout, when dry dark brown, 2–3 × 1.5 mm; calyx firm greenish brown, deeply campanulate 8–10 × 3–5 mm, the ovate-triangular acute lobes 2.5–3 mm; corolla tubular very slightly dilated upward, 26–30 × 3–4 mm, firm-textured greenish-ochroleucous, the lance-ovate nearly erect lobes 5–8.5 × 2.5–3 mm; filaments ±110–120, white, 4.5–6 cm, the tube 2–3 cm, not or scarcely exserted from corolla, anthers pallid ±0.25 × 0.3 mm; style a little longer than stamens, the funnelform stigma ±0.3 mm diam; ovary glabrous. *Pods* unknown.

In seasonally flooded white sand savanna and on sandy river banks, along and near Río Aracá at points 0°2′–0°30′N, 63°8′–30′W in N Amazonas, Brazil. — Map 50. — Fl. VII– VIII.

Macrosamanea prancei, notable for relatively few and large, glossy coriaceous leaflets and far-exserted bundles of whitish filaments, combines in one plant the leaf-formula (one pair of pinnae and about 2–4 pairs of leaflets) of *M. simabifolia* with the very large, stoutly pedicellate flowers of multifoliolate *M. spruceana*. The leaflets of *M. simabifolia* are prevailingly much narrower, the furthest pair on each pinna to 18–23, not 25–40 mm wide; when almost as wide (e.g., *Wurdack & Adderley 42992*, NY, from Río Atabapo in Amazonas, Venezuela) they number no more than 2 not 3–4 pairs. Such broad-leaved forms of *M. simabifolia* may suggest *M. prancei* vegetatively, but have the small narrow sessile calyx of their species. A recent collector (*Sette Silva 183*, NY) describes the plant as attaining about 10 m in stature, in contrast to the shrubby habit reported at the type locality; this observation requires confirmation.

11. Macrosamanea kegelii (Meisner) Kleinhoonte *in* Pulle, Fl. Suriname **2(2)** [= Vereen. Kol. Inst. Amsterdam Meded. **30**]: 331. 1940. *Pithecolobium kegelii* Meisner, Linnaea **21:** 249. 1848. — "[Surinam:] ad ripas fl. Cassepoerkreek, m. Nov. 1845, legit [Hermann Aribert Heinrich] Kegel n. 1173." — Holotypus, NY (herb. meisner.)!; isotypus, GOET!.

Pithecolobium stipellatum Bernhardi in sched.: Bentham, Trans. Linn. Soc. London **30:** 590. 1875, pro syn. — *Kappler 1341E* in herb. bernhardi., GOET!, MO!.

Amply leafy, potentially arborescent riparian *shrubs* attaining 8(–?) m, the terete older branches gray, lenticellate, the hornotinous ones and all lf-axes densely sordid-pilosulous with short ascending hairs, the thinly papery lfts remaining olivaceous when dry, sublustrous above, a little paler beneath, the fls disposed in short crowded pseudoracemes of few-fld umbelliform capitula arising either from knots on old branches or from lf-axils of slender hornotinous branchlets. *Stipules* erect, firm lanceolate 1.5–3 mm, dorsally glabrate and coarsely 3–6-nerved, tardily deciduous. *Lf-formula* v–viii/12–16; lf-stks 10–19 cm, shallowly or obscurely sulcate ventrally, the true petiole 6–12 mm, the longer interpinnal segments 23–35 mm; a depressed saucer-shaped or scutiform nectary 2–3.5 mm in long diameter shortly below first pair of pinnae, and progressively shorter ones between distal pairs of pinnae and lfts; pinna-pairs strongly decrescent proximally and less so distally (expanded lf-outline obovate), the rachis of longer ones (6.5–)7–12 cm, the interfoliolar segments about as long as greatest width of lft-blades, these subequilong except for a few abruptly decrescent proximally and for the longer, heteromorphic furthest pair; a pair of subulate or filiform paraphyllidia 0.3–0.9 mm at base of each pinna-rachis subcontiguous to pulvinus; lft-pulvinules transversely elliptic 0.4–0.5 × 1.2–1.6 mm; lfts at and near midrachis rhombic around an almost straight diagonal midrib, broadly cuneate at base, at apex triangular- or deltate-mucronulate, the basal and adaxial sides nearly straight, the distal side low-convex, the larger blades 21–28 × 5.5–9 mm, 2.3–3.1 times as long as wide; venation of midrib and 1 strong (sometimes few outer weak) intramarginal nerve produced well beyond mid blade, the midrib giving rise on each side to ±7–10 major and to several intercalated minor secondary nerves camptodrome well within the narrowly revolute margin, the tertiary venulation sinuous, all nerves sharply prominulous on both faces. *Inflorescence* little known, the primary axis apparently only 5–15 mm and the individual peduncles ±3–4 mm; rudimentary lf-stk subtending peduncles ±2 mm, charged ventrally with a horny nectary; capitula ±4–6-fld; bracts 1.5–2.5 mm, becoming dry, papery, few-nerved, lacking nectary, quickly deciduous; pedicels stout (0.5–)1–2.5 mm; calyx cylindric, slightly widened distally, 14–20 × 3.5–6 mm, prominulously many-nerved, densely puberulent below middle but more thinly so toward the teeth, the tube 10–13.5 mm, the lance-ovate (often unequal) teeth 3–5.5 mm; corolla at least when young more densely silky-puberulent with paler hairs, 24–31 mm, the usually 5 lanceolate lobes 4–7 mm; androecium 4–5 cm, the staminal sheath included in the corolla, the filaments creamy-white with pink tips; style to 6 cm, well exerted from the stamens, white, the dilated stigma ±0.2 mm diam. *Pods* unknown.

On river and creek banks, below 200 m, apparently localized in NE Surinam and immediately adjacent French Guiana, on the Surinam, Maroni, and Tapahoni rivers. — Map 52. — Fl. XI–XII. — *Watertamarinde*; *kabanfo* (Arawak).

When referring *M. kegelii* to his ser. *Coriacea*, Bentham remarked on the thin texture and pallid coloration of the foliage, which are reminiscent of some species of *Zygia* sect. *Zygiopsis*, and the relatively short androecial tube, at that time not otherwise known in the series. The species remains somewhat anomalous in *Macrosamanea* in two respects: the cauline inflorescence of shortly pedunculate few-flowered capitula, in which the peduncles are subtended by a rudimentary leaf-stalk and nectary, and the lack of nectary on the floral bracts. These two characters occur in species of *Zygia*, but coincide there with the small, short-calyxed *Zygia* flower. In proportionate lengths of calyx and corollas and in naked floral bracts *M. kegelii* is technically similar to the isolated *M. macrocalyx*, but differs significantly and obviously in leaf-formula, shape of leaflets, and cauliflory. The fruit of *M. kegelii*, when discovered, may contribute to our understanding of its affinities within *Macrosamanea*, which Kleinhoonte tentatively questioned, or to related zygioid genera.

XV. ALBIZIA DURAZZINI

Albizia Durazzini, Mag. Tosc. **3(4):** 10. 1772. — Sp. typica: *A. julibrissin* Durazzini. — Etymology and orthography: named (Nielsen, Adansonia, n. ser. **19:** 206. 1979) in honor of Filippo degli Albizzi, who brought seeds from Constantinople to Florence in 1749. The spelling *Albizzia*, adopted by Bentham in 1844 (London J. Bot. **3:** 84) and much copied, is incorrect.

Pseudalbizzia Britton & Rose, N. Amer. Fl. **23:** 48. 1928. — Sp. typica (unica): *Ps. berteriana* (de Candolle) Britton & Rose ≡ *Albizia berteriana* Balbis *ex* de Candolle.

Ours (whether native or naturalized) micro- and (few) macrophyllidious, unarmed, mostly deciduous *trees* of sympodial growth, the branches of any season terminating either in an efoliate or distally efoliate, simple pseudoraceme of capitula or in complex panicles of pseudoracemes. *Stipules* mostly small (<5 mm), but in *A. chinensis* enlarged foliaceous, always caducous, absent from fruiting specimens and perhaps lacking in some species. *Lf-formula* i–xiv(–xix)/3–63; lft-venation mostly palmate, in *A. burkartiana* reduced to simple midrib, in 2 Asiatic species palmate-pinnate and in anomalous *A. procera* pinnate. *Fls* dimorphic (obscurely so in some capitula of *A. niopoides*), homomorphic in *A. procera* only; the peripheral ones either sessile or pedicellate, in native species always small (calyx <3.5 mm, corolla <6.5 mm) but in naturalized ones larger (calyx to 5.5 and corolla ±7–12 mm); androecium 10–40(–46)-merous, <2 cm in native species and *A. procera* but well over 2 cm in other exotic species; intrastaminal disc known only in the terminal flower of *A. lebbeck* (Tomlinson, 1980). *Pods* sessile or shortly stipitate, in profile broad-linear, straight or nearly so, compressed or plano-compressed, variable in texture of valves and in mode of dehiscence: (a) papery and inertly dehiscent through both sutures, or (b) as the preceding but indehiscent, or (c) like the last but the valves segmented after fall of fruit; or (d) leathery, the valves segmented between seeds but retained by wiry sutures (resembling pods of *Hydrochorea*); or (e) pithy-lignescent, tardily segmented alike through valves and sutures (hence conventionally lomentiform), or (f) stiffly papery, the valves finally differentiated into continuous, inertly dehiscent exocarp and detached segmented endocarp encapsulating each seed in a buoyant envelope; funicle filiform or scarcely dilated; seed attached either basally or laterally above base (3.5–)4–11 mm, the exotesta mostly light brown or ivory-white or yellowish, occasionally brown, usually translucent when soaked, areolate, the pleurogram at middle (in *A. chinensis* below middle) of seed face, either U-shaped or less often complete. — A discontinuously circumtropical genus, most highly diversified in tropical America, Africa (including Madagascar), southeast Asia and Malesia, variously defined and its species consequently of indefinite number, represented in the Americas by 19 native species widely dispersed, many in monsoon climates or in riparian habitats within the lowland and low-montane tropics, one extending N in Mexico to desert foothills of S Sonora, three S into extratropical Brazil, N Argentina, Paraguay, and Amazonian Bolivia, two endemic to the Greater Antilles; four Asiatic species cultivated and two of these widely naturalized, one (*A. julibrissin*) frost-hardy northward in eastern United States to ±41° N.

The foregoing generic description of *Albizia* is deliberately tailored to those elements of the genus either native, or naturalized, or commonly cultivated in the Americas and should not be interpreted as definitive of the whole genus. Until a comprehensive worldwide study of *Albizia* can be undertaken, we adopt for it a provisional diagnostic syndrome of: sympodial growth; lack of armament; palmate or palmate-pinnate (not simply pinnate) venation of leaflets; inflorescence either: (a) composed in part of axillary, annual, sylleptic, strictly reproductive branches, (b) with ephemeral axes that do not persist as vegetative ones; or (c) both; dimorphic flowers, either capitulate or corymbose-umbellulate; and a straight or almost straight pod, never elastically dehiscent but diverse in texture and in release of the seeds. Our definition of *Albizia* is materially narrower than that proposed by Nielsen (1981: 180), which included among others our genera *Chloroleucon* (armed) and *Pseudosamanea* (indeterminate branches, pinnately veined leaflets). Our diagnostic syndrome assembled above purposely excludes the Asiatic *A. procera*, which is aberrant in *Albizia* in our sense by a combination of pinnately venulose leaflets and homomorphic flowers. The generic affiliation of *A. procera* is here left in abeyance, though we think it should be excluded from the genus.

The American native *Albizia* form a group that is homogeneous in most respects, but diverse in the late developmental states of the fruit. A similar but in detail different radiation has occurred in Madagascar (Capuron, 1970), and it is obviously not possible on a worldwide scale to define either *Albizia* itself, or any segregate from it, exclusively by characters of the legume. A telling example is offered by the American *Aa. subdimidiata, multiflora*, and *inundata*. Bentham knew all three from flowering specimens, which are so deceptively similar that he could not distinguish them even as species. Their pods, however, turned out to be so different at maturity (see description in the conspectus, spp. 16, 17, 19) that they have been transferred by subsequent authors to the genera *Samanea*, *Arthrosamanea*, and *Cathormion*, depending on relative importance attached to texture and dehiscence of the fruit. It now seems pretty certain that a more or less perfectly segmented fruit must have developed independently in several evolutionary lines of Ingeae, just as it has in *Acacia* (cf. *Dugandia rostrata* (Humboldt & Bonpland) Britton & Killip; *Manganaroa articulata* (Ducke) Spegazzini). The lomentiform (or in the case of *A. inundata* cryptolomentiform) fruits of African and Indian *Cathormion* and of American *Hydrochorea* and *Albizia* sect. *Arthrosamanea* are here interpreted as parallel but not homologous, and belong to taxonomic groups that can be readily separated and recognized by characters of inflorescence-architecture and leaflet-venation. Dispersal of papery pods is at least potentially anemochorous, whereas that of segmented pods is prevailingly hydrochorous. There is a strong correlation between method of seed-dispersal and microhabitat.

In summary, we recognize in the American flora two sections of *Albizia*: the native sect. *Arthrosamanea* (often a misnomer) distinguished by small flowers (corolla <6.5 mm) and short androecium (<2 cm); and a sect. *Albizia*, consisting of paleotropical species cultivated and naturalized, differing consistently only in relatively large flowers (corolla 7–12 mm) and elongate androecium (well over 2 cm). The first of these sections corresponds more or less neatly with *Pithecellobium* sect. *Samanea* ser. *Subarticulatae* (Americanae) and ser. *Parviflorae* of Bentham's *Revision of Mimoseae*; the second to elements of his *Albizzia* sect. *Eualbizzia* ser. *Obtusifoliae* and *Falcifoliae*. In the pages that follow we divide sect. *Arthrosamaea* into four series, but introduce serial nomenclature of the Old World group, for purposes of orientation, only into the Conspectus following.

Cassens and Miller (1981) noted that *Albizia* sects. *Albizia* and *Parviflorae* share, along with *Pithecellobium* sect. *Pithecellobium*, *Havardia*, and *Sphinga*, the distinction of being the only members of the Ingeae that have septate fibers and nonconfluent parenchyma. They distinguished the wood of *Albizia*, though, on the basis of ray width, pore diameter, vessel element length, and claimed that the two types do not appear closely related. Baretta-Kuipers (1981) also stated that *Albizia* has septate fibers in the wood. This is apparently true for the American species; however, Chauhan and Dayal (1985) found several Indian species for which this is not true.

Conspectus of American *Albizia*, native, adventive, and cultivated

¶ Fls of each unit of inflorescence dimorphic (except *A. barinensis*).

§ Inflorescence a compound panicle of pseudoracemose capitula, or of capitula axillary to coeval lvs; lft-venation palmate or exceptionally simple; fls small, the corolla of peripheral ones (2–)2.5–5.5 (in *A. carbonaria* –6.4) mm and their androecium 6–17 mm; native American; sect. *Arthrosamanea*:

* Pod plano-compressed papery, the valves continuous, inertly dehiscent; ser. *Paniculatae*.

† Sutures of pod not winged; seed straight.

1. *A. polycephala* (Bentham) Killip
2. *A. edwallii* (Hoehne) Barneby & Grimes
3. *A. burkartiana* Barneby & Grimes
4. *A. coripatensis* (Rusby) Schery
5. *A. tomentosa* (M. Micheli) Standley

6. *A. sinaloënsis* Britton & Rose
7. *A. leonardii* Barneby & Grimes

†† Sutures of pod not winged; seed oblique.

8. *A. barinensis* Cárdenas

††† Sutures of pod winged; seed straight; [macrophyllidious].

9. *A. adinocephala* (Donnell Smith) Britton & Rose

†††† Sutures of pod winged; seed oblique.

10. *A. niopoides* (Bentham) Burkart
  a. var. *niopoides*
  b. var. *colombiana* (Britton & Killip) Barneby & Grimes
11. *A. buntingii* Barneby & Grimes

** Pod plano-compressed papery, the valves continuous, indehiscent.

12. *A. carbonaria* Britton

*** Pod of **, but the valves developing fracture lines between seeds.

13. *A. berteriana* (de Candolle) Fawcett & Rendle

**** Pod lomentiform, plano-compressed, the ripe valves cracking transversely between seeds but the wiry sutures persisting; ser. *Arthrosamanea*.

14. *A. glabripetala* (Irwin) G. P. Lewis & P. E. Owen
15. *A. pistaciifolia* (Willdenow) Barneby & Grimes
16. *A. subdimidiata* (Splitgerber) Barneby & Grimes
  a. var. *subdimidiata*
  b. var. *minor* Barneby & Grimes

***** Pod lomentiform but only reluctantly separating into articles, the thick-textured valves and sutural keels breaking transversely under pressure; ser. *Multiflorae*.

17. *A. multiflora* (Kunth) Barneby & Grimes
18. *A. decandra* (Ducke) Barneby & Grimes

****** Pod a cryptoloment, dehiscent through the sutures, the valves differentiating into a continuous exocarp and a segmented endocarp separating into 1-seeded propagules; ser. *Inundatae*.

19. *A. inundata* (Martius) Barneby & Grimes

§§ Inflorescence a compound panicle of pseudoracemose capitula; lft-venation of §, but stipules herbaceous dilated; fls larger, the corolla of peripheral ones 7–8.3 mm and the androecium 26–34 mm; Asiatic; ser. not assigned.

20. *A. chinensis* (Osbeck) Merrill

§§§ Inflorescence reduced to a simple terminal pseudoraceme of capitula; venation palmate-pinnate; fls relatively large, the corolla of peripheral ones 7–12 mm, the androecium 24–38 mm; sect. *Albizia*.

* Inflorescence foliate, throughout or in part; lfts obtuse; ser. *Obtusifoliae*.

21. *A. lebbeck* (Linnaeus) Bentham

** Inflorescence efoliate, bracteate; lfts acute; ser. *Albizia*.

22. *A. julibrissin* Durazzini

¶¶ Fls of each unit of inflorescence homomorphic; lft-venation pinnate; Asiatic, planted in neotropics; sect. not assigned.

23. *A. procera* (Roxburgh) Bentham

Key to the native American series and to adventive and cultivated species of *Albizia*

1. Inflorescence: (a) a compound panicle, (b) composed largely of efoliolate pseudoracemose capitula, or (c) axillary to coeval leaves.
  2. Inflorescence of capitula axillary to coeval leaves . . . . 6. *A. sinaloënsis* (p. 215)
  2. Inflorescence a compound panicle or composed largely of efoliate pseudoracemose capitula.
    3. Fls of each unit of inflorescence dimorphic; lft-venation palmate or rarely simple.
      4. Stipules and bracts of inflorescence linear-lanceolate, or obscure, never dilated or foliaceous; areole of seed-coat at middle of seed-face.
        5. Pod plano-compressed, papery, dehiscent or not, the valves continuous, or exceptionally and very tardily developing fracture lines between seeds. . . . . (i) ser. *Paniculatae* (p. 208)
        5. Pod lomentiform or cryptolomentiform, the leathery or pithy valves articulate in some fashion.
          6. Exocarp of ripe pod transversely fractured between seeds, but the wiry sutures persistent entire . . . . (ii) ser. *Arthrosamanea* (p. 229)
          6. Exocarp of ripe pod either continuous and only the endocarp articulate, or the pod transversely articulate at once through valves and sutures.
            7. Valves of pod pithy-ligneous, articulate by simultaneous fracture through both valves and sutures . . . . (iii) ser. *Multiflorae* (p. 234)
            7. Valves of pod differentiated into continuous exocarp and articulate endocarp, the latter enclosing the seeds in an independently shed envelope. . . . (iv) ser. *Inundatae* (p. 238)
      4. Stipules (caducous) and bracts of inflorescence foliaceous, dilated; areole of seed-coat small, displaced toward base of seed-face . . . . 20. *A. chinensis* (p. 241)
    3. Fls of each unit of inflorescence homomorphic; lft-venation pinnate . . . . 23. *A. procera* (p. 244)
1. Inflorescence reduced to a simple terminal pseudoraceme.
  8. Peduncles all or mostly axillary to a coeval lf; lfts obtuse, their venation palmate-pinnate. . . 21. *A. lebbeck* (p. 242)
  8. Peduncles all bracteolate, the pseudoracemose inflorescence efoliate; lfts acute, their venation palmate. . . . . 22. *A. julibrissin* p. 243)

XVA. Sectio Arthrosamanea (Britton & Rose) Barneby & Grimes

Albizia sect. **Arthrosamanea** (Brittion & Rose) Barneby & Grimes, stat. nov. *Arthrosamanea* Britton & Rose *ex* Britton & Killip, Ann. New York Acad. Sci. **35:** 128, pro gen. 1936. — Sp. typica: *Arthrosamanea pistaciaefolia* (Willdenow) Britton & Rose ≡ *Mimosa pistaciaefolia* Willdenow ≡ *Albizia pistaciifolia* (Willdenow) Barneby & Grimes.

Fls small, either sessile or pedicellate, the corolla of peripheral ones 2.5–6.5 mm and the androecium 6–17 mm. *Pod* papery, leathery, or pithy-ligneous, the sutures either prominulous, or immersed, or narrowly winged; dehiscence inert, through one or both sutures, or none, or the valves breaking tardily into monospermous articles, the fruit in one species a cryptoloment, the exocarp continuous but the endocarp segmented. — Spp. 19, all American.

The sect. *Arthrosamanea* is distinguished from most American Ingeae similar in general aspect by venation of leaflets alone, but when this distinction fails, as in *Enterolobium* and some species of *Abarema*, the determinate stem lacking sylleptic buds giving rise to vegetative branches is decisive. The genera *Pseudosamanea*, *Samanea*, and *Balizia* sect. *Leucosamanea* differ further in the larger, often loosely umbellate peripheral flowers (corolla 7–15 mm) borne on mostly axillary peduncles, not in compound pseudoracemes, and in their characteristic fruits; whereas the genus *Hydrochorea*, which mimics members of ser. *Arthrosamanea* in lomentiform fruits, differs not only in pinnate venation of leaflets but also in truncate ovary.

Key to the species of genus *Albizia* sect. *Arthrosamanea*

1. Lfts at once few and large, those of longer pinnae (2–)3–5(–6) pairs, and the largest of them 9–30 mm wide.
    2. Larger lfts obovate or oblong, palmately 6–7-nerved from pulvinule; capitula 5–9-fld; native in N Colombia, N-centr. Venezuela, and Pacific Ecuador; valves of pod indehiscent, but under pressure breaking transversely between seeds, hence lomentiform. . . . . 15. *A. pistaciifolia* (p. 230)
    2. Larger lfts elliptic or ovate-elliptic, 3-nerved from pulvinule; capitula 16–36-fld; native in S Mexico and Central America, cult. in Greater Antilles; valves of pod papery, inertly dehiscent through both sutures . . . . 9. *A. adinocephala* (p. 218)
1. Lfts more numerous and smaller or narrower, in longer pinnae 6(–8) pairs upward and the largest rarely more than 9 mm wide.
    3. Dispersed in continental North America.
        4. Pedicel of outer peripheral fls 1.5–4 mm. . . . . 12. *A. carbonaria* (p. 224)
        4. Pedicel of outer peripheral fls <1 mm.
            5. Fls homomorphic.
                6. Lfts of longer pinnae ±28–60-jug., the largest 5–10 × 0.9–1.4 mm; lowland tropical Mexico and W Central America . . . . *A. niopoides* (p. 220)
                6. Lfts of longer pinnae ±16–20-jug., the largest 10–14 × 3–4 mm; of warm temperate NW Mexico . . . . *A. sinaloënsis* (p. 215)
            5. Of lowland tropical Mexico and W Central America; Panama.
                7. Lfts of longer pinnae 6–22(–28)-jug., the largest (10–)12–26 × 3–11(–14) mm; terminal fl highly modified.
                    8. Of Belize and NE Guatemala, N in Mexico to Yucatán, Veracruz, Chiapas, and Jalisco; pinnae 3–13-jug.; pod plano-compressed, the papery valves continuous, inertly dehiscent through sutures . . . . 5. *A. tomentosa* (p.213)
                    8. Of Panama and southward; pinnae of larger lvs 2–4-jug.; pod compressed but plump, indehiscent, the pithy-lignescent valves tardily lomentiform . . . 17. *A. multiflora* (p. 234)
                7. Lfts of longer pinnae ±28–60-jug., the largest 5–10 × 0.9–1.4 mm; terminal fl scarcely differentiated; Yucatán and Chiapas, Mexico, SE through Central America . . . . 10. *A. niopoides* (p. 220)
    3. Dispersed in West Indies and South America.
        9. Confined to the Greater Antilles.
            10. Lf-formula iii–v/9–12; Haiti . . . . 7. *A. leonardii* (p. 216)
            10. Lf-formula vii–xiii(–xv)/18–57.
                11. Pedicel of outer peripheral fls 1.5–4 mm; calyx 2.3–3.5 mm and corolla 4–6.4 mm, golden-silky externally; planted and perhaps weakly naturalized. . . . . . . 12. *A. carbonaria* (p. 224)
                11. Pedicels 0 or <1 mm; calyx 1.2–1.5 mm and corolla ±3–4 mm, glabrous or minutely puberulent; native . . . . 13. *A. berteriana* (p. 226)
        9. Dispersed in continental South America and S Lesser Antilles.
            12. Lfts of longer pinnae 8–14-jug. (occasionally to 17-jug. in extratropical *A. inundata*, but pinnae then only 2–4-jug.), the longest 11–28 mm.

13. Lfts subequilateral around the centric midrib; stamens 10; local in lower Amazon valley in Pará, Brazil; pod thick-textured indehiscent, tardily breaking transversely between seeds . . . . 18. *A. decandra* (p. 237)
13. Lfts inequilateral, the midrib substantially displaced forward from midblade; stamens 16–30; widespread over much of tropical and warm-temperate South America; pods various.
    14. Pedicel of outer peripheral fls 0.9–1.4 mm; pinnae of larger lvs 6–7-jug.; known only from Maracaibo basin, NW Venezuela . . . . 11. *A. buntingii* (p. 223)
    14. Pedicel of outer peripheral fls at most 0.6 mm, commonly obsolete; widespread in South America, but only *A. subdimidiata*, with 2–4-jug. pinnae, sympatric with *A. buntingii* in NW Venezuela.
        15. Pinnae of larger lvs (4–)5–8-jug.; lfts palmately 3-nerved from pulvinule; corolla silky-puberulent externally; pod plano-compressed, the papery valves inertly dehiscent through both sutures; Bolivia (Nor and Sur Yungas). . . . . 4. *A. coripatensis* (p. 212)
        15. Pinnae of larger lvs 2–4(–5)-jug.; lfts (4–)5–7-nerved from pulvinule; corolla either glabrous or its lobes thinly puberulent dorsally above middle; widespread species, much alike at anthesis, notably disparate in fruit.
            16. Lfts thinly puberulent at least dorsally, often on both faces; pod 16–24 mm wide, scarcely constricted (unless ovules abort), essentially indehiscent but tardily lomentiform, the pithy-ligneous valves ultimately breaking by transverse fracture between seeds into individually indehiscent 1-seeded segments; SW Ecuador and N Peru, in Peru extending E to the head of Río Maranón in N Cajamarca, but wholly extra-Hylaean; disjunctly to centr. Panama . . . . 17. *A. multiflora* (p. 234)
            16. Lfts glabrous facially (sometimes minutely ciliolate), or barbellate dorsally at very base; pod 6–14 mm wide, either dehiscent or not; widespread E of the Andes, in Ecuador and Peru found only in Amazonian lowland forest; 2 spp., deceptively similar at anthesis, sharply distinct in fruit.
                17. Lfts obtuse or obtusely deltate-apiculate and almost straight; pod indehiscent, the valves tardily separating by transverse cracks into 1-seeded segments, but neither seed nor segment released from wiry persistent sutures; Orinoco and Amazon basins from Colombia and Venezuela S to NE Bolivia, SE to the Guianas and in Brazil to N Maranhão, NW in Venezuela to the Maracaibo basin . . . . 16. *A. subdimidiata* (p. 232)
                17. Lfts nearly always porrect and deltately acute at apex; pod (cryptoloment) dehiscent inertly through the sutures, the valves differentiating when ripe into a continuous exocarp and a detached papery endocarp segmented between seeds, these shed in an individually indehiscent packet; basins of ríos Paraguai, Paraná, and Uruguai northward from 33°S in Argentina to Paraguay and adjacent Brazil and Bolivia, in Brazil extending discontinuously N to the São Francisco valley in Bahia and to the Amazon River in Pará. . . . . 19. *A. inundata* (p. 238)

12. Lfts of longer pinnae 16–63-jug., the longest 4.5–14 mm; but if extratropical in dispersal (*A. edwallii*) then pinnae of larger lvs 5–9-jug. and some lfts to 16 mm.
    18. Pedicel of outer peripheral fls 1.5–4 mm; longest lfts 2.7–4(–4.4) mm; N Colombia and SE Venezuela (cult. elsewhere). . . . . 12. *A. carbonaria* (p. 224)
    18. Pedicel of outer peripheral fls <1.5 mm, commonly obsolete.
        19. Corolla glabrous externally or only very thinly minutely puberulent toward the tip of lobes; pod various.
            20. Lfts 1-nerved from pulvinule; local in extratropical S Brazil . . . 3. *A. burkartiana* (p. 211)
            20. Lfts 2–4(–5)-nerved from pulvinule; widespread from SE Brazil and NE Argentina northward.
                21. Androecium 20–40-merous; pod plano-compressed, the ripe valves papery continuous, inertly dehiscent along both sutures, these narrowly winged. . . . . 10. *A. niopoides* (p. 220)
                21. Andreocium ±16-merous; pod lomentiform indehiscent, the ripe valves transversely segemented between seeds, sutures not winged; localized in N Brazil (Roraima), Venezuelan Guayana, and Guyana . . . . 14. *A. glabripetala* (p. 229)
        19. Corolla externally sericeous or pilosulous, either overall, or at least beyond the tube; pod plano-compressed, dehiscent through the sutures.

22. Inflorescence paniculate and exserted beyond foliage; longest peripheral fls 2.8–3.5 mm; tropical E Brazil in lat. 8°–23°S. . . . . 1. *A. polycephala* (p. 208)
22. Inflorescence more shortly paniculate, immersed in foliage; longest peripheral fls 4–6 mm; allopatric.
    23. Lfts inequilaterally lance-oblong, the larger ones 3–4 times as long as wide; corolla appressed-silky; androecium mostly 15–22-merous; extratropical Brazil in lat. 23°–30°S . . . . 2. *A. edwallii* (p. 209)
    23. Lfts oblong, the larger ones 2.3–3 times as long as wide; corolla finely gray-pilosulous; androecium 26–30-merous; Orinoco basin in Venezuela, and adjacent Guyana . . . . 8. *A. barinensis* (p. 217)

XVA(I). SERIES PANICULATAE (BENTHAM) BARNEBY & GRIMES

Albizia sect. **Arthrosamanea** ser. **Paniculatae** (Bentham) Barneby & Grimes, comb. nov. *Pithecolobium* sect. *Samanea* ser. *Paniculatae* Bentham, London J. Bot. **3:** 219. 1844. — Sp. lectotypica: *P. polycephalum* Bentham ≡ *Albizia polycephala* (Bentham) Killip.

Pithecolobium sect. *Samanea* ser. *Parviflorae* [*sic*] Bentham, Trans. Linn. Soc. London **30:** 591 (exclus. sp. 77). 1875 & *in* Martius, Fl. Bras. **15(2):** 445. 1876. — Sp. lectotypica: *P. polycephalum* ≡ *Albizia polycephala* (Bentham) Killip.
Pseudalbizzia Britton & Rose, N. Amer. Fl. **23:** 48. 1928. — Sp. typica (unica): *Ps. berteriana* (de Candolle) Britton & Rose ≡ *Acacia berteriana* Balbis *ex* de Candolle ≡ *Albizia berteriana* (de Candolle) Fawcett & Rendle. — Equated with *Albizia* by Nielsen, 1981: 180.
Albizzia berteriana sensu Bisse, 1988: 226.

Pods broad-linear, the ripe valves papery, continuous or developing transverse fracture lines at late maturity, but not truly lomentiform, the sutures sometimes narrowly winged; dehiscence inert, through one or both sutures, or the seeds released only by weathering on the ground. — Spp. 13, with range of the section.

1. Albizia polycephala (Bentham) Killip *ex* Record, Trop. Woods **63:** 6. 1940. *Pithecolobium polycephalum* Bentham, London J. Bot. **3:** 219, syn. Velloziano dubio excluso. 1844.– "Tropical Brazil, *Sello, Pohl*; near Rio de Janeiro, *Miers*, near Ilheos, *Blanchet*, n. 1848." — Lectoholotypus, *Pohl d. 1420*, K(herb. bentham.)! ≡ NY Neg. *2022*; isotypi, F!, K(herb. hooker.)!, NY!, US!; paratypi, *Sello 219*, *Miers*, *Blanchet 1848*, all K(herb. bentham.)!. — *Feuilleea terminalis* O. Kuntze, Revis. Gen. Pl. **1:** 187. 1891. *Samanea polycephala* (Bentham) Pittier, Arb. Arbust. Venez. 55, in clave, basionymo imperfecte citato. 1925.

Amply microphyllidious *trees* 4–10 m with smoothish mottled trunk potentially 5 (but usually 1–3) dm dbh, sometimes flowering precociously as a shrub 2 m, the young stems and all lf- and inflorescence-axes densely tomentellous and sometimes also pilosulous with incumbent or straighter spreading, golden-brown or sordidly yellowish hairs to 0.15–0.8 mm, the lvs bicolored, on upper face dark green (brunnescent when dry) sublustrous and either glabrous or thinly minutely puberulent, beneath paler dull (usually tan or orange when dry) and pilosulous either thinly overall or only along principal nerves, the inflorescence a pyramidal, efoliate or only proximally few-foliate panicle of small whitish capitula at least equaling and more often overtopping the distal lvs. *Stipules* narrowly triangular or linear-lanceolate 0.8–1.8 mm, caducous, the externally nerveless blade puberulent like adjacent stem. *Lf-formula* vi–xi(–xii)/16–22; lf-stks of longer lvs (8–)10–19(–24) cm, the petiole 2–4(–5) cm, at middle 1–1.8 mm diam, the longer interpinnal segments 9–19 mm, the ventral groove continuous between pinna-pairs; a nectary well below midpetiole round or vertically elongate 0.8–4 mm diam, either shallow-cupular or almost plane, in profile not over 0.4 mm tall, sometimes immersed in petiolar groove or even obsolete, a much smaller nectary at tip of most pinnae; pinnae a little decrescent proximally, the rachis of longer ones 6–9(–10.5) cm, the longer interfoliolar segments 2–4(–4.7) mm; lft-pulvinules 0.1–0.3 × 0.3–0.6 mm, the lfts sessile against the rachis; lfts gently decrescent toward each end of rachis, the first pair often reduced to paraphyllidia or its anterior member lacking, the blades of the rest oblong from obliquely truncate base, deltately subacute, the larger ones (4.5–)5–14 × (1.9–)2.1–4(–4.3) mm, 2.2–3.1 times as long as wide, the margin of all strongly revolute; venation of 3–4(–5) nerves from pulvinule, the nearly straight midrib a little forwardly displaced and giving rise on each side to 2–4 slender secondary nerves, the inner of 2(–3) posterior primary nerves incurved-ascending to anastomosis a little beyond midblade, the outer posterior nerve and sometimes a faint anterior one very short and weak, the whole venation immersed on upper face. *Primary axis* of inflorescence ±(1–)1.5–3 dm; peduncles 3–8(–10) per node of inflorescence, biseriate upward from node, the longest of each fascicle 10–22 mm, the rest progressively shorter proximally; capitula 10–22-fld, without filaments 6–7.5 (–8) mm diam, the receptacle including short terminal pedestal 1.5–2.5 mm; bracts oblong-obovate or spatulate 0.6–1 mm, tardily deciduous; fls dimorphic, the perianth of all 5-merous and externally yellowish-silky-strigulose overall; PERIPHERAL FLS: calyx turbinate-campanulate (1–)1.2–1.8 mm, either sessile or contracted at base into a pedicel 0.1–0.5 mm, the depressed-deltate teeth 0.1–0.4 mm; corolla 2.8–3.5 mm, the erect ovate lobes 0.7–1.4 × 0.5–1 mm; androecium 9–14-merous 11–14.5 mm, the stemonozone ±0.45 mm, the tube 1–2.2(–2.5) mm; ovary at anthesis glabrous, conical at apex, following fertilization becoming densely yellowish tomentulose; style a little longer than longer stamens, slightly dilated at the stigma; TERMINAL FL: sessile, the calyx 2.1–2.9 × 1.2–2 mm, the corolla 4–5.5 mm; androecium 16–25-merous, the stemonozone 1–1.4 mm, the dilated tube 5–7 mm, shortly but evidently exserted. *Pods* solitary, sessile or cuneately contracted at base into a pseudostipe to 4 mm, the broad-linear, straight or nearly straight, plano-compressed body (8–)9–15 × (1.5–)1.7–2.7(–3.2) cm, rounded at apex but minutely apiculate, 8–12-seeded, the papery, early glabrate, sordid-olivaceous or castaneous, transversely venulose valves becoming low-bullate over each seed, framed by almost straight, simple sutures 1.2–1.4 mm wide, the pallid endocarp not adherent between seeds; dehiscence tardy, through both sutures, inert; funicles ribbonlike, sigmoid at apex; seeds (few seen) at middle of valves, obliquely ascending, disciform, in broad view 5.5–9.5 mm diam, ±1.5 mm thick, the translucent, brownish or grayish testa produced as a peripheral wing 0.2–0.4 mm wide, adherent to the embryo, which does not fill the testa-cavity, the pleurogram small, U-shaped.

In caatinga and on creek banks in cerrado climax, ascending from the coastal lowlands, where surviving in and around cocoa plantations, to 1000 m in Bahia and to 1600 m in S-centr. Minas Gerais, and coming out to coastal restinga southward, endemic to E Brazil: frequent on the Atlantic slope in Bahia between 10° and 16°S; Sa do Espinhaço in S-centr. Minas Gerais; coast of Rio de Janeiro; N in scattered stations to Sa Baturité in Ceará; and known by a single collection from SW Goiás (Jataí). — Map 53. — Fl. XII–III; fr. IV–VII(–?). — *Angico monjollo, cabuir, camunzé, comondongo, farinha seca; monzê.*

Albizia polycephala and its North American analogue *A. tomentosa* provide a blueprint, as it were, for the sect. *Arthrosamanea*, having the essential characters of the group but without any striking individual one: elaborate leaves, effuse panicles of small white capitula, and plano-compressed papery pods hardly different from those of *A. julibrissin*. They are very closely related and far from easy to separate in clearly contrasting terms, but *A. polycephala* is on the average a smaller tree when full grown, has a further exserted panicle of more densely pubescent and on the average slightly smaller flowers, and only 9–14, not 18–32 stamens in the peripheral flowers. Their remotely vicariant ranges of dispersal furnish the most accessible criterion.

2. Albizia edwallii (Hoehne) Barneby & Grimes, comb. nov. *Pithecolobium edwallii* Hoehne, Bol. Inst. Brasil. Sci. **2:** 243, t. 1. 1926. — "São Paulo, n° 13077 (no. 5608, *Dr. Gustavo Edwall*, Fazenda da Cascata, Est. Alfredo Rodrigues, out. 1901. Ex Erv. Commissão Geogr. e Geologica do S. Paulo). Butantan n° 414 correspondente ao n° 8 no Serviço Florestal da Companha Paulista de Estradas de Ferro, *Navarro de Andrade*, Rio Claro, Julho de 1917, mais n° 1920 idem, Batovi, em 6/918 sob n° 172 do Serviço mencionado." — Lectotypus, SP *13077 = Edwall 5608*, SP!; paratypi, *Andrade 8*, RB!, SP!; *O. Vecchi 172*, SP!.

Albizia austrobrasilica Burkart *in* Reitz, Fl. Ilustr. Catar., Legum. Mimos. 109, estampas 14, 15. 1979. — "Santa Catarina: Lontras: Riachuelo, beira de rio, 350 m . . . Reitz & Klein 8.649 (14.03.1959) . . ." — Holotypus, SI, not seen; paratypi, *Smith & Klein 13114, 13183, 13255*, all NY!, US!.

Amply microphyllidious *trees* (4–)6–15 m with trunk to 3.5 dm dbh, the young stems and all lf- and inflorescence-axes more or less densely tomentellous and pilosulous with incumbent and straighter spreading, golden, bronze or sordid-pallid hairs to 0.2–0.5 mm, the lfts bicolored, on upper face olivaceous (brunnescent when dry) sublustrous glabrous, beneath paler dull, barbellate (sometimes only minutely) at base in anterior angle of midrib but otherwise glabrous, thinly ciliolate all around or sometimes only proximally, the inflorescence a relatively compact, efoliate or only proximally few-foliate panicle of small whitish capitula surpassed by distal lvs. *Stipules* narrowly ovate or lanceolate (1–)2–4.5 mm, caducous

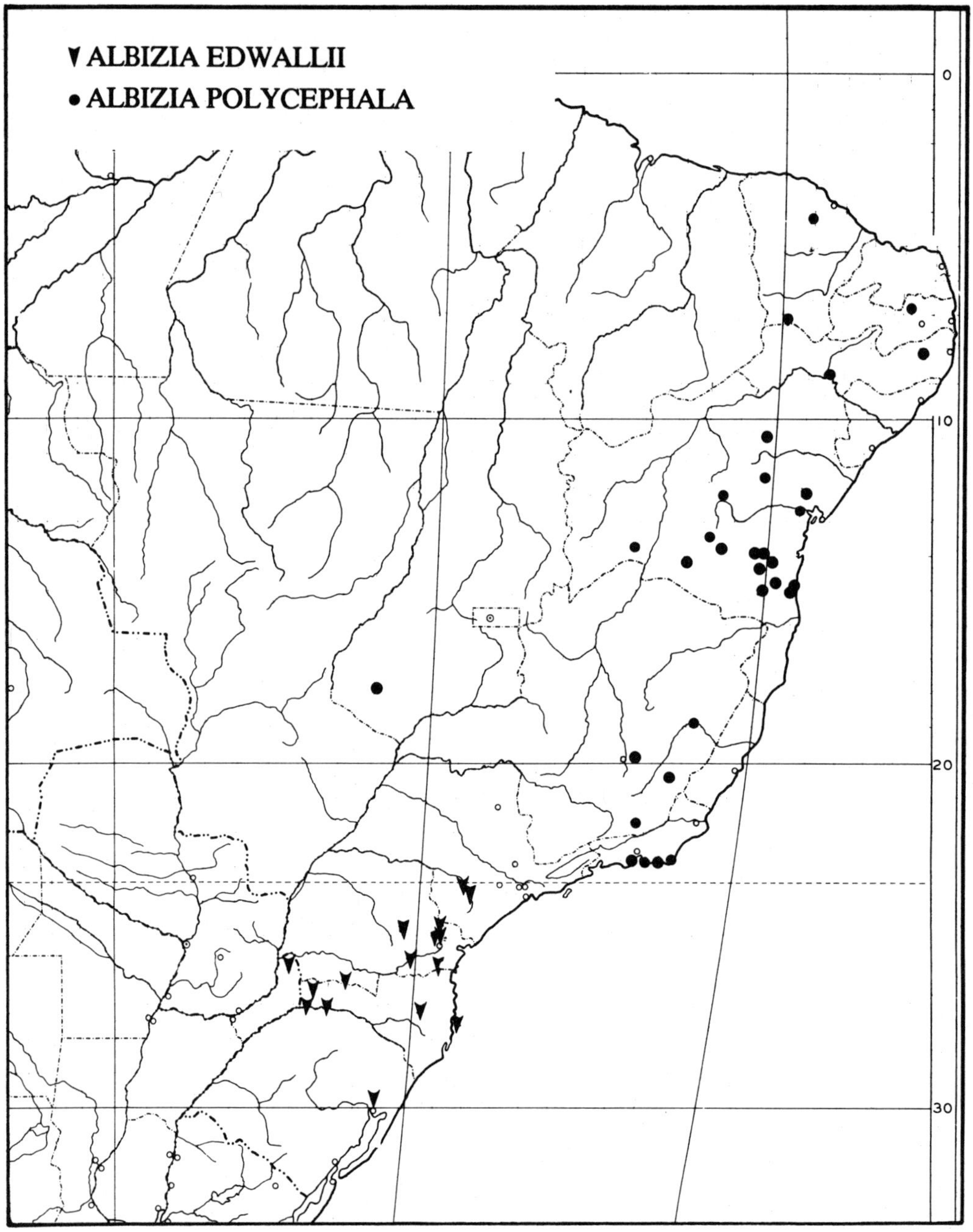

MAP 53. Eastern and southern Brazil. Range of *Albizia polycephala* (Bentham) Killip and *A. edwallii* (Hoehne) Barneby & Grimes.

from a small pallid scar, the externally nerveless blade tomentulose like adjacent stem. *Lf-formula* (iv–) v–viii(–xi)/17–26 (in saplings, no further described, to xiv/30) lf-stks of longer lvs (8–)10–21(–25) cm, the petiole 3.5–5 cm, at middle 1.3–2.2 mm diam, the longer interpinnal segments 1.5–3 cm, the ventral groove continuous between pinna-pairs; a livid shallow-cupular crumpled nectary near or below midpetiole either round or vertically elongate 1–2.4 mm diam, in profile 0.3–0.6 mm tall, no nectary between pinna-pairs but much smaller ones between 1–2 furthest pairs of lfts; pinnae a little decrescent at each end of lf-stk, the rachis of longer ones 7–14 cm, the longer interfoliolar segments 3–5 mm; lft-pulvinules 0.3–0.7 × 0.6–0.9 mm, the lfts almost sessile; lfts decrescent at each end of rachis, the first pair commonly reduced to paraphyllidia or the first anterior one lacking, the blade of the rest inequilaterally lance-oblong from shallowly semicordate base, deltately acute, the larger ones 10.5–16 × (2.7–)2.9–4.5 mm, (2.9–)3–4 times as long as wide, the margin of all loosely revolute; venation of 3–4 nerves from pulvinule, the nearly straight midrib forwardly displaced, strongly so at base, at middle dividing blade 1:2–3, giving rise on each side to 3–5 ascending secondary nerves and to a random tertiary reticulum, the inner posterior primary nerve incurved to anastomosis near or a little beyond midblade, the outer posterior ones much shorter, the whole venation weakly elevated on upper face, sharply so beneath. *Primary axis* of panicle becoming at full anthesis 1–1.5 dm; peduncles 2–6 per node of inflorescence, the longest of each fascicle 12–22 mm, the proximal ones progressively shorter; capitula at full anthesis hemispherical, 12–26-fld, without filaments mostly 8–10 mm diam, the receptacle including very short terminal pedestal 1.5–2.5(–3) mm; bracts ovate, oblong-obovate or spatulate 0.6–1.2 mm, tardily deciduous (sometimes a larger bract shortly below lowest fls); fls dimorphic, the greenish-white perianth of all 5(–6)-merous, silky-puberulent overall or the tube, alike of calyx and corolla, often glabrate proximally; PERIPHERAL FLS: calyx narrowly campanulate 1.7–2.5 × 1–1.3 mm, either sessile or contracted into a pedicel <0.6 mm, the depressed-deltate teeth 0.1–0.3 mm; corolla 4–6 mm, the erect or ascending, ovate lobes 1.3–1.8 × 0.9–1.2 mm; androecium (11–)15–22-merous, (12–)13–19 mm, the stemonozone 0.55–0.8 mm, the tube 1.2–2.5 mm; ovary shortly stipitate, at anthesis glabrous, conical at apex; style a little longer than longest filaments, a trifle dilated at apex; TERMINAL FL: sessile, the calyx campanulate 2.7–3 × 1.7–2.3 mm, the corolla 5.5–6.5 mm; androecium 18–30-merous, the dilated tube 4–6.5 mm, nearly as long as or a trifle longer than corolla, the free part of filaments thickened proximally. *Pods* solitary, subsessile or contracted into a pseudostipe to ±10 mm, the broad-linear, straight, plano-compressed body 9–18 × 2–3 cm, ±10-seeded, in other respects like that of *A. polycephala*, but the sutures tending to separate from the valves in age; seeds in broad view oval ±11 × 6 mm.

In Atlantic rain forest, araucaria forest, and mata branca, 200–600(–?) m, extratropical E Brazil between ±23°30′ and 30°S latitude: along and near the Atlantic-Uruguay River watershed in S São Paulo, Paraná, and Santa Catarina, there extending to the coast near Florianópolis, and westward along the lower valleys of ríos Iguaçú and Pelotas to immediately adjacent Missiones, Argentina, border; S to NE Río Grande do Sul (S. Leopoldo). — Map 53. — Fl. XI–I. —*farinha seca, Pau gambá.*

Albizia edwallii is closely related to *A. polycephala*, from which it differs in more coarsely divided leaves, in larger, more evidently venulose leaflets, and in more contracted panicles of pseudoracemes immersed in contemporary leaves. The individual flowers are a little larger and tend to be less densely silky-pubescent, while the stamens of peripheral flowers are slightly more numerous and longer. Burkart, who overlooked *P. edwallii*, contrasted the glabrous pod of his *A. austrobrasilica* with the supposedly rufous-tomentellous pod of *Pithecellobium polycephalum* as described by Bentham. It is, however, only in a very immature stage of development that the pod-valves of *A. polycephala* are tomentulose; at maturity they become glabrous, lustrous and venulose like those of *A. edwallii*. Although the morphological differences (summarized in the key that follows) are weak, they are reinforced by the allopatry of the species, one north and the other south of the Tropic line. The following key contrasts *Albizia polycephala* and *A. edwallii*.

1. Central axis of paniculate inflorescence (1–) 1.5–3 dm, this exserted from foliage; capitula without filaments 6–7.5(–8) mm diam; corolla of peripheral fls 2.8–3.5 mm, densely sericeous overall; androecium of peripheral fls 9–14-merous; tropical E Brazil, from Pernambuco S to Rio de Janeiro .... *A. polycephala*
1. Central axis of paniculate inflorescence 1–1.5 dm, immersed in foliage; capitula without filaments 8–10 mm diam; corolla of peripheral fls 4–6 mm, the lobes silky-puberulent but the tube glabrous or glabrescent; androecium of peripheral fls (11–)15–22-merous; extratropical SE Brazil, from E São Paulo to N Río Grande do Sul .... *A. edwallii*

3. Albizia burkartiana Barneby & Grimes, sp. nov., *A. edwallii* et *A. niopoide* arcte afinis, sed foliolis 1-nerviis, costa mediam laminam versus subcentrica, diversa. — BRAZIL. Santa Catarina: in disturbed woodland, 700 m, Capinzal, on upper Río Uruguai near 52°W, 21.XII.1973 (fl), *P. R. Reitz & R. M. Klein 14359.* — Holotypus, NY.

Albizia polycephala sensu Burkart *in* Reitz, Fl. Ilustr. Catar. Legum. Mimos. 115 + map. 1979; non *Pithecolobium polycephalum* Bentham.

Microphyllidious *trees* 8–10 m, the strongly ribbed and sulcate young branchlets, the ventral face of lf-axes, and the pedicels minutely loosely gray-puberulent, the narrow and crowded, olivaceous (brunnescent) lfts bicolored, dull on both faces, paler beneath, except for a small tuft of hairs in the proximal angle of midrib on lower face glabrous facially, minutely ciliolate, the inflorescence a pyramidal, proximally foliate panicle of pseudoracemes not or only shortly exserted from foliage, the small whitish or ochroleucous capitula almost glabrous. *Stipules* narrowly lanceolate 2.5–4 mm, early caducous; a cone 1–1.5 mm long of loosely imbricate 1-nerved stipules in the axil of most leaves proximal to any axillary branchlet or fascicle of peduncles. *Lf-formula* viii–xii/28–33; lf-stks mostly 8–14 cm, the petiole 3–4.5 cm, at middle 0.7–0.9 mm diam, the ventral groove continuous between pinna-pairs, the longer interpinnal segments 6–15 mm; nectary below midpetiole sessile fuscous, shallowly cupular and thick-rimmed, round or vertically elongate 1.6–2.5 mm diam, in profile 0.4–0.6 mm tall, no further nectaries either between or at tip of pinnae; pinnae a little decrescent toward each end of lf-stk, the rachis of longer ones 3.5–5 cm, the longer interfoliolar segments 0.8–1.3 mm; pulvinules 0.25–0.4 × 0.2–0.25 mm; lfts decrescent at each end of rachis, otherwise subequilong, the first pair represented by linear paraphyllidia 0.6–1.5 mm, the blade linear-lanceolate from shallowly semicordate base, subacute, either straight or very slightly arched forward, those near midrachis 4.5–5.5 × 1–1.3 mm; midrib slender, a little forwardly displaced from mid-blade, simple, on upper face immersed but discolored, on lower face weakly prominulous, sometimes an extremely short and faint posterior primary nerve barely perceptible. *Primary axis* of panicle ±1–2 dm, that of pseudoracemes 2–10 cm; capitula 2–4 per node, ±8–14-fld, without filaments (4.5–)5–7 mm diam; the peduncle of distal one at each node 4–9 mm, the receptacle including squat terminal pedestal ±1–1.5 mm long; flowers sessile or the lowest of a capitulum elevated on a pedicel not over 0.3 mm, scarcely dimorphic except for longer staminal tube in the terminal fl (this sometimes abortive), the greenish white perianth 5(or in terminal fl 6)-merous and glabrous except for microscopically ciliolate calyx-teeth and sometimes sparsely minutely puberulent corolla-lobes; calyx widely turbinate-campanulate 1–1.3 mm, the deltate teeth 0.15–0.4 mm; corolla 2.5–3.5 mm, campanulate but somewhat dilated distally, the ovate lobes 0.7–1 mm; androecium 25–32-merous 8–9 mm, the stemonozone scarcely 0.5 mm, the tube of peripheral ones 1.3–1.8 mm, the tube of terminal one 1.8–3.5 mm; ovary at anthesis glabrous, symmetrically conical at apex; style a trifle longer than longest fialments, barely dilated at the stigma. *Pods* not yet known.

In Paraná pine woodland near 600–700 m, known only from the basins of ríos Iguaçú and Uruguai between 52°–53°W and 26°–27°S in far SW Paraná (mun. Clevelândia) and W Sta Catarina, Brazil. — Fl. XII–I(–?). — *Angico-branco.*

This still imperfectly known albizia was mistaken by Burkart for *A. polycephala*, a species of tropical dispersal different in palmately nerved leaflets and sericeous perianth, although similar in general facies. It seems more nearly related to extratropical *A. edwallii* and to the southern populations of *A. niopoides* that have been distinguished as *A. hassleri.* These have similar glabrescent perianth but differ in palmately nerved leaflets with evidently, often strongly displaced costa.

4. Albizia coripatensis (Rusby) Schery, Ann. Missouri Bot. Gard. **37:** 400. 1950, basionymo imperfecte citato. *Pithecolobium coripatense* Rusby, Bull. New York Bot. Gard. **4:** 349. 1907. — "[BOLIVIA. Sud Yungas:] . . . at Coripata, May 6, 1894 [fl. jun.] . . . [*M. Bang*] *No. 2176.*" — Holotypus (3 sheets) NY!; isotypi, BM!, MINN!, US *350066*!. — *Samanea coripatensis* (Rusby) Killip *ex* Record, Trop. Woods **63:** 6. 1940.

Amply mesophyllidious *trees* 5–15 m with trunk attaining ±1(–?) dm dbh, the sulcate new stems, foliage and all axes of inflorescence densely pilosulous with erect and forwardly incurved, at first yellowish but soon whitish hairs to 0.1–0.25(–0.35) mm, the lfts moderately bicolored, dull olivaceous above, hardly paler but more yellowish beneath, the inflorescence composed of short efoliate pseudoracemes of very small whitish capitula both terminal to branchlets of current year and axillary to distal lvs, immersed in foliage. *Stipules* subulate 1.2–1.8 mm, puberulent dorsally like the stem, tardily deciduous. *Lf-formula* (iv–)v–viii/8–12(–13); lf-stks 9–18 cm, the petiole including firm, little-swollen pulvinus, 2.2–3.5 cm, at middle 1.3–1.7 mm diam, the longer interpinnal segments 1.5–2.8 cm; nectary near mid-petiole sessile, low-cupular thick-rimmed, 0.3–0.5 mm tall and in vertical view 0.8–1.6 mm diam, similar but smaller nectaries at insertion of some distal pinnae and at tip of pinna-rachises; pinnae decrescent proximally, otherwise subequilong, the rachis of distal pairs (6–)7–9.5 cm, the longer interfoliolar segments 6–11 mm; lf-pulvinules in dorsal view 0.4–0.5 mm and about as wide; lfts decrescent toward base of rachis, the first pair, contiguous to pinna-pulvinus, represented by paraphyllidia 0.4–1 mm, the rest (disregarding the broader furthest pair) subequiform, in outline inequilaterally oblong-ovate from obliquely truncate, bluntly semicordate base, broadly obtuse at apex, the larger ones 15–23 × 5–10 mm, 2.2–2.8(–3) times as long as wide; primary venation of 3 nerves from pulvinule, the midrib a little forwardly displaced and gently incurved distally, giving rise to 5–7 pairs of major secondaries, the inner posterior primary nerve ascending-incurved to anastomosis well within the revolute margin either close to or well below mid-blade, the outer primary nerve very short, a few random tertiary venules sometimes faintly visible externally, the whole venation pallidly discolored above, finely sharply prominulous beneath. *Primary axis* of pseudoracemes 2–5 cm; peduncles fasciculate by 2–3, at anthesis mostly <1 cm, becoming 6–17 mm in fruit; capitula hemispherical ±10–17-fld, without filaments ±7 mm diam, the receptacle, including a short terminal pedestal, 1.2–2 mm; bracts ovate 0.4–0.6 mm, persistent; fls dimorphic, the peripheral ones progressively a little larger distally, either sessile or raised on a pedicel to 0.3 mm, the terminal one largest and sessile, the perianth of all densely puberulent; calyx campanulate 1–1.5 mm, the depressed-deltate teeth 0.1–0.3 mm; corolla 2.5–3.7 (of terminal fl to 4) mm, the ovate lobes 1–1.3 mm; androecium 7–8.5 mm, the stemonozone ±0.5 mm, the tube progressively longer and wider upward through the capitulum, that of lowest fls <1.5 mm and of distal fls to 4 mm, the white filaments exactly 16 in all fls examined but in the protologue reported as "about 20–25"; ovary at anthesis glabrous, conical at apex; style about as long as stamens, the stigma poriform. *Pods* solitary or rarely 2 per capitulum, sessile but cuneately contracted at base into a pseudostipe 1.5–3 mm, the body broad-linear straight plano-compressed 9.5–16.5 × 2.1–2.6 cm, 9–12-seeded, abruptly recurved-apiculate at obtuse apex, framed by straight or shallowly undulate, 3-angulate sutures ±1 mm wide (the central angle sharp-edged and 0.3–0.5 mm wide), the papery valves fuscous-greenish becoming tan-brown, closely transverse-venulose and densely puberulent overall, the endocarp pallid but not satiny within, not septiferous; dehiscence inert, through both sutures; seeds transverse at middle of pod, strongly compressed, in broad profile ±7 × 6 mm, the translucent testa 0.1–0.2 mm thick, loosely investing the thin-textured greenish embryo, the finely incised pleurogram inversely U-shaped; cotyledons thinly foliaceous, not concealing the radicle, acutely lobed at base.

In moist woodland of the Yungas valleys, now surviving along road banks and in pasture-thickets, 800–1390 m, narrowly endemic (at 15°50′–16°40′S, 66°40′–67°30′W) to prov. Nor and Sud Yungas of dept. La Paz, Bolivia. — Map 57B. — Fl. III–V(–?).

Gray indumentum of the mature foliage, relatively few and broad, marginally revolute leaflets, a contracted panicle of pseudoracemose capitula, and dispersal restricted to middle elevations of the Bolivian Yungas effectively characterize *A. coripatensis* among albizias akin to *A. polycephala.*

5. Albizia tomentosa (M. Micheli) Standley, J. Wash. Acad. Sci. **13:** 6. 1923. *Pithecolobium tomentosum* M. Micheli, Mém. Soc. Phys. Genève **34:** 285, t. 28. 1903. — "[MEXICO. Michoacán:] . . . rives de l'Espiritu Santo [near 18°50′N, 101°75′W — see McVaugh, Candollea **13:** 197. 1951], 600 m, [19] avril, 1898, [*Langlassé*] n. 107; Zilmatango [on original label Zihuatanejo, Guerrero 6] août 1898, n. 280." — Lectotypus (Standley, 1922: 396): *Langlassé 107*, G!; isotypi, P!, NY (fragm.)!, US *385667*!; paratypi, *Langlassé 280*, G!, P!.

Albizzia (?) *purpusii* Britton & Rose, N. Amer. Fl. **23:** 45. 1928. — "Remulatero [= Rancho Remudadero, 19°15′N, 96°34′W], Veracruz, April, 1922, [*C.A.*] *Purpus 8723*." — Holotypus, NY!; isotypi, UC *214372*!, US *1169664*!.

Albizzia hummeliana Britton & Rose, N. Amer. Fl. **23:** 46. 1928. — "Near Hill Bank, New River Lagoon, British Honduras, January 26, 1926, *Samuel J. Record 27*." — Holotypus, US *1209912*!; clastotypus (fragm.) + photo, NY!. — Equated with *A. tomentosa* by Standley & Steyermark, 1946: 18.

Albizzia tomentosa sensu Standley, Contr. U.S. Natl. Herb. **23:** 1659. 1926; Britton & Rose, 1928: 45; Standley & Steyermark 1946: 18; McVaugh, 1987: 46.

Amply foliate *trees* 5–25(–30) m with trunk attaining 7(–8) dm dbh, sometimes fertile as treelets 2–4 m, the bluntly angulate young stems and all axes of lvs and inflorescences pilosulous with widely spreading-incurved or forwardly incumbent, pallid or lutescent hairs to 0.1–0.25(–0.35) mm, the lvs moderately bicolored, the dull olivaceous lfts somewhat paler beneath, either puberulent on both faces (but often sparsely inconspicuously so) or pilosulous along midrib beneath, the inflorescence composed of efoliate pseudoracemes of small fasciculate capitula, the lowest of these usually subtended by a developed lf but the distal ones forming an immersed or shortly exserted, either efoliate or hysteranthously foliate panicle (random pseudoracemes reduced to a single axillary fascicle), the axis of each element of inflorescence determinate. *Stipules* subulate or narrowly triangular 0.8–2.2 mm, caducous. *Lf-formula* iii–xiii (–xv)/6–22(–28); lf-stk of lvs at and near base of panicle 4.5–22(–25) cm, the petiole 1.2–3.5(–5) cm, at middle (0.8–)1.2–2.4 mm diam, the ventral groove continuous between pinnae, the longer interpinnal segments 11–21(–25) mm; nectary below midpetiole, often adjacent to lf-pulvinus, sessile, shallowly patelliform, in outline narrowly elliptic (1.5–)1.8–5.5(–8) mm, no further nectaries on lf-stk but small ones on pinna-rachises at furthest 1–3 pairs of lfts; pinnae distally accrescent, the rachis of longer pairs 4.5–10.5(12) cm, their longer interfoliolar segments 4–9(–11) mm; lfts a little decrescent at one or both ends of rachis, otherwise subequilong, in outline symmetrically or asymmetrically oblong, broad-linear, or oblong-elliptic from an either obliquely truncate, or shallowly cordate, or semicordate base, at apex either obtuse or obtuse and apiculate, those near midrachis (11–)12–22(–24) × 3–8(–11.5) mm, (2–)2.2–4 times as long as wide; venation initially palmate, the slender, dorsally prominulous, straight or beyond middle gently incurved midrib either nearly centric or forwardly displaced to divide blade 1:1.5–2.5, the inner posterior primary nerve incurved-ascending beyond midblade, the outer 2–3 progressively shorter, the anterior primary faint or 0, the midrib giving rise on each side to 5–8 weak secondary nerves brochidodrome shortly within the revolute margin, tertiary venulation weak and random, the whole venulation immersed or almost so on ventral face. *Primary axis* of pseudoracemes (2–)4–14 cm; longer peduncles 12–26 mm; capitula (14–)20–40-fld, the receptacle not over 2 mm, terminal pedestal 0; bracts persistent through anthesis, spatulate (or the lowest obovate-cuneate) (0.4–)0.7–1.3 mm, the incurved tip pilosulous externally; fls dimorphic, the peripheral ones either shortly pedicellate or subsessile, the terminal one sessile, stouter but scarcely longer, its androecium modified; perianth 5 (or in random terminal fls 6)-merous, the calyx usually pilosulous externally around the orifice, sometimes only ciliolate, sometimes puberulent overall, the greenish corolla either glabrous or distally puberulent, the filaments white; PERIPHERAL FLS: pedicel (0–)0.3–0.8 × 0.3–0.6 mm; calyx turbinate, 1.4–2.4 × 1.1–1.6 mm, the tube bluntly pentagonal, the depressed-deltate or -ovate teeth 0.2–0.5 mm; corolla 3–5(–5.7) mm, the lance-ovate lobes 1.1–1.8 × 0.65–1.1 mm; androecium 18–32-merous, 9.5–17 mm long, the stemonozone 0.1–0.8 mm, the tube 1.6–3.5(–4.4) mm; ovary slenderly ellipsoid, attenuate into style, glabrous: stigma scarcely dilated, not more than 0.2 mm diam; TERMINAL FL: calyx broadly campanulate 2–3 × 2–2.6 mm, the teeth 0.2–0.6 mm; corolla 4.2–5.5 mm; androecium 26–38-merous, the tube (2.5–)3–6(–9) mm, at apex 1–2 mm diam, the filaments thence either basally thickened or filiform. *Pods* 1(–2) per capitulum, sessile or almost so, in profile broad-linear straight, plano-compressed, cuneately attenuate at base, abruptly apiculate, when well fertilized 10–14 × 1.5–2.7 cm, 8–12-seeded, the stiffly papery, fuscous-green or ultimately stramineous, closely transverse-venulose, either glabrous, or minutely puberulous, or subvelutinous valves framed by dilated, dorsally carinate (but not winged), puberulent sutures (0.7–)1.1–1.6 mm wide, the endocarp smooth but not satiny, tan overall; dehiscence tardy, inert, through both sutures; seeds transverse at middle of pod, basifixed on compressed, apically sigmoid funicle, oblong-ellipsoid, strongly compressed, in broad profile 6.5–8 × 4.5–5.3 mm, the smooth tan testa translucent when soaked (resembling grapeskin) loosely investing the embryo, commonly with a white enlarged ridge produced from hilum to just below the U-shaped pleurogram, this ±3 × 1.7–2.5 mm.

In drought-deciduous woodland and at edge of evergreen forest or in savanna tree islands, from sea level to 450 m, scattered in lowland tropical Mexico and adjacent Belize and Guatemala: in Mexico extending on the Pacific slope from Jalisco through the Balsas basin and along the coast to Chiapas, and (in slightly modified form) on the Gulf-Caribbean slope from centr. Veracruz to Yucatán, S into centr. Belize and NW Guatemala (Huehuetenango, Petén). — Map 54. — Fl. III–VII. — *Guejillo* (*guajillo*), *parotillo* (SW Mexico); *tepozonte* (Veracruz); *cantemoc* (Guatemala).

As defined by the foregoing description and synonymy, *A. tomentosa* is polymorphic in foliage, though nearly uniform, except for length of filaments, in the flower, and essentially so in form of the fruit. On the Pacific slope the leaf-formula is mostly ii–v/6–9(–10) and the androecium is 12–17 mm long; on the Gulf-Caribbean slope the leaf-formula is v–xii (–xv)/(10–)12–28 and the androecium of peripheral

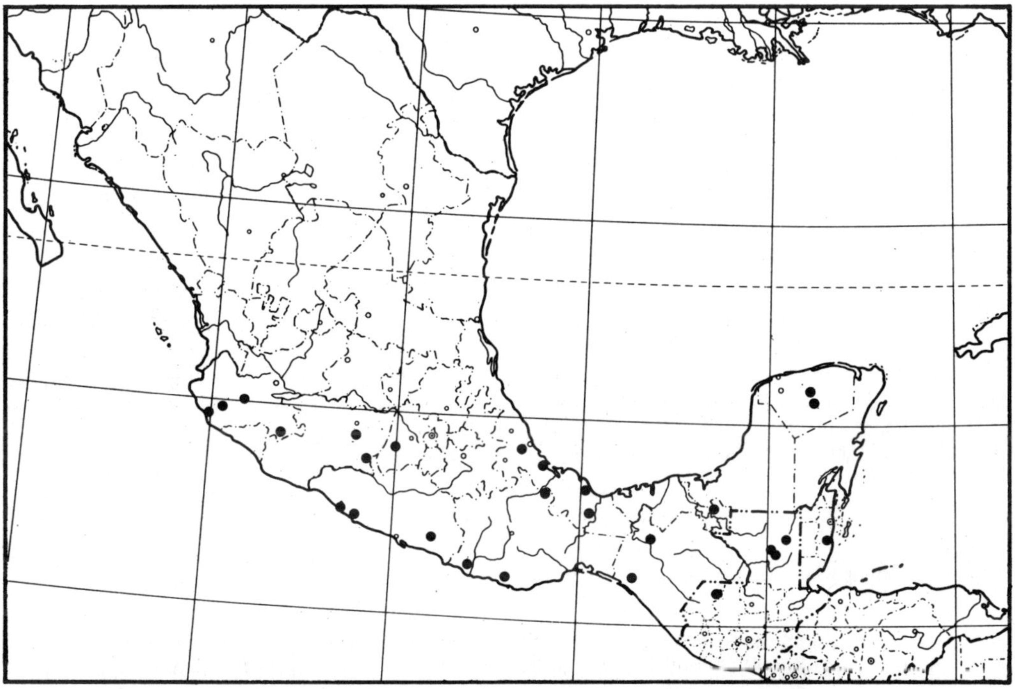

MAP 54. Mexico, Guatemala, and Belize. Range of *Albizia tomentosa* (M. Micheli) Standley.

flowers is only 9.5–12 mm long. The leaflets vary in outline and proportions, being generally broader when fewer, and more inequilateral when more numerous and narrower, but the variation is far from stabilized on either slope of the continent. The eastern form is confined to the *selva mediana caducifolia* of the coastal plain and lower interior foothills. We suspect that *A. purpusii*, corresponding with the eastern phase of the species, may deserve infraspecific rank. *Albizzia hummeliana* was described from a fruiting specimen of this phase of *A. tomentosa*.

6. Albizia sinaloënsis Britton & Rose, N. Amer. Fl. **23:** 45. 1928. — "Sinaloa. Type from the vicinity of Fuerte, March 26, 1910, *Rose, Standley & Russell 13559*." — Holotypus, NY!; isotypus, US *636383*!.

Albizzia sinaloënsis sensu Wiggins, 1964: 591.

Amply microphyllidious *trees* attaining 20 m, with yellowish bark, the young branchlets and all lf-axes and peduncles densely pilosulous with spreading whitish hairs to 0.3–0.4 mm, the pallidly olivaceous lvs moderately bicolored, the lfts inconspicuously puberulent on upper face, paler and thinly pilosulous beneath at least along the prominulous midrib, the capitula solitary and paired in lf-axils, the annual leafy shoot continuous beyond them. *Stipules* caducous, not seen. *Lf-formula* iv–ix/16–20; lf-stks 8–19 cm, the petiole 2.5–5.5 cm, at middle 0.9–1.3 mm diam, the longer interpinnal segments 1.4–2.3 cm; nectary below midpetiole, often close to lf-pulvinus, sessile or immersed, elliptic shallowly concave 1.4–2.6 mm in long diameter, smaller cupular nectaries between 1–2 furthest pinna-pairs, and yet smaller ones between 1–2 furthest pairs of lfts; pinnae often not exactly opposite, proximally decrescent, the rachis of longer ones ±5–7 cm, the longer interfoliolar segments 2.5–4 mm; lft-pulvinules 0.4–0.5 × 0.4 mm; lfts little graduated, narrowly oblong-elliptic or broadly linear from inequilateral, shallowly semicordate base, either obtuse or obscurely apiculate, those near and beyond midrachis 10–14 × 3–4 mm, 3–4 times as long as wide; venation of 2–3 primary nerves from pulvinule, the slender, dorsally cariniform midrib subcentric, straight or nearly so, the inner (or only) posterior primary nerve incurved-ascending nearly to or shortly beyond midblade, the outer posterior one short and weak, the midrib giving rise on anterior side to 4–6 and on posterior side to 2–4 simple secondary nerves brochidodrome shortly within the revolute and ciliolate margin, the whole venulation immersed or almost so on upper face, slenderly prominulous beneath. *Peduncles* ±10–35 mm, the subhemispherical receptacle ±1.5 mm diam; inflorescence known from one not quite mature specimen: bracts spatulate ±2 mm; fls homomorphic, all subsessile; perianth 5-merous, the calyx and greenish corolla pilosulous externally on lobes and near orifice, calyx turbinate-campanulate ±2.0 mm, the depressed-deltate or -ovate teeth ±0.5 mm; corolla ±3.8 mm, the lance-triangular lobes ±0.9–1 × 0.7 mm; androecium ±36-merous, 5–5.5 mm long, the tube 2–3 mm; ovary shortly ellipsoid, abruptly attenuate into style, glabrous. *Pods* 1–2 per capitulum, sessile, broad-linear, straight, 9–15 × 1.6–2.4 cm, cuneately or abruptly contracted at base, abruptly apiculate, planocompressed, 8–12-seeded, the stiffly papery, brownish or fuscous-stramineous valves framed by dilated, 3-angulate but not winged sutures 1.6–2 mm wide, low-convex over seeds, the exocarp minutely puberulent (glabrescent) and transversely venulose, the endocarp cream-buff, smooth but not satiny, the cavity continuous; dehiscence inert, through both sutures; seeds transverse at middle of pod, basifixed on an apically sigmoid, compressed funicle, compressed-lentiform, in broad view 6–8 × 5–6.2 mm, the smooth buff testa becoming translucent when wetted (grapeskin-like), closely investing the embryo, commonly with a horny white ridge produced from hilum to just below pleurogram, this finely engraved, U-shaped ±2 × 1.5 mm.

In deep alluvial soils of river valleys below 200 m, along the W piedmont of Sa Madre Occidental between the Yaqui and Fuerte rivers in S Sonora and N Sinaloa (lat. 26°20′–27°30′N). — Fl. period unknown; fr. ripe X–III. — *Palo Joso.*

For lack of adequate materials we have but an imprecise perception of *A. sinaloënsis*, which seems in most respects a form of *A. tomentosa* adapted to the severe climate around the south margin of the Sonoran Desert. The capitula of homomorphic flowers axillary to coeval leaves set the species apart from all other New World *Albiziae*, and recall species of *Havardia*. On the basis of overall similarity of fruit and seeds with small pleurogram and (commonly) horny ridge produced from hilum to near the pleurogram (the latter unknown to us in other New World Ingeae), we assume a close relationship to *A. tomentosa*. The descriptions by Britton and Rose (1928: 45) suggest a difference in leaf-formula (*A. tomentosa* iii–iv/6–11; *A. sinaloënsis* vi–ix/12–15), but this has proved ineffective.

7. Albizia leonardii Britton & Rose *ex* Barneby & Grimes, sp. nov., florum (nisi alabastra valde immatura) defectu affinitatis incertae ulterius inquirendae, inter congeneres antillanos his signis autem recognoscenda: foliorum formula iii–v/9–12; pinnae longiores 3–9 cm longae, foliolis oblongis subretusis 6–16 × 2.8–7 mm usque e pulvinulo palmatim 3–4-nerviis et reticulatis; legumina lato-linearia planocompressa ±13–14 × 2.2–2.5 cm, basi in stipitem ±1 cm contracta 14-sperma, valvulis tenuiter chartaceis glabris lucidis, secus utramque suturam dehiscentibus. — HAITI. Dépt. du Nord: in dry thicket, 350 m, Habitation Baille n. of Atalaye plantation, S. Michel de l'Atalaye, 26 nov 1925, *E. C. Leonard 7490.* — Holotypus, US; clastotypus + photo, NY. — Dépt. Artibonite: zona árida, 130 m, Dubedou (de Gonaives) ±20 km al N. de Gonaives en la carretera a Port-de-Paix, 8 jun 1985 (young bud), *T. Zanoni & al. 34986.* — Paratypi, JBSD, NY.

Unarmed *treelets* 3(–?) m with rough gray annotinous and older branchlets, the hornotinous ones together with lf- and inflorescence-axes pilosulous with pallid incurved hairs to ±0.1–0.2 mm, the plane papery olivaceous lfts moderately bicolored, glabrous above, paler and remotely strigulose beneath, the inflorescence (known only in bud and in mature fruit) apparently consisting of efoliate (possibly hysteranthously foliate) pseudoracemes of subglobose capitula axillary to and surpassed by fully expanded primary lvs. *Stipules* subulate 1–3 mm, either deciduous or subpersistent. *Lf-formula* iii–v/9–12; lf-stks of larger lvs 3–7.5 cm, the petiole 7–20 mm, at middle 0.8–1.5 mm diam, the longer interpinnal segments 8–16 mm, the ventral groove continuous between pinna-pairs; nectary immediately below first pair of pinnae sessile cupulate thick-rimmed 0.6–1 mm diam, in profile 0.2–0.5 mm tall, a similar, scarcely smaller nectary at tip of lf-stk and a much smaller one at tip of most pinna-rachises; pinnae distally accrescent, the rachis of longer ones 3–9 cm, the longer interfoliolar segments 3–8 mm; paraphyllidia minute subulate caducous; lft-pulvinules 0.4–0.9 × 0.4–0.5 mm, not wrinkled; lfts proximally decrescent, otherwise subequilong, oblong from shallowly semicordate base, at apex broadly shallowly retuse and minutely apiculate, the larger ones 6–16 × 2.8–7 mm, 1.9–2.3 times as long as wide; venation of 3–4 nerves from pulvinule, the straight midrib forwardly displaced to divide blade ±1:2, the inner posterior primary nerve produced to anastomosis well within the plane margin, near or little beyond midblade, the outer posterior ones much shorter, the midrib giving rise on broad side of blade to 2–4 major (and random minor) secondary nerves and to a lax reticulum of veinlets, the whole venation finely prominulous on both faces. *Capitula* fasciculate by 2–5 in the axil of an incipient lf, ±12–20-fld, in young bud globose gray-strigulose, the floral bracts minute. *Pods* broad-linear straight, plano-compressed, cuneately contracted at base into a stipe ±1 cm, obtuse at apex, ±13–14 × 2.2–2.5 cm, ±14-seeded, the thinly papery, brownish-stramineous, glabrous sublustrous valves framed by straight, dorsally plane or shallow-sulcate sutures 1.2–1.5 mm wide, low-convex over each seed, delicately cross-nerved, within pallid but scarcely lustrous, the cavity continuous; dehiscence through the length of both sutures, inert; seeds horizontal at middle of pod, basifixed on almost filiform funicle, in broad view obovate or nearly round 5.2–6.5 × 4.8–6 mm, strongly compressed, the coriaceous, smooth castaneous testa ±0.2 mm thick, only loosely investing the embryo, the deeply U-shaped pleurogram 2.3–2.5 × ±1.5 mm.

In dry thickets, 130–350 m, apparently rare and local, known only from NW Haiti, in depts. Artibonite and Nord. — Fl. ±VII; fr. ripe in XI.

Until the flowers are obtained, the affinities, even the genus, of *A. leonardii* will remain controversial. Assuming that it is correctly placed in *Albizia* sect. *Arthrosamanea*, as the ripe fruit and incipient inflorescence suggest, and that it is native in the chaparral of remote northwest Haiti, the chances of its having been described previously are extremely small. In the setting of Hispaniola, *A. leonardii* is obviously different from the native *A. berteriana* and the introduced *A. carbonaria* in leaf-formula alone; and from the common weedy *A. lebbeck* in smaller leaves and leaflets, and narrower, thinly papery pod not raised as a dome over each seed (the seeds moreover one-half smaller). For ecogeographic reasons we have compared *A. leonardii* also with *Pseudosamanea cubana*, but this not only has the characteristic leaflet-venation of its genus but densely tomentulose pods containing 24–30 (not about 14) seeds.

The pod of *A. leonardii*, detached from the inflorescence, might pass for that of *Acacia*, *Leucaena*, or *Lysiloma*. The plant is unarmed, which is unusual in American *Acacia*, and does not resemble in habit or foliage any species of the genus known from Hispaniola (see Barneby & Zanoni, Moscosoa **5:** 4–27. 1989). *Leucaena* is represented on Hispaniola by *L. leucocephala* (Lamarck) DeWit and *L. pseudotrichodes* de Candolle, unmistakably different in foliage. And the two species of *Lysiloma* that reach Hispaniola from the west are so distinct in leaflet-venation, membranous stipules, and livid exfoliating epidermis of the pod-valves that they can be set aside as irrelevant.

We accept in principle that description of a new species from imperfect or litigious material is to be discouraged. Here, however, we have a tree, forming part of a highly endangered flora, that has been known for 65 years, but has only recently and only once been seen again. Presumably it is rare, local, and a candidate for protection in its native habitat and for propagation in horticulture. We feel justified in these circumstances in placing its existence on record, in the hope that doing so will encourage further observation of its morphology and relationships.

8. Albizia barinensis Cárdenas, Ernstia **21:** 5 + fig. 1983. — ". . . muy cerca de Punta de Piedra, Edo. Barinas [Venezuela] . . . 3–IV–1976 [fl, fr], *Lourdes Cárdenas de Guevara 2273*." — Holotypus, MY (not seen); isotypi, BM!, NY!, US!, VEN!.

Broad-crowned microphyllidious *trees* 6–20 m with trunk attaining 3–5 dm dbh, the young stems, leaves (except for glabrescent upper face of lfts) and inflorescence softly pilosulous throughout with ascending or forwardly incurved hairs to 0.15–0.3 mm, the brownish olivaceous lvs moderately bicolored, dull on both faces, a little paler beneath, the inflorescence an efoliate simple pseudoraceme or small efoliate panicle of pseudoracemes terminal to branchlets and surpassed by upper lvs, the capitula of small greenish white fls globose in bud, subhemispherical at anthesis. *Stipules* linear-subulate 1–2.5 mm, caducous, leaving a minute or barely perceptible scar. *Lf-formula* (vi–)viii–xii(–xiv)/16–24(–"28"); lf-stks 10–16(–21) cm, the petiole ±3–4.5 cm, at middle 1–1.8 mm diam, the longer interpinnal segments 12–16 mm, the ventral groove continuous between pinna-pairs except where interrupted by a nectary; nectary either beyond or below midpetiole sessile, depressed-cupular 0.8–1.4 mm diam, in profile 0.3–0.4 mm tall, the thickened margin densely puberulent, similar but smaller nectaries at insertion of 1–4 furthest pairs of pinnae and yet smaller ones between several distal pairs of lfts; pinnae decrescent proximally, thence subequilong, the rachis of longer ones 7–9(–10) cm, the longer interfoliolar segments 3.5–5 mm; lft-pulvinules in dorsal view 0.3–0.5 mm, the lfts subsessile against rachis; lfts subequilong except at very ends of rachis, the first pair degraded to linear-subulate paraphyllidia 0.6–1 mm, the rest oblong from obliquely truncate base, depressed-deltate at apex, the larger ones 8.5–11.5(–12) × 3.2–4.8 mm, 2.3–3 times as long as wide; venation of 3(–4) nerves from pulvinule, the straight midrib a little forwardly displaced, giving rise on one or both sides, near and beyond middle, to 2–4 faint secondary nerves looping or expiring close within the revolute margin, the inner posterior primary nerve incurved to anastomosis at or well beyond midblade, a faint random tertiary venulation perceptible on upper face, the primary and secondary nerves slenderly prominulous on both faces. *Primary axis* of pseudoracemes 6–16 cm, that of branches, when present to ±2.5 cm; peduncles (1–)2–6 per node of inflorescence, the longest (distal) peduncle 9–16 mm; capitula (12–)16–26-fld, without filaments ±8–10 mm diam, the globose receptacle 0.6–1 mm diam (terminal pedestal lacking); bracts linear-spatulate or obovate 0.5–0.8 mm, deciduous; fls homomorphic, subsessile or the lowest contracted into pedicel 0.2–0.5 mm; perianth 5(–6)-merous, finely pilosulous overall; ovary symmetrically conical at apex, at anthesis glabrous but pilosulous soon after fertilization; calyx narrowly campanulate, faintly 5-nerved, the tube 1.4–1.8 × 1–1.2 mm, the depressed-deltate teeth ±0.2 mm; corolla 4–4.5 mm, the sinuses between lobes variably cut back, 0.5–2 mm deep; androecium 26–30-merous, 7–9 mm, the stemonozone a little less than 1 mm, the tube 3.8–4.8 mm, a little shorter or longer than corolla. *Pods* subsessile, broad-linear or linear-elliptic in profile, planocompressed, contracted at base into a very short stipe-like neck, abruptly apiculate, the body (7–)10–14 × 1.4–2.5 cm, 7–10-seeded, the thin-textured valves early glabrescent, glaucous-green at length becoming papery and brownish, finely cross-venulose, low-convex over each seed, internally pale buff but not satiny, framed by straight simple sutures ±1 mm side; dehiscence tardy, inert, through both sutures; funicle ribbonlike ±1 mm wide, not sigmoid, appearing to enter seed well above middle; seeds at middle of valves, appearing horizontal but in reality pitched backward and broader than long, in broad profile elliptic 4.5–5 × 7–9 mm, strongly compressed, the hard bony testa pale fawn, smooth but not lustrous, ±0.15 mm thick in section, the small pleurogram obliquely C-shaped, the obtuse periphery girdled with a detachable, very narrow fringe of cellular tissue.

In seasonally dry woodland and surviving in pastures, ±200–300 m, scattered over the middle and lower Orinoco valley in Venezuela, from Barinas and Portuguesa E to Anzoátegui and N Bolívar, extending N to the Caribbean coast in Sucre; apparently isolated on the Kanuku Mts in SW Guyana. — Map 55. — Fl. III–VIII(–XI), the fruits long persisting and sometimes contemporary with new fls.

Oblong obtuse leaflets and densely gray-pilosulous flowers are distinctive features of *A. barinensis*. The similar, but remotely allopatric, *A. coripatensis* has lower leaf-formula and larger individual leaflets.

9. Albizia adinocephala (Donnell Smith) Britton & Rose [*ex* Record, Trop. Woods **10:** 22. 1927, basionymo haud citato], N. Amer. Fl. **23:** 47. 1928. *Pithecellobium adinocephalum* Donnell Smith, Bot. Gaz. (Crawfordsville) **57:** 419. 1914. — "Ad fundum *La Verbena* prope Alajuelita, San Jose, Costa Rica, alt. 1000 m, Ad. Tonduz; Aug. 1894 [fr.], no. 8932; Dec. 1894 [fl. + fr.], no. 9077 . . . Peninsula Nicoya, Costa Rica, Jan 1900, Ad. Tonduz, n. 13531." — Lectotypus, *Tonduz 9077*, US!; isotypi, G!, NY!, US (see next entry)!; paratypus, *Tonduz 8932*, US *471616* (fr.), *942455* (fl.).

Pithecollobium discolor Pittier, Contr. U.S. Natl. Herb. **20:** 464. 1922. — "Type in U.S. National Herbarium, no. 471907, collected at La Verbena, near San José, Costa Rica . . . A. Tonduz (Inst. Fis. Geogr. Costa Rica no. 9077)." — Holotypus (fl.) US *471907*!; isotypi (fr.), US *354245* + *Tonduz 13531* and *Pittier 5737*, all US!. — Based on an isotypus of the preceding and equated with it by Britton & Rose, 1928, l. c. — Non *P. discolor* Britton, 1914.

Macrophyllidious broad-crowned semideciduous *trees* 10–30 m with smooth gray trunk attaining 3–5 dm dbh, the annotinous branches gray, finely sulcate lengthwise, the young branchlets and all axes of lvs and inflorescence strigulose with subappressed whitish hairs <0.2 mm, the papery lfts moderately bicolored, above olivaceous sublustrous and often glabrous except for cilolate midrib but sometimes remotely strigulose overall, beneath paler dull and always at least thinly or remotely puberulent overall, the inflorescence composed of mostly simple, rarely few-branched, efoliate pseudoracemes of small dense globose capitula terminal either to primary or occasionally to some lateral hornotinous branchlets (rarely a few axillary capitula also), mostly surpassed by foliage. *Stipules* triangular-subulate 0.5 mm or less, early caducous and lacking from many specimens, perhaps sometimes obsolete. *Lf-formula* i–iii/(2–)3–4(–5); lf-stks (1.5–)2.5–9(–10.5, in some juvenile lvs, no further mentioned, to 20) cm, the interpinnal segments (1–)1.5–3.3(–4.5) cm; nectary situated below midpetiole, sometimes close to lf-pulvinus, sessile or shallowly recessed into groove of petiole, shallowly cupular with smooth rim, vertically dilated, <0.5 mm

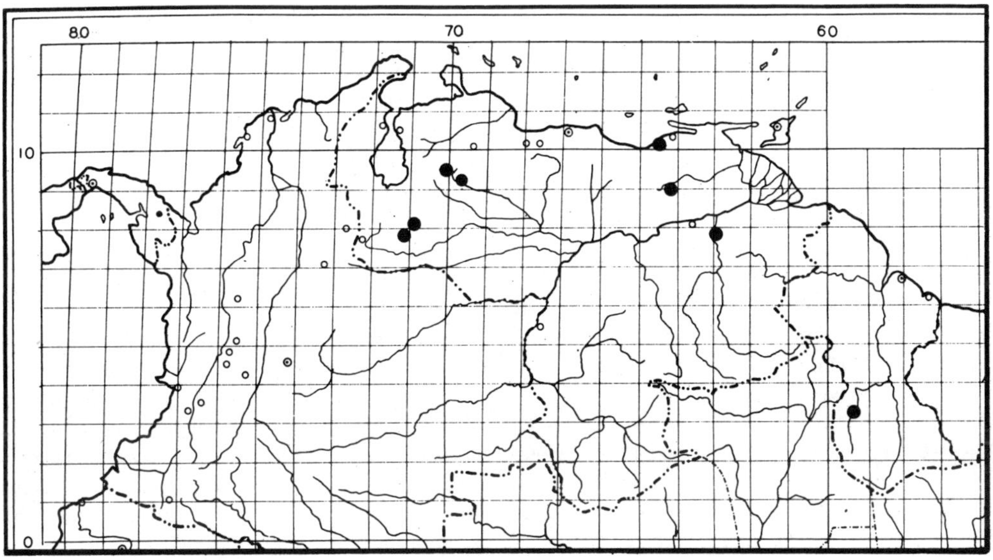

MAP 55. Venezuela and Guyana. Range of *Albizia barinensis* Cárdenas.

tall in profile but in vertical view (1–)1.6–3.2(–4) mm in long diameter, mostly 3–6 times as long as wide, similar but smaller nectaries between 1–2 furthest pairs of lfts but none between pinna-pairs; pinnae when 2 pairs or more slightly accrescent distally, the rachis of furthest or of the single pair 3–8 cm, the longer interfoliolar segments 9–21(–23) mm; lft-pulvinules 1–1.7(–2.3) × 0.6–0.8 mm; lfts accrescent distally but otherwise subequiform, subsymmetrically elliptic or elliptic-ovate from cuneate or less commonly flabellate base, very shortly and bluntly acuminate, the longer ones 3–5.5(–6) × 1.2–2.6(–3) cm, (1.8–)2–2.6(–3) times as long as wide; primary venation of 3 nerves from pulvinule, the posterior and anterior ones both short, weak and submarginal, the much stronger subcentric, straight or very gently incurved midrib giving rise on each side to 8–13 major secondary nerves brochidodrome shortly within the revolute margin and giving rise to an open reticulum of venules, the whole venation prominulous on both faces of blade and often pallidly discolored on the upper. *Primary axis* of panicle or of individual pseudoracemes (3–)4–10(–15) cm; peduncles mostly 2–5 (–6) per node of inflorescence, randomly solitary, the longer peduncles 9–27 mm; capitula globose, 16–36-fld, before anthesis gray-puberulent, at anthesis (without filaments) 6–8(–9) mm diam, the receptacle including pedestal of terminal fl (sometimes lacking) ±1–2 mm; bracts heteromorphic, the lowest ovate, 5 mm or less, the interfloral ones linear-spatulate 6–9 mm, the tiny incurved blade puberulent dorsally, all commonly persistent through and some time after anthesis; fls dimorphic but not greatly so, the terminal flower sometimes abortive; perianth 5-merous, the corolla greenish white, fragrant, the filaments white; PERIPHERAL FLS: calyx turbinate 1.3–1.7 mm, contracted at base into a pedicel 0.2–0.9 mm, the tube 5-grooved below the sinuses of the teeth, these 0.1–0.3 mm, obtuse, incurved and dorsally convex at apex, puberulent externally; corolla narrowly trumpet-shaped 3–4.3 mm, the lance-ovate subacute lobes 1.2–1.5 × 0.7–0.8 mm, glabrous or almost so; androecium 8–11 mm, the stemonozone 0.2–0.5 mm, the tube 2–3.3 mm; ovary glabrous, symmetrically tapering at tip, the style about as long as androecium, the stigma poriform; TERMINAL FL: sessile or almost so, calyx campanulate 1.8–2.2 × 1.7–2 mm, the teeth 0.2–0.3 mm; corolla ±4.5–4.8 mm; androecium 8.5–11.5 mm, the tube 4–6 mm. *Pods* often solitary, sometimes 2–3 per capitulum, sessile but cuneately contracted at base into a short pseudostipe, the body broad-linear straight, plano-compressed, when fully fertile 10–14.5 × 1.6–2.7 (locally in Honduras –3.5) cm, 10–13-seeded, the glabrous valves scarcely elevated over seeds, framed by 3-carinate sutures (the terminal rib produced as an incipient wing 0.1–0.5 mm wide), green or fuscous becoming light tan and discolored over seeds, densely sharply transverse-venulose, pallid but scarcely lustrous internally; dehiscence inert, through both sutures; funicle sigmoid apically, straight below; seeds (heavily predated by bruchid larvae) transverse at midpod, attached at very base to funicle, compressed-lentiform 6.5–7 × 5–5.5 × 1–1.5 mm, narrowly obtusely marginate, the testa smooth tan translucent, loosely investing the thin green embryo, the pleurogram inversely U-shaped, 4.5–6 × 1.5–1.75 mm.

In upland forest and in hill country or on river banks in the lowlands, attaining 1100 m in Costa Rica, widespread from S Mexico (Oaxaca, Chiapas), N Guatemala (Petén) and Belize SE to centr. Panama (Coclé), planted in the canal zone of Panama, on Puerto Rico, and perhaps elsewhere. — Map 56. — Fl. VI–XII, the papery pods long persistent on annotinous branchlets. — *Gavilancillo, palometa* (Guatemala); *chaculatapa, chilpite* (*chipilse*), *conacaste blanco* (El Salvador); *barbona* (Nicaragua).

The few (3–4, exceptionally 5 pairs) of ample, ovate-elliptic leaflets, the small, crowded capitula (to which the epithet refers), the dorsally convex calyx-teeth, and the sharply closely cross-venulose fruit contribute to an unmistakable facies.

10. Albizia niopoides (Bentham) Burkart, Las Legum. Argent., ed. 2, 542. 1952. *Pithecolobium niopoides* Spruce *ex* Bentham, Trans. Linn. Soc. London **30:** 591. 1875 & *in* Martius, Fl. Bras. **15(2):** 447. 1876. — Typus infra sub var. *niopoide* indicatur.

Amply microphyllidious *trees* flowering when 4–7 m tall but in favorable conditions attaining a stature of 25–40 m and trunk 3.5–12(–15) dm dbh, the bark pallid smooth, peeling in flakes, the foliage bicolored, the narrow crowded lfts olivaceous and somewhat lustrous above, paler dull beneath, the young stems, ventral face of all lf-axes, and all axes of inflorescence either thinly or densely villosulous with fine, erect and incurved, pallid or yellowish hairs to 0.1–0.3 mm (but in var. *colombiana* the indumentum reduced to a rudiment along the ventral rib of pinnae), the lfts mostly glabrous on both faces (rarely puberulent ventrally near base) and minutely ciliolate, the inflorescence composed mostly or entirely of efoliate pseudoracemes of small globose capitula either all axillary to and shorter than the suffulcrant lf or less often partly or wholly terminal and paniculate (but not surpassing the foliage), rarely a few peduncles axillary to fully developed lvs. *Stipules* linear-subulate 0.7–2.5 mm, very early dry fugacious, absent from many flowering and from all fruiting specimens. *Lf-formula* ix–xix (seldom over xii)/28–57(–63); lf-stk of major lvs (4.5–)6–16(–21) cm, the petiole together with firm livid pulvinus (1.5–)2–4.5 cm, the longer interpinnal segments 8–24 cm; a round or elliptic, cupular thick-rimmed nectary near midpetiole, either sessile or obconic, in profile 0.2–0.5(–0.6) mm tall and in vertical view 0.6–1.5(–1.8) mm diam, a similar but smaller nectary sometimes between 1–2 furthest pinna-pairs and a yet smaller one occasional at tip of pinna-rachises; pinnae decrescent proximally, either the furthest or the antepenultimate pair longest, the rachis of longer ones 5–11.5 cm, the longer interfoliolar segments 0.6–2.5 mm; lft-pulvinules 0.2–0.4 × 0.3–0.55 mm; lfts decrescent at base and sometimes also at extreme apex of pinna-rachis, otherwise subequilong, the first pair commonly reduced to subulate paraphyllidia, the blades linear or linear-lanceolate from bluntly auriculate base, obtuse or deltately subacute at apex, commonly arched forward and low-convex ventrally, the larger ones 5–13 × 0.9–2.5 mm; primary venation palmate, the asymmetric midrib submarginal at base of lft but becoming subcentric at its apex, at middle dividing blade 1:2–5, either simple or 1–2-branched on proximal side, the inner posterior primary nerve ascending, close to the almost plane margin, to or beyond midblade, the 1–2 outer posterior primaries short or obscure, the whole venation immersed or weakly elevated on upper face, prominulous beneath. *Primary axis* of pseudoracemes (or of panicle-branches) 1.5–6(–10) cm; peduncles fasciculate by 2–3, 4–14 (or random axillary capitula to 25) mm; capitula without filaments 5–8 mm diam; bracts spatulate, <1 mm, persistent; fls sessile or cuneately contracted into a very short pedicel, homomorphic or nearly so, the terminal fl sometimes a trifle longer than the peripheral ones but its androecium not modified except for slightly longer filament-tube, the perianth greenish white, the calyx most commonly puberulent overall, or sometimes glabrescent externally but ciliolate, or sometimes quite glabrous, the tube 0.9–1.5(–1.7) mm; corolla (2.7–)3–4.5 mm, the often unequal ovate lobes 0.4–1.2 mm, glabrous except for often thinly puberulent lobes or wholly glabrous; androecium 20–34(–40)-merous, (5.5–)6.5–9 mm long, the stemonozone not over 0.6 mm, sometimes obscure, the tube (1.8–)2.4–4.2 mm, usually no longer than the corolla, rarely up to 0.6 mm longer; ovary glabrous, attenuate into style, this up to 1 mm longer than the longest stamens, the stigma poriform. *Pods* solitary or rarely 2–3 per capitulum, sessile but abruptly contracted at base into a pseudostipe 2–8 mm, in profile broad-linear 8.5–16.5 × 1.4–2.6 cm,

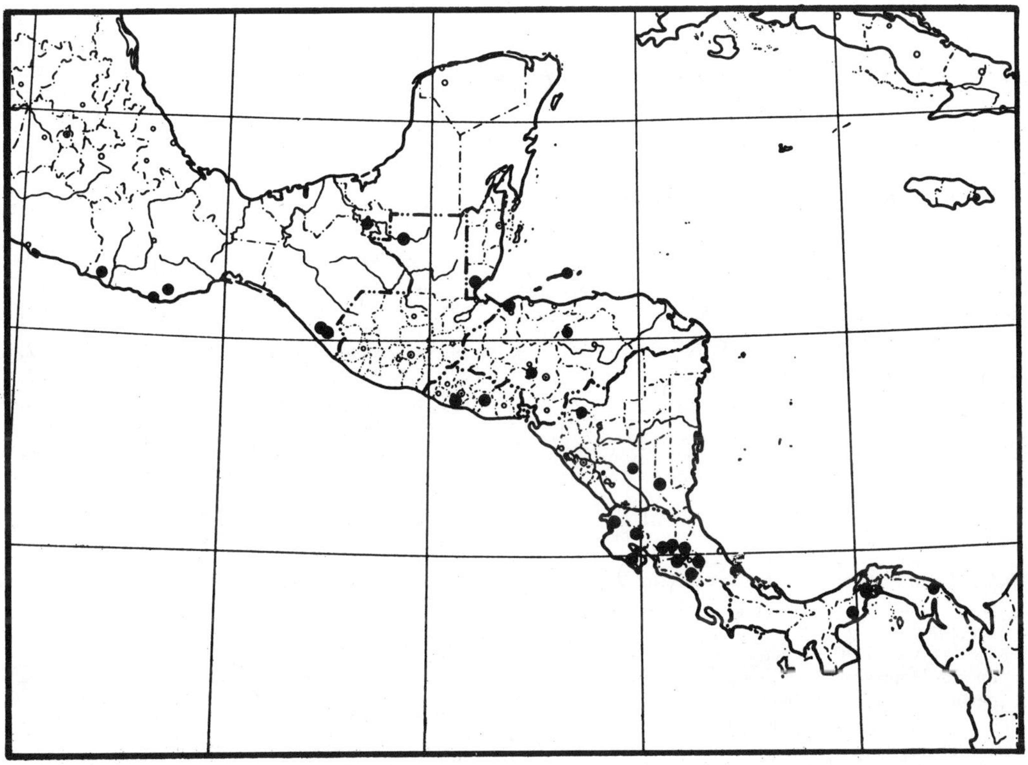

MAP 56. Southern Mexico and Central America. Range of *Albizia adinocephala* (Donnell Smith) Britton & Rose.

straight or nearly so, commonly apiculate, when well fertilized 9–13-seeded, the body plano-compressed, framed by slightly elevated sutures produced as a sharp-edged wing to (0.8–)1–2.3 mm wide, the glabrous valves at first either green, or brown with fuscous transverse bands between the seeds, or nigrescent and pruinose, when ripe stiffly papery, scarcely elevated over the seed-cavities, indistinctly venulose, satiny within; dehiscence inert, through both sutures; funicles linear, contorted at apex; seeds horizontal at middle of pod, obliquely attached to funicle, strongly compressed, in broad profile 5.5–7.6 × 4–5 mm, the smooth translucent testa scarcely 0.1 mm thick, the pleurogram delicately incised, inversely U-shaped; cotyledons thinly foliaceous, concealing the radicle.

Albizia niopoides is most reliably characterized in the context of its immediate relatives by the combination of numerous narrow leaflets and a glabrous, sharply wing-marginate pod that shows no sign, even after dehiscence, of transverse fracture between seeds. Its discontinuous range has encouraged local variations in leaf-formula, indumentum, and size of flowers, but these are, with one exception, poorly correlated with each other or with dispersal patterns, and some specimens of what has been called *Pithecolobium caribaeum* in Mexico and the Antilles are identical with some that have passed in Paraguay and one on the Brazilian Planalto as *P. hassleri*. The type of *P. niopoides* has the smallest flowers so far known in the species, but no other distinctive quality. However, no fruit of *P. niopoides* from the Amazon valley is available for comparison, and this might possibly provide a surprise. At the western edge of Amazonia, in Ecuador and Peru, the leaf-formula rises from an average iv–x/35–50 to ix–xix/45–60, and the leafstalks become relatively much longer than elsewhere. However, random occurrence of high leaflet-numbers are known to coincide with relatively few pairs of pinnae, which weakens the contrast. A single individual from the northwest corner of Panama (Pto Obaldia, *Pittier 4326*, NY) has the high leaf-formula of the sub-Andean populations, but larger and glabrate perianth, suggesting the vicariant *A. colombiana*, mentioned in the next paragraph. Immature specimens collected at Tarapoto (dept. and prov. San Martín), Peru by Spruce (*4407*, NY) and misidentified by

Bentham (1875: 591) as *Pithecolobium polycephalum* suggest an aberrant form of *A. niopoides* with relatively few (to 22) pairs of relatively broad leaflets (to 2.5 mm wide). Complementary fruits from the same region are necessary for interpretation of these enigmatic plants.

In northern Colombia and northwest Venezuela the commonest form of *A. niopoides* is almost glabrous, has relatively low leaf-formula, and leaflets wider and further distant along the pinna-rachis than elsewhere, and the pod tends to be very dark-colored and pruinose (as though glaucous) when ripe. This has been described as *Albizia colombiana* and is retained here at the rank of variety.

Key to the varieties of *A. niopoides*

1. Axes of lvs and inflorescence villosulous; lfts relatively crowded, short, and narrow, the longer interfoliolar segments of pinna-rachis 0.6–1.8 (–1.9) mm, and the largest lfts 5–8.5(–10) × 0.9–1.3(–1.4) mm; ripe pod tan or greenish fuscous, not pruinose; range of the species ..... 10a. var. *niopoides*
1. Axes of lvs and inflorescence glabrous except for rudimentary puberulence of ventral ridge of pinnae; lfts relatively distant and wide, the longer interfoliolar segments of pinna-rachis 1.4–2.5 mm, and the largest lfts 8.5–13 × 1.5–2.5 mm; ripe pod commonly nigrescent and pruinose; local in N Colombia and in Maracaibo basin of est. Zulia in Venezuela .... 10b. var. *colombiana*

10a. Albizia niopoides (Bentham) Burkart var. **niopoides.** *Pithecolobium niopoides* Spruce *ex* Bentham, 11. cc. 1875, 1876, sens. str. — "Habitat prope Santarem prov. Para: Spruce n. 1088 [IX.1851, fl.]." — Holotypus, K(herb. bentham.)! = NY Neg. *2023*; isotypi, †B = F Neg. *1213*, BM!, P!. — *Feuilleea niopoides* (Bentham) O. Kuntze, Revis. Gen. Pl. **1:** 188. 1891.

Pithecolobium caribaeum Urban, Symb. Ant. **2:** 260. 1900. — "Hab. in St. Vincent supra Kingston prope Lowman's Village . . . 170 m. Jul. flor. (an indigena?): [H. H.] Smith n. 902; Grenada inter St. George et La Pointe . . . Dec. fruct.: Eggers n. 6431; Tobago prope Aukenskeoch . . . Nov. fruct.: Eggers n. 5920; Trinidad in horto botanico cultum: Broadway n. 3800, Hart. n. 6361, [*Broadway*] 6362." — All the original material at B destroyed; lecto- and neotypus; Tobago, fruct. in nov. 1889, *Eggers 5920*, K!; isolecto- and isoneotypus, P!; paratypi, *Hart 6362*, NY!, *Broadway 3800*, K!, *H. H. Smith 902*, K!. — *Albizzia caribaea* (Urban) Britton & Rose, N. Amer. Fl. **23:** 44. 1928. *A. caribaea* sensu Witsberger, 1982: 30, fig. 30. — Equated with *A. niopoides* by Rico, 1992: 700.

Pithecolobium hassleri Chodat *in* Chodat & Hassler, Bull. Herb. Boissier, ser. 2, **4:** 483. 1904. — "[Paraguay.] pr. San Bernardino, Jan., [*Hassler*] n. 3719; ad marginem silvae in regione cursus superioris fluminis Apa, Nov., [*Hassler*] no. 8006." — Syntypi, *Hassler 3719, 8006*, G!; isosyntypi, NY! *3719*, P!. — *Albizzia hassleri* (Chodat) Burkart, Darwiniana **7:** 517. 1947. — Equated with *A. niopoides* by Rico, 1992: 700.

Albizia richardiana King & Prain, Ann. Roy. Bot. Gard. (Calcutta) **9:** 32. 1906. — Described from plants cultivated at Calcutta. — Equated by Rico, 1992: 700, with *A. niopoides*, on basis of material at K, not verified by us.

Senegalia liebmannii Britton & Rose, N. Amer. Fl. **23:** 116. 1928. — "Pacific Coast of Oaxaca, 1841–3, *Liebmann 4380*." — Holotypus, US!; isotypus (fragm.), NY! — Equated with *A. niopoides* by Rico, 1992: 700, who mistakenly attributed the holotype to NY.

Senegalia guacamayo Britton & Killip, Ann. New York Acad. Sci. **35:** 142. 1936. — "Prado, Santa Marta, Magdalena, Colombia, January 13, 1930, [*S. J.*] *Record 7* . . ." — Holotypus, NY!. — *Acacia guacamayo* (Britton & Killip) Standley, Trop. Woods **52:** 26. 1937. *Albizia guacamayo* (Britton & Killip) Cárdenas, Ernstia **1(4):** 133. 1992.

Albizia hassleri sensu Bernardi, 1984: 171; Ortega T., et al. *in* Spichiger, Fl. Paraguay, ser. espec. **3:** 90, vi (fig. 32, slash). 1989.

Albizia niopoides sensu Rico, 1992: map 1.

Albizia guacamayo sensu Cárdenas, Ernstia **1(4):** 147, fig. 1. 1992.

Lf-formula iv–xix/33–57(–63); otherwise as given in description of species and in key to varieties.

In seasonally deciduous or semideciduous woodlands, or in moister gallery and other tropical evergreen forest, from near sea level up to 700 m in Central America, up to 1100 m on the Brazilian Planalto, and 1300 m in Peru, widely dispersed around, and very locally within, the Amazon Basin northward from N Paraná in Brazil, Paraguay and E Bolivia to N Colombia, N Venezuela, Trinidad and Tobago, thence sparingly N into the Lesser Antilles and discontinuously (perhaps sometimes planted) NW into S Mexico (S Oaxaca, Chiapas, and Yucatán). — Maps 57A, B. — Fl. in Orinoco basin and N Colombia northward mostly II–VI, from the Equator southward mostly X–II. — *Tantacayo* (Panama); *wild tamarind* (Grenada); *guacamayo, tribunal* (Colombia); *carabelí, caro, hueso de pescado, turupía camburi chiquito* (Venezuela); *llambo paspaco, paspaco blanco* (Peru); *angico branco, gurujuba* (Brazil); *jebió* (Bolivia); *yvyrá-ju* (*yvirá-yú*) (Paraguay).

10b. Albizia niopoides var. **colombiana** (Britton & Killip) Barneby & Grimes, comb. et stat. nov. *Albizzia colombiana* Britton *ex* Britton & Killip, Ann. New York Acad. Sci. **35:** 131. 1936. — "Near Bonda, Santa Marta, Magdalena, Colombia, August 3, 1899, *H. H. Smith 38*, type . . .; *38a, 2067*; *Record 75* . . ." — Holotypus + all paratypi, NY!.

Lf-formula iv–vi/28–39; otherwise as given in description of species and in key to varieties.

In semideciduous dry forest communities below 150 m, locally plentiful in the Caribbean lowlands of N Colombia (Bolívar, Atlántico, Magdalena) and adjacent NW Venezuela (Maracaibo Basin in Zulia). — Map 57B. — Fl. VIII–IX(–?). — *Guacamayo.*

A collection of var. *colombiana* from the roadside near La Victoria, state of Aragua, Venezuela (*Cárdenas 304*, GH), is omitted from the range statement and map until its native status can be confirmed.

11. Albizia buntingii Barneby & Grimes, sp. nov., *A. niopoidi*, *A. barinensi*, *A. glabripetalae*, et *A. subdimidiatae* notulis multis similis, affinitatis verae leguminis defectu incertae, a prima pinnarum longiorum foliolis 11–12 (nec 28–63)-jugis usque, majoribus 4–5 (nec 0.9–2.5) mm latis et e pulvinulo palmatim 5–6 (nec 3–4)-nerviis, flosculisque peripthericis pedicello 0.9–1.4 mm usque longo elevatis (nec subsessilibus); a secunda foliolorum numero minori, pedicellis flosculisque glabris; a tertia foliolorum numero minori sed ambitu multo majori (13–14 × 4–5, nec 5–8 × 1.4–2.3 mm usque) androecioque 28–30 (nec 16)-mero; a quarta, quoad foliolorum numerum comparabili, pinnis 6–7 (nec 2–4, raro 5)-jugis, flosculisque et pedicellatis et majoribus (4–4.5, nec 2–3.3 mm longis) diversa. — VENEZUELA. Zulia: alrededores de Casigua El Cubo, 100 m, al este del empalme de la vía hacia Casigua con la carretera Machiques-La Fría. Alt. ca. 125 m, 25.II.1985 (fl.), *George S. Bunting 13370*. — Holotypus, NY.

Rather coarsely microphyllidious *trees*, so far as known ±3 m with trunk 1 dm dbh and dark brown, angulately ribbed branchlets, except for glabrous or subglabrous upper face of dull olivaceous, thin-textured lfts puberulent throughout, the branchlets microscopically so and early glabrate, the axes of lvs and inflorescences pilosulous with incurved hairs to 0.2 mm, the umbelliform capitula of greenish fls borne singly and geminate at nodes of short efoliate pseudoracemes arising singly or 2–3 together from the axil of fully expanded or lately shed lvs, the inflorescence consequently immersed in foliage. *Stipules* not seen, probably very small and evanescent, the scar minute. *Lf-formula* vi–vii/11–12; lf-stks 11–16 cm, the petiole 5–6 cm, at middle 1.2–1.5 mm diam, the ventral sulcus continuous between pinnae, the longer interpinnal segments ±15–18 mm; a sessile, shallow-cupular nectary ±1 mm diam a little below midpetiole, a smaller one at tip of lf-stk, and yet smaller ones between 1–2 furthest pairs of lfts; pinnae gently accrescent distally, the rachis of furthest and penultimate pairs 6–7 cm, the longer interpinnal segments 5–5.5 mm; pulvinules 0.5–0.6 mm diam, blackish; lfts decrescent near base of rachis, the first pair represented by subulate or falcate paraphyllidia 0.3–0.7 mm, the rest inequilaterally oblong from obliquely subtruncate base (the posterior angle acutely rectangular), depressed-deltate-apiculate, the larger ones 13–14 × 4–5 mm, 2.8–3.3 times as long as wide; venation of 5–6 nerves from pulvinule, the straight midrib giving rise on each side to 4–5 weak secondary nerves, the inner posterior primary produced well beyond midblade to anastomosis well within the loosely revolute margin, the outer posterior primaries progressively short, the one anterior primary nerve either very short or obsolete, all these and a random oblique or irregular tertiary venulation bluntly prominulous above. *Primary axis* of pseudoracemes ±6–7 cm; longer peduncles 9–14 mm; capitula 12–26-fld, without the cream-white filaments 9–10 mm diam; bracts linear-oblanceolate, to 2 mm, caducous long before anthesis; fls moderately or strongly dimorphic, all at least shortly pedicellate, 1–3 terminal ones at least a little and sometimes much longer than the rest and their androecial tube much longer; perianth of all fls 5-merous and glabrous except for a few minute pallid hairs on calyx-teeth and toward tip of corolla-lobes; PERIPHERAL FLS: pedicel 0.9–1.4 × 0.2 mm; calyx campanulate 1.2–1.6 × 1–1.2 mm, the triangular teeth ±0.4 mm; corolla 4–4.5 mm, the ovate lobes 1.2–2 × 0.8–1.3 mm; androecium 28–30-merous ±10 mm, the stemonozone ±0.5 mm, the tube 3.5–4.5 mm; ovary glabrous, conical at apex; style as long as or a little longer than longer filaments, scarcely dilated at the stigma; TERMINAL FL(S): pedicel 0.5 mm or less; calyx 1.6–1.7 × 1.2–1.7 mm; corolla 4–6 mm; androecium 9.5–13 mm, the tube 4.5–8 mm. *Pods* unknown.

In unrecorded habitat at ±125 m, to be expected in seasonally dry brush-woodland, known only from the type locality in the Maracaibo basin, Venezuela. — Fl. II–III.

Albizia buntingii belongs to a group of species that are separable at anthesis by rather subtle characters of pubescence, leaflet-number, -size, and -venation, flower-size, and number of filaments in peripheral flowers of the capitulum, but are suprisingly diverse in ultimate form of the pod. In absence of this, the true close affinity of *A. buntingii* cannot be assessed, and we have contrasted it in the diagnosis with

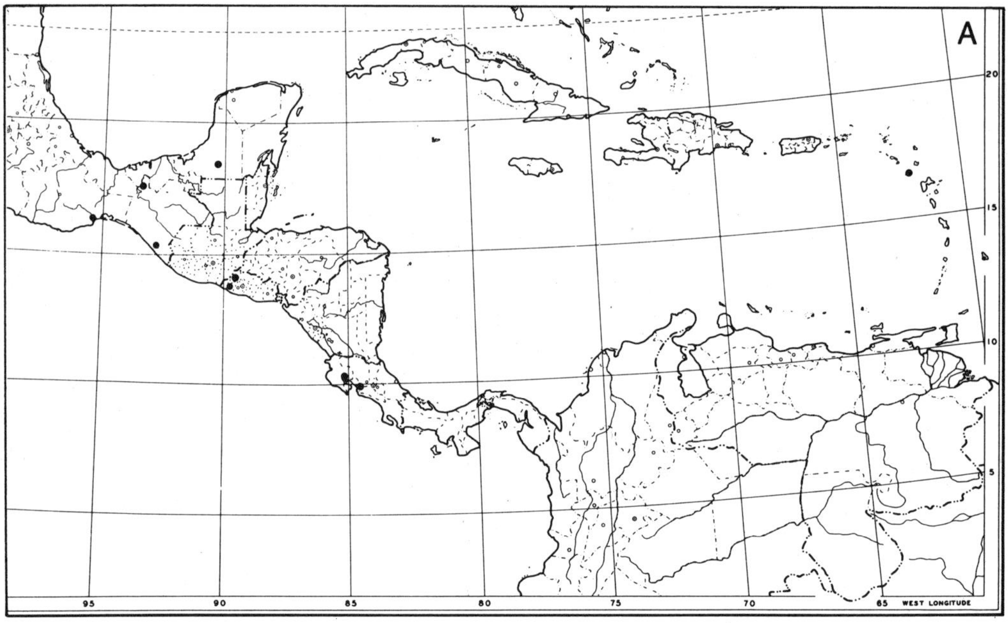

MAP 57A. Central America and Montserrat. Range of *Albizia niopoides* (Bentham) Burkart var. *niopoides*.

species that are known to occur nearest to it, and with which in practice it is most likely to be confused. *Albizia niopoides*, the one of these that is known to occur also, and in two varieties, in the Maracaibo basin, differs most obviously in that its longer pinnae are composed of 30–40 (not 11–12) pairs of leaflets at most 1–2 (not 4–5) mm wide. *Albizia barinensis*, primarily of the Orinoco basin in Venezuela, has both pinnae and leaflets rather more numerous and peripheral flowers at once subsessile and pilosulous. *Albizia glabripetala*, from the periphery of Guayana far to the southeast of the Maracaibo basin, resembles *A. buntingii* in often pedicellate flowers, but has narrower, more numerous leaflets and only half as many (±16, not 28–30) stamens per flower. Finally, the widely dispersed *A. subdimidiata* differs in fewer, mostly 2–4 pairs of pinnae, somewhat broader leaflets, almost or truly sessile peripheral flowers, and substantially smaller corolla about 2–3.3 (not 4–4.5) mm long. Of the four species just compared with *A. buntingii*, the first two have plano-compressed acacia-like pods inertly dehiscent through the sutures, whereas the pods of the last two are lomentiform, indehiscent in the strict sense, but tend to split transversely between seeds. It is not possible to foretell to which type, if either, the pod of *A. buntingii* will conform.

12. Albizia carbonaria Britton, Sci. Surv. Porto Rico Virgin Is. **6(3):** 348. 1926. — "Grown at the forest station Río Piedras, Porto Rico, from seedlings received in 1921, from the Bureau of Plant Industry; the seed was procured from Palmira [Valle de Cauca], Colombia, in 1920. Type collected by *C. L. Bates*, October, 1924, at the Forest Station, Río Piedras . . ." — Holotypus, NY!; paratypi, *Britton & al. in 1923, 1924 & 1925*, NY!. — *Pithecellobium carbonarium* (Britton) Niezgoda & Nevling, Phytologia **44:** 310. 1979. — *Acacia* sp. seu *Carbonero* Britton, Sci. Surv. Porto Rico Virgin Is. **5(3):** 355 (in nota). 1924.

Albizia malacocarpa Standley [*in* Standley & Calderón, Lista Pl. Salvador 96. 1925, nom. nud.] *ex* Britton & Rose, N. Amer. Fl. **23:** 44. 1928. — "Type from San Salvador, El Salvador, *Calderón 2042*, a fruiting specimen; description of flowers taken from [*R.S.*] *Williams 952*, Cana [= Pirre, Darién], Panama." — Syntypi, NY!; isosyntypus, (*2042*), US *1169954*!. — Equated with *Pithecellobium carbonarium* by Britton & Killip, 1936: 131; and by Nielsen, 1985: 47.

Albizia sumatrana v. Steenis, Encycl. Ned.-Ind., ed. 2, Suppl. **6:** 864. 1931 & Blumea **5:** 509, pl. 1. 1945. Equated with *A. carbonaria* by Nielsen, 1985: 47.

Trees of rapid growth flowering at 4–6 m but potentially attaining 30 m and with a long-unbranched trunk to 9 dm dbh, the nearly smooth bark exfoliating

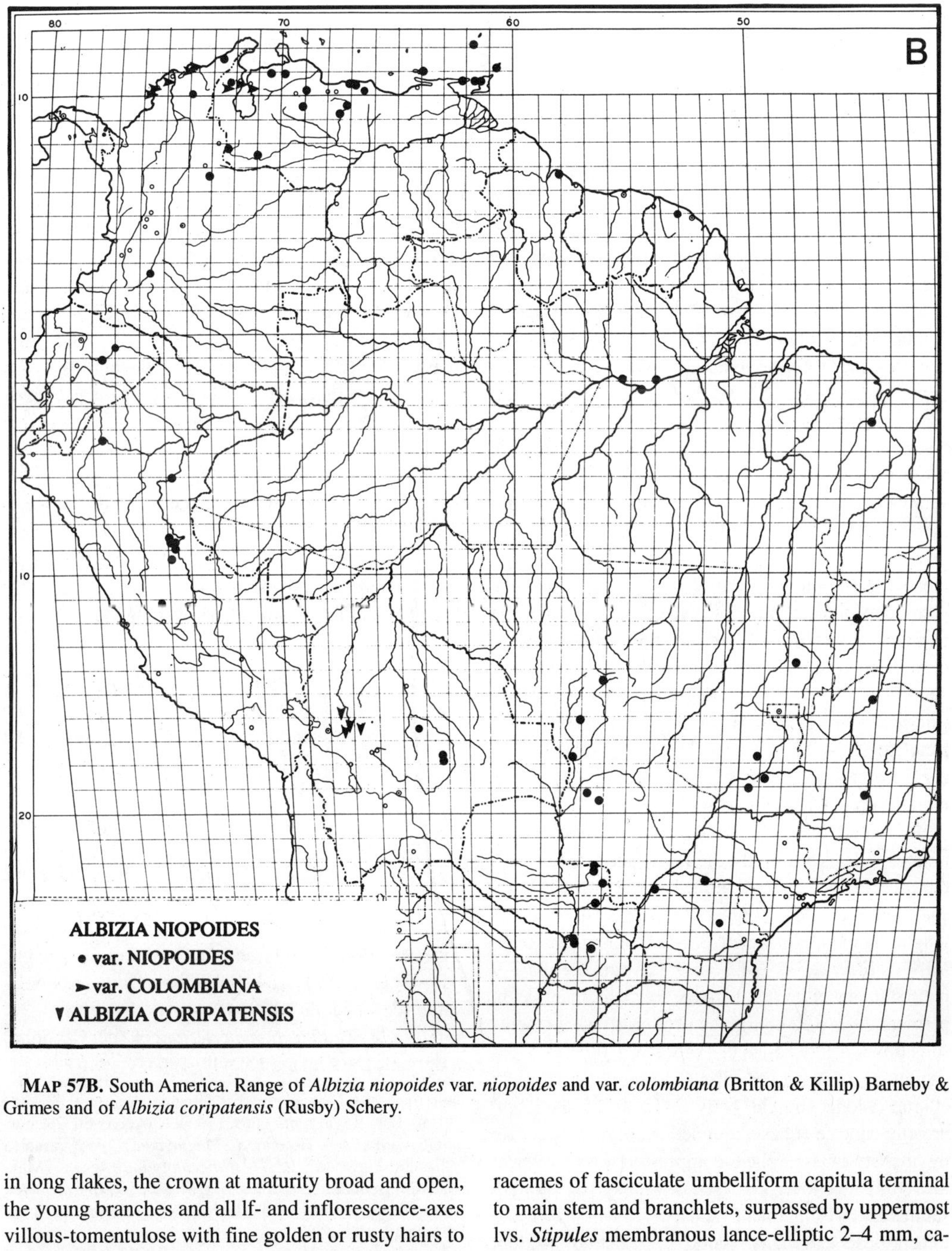

MAP 57B. South America. Range of *Albizia niopoides* var. *niopoides* and var. *colombiana* (Britton & Killip) Barneby & Grimes and of *Albizia coripatensis* (Rusby) Schery.

in long flakes, the crown at maturity broad and open, the young branches and all lf- and inflorescence-axes villous-tomentulose with fine golden or rusty hairs to 0.25–0.6(–0.75) mm, the lvs prominently bicolored, the membranous plane lfts above dull dark green (brittle and brunnescent when dried) and minutely puberulent, beneath pale green, puberulent and barbellate on anterior side of midrib, the inflorescence a panicle of proximally foliate, distally efoliate pseudoracemes of fasciculate umbelliform capitula terminal to main stem and branchlets, surpassed by uppermost lvs. *Stipules* membranous lance-elliptic 2–4 mm, caducous at earliest expansion of associated lf, absent from mature specimens. *Lf-formula* viii–xiii/(18–)20–30; lf-stk of major lvs of fertile branches 8–17 (of seedling lvs, no further mentioned, –28) cm, the petiole including poorly differentiated pulvinus 2–4(–4.5) cm, at middle 0.7–1.4 mm diam, the longer interpinnal segments 7–13 mm, the ventral groove continuous between pinna-pairs; petiolar nectary often lacking or reduced to a minute pore recessed into petiolar groove, but erratically present, situated near or somewhat below midpetiole, shallow-cupular, grossly marginate, in profile 0.5–1 mm tall and 1–4 mm diam, fuscous; pinnae proximally decrescent, otherwise subequilong, the rachis of longer ones (3–)3.5–7.5(–8) cm, the longer interfoliolar segments (1.3–)1.5–2.8 (–3) mm; pulvinules in dorsal view 0.2–0.4(–0.5) mm; lfts decrescent at very ends of rachis, elsewhere equilong, the blades asymmetrically linear-oblong, lance-elliptic or narrowly ovate-oblong from shallowly semicordate base, abruptly minutely deltate-apiculate, the larger ones 4.5–8 × (–1.3)1.5–3.2 mm, 2.7–3.8(–4.4) times as long as wide; midrib displaced to divide blade 1:2–4(–5), slightly or scarcely falcate, immersed above, sharply prominulous beneath, a faint secondary venulation visible in some larger lfts but wholly immersed. *Peduncles* fasciculate by 3–6(–7), the longest of each fascicle 1.8–4.6 cm; capitula without filaments 11–16 mm diam, the whitish fls dimorphic, the peripheral ones elevated on pedicel ±1.5–4 mm, the terminal fl sessile, the receptacle including short drum-shaped pediment of terminal fl 1–2 mm, the perianth of all fls 5-merous and golden-silky-puberulent externally; PERIPHERAL FLS: calyx narrowly campanulate (2.3–)2.5–3.5 × (1.2–)1.5–1.8 mm, the incurved obtuse teeth 0.6–1.2 mm; corolla (4–)4.4–6.4 mm, the erect lobes (1.1–)1.4–2.1 mm, the 5 segments in age separating to base; androecium white 19–24-merous, 13–16.5 mm, the stemonozone 0.5–0.8 mm, the tube (1.7–)2–3.3(–4) mm; ovary slenderly ellipsoid, symmetrically conical at apex; style 1–3 mm longer than longest stamen, 0.1–0.15 mm diam; TERMINAL FL: calyx 3.7–4.5 × 1.7–2.1 mm; corolla 7–9 mm; stamens either as many or twice as many as in peripheral fls (±24–45), the free part of filaments sometimes a little dilated. *Pods* solitary or to 3 per capitulum, stipitate, the stipe 4–14 mm, the narrowly elliptic-oblong straight body 7–12 × 1.5–2.35 cm, broadly cuneate at base, rounded at apex but apiculate by the style-base, plano-compressed except where low-colliculate over ripe seeds, framed by straight sutures ±0.8–1 mm thick, the papery valves brown and tomentulose externally, pallid and lustrous within; dehiscence very tardy, the pod long-persistent on branches, falling unopened, the seeds apparently released only by weathering; funicle filiform; seeds of well-fertilized fruits 20–26, transverse, broadside up, at middle of valves, in broad outline elliptic 3.6–4.2 × 1.6–1.9 mm, the testa firm smooth, pale tan but a little darker within the narrowly inverse-U-shaped pleurogram; a pallid lens distal to the funicle-attachment.

A species of debatable aboriginal range, but seemingly native in moist upland tropical forest, mostly between 700 and 1800 m, in centr. and N Colombia, E Panama (Darién), and the eastern cordillera in Venezuela (Trujillo and Mérida), within this range also preserved and planted for coffee-shade and as an ornamental shade tree; introduced in cafetales and occasionally wild in the lowlands of Central America (Costa Rica; El Salvador), the Greater Antilles (Cuba; Dominican Republic; Puerto Rico), centr. Peru, and SE Brazil (São Paulo). — Map 58A, B. — Fl. primarily V–XII but intermittently throughout the year, the ripe frs. often coeval with new fls. — *Carbonero, muche, piñón, pisquín* (Colombian).

Among related albizias with multifoliolate leaves, *A. carbonaria* is notable for the dorsally pallid leaflets, the dense golden indumentum of the inflorescence, the distinctly pediceled flowers, and the papery tomentulose pod. Its rapid growth and open crown recommend it as a shade tree in cafetales, where it has been planted far outside its native range.

13. Albizia berteriana (de Candolle) Fawcett & Rendle, Fl. Jamaica **4:** 145. 1920. *Acacia berteriana* Balbis *ex* de Candolle, Prodr. **2:** 470. 1825. — "... in Sancto Domingo (Bertero)." — Holotypus, labeled: "St. Domingue, Bertero, misit Balbis, 1821" G-DC (fl, fr)! = NY Neg. 33453. — *Pithecolobium berterianum* (de Candolle) Bentham, London J. Bot. **3:** 220. 1844. *Albizzia berteriana* (de Candolle) Gómez de la Maza, Dic. bot. nom. vulg. cubanos y puertoriquenos 10. 1889, nomen, basionymo imperfecte cit. *Feuilleea berteriana* (de Candolle) O. Kuntze, Revis. Gen. Pl. **1:** 184. 1891. *Pseudalbizzia berteriana* (de Candolle) Britton & Rose, N. Amer. Fl. **23:** 48. 1928. *Cathormion berterianum* (de Candolle) Burkart, Darwiniana **13:** 447. 1964.

Inga fragrans Macfadyen, Fl. Jamaica **1:** 309. 1837. — "Hab. Port-Royal mountains. Below Moccha and Chesterfield works." — Holotypus, *Macfayden s. n.* e Jamaica "bastard logwood", K!. — *Pithecolobium fragrans* (Macfadyen) Bentham, London J. Bot. **3:** 220. 1844. — Equated with the preceding and illegitimately referred to it by Bentham, 1875: 592.

Acacia litoralis A. Richard *in* Sagra, Hist. Phys. Cuba, Bot. Pl. Vasc. 467. 1846. — "Crescit in *Vuelta de Abajo* ... Jose Maria Valenzuela ... *moruro de costa*." — Holotypus, not seen. — Equated with *Pithecolobium berterianum* by Urban, Symb. Ant. **2:** 261. 1900.

A. sarmentosa sensu Combs, Trans. Acad. Sci. St. Louis **7(17):** 418. 1897; non Desv. (Cieneguita, Santa Clara, *Combs 237*, NY!).

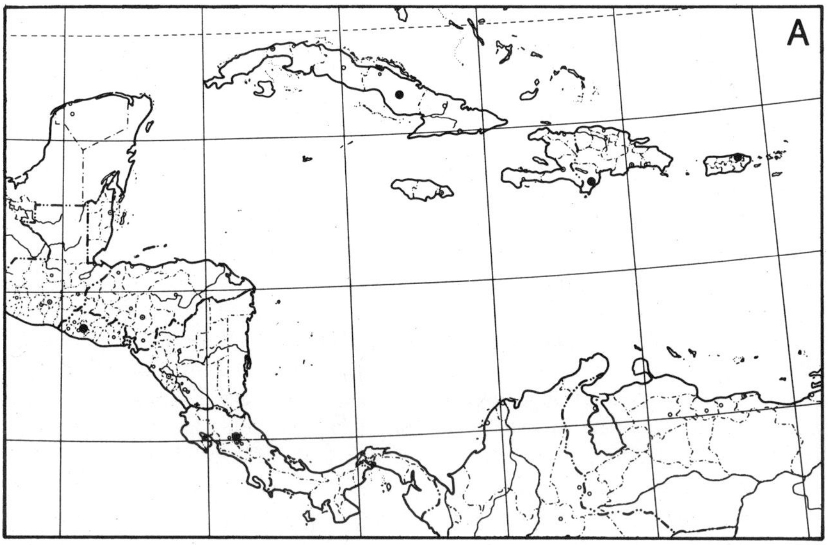

Map 58A. Central America and the West Indies. Range of *Albizia carbonaria* Britton.

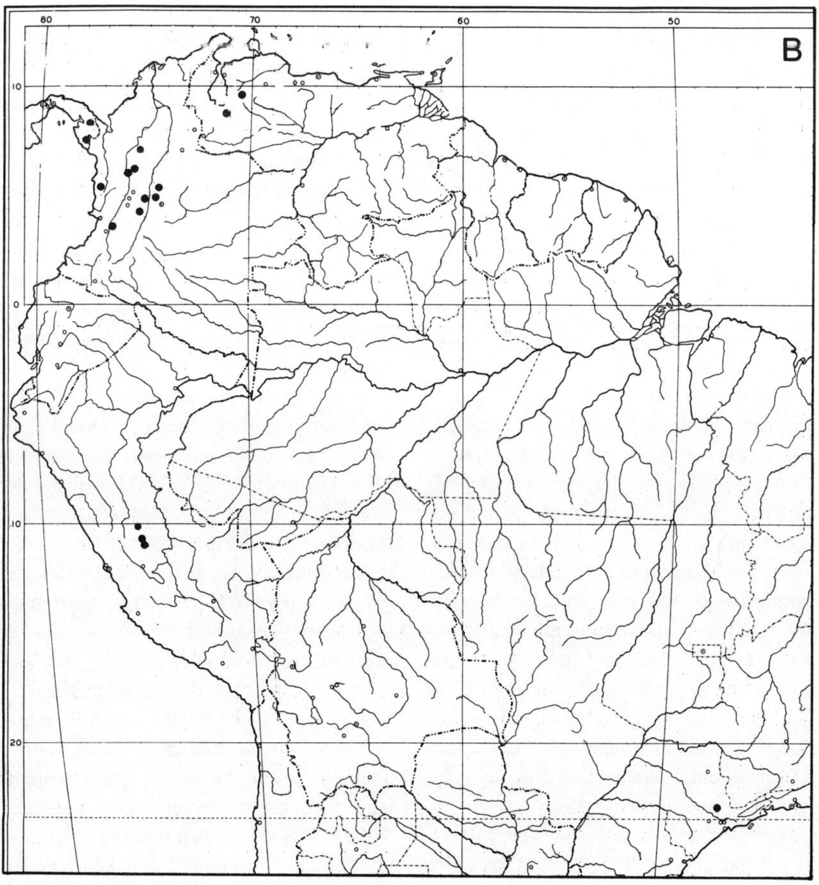

Map 58B. South America. Range of *Albizia carbonaria*.

Amply microphyllidious *trees* with smooth or finely cracked bark, flowering when 2.5–4 m but potentially attaining 14 m, the young branchlets, the lf-axes (at least ventrally) and all axes of inflorescence either thinly or densely puberulent with fine pallid, forwardly incurved hairs to 0.1–0.2 mm and often in addition minutely livid-granular, the lvs bicolored, the small firm crowded lfts lustrous dark green (when dry brunnescent) and glabrous on upper face, beneath dull pale green (when dry tan) and proximally pilosulous on anterior side of midrib, the inflorescence a terminal, more or less pyramidal, proximally foliate but distally efoliate panicle of pseudoracemes, the small whitish capitula globose in bud, subhemispherical at anthesis, the ascending or randomly spreading pods long-persistent. *Stipules* deltate, <0.5 mm, very early deciduous or perhaps sometimes obsolete, lacking from fruiting specimens. *Lf-formula* vii–xii (–xv)/(32–)35–52(–57); lft-stk of longer lvs (8.5–) 10–22 cm, the ventral groove bridged between pinna-pairs, the petiole (2–)3–7 cm, the longer interpinnal segments 8–16(–18) mm; petiolar nectary well below midpetiole but not contiguous to the firm pulvinus, sessile elliptic, shallowly patelliform, in profile commonly 0.2–0.5 mm tall and 1.5–3.2 mm in long diameter (but lacking in random lvs), smaller cupular nectaries between 1–2 distal pairs of pinnae and at tip of each pinna-rachis; pinnae decrescent at base and often toward apex of lf-stk, the rachis of long ones (4.5–) 5.5–9(–9.5) cm, the longer interfoliolar segments 0.7–2(–2.4) mm; lft-pulvinules in dorsal view 0.2–0.45 mm; lfts gradually decrescent toward each end of rachis, linear or linear-lanceolate from bluntly auriculate base, rounded or obtusely deltate at apex, either straight or very gently either forwardly or backwardly falcate, those near midpinna 5.5–9.5 × 1–2.1(–2.3) mm, (4–)4.2–6.1 times as long as wide; venation of 2–3 nerves from pulvinule, the dorsally prominulous midrib forwardly displaced at base becoming subcentric at and beyond midblade, giving rise on each side to 2–4(–5) weak subhorizontal secondary venules faintly anastomosing within the plane or dorsally thickened, nearly always ciliolate margin, the posterior primary nerves very short and faint, the whole venation immersed on upper face. *Peduncles* fasciculate by (2–)3–6, biseriate above each node of pseudoraceme, the longest of each fascicle 3–11 mm; capitula to 12–24-fld, without filaments 6–8 mm diam, the fragrant whitish fls obscurely dimorphic, the perianth of all 5-merous and externally glabrous except for always ciliolate calyx-teeth and sometimes thinly ciliolate corolla-lobes; receptacle subglobose ±1–1.5 mm diam, terminal pedestal 0; bracts obovate or spatulate 0.3–0.7 mm, ciliolate, tardily deciduous; PERIPHERAL FLS: sessile or contracted at cuneate base into a thick pedicel <1 mm, the campanulate or turbinate-campanulate calyx 1.2–1.5 mm, the depressed-deltate teeth 0.1–0.2 mm; corolla 3.2–4 mm, the erect or recurved, lanceolate lobes ±1.1–1.3 mm; androecium 18–24-merous, (5–)6–8.5 mm, the stemonozone 0.3–0.5 mm, the tube 1.4–2.7 mm; ovary symmetrically conic at apex; style a trifle longer than androecium, the stigma not dilated; TERMINAL FL: sessile, the calyx 1.1–2 × 1.4–1.8 mm; corolla of peripheral fls; stamens as many as in peripheral fls or up to 33, the tube dilated and 3–4 mm long. *Pods* 1–3(–4) per capitulum, in profile broad-linear contracted at base into a stipe or 2-ribbed pseudostipe 3–11 mm, the body when well fertilized (6–)8–13(–14) × (0.8–) 0.9–1.5(–1.65) cm, straight, plano-compressed becoming low-convex and somewhat discolored over each seed, the papery or toward the sutures thinly pithy valves framed by straight or obscurely undulate sutures ±0.4–0.65 mm wide, at first green becoming light brown, puberulent and livid-granular overall, when ripe developing horizontal fracture-lines between seeds; dehiscence 0, the pod commonly falling entire and breaking on the ground into 8–12 individually indehiscent segments; funicle filiform; seeds transverse at middle of valves, plumply lentiform or discoid 4–6.8 × 3.5–5.5 mm, the smooth, pale tan testa of grapeskin consistency when moistened, very narrowly winged or ridged around the periphery, the pleurogram U-shaped ±2–2.5 mm diam; cotyledons pale green or yellowish tan, concealing the plumule; no endosperm.

In semideciduous thorn-scrub and open woodland, commonly on limestone, occasionally on serpentine and perhaps other substrates, ascending from coastal lowlands up to 660 m in Jamaica and 700 m in Dominican Republic, widespread and locally plentiful in the Greater Antilles: centr. and E Cuba; Jamaica (mostly on S slope); Hispaniola (widespread). — Map 59. — Fl. (IV)V–IX; fr. long-persistent on branches, sometimes coeval with new fls. — *Abey blanco, hoja menuda, moruro blanco* (Cuba); *candelón, corbano, corbano blanco* (Domincan Republic).

When fresh, the amply multifoliolate leaves of *A. berteriana* are dark lustrous green on the upper face, a little paler and dull below. The prolific terminal panicles of papery stramineous pods persist at tip of dead leafless branchlets well into the dry season, falling at random, and piling up in windrows under the tree. In drying, the pod valves develop a transverse fracture

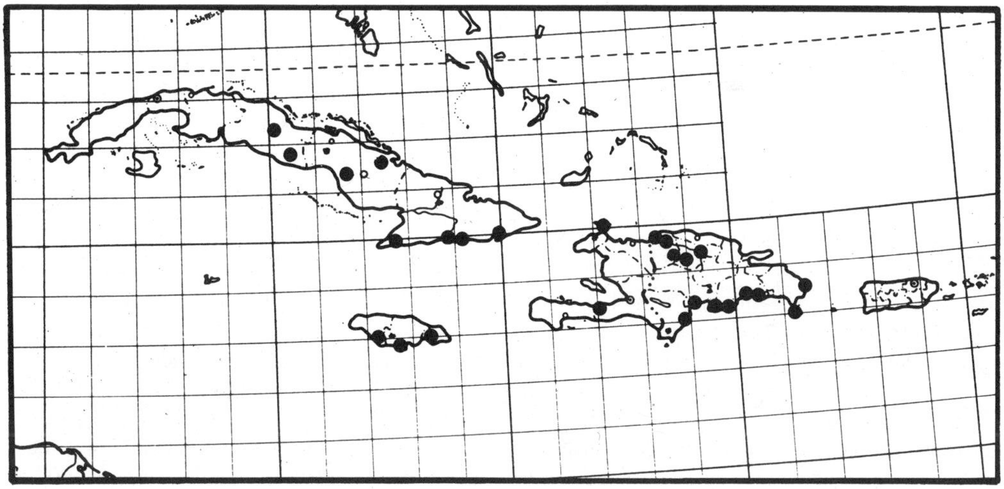

Map 59. The Greater Antilles. Range of *Albizia berteriana* (de Candolle) Fawcett & Rendle.

line between seeds, but disarticulate only reluctantly, and hardly deserve the term lomentiform, less so, in any case, than those of *A. subdimidiata* and kindred. The inconsiderable genus *Pseudalbizzia* Britton & Rose was based exclusively on this feature of the fruit.

Seeds of *A. berteriana* collected at JBSD (*Barneby s.n.*) had phanerocotylar germination and two, opposite, simply pinnate (10–12-foliolate) eophylls.

XVA(II). Series ARTHROSAMANEA (Britton & Rose) Barneby & Grimes

Albizia sect. **Arthrosamanea** (Britton & Rose) Barneby & Grimes ser. **Arthrosamanea.** *Arthrosamanea* Britton & Rose *ex* Britton & Killip, Ann. New York Acad. Sci. **35:** 128, pro gen. 1936, sens. str. — Typus supra (p. 206) sub sectionis titulo indicatur.

Pods linear or broad-linear, straight or subdecurved, plano-compressed, at maturity lomentiform, the valves transversely cracking between seeds, but the monospermous, indehiscent articles long retained by wiry continuous sutures. — Spp. 3, of N South America, Amazonia, and Pacific Ecuador.

14. Albizia glabripetala (Irwin) G. P. Lewis & P. E. Owen, Leg. Ilha de Maracá 42. 1989. *Pithecellobium glabripetalum* Irwin, Mem. New York Bot. Gard. **15:** 109, fig. 5. 1966. — "BRITISH GUIANA. "Orealla," Corantyne R., Oct. 1879, *G. S. Jenman 364* . . ." — Holotypus, NY!; isotypi, BM!, P! (not found at K in 1990); 4 paratypi: *Forest Dept. 5397, 5642, Schomburgk 525, Wurdack & Monachino 41263*, all NY!.

Pithecolobium polycephalum sensu Bentham, 1875: 591, 644, quoad *Rich. & Rob. Schomburgk 525* e Guyana; non aliter.

Amply microphllyidious *trees* 5–15(–18) m with almost smooth trunk attaining 2–3 dm dbh and sulcate branchlets, except for sometimes glabrate (never densely pubescent) upper face of lfts finely minutely puberulent throughout with loose pallid hairs 0.1–0.3 mm, the lvs bicolored, the ventrally convex lfts dark olivaceous (brunnescent) and sublustrous above, paler dull beneath, the small hemispherical capitula of whitish fls borne in a simple terminal pseudoraceme or a small few-branched panicle terminal to hornotinous leafy branchlets, not or scarcely exserted from foliage. *Stipules* (few seen) triangular-subulate 0.4–1 mm, very early caducous. *Lf-formula* (iv–)v–x/20–35; lf-stks 6–16.5 cm, the petiole 2.5–4.5 cm, at middle 0.9–1.6 mm diam, the ventral groove continuous between pinnae, the longer interpinnal segments 6–12 mm; an elliptic, depressed-cupular nectary near or commonly well below midpetiole, ±1–1.5 mm diam and <0.25 mm tall, rarely also a minute one at tip of some pinna-rachises; pinnae little graduated in length, the rachis of longer ones 5–8.5 cm, the longer interfoliolar segments 1.2–2.2 mm; pulvinules 0.2–0.4 mm, the lfts appearing sessile against the rachis; lfts equilong except at extreme ends of rachis and the first pair commonly represented by minute subulate paraphyllidia, the blades inequilaterally lance-oblong from postically truncate or semicordate base, at apex deltately subacute and often incipiently porrect, the longer ones 5–8 × 1.4–2.3 mm, 3–3.9 times as long as wide; venation of 3–4 nerves from pulvinule, the straight or almost straight, unbranched, dorsally prominulous and there (on anterior side) barbellate midrib forwardly displaced to divide blade 1:2.3–3.2, the inner posterior nerve produced well beyond midblade before anastomosis, the outer one or two much shorter, tertiary venulation 0, the whole venation weakly bluntly prominulous on upper face, more sharply so beneath. *Primary axis* of simple pseudoraceme and secondary divisions of panicle ±8–12 cm; peduncles (1–)2–3 per node, the longest of each fascicle mostly 15–25 (at distal nodes somewhat shorter), the capitula 9–22-fld, without filaments 6–7.5 mm diam, the fls dimorphic, the peripheral ones at least shortly pedicellate, the much larger terminal one sessile on a short pedestal, the whole receptacle 1–2 mm; bracts minute, ciliolate; perianth of all fls 5-merous, or the terminal fl 6-merous, glabrous except for microscopically ciliolate calyx-teeth, or for a few minute scattered hairs on corolla-lobes; PERIPHERAL FLS: pedicel 0.2–1.2 mm; calyx campanulate 0.7–1 × 0.8–0.9 mm, the depressed-deltate teeth 0.1–0.2 mm; corolla greenish white or faintly anthocyanic 2.8–3.2 mm, the ovate lobes 1.2–1.5 × 0.8–0.9 mm; androecium 15–16-merous, white, the longer filaments 10.5–12 mm, the stemonozone ±0.5 mm, the tube ±1.5 mm; ovary glabrous, conical at apex; style a little longer than longer filaments, scarcely dilated at the stigma; TERMINAL FL: sessile; calyx 1.4–1.6 × 1.4 mm; corolla 3.4–4.7 mm; androecium 20–22-merous, the stemonozone ±1 mm, the tube 3.5–7.5 × 1.2–1.3 mm, the free portion of filaments dilated and arched outward. *Pods* 1 or 2 per peduncle, sessile but often contracted basally, due to abortion of first seed, into a short sterile neck, in profile broad-linear, at apex broadly obtuse-apiculate, when well fertilized 11–18 × 1.5–2.2 cm, 10–14-seeded, plano-compressed becoming low-convex over each seed and depressed between them, the stiffly papery, dark brown, when young venulose valves framed by obtuse, scarcely or not constricted sutures ±1 mm wide, either glabrous (in Guyana) or finely minutely strigulose (in Brazil), when mature transversely cracking between seeds; dehiscence not seen, much delayed or the seeds perhaps released only by weathering, the apparent segments not liberated from the wiry sutures; seeds (few seen) transverse, basifixed on a filiform funicle, in broad profile 7 × 5 mm, the hard testa bone-white, smooth but dull, closely investing the horny biconvex embryo, the inversely U-shaped pleurogram ±2 mm diam.

In low, seasonally flooded woodland, sometimes with *Mauritia* palms, on creek and river banks, and at edge of white sand savanna, below 200 m, seemingly uncommon about the periphery of the Guayana Highland in T. F. Amazonas (Esmeralda) and state of Bolívar (Río Suapure; Foráneo La Urbana), Venezuela, on the upper Rupununi and middle Courantyne rivers in Guyana (and presumably adjoining Surinam), and on upper Río Branco (Ilha de Maracá) in N state of Roraima, Brazil. — Map 60. — Fl. IX–I. — *Angico* (Brazil); *manariballi* (Guyana).

Albizia glabripetala has been recognized from the first as closely related to the widespread and rather variable *A. niopoides* from which it differs most clearly in the tardily segmented valves of the pod and lack of wing along its ventral suture. Slightly fewer stamens and, on the average, fewer leaflets also, are weak supporting characters. It is more readily distinguished from allopatric *A. polycephala*, to which Bentham referred the earliest collection (*Schomburgk 525*), by the glabrate, not silky-pilosulous perianth.

A single large tree (±15 m, 8 dm dbh) in Reserva Ecolôgica de Maracá (Roraima), sampled both in flower (*Lewis 1652*, NY) and in fruit (*Lewis 1533*, NY), differs from typical *A. glabripetala*, which is also recorded from Ilha de Maracá (*Milliken 88*, NY), in these characters and is provisionally excluded from our description of the species: relatively ample, oblong-elliptic leaflets to 10 × 4 mm; capitula to 7 per node of inflorescence; peripheral flowers subsessile and a trifle larger (calyx ±2 mm and corolla to 4.2 mm); androecium of peripheral flowers to 22-merous; pod smaller and narrower (7–8 × 1.1 cm). It is perhaps a distinct taxon, needing further observation.

15. Albizia pistaciifolia (Willdenow) Barneby & Grimes, comb. nov. *Mimosa pistaciaefolia* Willdenow, Sp. Pl. **4:** 1028. 1806. — "*Habitat ad* Caracas . . . *Bredemeyer*." — Holotypus, B-WILLD, seen in Microform IDC 86. 1382: 111. 9!. — *Samanea pistaciaefolia* (Willdenow) Dugand, Apunt. Hist. Nat. Dept. Atlántico 31. 1933. *Arthrosamanea pistaciaefolia* (Willdenow) Britton & Rose ex Britton & Killip, Ann. New York Acad. Sci. **35:** 128. 1936. *Cathormion pistaciaefolium* (Willdenow) Burkart, Darwiniana **13:** 446. 1946. — Non *Pithecolobium pistaciifolium* Standley, 1935.

Inga triflora G. Don, Gen. hist. **2:** 394. 1832. — "Native of Guayaquil. *Mimosa triflora* Ruiz et Pav[ón] in herb. Lamb[ert]." — Holotypus, not found; isotypus, *Tafalla* ex hb. Pavon., G(herb. boissier.)!. — *Pithecolobium triflorum* (G. Don) Bentham, Trans. Linn. Soc. London **30:** 591. 1875.

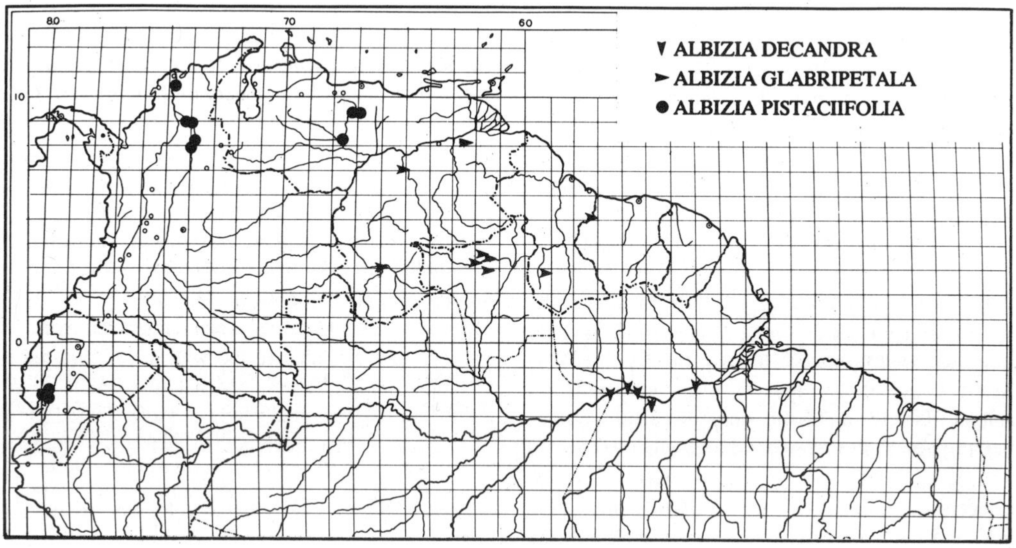

MAP 60. Northern South America. Range of *Albizia pistaciifolia* (Willdenow) Barneby & Grimes, *A. decandra* (Ducke) Barneby & Grimes, and *A. glabripetala* (Irwin) G. P. Lewis & P. E. Owen.

Pithecolobium daulense Spruce *ex* Bentham, Trans. Linn. Soc. London **30:** 586. 1875. — "Tropical America: savannahs on the Daule river near Guayaquil [Ecuador], *Spruce*, n. 6330." — Holotypus, K(herb. bentham.)! = NY Neg. *2041*; isotypi, †B = F Neg. *1190*!, BM!, NY!, P!. — *Feuilleea daulensis* (Bentham) O. Kuntze, Rev. Gen. Pl. **1:** 187. 1891. *Arthrosamanea daulensis* (Bentham) Killip *ex* Record, Trop. Woods **63:** 6. 1940. *Cathormion daulense* (Bentham) Burkart, Darwiniana **13:** 447. 1964.

Samanea guajacifolia Pittier, Arb. Arbust. Venez. **4/5:** 53. 1926. — "[Venezuela]. Selvas xerófilas de el Sombrero, Guárico, con frutos Febrero 19 1924 (H. Pittier 11441, tipo)." — Holotypus, US *1193250*!; isotypi G!, NY!.

Broad-crowned semideciduous macrophyllidious *trees* (3–)4–18 m with trunk potentially attaining 5 m dbh or more and angulate young branches clothed with smooth, vertically splitting epidermis, appearing glabrous but the ventral face of lf- and all inflorescence-axes minutely thinly puberulent or silky-strigulose, the ample plane papery venulose lfts olivaceous brunnescent, a little paler beneath, the slender simple efoliate pseudoracemes of loosely few-fld capitula arising singly or geminate from coeval lf-axils, the fruits often long persisting on defoliate branchlets. *Stipules* (seldom seen) ovate-triangular, <1 mm, fugacious before maturity of associated lf. *Lf-formula* i–ii/3–5 (–6); lf-stks 1–5.5 cm, the petiole 1–3.5 cm, at middle 0.7–1.3 mm diam, the 1 interpinnal segment, when present, 1.5–2.5 cm; a sessile shallow-cupular nectary erratically situated either near middle, or close to either end, of petiole, 0.6–1.6 mm diam <0.25 mm tall, or sometimes immersed in petiole-groove, or in some lvs lacking, and often small or obscure nectaries at tip of pinna-rachises; rachis of pinnae 3.5–7.5 cm, the longer interfoliolar segments 1–2(–2.5) cm; lft-pulvinules strongly cross-wrinkled, 0.9–1.4 × 0.7–1.1 mm; lfts just perceptibly or quite strongly accrescent distally, in outline oblong or oblong-elliptic to inequilaterally ovate or obovate, at base asymmetrically rounded, semicordate, or rarely flabellate, broadly obtuse at apex, the penultimate pair 19–32(–48) × 9–18(–26) mm, 1.5–2 times as long as wide; venation of (5–)6–7 primary nerves from pulvinule, the straight midrib a little forwardly displaced, giving rise on each side to 2–4 narrowly ascending major secondaries, the inner posterior primary nerve incurved-ascending to anastomosis well beyond mid-blade, the outer posterior ones progressively shorter, these all prominulous on both faces of blade and generating an irregular, rather open reticulum of venules. *Primary axis* of pseudoracemes 7–14 cm; peduncles 2–6 per node, the longest of each fascicle 3–8 mm; capitula hemispherical (3–)5–9-fld, without filaments 5–6 mm diam, the receptacle 1–1.5 mm; bracts rhombic-ovate 0.3–0.6 mm, persistent; fls all sessile but dimorphic, the perianth of all greenish, glabrous except for microscopically ciliolate calyx-teeth, 5-merous or that of the terminal fl 6-merous, the filaments white; PERIPHERAL FLS: calyx turbinate-campanulate ±1.2 × 1–1.4 mm, the depressed-deltate teeth 0.2 mm; corolla 3–3.3 mm, the lobes 1.2–1.4 × 0.8 mm; androecium 28–36-merous, 9–10 mm, the stemonozone 0.3–0.45 mm, the tube 1.7–2.2 mm; ovary glabrous, symmetrically conic at apex; style as long as or a little longer than longest stamens, usually little dilated at apex but randomly to 0.4 mm diam; TERMINAL FL: scarcely longer but stouter than the peripheral ones, the calyx 1.1–1.4 × 1.3–2 mm, the corolla ±3.5 mm; androecium 32–42-merous, the tube 3.3–3.5 × 1.3–1.7 mm. *Pods* solitary, in profile undulately linear, straight or nearly so, contracted at base into a stipelike neck 2–8 mm, at apex obtuse apiculate, when well fertilized 10–16 × 1–1.5 cm, 11–15-seeded, framed by narrowly 2-sulcate, wiry and continuous, evenly shallowly (deeply where ovules abort) constricted sutures 0.7–1 mm wide, the stiffly leathery or subligneous, light brown, rough but scarcely venulose valves becoming low-convex over each seed and transversely sulcate between them, the individual seed-cavities (7–)9–12 mm long, the whole fruit lomentiform in appearance but not separating into 1-seeded propagules; dehiscence in strict sense 0, the mature valves reluctantly separating along the transverse sulci but retained by the sutures, the whole fruit long persistent on the tree; seeds transverse on filiform funicle, compressed but plump, in broad view round or oblong-elliptic ±6.5–7.5 mm in greatest diameter, the smooth crustaceous yellowish testa closely investing the plano-convex cotyledons, the centric pleurogram complete, ±3 mm diam.

Restricted to wet or annually flooded habitats within seasonally dry woodland communities below 250 m, locally plentiful in three small and widely disjunct areas in NW South America: lower Magdalena valley in Atlántico, Bolívar and N Magdalena, Colombia; Orinoco basin, along ríos Guanare, Guárico, Orituco and (lower) Apure in Portuguesa, Guárico, and Apure, Venezuela; and collected frequently, from late eighteenth century onward, within a radius of ±30 km of Guayaquil, in Guayas, Ecuador. — Map 60. — Fl. IX–II(–?). — *Guayacán cenega, guayacán chaparro, guayacán hobo* (Colombia); *carabalí, quiebrahacho, vera macho* (Venezuela); *nance, tinto de bajos* (Ecuador).

In context of its immediate relatives *A. pistaciifolia* is instantly recognized by leaf-formula of 1 or less often 2 pairs of pinnae and not more than 6 pairs of relatively ample, venulose leaflets. The foliage in gross aspect recalls that of the Hispaniolan *Cojoba filipes* (Ventenat) Barneby & Grimes, which differs in organization of the inflorescence and the entirely different, subterete fleshy fruit.

The proliferation of names applied to *A. pistaciifolia* is accidental and not the result of critical differentiation. When Bentham recognized *Pithecolobium triflorum*, he was ignorant of the true nature of the prior *Mimosa pistaciifolia* and also of the seemingly lomentiform pod, which, on discovery, was described as *Pithecolobium daulense*. Pittier compared *Samanea guajacifolia* to *Pithecellobium marginatum*, comparable indeed in superficial aspects of the pod but fundamentally different in the corymbiform capitula elevated on a long axillary peduncle.

16. Albizia subdimidiata (Splitgerber) Barneby & Grimes, comb. nov. *Acacia subdimidiata* Splitgerber, Tijdschr. Natuurl. Gesch. Physiol. **9:** 112. 1842. — Typus infra sub var. *subdimidiata* indicatur.

Amply, often rather coarsely foliate *trees* 5–20 (–25) m with single or multiple trunk attaining 1–3 (–4) dm dbh, the hornotinous branches ridged and sulcate lengthwise, buttressed under each lf, with smooth, castaneous or atrocastaneous, at length vertically splitting and striped epidermis, the lf-stks (at least ventrally) and axes of inflorescence always, the branchlets sometimes, pilosulous with pallid hairs to 0.1–0.25 mm, the lfts glabrous facially except for midrib proximally barbellate on its anterior side beneath, bicolored, olivaceous (brunnescent) and dull or sublustrous above, pallid and usually microscopically pallid-papillate beneath, the simple pseudoracemes of very small, fragrant, greenish white or ochroleucous capitula arising singly or 2–3 together from the axil of coeval (or newly shed) lvs and much surpassed by them, the fruits long persistent on defoliate annotinous branches. *Stipules* deltate or triangular-ovate 0.8–1.8 mm, very early dry caducous, lacking from mature lvs. *Lf-formula* ii–iv(–v)/(7–)8–13(–14), in saplings the pinnae sometimes to 5-jug.; lf-stks of larger lvs mostly (5–)6–16(–20), in var. *minor* only 1.5–5 cm, the petiole (3–)3.5–7, in var. *minor* 1.5–3.2 cm, at middle 0.7–1.9 mm diam, the ventral groove continuous between the often not exactly opposite pinna-pairs, the longest interpinnal segments (1.5–) 2–4, in var. *minor* 0.6–1.6 cm; near or below mid-petiole (exceptionally near first pinna-pair) a shallow-cupular or rarely almost plane, round or elliptic, glabrous nectary (0.7–)1–2.6 mm diam, in profile not over 0.4 mm tall, rarely a second nectary at tip of lf-stk and nearly always yet smaller ones between furthest 1–2 pairs of lfts; pinnae a little decrescent proximally, the rachis of longer ones (5–)6–9(–11) or in var. *minor* only 3–5.5 cm, the longer interfoliolar segments (5–)6–11 or in var. *minor* only 3–6 mm; lft-pulvinules 0.7–1 × 0.5–0.8 mm; lfts a little decrescent proximally, thence subequilong, in outline lance-oblong, oblong, semi-ovate, or asymmetrically ovate-elliptic, from inequilaterally rounded or semicordate base, either acutely or obtusely deltate-apiculate, straight or gently incurved near apex, the larger ones 16–25(–29) × 6–9(–12.5) or in var. *minor* only 11–14 × 3–5 mm, (2–)2.2–3.3(–3.6) times as long as wide; venation of (4)5–6 primary nerves from pulvinule, the straight or near apex incurved midrib forwardly displaced to divide blade 1:2–3 and giving rise on anterior side to ±5–7 widely ascending major secondary venules and on posterior side to fewer, more narrowly ascending ones, the inner posterior primary nerve produced far beyond midblade to anastomosis well within the plane or loosely revolute margin, the outer posterior ones progressively shorter, the 1 anterior primary nerve very short or subobsolete, all these, together with random tertiary venulation, sharply prominulous dorsally, less so but commonly pallid ventrally. *Primary axis* of pseudoraceme 7–21 cm; peduncles fasciculate by 2–6, the longer of each fascicle 3–9(–11) mm; capitula (8–)11–22-fld, without filaments 5–8.5 mm diam, the globose or claviform receptacle 0.6–2 mm; bracts rhombic-ovate or spatulate 0.4–0.7 mm, glabrate dorsally but ciliolate, persistent; fls dimorphic, the terminal one of each head hardly longer than the rest but wider and with modified androecium, the perianth 5(or that of terminal fl 6–7)-merous, all greenish white and either glabrous except for minutely ciliolate calyx-teeth or the calyx-tube and tip of corolla-lobes microscopically puberulent; PERIPHERAL FLS: sessile or raised on an obscure pedicel <0.6 mm, the calyx campanulate or turbinate-campanulate (0.65–)0.7–1.2 × 0.7–1 mm, the depressed-deltate or deltate-ovate teeth 0.1–0.25 mm; corolla (2–)2.2–3.3 mm, the lobes 0.9–1.5 × 0.65–0.9 mm; androecium (12–)18–25-merous, 6–9 mm long, the stemonozone 0.2–0.5 mm, the tube (0.5–)1.5–2.2 mm; ovary glabrous, symmetrically conical at apex; style a little surpassing the longest stamens, slightly dilated at tip; TERMINAL FL: always sessile, the shallowly campanulate calyx (0.75–)0.8–1.3 × 1.3–2.1 mm, the corolla 2–3.7 mm; tube (1.2–)1.8–3 × 1.2–3 mm, dilated upward; filaments (20–)24–32, irregularly 2-ranked, the free part of stronger ones proximally dilated. *Pods* solitary, subsessile or commonly contracted at base into a laterally sulcate pseudostipe 1.5–5 mm, the body undulately linear in profile, either straight or decurved, (8–)9–14(–15) or in var. *minor* only 6–8 mm wide, when well fertilized (6–)8–17 cm long and (8–)10–15-seeded, the leathery, at first plano-compressed, atrocastaneous and lustrous, and elaborately venulose valves at length sordid-brown or dull buff, low-convex over each seed, and transversely sulcate between them, the obtuse wiry sutures 0.6–1.1 mm wide, evenly crenately indented their whole length, each seed-chamber (7–)8–15 mm long; fruit lomentiform but not truly dehiscent, the valves often cracking along sulcus between seeds but the sutures continuous and tenaciously attached to the 1-seeded segments, the seeds not individually dispersed; seeds nearly round in broad view, or obtusely quadrate or broad-ellipsoid, compressed but plump, in broad profile 5.3–7(8) mm diam, the bone-white, putty-colored or fawn-brown testa ±0.1–0.15 mm thick, closely investing the hard embryo and charged on each face with a complete pleurogram 2.7–4.5 mm diam.

This is the first of three species, only cryptically different at anthesis but decisively different in fruit, that Bentham included together in his concept of *Pithecolobium multiflorum*. The carpological differences of *A. subdimidiata*, *A. inundata*, and the genuine *A. multiflora* are set out in the conspectus to the genus (p. 205). In practice the accompanying dispersal map should be most helpful in placing flowering specimens of the three species, only two of which are narrowly sympatric, and those only in the lower Amazon valley.

Albizia subdimidiata is variable in pubescence of young stems, in amplitude of foliage, in several small features of the individual flower, and in width of pod, but we can discern only one localized race sufficiently modified to deserve taxonomic status.

Key to the varieties of *A. subdimidiata*

1. Lvs relatively ample, the longer lf-stks 6–16(–20) cm, the petiole (3–)3.5–7 cm × 0.9–1.9 mm, the rachis of longer pinnae (5–)6–9(–11) cm, and its interfoliolar segments to (5–)6–11 mm; larger lfts 16–25(–29) × 6–9(–12.5) mm; range of the species .... 16a. var. *subdimidiata*
1. Lvs smaller in all parts, the longer lf-stks 1.5–5 cm, the petiole 1.5–3.2 cm × 0.7–0.9 mm, the rachis of longer pinnae 3–5.5 cm, the longer interfoliolar segments 3–6 mm; larger lfts 11–14 × 3–5 mm; local in N state of Roraima, Brazil, and in adjacent S Guyana........................16b. var. *minor*

16a. Albizia subdimidiata (Splitgerber) Barneby & Grimes, var. **subdimidiata.** *Acacia subdimidiata* Splitgerber, 1842, l. c., sens. str. — "Habitat ad ripas fluminis Surinami superioris, ubi lege [*sic*] florentem mense Aprili." — Holotypus, *Splitgerber 917*, "prope Goode Savanne," 27.IV.1838, L! = NY Neg. *12075*. — Equated with *Pithecolobium multiflorum* by Bentham, London. J. Bot. **3:** 220. 1844.

Pithecolobium multiflorum sensu Bentham, 1875: 591 & 1876: 445, ex parte, typo excluso; Lindeman & Mennega, Bomenboek Surinam 191. 1963.
Arthrosamanea multiflora sensu Kleinhoonte, 1940: 326, ex parte, typo et synon. nonnullis exclusis.

Characters of the species, as modified by the foregoing key.

On river banks, in várzea forest, or at forest margins in seasonally inundated campo, 5–350 m, widely dispersed between 9°N and 15°S in South America: Orinoco and Maracaibo basins in Venezuela and far NE Colombia (Boyacá); lowland Guyana and Surinam; around the W and SW periphery of the Amazon basin from Peru (Loreto) to Bolivia (NE La Paz, Pando, and Beni) and in scattered localities along the Ríos Solimões and Amazonas to W Pará, Brazil; somewhat isolated on upper Río Guama in Pará and on Río Marucassumé in Maranhão. — Map 61. — Fl. (VII–)VIII–?. — *Hueso de pescado* (Venezuela, a name applied also to other, habitally similar mimosoid trees); *swamp manariballi* (Guyana); *pashaca* (Peru); *paracaxirana, paricarana* (Amazonas); *acuhy do campo* (Maranhão, used for tanning); *coronha, marimari bravo* (Pará).

The var. *subdimidiata* is somewhat variable in pubescence and in width of the pod. Most populations in the Orinoco basin and the Guianas have minutely puberulent young branchlets, and the pods often transiently so, whereas in Amazonia the branchlets are glabrous (and commonly atrocastaneous in color), and the pods are glabrate at a very early stage. The differences are slight and seem taxonomically unimportant. No geographic correlation with width of pod has been detected.

16b. Albizia subdimidiata var. **minor** Barneby & Grimes, var. nov., a var. *subdimidiata* foliis deminutis, petiolo communi 1.5–5 (nec 6–20) cm usque, petiolo vero vix 3 (nec 3–7) cm, pinnarum (2–3-jugarum) longiorum rachi 3–5.5 (nec 5–11) cm, foliolisque usque 11–14 (nec 16–29) mm longis diversa. — GUYANA: at edge of forest, among rocks, basin of Essequibo river near 4°53′N, Kuyaliwak Falls, 1.X.1937 (fl.), *A. C. Smith 2156*. — Holotypus, NY; isotypi, K, P.

Trees 5–8 m; other characters of the species, except as modified in key and protologue.

In seasonally flooded savanna and at forest margins or on stream banks, 100–450 m, apparently localized on the upper Río Branco in N state of Roraima, Brazil (Maracá), and in the middle and upper Essequibo basin in W-centr and SW Guyana. — Map 61. — Fl. IX–X(–?).

The typus of var. *minor* was referred by Sandwith to *Pithecolobium multiflorum* (sensu Bentham), and in isolation might pass for an incidental small-leaved form of it. But three further collections from the same general region confirm that there is something more to var. *minor* than fortuitous dwarfing of the foliage.

XVA(III). SERIES MULTIFLORAE BARNEBY & GRIMES

Albizia sect. **Arthrosamanea** ser. **Multiflorae** Barneby & Grimes, ser. nov., a ser. *Arthrosamanea* caeterius haud diversa leguminis valvulis crasse lignoso-fibrosis indehiscentibus, sed maturis lomentiformibus in articulos monospermos tarde solutis abiens. — Sp. typica: *A. multiflora* (Kunth) Barneby & Grimes ≡ *Acacia multiflora* Kunth ≡ *Pithecolobium multiflorum* (Kunth) Bentham.

Pods broad-linear, straight, compressed but plump, the pithy-ligneous valves ±2–3 mm thick in section, at maturity lomentiform, transversely septate and, under slight pressure or after falling, breaking apart, through valves and suture alike, into monspermous articles. — Spp. 2, one of lower Amazonian Brazil, one of Pacific Ecuador and NW Peru.

17. Albizia multiflora (Kunth) Barneby & Grimes, comb. nov. *Acacia multiflora* Kunth *in* Humboldt, Bonpland & Kunth, Nov. Gen. Sp. **6(qu):** 277. 1824. — *Typus sub var. multiflora* infra indicatur.

Slender, amply leafy *trees* 5–15 m, with smooth glabrous, in second year vertically striped branchlets deeply sulcate under the buttressed lf-pedestals, the lf- and inflorescence-axes puberulent with incurved hairs <0.15 mm, the broad plane lfts bicolored, dull olivaceous above, pallid beneath, the lfts minutely inconspicuously puberulent on both faces or glabrescent on upper one, the simple efoliate pseudoracemes of small greenish white capitula arising singly or geminate from coeval lf-axils, the firm broad-linear pods maturing at defoliate nodes below the hornotinous lvs. *Stipules* either obsolete or very early caducous, none seen. *Lf-formula* ii–iv(or v, fide Schery)/5–11; lf-stks 7.5–18 cm, the petiole including wrinkled

MAP 61. South America. Range of *Albizia multiflora* (Kunth) Barneby & Grimes, *A. inundata* (Martius) Barneby & Grimes, and *A. subdimidiata* (Splitgerber) Barneby & Grimes var. *subdimidiata* and var. *minor* Barneby & Grimes.

pulvinus (3–)4.5–9 cm, at middle 0.9–1.5 mm diam, the ventral groove continuous between often not quite opposite pinna-pairs, the one or the longest interpinnal segment (2–)2.5–4 cm; a sessile glabrous cupular nectary well below midpetiole or contiguous to lf-pulvinus 0.7–1.5 mm diam, <0.3 mm tall or sometimes almost immersed in petiolar groove, a smaller nectary at tip of each pinna-rachis; pinnae little graduated, the rachis of longer ones 6–11.5 cm, the longer interfoliolar segments 7–18 mm; pulvinules 0.5–1.1 × 0.5–0.9 mm; lfts a little decrescent near base of rachis, otherwise subequilong, the blade oblong or semi-ovate from inequilaterally rounded or shallowly semicordate base, broadly obtuse but sometimes also minutely bluntly apiculate, the longest 17–37 × 6–17 mm; venation of 5–6 primary nerves from pulvinule, the almost straight midrib forwardly displaced to divide blade ±1:2–2.5, giving rise on anterior side to 2–4 and on the posterior to 1–2 major secondary venules, the inner posterior nerve incurved-ascending well beyond midblade, the outer ones progressively shorter, the tertiary venules few and random, the whole venation finely prominulous dorsally, little elevated but pallidly discolored above. *Primary axis* of pseudoracemes 4–19 cm; longer peduncles 5–11 mm; capitula without filaments 6–8 mm diam, the receptacle, including short terminal pedestal, not over 1.5 mm; bracts ovate or spatulate 0.4–0.7 mm, persistent; flowers sessile, dimorphic, externally either microscopically puberulent overall or only so near orifice of calyx and at tip of corolla-lobes, the perianth of all 5-merous or that of one or more distal ones 6-merous; PERIPHERAL FLS: calyx campanulate 1–1.8 × 0.7–1.4 mm, the depressed-deltate teeth 0.1–0.25 mm; corolla 2.8–4.2 mm, the lobes 0.9–1.2 × 0.7–1 mm; androecium 24–30-merous, 6.5–9 mm, the stemonozone ±0.5 mm, the tube 1.4–3.4 mm; ovary glabrous, subsymmetrically conic at apex; style as long as or a trifle longer than stamens, not dilated at apex; TERMINAL FL: no longer or only a little longer but stouter than peripheral ones, the calyx 0.8–1.6 × 1–1.8 mm, the corolla 2.5–4.1 mm, its lobes 1.1–1.6 mm; androecium 26–38(–40)-merous, the tube 2.5–3.8 mm, the free part of filaments dilated at base. *Pods* in profile broad-linear, straight or almost so, 11–17 × 1.6–2.4 cm, abruptly contracted at base into a short necklike stipe and abruptly apiculate, compressed but plump and facially corrugate by shallow transverse interseminal sulci, the pithy-ligneous, glabrous, purplish-castaneous, scarcely venulose [but reportedly resinous] valves framed by almost straight (except where ovules abort) blunt sutures, at maturity 2–2.5 mm thick; dehiscence none or greatly delayed and incipiently lomentiform, the fruit (when dry) under slight pressure breaking through the interseminal grooves and through both sutures into closed, 1-seeded segments; seeds (few seen) transverse on filiform funicle, ±9.5–10 × 5.5–7 × 3 mm, the complete pleurogram ±6–6.5 × 3.5 mm.

Following Bentham, who for lack of the telltale fruits confused the relatively local *A. multiflora* of the Pacific slope with the much more widely dispersed Amazonian *A. subdimidiata*, botanists have universally misapplied the epithet *multiflora*. The fruit of genuine *A. multiflora*, well described by A. Gentry (1988, l. c., as *Pithecellobium paucipinnatum*), is a plumply compressed, pithy-lignescent pod more than 1.5 cm wide that has almost straight sutural margins and only cryptically segmented valves, whereas that of *A. subdimidiata* is leathery plano-compressed and only 8–14 mm wide, with margins undulately pinched between seeds and transverse fracture lines externally evident at an early stage of maturity. The prolonged eclipse of genuine *A. multiflora* in the shadow if *A. subdimidiata* induced the superfluous description of *Pithecolobium weberbaueri* and *Albizia paucipinnata*.

Key to the varieties of *A. multiflora*

1. Lfts of longer pinnae 8–11 pairs, separated by interfoliolar segments 7–11 mm and the longer blades 17–26 mm; ranging northward from centr. and N dept. Cajamarca (prov. Cajamarca near 7°20′S; prov. Jaén near 5°45′S) ....17a. var. *multiflora*
1. Lfts of longer pinnae 5–6 pairs, separated by interfoliolar segments to 14–18 mm and the longer blades 30–37 mm; known only from Cajamarca prov. Contumazá near lat. 7°35′S ....17b. var. *sagasteguii*

17a. Albizia multiflora (Kunth) Barneby & Grimes var. **multiflora.** *Acacia multiflora* Kunth, 1824, l. c., sens str. — "Crescit prope San Felipe, in declivitate orientali Andium Bracamorensium [±5°45′S, 79°20′W in SW prov. Jaén, dept. Cajamarca, Peru], alt. 980 hex [±1800 m]." — Holotypus, *Humboldt & Bonpland 3562* collected in Aug. 1802, P(HBK)!; isotypus, P(herb. bonpland.)!. — *Pithecolobium multiflorum* (Kunth) Bentham, London J. Bot. **3:** 220. 1844. *Feuilleea multiflora* (Kunth) O. Kuntze, Revis. Gen. Pl. **1:** 188. 1891. *Samanea multiflora* (Kunth) Pittier, Arb. Legum. **1:** 61. 1927 (citatione manca). *Senegalia multiflora* (Kunth) Killip, Trop. Woods **63:** 6. 1927. *Arthrosamanea multiflora* (Kunth) Kleinhoonte in Pulle, Fl. Suriname **2(2):** 327. 1940 (quoad basionym. solum). *Cathormion multiflorum* (Kunth) Burkart, Darwiniana **13:** 446. 1964 (quoad basionym., haud conceptum).

Pithecolobium weberbaueri Harms, Repert. Spec. Nov. Regni Veg. **16:** 350. 1919. — "Peru: Depart. Cajamarca, Prov. Jaen, Tal des Maranon zwischen Jaen u. Bellavista, 5°30′–5°40′S Br., 600 m . . . (A. Weberbauer no. 6207. — April 1912 . . .)." — Holotypus, †B = F Neg. 1221!; isotypus not found at GH in 1992. — Equated with *Pithecolobium multiflorum* by Macbride, 1943: 59.

Albizia paucipinnata Schery, Ann. Missouri Bot. Gard. **37:** 400. 1950. — "Ecuador: Prov. El Oro, along Rio Amarillo upstream from Portovelo [Portoviejo], alt. 640–760 m, *J. A. Steyermark 54035*." — Holotypus, MO, n. v.; isotypi, F!

= F Neg. *53643*, NY!. — *Pithecellobium paucipinnatum* (Schery) A. Gentry & C. Dodson *in* A. Gentry, Ann. Missouri Bot. Gard. **75:** 1438. 1988.

Pithecolobium multiflorum sensu Bentham, 1875; 591 & 1876; 445, minore ex parte, nomen his locis illegitimum, ob synonymum prius '*Acacia inundata*' inclusum; sensu Macbride, 1943: 59.

As described in key to varieties.

In dry forest and disturbed brush-woodland, extending in Ecuador from near sea level up to 700 m and in inter-Andean valleys of Peru up to 1800 m, locally plentiful in SW Ecuador (Guayas, El Oro, S Manabí, S Loja) and NW Peru (Tumbes, Piura), thence passing E over the Andean crest to the sources of Río Maranón in centr. and N dept. Cajamarca (prov. Cajamarca and Jaén); disjunct in centr. Panama (Sardinilla, prov. Colón), but collected in Colombia (without locality, not mapped) by J. C. Mutis before 1808. — Map 61. — Fl. VIII–X(–IV). — *Compoño* (Guayas); *angol, angolo* (Piura, Lambayeque).

Albizia multiflora was collected first at Guayaquil, probably by Juan Tafalla (s.n., OXF!).

17b. Albizia multiflora var. **sagasteguii** Barneby & Grimes, var. nov., a var. *multiflora* foliolis minus numerosis atque majoribus, pinnarum majorum 5–6-jugis, lamina 30–37 (nec 17–26) mm longa abiens. PERU. Depto Cajamarca prov. Contumazá: in a quebrada near San Benito, ±78°55′W, 7°35′S, *A. Sagástegui A. 15410.* — Holotypus, F; isotypus, NY.

As described in key to varieties; tree 10 m with whitish fls.

Quebrada at 1200 m, known only from the type collection. Fl. X–XI.

18. Albizia decandra (Ducke) Barneby & Grimes, comb. nov. *Pithecolobium decandrum* Ducke, Arch. Jard. Bot. Rio de Janeiro **5:** 121. 1930. — "... Para circa Obidos (Herb. [MG] n. 15.724, et herb. [RB] n. 10174) et ... ad lacum Faro ([RB] n. 20.198), ubi florebat januario 1927 (legit A. Ducke). Arbores fructiferas vidi etiam prope Oriximiná . . . et in regione Jutahy inter Almeirim et Prainha." — Lectotypus, *Ducke s.n.*, 19.I.27 (fl.) in RB *10174* (8 sheets), RB!; isotypi, K!, NY!, U!, US *1441564*!; paratypi, *Ducke s.n.*, in MG *15724*, MG!, 10.XI.19 (fr.), cum RB *10174* associatum, K!, NY!, P!, RB!, *Ducke s.n.*, 27.I.1927, K!, RB *20198*!.

Pithecellobium decandrum sensu Ducke, Bol. Técn. Inst. Agron. N. **18:** 38. 1949.

Trees 20–30 m, appearing glabrous but the very young stems and the axes of lvs and inflorescence thinly puberulent with incurved pallid hairs <0.3 mm, the lvs bicolored, the ventrally low-convex, papery lfts dull olivaceous on both faces but paler beneath, the hemispherical or subglobose capitula of whitish fls borne on short simple efoliate pseudoracemes axillary to coeval lvs and either immersed in, or barely surpassing, the foliage. *Stipules* not seen, fide Ducke small linear, early caducous. *Lf-formula* (i–)ii–iv/9–12(–13); lf-stks 3–12 cm, the petiole 2–5 cm, at middle 0.7–1.3 mm diam, the ventral sulcus continuous between pinna-pairs, the longer (or only) interpinnal segments 1.3–2.5 cm; a round sessile glabrous, shallowly cupular nectary 0.5–1.3 mm diam, <0.3 mm tall, below midpetiole of most lvs, but this lacking in some lvs or displaced to very base of petiole, sometimes a smaller nectary at apex of lf-stk and a yet smaller one at tip of pinna-rachises; pinnae accrescent distally, the rachis of further pairs 5.5–8.5 cm, the longer interfoliolar segments 5–7.5 mm; pulvinules 0.4–0.6 mm diam; lfts decrescent proximally, scarcely so distally, the first pair often represented by linear paraphyllidia arising next to pinna-pulvinus, the rest oblong-elliptic from asymmetrically rounded or shallowly semicordate base, broadly obtuse at apex, those near midrachis 12–18 × 5–8 mm, 2.2–2.5 times as long as wide; venation of 4(–5) primary nerves from pulvinule, the almost centric, straight midrib giving rise on each side to 3–5 pairs of secondary nerves brochidodrome shortly within the margin, the inner posterior and the one anterior primary shorter, the connecting tertiary venules weak and random, the whole venation finely sharply prominulous beneath, scarcely, or in age sharply so and pallidly discolored above. *Primary axis* of pseudoracemes 6–15 cm, floriferous from near or beyond middle, the peduncles mostly geminate at the nodes, the longer ones 11–19 mm, the dimorphic fls all contiguous or 1–2 displaced onto the peduncle, the receptacle including squat terminal pedestal ±2–4 mm; bracts ovate-lanceolate ±1–1.5 mm, persistent; perianth 5- or of some terminal fls 6-merous, submembranous glabrous, whitish or greenish; PERIPHERAL FLS: subsessile or elevated on slender pedicel to 0.1–1.2 mm; calyx campanulate 1–1.2 × 0.9–1 mm, the depressed-deltate teeth 0.1–0.2 mm; corolla 3.6–3.8 mm, the lobes 1.3–2 × 0.8–1 mm; androecium 10-merous, 11–17 mm, the stemonozone ±0.5 mm, the tube 1–1.4 mm; ovary glabrous, symmetrically conical at apex; style 1–3 mm longer than longest stamens, slightly dilated at the stigma; TERMINAL FL: sessile, the broadly shortly campanulate calyx 1.5–2 × 1.7–2.2 mm, the corolla 5–5.5 mm; androecium to 15(–?)-merous, the tube 4.5–7 mm, the free part of filaments dilated at base. *Pods* solitary, sessile, broad-linear straight or nearly so, when well fertilized 8–12 × 1.7–2.1 cm, compressed but to ±6 mm thick, the obtuse sutures ±1.6 mm wide, straight or almost imperceptibly recessed between seeds, the valves low-convex over each seed and transversely shallow-sulcate between them, the castaneous exocarp 1.2–3 mm thick directly over each seed, the crustaceous endocarp ±0.2 mm thick, each seed-locule ±8–10 mm long; dehiscence 0, but the valves, under moderate pressure, breaking transversely through the interseminal sulci and through the sutures into individually indehiscent 1-seeded locules; seeds transverse on filiform funicle, in broad view oblong ±8–8.5 × 4.5–5 mm, the hard smooth, yellowish brown testa charged on each face with a deep, inversely U-shaped pleurogram 3.5–4 × 2 mm; embryo clothed in thin translucent endosperm.

In forest on terra firme, below 100 m, apparently confined to the lower valley of the Amazon and its immediate tributaries in Pará and extreme E Amazonas, Brazil, between the mouths of ríos Nhamundá (Faro) and Parú (near Almeirim). — Map 60. — Fl. XII–II.

Albizia decandra is well characterized by leaf-formula, subequilateral obtuse leaflets, decandrous peripheral flowers, and thick-walled, indehiscent fruit. The pod bears a strong likeness to that of the Ecuadorian and north Peruvian *A. multiflora.*

XVA(IV). SERIES INUNDATAE BARNEBY & GRIMES

Albizia sect. **Arthrosamanea** ser. **Inundatae** Barneby & Grimes, ser. nov., a ser. *Multifloris* caeterius congruis legumine cryptolomentiformi, valvarum maturarum exocarpio continuo per suturas pigre dehiscenti sed endocarpio in articulos monospermos more *Plathymeniae* Bentham secedenti diversa. — Sp. typica (unica): *Albizia inundata* (Martius) Barneby & Grimes ≡ *Acacia inundata* Martius = *Arthrosamanea polycephala* (Grisebach) Burkart (non *Arthrosamanea polycephala* (Bentham) Killip). — Sp. 1, discontinuously widespread from lower Amazon valley to E Brazil, N Argentina, E Bolivia.

At anthesis simulating ser. *Multiflorae*, but differing in the cryptolomentiform fruit, described for the species following.

19. Albizia inundata (Martius) Barneby & Grimes, comb. nov. *Acacia inundata* Martius *in* Spix & Martius, Reise Bras. **1:** 555. 1823. — ". . . im Alagadisso [a tract of seasonally inundated riparian forest along Río São Francisco in N Minas Gerais, Brazil]." — Holotypus, labeled "in virgultis ad fl. S. Francisco prope Salgado . . . Aug.[1818]" flor., M! = NY Neg. *12583.*

Enterolobium polycephalum Grisebach, Symb. Fl. Argent. 123. 1879. — "O[rán]: Gr. [= Gran] Chaco, in palmetis pr. Laguna del Palmar." — Typus (Burkart, Darwiniana **9:** 69. 1949), *Lorentz & Hieronymus 572*, coll. Aug. 1873 (fl.); holotypus, GOET (not seen); isotypi, CORD (fide Burkart, l. c.), NY!, S!. — *Feuilleea polycephala* (Grisebach) O. Kuntze, Revis. Gen. Pl. **1:** 188. 1891. *Arthrosamanea polycephala* (Grisebach) Burkart, Darwiniana **9:** 69. 1949. *Cathormion polycephalum* Burkart, Darwiniana **13:** 448. 1964. — Non *Albizia polycephala* Killip, 1940.

Pithecolobium pendulum Lindman, Bih. Kongl. Svenska Vetensk.-Akad. Handl. **24,** Afd. **III(7):** 54, fig. 14. 1898. — "Paraguay: El Chaco, in ripis amnium nunc silvaticis (Río Pilcomayo), nunc apertis (Río Negro) copiose (A[nisits] 2005)." — Holotypus, S (not seen). — Described from sterile material; equated with *Arthrosamanea polyantha* by Burkart, Darwiniana **9:** 66. 1949.

Pithecolobium multiflorum var. *brevipedunculata* [sic] Chodat & Hassler, Bull. Herb. Boissier, ser. 2, **4:** 483. 1904. — "[Paraguay] . . . in humidis insulae Chaco-y pr. Concepcion. Aug., [*Hassler*] no. 7217." — Holotypus, G (not seen); isotypi, NY!, P!, UC *940803*!, *950950*! US!.

Pithecolobium multiflorum sensu Bentham, 1875; 591 & 1876: 445, ex parte, the mixture astutely analyzed by Burkart, Darwiniana **9:** 69, 1949.

Arthrosamanea polyantha [sic] sensu Burkart, Darwiniana **9:** 66. 1949, exclus. basionymo *Acacia polyanthes* K. Sprengel, Syst. Veg. **5:** 3. 1826, substituto pro *A. multiflora* K. Sprengel, Syst. Veg. **3:** 142. 1828, homonymo posteriori ac nom. dub. (typo perdito); non *Acacia multiflora* Kunth in Humboldt, Bonpland & Kunth, 1824. — *Cathormion polyanthum* [*sic*] Burkart, Darwiniana **13:** 447. 1964. *Albizia polyantha* G. P. Lewis, Leg. Bahia 164. 1987.

Arthrosamanea polyantha sensu Burkart, Las Leguminosas Argentinas 2da ed., 108, fig. 11. 1952. *Cathormion polyanthum* sensu Burkart, Fl. Ilustr. Sta. Catarina, Legum. Mimos. 91. 1979; Fl. Ilustr. Entre Ríos **3:** 467, fig. 205. 1987; sensu Bernardi, 1984: 179, fig. 45; sensu Hoc, 1992b: 260, sequ.

Trees 3–12(–14) m with trunk attaining ±3 dm dbh, or sometimes (see discussion below of *Enterolobium polycephalum* Grisebach) a bushy treelet fertile at 1–3 m and propagating by suckers, the young branches strongly ribbed and sulcate, buttressed at each node, becoming vertically striped in second season, the new growth except for gray-puberulent axes of inflorescence glabrous or only minutely or thinly puberulent on ventral margins of lf-stk and pinna-rachis, the bicolored lfts glabrous facially except sometimes for a few hairs in proximal basal angle of midrib or along its proximal half, on upper face green or dark brownish green and moderately to highly lustrous, paler dull, sometimes glaucescent but only rarely pallid-papillate beneath, the narrow, simple or feebly branched, efoliate pseudoracemes of small greenish white capitula arising singly or paired in coeval lf-axils, the furthest often appearing terminal, all shorter than the subtending lf. *Stipules* papery, triangular or ovate ±1–1.5 mm, nerveless or faintly 3-nerved, early caducous, absent from fruiting spms. *Lf-formula* ii–iv/(8–)9–14(fide Burkart –17); lf-stks 6–15 cm, the petiole including wrinkled pulvinus 3–6.5(–8) cm, at middle 0.9–1.5 mm diam, the one or the longest interpinnal segment 1.8–3.2(–3.5) cm, the ventral groove continuous between pinnae; a sessile, shallow-cupular, almost round or vertically elliptic, brown glabrous nectary normally 0.8–2.1 × 0.8–2.1 mm close to or somewhat below midpetiole, exceptionally vestigial, a smaller one rarely just below first pinna-pair, and yet smaller ones at tip of lf-stk and each pinna; pinnae of nearly equal length or the first and last a little shorter, the rachis of longer ones 5.5–9.5 cm, the longer interfoliolar segments (4–)5–10 mm; pulvinules 0.6–1.4 × 0.5–0.9 mm; lfts a little decrescent at each end of rachis, the firm, when dry ventrally convex blade narrowly oblong or more commonly lanceolate from inequilaterally rounded base, acutely triangular or acutely deltate and almost always falcately porrect at very apex, the longer ones (15–)16–28 × (3.5–)4–9 mm, (2.7–) 3.1–4(–5) times as long as wide; venation of 5–7 primary nerves from pulvinule, the midrib forwardly displaced to divide blade ±1:2–4, straight proximally, near apex falcately porrect, the inner posterior nerve incurved-ascending almost to blade's apex, the outer posterior ones progressively shorter, the anterior ones(s) weak and short, the primary nerves giving rise to few random, narrowly ascending connecting venules, the whole venation finely but sharply prominulous beneath, less so above. *Primary axis* of inflorescence 3–14 cm, that of rare secondary branchlets <2 cm; peduncles at most nodes 2–5, the longest of each fascicle (2–)3–8 mm; capitula 4–16-fld, without filaments ±5–6.5 mm diam, the axis including a sometimes differentiated terminal pedestal 1 mm or less, the fls sessile or almost so, of nearly the same length but the terminal one (sometimes abortive) a little broader and with slightly more numerous stamens; bracts deltate-ovate 0.3–0.5 mm, deciduous; perianth 5-merous greenish, glabrous except for microscopically ciliolate calyx-teeth; calyx campanulate or turbinate-campanulate 0.85–1 mm and about as wide, the depressed-deltate teeth 0.15–0.3 mm; corolla 2.6–3.3(3.5) mm, the ovate lobes ±1.3 mm; filaments of peripheral fls 22–46 (of some terminal fls to 64), when straightened out 6–7 mm, the stemonozone ±0.5 mm, the tube 1.1–1.7 mm; ovary glabrous, conical at apex; style slightly surpassing the longer stamens, scarcely dilated at apex. *Pods* (cryptoloments) one or exceptionally 2 per capitulum, either sessile or more often contracted at base into a laterally grooved pseudostipe 1.5–8 mm, in profile broad-linear, straight or gently arched downward, abruptly apiculate by the style-base, when well-fertilized (9–)11–18 × 0.9–1.4 cm, (8–)11–18-seeded, the stiffly papery, plano-compressed, dull brown or livid-brown, coarsely venulose, glabrous valves framed by obtuse, straight or shallowly undulate (only where ovules abort more deeply constricted) sutures 1–1.5 mm wide, when ripe shallowly transversely sulcate between seeds, finally separating into continuous exocarp and pale buff endocarp, this permanently clothing the seeds and disjointing between them; dehiscence through both sutures, the exocarp narrowly gaping to emit the seeds, these each wrapped in a square, rhombic, or rhombic oblong envelope; seeds oblong or elliptic in broad view, ±5.5–7 × 4.2–5.2 mm, the smooth sublustrous testa fawn or putty-colored, ±0.2 mm thick, closely investing the embryo, the inversely rounded or elliptic, U-shaped pleurogram 2.5–3 mm diam.

In riparian woodland and seasonally flooded varzea forest, on river-islands and in low-lying palm savanna, 5–450 m, locally common or even dominant, frequent in the Paraná–Paraguai–Uruguai basin northward from ±33°S in Río Negro, Uruguay, and in Corrientes and adjacent Buenos Aires in Argentina to the Pantanal in Mato Grosso, Brazil, and the headwaters of Río Beni in Bolivia, NW in Argentina to Salta (Orán) and Jujuy (*Schuel 103*, W); disjunct in E Brazil along and near Río São Francisco in extreme N Minas Gerais and interior Bahia, and in the upper Tocantins valley in the same latitude in Goiás, in S Piauí and Ceará; and reappearing in Pará, Brazil, on the Amazon and immediate tributaries downstream from mouth of Río Trombetas and on the lower Río Tocantins. — Map 61. — Fl. VII–XII. *Maloxo, palo-flojo, timbó-atá, timbó blanco.*

While fully endorsing Burkart's definition of this species, we disagree with his proposed nomenclature. *Acacia multiflora* K. Sprengel was described from a Sello specimen, now lost, in a phrase of only 14 words and in terms so generalized that its identity, in absence of a type, is essentially unknowable. Bentham (1875: 591) thought it "very probable" that Sprengel's name might be equated with *A. multiflora*

Kunth, but Bentham took an extremely broad and now indefensible view of the latter species. Burkart promoted Bentham's probability into virtual certainty, adding that the species might be recognized, in the context of Sello's itineraries, by the falcate leaflets. There is nothing in the diagnosis, however, to show that the type actually came from *southern* Brazil (though this is probable) and not a word on structure of the androecium that might exclude *A. multiflora* Sprengel from the genus *Acacia* as now constituted, or from other genera of Mimosoideae. If it were indeed an *Albizia* or a *Pithecellobium*, it might as well, from the scant morphological and geographic data in the protologue, be applicable to *Albizia edwallii*, found commonly on Sello's route through Paraná and Santa Catarina. We consider *A. multiflora* Sprengel a nomen dubium. Furthermore, it has not been noticed that when Sprengel substituted for his posterior *multiflora* a new epithet, this was spelled *polyanthes* (in accusative case *polyanthen*, not *polyantham*). When transferred by Burkart to *Arthrosamanea* the epithet should have been *A. polyanthes*. The spelling and use of this combination are, however, of no more than academic interest now that a yet earlier name for the taxon involved has been found.

It is not without misgiving that we here reduce *Enterolobium polycephalum* Grisebach to synonymy of *A. inundata*. Burkart regarded them, first in genus *Arthrosamanea* and later in genus *Cathormion*, as closely related species distinguished by habit of growth, length of pseudoracemes (vaguely precised), chromosome-number, and (indecisively) stamen-number. When the stamens fall within the range of 32–37, numbers often encountered in peripheral flowers of the capitulum, and no data on habit or chromosomes are available, the two taxa are practically indistinguishable. The dwarf, stooling form described as *polycephala* by Burkart may be distinctive in the field, but we can find nothing by which to recognize it in the herbarium. The polyploidy ($2n = 52$) ascribed to *A. inundata* by Burkart (1952: 108, sub *A. polycephala*) requires testing not only in Argentina but in the disjunct populations in the São Francisco and Amazon basins in Brazil. Compare Hoc, 1992b: 260, sequ.

XVB. Sectio ALBIZIA

Albizia Durazzini sect. **Albizia.** — Typum sub generis titulo indicavimus.

Differing from native American species either in larger flowers (corolla 7 mm upward) with longer androecium (>2 cm), or by lack of a dimorphic fl in each capitulum, or by pinnate leaflet-venation, or by some combination of these characters. — An Old World group, of imprecise definition.

The four species of *Albizia*, all of Old World origin, that are naturalized in the Americas or cultivated there in significant numbers, were referred by Bentham (1875: 562, 564, 568) to three subdivisions of a typical sect. *Eualbizzia*. We lack detailed knowledge of the genus in Asia and Africa necessary for judicious evaluation of these supraspecific categories, which Nielsen, in his monographic study (1979b: 201) of mainland southeast Asian *Albizia*, has provisionally abandoned, though they were maintained as lately as 1983 by Huang. For the sake of orientation, however, we have introduced Bentham's serial names into the key that follows. It will be noticed that while the first three species form a relatively homogeneous cluster, the fourth, *A. procera*, stands well apart, and could perhaps with advantage be excluded from the genus.

In addition to those described below, *A. lebbekoides* de Candolle, a Melanesian species resembling *A. procera* in terminally paniculate inflorescence and homomorphic flowers but differing in strong forward displacement of the leaflet's costa, is reported by Isely (Vasc. Fl. SE. U.S. 3(2): 12. 1990) as "slightly established" in urban waste places in southern Florida. We have seen no American material of *A. lebbeckoides* and omit it from our key. The African *A. gummifera* (J. F. Gmelin) C. A. Smith, a member of *Albizia* sect. *Zygia* Bentham (not genus *Zygia* P. Browne), has been collected along the Campinas-Mogi-mirim highway in São Paulo (*Gabrielli & al. 11573*), but is not known to be established outside cultivation. The African section *Zygia*, which if raised to generic level would assume the name *Sassa* Bruce, is instantly recognized among Ingeae with heteranthous capitula by an extremely long, far-exserted androecial tube divided into a tassel less than one-third as long. The Australian *Albizia lophantha* and the Malesian *A. falcataria* are planted and also weakly adventive in the Americas. *Albizia lophantha* has been transferred by Nielsen to the segregate genus *Paraserianthes*; *A. falcataria*, also transferred to *Paraserianthes* by Nielsen, is herein treated in *Falcataria*. See both genera for description and comment.

Chauhan and Dayal (1985) found variation in the wood-anatomy of several Indian albizias. For instance, *A. procera* lacks septate fibers otherwise common in the New World members.

Key to the species of genus *Albizia* sect. *Albizia* planted or naturalized in the Americas

1. Venation of lfts palmate; fls of each capitulum dimorphic, and corolla of peripheral ones 7–12 mm; stipules present but often caducous, absent from many specimens at anthesis and subsequently.
   2. Lf-formula v–xv/19–43; larger lfts 6–13 × 1.5–3.7 mm, acute; peripheral fls (sub)sessile, the pedicel <1 mm, often 0; pods 9–17 × 1.6–2.6 cm [ser. *Falcifoliae* Bentham ≡ ser. *Albizia*].
      3. Stipules linear-lanceolate 1-nerved; lfts of longer pinnae 19–30-jug., bicolored, green above, pallid beneath; inflorescence simply pseudoracemose, the primary axis ±3–9 cm, abruptly flexuous; pod inertly dehiscent; pleurogram of seed elongately U-shaped, not downwardly displaced; warm-temperate, extensively naturalized in E United States and planted elsewhere . . . . 22. *A. julibrissin* (p. 243)
      3. Stipules foliaceous, ovate-flabellate or oblong-ovate from cordate base [caveat: caducous]; lfts of longer pinnae 33–43-jug., pallid on both faces; inflorescence paniculate 1–2 dm, the primary and secondary axes not abruptly flexuous; pod falling indehisced; pleurogram downwardly displaced to below midface of seed; tropical, sparingly planted, scarcely naturalized . . . . 20. *A. chinensis* (p. 241)
   2. Lf-formula ii–iv/4–10; larger lfts at least 3 × 1 cm, broadly obtuse; capitula umbelliform, the pedicel of outer peripheral fls 1.5–5.5 mm; pods 13–32 × 2.5–5 cm; tropical only [ser. *Obtusifoliae* Bentham] . . . . 21. *A. lebbeck* (p. 242)
1. Venation of lfts pinnate; fls of each capitulum homomorphic, and corolla ±5.5–6.5 mm; stipules 0; tropical only [ser. *Obtusifoliae* "Parviflorae" Bentham] . . . . 23. *A. procera* (p. 244)

20. Albizia chinensis (Osbeck) Merrill, Amer. J. Bot. **3:** 575. 1916. *Mimosa chinensis* Osbeck, Dagb. Ostind. Resa 233. 1757. — "[Translated from Danish by J. R. Forster, Voy. China W. India **1:** 379. 1771:] . . . the French Island [near Canton, China]." — No type found in Scandinavian herbaria by Nielsen (1979b: 221), but the protologue decisive.

Mimosa stipulacea Roxburgh, Fl. Ind. ed. 1832, **2:** 549. 1783. — "A native of the mountains north of Bengal." — Type to be sought at K.
Pithecolobium serronii Glaziou, Bull. Soc. Bot. France, Mém. **3:** 190. 1905, nom. nud. — "Entre Gavea et Floresta da Tijuca, Rio-Jan[eiro], [*Glaziou*] n° *13772*." — Spm. authent., P!
Albizzia stipulata [*sic*] sensu Bentham, 1875: 568 (extensive synonymy); Britton, Sci. Surv. Porto Rico Virgin Is. **5(3):** 351. 1924.
Albizia chinensis sensu Nielsen, Opera Bot. **81:** 40. 1985; Adansonia **19:** 221. 1970; *in* Aubréville, Fl. Cambodge, Laos & Viêt-nam **19:** 84, pl. 15, figs. 1–5. 1981.

Microphyllidious *trees* (4–)6–40 m with trunk attaining 1–2(–?) dm dbh and deeply fissured, dark gray bark, the young stems and all axes of the terminal, efoliate or proximally foliate panicle of pseudoracemose fasciculate capitula pilosulous with fine silky, white or yellowish hairs to 0.25–0.55 mm, the pallid or subglaucescent lfts nearly concolorous, always ciliolate but either thinly silky or glabrous facially. *Stipules* foliaceous, ovate-flabellate or oblong-ovate from cordate-amplexicaul base, ±1–2.5 cm, acute or shortly caudate-acuminate, weakly palmately nerved, very early caducous, lacking from most fully flowering and from all fruiting specimens. *Lf-formula* vii–xv/33–43; lf-stk of longer lvs 13–26 cm, the petiole ±2.5–5 cm, at middle 1.5–3 mm diam, the longer interpinnal segments 12–22 mm; nectary near or below midpetiole sessile, depressed-verruciform crumpled, or shallowly concave, in outline more or less vertically elongated and (1–)1.5–4 mm diam, similar but smaller nectaries between a few distal pairs, rarely between all pairs, and much smaller ones on pinna-rachises between 1–3 furthest pairs of lfts; pinnae decrescent proximally, then subequilong, those at and beyond mid-lf (6–)7–13 cm, the longer interfoliolar segments (1.5–)2–3.5 mm; lft-pulvinules 0.3–0.6 mm; lfts subequilong except at very ends of rachis, the blades inequilaterally lanceolate or dimidiately narrow-ovate from postically obtusangulate base, abruptly acute, straight or incipiently falcate, the longer ones (6–)6.5–10 × (1.3–)1.6–3 mm, 3.2–4.5 times as long as wide; venation palmate, of 3–4 primary nerves from pulvinule, the midrib displaced to within 0.1–0.4 mm of the plane anterior margin, the inner posterior primary nerve incurved-ascending about to midblade, the midrib on posterior side 1–3-branched beyond middle, the whole venation prominulous only on dorsal face, imperceptible or almost so on ventral face. Primary *axis of panicle* ±1–2 dm; peduncles fasciculate by 2–8, each fascicle subtended in vernation by a very early caducous, foliaceous bract resembling a stipule but smaller, and occasionally also by a rudimentary lf-stk with nectary; longer (distal) peduncle of each fascicle ±15–35 mm, this often not fertile, the fruiting peduncles therefore often shorter; capitula (7–)10–23-fld, the receptacle including short terminal pedestal 1.5–3.5 mm; bracts linear or linear-lanceolate, those subtending proximal fls 1.5–3.5 mm, all caducous before anthesis; fls sessile or almost so, 5-merous, dimorphic, the perianth of all whitish, finely yellowish pilosulous; PERIPHERAL FLS: calyx deeply campanulate, slightly enlarged distally, 3.2–4.4 × 1.3–1.8 mm, the deltate teeth 0.15–0.3 mm; corolla tubular-infundibuliform 7–8.3 mm, the ovate-triangular lobes 1.3–1.7 mm, gently recurved in age; androecium 18–22-merous, 26–34 mm, the stemonozone almost 0, the tube 4.5–7.5 mm, the free filaments usually whitish, sometimes pink-tinged; no intrastaminal disc; ovary in profile very slenderly lance-elliptic, tapering into the style; stigma poriform; TERMINAL FL (abortive or poorly differentiated in some smaller capitula): ordinarily stouter than peripheral fls but hardly longer; calyx broadly campanulate 2.5–3.3 × 1.8–2.2 mm; corolla-tube ±1–1.5 mm diam. *Pods* 1(–2) per capitulum, sessile, in profile broad-linear, straight, plano-compressed, cuneate at base, either obtuse or abruptly short-acuminate at apex, when well fertilized 9–14 × 1.6–2.6 cm, 9–14-seeded, the stiffly papery, dull tan, glabrous, minutely pitted valves framed by slender, often micropuberulent sutures, composed of thin brittle, externally evenulose exocarp and thinly papery whitish endocarp, not or scarcely elevated over seeds; dehiscence 0, the seeds released by weathering on the ground; seeds compressed-obovoid-ellipsoid ±3.5–6 × 3–4.5 mm, the testa fuscous dull, smooth or finely pitted, the pleurogram close above the hilum, 0.8–1.5 mm diam.

Native in deciduous and semideciduous forest but surviving disturbance and becoming weedy in waste places, widespread over India, Sri Lanka, S China, Indochina, Java, and Borneo, commonly planted for shade in coffee and tea orchards; introduced in the New World at agricultural experiment stations and in botanical gardens (Hawai'i, Puerto Rico, Guadeloupe, Trinidad, Rio de Janeiro, doubtless elsewhere), and occasionally subspontaneous, but nowhere an aggressive weed like *A. lebbeck* or *A. julibrissin*. — Fl. following rains, the fruit long persistent.

Albizia chinensis is one of a small group of southeast Asian species characterized by dimidiate leaflets. In this group, as shown by Nielsen (1979b: 221), it is readily distinguished by the combination of enlarged foliaceous stipules, indehiscent pods, and small pleurogram situated close above the hilum of the seed. From *A. julibrissin* it differs further in the paniculately branched inflorescence with straight or at least not abruptly zigzag primary axis and less clearly differentiated terminal flower in the capitula.

21. Albizia lebbeck (Linnaeus) Bentham, London J. Bot. **3:** 87 ("*Lebbek,*" a variant spelling often subsequently copied). 1844. *Mimosa lebbeck* Linnaeus, Sp. Pl. 516. 1753. — "*Habitat in Ægypto superiore.*" — Holotypus (Brenan, Fl. Trop. E. Afr., Leguminosae subfam. Mimosoideae 147. 1959): LINN *1228.16.* — *Acacia lebbeck* (Linnaeus) Willdenow, Sp. Pl. **4:** 1066. 1806. *Feuilleea lebbek* (Linneaus) O. Kuntze, Revis. Gen. Pl. **1:** 184. 1891.

Mimosa speciosa Jacquin, Icon. Pl. Rar. **1:** 19, t. 198. 1786 & Collectanea **1:** 47. 1787. — ". . . ex semine sine titulo ex India Orientali transmisso enata . . ." — No typus seen, but the cited plate unequivocal; equated with *A. lebbeck* by Bentham, 1844, l. c. — *Acacia speciosa* (Jacquin) Willdenow, Sp. Pl. **4:** 1066. 1806.

Albizia lebbeck var. *pubescens* Bentham, London. J. Bot. **3:** 87. 1844. — ". . . with the smooth form in East India . . ." — No typus identified at K in 1990.

Pithecolobium splitgerberianum Miquel, Stirp. Surinam. Select. **5:** 1851. — "[Surinam] Herb. Hostm[ann.] n. 459." — Holotypus, U!; isotypus, G!.

Albizia lebbeck sensu Bentham, 1875: 562 ("*Lebbek*"); Rock, Leg. Pl. Hawaii, pl. 5, 6. 1920; Kirtikar & Basu, Ind. Med. Pl., ed. 2, **2:** 936, t. 383. 1935; T'ang-jui Liu, Illust. Lign. Pl. Taiwan **1:** t. 424. 1960; Little, Common Trees Puerto Rico & Virgin Is. fig. 60. 1964; Isely, 1973: 76, map 15 (Florida); Verdcourt, Man. New Guinea Leg. fig. 49. 1979; Elias, Trees N. Amer. fig. 637. 1980; Bisse, 1988: 227; Wagner & al., Man. Fl. Pl. Hawai'i **1:** 644, pl. 87. 1990.

Macrophyllidious, deciduous *trees* of rapid growth mostly 3–15(–18) m but sometimes fertile as saplings 1.5 m upward, the simple or branched trunk attaining 2–6 dm dbh, the young branchlets and all axes of lvs and inflorescence usually finely pilosulous with straight erect or with incurved hairs to ±0.1–0.4 mm, sometimes glabrescent, the lvs bicolored, the papery-membranous lfts above bright green, sublustrous and most often glabrous but sometimes sparsely strigulose, beneath paler, often glaucescent, and commonly puberulent at least along some nerves but occasionally glabrous, the inflorescence either a short terminal, proximally foliate pseudoraceme of compact umbelliform capitula immersed in foliage, or composed of capitula axillary to a succession of lvs, the primary axis rather abruptly flexuous. *Stipules* (few seen) submembranous linear-oblong 2–4 mm obtuse, caducous long before expansion of lf, absent from most fl and from all fruiting specimens. *Lf-formula* ii–iii(–iv)/4–9(–10); lf-stks (6–)7–17(–22) cm, the petiole (3.5–) 4–11(–12.5) cm, at middle 1.4–2.6 mm diam, subterete proximally, narrowly sulcate distally, the sulcus continuous or randomly interrupted at insertion of pinnae by a nectary, the first (only) interpinnal segment 2–5 cm; first petiolar nectary inserted usually below, sometimes near midpetiole, or close to lf-pulvinus, sessile, plane or shallowly concave at middle, 1.8–4.5 mm diam, in profile 0.25–0.6 mm tall (exceptionally lacking), a similar but smaller nectary sometimes below all (or only the distal) pinna-pairs, and yet smaller ones on 2–several distal segments of pinna-rachises; pinnae subdecrescent proximally or subequilong, the rachis of longest pair (4–)6–14.5 cm, the longer interfoliolar segments (1–)1.2–2.2 cm; lft-pulvinules (1–)1.2–2.5 mm, cuneately widened upward to (0.5–)0.6–1.1 mm; lfts decrescent proximally, thence subequilong, the blades oblong or ovate-oblong from inequilateral base, broadly obtuse or obscurely emarginate, the largest 30–48 × (8.5–)11–17 mm, 1.8–3.1 times as long as wide; venation palmate and then pinnate, of 4–6 primary nerves from pulvinule, the straight or gently incurved midrib displaced to divide blade 1:1.5–2.2, the inner posterior primary nerve produced nearly to midblade or well short of it, the midrib giving rise on each side to 5–8 (and random intercalary minor) secondary nerves brochidodrome well within the slenderly nerved margin, these further generating a reticulum of fine veinlets, the whole venation prominulous on both faces of lft. *Peduncles* (1–)2–4-nate, 3–12(–13) cm; capitula 18–45-fld, prior to anthesis globose, becoming hemispherical, the fls dimorphic, the peripheral ones pedicellate, the terminal one larger and sessile, the scarcely dilated receptacle ±2.5–6 mm; bracts oblanceolate, linear-spatulate, or subrhombic 1.2–4 mm, puberulent dorsally near apex, fugacious from below young fl-buds; perianth 5(–6)-merous, the calyx finely pilosulous, the greenish corolla strigulose on lobes only; PERIPHERAL FLS: pedicels (1–)1.5–5.5 mm; fl-buds slenderly or plumply pyriform; calyx narrowly campanulate and often a little expanded beyond middle, 3.7–5.8 × 1.9–2.5 mm, toward base weakly 5-angulate, the ovate teeth 0.3–1.1 mm; corolla slenderly vase-shaped 7.7–12.3 mm, the ovate-acute lobes 1.6–3.3 mm; androecium (22–)26–42-merous, (18–)24–36(–44) mm, the stemonozone 0.7–1.6 mm, the tube (3–)3.5–5.5 mm; ovary sessile, linear-ellipsoid, glabrous; TERMINAL FL: calyx 4.8–6 × 2.2–3.5 mm; corolla 9–14.5 mm; androecium mostly 40–52-merous, the free filaments partly filiform and partly ribbonlike, or sometimes fewer and all dilated, the tube 11.5–19 mm. *Pods* randomly ascending, 1–5 per capitulum, sessile or almost so, in profile broad-linear or narrowly oblong-elliptic 13.0–32 × 2.5–5.2 cm, broad-cuneate at base, rostellate or apiculate at apex, framed by essentially straight, dorsally plane or shallowly bisulcate sutures 1.2–2.2 mm wide, 8–12-seeded, the stiffly papery, stramineous and glabrous valves bullately dilated over each seed (the bullae ±1.5–2.5 cm diam and alternately convex and concave along each valve), pallid within, the cavity continuous; dehiscence tardy, often after fall of pod, inert, through the ventral suture only, this narrowly gaping to release seeds; seeds transverse at middle of pod, basifixed on a filiform funicle, compressed but plump, in broad profile oblong-elliptic or round, emarginate at hilum, ±9–11 mm in greatest diameter and ±3 mm thick, the smooth but hardly lustrous, khaki or light reddish brown testa consisting of a thin outer integument and a hard crustaceous inner shell, the inversely U-shaped pleurogram occupying about half of each face; cotyledons plump, surrounded peripherally by a band of endosperm.

Native of tropical E Asia, by mid-nineteenth century randomly circumtropical, both as a cultivated tree (occasionally used for coffee-shade or about dwellings) and a living fence, becoming weedy in disturbed woodland and in pastureland, from sea level to ±1000 m, in the New World known from W and Gulf lowland Mexico (Sinaloa; Tamaulipas to Yucatán Peninsula), thence S through Chiapas and Central America to Panama; Bermuda and Bahamas; Greater and Lesser Antilles (Cuba, Jamaica, Hispaniola, Puerto Rico and Virgin Is., S to the Grenadines); and in South America from N Colombia E through N and centr. Venezuela to coastal Guyana and Surinam, thence S, mostly as a shade and street tree, from Maranhão, Brazil southward to Minas Gerais, Brasília, and Guanabara. — Fl. irregular, but usually at onset of rainy seasons. — *Acacia* (Mexico and Central America); *mata-ratón* (Panama); *lengua de mujer* (Latin America) or *woman's tongue* (former British colonies, in reference to the rattling of seed-pods); *bois noir* (Fr. Antilles); *baile de caballero* (NW Venezuela).

22. Albizia julibrissin Durazzini, Mag. Tosc. **3(4):** 11. 1772. — Described from a plant grown at Florence, raised from seed brought from Constantinople. — Neotypus (Nielsen, Adansonia **19:** 206, 213, pl. 2. 1979): *Herb. Micheli. s.n.*, FI (not seen).

A. julibrissin sensu Bentham, 1875: 568; Britton, N. Amer. Trees 519, fig. 480. 1908; Fiori & Paoletti, Iconogr. Fl. Ital., ed. 3, fig. 1838. 1933; Palibin *in* Fl. S. S. S. R. **11:** 10, t. 1(3). 1945; Prilipko *in* Fl. Azerbaijana **5:** 205, t. 18. 1954; T'ang-jui Liu, Illust. Lign. Pl. Taiwan **1:** t. 423. 1960; Wilbur, Leg. Pl. N. Carolina, fig. 1. 1963; Gleason, Illust. Fl. NE U.S. & Canada **2:** 381. 1963; Isely, 1973: 75, map 15 (U.S.A.); Elias, Trees N. Amer. fig. 638. 1980;

Cronquist, Integr. Syst. 593, fig. 5.15 (optima!). 1981; Nielsen, 1981: fig. 1(12a, 12b); *in* Aubréville, Fl. Cambodge, Laos & Viêt-nam **19:** 82, pl. 14, figs. 9–14. 1981.

Amply microphyllidious, broad-crowned, unarmed, deciduous *trees* of rapid growth and early maturity, attaining 6–12 m but often precociously fertile, appearing glabrous but the lf- and some inflorescence-axes thinly (in var. *mollis* (Wallich) Bentham more densely) pilosulous with loose pallid hairs <0.5 mm, the foliage bicolored, the thin-textured lfts green and glabrous above, beneath paler and often thinly silky at least on anterior side of midrib, often ciliolate, the inflorescence a simple, abruptly flexuous, mostly efoliate, often at anthesis corymbiform pseudoraceme of capitula terminal to homotinous branchlets and surpassed by uppermost lvs, the overall branching-pattern sympodial. *Stipules* submembranous linear-lanceolate, caducous as the associated lf expands, absent from most specimens. *Lf-formula* (iv–)v–x (–xii)/19–30; lf-stks ±1–2 (in saplings –3.5) dm, petiole ±3–7 cm, the longer interpinnal segments 1–1.8 (–2.5) cm; petiolar nectary well below midpetiole sessile, round or elliptic, scutiform or shallowly patelliform 0.8–2.5 mm diam, a smaller nectary at tip of most lf-stks and yet smaller ones at tip of most pinna-rachises; pinnae decrescent proximally, scarcely so distally, the rachis of the penultimate pair 5.5–11 cm, the longer interfoliolar segments 2–4 mm; pulvinules 0.4–0.7 mm, not or scarcely wrinkled; lfts subequilong except at far ends of rachis, one or both of the first pair usually reduced to a paraphyllidium, the rest dimidiately narrow-oblong or lance-oblong from postically subrectangular base, abruptly acuminulate or apiculate at erect or often incurved apex, those near midrachis 8–13.5 × 2.2–3.7 mm, 2.8–4.3 times as long as wide; venation palmate and then pinnate, the midrib forwardly displaced to divide blade 1:(5–)6–10, the innermost of 2–3 posterior primary nerves incurved-ascending to anastomosis a little short of midblade, the outer one(s) shorter, the midrib giving rise on posterior side only to 2–4 weak secondary nerves and a faint reticulum of veinlets, the whole venulation either immersed or faintly prominulous on upper face, more sharply but finely raised beneath. *Primary axis* of inflorescence ±3–9 cm, efoliate throughout or sometimes foliate at the first 1–3 nodes, the peduncles solitary or 2–3 per node, those at proximal nodes longest, ±(2.5–)3–5 cm; capitula 12–20-fld, the fls dimorphic, the peripheral ones either sessile or contracted at base into a stout pedicel to 0.8 mm, all bisexual, the terminal ones always sessile, longer, functionally staminate, and the filament-tube far-exserted, sometimes one fl (or flowerless bract) downwardly displaced onto peduncle, the hemispherical receptacle commonly 1.5–2.5 mm diam; bracts linear-spatulate ±1 mm, quickly deciduous; the calyx of all fls submembranous, pale green, either glabrous ciliolate or thinly pilosulous, the greenish corolla externally silky-strigulose or pilosulous, at least on lobes; PERIPHERAL FLS: calyx campanulate or deeply campanulate 2.2–4.2 × 1.1–1.7 mm, the incurved, ovate obtuse or deltate teeth 0.3–0.7 mm; corolla slenderly trumpet-shaped 7–11 mm, the erect ovate lobes 1.3–2.6 × 0.9–1.2 mm; androecium 24–34-merous, 24–38 mm, the stemonozone 1 mm or less, the tube 6.5–13 mm (a little shorter to 2 mm longer than corolla), at separation of the filaments 1 mm diam or less, the free part of filaments pink or pinkish orange; TERMINAL FL: calyx 3.5–4.5 × 2.2–2.8 mm; corolla ±9–11 mm; filament-tube 11–16.5 mm, at separation ±2 mm diam, internally thick-ribbed at base but without disc; ovary slenderly ellipsoid, compressed, attenuate at both ends, glabrous at anthesis; style filiform; ovules 10–14. *Pods* 1–3 per capitulum, in profile broad-linear straight 10–17.5 × 1.7–2.6 cm, attenuate at both ends, at base into a short or obscurely differentiated stipe, at apex into an erect cusp, plano-compressed, the papery, at first green, when ripe stramineous or fuscous-stramineous valves framed by obtuse sutures ±1 mm wide, bullate on each face over each seed, the cavity continuous; dehiscence inert, through both sutures, tardy, often delayed till after fall of fruit; seeds transverse, basifixed on sinuously filiform funicle, in broad view elliptic, compressed but plump, the brittle exotesta dull brown, charged on each face with a long U-shaped pleurogram, the lens convex, pallid, conspicuous.

Native in warm-temperate and subtropical Asia, widely planted and naturalized elsewhere in cool-winter, hot-summer climates. — Fl. midsummer. — *Mimosa, silk-tree.*

Albizia julibrissin lacks septate fibers in the wood (Chauhan & Dayal, 1985), a feature common in New World species.

23. Albizia procera (Roxburgh) Bentham, London J. Bot. **3:** 89. 1844. *Mimosa procera* Roxburgh, Pl. Coromandel **2:** 12, pl. 121. 1798. — ". . . native of the mountainous parts of the [Coromandel] coast [Madras, SE peninsular India]." — Holotypus (Nielsen, 1985: 38): Roxburgh, 1798, pl. 121, supra cit. — *Acacia procera* (Roxburgh) Willdenow, Sp. Pl. **4:** 1063. 1805. *Feuilleea procera* (Roxburgh) O. Kuntze, Revis. Gen. Pl. **1:** 184. 1891.

A. procera sensu Bentham, 1875: 564; Pierre, Fl. Forest. Cochinch. **4:** t. 398B. 1899; Rock, Legum Pl. Hawaii, pl. 4. 1920; Little & Wadsworth, Common Trees Puerto Rico 148, fig. 61. 1964; Isely, 1973: 77; Verdcourt, Man. New Guinea Leg. fig. 50. 1979; Nielsen *in* Aubréville, Fl. Cambodge, Laos & Viêt-nam **19:** 89, pl. 16, figs. 1–4. 1981; A. C. Smith, Fl. Viti. Nova **3:** 78. 1985.

Amply plurifoliolate *trees* of rapid growth and early maturity, flowering when 2–4 m but attaining 25 m with smooth gray trunk 3–6 (in old age reportedly to 20) dm dbh, appearing glabrous and the calyces and pods in fact always so but the lf-axes, lft-pulvinules, and one or both faces of the conspicuously bicolored lfts commonly strigulose with fine whitish hairs <0.35 mm, the inflorescence composed of efoliate pseudoracemes of small fasciculate capitula axillary to uppermost lvs and an efoliate terminal panicle, this not or only shortly exserted from foliage. *Stipules* 0. *Lf-formula* (ii–)iii–vi/(7–)8–12(–13); lf-stks (9)10–28(36) cm, the petiole including livid pulvinus (5.5–)7–13 cm, the narrow shallow ventral groove continuous between pinna-pairs; nectary below midpetiole sessile, shallowly patelliform or almost plane, in outline elliptic or linear-elliptic 3.5–9 mm in long diameter, no nectary elsewhere on lf-stk, but small ones on rachis of some pinnae close below 1–3 furthest lft-pairs; pinnae ordinarily decrescent at each end of lf-stk, the rachis of longer ones 11–22 cm, the longer interfoliolar segments 11–22 (–26) mm; pulvinus of lfts (1.2–)1.5–2.7(–3) mm, discolored or not but not deeply wrinkled; lfts decrescent at each end of rachis, otherwise subequilong, in outline ovate-, oblong-, or obovate-elliptic from inequilaterally cuneate or flabellate base, at apex obtuse, deltately subacute, or shallowly emarginate, minutely callous-apiculate, those near midrachis (21–)25–50 × (10–)11–22 mm; venation pinnate, the straight midrib subcentric at and beyond midblade, giving rise on each side to 5–9 major (and random intercalary) secondary nerves brochidodrome shortly within the very narrowly revolute margin, the tertiary venulation often immersed, scarcely prominulous when raised, the midrib and secondary nerves usually pallidly discolored on ventral face of blade, the color contrasting with the dull dark green intervenium. Primary *axis of pseudoracemes* or panicle 7–22 cm, the capitula mostly fasciculate by 2–5(–6, randomly solitary), the longest peduncle of each fascicle 7–19 mm; bracts membranous 0.5–1.2 mm, caducous from beneath the immature fl-buds, one bract at base of capitulum sometimes a little larger; capitula 12–23-fld, the flowers radiately spreading in a subglobose cluster, the globose or pyriform receptacle 1–2 mm; fls homomorphic, sessile, 5-merous, the calyx glabrous, the corolla-lobes usually pallid-strigulose overall or only at the tip; calyx membranous, greenish, veinless, deeply campanulate 2.6–3.3 × 1.1–1.4 mm, the sometimes unequal teeth 0.2–0.6 mm, the whole becoming dry and often circumscissile at base; corolla ochroleucous 5.5–6.7 mm, the ascending lobes 1.2–2.2 mm; androecium 30–40-merous, 9–11 mm, the stemonozone 1.1–1.5 mm, the tube 4.5–5.7 mm, a trifle shorter than the corolla, the free filaments whitish; ovary in profile narrowly elliptic ±1 mm, contracted into a stipe ±0.2 mm, glabrous; style filiform, scarcely thickened at apex, the stigma poriform. *Pods* 1–3 per capitulum, in profile broad-linear 12–18 × 1.8–2.5 cm, contracted at base into stout stipe 1–4 mm, shortly tapering or abruptly obtuse at apex, plano-compressed, the stiffly papery, reddish brown, smooth glabrous valves framed by straight sutures, low-convex and fuscous over the seeds; dehiscence tardy, through the ventral suture, the valves narrowly gaping to release the seeds; funicles filiform, straight, seeds transverse, well spaced along the continuous cavity, compressed-ellipsoid, in broad profile ±6–7 × 4.5–5 mm, the brittle testa ±0.1 mm thick, externally dull brown, the pleurogram at middle of broad faces U-shaped 4.5–5 × 2.5–3 mm; endosperm present.

In monsoon woodland, 1–1200 m, surviving disturbance in hedges and pastures, widespread in SE Asia (India, Sri Lanka, Indochina, S China, Taiwan), the Philippines, New Guinea, N Australia (Queensland), introduced at experiment stations and in botanical gardens of the neotropics as a shade tree and source of fuel, feebly spontaneous in hedges and disturbed places, fairly frequent on Cuba, Jamaica, Puerto Rico, and Hispaniola, to be expected in the Lesser Antilles; Rio de Janeiro, Brazil. — Fl. following rains, the frs. long persistent, sometimes shed with the inflorescence-axis and dehiscent on the ground. — *White siris.*

XVI. ENTEROLOBIUM MARTIUS

Enterolobium Martius, Flora **20(2,** Beibl. **8** [Herb. Fl. Bras.]): 116. 1837. — Sp. typica: *E. timbouva* Martius.

Enterolobium sensu Bentham, 1875: 598; Mesquita, Rev. Tax. Gen. *Enterolobium* 70 (antecedent literature). 1991.

Unarmed, deciduous, either macro-or microphyllidious *trees,* the young growth more or less densely pubescent with simple, pallid or brownish hairs. *Stipules* small, mostly caducous. *Lf-formula* ii–xxx/4–80; petiolar nectary either cupular-patelliform, or mounded, or sunk in petiolar groove; lft-pulvinules <1 mm; venation of lfts palmate, palmate-pinnate, or (in narrowest blades) reduced to midrib. *Inflorescence* of either homo- or dimorphic, compact or sometimes umbelliform capitula borne either singly, or in fascicles, or in efoliate pseudoracemes either (a) at efoliate nodes below current year's foliage and axillary to some early lvs of the annual cycle or (b) wholly in contemporary lf-axils, but not terminally paniculate, the meristem of major branches inhibited beyond the last unit of inflorescence and the whole ramification sympodial; fls 5–8-merous, the calyx of peripheral ones campanulate, turbinate- or cylindro-campanulate; corolla gamopetalous, twice or less than twice as long as calyx; androecium ("8"–)10–70-merous, the tube variable in length, the intrastaminal disc often well developed; ovary either tapering or truncate at apex; stigma minute, poriform; ovules 22–32. *Pods* sessile, in profile oblong or broad-linear decurved through ½ to 2 circles into a reniform-auriculiform, or annular, or compressed-helicoid figure, the sutures immersed, the outer (seminiferous) one either evenly arcuate or undulately recessed between seeds, the valves composed of (a) thin, fuscous or blackish exocarp, (b) thick, either dry mealy-fibrous or resinous-pulpy mesocarp, and (c) papery pallid endocarp inflexed to form complete, often partly membranous interseminal septa; dehiscence 0, the seeds released by weathering on the ground or excreted by herbivores; seeds transverse, uniseriate or (*E. contortisiliquum*) irregularly biseriate, borne on crumpled filiform funicle, plumply compressed-ovoid-ellipsoid, the hard testa castaneous or fuscous, some with pallid halo following the line of the complete or incomplete pleurogram, or pleurogram lacking; endosperm 0. — Spp. (at last revision) 10, but (see note following) perhaps a few more, widespread over the lowland and submontane American tropics from S Mexico and Greater Antilles S through the Magdalena valley in Colombia, the Orinoco basin in Venezuela, and Guyana almost throughout Brazil to E Bolivia, Paraguay, N Argentina, and Uruguay, 6 in Amazonian Brazil.

No account of American Ingeae would be complete without *Enterolobium*, but as this genus has been revised in detail by Mesquita (1990) we provide here no more than an outine of its taxonomy, in order to circumvent confusion between it and compatriot species of other genera. Below will be found a conspectus of the genus that may function as a multiple-entry key to the species, remarks on affinities in the tribe, and random reflections on taxonomic points still in need of clarification. For synonymy, typifications, and other relevant matters the reader is referred to Mesquita's monograph.

Enterolobium was placed by Bentham (1875: 598) next following *Pithecolobium* sect. *Chloroleucon* (genus *Chloroleucon* of these pages), with the comments that it differed from *Pithecolobium* only in the broad, thick-walled, indehiscent fruit, and that it might as well be rated as a mere section of it. Mesquita (1990: 43) has suggested that *Enterolobium* is most closely related to *Albizia*, and that these genera might both be derived from *Pithecellobium*. Because he did not, however, define *Albizia*, his estimate of affinity lacks precision; and we think that *Pithecellobium* sens. str., an American group characterized by a specialized fruit and seeds, is an improbable ancestor for a genus, however circumscribed, that contains the type-species of *Albizia*, the Asiatic *A. julibrissin*.

Enterolobium exhibits more variation in inflorescence structure than any other genus of New World Ingeae, and in pattern of inflorescence is intermediate between *Chloroleucon*, which differs in striate resting-buds and in narrower dehiscent (but sometimes spirally coiled) fruit, and *Albizia* ser. *Arthrosamanea*. The inflorescences of *Chloroleucon* and ser. *Arthrosamanea* are borne on anauxotelic (sympodial) axes. In *Chloroleucon* the axis continues to elongate beyond the furthest axillary inflorescence-unit (though determinate); the inflorescence at anthesis is never the terminal-paniculate inflorescence of ser. *Arthrosamanea*, but in contrast becomes a terminal panicle of capitula by suppression of the leaves (the capitula are subtended by bracts). Some species of *Enterolobium* (e.g., *E. barnebianum*), and individuals of other species, often have precocious inflorescences (i.e., the leaves are suppressed at the oldest nodes) and simple axillary capitula or fascicles of capitula at younger nodes, thus resembling *Chloroleucon*. Other species (e.g., *E. contortisiliquum*) or individuals will often have leaves suppressed at the lowest oldest nodes of some proleptic branch systems and also at the terminus of the meristem, thus forming at anthesis a terminal pseudoraceme of capitula. In *Enterolobium*, *Albizia* ser. *Arthrosamanea* and *Chloroleucon*, vegetative branching is by development of proleptic buds. In *Chloroleucon*, sylleptic branch-formation has never been observed. It is usual in ser. *Arthrosamanea* and common in some species of *Enterolobium* (e.g., *E. contortosiliquum*, *barinense*, and *cyclocarpum*) for sylleptic buds to develop, but they then form annual, axillary strictly reproductive branch systems with ephemeral axes; that is, the axes never contribute to the vegetative architecture of the tree.

The genus remains best characterized by the proportionately broad, thick-walled, often resinous, internally septate indehiscent pod decurved around a very short, annular or C-shaped dorsal suture into a ring or compressed spiral, but no one attribute of the fruit is peculiar to the genus. The fruit of *Samanea* is, for example, similarly thick-walled, resinous, septate, and indehiscent (but straight), whereas that of *Abarema microcalyx* var. *enterolobioides* is externally similar to that of *Enterolobium* but tardily dehiscent. Niezgoda et al. (1983) included *E. contortisiliquum* in a study of the ultrastructure of the pollen of Ingeae. They found that the pollen of *E. contortisiliquum*, unlike that of any other taxon they studied, is characterized by an ektexine-free interstitium. The tectum is imperforate, as is that of all species of *Chloroleucon*, and in contrast to all species of ser. *Arthrosamanea* except *Albizia carbonaria*.

Enterolobium contortisiliquum at anthesis closely resembles *Blanchetiodendron blanchetii*, which was, in fact, first described in *Enterolobium*, but the straight papery, dehiscent fruits and thin-skinned discoid seeds of the latter are decisively different. *Enterolobium schomburgkii*, on the other hand, is more similar to some African species of *Cathormion* (as delimited by Brenan & Brummit, 1965: 192, sequ.), especially to *C. dinklagei* (Harms) Hutchinson & Dandy, which scarcely differs except in the more nearly lomentiform fruit. Whether African *Cathormion* is congeneric with the Asiatic *Cathormion* Hasskarl (sens. str.), which is armed with indefinite axillary thorns and by spinescent stipules on sucker shoots, is a question that cannot be addressed in this context. It is worth notice, however, that while truly indehiscent the pod of most enterolobiums breaks apart under no great pressure along the line of the interseminal septa; it is, as it were, cryptically lomentoid, thus not fundamentally different from that of *Cathormion*.

The expansion of *Enterolobium* to embrace *Samanea*, proposed by Prain in 1897 (J. Asiatic Soc. Bengal, pt. 2 Nat. Hist. 66. 352), acknowledges an undeniable similarity in resinous-pithy texture of the pods' valves but takes no account of the profound differences in inflorescence-architecture and leaflet-venation; it has found no acceptance in the literature.

Eight of the 10 species of *Enterolobium* recognized by Mesquita, including the generitype, are uniform in essential structure, except that the flowers of each capitulum are dimorphic in *E. barnebianum* (? always) and reported to be so in *E. barinense*, whereas they are homomorphic in the remainder. The three species *E. oldemanii*, *E. schomburgkii*, and *E. glaziovii* differ from the remainder in multiple characteristics emphasized in the conspectus that follows, and are segregated as a sect. *Robrichia*.

Conspectus of genus *Enterolobium*

¶ Indumentum of pallid hairs; lf-formula (i–)ii–xii/4–21, the larger lfts at least (1.5–)2 mm wide; androecium 20–68-merous; terminal fl of capitula lacking a well-defined intrastaminal disc; ovary at anthesis glabrous (but often pubescent after fertilization), tapering or conic at apex; valves of pod in profile 3–6.5 cm wide; seeds 12–18 mm long; sect. *Enterolobium.*

* Lfts inequilaterally or dimidiately lanceolate, acute or acuminulate; fls of each capitulum homomorphic.

1. *E. cyclocarpum* (Jacquin) Grisebach. Lf-formula v–xii/21–31; valves of pod raised over each individual seed and depressed between them; seeds uniseriate; mesocarp of pod resinous. — Venezuela and closely adjacent Brazil, NW to Mexico and West Indies.
2. *E. timbouva* Martius. Lf-formula ii–iv/8–12; valves of pod not raised over each seed; seeds 1-seriate; mesocarp fibrous-mealy. — Brazil southward from lower Amazon valley to Mato Grosso do Sul and Paraguay.
3. *E. contortisiliquum* (Vellozo) Morong. Lf-formula v–vii/12–20; valves of pod not raised over each seed; seeds biseriate; mesocarp resinous. — Brazil southward from S periphery of Amazon basin to Bolivia, Paraguay, N. Argentina, and Uruguay.

** Lfts sudimidiately lance-oblong, obtuse; fls of each capitulum dimorphic.

4. *E. barnebianum* Mesquita & da Silva. Lf-formula v–vii/12–20; lfts relatively small, obtuse at apex, and the strongly displaced midrib simple. — N and W Amazonia, N to the headwaters of Río Paragua in S state of Bolívar, Venezuela.

*** Lfts oblong or oblong-elliptic and broadly obtuse at apex, the midrib subcentric or only moderately displaced, branched on each side; fls (described as) dimorphic; pod glabrous, the mesocarp resinous.

5. *E. barinense* Cárdenas & Rodríguez. Lf-formula v–xiii/11–20. — S Venezuela.

**** Lfts oblong or oblong-elliptic, broadly obtuse; fls homomorphic; pod velutinous, the mesocarp (so far as known) dry mealy-fibrous.

6. *E. maximum* Ducke. Branches not corky; lf-formula (ii–)iii–vi/6–11; pod well developed, coiled into a full circle or further. — Amazonian forest.
7. *E. monjollo* (Vellozo) Martius. Branches not corky; lf-formula iv–vii/7–12; pod decurved through about ½ circle. — Atlantic forest of Brazil (Bahia to Rio de Janeiro).

8. *E. gummiferum* (Martius) Macbride. Branches corky; lf-formula (i–)ii–iii/4–6(–7, exceptionally more); pod incurved through a full circle or more. — Campos cerrados of the Brazilian planalto and its periphery.

¶¶ Indumentum of brown or sordid hairs; lf-formula ix–xxiii/(32–)40–83, the lfts inequilaterally lance-linear 2.5–6 × 0.5–1.2, mostly 1-nerved; androecium (8–)10–14-merous, that of terminal fls surrounded at base by a lobed disc; ovary at anthesis strigulose overall, obliquely truncate at apex; valves of pod in profile 1.8–3(–3.3) cm wide, incurved into an ear-shaped or helicoid figure 3.5–6 dm diam; seeds uniseriate ±6–8 mm long; sect. *Robrichia* Barneby & Grimes.

* Fls of each capitulum dimorphic.

9. *E. oldemanii* Barneby & Grimes. Calyx of peripheral fls 5–6 mm; corolla 7–8.2 mm. — French Guiana.

10. *E. schomburgkii* Bentham. Calyx of peripheral fls (2.3–)2.5–3.3(–3.5) mm, corolla (3.2–)3.5–5.2 mm. — Amazon basin and the Guianas NW to S Mexico.

** Fls of each capitulum homomorphic.

11. *E. glaziovii* (Bentham) Mesquita. Fls of each capitulum homomorphic; exocarp of pod microtumulose overall. — Atlantic forest of Brazil (Bahia to Rio de Janeiro).

NOTES ON THE SPECIES OF *ENTEROLOBIUM* MARTIUS

1. *E. cyclocarpum* (Mesquita, 1990: 112, figs. 19–22).
The exterior separation of successive seed-locules by shallow grooves radiating from the center of the wheel-like fruit distinguishes the pod from that of congeners. Add synonym: *Albizzia* (?) *longipes* Britton & Killip, Ann. New York Acad. Sci. **35:** 132. 1936. — ". . . vicinity of Barranquilla, Atlantico, Colombia, February, 1928, *Brother Elias 499* . . ." — Holotypus, NY!; isotypus, US *1420061*!, this correctly equated (in adnot.) by Dugand with *A. cyclocarpum*.

2. *E. timbouva* (Mesquita, 1990: 184, figs. 38–41).
Mesquita's record of *M. timbouva* from Santa Marta, Colombia (*H. H. Smith 5285*, cited but not mapped) and his mapped record from near Pucallpa, Peru, are probably mistaken.

3. *E. contortisiliquum* (Mesquita, 1990: 94, figs. 15–18); sensu Bernardi, 1984: 188, fig. 45.
Where their ranges overlap, *E. contortisiliquum* may be distinguished from *E. timbouva* by higher leaf-formula and by displacement of seeds into two series.

4. *E. barnebianum* (Mesquita, 1990: 86, figs. 13, 14).
The pod of Brazilian material of *E. barnebianum* that we have seen has mealy endocarp, as described in the protologue. Specimens from Ecuador (*Zak & Rubio 4275, 4298*; *S. King 962*; *Rubio 113*, all NY) and Peru (*R. Vásquez 5422*, NY) that resemble *E. barnebianum* in number, size, and venation of leaflets and are in these respects unlike any other described species of the genus, have internally resinous pods recalling *E. barinense*. We lack correlated flowers and fruits needed for solution of this puzzle.

5. *E. barinense* (Mesquita, 1990: 79, figs. 83–85).
Mesquita distinguished the "circinate" fruit of *E. barinense* from the ear-shaped, decurved, or helically coiled fruits of all congeners. The term *circinate* was used in the protologue, but erroneously, for the pod is coiled into a compressed spiral nowise different from that of *E. maximum* or *E. cyclocarpum*, not incurved on itself like a bishop's crozier or a watchspring. The record from Villavicencio, "*Karsira*" [probably *Karsten*] requires confirmation.

6. *E. maximum* Ducke (Mesquita, 1990: 150, figs. 28–30).
The leaves of *E. maximum* closely resemble those of *E. barinense*, but the homomorphic flowers and dry-walled, externally velutinous pod are distinctive.

7. *E. monjollo* (Vellozo) Bentham (Mesquita, 1990: 159, figs. 31–33).
Mesquita separated this species from the preceding by characters of puberulent, not glabrous leaflets, and auriculiform, not circular or recurved pods. We find that the leaflets of *E. maximum* are as often as not finely puberulent beneath, and it is embarrassing that Vellozo illustrated the pods of *Mimosa monjollo* as recurved through more than a full circle. For the present the disjunction in range between Amazonian and Atlantic forest seems the most practical criterion.

8. *E. gummiferum* (Martius) Macbride (Mesquita, 1990: 137, figs. 25–27).
In the field, *E. gummiferum* is notable for corky, deeply fissured trunks and branches, a feature common in cerrado climax. Almost all specimens are instantly recognizable by the low leaf-formula (i–)ii–iii/4–6(–7), or by the massive velutinous pod with mealy-fibrous, nonresinous valves, or by both. We have, however, two ambiguous collections: from Xavantina, Mato Grosso, *Irwin & Soderstrom 6369* (fl., NY), with leaf-formula to iii/12, suggesting passage into *E. maximum*; and from BR-364, km 80 NE of Cuiabá, Mato Grosso, *Cid & al. 4449* (fl., fr.,

NY), with leaves of typical *E. gummiferum* but a glabrescent pod heavy with resin. These obviously need further observation.

Enterolobium sect. **Robrichia** Barneby & Grimes, sect. nov., a sect. *Enterolobio* indumento e pilis sordidis (nec albidis), foliolis numerosissimis minutis (2.5–6 × 0.5–1.2 mm), ovario florum periphericorum oblique truncato ubique pubescenti et disco lobulato circumdato, androecio 10–14 (nec 20–68)-mero, necnon legumine parvulo (valvulis plerumque 18–30, nec 30–65 mm latis) seminibusque dimidio minoribus (6–8, nec 12–18 mm longis) diversa. — Sp. typica: *E. schomburgkii* Bentham. — Etymology: commemorating the brothers Robert and Richard Schomburgk.

The section consists of a group of three species, closely resembling the African *Cathormion dinklagei* (Harms) Keay, which differs in the rather narrower, less incurved, sinuately marginate, incipiently lomentiform rather than continuous valves of the pod.

Record (1924) stated that the wood of *E. schomburgkii* is considerably different from that of all species of sect. *Enterolobium*, but he did not state by which characters. He did not examine the other species of sect. *Robrichia*.

9. Enterolobium oldemanii Barneby & Grimes, sp. nov. *E. schomburgkii* floribus terminalibus disco lobulato provisis et staminibus 10-meris simulans sed ab eo foliolis plurumque latioribus (0.7–1.2 nec 0.5–0.8 mm) floribus fere duplo longioribus et ovario minus pubescenti minus oblique tamen valde truncato differt. — FRENCH GUIANA. Ancienne réserve forestière du Matoury, Ile de Cayenne, 30 aout 1966 (fl.), *Oldeman 2242*. — Holotypus, P!; isotypi, CAY (2 sheets)!, NY (fragm.)!. Fig. 18H.

Microphyllidious *trees* to 30 m, resembling *E. schomburgkii* in indumentum and in inflorescence architecture, but the lfts somewhat wider, the fls nearly twice as large, and the ovary less pubescent and less obliquely yet strongly truncate; perulate buds not seen, but the growth obviously seasonal as in *E. schomburgkii*. *Stipules* on new growth thick-coriaceous, inequilaterally triangular-lanceolate or lanceolate 4.0–4.75 × 2.0–2.25 mm, quickly deciduous from a transverse scar. *Lf-formula* xvi–xxiii/35–49; lf-stks 6.5–14 cm; petiole 9–20 mm, tapering from an enlarged but scarcely differentiated pulvinus, at middle 0.8–1.8 mm diam; petiolar nectary directly adjacent to pulvinus, at midpetiole, or close to first pinna-pair, sessile, cupular thick-rimmed 1.0–1.4 mm diam, similar but smaller nectaries between ultimate 3–5 pinna-pairs; interpinnal segments 3–6 mm; pinnae a little decrescent at both ends of lf-stk, the rachis of longer ones (2.6–)3.5–5.7 cm; interfoliolar segments 0.5–0.85(–1) mm; lfts attached on the ventral side of the lf-stk, adjacent to a centric rib, appearing to the naked eye sessile against rachis, the darkened wrinkled lft-pulvinules ±0.3–0.35 mm; lfts decrescent at each end of pinna, linear to linear-elliptic from an inequilaterally truncate base, at apex broadly acuminate, straight, plane, commonly ciliolate, the longer ones 5–6 × 0.7–1.2 mm, venation reduced to a forwardly displaced midrib dividing blade 1:2–3. *Peduncles* 1–2(–3) per node, 1.2–3.6 cm; capitula 32–38-fld, the clavate or globose receptacle ±3.0–5.25 mm; bracts persistent into anthesis, that of peripheral fls linear-spatulate to oblanceolate ±1 mm, hyaline-margined, pubescent at apex, that of terminal fl about twice as large, not hyaline-margined; PERIPHERAL FLS: calyx tubular 5–6 mm, contracted at base into an obconic pedicel 0.5–1 mm, the broad-deltate teeth ±0.9 mm; corolla 7–8.2 mm, the lanceolate lobes 2.3–2.4 × 1 mm; androecium 10-merous, 20–29 mm, the stemonozone 1–1.6 mm, the tube 3–5 mm; ovary 2.6–3.5 mm, obovoid truncate, densely puberulent or puberulent only along the ventral suture; style slightly longer than stamens; TERMINAL FL: calyx sessile, deeply campanulate 5.5–6 × 2.8–3.3 mm, the teeth ±1 mm; corolla ±11 mm, the lobes 2.5 mm; androecial tube exserted 1.5–2 mm; ovary rudimentary or similar in shape and size to that of peripheral fl, surrounded by a shallow disc. *Pods* to 3 per capitulum, curved through ½–1½ circles 9–11 cm diam, the sutural rib slightly enlarged, the finely venulose valves 3.4–4.5 cm broad, to 6 mm thick over seeds, the mesocarp brown-ligneous, not at all granular or mealy; seeds (not seen fully mature) in 2 or 3 rows, ovate-elliptic in outline ±9 × 5 mm, the seed-coat apparently shiny brown.

Known only from French Guiana, in undisturbed forest near Cayenne (Matoury; Tonnegrande), and between the Sinnamary and Counamama rivers. — Fl. VIII–IX. — *Préfontaine* (used in cabinetwork and joinery), *St Martin jaune*.

Enterolobium oldemanii is closely related to *E. schomburgkii*, with which it shares persistent golden-brown indumentum, dimorphic flowers with a nectary surrounding the truncate ovary, and an androecium of mostly 10 stamens. It differs most noticeably in the size of the flowers, which are about twice as

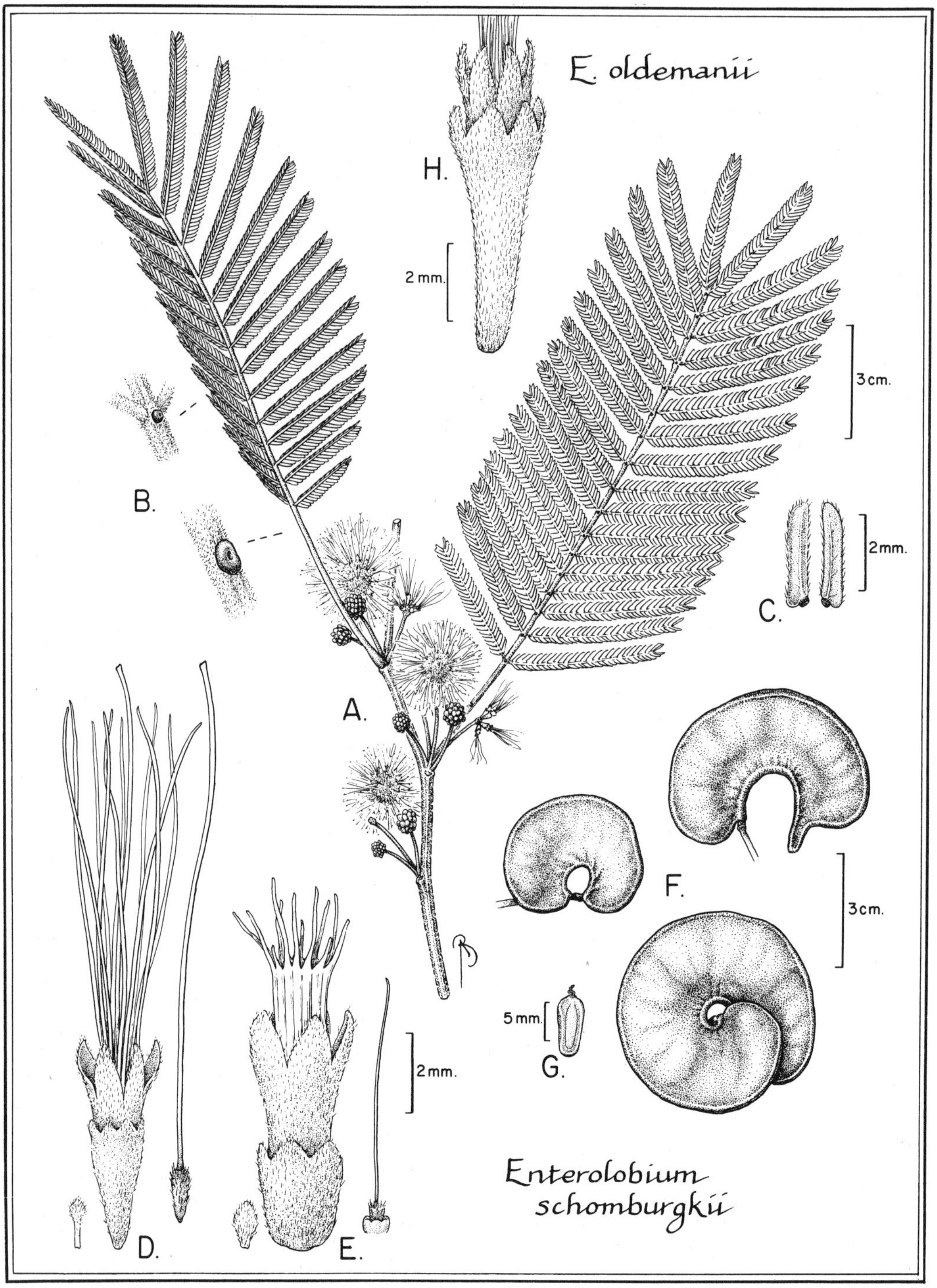

FIG. 18. Illustration of *Enterolobium schomburgkii* (A–G) and *E. oldemanii* (H). **A.** Habit. (Drawn from *Silva 4781.*) **B.** Leaf-nectaries. (*Silva 4781.*) **C.** Leaflets. (*Silva 4781.*) **D.** Peripheral flower. (*Silva 1387.*) **E.** Dimorphic terminal flower. (*Silva 1387.*) **F.** Fruits. (*Silva 4781*, left ft; *Smith 41781*, right ft; *Silva 2474*, lower ft.) **G.** Seed. (*Silva 4781.*) **H.** Dimorphic terminal flower of *Enterolobium oldemanii*. (*Oldeman 2242*; the largest fruit, lowermost of the three, is drawn from *Silva 2474* (NY), collected in Brazil.) The largest fruit matches closely those found on *E. oldemanii* in French Guiana, and it may be that *E. oldemanii* extends into Brazil. However, until flowers are collected this possibility remains unconfirmed.

long as those of *E. schomburgkii*, but also in thick coriaceous stipules, wider leaflets, and overall fewer peduncles per node. *Enterolobium schomburgkii* is notable for possessing two morphs of fruit, described under the following. The systematic significance of these two morphs is unknown.

It is with pleasure that we name this species in honor of Dr. R. A. A. Oldeman, indefatigable collector of the flora of French Guiana, who is best known for his contribution to the understanding of the architecture of forest trees.

10. E. schomburgkii Bentham (Mesquita, 1990: 168, figs. 34–37). Fig. 18A–G.

Throughout its discontinuously far-flung range *E. schomburgkii* is remarkably stable in characters of the foliage and consequently in general aspect, and over most of that range equally so in characters of flower and fruit. There are, nevertheless, some anomalous collections that foreshadow, as yet only tentatively, one or more cryptic taxa at least as distinct from typical *E. schomburgkii* as *E. glaziovii* has turned out to be (Mesquita, l.c., infra). The pod of typical *E. schomburgkii*, invariably, so far as known, outside the Amazon basin and prevalently within it, has ligneous valves 3–6 mm thick in section at their thickest point, and immersed sutures (hence a blunt outer margin). Occasionally, in French Guiana (*Serv. Forest. 202M, 7695*, NY) and again in western Amazonian Brazil (ríos Solimões and Purús, *Krukoff 8816*, *Prance 13652*, both NY) the valves are only about 1–2.5 mm thick at their thickest point and the sutures are externally evident, sometimes shallowly sulcate lengthwise. The individual seed-chambers of the thicker pod are not differentiated externally, but in the discoid thinner pod the valves are at least shallowly depressed between seeds, as in the pods of *E. cyclocarpum*. It is tempting to associate the smaller flower with the woodier pod, and the larger flower of *E. oldemanii* with the thinner pod, especially since the latter have both been found in the neighborhood of Cayenne; but the connection can be made only by observing one and the same tree at anthesis and in fruit, states that never coincide. As against this hypothesis it must be recorded that small-flowered *E. schomburgkii* also occurs in typical form in Atlantic French Guiana, and no instance of large flowers has come to light as yet in upper Amazonia.

11. E. glaziovii (Bentham) Mesquita (1999: 127, figs. 23, 24).

The pod of *E. glaziovii* differs from that of *E. schomburgkii* only slightly in superficial rugulosity of the epicarp, but the capitulate flowers are indeed homomorphic, as Mesquita discovered. Mesquita described the pod of *E. glaziovii* as glabrous, but every sample of it that we have seen is brown-strigulose, especially when young and toward the inner (dorsal) suture.

XVII. CEDRELINGA Ducke

Cedrelinga Ducke, Arch. Jard. Bot. Rio de Janeiro **3:** 70. 1922. — Sp. typica: *C. cateniformis* (Ducke) Ducke ≡ *Piptadenia catenaeformis* Ducke.

Cedrelinga is a monotypic genus without obvious close kindred in Ingeae, but seems most closely related to *Albizia* and *Enterolobium* sect. *Enterolobium* on the basis of the sylleptic, annual, axillary, strictly reproductive branches. Apart from its great stature, it may be known at anthesis by the unusual asymmetry of the ample, finely reticulate leaflets (the costa at once arched backward and postically displaced from midblade), by the terminal panicle of pseudoracemose capitula, and by corolla lobes separating downward as far as the stemonozone, therefore at maturity lacking a tube free from the androecium. The very large, papery lomentiform fruit twisted through about 90° at each interseminal isthmus is unique and unmistakable. The thin, exareolate, eventually fragile seed-testa suggests affinity with the *Zygia* group of Ingeae, but this hypothesis requires confirmation from seed morphology. Such anemochorous seeds could have evolved independently.

In his key to Mimosoideae of Brazilian Amazonia, Ducke (1949: 18) described the peduncles of *Cedrelinga* as arising from swollen nodosities resembling those bearing the individual flowers of many phaseoloid Fabaceae. The inflorescence nodes of *Cedrelinga* are indeed a trifle thickened under the serially bifarious peduncles, which are otherwise not different from those of other Ingeae, but the swelling is of course not homologous with those of Phaseoleae.

Cárdenas (1974: 124, 129) without explanation referred *Cedrelinga* to a comprehensive *Pithecellobium* of vast morphological diversity, but simultaneously maintained *Enterolobium* as a distinct genus. Such inequitable character-weighting seems capricious and inadmissible.

1. Cedrelinga cateniformis (Ducke) Ducke, Arch. Jard. Bot. Rio de Janeiro **3:** 70 estampa 6. 1922 "*catenaeformis*." *Piptadenia catenaeformis* Ducke, Arch. Jard. Bot. Rio de Janeiro **1:** 17. 1915. — "Ad ripas paludosas rivulorum silvestrium prope

Oriximina ad flumen Trombetas inferius, 2-III-1915, H[erb]. A[maz]. M[us]. P[araense] 15.704, et prope Obidos, 6-III-1915, n. 15.710; specimina omnia fructifera, ab A. Ducke lecta." — Syntypi, P!, RB!; isosyntypus, (*15710*), BM!. — *Pithecellobium catenaeformis* [*sic*] (Ducke) Cárdenas, Revista Fac. Agron. Maracay **7(3):** 124, 129 (lacking full citation of protologue). 1974. Fig. 19.

Cedrelinga catenaeformis sensu Macbride, 1943: 65; Ducke, 1949: 48; Mohlenbrock, 1963: 438; Hutchinson, 1964: 295; Nielsen 1981: 188, fig. 5(3); Spichiger, Boissiera **43:** 322, fig. 160. 1989.

Macrophyllidious *trees* of potentially great stature, fertile at 30–60(–66) m, with long simple, smooth but vertically furrowed, reddish trunk buttressed at base and to 1–2, exceptionally somewhat >3 m dbh, the emergent crown relatively narrow, the young branchlets and axes of inflorescence densely minutely brown-puberulent, the foliage glabrous subconcolorous, the broad, stiffly papery lfts lustrous olivaceous above, duller but scarcely paler beneath, reticulate on both faces but less sharply so dorsally, the inflorescence a terminal, seemingly efoliate panicle of pseudoracemose capitula biseriately fasciculate in the axil of inhibited lvs, the whole panicle shorter than the developed upper lvs. *Stipules* 0 (no scars). *Lf-formula* ii–iii/2–4(–5), the lfts 20–36 per lf; lf-stks (6.5–)8–16(–22) cm, obscurely sulcate ventrally, the petiole including firm pulvinus (3–)4–8 cm, at middle ±2–3 mm diam, the 1 or the longer of 2 interpinnal segments (3–)4–7.5 cm; petiolar nectaries at or close below insertion of each pinna-pair, low-convex and wrinkled or sunk into epidermis and amorphous, 2.5–5 mm diam, similar but smaller nectaries between furthest 1–2 pairs of lfts; pinnae a little accrescent distally, the rachis of longer ones 6–12(–16.5) cm, the 1 or the longer of 2–3 interfoliolar segments (2–)2.5–5 cm; lf-pulvinules livid wrinkled (3.5–)4–7 × 0.8–1.5 mm; lfts a little or distinctly accrescent distally, asymmetrically ovate or ovate-elliptic from postically cuneate, antically broader and rounded base, shortly bluntly acuminate, the distal pair 7–12(–13) × 3.5–5.7 cm, 1.8–2.6 times as long as wide; venation pinnate, the gently retro-arcuate midrib slightly displaced backward from midblade, giving rise on each side to (4–)5–6 major secondary nerves widely incurved-ascending to weak anastomosis close within the plane, somewhat thickened margin and to numerous crowded, more widely divergent and much finer intercalary nerves, these in turn generating a fine polygonal reticulum of veinlets. Primary *axis of panicle* or of its longer pseudoracemose branches ±1–1.5 dm; peduncles 2–6 per node, the longest of a fascicle 5–16 mm; capitula hemispherical, densely 8–20-fld, the receptacle 1–2 mm; bracts ovate or obovate-spatulate 0.3–0.7 mm, puberulent, persistent; fls all sessile, homomorphic, the 5-merous perianth greenish white (brunnescent when dried), glabrous except for minutely ciliolate calyx-teeth and sometimes papillate tip of corolla-lobes; calyx short-campanulate 1–1.4 × 1.1–1.3 mm, the depressed-deltate teeth 0.2–0.25 mm; corolla 4–4.6 mm, the lobes ovate ±1 × 0.9 mm at early anthesis, later separating more deeply, finally free beyond the stemonozone; androecium whitish, 24–30-merous, 9–11 mm, the stemonozone ±0.5 mm, the tube 1.4–2.4 mm; no intrastaminal disc; ovary shortly stipitate, ellipsoid, abruptly conic at apex; style (in random fls short, nonfunctional) a little longer than longest stamens, dilated to ±0.2 mm diam at the stigma. *Pods* 1 per capitulum, pendulous, lomentiform, in profile broad-linear but deeply constricted between each of the 2–5(–6) very large plano-compressed 1-seeded articles, the whole fruit 2–7 dm long, each segment oblong-elliptic 10–15(–17) × (3.3–)3.5–6 cm, and the isthmi 3–18 mm wide, the whole body twisted through ±90° at each isthmus but plane and straight between them, the pale green, when ripe papery and brownish stramineous, glabrous, sinuously venulose valves framed by relatively slender sutures, adherent to each other except where low-convex (and often discolored and more emphatically venulose) over the seeds; dehiscence 0, the 1-seeded segments (resembling whole pods of *Platymiscium*) breaking apart by transverse fission at the isthmi and shed individually; seeds descending at middle of pod, disciform, in broad outline elliptic 25–32 × 14–18 mm (orbicular 10–11 mm diam), ±1.5 mm thick, the thin dry translucent, when ripe fragile testa loosely investing the livid embryo; pleurogram 0.

In either wet or seasonally dry, primeval forest, especially along streams, 50–250m, and in prae-Andean Peru, up to 800 m, scattered through equatorial latitudes from the Amazon delta region and lower reaches of major tributary streams in Pará, Brazil, westward to the upper Río Napo in Ecuador, the upper Huallaga and middle Ucayali in Peru, in Brazil S to W Acre and NW Mato Grosso, and N into Surinam (Bigi Poika, Maripston road, not mapped), French Guiana (Saül), and Venezuelan Guayana (on Río Negro just into Amazonas, and on Río Canaracuni near 4°30'N in state of Bolívar). — Map 62. — Fl. IV–V; X–XII, the record very likely incomplete.

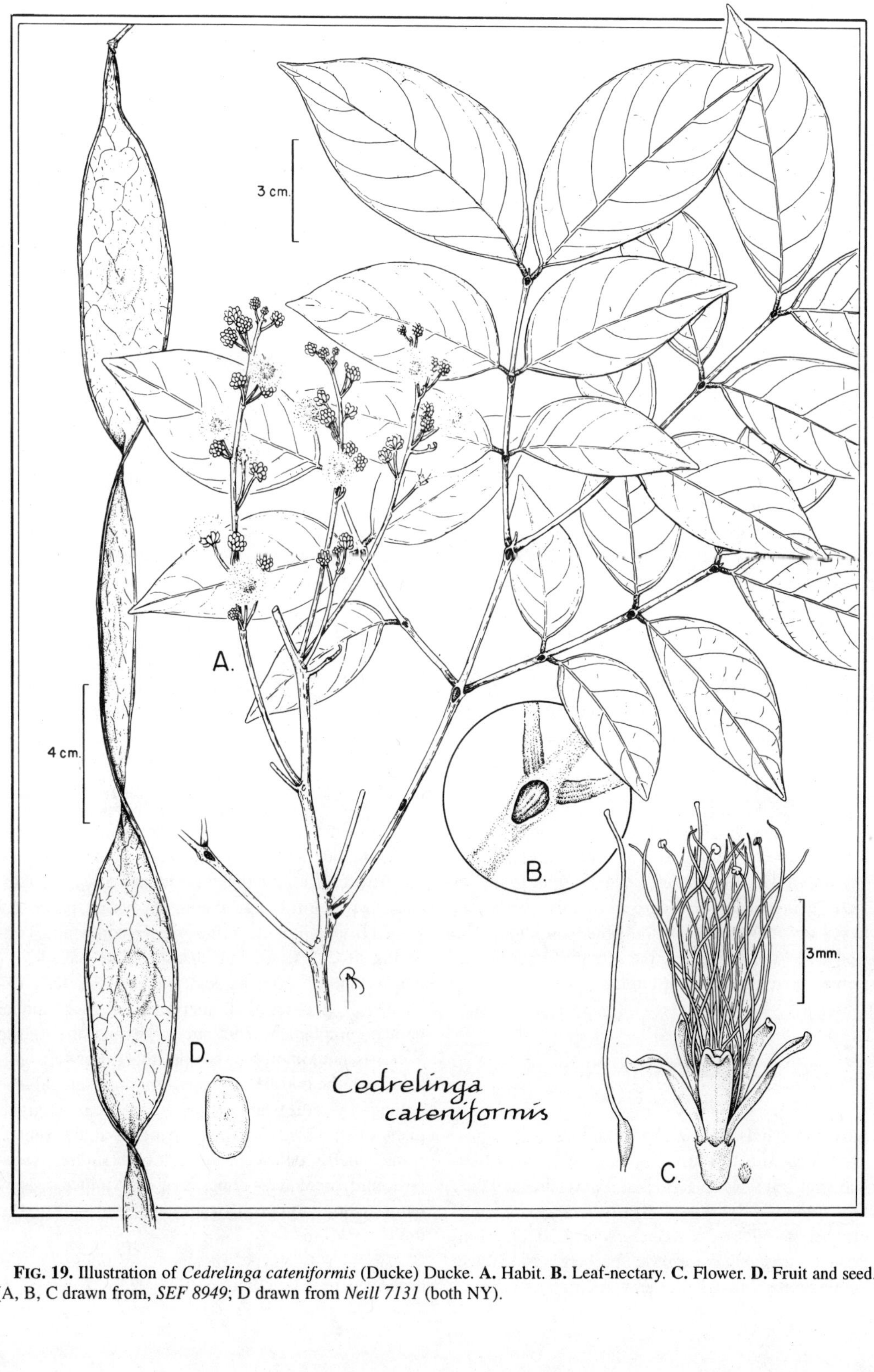

FIG. 19. Illustration of *Cedrelinga cateniformis* (Ducke) Ducke. **A.** Habit. **B.** Leaf-nectary. **C.** Flower. **D.** Fruit and seed. (A, B, C drawn from, *SEF 8949*; D drawn from *Neill 7131* (both NY).

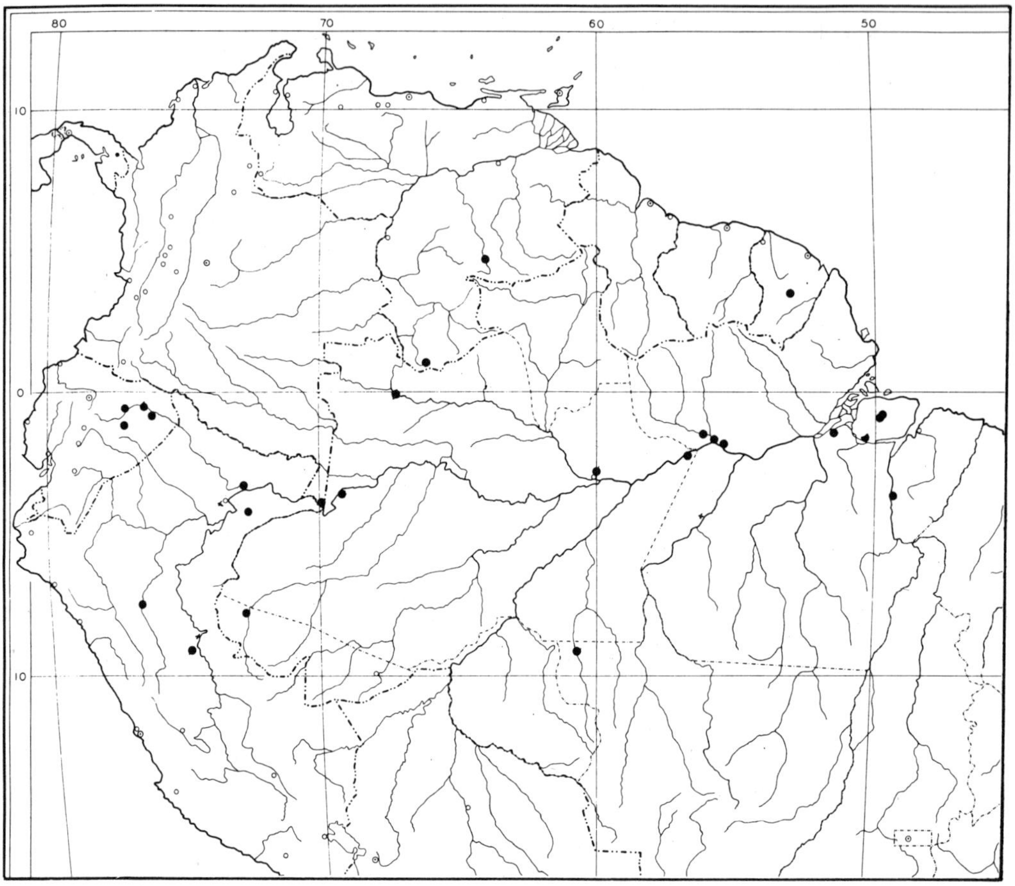

MAP 62. The Amazon basin. Range of *Cedrelinga cateniformis* (Ducke) Ducke.

— *Chuncho* (Quichua of Napo); *huayracaspi*, *tornillo*, *tornillo rosado* (Peru); *cedrorana* (Brazil, from resemblance of smooth reddish trunk to that of *Cedrela*); *iacaiacá* (Río Negro in Brazil); *cachicana*, *mure*, *guaura* (Venezuela); *don-ceder* (Surinam). The wood is like that of *Cedrela* but spongier, malodorous when worked, used in boat-building.

XVIII. FALCATARIA (NIELSEN) BARNEBY & GRIMES

Falcataria (Nielsen) Barneby & Grimes, gen. et stat. nov. *Paraserianthes* sect. *Falcataria* Nielsen *in* Nielsen et al., Bull. Mus. Hist. Nat. B, Adansonia IV, **5:** 327. 1983. — Sp. typica: *Paraserianthes falcataria* (Linnaeus) Nielsen ≡ *Adenanthera falcataria* Linnaeus ≡ *Albizia falcataria* (L.) Fosberg = **Falcataria molluccana** (Miquel) Barneby & Grimes.

Unarmed *trees,* at anthesis resembling *Serianthes* Bentham, but the lfts opposite along pinna-rachis and the fruit essentially that of *Albizia*. *Stipules* caducous. *Lvs* multifoliolate, the lfts subsessile, palmately veined from pulvinule. *Inflorescence* compound, consisting of 2–3 times branched, efoliate panicles of few-fld spikes, each panicle axillary to a fully expanded lf, the terminal meristem of each annual branch-complement continuous beyond the fertile axes; fls homomorphic, sessile, the calyx subhemispherical, the corolla sericeous, an intrastaminal disc around base of ovary. *Pods* broad-linear, straight, plano-compressed, narrowly winged along ventral suture, inertly dehiscent through both sutures; seed-coat hard, areolate. — Spp. 3, of the Moluccas, New Guinea, Solomon Is., Queensland, one cultivated in the New World.

Although the type species of *Falcataria* was described by Rumphius in 1743 (Herb. Amboin. **3:**

176), it was not perceived by Bentham (1875) as a distinct taxon. In the present century it has appeared in the literature (summarized by Nielsen et al., 1983a: 357) under the names *Albizia falcata* (the epithet misapplied), *Albizia moluccana*, and *A. falcataria* Fosberg. While it has a straight plano-compressed, stiffly papyraceous pod and areolate seeds that would not of themselves be out of place in *Albizia*, it is clearly not congeneric with the core of that genus as now defined, the indeterminate branching, spicate units of inflorescence and homomorphic flowers being incompatible with *A. julibrissin* and its close relatives. We fully endorse the expulsion of *A. falcataria* proposed by Nielsen, but we dissent from its association in one genus with *Paraserianthes lophantha*. The differences between them in inflorescence-architecture, and consequently in overall aspect both at anthesis and in fruit, are at least of equal value as those admitted elsewhere in contemporary classification of tribe Ingeae. Nielsen et al. (1983b: 351) found further differences in pollen and in the eophylls of seedlings.

The unnaturally close association, in some supraspecific category, of *Albizia lophantha* and *A. falcataria* goes back no further than Fosberg's revision of *Albizia* sect. *Pachysperma*, a name substituted (Fosberg, 1965: 74) for illegitimate *A.* sect. *Lophantha* ser. *Pachyspermae* Bentham (1875: 559). Bentham had divided his sect. *Lophantha*, which incorporated all spicate-racemose albizias then known to him, into two series; these were raised by Fosberg to sectional rank within *Albizia* and by Nielsen to independent genera, the ser. *Platyspermae* Bentham corresponding to *Archidendropsis* Nielsen, and the ser. *Pachyspermae* Bentham to *Paraserianthes* Nielsen. With this we have no quarrel, for they are substantially differentiated taxa. *Paraserianthes* was further divided by Nielsen into a monotypic sect. *Paraserianthes*, consisting of *P. lophantha* and its varieties, and sect. *Falcataria* Nielsen. It is this last group of three species, which collectively resemble *Serianthes* in characters of the inflorescence but differ in opposite leaflets and stiffly papery, dehiscent, not lignescent pod, that we here propose to treat as a discrete genus. Because the earliest epithet *falcataria* would be tautonymic in the genus *Falcataria* we are obliged to fall back on the next oldest epithet *moluccana*. It is a matter of regret that *Paraserianthes* Nielsen should have been typified by *P. lophantha*, which has only a remote resemblance to *Serianthes* Bentham, rather than by *Albizia falcataria*, which really does suggest *Serianthes* in most characters; this of course cannot be remedied.

1. Falcataria moluccana (Miquel) Barneby & Grimes, comb. nov. *Albizzia moluccana* Miquel, Fl. Ned. Ind. **1:** 26. 1855. — Described from sterile material, seen at L and equated with *Albizia falcataria* by Nielsen et al, 1983b: 358.

Adenanthera falcataria Linnaeus, Sp. Pl., Ed. 2, 550. 1762, based wholly on *Clypearia alba* Rumphius, Herb. Amboin. **3:** 176, t. 111. 1743. — Holotypus, the Rumphian protologue!. — *Albizzia falcataria* (Linnaeus) Fosberg, Reinwardtia **7:** 88. 1965. *Paraserianthes falcataria* (Linnaeus) I. Nielsen *in* Nielsen et al., Bull. Mus. Hist. Nat. B, Adansonia IV, **5:** 327. 1983.

Paraserianthes falcataria subsp. *falcataria* sensu I. Nielsen et al., Bull. Mus. Hist. Nat. B, Adansonia IV, **5:** 357, pl. 10 (map). 1983, q.v. for antecedent synonymy, literature, etc.

Albizia falcataria sensu Cárdenas, Ernstia **1(4):** 139, fig. 1. 1992.

Broad-crowned microphyllidious *trees* of rapid growth attaining 15–35 m, with smooth gray trunk to 2 m diam and lenticellate branches, the young stems and lf-axes pilosulous or tomentulose with pallid or sordid hairs to 0.15–0.3 mm, the ample lvs bicolored, the plane papery lfts dark dull green on upper face, pale beneath, minutely puberulent on both faces or sometimes glabrate ventrally, often pilosulous along midrib beneath, the inflorescence consisting of efoliate, twice or thrice branched panicles of loose, few-fld spikes arising from the axil of and mostly shorter than a few distal lvs, the terminal meristem continuous beyond them. *Stipules* thinly herbaceous, narrowly lanceolate 3–5.5 mm, covering the lf-primordium but caducous as this starts to develop. *Lf-formula* (vi–) vii–xi(–xii)/15–25; lf-stks of larger lvs ±1–2.5 dm, tapering distally, the petiole 2–6.5 mm and the longer interpinnal segments 12–16 mm, the ventral groove continuous; nectary either near mid-petiole or close below first pinna-pair sessile, oblong-ellipsoid, convex but dimpled, 2–7 mm diam, small round button-like nectaries at insertion of furthest 2–4(–5) pinna-pairs and yet smaller cupular ones on 1–3 furthest segments of pinna-rachises; pinnae conspicuously decrescent proximally, the lowest very short, the rachis of longer distal ones 7–11 cm, the longer interfoliolar segments 4–7.5 mm; lft-pulvinules 0.5–0.75 × 0.45–0.7 mm; lfts obliquely oblong or ovate-oblong from asymmetrically wide-cuneate base, abruptly deltate-apiculate, the larger ones 11–17.5 × 4.3–6 mm, ±2–3.2 times as long as wide; venation of 3–4 nerves from pulvinule, the straight or distally incurved, dorsally prominulous midrib forwardly displaced to divide blade 1:2.2–3.5, the much weaker, inner posterior primary nerve incurved-ascending to anastomosis at or near mid-blade, the outer posterior one(s) shorter, the midrib giving rise on blade's broad side to 2–4 feeble secondary nerves, the tertiary venulation scarcely or not perceptible, the whole venation immersed or almost so on upper face of blade. *Primary axis* of panicles 8–22 cm, that of ultimate flowering spikes 1.5–4 cm, the spikes 6–21-fld, the flowers maturing in acropetal series, the unfertilized ones quickly shed; bracts ovate or spatulate ±1–2.5 mm, covering the very young fl-buds but caducous as these begin to swell; perianth 5(–6)-merous, densely appressed-silky externally, the vesture of the corolla a little denser and paler than that of calyx; calyx widely campanulate or hemispherical 2.2–2.9 mm diam, the low-deltate or broad-triangular teeth 0.3–0.7 mm; corolla 6–7.2 mm, the ovate lobes 2.4–4 × 1.4–2.1 mm; androecium 84–148-merous, 14–18 mm long, the stemonozone ±1–1.3 mm, the tube 3.6–4.5 mm, internally at base thickened into a disc surrounding the very short stipe of the glabrous ovary, this abruptly contracted into the style; stigma poriform. *Pods* seldom more than 1 per spike, broad-linear, planocompressed, cuneately contracted at base into a pseudostipe 4–8 mm, abruptly apiculate, the body in profile 7.5–11.5 × 1.6–2.1 cm, the thin, greenish or fuscous, finally stramineous and papery valves either glabrous or remotely puberulent (especially near and along sutures), 10–15-seeded, framed by moderately dilated, straight sutures, at least the ventral (seminiferous) suture produced dorsally as a sharp-edged wing 1.5–2.5 mm wide; dehiscence through both sutures, inert; seeds horizontal near middle of pod, basifixed on a tapelike funicle, compressed-ellipsoid ±6 × 3 mm, the smooth, dull brown testa ±0.1–0.15 mm thick, clothed in a membranous outer coat peeling when soaked, the narrowly U-shaped pleurogram ±4–4.5 × 1–1.5 mm; endosperm 0.

Native (Nielsen, l.c.) in primary and disturbed rain forest, from sea level up to 2300 m, in the Moluccas and New Guinea, the Bismark Archipelago, and (subsp. *solomonensis* Nielsen) the Solomon Is.; naturalized on Fiji (A. C. Smith, Fl. Viti. Nova 3: 77. 1985); the typical subspecies, as described herein, locally planted for shade and reforestation in the Neotropics (Cuba, Trinidad, Venezuela, Ecuador, Brazil, Hawaii, doubtless elsewhere), not recorded as weedy. — Fl. with new foliage. — *Morrocoy* (Venezuela, perhaps transferred from another species); *luminosa* (Ecuador).

Note that the foregoing description refers exclusively to the typical subspecies as represented in cultivation or naturalized in the New World.

XIX. PARASERIANTHES NIELSEN

Paraserianthes Nielsen, Bull. Mus. Hist. Nat. B, Adansonia, ser. 4, **5:** 326, quoad solam sect. *Paraserianthem*, caeteris exclusis. 1983, based on *Albizzia* sect. *Lophantha* Miquel, Fl. Ned. Ind. **1:** 29. 1855. — Sp. typica (unica): *P. lophantha* (Willdenow) Nielsen ≡ *Acacia lophantha* Willdenow ≡ *Albizzia lophantha* Bentham.

Albizzia sect. *Spiciflorae* Bentham, London J. Bot. **3:** 85. 1844, parva ex parte (sp. 3), exclus. sp. lectotypica *A. fulgens* Labillardière (Fosberg, 1965: 73) quae ≡ *Archidendropsis fulgens* (Labillardière) Nielsen.

Albizzia sect. *Spiciflorae* ser. *Pachyspermae* Bentham, Trans. Linn. Soc. London **30:** 559. 1875. — Sp. typica: *A. lophantha* (Willdenow) Bentham ≡ *Paraserianthes lophantha* (Willdenow) Nielsen — *Albizia* sect. *Pachysperma* (Bentham) Fosberg, Reinwardtia **7:** 74, ex parte, omnibus praeter typum exclusis. 1965.

With the exclusion of *Falcataria*, as proposed above, *Paraserianthes* becomes monotypic, consisting however of four varieties (Nielsen et al., 1983b: 351, pl. 9). The description that follows incorporates the generic characters but applies precisely only to var. *lophantha*, native to southwestern Australia, occasionally cultivated and naturalized in the Americas.

Paraserianthes seems taxonomically isolated in Ingeae. The very numerous stamens of the individual flower, short in proportion to the calyx and only shortly united at base, and the minute anthers (0.2–0.25 mm long, 1.1–1.3 times wider than long) are reminiscent of some species of *Acacia*.

1. Paraserianthes lophantha (Willdenow) Nielsen, *in* Nielsen, Guinet & Baretta-Kuipers, Bull. Mus. Hist. Nat. B, Adansonia, ser. 4, **5:** 326. 1983. *Acacia lophantha* Willdenow, Sp. Pl. **4:** 1070. 1806, technically a substitute for *Mimosa distachya* Ventenat, Descr. Pl. Nouv. **5:** t. 20. 1801. — ". . . découvert par Lahaye sur les côtes occidentales de la Nouvelle Hollande, à 34 degrés de latitude, introduit chez Cels en l'an 5 [1797]." — Non *Mimosa distachya* Cavanilles, 1795. — Lectotypus (Nielsen, et al. op. cit. **5(4):** 353. 1983): B-WILLD *19419*, received from J. H. Seidel, presumably a plant cultivated at Dresden, seen in Microform 86. 1388: III. 6!. Since Willdenow quoted *M. distachya* Ventenat as a synonym of *A. lophantha* a more appropriate typus might be sought in herb. ventenat. at G. — *Mimosa lophantha* (Willdenow) Persoon, Syn. Pl. **2:** 264. 1807. *Albizzia lophantha* (Willdenow) Bentham, London J. Bot. **3:** 86. 1844. *Feuilleea distachya* (Ventenat) O. Kuntze, Revis. Gen. Pl. **1:** 185. 1891.

Albizzia distachya (Ventenat) Macbride, Contr. Gray Herb., n. ser. **59:** 3. 1919.

Pithecolobium lophanthum sensu Bentham, 1875: 559; *Paraserianthes lophantha* subsp. *lophantha* Nielsen *in* Nielsen, Guinet & Barreta-Kuipers, Bull. Mus. Hist. Nat. B. Adansonia, ser. 4, **5:** 352. 1983, q.v. for further synonymy and literature.
Albizia distachya sensu da Costa & Franquinho, Madeira Pl. & Flores, ed. 11, 37. 1990.

Microphyllidious *trees* of rapid growth and early maturity, attaining 8 m, the young branchlets and lf-axes densely tomentellous with sordid hairs to ±0.15–0.25 mm, the bicolored lfts dull dark olivaceous above, beneath paler and either glabrous or thinly strigulose, the fls borne in shortly pedunculate racemes either solitary or 2–4 together in the axil of coevally expanding lvs and greatly surpassed by them, the terminal meristem continuous. *Stipules* narrowly lanceolate or subulate ±1.5–3 mm, early caducous from small scar, absent from fruiting stems. *Lf-formula* vii–xi(–xiii)/24–38(–40); lf-stks 11–23 cm, tapering distally, the petiole ±3–6 cm, at middle 1.4–3.2 mm diam, the ventral groove continuous between pinna-pairs, the longer interpinnal segments 11–19 mm; petiolar nectary near or shortly below midpetiole sessile, elliptic-oblong, plane or shallowly dimpled or, when old, low-convex, 2–7 mm diam, in profile 0.3–1 mm tall, sometimes a second, smaller nectary above midpetiole and small round ones between the 1–2 furthest pinna-pairs, rarely a very small nectary at tip of some pinna-rachises; pinnae decrescent either proximally, or distally, or toward each end of lf-stk, the rachis of longer ones 7–11.5 cm, the longer interfoliolar segments ±2–4 mm; lft-pulvinules 0.2–0.6 × 0.4–0.5 mm, not wrinkled; lfts subequilong except at very ends of rachis, in outline broad-linear, linear-lanceolate or narrowly oblong from shallowly semicordate base, either acute, or deltately subacute, or obtuse apiculate, the longer ones 6.5–12 × 1.3–3(–3.8) mm, 3.2–5(–5.5) times as long as wide; venation weak, when perceptible palmate, the midrib forwardly displaced to divide blade ±1:2, the inner of 2–3 posterior primary nerves incurved to weak anastomosis beyond midblade, the outer 1–2 weak and short, the midrib either simple or weakly 1–2 branched on each side beyond middle, the whole venation immersed or almost so above, only the midrib prominulous beneath. *Peduncles* 1–2 cm, bracteolate near middle, the bracteole caducous; racemes ±45–70-fld, the axis 3–6 cm, the fls homomorphic; bracts rhombic-obovate ±1 mm, caducous as the fl-buds begin to swell; pedicels subhorizontal or a trifle declined 1–2.5 mm; perianth 5-merous, finely strigulose externally; calyx shallowly campanulate 2.2–3 mm, nearly as wide, the deltate or triangular teeth 0.6–1 mm, the tube sometimes circumscissile at base when past anthesis; corolla greenish 5–7.2 mm, the ovate lobes 1.4–3 × 1.1–1.8 mm; androecium greenish- or lemon-yellow, 96–154-merous, 13–16.5 mm, the stemonozone obscure or to 0.5 mm, the tube 3–5.5 mm, sometimes thickened at base around stipe of ovary, this glabrous, the stipe <1 mm, the body compressed, shallowly sulcate laterally, tapering into the style; stigma poriform. *Pods* usually several per spike, stipitate, the stipe (including transitional pseudostipe) ±5–9 mm, the body in profile broad-linear or narrowly oblong, straight or nearly so, when well fertilized 8–11(–12) × (1.3–)1.5–2.3(–2.4) cm, 8–11-seeded, strongly compressed but bullately elevated over each seed, the stiffly papery, reddish brown or fuscous, sometimes finally pruinose, always glabrous valves framed by sutures ±1–1.3 mm wide (the ventral suture sharp-edged, the dorsal one plane or shallowly sulcate), the ripe valves either weakly cross-venulose especially near the sutures or reticulately so over seeds, in section ±0.2 mm thick, not septiferous between seeds; dehiscence tardy, inert, through both sutures; funicles compressed, sigmoid distally; seeds horizontal at middle of pod, plumply ellipsoid, in broad view 6–8.2 × 4–5.2 mm, the hard testa dull black or fuscous, the incomplete pleurogram deeply U-shaped.

A species of bicentric range in East Indies and SW Australia (Nielsen et al., 1983b, pl. 9) of which the typical subspecies, endemic to SW Australia, has been cultivated in tropical and warm-temperate regions of both hemispheres, in the Americas specifically in W-centr. and S California, S Florida, Colombia, N Venezuela, Ecuador, E Bolivia, seldom and nowhere extensively naturalized. — *Chipo* (Colombia, where also *bracatinga*, properly *Mimosa scabrella* Bentham, is recorded).

XX. LYSILOMA BENTHAM

Lysiloma Bentham, London J. Bot. **3:** 82. 1944. — Generitypus (Hutchinson, Genera Fl. Pl. **1:** 296. 1964): *L. schiedeana* [*sic*] Bentham = *L. divaricatum* (Jacquin) Macbride, the only species that could at the time have provided the generic character of "the pod of a *Mimosa* with . . . monadelphous stamens." The earlier lectotypification (Britton & Rose, 1928: 76) by *L. bahamense* Bentham = *L.*

latisiliquum (Linnaeus) Bentham was an arbitrary selection based on page-priority that contradicts the essential differential character of the genus by having indehiscent fruits. The lectotype was, nevertheless, confirmed by Isely (Castanea **35:** 249. 1970) on the grounds that when *L. bahamense* turned out, by taxonomic opinion, to belong to *Lysiloma* Bentham, it became an appropriate generitype.

Lysiloma sensu Bentham & Hooker, Gen. Pl. **1(2):** 595. 1865; Bentham, 1875: 533–536, exclus. spp. 1 (= *Pseudosamanea*), 5 (= *Acacia*), 10 (= *Acacia*); Standley, 1922: 366–391; Britton & Rose, 1928: 76–84; Mohlenbrock, 1963: 436, 441; Isely, 1970: 249–252; 1973: 96–99; R. Thompson, 1980; Nielsen, 1981: 182; McVaugh, 1987: 188–193, fig. 26.

Unarmed, deciduous or semidedicuous *trees* with one or more trunks and often flaking bark, the sympodial branchlets thickened at the nodes by a prominent buttress under each lf-scar, commonly lenticellate, sometimes canescent in age; indumentum, when present, of white or sordid white (never brown or golden) hairs. *Stipules* evanescent, absent from all fruiting and most mature flowering specimens, commonly submembranous, varying (sometimes within a species, sometimes between lower and upper nodes of a young branchlet) from linear to obliquely dilated, semicordate, or flabellate. *Lf-formula* i/1.5 to ±xxx/50, the lfts consequently from exactly 6 to 1000+ per leaf; petiolar nectary near or well above midpetiole, sessile, either obesely cupular or mounded small-pored, and smaller nectaries often between some distal pinna-pairs and toward tip of pinna-rachises; lfts linear to elliptic or obovate, the size reciprocally adjusted to number, the venation simple, or weakly pinnate, or palmate-pinnate. Inflorescence of two sorts: (a) precocious, the usually bracteolate peduncles solitary or fasciculate at earliest efoliate nodes of flush-growth and often also axillary to earliest lvs; and (b) axillary to distal lvs of flush-growth and assembled beyond these into a short efoliate (bracteate) pseudoraceme; units of inflorescence either spicate-racemose (amentiform), or spherical capitate, or hemispherical-umbellulate, the individual fls bracteate in all but *L. sabicu*; fls homomorphic or almost so, the central fl(s) of some capitula sometimes a little thicker, but no longer, than the rest, its androecium not modified; calyx campanulate short-toothed; corolla ±0.5–1.5 times as long as calyx; androecium whitish 12–32-merous, 7–16(–20) mm, the tube 1.5–7 mm, the anthers 0.15–0.25 mm; gynoecium sessile or almost so, glabrous at anthesis (but the fruit occasionally puberulent). *Fruit* either a craspedium or an indehiscent legume, long persistent on tree (often into anthesis of second season), stipitate, in profile broad-linear or oblong-elliptic, acute or obtuse, framed by prominent sutures, the papery-crustaceous valves plano-compressed except where crumpled or low-bullate over seeds, straight but sometimes twisted through ±180° either above or below middle, the exocarp usually livid-purple-castaneous but when ripe nigrescent and exfoliating to reveal the stramineous endocarp, less often green ripening pale-brown and not exfoliating, the two endocarpic panels sealed around the seeds and, following craspedial dehiscence, falling together as a unit; dehiscence in most species craspedial, the valves then separating from the sutures and falling together out of the marginal frame, the seeds then released on the ground either through gaping margins of the endocarpic panels, or, when dehiscence none, through decay of the panels themselves; funicles filiform; seeds transverse, compressed-ellipsoid, rather lustrous brown, the pleurogram on middle of each seed-face commonly U-shaped, in *L. candidum* very small and semicircular, in *L. sabicu* complete. — Spp. 8, adapted to monsoon and desert climates at low and submontane elevations in Mexico and Greater Antilles, one of these reaching subtropical Florida, one western species extending N into SE Arizona; records from South America are based on misidentified *Acacia*.

Contrary to a tradition that goes back to the first description in 1844 and persists to this day, *Lysiloma* is not characterized as a genus by craspedial dehiscence of the fruit. Bentham correctly observed and described craspedia of two species, but cautiously remarked that the generic affinity of the rest, and the status of the genus itself, required confirmation from further study. The myth of uniformly craspedial dehiscence in *Lysiloma*, like that of the arillate seed in *Klugiodendron*, illustrates the vitality of misconceptions that accumulate verisimilitude by reiteration, and thus resist correction.

This specialized craspedial dehiscence, whereby the pods' valves are detached as a unit, together with the seeds, from the sutural frame that often persists in the form of threads on the parent branch, is indeed characteristic of all the microphyll lysilomas widespread over continental Mexico and Central America, but in the two primarily West Indian species that are described at length on the following pages the entire pod falls indehisced, and the seeds are released only by decay of the valves on the ground. The carpological character that in reality distinguishes *Lysiloma* from other New World ingeoid Mimosaceae is the permanent adherence of the two parallel sutural ribs around the whole periphery of the pod, ribs that together form a fused keel that may or may not, depending on the species, separate eventually from the valves. The indehiscent fruits of *L. latisiliquum* and *L. sabicu* have been correctly described until now only by Ralph Thompson in his unpublished monograph of the genus. Thompson (1980) proposed a division of the genus into two subgenera based on this property of the fruit, but having accepted *L. bahamense* (which is *L. latisiliquum*) as generitype he was obliged to refer the genuine, craspedial species, which ought to constitute the subgen. *Lysiloma*, as a newly named subgenus. This we are unable to accept. Nomenclature and typification aside, we question the need for subgeneric division of the genus. The two West Indian lysilomas, while alike in almost all features of the fruit, differ from one another in several characters, some of which each shares with some member of the typical continental group, suggesting independent derivation. Thus *L. latisiliquum* shares a distally efoliate inflorescence only with *L. candidum*; while *L. sabicu* resembles *L. tergeminum*, and no other, in ebracteolate peduncles.

The phylogenetic relationships of *Lysiloma* can only be guessed at. The genus was thought by Nielsen (1981: 182) to be most nearly allied to *Albizia*, and perhaps better reduced to a section of that genus. It differs, however, from *Albizia* as defined herein, not only in the permanently sealed sutures of the pod but also in inflorescence architecture, in which it resembles *Chloroleucon*, *Enterolobium*, and *Hesperalbizia*. Its resemblance to *Acacia* is in most respects deceptively close, but *Lysiloma* is nevertheless effectively separable from habitally similar, unarmed *Acacia* by a syndrome of longer staminal tube (at least twice as long as diameter, generally longer) and the pods' peculiar sealed frame.

Taxonomic opinion is contradictory about the size of *Lysiloma*, Hutchinson (1964) claiming 50 species, Nielsen (1981) about 35, and Thompson (following unpublished analysis, 1980) finding only nine major taxa. The common microphyll species are plastic in leaf-formula, pubescence, and width of pod, and have provided types of numerous microspecies based on single collections. Our perception of *Lysiloma*, which differs taxonomically from that of Thompson only in combining *L. divaricatum* and *L. microphyllum*, which are completely confluent in southwestern Mexico, is presented in the conspectus following. Only two species, both primarily West Indian, are here described and mapped.

Five taxa based on West Indian or South American types are excluded from *Lysiloma*:

1. *Lysiloma ambiguum* (Vogel) Urban, Ark. Bot. 22A, **8:** 28. 1929 "*ambigua!*". *1929.* ≡ *Acacia ambigua* Vogel, Linnaea **10:** 600. 1836, homon. poster. ≡ *Acacia vogeliana* Steudel, Nomencl. ed. 2, **1:** 9. 1840. ≡ *Lysiloma vogelianum* Stehlé, Bull. Mus. Hist. Nat. Paris, ser. 2, **18:** 193. 1946 "*vogeliana*".
2. *Lysiloma guachapele* Bentham, 1875: 533. — Treated herein as *Pseudosamanea guachapele* Harms.
3. *Lysiloma marchianum* Grisebach, Fl. Brit. W. I. 223, 1860 "*marchiana*". — Equated with *Calliandra portoricensis* (Jacquin) Bentham by Bentham, 1875: 543, which ≡ *Zapoteca portoricensis* H. Hernández.
4. *Lysiloma polyphyllum* (Clos) Bentham, 1875: 535 ("*polyphylla*"), based on *Acacia polyphylla* Clos *in* Gay, Fl. Chil. **2:** 254. 1847, homon. poster. — ". . . proviene probablemente del Tucumán. Se cultiva en San Isidro y otros varios puntos de Coquimbo." — Holotypus: *C. Gay s.n.*, P (annotated by Bentham)!; isotypus, *Gay 366*, K!. — We equate this with *Acacia visco* Grisebach.
5. *Lysiloma rostratum* (Willdenow) Bentham, London J. Bot. **3:** 84 1844 "*rostrata*". ≡ *Acacia rostrata* Humboldt & Bonpland *ex* Willdenow ≡ *Dugandia rostrata* (Willdenow) Britton & Killip, Ann. New York Acad. Sci. **35:** 138. 1936.

Conspectus of genus *Lysiloma*

¶ Fruit a craspedium, at least tardily dehiscent, the continuous valves separating from the sutural keels and falling as a unit, the livid or purplish-brown or purplish-green exocarp peeling in flakes, the stramineous, stiffly papery-crustaceous endocarp narrowly gaping along each margin to release the seeds; inflorescence precocious, the bracteolate peduncles arising either from earliest efoliate nodes of flush-growth, or from lower lf-axils, or from both, but none pseudoracemose; lvs multifoliolate, the lf-formula not less than iv/25 and the largest lfts 2–9 × 0.5–2.4 mm; units of inflorescence various:

* Units of inflorescence amentiform, narrowly cylindric, the floral axis 3.5–6.5 cm; lf-formula vii–xvi(–xviii)/33–50(–54), the largest lfts 4–9 × (1–)1.3–2.4 mm; Mexico (southward from Sinaloa and adjacent Chihuahua on the Pacific slope and from S Tamaulipas on the Gulf slope) SE to El Salvador and W Honduras. *L. acapulcense* (Kunth) Bentham

** Units of inflorescence capituliform, ellipsoid, densely racemose, the floral axis 8–20 mm; lf-formula (xiv–)xvii–xxx/30–35, the largest lfts 2–4.5(6) × 0.5–1.4 mm; Mexico (Guerrero and

Veracruz) Se to W Costa Rica. *L. auritum* (Jacquin) Macbride

*** Units of inflorescence capituliform, suborbicular, the floral axis 3–8 mm, some fls distinctly though shortly pedicellate; lf-formula v–ix/30–50, the largest lfts 5–7.5(–9) × 1.7–3 mm; Sonoran Desert and surrounding foothills, SE Arizona through Sonora to SW Chihuahua and N Sinaloa, sympatric there with *L. divaricatum*, but differing in broader, strongly bicolored (not bright green) lfts drying brown below, dove-gray above, as well as in longer floral receptacle: *L. watsonii* Rose

**** Units of inflorescence truly capitate, the floral receptacle not over 2.5 mm and the fls all sessile; lf-formula iv–x(–xvii)/26–46, the largest lfts 3–8 × 0.7–2.2 mm, either faintly 1-nerved (syn: *L. microphyllum* Bentham) or very finely and weakly penninerved; widely dispersed in Mexico around and S of the Meseta Central and Cordillera Neovolcánica, from centr. and S Baja California Sur and N Sonora on the Pacific slope, and from S Tamaulipas and SE San Luis Potosí on the Gulf slope, SE to NW Costa Rica. *L. divaricatum* (Jacquin) Macbride

¶¶ Fruit craspedial, as in ¶, with nigrescent exocarp peeling in flakes, but the inflorescence of *L. latisiliquum*, the capitula arising both from lf-axils and from efoliate distal nodes beyond developed lvs, thus assembled into short bracteate pseudoracemes; lf-formula i–ii/4–7(–9), the larger lfts 6–12 mm wide, pinnately nerved; Baja California Sur and extreme SE Baja California, one station in coastal Sonora. *L. candidum* T. S. Brandegee

¶¶¶ Fruit a craspedium, but the pale green or light tan, transversely striate exocarp never peeling; inflorescence as in ¶, the units of inflorescence densely capitulate, the fls sessile; lf-formula (of *Mimosa albida* Kunth) i/1.5, the ample, palmately nerved lfts exactly 6 per lf, the distal pair of each pinna ±3.5–7 × 2–4 cm; peduncles ebracteolate (as in *L. sabicu*); SW continental Mexico, from S Nayarit to Morelos and Guerrero. *L. tergeminum* Bentham

¶¶¶¶ Fruit an indehiscent pod, the valves never separating from the sutural keels, the nigrescent exocarp peeling in small tesserae to reveal the stramineous endocarp, this sealed around the seeds, which are released only by decay on the ground.

* Inflorescence developing with and beyond the lvs, the bracteolate peduncles arising from later lf-axils and distally assembled into short bracteate pseudoracemes; capitula globose, the fls sessile, bracteate; lf-formula ii–v(–vi)/(11–)12–26, the largest lfts 3–6.6 mm wide; pod acute or acuminate at apex; S Florida, Bahamas, Cuba, SW Hispaniola; Yucatán Peninsula westward from Tabasco and S into adjacent Guatemala (Petén) and N Belize: *L. latisiliquum* (Linnaeus) Bentham

** Inflorescence precocious, the ebracteolate peduncles arising from earliest efoliate nodes of flush growth and from lower lf-axils, none pseudoracemose; capitula hemispherical umbelliform, the fls pedicellate ebracteate; lf-formula ii–iv/(3–)4–6 (–7), the largest lfts 8–17 mm wide; pod obtuse at apex; Bahamas, Cuba, NW Hispaniola, and one station in N-centr. Puerto Rico: *L. sabicu* Bentham

Key to the West Indian species of genus *Lysiloma*

1. Stipules semicordate-auriculate at base; distal capitula axillary to bracts, pseudoracemose; petiolar nectary verruciform, obtuse; lfts of longer pinnae (11–)13–36-jug., the largest 2.4–4 mm wide, the pulvinules 0.3–0.6 mm; peduncles bracteate above middle; capitula spherical, the fls sessile, the fl-buds densely whitish-puberulent with crimped hairs, the interfloral bracts linear or linear-spatulate, persistent; pod acuminate at both ends; pleurogram of seed-face U-shaped . . . . 1. *L. latisiliquum* (p. 260)
1. Stipules cuneate-flabellate at base; all but the earliest capitula axillary to a fully expanded lf; petiolar nectary shallowly cupular, depressed or subimmersed; lfts of longer pinnae 4–6(–7)-jug., the largest 8–17(–30) mm wide, the pulvinules 1–2.2 mm; peduncles ebracteate; capitula hemispherical, all or at least the lower fls pedicellate, ebracteolate, the fl-buds only minutely puberulent at very apex; pod broadly obtuse at apex; pleurogram of seed-face elliptic complete . . . . 2. *L. sabicu* (p. 261)

1. Lysiloma latisiliquum (Linnaeus) Bentham, Trans. Linn. Soc. London 30: 534. 1875 "*latisiliqua*". 1875. *Mimosa latisiliqua* Linnaeus, Sp. Pl. 519. 1753. — "*Habitat in* America *calidiore*." — Holotypus (De Wit, Taxon **10:** 52. 1961): *Acacia non spinosa, siliquis latis compressis, flore albo* Plumier [Nova Pl. Amer. 17, phrase-name only quoted verbatim by Linnaeus] *ex* Burman, Pl. Amer. fasc. prim. 2, Pl. 6 (reproduced in Taxon **24:** 350, fig 1. 1975). 1755. — The alternative typification by Gillis & Stearn, Taxon **23:** 185–191. 1974, which identified *Mimosa latisiliqua* with *Leucaena leucocephala* (Lamarck) De Wit, was definitively rejected by De Wit, Taxon **24:** 349–352. 1975, and by Shaw and Schubert, J. Arnold Arb. **57:** 113–118. 1976. — *Acacia latisiliqua* (Linneaus) Willdenow, Sp. Pl. **4:** 1067. 1805. *Leucaena latisiliqua* (Linnaeus) Gillis *in* Gillis & Stearn, Taxon **23:** 190. 1974.

Lysiloma bahamensis [*sic*] Bentham, London J. Bot. **3:** 82. 1844. — "Bahamas, [*William*] *Swainson*." — Holotypus, K (photo: Taxon **23:** 189, fig. 1, bottom. 1974)!. — *Acacia bahamensis* (Bentham) Grisebach, Fl. Brit. W. I. 221. 1860.
Lysiloma latisiliquum subsp. *lataefoliolatum* [*sic*] Barreto & Yakovlev, Acta Bot. Cubana **41:** 6. 1987. — "[Cuba.] Guantánamo . . . Maisí, Bisse & Kohler 8027." — Holotypus, HAJB, not seen; isotypus (Bässler, Gleditschia **17:** 67. 1989), BHU, not seen; isoparatypus, *León 11673*, NY!.

Lysiloma latisiliquum sensu Sargent, Silva **3:** 129, pl. 144 1892 *"latisiliqua"*. Urban, Symb. Ant. **2:** 264. 1900; **8:** 255. 1920; Isely, 1973: 97, map 26 (Florida only); R. Thompson, 1980: 26, fig. 4 (map); Correll & Correll: 1982: 656, fig. 269; Barreto & Yakovlev, Acta Bot. Cubana **41:** 4, fig. 5. 1987; Isely, Vasc. Fl. SE U.S. **3(2):** 12. 1990.

Lysiloma bahamense sensu Britton & Millspaugh, Bahama Fl. 158. 1920; Sargent, Man. Trees, ed. 2, 590, fig. 541. 1922; Standley, 1922: 390; Britton & Rose, 1928: 81; Standley & Steyermark, 1946: 50; León & Alain, Fl. Cuba **2:** 239. 1951; Pennington & Sarukhán, 1968: 178 + fig.; Liogier, 1985: 46.

Unarmed *trees* 5–13(–20) m, some flowering as arborescent shrubs, with smooth or scaley trunk attaining 1+ m dbh and fuscous pallid-lenticellate branchlets, appearing glabrous but the lf-axes often and those of inflorescence always at least minutely puberulent or strigillose with fine whitish hairs <0.3 mm, the bicolored lfts either glabrous on both faces, or minutely puberulent on one or both faces, or sometimes in addition weakly barbellate dorsally in anterior basal angle of midrib, the dense globose capitula fasciculate (solitary) at nodes of new shoots, the early ones subtended by a lf, the later by a pair of stipular bracts progressively smaller distally, then forming a terminal pseudoraceme, the terminal meristem of each flowering stem determinate and the whole branching pattern sympodial; resting buds not developed. *Stipules* green but submembranous, caducous, those at early nodes of new branchlets inequilaterally (ob)ovate to lance-elliptic from auriculate or semicordate base, acute, 10–26 × 3–12(–15) mm, those subtending peduncles progressively smaller and narrower distally, lanceolate 2–8 × 0.5–4.5 mm. *Lf-formula* ii–v(–vi)/(11–)13–26(–28); lf-stk of longer lvs (3–)4.5–10.5(–12) cm, the petiole (1)1.7–4(4.5) cm, usually a little longer than the interpinnal segments, these to 0.8–2.2(–3) cm; nectary near or above midpetiole, rarely next to first pair of pulvinules, plumply verruciform, obovoid or obliquely hemispherical, obtuse (0.4–)0.6–1.5 mm tall, at base 0.9–1.8 mm diam, fuscous or nigrescent, a similar nectary often at tip of lf-stk and randomly between pinna-pairs, a much smaller one at tip of most pinna-rachises; rachis of longer pinnae (4–)5–10 cm, the longer interfoliolar segments 3–5.5(–6.5) mm; pulvinules 0.3–0.6 mm; lfts decrescent at each end of rachis but otherwise subequilong, the blade narrowly oblong or oblong-elliptic from obliquely truncate or shallowly semicordate base, obtuse or obtusely triangular at apex, the larger ones 9–16 × 3–6.6 mm, 2.4–3.6(–4) times as long as wide; venation slenderly palmate-pinnate, the midrib centric or only a little forwardly displaced, prominulous only dorsally, the 1(–2) anterior primary nerves produced about to midblade, the 3–4 pairs of feeble immersed secondary ones sometimes visible only dorsally. *Peduncles* 1–5 per node, 10–34 mm, bracteolate shortly below capitulum, the bracteole 1–2.5 mm; capitula very densely (32–)40–54(–88)-fld, the fls sessile, subequilong, but the terminal one sometimes stouter and occasionally with 2 gynoecia, the perianth submembranous with thicker and crisply gray-puberulent lobes; bracts linear-oblanceolate or spatulate 0.9–1.8 mm, persistent; calyx turbinate-campanulate 1.6–2.6 mm, the obtuse teeth 0.4–0.7 mm; corolla 2.7–3.7(–4) mm, the ovate-triangular lobes 0.7–1.3 mm; androecium 12–24(–30)-merous, 11.5–14(–15) mm, the tube 0 9–1.5(–1.7) mm, about as long as the glabrous ovary, this obliquely conic at apex. *Pods* geotropic, glabrous, the stipe 14–43 mm, the straight (and not twisted), plano-compressed body oblong-elliptic in profile (5.5–)7.5–16(–18) × 2–4.5(–6.2) cm, acuminate or attentuate at each end, when well fertilized 9–13(–14)-seeded, the stiffly chartaceous valves framed by dorsally flattened or shallowly sulcate sutures, becoming low-bullate or crumpled over the seeds, the thin exocarp lustrous dark reddish-brown becoming dull black in age, deciduous in tesserae to expose the light tan or stramineous endocarp; dehiscence 0, the whole fruit tardily deciduous and the seeds released by weathering of the valves on the ground; funicles filiform; seeds narrowly elliptic in broad view, 6–7 × 3–4 mm, the testa dark brown sublustrous, the narrowly U-shaped pleurogram at middle of seed-faces.

In hammock, coppice, and scrub-woodland, mostly on limestone below 50 m, a vigorous pioneer following fire or cutting, locally plentiful and sometimes weedy in wasteland, in Florida locally cultivated; S peninsular Florida and the Keys; common nearly throughout the Bahamas (unknown from Andros) southeastward from Grand Bahama to the Caicos Is. and NW coast of Haiti; scattered through the length of Cuba; and plentiful in deciduous woodland of Yucatán Peninsula eastward from Tabasco, Mexico, S just into Guatemala (Petén) and Belize (distr. Cayo). — Map 63. — Fl. most prolifically II–V(VI), sporadically in other months. — *Singing beans*, *Wild tamarind* (Florida, Bahamas); *dormido*, *soplillo*, *tengüe*, (Cuba); *salam*, *salem*, *tzalam*, (Yucatán, where the bark is used for tanning).

2. Lysiloma sabicu Bentham, Hooker's J. Bot. Kew Gard. Misc. **6:** 236. 1854. — Described from

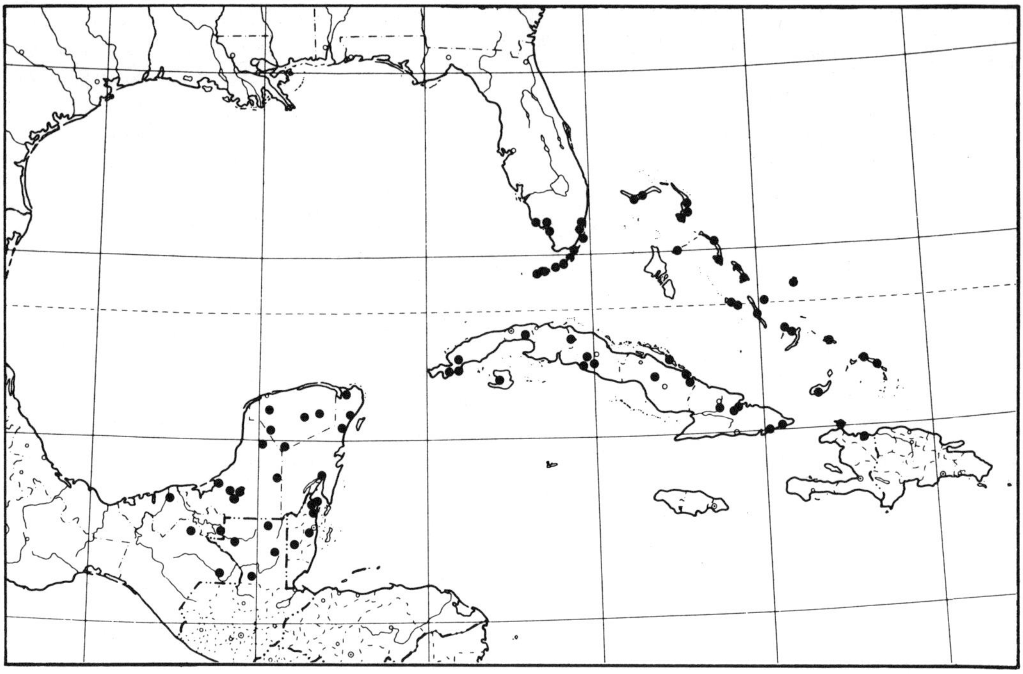

MAP 63. Central America and the West Indies. Range of *Lysiloma latisiliquum* (Linnaeus) Bentham.

specimens collected in Cuba by Ramón de la Sagra, K (photo: Taxon **23:** 190, fig. 1, top. 1974)!.

Acacia latisiliqua (ß) *paucifoliola* de Candolle, Prodr. **2:** 467. 1825. — " . . . in Sancto-Domingo (Bertero . . . in h[erb.] Balb[is.]." — Holotypus, G-DC, not seen. — *Lysiloma paucifoliola* (de Candolle) A. S. Hitchcock *ex* Northrop, Mem. Torrey Bot. Club **12:** 38. 1902, nom. nud.
Lysiloma sabicu sensu Bentham, 1875: 534; Oliver *in* Hooker, Icon. pl. 1722. 1887; Urban, Symb. Ant. **2:** 264. 1900; Britton & Millspaugh, Bahama Fl. 158. 1920: Isely, 1973: 98; Gillis & Stearn, Taxon **23:** 190. 1974; Shaw & Schubert, J. Arnold Arb. **57:** 113. 1976; R. Thompson, 1980: 17, fig. 2 (map); Correll & Correll, 1982: 658; Liogier, 1985: 46; Barreto & Yakovlev, Acta Bot. Cubana **41:** 2, fig. 5 (map, Cuba). 1985.
Acacia formosa sensu A. Richard *in* Sagra, Hist. Phys. Cuba, Bot. pl. vasc. 463. 1845, typ. exclus.; non *Acacia formosa* Kunth. *Lysiloma formosa* A. S. Hitchcock, Ann. Rep. Missouri Bot. Gard. **4:** 83. 1893, nomen.
Lysiloma latisiliqua [*sic*] Britton & Rose, 1928: 80, basionym. exclus. León & Alain, Fl. Cuba **2:** 239. 1951; non *L. latisiliqua* (Linnaeus) Bentham.

Semideciduous unarmed *arborescent shrubs*, in undisturbed woodland attaining 12 m but potentially flowering when 1 m tall, the annotinous branchlets purplish-brown pallid-lenticellate, prominently buttressed at defoliate nodes, glabrous, the lvs and fls also glabrous throughout except for distally puberulent corolla-lobes and for occasional lfts minutely barbellate dorsally in anterior basal angle of midrib, the plane, thin-textured lfts more or less bicolored, paler beneath, the hemispherical capitula of whitish fls arising singly or geminate from the earliest nodes of short-shoots lateral to indeterminate barren long-shoots, subtended by dilated, green but submembranous, caducous stipules and sometimes also a developed lf; resting buds plumply ovoid or subglobose 1–2.5(–3) mm, the perules firm, lustrous brown, not striate externally. *Stipules* obovate-cuneate obtuse or broadly oblanceolate subacute 7–16 × 4–10 mm, flabellately several-nerved from base. *Lf-formula* ii–iv/(3–)4–6(–7); lf-stks 3–12 cm, the slender petiole scarcely longer than the longest interpinnal segment, this (1.2–)1.5–3.3 cm; a sessile shallow-cupular or sunken nectary either near, or below, or well above midpetiole 0.6–1.3 mm diam, rarely a minute one at apex of lf-stk; pinnae subequilong or shorter proximally, the rachis of longer ones 3.5–7 cm, the longer interfoliolar segments (6–)8–18(–20) mm; lft-pulvinules subterete 1–2.2 mm; lfts moderately accrescent upward, obovate or broadly oblong-elliptic obtuse, the larger ones (1.2–)1.4–2.7(–4.7) × 0.8–1.7 (–3) cm, 1.3–1.9(–2.4) times as long as wide; venation pinnate, the subcentric, straight or faintly incurved

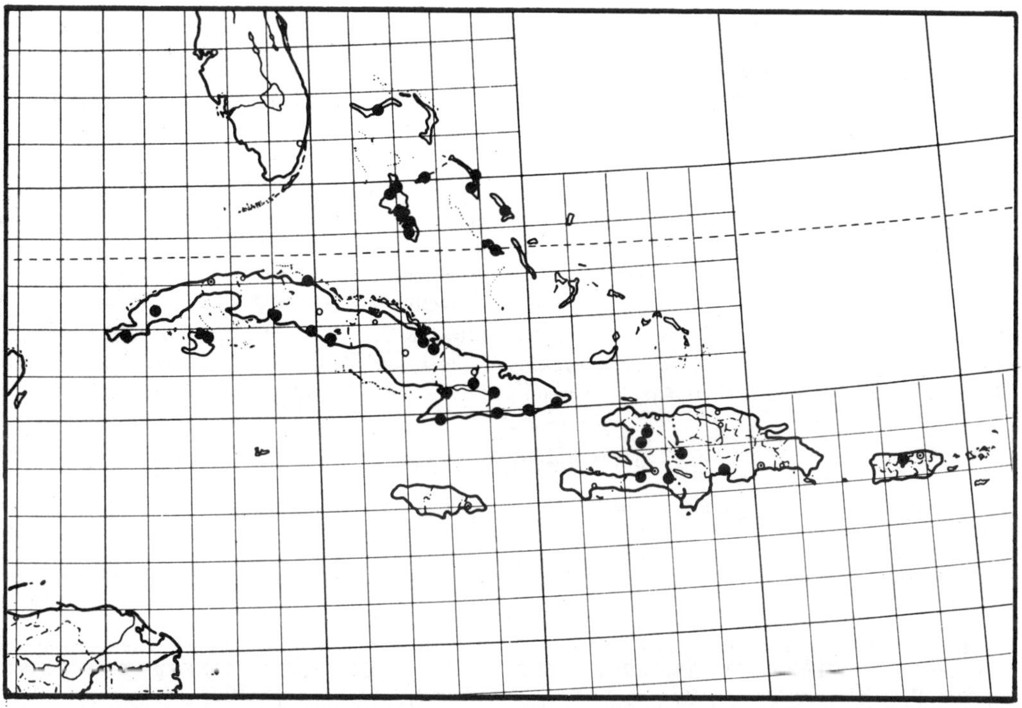

MAP 64. Greater Antilles and Bahamas. Range of *Lysiloma sabicu* Bentham.

midrib giving rise on each side to (4–)5–8(–9) secondary venules brochidodrome well within the margin, a weak tertiary venulation at times elevated on one or both faces. *Peduncles* 3–5 cm, ebracteolate; capitula 22–45-fld, the receptacle including short pedestal 1.5–3.5 mm; bracts obsolete; fls subhomomorphic in perianth but the androecial tube of 1–4 distal ones longer and more dilated upward; pedicel of lowest fls 0.4–1.4 mm, of distal fls progressively shorter or 0; calyx narrowly campanulate, bluntly angulate, 2.5–3.6(–4) × 1.4–1.9 mm, the often unequal teeth 0.2–0.9 mm; corolla 3.8–5 mm, the often unequal lobes 1–1.6 mm; androecia 14–22-merous, 10.5–16.5 mm, the tube in peripheral fls 2.2–4.5, in terminal fls 4.3–7 mm, but these occasionally abortive; ovary glabrous, abruptly contracted into the style; stigma scarcely dilated, ±0.12–0.15 mm diam. *Pods* 1–5 per capitulum, geotropic, the stipe (1.6–) 1.8–3.6 cm, the plano-compressed body in profile oblong, elliptic-oblong or oblanceolate (7–)8–14 (–15.5) × (1.9–)2.2–4 cm, often attenuate toward base and there twisted through 180°, flat thereafter and framed by slender, dorsally plane, narrowly prominulous sutures, the thin-textured valves consisting of livid-nigrescent glabrous exocarp breaking up at maturity by transverse and vertical cracks into readily deciduous tesserae and thereby exposing the smooth papery stramineous endocarp, this not spontaneously separating from the sutural ribs, the 2 valves closely appressed toward the margin and between successive seeds, low-bullate over them; dehiscence 0, the fruit sometimes retained unopened on the tree into anthesis of second year; funicles filiform, straight; seeds transverse, compressed-ellipsoid, in broad view 5.8–8.5 × 3.6–4.2 mm, the testa fuscous-castaneous, microscopically roughened, the pleurogram at middle of seed-face elliptic complete 3–4 × 1.3–2.2 mm.

In coppice on coral-limestone near sea level and in Hispaniola inland and up to 600 m in xeromorphic woodland, local but gregarious: NW Bahamas, (Grand Bahama, Andros, New Providence, Eleuthera, Cat and the Exumas); Cuba (widespread from W Pinar del Río and Isla de la Juventud to E Oriente); Hispaniola (SE Haiti and SW Dominican Republic); and Puerto Rico (Cambalache Forest); cultivated and perhaps locally naturalized in S Penisular Florida; reported (Urban, Symb. Ant. **2:** 264. 1920; Standley 1922: 390) from Yucatán (*xiaxek*), but no Mexican specimens seen. — Map 64. — Fl. II–V, sporadically later. — *Horseflesh* (Bahamas); *jigüe*, *jiqui*, *sabicú* (Cuba); *caracolí* (Dominican Republic, where also recorded for *Pithecellobium unguis-cati*).

Acknowledgments

We thank Ivan Nielsen and Melissa Luckow for critical review of the manuscript; M. Luckow, furthermore, for continuing discussion on characters and taxonomy of the Mimosaceae. We are indebted to William Buck for meticulous and prolonged editorial guidance and to Bobbi Angell for 19 skillfully executed illustrations. The following individuals deserve special thanks for assistance in the field: Maria Lúcia Kawasaki, Bruce and Sherri Nelson, Basil Stergios, Daniel Sabatier and M.-F. Prévost. For assistance or discussion on various aspects of this work we thank H. C. de Lima, Stéphan Maumont, Kevin Nixon, and Dennis Stevenson. Gwilym Lewis kindly provided the illustration of *Leucochloron limae*, superbly executed by Eleanor Catherine. We appreciate the assistance and encouragement of the staff of the Institute of Systematic Botany of The New York Botanical Garden, and of Patricia Holmgren, Jacquelyn Kallunki, Sat Misir, Robin De Luise, and Gina Lindsay of the NYBG Herbarium for help with procurement and processing of numerous loans involved in this study. Likewise we acknowledge the curators of the following herbaria for loans and/or hospitality: A-GH, BM, BR, CAS, CAY, F, IAN, K, M, MBM, MG, MO, P, PORT, RB, SP, SPF, U, US, and VEN. Financial support for this study was provided by NSF grants BSR–9001375 and BSR–8715778.

Literature Cited

Adams, C. D. 1972. Flowering plants of Jamaica. Univ. West Indies, Mona, Jamaica.

Baretta-Kuipers, T. 1981. Wood anatomy of Leguminosae: its relevance to taxonomy. Pages 677–705. *in*: R. M. Polhill & P. H. Raven (eds.), Advances in legume systematics, pt. 2. Royal Botanic Gardens, Kew.

Barneby, R. C. 1991. Sensitivae censitae. A description of the genus *Mimosa* Linnaeus (Mimosaceae) in the New World. Mem. New York Bot. Gard. **65:** 1–835.

Barneby, R. C. & J. W. Grimes. 1984. Two leguminous forest trees new to the flora of French Guiana. Brittonia **36:** 45–48.

Bentham, G. 1844. Notes on Mimoseae. London J. Bot. **3:** 82–112, 195–226.

———. 1875. Revision of the suborder Mimoseae. Trans. Linn. Soc. London **30:** 335–664, pls. LXVI–LXX.

———. 1876. Leguminosae subordo Mimoseae. Pages 258–527, *in*: C. F. P. Martius, Flora Brasiliensis **10(2):** pls. 67–138.

Bentley, B. L. 1977. Extrafloral nectaries and protection by pugnacious bodyguards. Annual Rev. Ecol. Syst. **8:** 407–427.

Bernardi, L. 1984. Leguminosae Mimosoideae-Ingeae *in* Contribución a la dendrología paraguaya. Boissiera **35:** 168–203.

Bisse, J. 1988. Arboles de Cuba. Editorial Científico-Técnica, La Habana.

Brenan, J. P. M. & R. K. Brummitt. 1965. New and little known species from the Flora Zambesiaca area. XIX. Leguminosae–Mimosoideae. Bol. Soc. Brot., ser. 2, **39:** 189–204.

Briggs, B. G. & L. A. S. Johnson. 1979. Evolution in the Myrtaceae — evidence from inflorescence structure. Proc. Linn. Soc. New South Wales **102:** 157–256.

Britton, N. L. & E. P. Killip. 1936. Mimosaceae and Caesalpiniaceae of Colombia. Ann. New York Acad. Sci. **35:** 101–208.

Britton, N. L. & J. N. Rose. 1928. Mimosaceae. N. Amer. Fl. **23:** 1–194.

Burkart, A. E. 1952. Las Leguminosas argentinas, silvestres y cultivadas, 2da ed., pp. 1–569, figs. 1–130, láms. I–XXIII. ed. Acme Agency, Buenos Aires.

———. 1979. Leguminosas Mimosoideas. Pages 1–299 *in* P. R. Reitz, Flora ilustrada catarinense, fasciculo LEGU(1). Herbario "Barbosa Rodrigues," Itajai, Brazil.

Capuron, R. 1970. Contribution à l'étude de la flore forestière de Madagascar: notes sur les *Albizia* Duraz. Adansonia, **11:** 357–382, pls. 1–4.

Cárdenas de Guevara, L. 1974. Los géneros venezolanos de las Mimosoideae (Leguminosae). Revista. Fac. Agron. Maracay **7(3):** 109–194, láms. 1–24.

Carr, D. J. & S. G. M. Carr. 1959. Developmental morphology of the floral organs of *Eucalyptus*. I. The inflorescence. Austral. J. Bot. **7:** 109–141.

Cassens, D. L. & R. B. Miller. 1981. Wood anatomy of the New World *Pithecellobium* (sensu lato). J. Arnold Arbor. **62:** 1–44.

Chauhan, L. & R. Dayal. 1985. Wood anatomy of Indian *Albizia*. IAWA Bull. **6:** 213–218.

Corner, E. J. H. 1951. The leguminous seed. Phytomorphology **1:** 117–150.

Correll, D. S. & H. B. Correll. 1982. Flora of the Bahama Archipelago. J. Kramer, Vaduz.

Coster, C. 1927–1928. Zur Anatomie und Physiologie der Zuwachszonen und Jahresringbidung in der Tropen. Ann. Jard. Bot. Buitenzorg **37:** 49–160; **38:** 1–114.

Derstine, K. S. & S. C. Tucker. 1991. Organ initiation and development of inflorescences and flowers of *Acacia baileyana*. Amer. J. Bot. **78:** 816–832.

Ducke, A. 1949. Notas sôbre a flora neotrópica. II. As leguminosas da Amazônia brasiliera, 2da ed. Bol. Técn. Inst. Agron. N. **18:** 1–248.

Elias, T. S. 1981. Mimosoideae. Pages 143–152, figs. 1–5, *in:* R. M. Polhill & P. H. Raven (eds.), Advances in legume systematics, pt 2. Royal Botanic Gardens, Kew.

Estrada C., A. E. & J. A. Marroquín de la Fuente. 1992. Leguminosas en el Centro-sur de Nuevo León. Univ. Autonoma de Nuevo León, Facultad de Ciencias Forestales, Linares.

Farris, J. S. 1988. HENNIG86. Distributed by the author.

Fosberg, F. R. 1965. Revision of *Albizia* sect. *Pachysperma* (Leguminosae–Mimosoideae). Reinwardtia **7(1):** 71–90.

Gentry, A. 1988. New species and a new combination for plants from trans-Andean South America. Ann. Missouri Bot. Gard. **75:** 1429–1439.

Grimes, J. W. 1992. Metamerism, heterochrony, and inflorescence morphology of the *Pithecellobium*-complex (Leguminosae: Mimosoideae: Ingeae). Brittonia **44:** 140–159.

———. 1995. Generic relationships of Mimosoideae tribe Ingeae, with emphasis on the New World *Pithecellobium*-complex. Pages 101–121 *in* M. D. Crisp & J. J. Doyle (eds.), Advances in legume systematics, pt. 7. Royal Botanic Gardens, Kew.

Guinet, P. 1981. Mimosoideae: the characters of their pollen grains. Pages 835–855 *in* R. M. Polhill & P. H. Raven (eds.), Advances in legume systematics, pt. 2. Royal Botanic Gardens, Kew.

Hernández, H. M. 1986. *Zapoteca*: a new genus of neotropical Mimosoideae. Ann. Missouri Bot. Gard. **73:** 755–763.

Hickey, L. J. 1973. Classification of the architecture of dicotyledonous leaves. Amer. J. Bot. **60:** 17–33.

Hoc, P. S. 1981. El género *Pithecellobium* en la Argentina. Darwiniana **23:** 523–558, figs. 1–10.

———. 1992a. Notulae ad floram paraquaiensem 34. Sinonimia en el género *Pithecellobium* C. Martius (Leguminosae, Ingeae). Candollea **47:** 89–94.

———. 1992b. Sinonomia en el genero *Cathormion* Benth. Hassk (Leguminosae Ingeae). Candollea **47:** 260–264.

Holltum, R. E. 1953. Evolutionary trends in an equatorial climate. Symp. Soc. Exp. Biol. **7:** 159–173.

Howard, R. A. 1988. Leguminosae. *In* Flora of the Lesser Antilles **4(1):** 334–538. Arnold Arboretum, Jamaica Plain.

Huang, Y. 1983. A taxonomic study of the genus *Albizia* Durazz. from China. Acta Bot. Yunnanica **5:** 127–139.

Hutchinson, J. 1964. The Genera of flowering plants, vol. 1. Clarendon Press, Oxford.

Irwin, H. S. & R. C. Barneby. 1982. The American Cassiinae: a synoptical revision of Leguminosae tribe Cassieae subtribe Cassiinae of the New World. Mem. New York Bot. Gard. **35:** 1–918.

Isely, D. 1970. Legumes of the United States V. Castanea **35:** 244–260.

———. 1973. Leguminosae of the United States: I. Subfamily Mimosoideae. Mem. New York Bot. Gard. **23(1):** 1–152, maps 1–48.

Kleinhoonte, A. 1940. Mimosaceae. *In* A. A. Pulle, Flora of Suriname **2(2):** 258–331. Brill, Leiden.

Koriba, K. 1958. On the periodicity of tree-growth in the tropics. Gard. Bull. Straits Settlem. **17:** 11–81.

Lewis, G. P. 1987. Legumes of Bahia. Royal Botanic Gardens, Kew.

———. **& P. E. Owen.** 1989. Legumes of the Ilha de Maracá. Royal Botanic Gardens, Kew.

Liogier, A. H. 1985. Leguminosae subfam. Mimosoideae. Pages 12–71 *in* La Flora de la Española, vol. 3. Univ. Central del Este, San Pedro de Macorís, R. D.

Macbride, J. F. 1943. Flora of Peru. Family Leguminosae. Field Mus. Nat. Hist., Bot. Ser. **13:** 83–96.

Maumont, S. 1993. Seed-coat anatomy of non-pleurogrammic seeds in the tribe Ingeae (Leguminosae, Mimosoideae). Brittonia **45:** 249–259.

McKey, D. 1989. Interactions between ants and leguminous plants. Advances in Legume Biology. Monogr. Syst. Bot. Missouri Bot. Gard. **26:** 673–718.

McVaugh, R. 1987. Flora Novo-Galiciana. Vol. 5, Leguminosae. Univ. Michigan, Ann Arbor.

Mesquita, A. de L. 1990. Revisão taxonômica do gênero *Enterolobium* Mart. (Mimosoideae) para a região neotropical. M.S. thesis. Univ. Fed. Rural de Pernambuco. 222 pp.

Mohlenbrock, R. H. 1963. Reorganization of genera within tribe Ingeae of the mimosoid Leguminosae. Reinwardtia **6:** 429–447.

Nielsen, I. 1979a. Notes on the genus *Archidendron* F. von Mueller and *Pithecellobium* Martius in mainland S.E. Asia. Adansonia **19:** 3–37, pl. 1, 2.

———. Notes on the genus *Albizia* Durazz. (Leguminosae–Mimosoideae) in mainland S.E. Asia. Adansonia **19:** 197–229, pl. 1–4.

———. 1981. Ingeae. Pages 173–190. *in* R. M. Polhill & P. H. Raven (eds.), Advances in legume systematics, pt. 2. Royal Botanic Gardens, Kew.

———. 1985. The Malesian species of *Acacia* and *Albizia*. Opera Bot. **81:** 5–50, figs. 1–26.

———**, P. Guinet & T. Baretta-Kuipers.** 1983a. Studies in the Malesian, Australian and Pacific Ingeae (Leguminosae–Mimosoideae): The genera *Archidendropsis*, *Wallaceodendron*, *Paraserianthes*, *Pararchidendron* and *Serianthes* (part 1). Bull. Mus. Natl. Hist. Nat. IV, sect. B. Adansonia **5:** 303–329, pls. 1–6.

———**,** ——— **&** ———. 1983b. Studies in the Malesian, Australian and Pacific Ingeae (Leguminosae–Mimosoideae): the genera *Archidendropsis*, *Wallaceodendron*, *Paraserianthes*, *Pararchidendron* and *Serianthes* (part 2). Bull. Mus. Natl. Hist. Nat. IV, sect. B, Adansonia **5:** 335–360, pls. 1–10.

Niezgoda, C. J., S. M. Feuer & L. I. Nevling. 1983. Pollen ultrastructure of the tribe Ingeae (Mimosoideae: Leguminosae). Amer. J. Bot. **70:** 650–667.

Nixon, K. 1992. CLADOS. Distributed by author.

Pennington, T. D. & J. Sarukhán. 1968. Manual para la identificacion de campo de los principales arboles tropicales de México. Inst. Nac. Invest. Forestales, Mexico.

Pittier, H. 1927. Arboles y arbustos del orden de las Leguminosas. I. Mimosaceas. Tipografia Americana, Caracas.

Record, S. J. & C. D. Mell. 1924. Timbers of Tropical America. Yale Univ. Press, New Haven.

Rico A., M. de L. 1991. New species, combinations and synonyms for *Zygia*, *Cojoba*, *Marmaroxylon* and *Pithecellobium* (Leguminosae—Mimosoideae—Ingeae). Kew Bull. **46:** 493–521, figs. 1–6.

———. 1992. Notes on *Albizia niopoides* (Spruce *ex* Bentham) Burkart (Leguminosae: Mimosoideae). Kew Bull. **47:** 699–702, map 1.

Standley, P. C. 1922 (jul). Mimosaceae. *In* Trees and shrubs of Mexico. Contrib. U.S. Natl. Herb. **23(2):** 348–400.

Standley, P. C. & J. A. Steyermark. 1946. Flora of Guatemala, Leguminosae. Fieldiana, Bot. **24:** 1–368.

Swofford, D. 1990, Phylogenetic analysis using parsimony, vers. 3.0. Illinois Natural History Survey, Urbana.

Thomlinson, P. B. & A. M. Gill. 1973. Growth habits of tropical trees: some guiding principles. Pages 129–143 *In* B. J. Meggers, E. S. Ayensu & W. D. Duckworth (eds)., Tropical forest ecosystems in Africa and South America: a comparative review. Smithsonian Institution Press, Washington, DC.

Thompson, R. L. 1980. A revision of *Lysiloma* (Leguminosae). Ph.D. dissertation, Southern Illinois Univ., Carbondale. 107 pp. + 18 figs.

Tomlinson, P. B. 1980. The biology of trees native to tropical Florida. Harvard Univ. Press, Cambridge.

Urban, I. 1900. Leguminosae novae vel minus incognitae. I. Symbolae Antillanae **2:** 257–335.

Wiggins, I. L. 1964. Mimosaceae. *In* F. Shreve & I. R. Wiggins, Vegetation and flora of the Sonoran Desert. Stanford Univ. Press, Stanford.

Witsberger, D. 1982. Arboles del Parque Deininger [El Salvador]. Min. Educación, San Salvador.

Numerical List of Taxa Listed as Exsiccatae

I. HYDROCHOREA Barneby & Grimes
1. *H. gonggrijpii* (Kleinhoonte) Barneby & Grimes
2. *H. corymbosa* (L. C. Richard) Barneby & Grimes
3. *H. marginata* (Bentham) Barneby & Grimes
a. var. *marginata*
b. var. *panurensis* (Bentham) Barneby & Grimes
c. var. *scheryi* Barneby & Grimes
4. *H. acreana* (Macbride) Barneby & Grimes

II. BALIZIA Barneby & Grimes
1. *B. leucocalyx* (Britton & Rose) Barneby & Grimes
2. *B. pedicellaris* (de Candolle) Barneby & Grimes
3. *B. elegans* (Ducke) Barneby & Grimes

III. ABAREMA Pittier
1. *A. floribunda* (Bentham) Barneby & Grimes
2. *A. ferruginea* (Bentham) Pittier
3. *A. gallorum* Barneby & Grimes
4. *A. longipedunculata* (Irwin) Barneby & Grimes
5. *A. idiopoda* (Blake) Barneby & Grimes
6. *A. curvicarpa* (Irwin) Barneby & Grimes
a. var. *curvicarpa*
b. var. *rodriguesii* Barneby & Grimes
7. *A. macradenia* (Pittier) Barneby & Grimes
8. *A. ganymedea* Barneby & Grimes
9. *A. piresii* Barneby & Grimes
10. *A. auriculata* (Bentham) Barneby & Grimes
11. *A. microcalyx* (Bentham) Barneby & Grimes
a. var. *microcalyx*
b. var. *enterolobioides* Barneby & Grimes
c. var. *parauaquarae* (Ducke) Barneby & Grimes
12. *A. commutata* Barneby & Grimes
13. *A. jupunba* (Willdenow) Britton & Killip
a. var. *jupunba*
b. var. *trapezifolia* (Vahl) Barneby & Grimes
14. *A. barbouriana* (Standley) Barneby & Grimes
a. var. *barbouriana*
b. var. *arenaria* (Ducke) Barneby & Grimes
15. *A. adenophora* (Ducke) Barneby & Grimes
16. *A. zollerana* (Standley & Steyermark) Barneby & Grimes
17. *A. leucophylla* (Bentham) Barneby & Grimes
a. var. *leucophylla*
b. var. *vaupesensis* Barneby & Grimes
18. *A. mataybifolia* (Sandwith) Barneby & Grimes
19. *A. laeta* (Bentham) Barneby & Grimes
20. *A. campestris* (Bentham) Barneby & Grimes
21. *A. cochleata* (Willdenow) Barneby & Grimes
a. var. *cochleata*
b. var. *moniliformis* (Ducke) Barneby & Grimes
22. *A. callejasii* Barneby & Grimes
23. *A. lehmannii* (Britton & Killip) Barneby & Grimes
24. *A. killipii* (Britton & Killip) Barneby & Grimes
25. *A. racemiflora* (Donnell Smith) Barneby & Grimes
26. *A. obovata* (Bentham) Barneby & Grimes
27. *A. brachystachya* (de Candolle) Barneby & Grimes
28. *A. filamentosa* (Bentham) Pittier
29. *A. turbinata* (Bentham) Barneby & Grimes
30. *A. cochliacarpos* (Gomes) Barneby & Grimes
31. *A. langsdorfii* (Bentham) Barneby & Grimes
32. *A. asplenifolia* (Grisebach) Barneby & Grimes
33. *A. glauca* (Urban) Barneby & Grimes
34. *A. obovalis* (A. Richard) Barneby & Grimes
35. *A. alexandri* (Urban) Barneby & Grimes
a. var. *alexandri*
b. var. *troyana* (Urban) Barneby & Grimes
36. *A. oppositifolia* (Urban) Barneby & Grimes
37. *A. nipensis* (Britton) Barneby & Grimes
38. *A. abbottii* (Rose & Leonard) Barneby & Grimes
39. *A. levelii* (Cowan) Barneby & Grimes
40. *A. oxyphyllidia* Barneby & Grimes
41. *A. villifera* (Ducke) Barneby & Grimes
42. *A. josephi* Barneby & Grimes
43. *A. centiflora* Barneby & Grimes
44. *A. ricoae* Barneby & Grimes, nom. provis. — Nicaragua
45. *A. agropecuaria* Barneby & Grimes, nom. provis. — Brazil

IV. HESPERALBIZIA Barneby & Grimes
1. *H. occidentalis* (T. S. Brandegee) Barneby & Grimes

V. PSEUDOSAMANEA Harms
1. *Ps. guachapele* (Kunth) Harms
2. *Ps. cubana* (Britton & Rose) Barneby & Grimes

VI. SAMANEA Merrill
1. *S. saman* (Jacquin) Merrill
2. *S. tubulosa* (Bentham) Barneby & Grimes
3. *S. inopinata* (Harms) Barneby & Grimes

VII. BLANCHETIODENDRON Barneby & Grimes
1. *B. blanchetii* (Bentham) Barneby & Grimes

VIII. LEUCOCHLORON Barneby & Grimes
1. *L. limae* Barneby & Grimes
2. *L. incuriale* (Vellozo) Barneby & Grimes
3. *L. minarum* (Harms) Barneby & Grimes
4. *L. foederale* (Barneby & Grimes) Barneby & Grimes

IX. CHLOROLEUCON (Bentham) Britton & Rose
1. *Ch. guantanamense* (Britton) Britton & Rose
2. *Ch. chacoënse* (Burkart) Barneby & Grimes
3. *Ch. eurycyclum* Barneby & Grimes
4. *Ch. acacioides* (Ducke) Barneby & Grimes
5. *Ch. extortum* Barneby & Grimes
6. *Ch. dumosum* (Bentham) G. P. Lewis
7. *Ch. tenuiflorum* (Bentham) Barneby & Grimes
8. *Ch. tortum* (Martius) Pittier ex Barneby & Grimes
9. *Ch. foliolosum* (Bentham) G. P. Lewis
10. *Ch. mangense* (Jacquin) Britton & Rose
a. var. *mangense*
b. var. *mathewsii* (Bentham) Barneby & Grimes
c. var. *vincentis* (Bentham) Barneby & Grimes
d. var. *leucospermum* (T. Brandegee) Barneby & Grimes
e. var. *lentiscifolium* (A. Richard) Barneby & Grimes
f. var. *tetrazyx* Barneby & Grimes

X. SPHINGA Barneby & Grimes
1. *S. prehensilis* (C. Wright) Barneby & Grimes
2. *S. acatlensis* (Bentham) Barneby & Grimes
3. *S. platyloba* (de Candolle) Barneby & Grimes

XI. HAVARDIA Small
1. *H. sonorae* (S. Watson) Britton & Rose
2. *H. campylacanthus* (L. Rico & M. Sousa) Barneby & Grimes
3. *H. pallens* (Bentham) Britton & Rose

4. *H. albicans* (Kunth) Britton & Rose
5. *H. mexicana* (Rose) Britton & Rose

XII. EBENOPSIS Britton & Rose
1. *E. caesalpinioides* (Standley) Britton & Rose
2. *E. ebano* (Berlandier) Barneby & Grimes
3. *E. confinis* (Standley) Barneby & Grimes

XIII. PAINTERIA Britton & Rose
1. *P. leptophylla* (de Candolle) Britton & Rose
2. *P. elachistophylla* (S. Watson) Britton & Rose
3. *P. revoluta* (Rose) Britton & Rose

XIV. MACROSAMANEA Britton & Killip
1. *M. macrocalyx* (Ducke) Barneby & Grimes
2. *M. spruceana* (Bentham) Killip
3. *M. consanguinea* (Cowan) Barneby & Grimes
4. *M. froesii* Barneby & Grimes
5. *M. discolor* (Willdenow) Britton & Killip *ex* Britton & Killip
   a. var. *discolor*
   b. var. *arenicola* (Cowan) Barneby & Grimes
6. *M. duckei* (Huber) Barneby & Grimes
7. *M. pubiramea* (Steudel) Barneby & Grimes
   a. var. *pubiramea*
   b. var. *lindsaeifolia* (Bentham) Barneby & Grimes
8. *M. amplissima* (Ducke) Barneby & Grimes
9. *M. simabifolia* (Bentham) Pittier
10. *M. prancei* (Barneby) Barneby & Grimes
11. *M. kegelii* (Meisner) Kleinhoonte

XV. ALBIZIA Durazzini
1. *A. polycephala* (Bentham) Killip *ex* Record
2. *A. edwallii* (Hoehne) Barneby & Grimes
3. *A. burkartiana* Barneby & Grimes
4. *A. coripatensis* (Rusby) Schery
5. *A. tomentosa* (M. Micheli) Standley
6. *A. sinaloënsis* Britton & Rose
7. *A. leonardii* Britton & Rose *ex* Barneby & Grimes
8. *A. barinensis* Cárdenas
9. *A. adinocephala* (Donnell Smith) Britton & Rose
10. *A. niopoides* (Bentham) Burkart
   a. var. *niopoides*
   b. var. *colombiana* (Britton & Killip) Barneby & Grimes
11. *A. buntingii* Barneby & Grimes
12. *A. carbonaria* Britton
13. *A. berteriana* (de Candolle) Fawcett & Rendle
14. *A. glabripetala* (Irwin) G. P. Lewis & P. E. Owen
15. *A. pistaciifolia* (Willdenow) Barneby & Grimes
16. *A. subdimidiata* (Splitgerber) Barneby & Grimes
   a. var. *subdimidiata*
   b. var. *minor* Barneby & Grimes
17. *A. multiflora* (Kunth) Barneby & Grimes
   a. var. *multiflora*
   b. var. *sagasteguii* Barneby & Grimes
18. *A. decandra* (Ducke) Barneby & Grimes
19. *A. inundata* (Martius) Barneby & Grimes
20. *A. chinensis* (Osbeck) Merrill
21. *A. lebbeck* (Linnaeus) Bentham
22. *A. julibrissin* Durazzini
23. *A. procera* (Roxburgh) Bentham

XVI. ENTEROLOBIUM Martius
1. *E. cyclocarpum* (Jacquin) Grisebach
2. *E. timbouva* Martius
3. *E. contortisiliquum* (Vellozo) Morong
4. *E. barnebianum* Mesquita & da Silva
5. *E. barinense* Cárdenas & Rodríguez
6. *E. maximum* Ducke
7. *E. monjollo* (Vellozo) Martius
8. *E. gummiferum* (Martius) Macbride
9. *E. oldemanii* Barneby & Grimes
10. *E. schomburgkii* Bentham
11. *E. glaziovii* (Bentham) Mesquita

XVII. CEDRELINGA Ducke
1. *C. cateniformis* (Ducke) Ducke

XVIII. FALCATARIA (Nielsen) Barneby & Grimes
1. *F. moluccana* (Miquel) Barneby & Grimes

XIX. PARASERIANTHES Nielsen
1. *P. lophantha* (Willdenow) Nielsen

XX. LYSILOMA Bentham
- *L. acapulcense* (Kunth) Bentham
- *L. auritum* (Jacquin) Macbride
- *L. watsonii* Rose
- *L. divaricatum* (Jacquin) Macbride
- *L. candidum* T. S. Brandegee
- *L. tergeminum* Bentham
1. *L. latisiliquum* (Linnaeus) Bentham
2. *L. sabicu* Bentham

List of Numbered Exsiccatae

The numbers in parentheses refer to those in the Numerical List of Taxa. For genus *Enterolobium*, only specimens of sect. *Robrichia* are cited; for *Lysiloma*, only those from the West Indies.

Aa, H. A. van der, 1089 (III-19).
Abbott, W. L., 2258 (III-38).
Abrams, L. R., 13264 (XI-5).
Acad. Rhenotraiect., *see* Bureau v. h. Boschwezen
Acevedo R., P., 2148 (XV-23); 2631 (VI-1); 3293 (I-2); 5029 (III-18).
Acuña, J., 12440 (III-32); 12441 (III-37); 12442 (III-34); 18679 (X-1); 19662 (III-37); 19663 (III-32).
Adams, C. D., 10931 (III-35b).
Agra, M. F., 681 (VI-3).
Aguilar, J. M., 48 (XV-5); 62 (IX-10d); 91 (XV-5); 335 (IX-7).
Ahshapanek, D., 299 (XV-21).
Alain, Bro., 4637 (III-32); 5800 (III-32); 6167 (IX-10e).
Albert de Escobar, L., 5054 (III-14a); 5150 (III-23); 7332 (III-23); 7333 (III-23); 7465 (III-23); 7532 (III-23); 7708 (III-14a); 7752 (III-23).
Albuquerque, B., 527 (XV-16a); 610 (XIV-7b); 688 (I-3a); 885 (III-19); 1039 (I-3a).
Alcorn, J. B., 2982 (XV-5).
Alencar, L., 109 (I-3b); 297 (XIV-5).
Alexandre, D. Y., 440 (IX-4).
Alexiades, M. N., 1116 (XVI-10).
Alfredo Pérez J., L., 1160 (IX-10d); 1204 (IX-10d).
Allart, 487 (XIX-1).
Allen, P. H., 3017 (III-17a); 3379 (III-1); 4433 (VI-1); 4437 (VI-1); 6042 (III-7); 6591 (XV-10a); 7007 (XV-12); 7212 (XV-10a); 7252 (VI-1); 7612 (XV-10a); 16042 (III-7).

Boom, B. M., 1635 (III-18); 1663 (III-14a); 2300 (XVII-1); 2314 (III-13a); 4371 (XVI-10); 4942 (XVI-10); 5459 (III-13a); 7323 (III-13b); 7859 (III-13a); 8715 (III-19); 14943 (III-18).

Bonpland, A., 415 (XV-19).

Bourgeau, E., 579 (XIII-1).

Brace, L. J. K., 1485 (XX-1); 1689 (XX-1); 1987 (XX-1); 2017 (III-33); 3936 (XX-1); 4307 (XX-1); 5083 (XX-2).

Brade, A. C., 15932 (VIII-2); 28010 (IX-6).

Braga, P. I. S., 2338 (III-31).

Brandegee, T. S., 37 (XIII-2); 199 (XII-3); 200 (IV-1).

Brazão, J. E. M., 56 (III-30); 215 (III-30).

Breedlove, D. E., 9615 (X-3); 9908 (III-16); 23573 (XV-21); 23700 (XV-21); 24433 (IX-10a); 30235 (V-1); 34350 (X-3); 34735 (XV-10a); 35280 (III-5); 39863 (XV-5); 42136 (IX-10d); 42141 (X-3); 42219 (IX-10d); 42357 (IV-1); 43154 (XII-3); 43158 (XII-3); 43199 (XII-3); 43217 (IX-10d); 43225 (IX-10d); 44027 (XIII-1); 46408 (X-3); 46477 (XI-3); 46990 (X-3); 46999 (X-3); 47131 (IX-10d); 47136 (IV-1); 47490 (XV-10a); 48492 (IV-1); 48538 (XV-10a); 50340 (IX-10d); 50356 (X-3); 50392 (X-3); 53730 (IX-10d).

Brenes, A. M., 6676 (XV-9); 17199 (XV-9).

Breteler, F. J., 4614 (XV-12); 4708 (XVI-10); 4795 (III-11a); 4968 (III-13b); 4989 (II-2).

Bristan, N., 1062 (XV-12).

Britton, N. L., 151 (XX-1); 162 (XX-1); 258 (XX-1); 355 (XX-2); 499 (VI-1); 649 (III 35b); 779 (III 13b); 1199 (III-13b); 1300 (XV-13); 1658 (XV-13); 1775 (III-13b); 1870 (XX-2); 1931 (XV-13); 2050 (V-2 - TYPUS); 2051 (IX-1); 2066 (XX-2); 2202 (XV-13); 2578 (XX-1); 2671 (XV-10a); 2734 (XV-10a); 2851 (XX-1); 2887 (XX-2); 2943 (XX-2); 3021 (XX-1); 4583 (XX-1); 4625 (IX-10e); 5079 (VI-1); 5346 (III-36); 5432 (XX-2); 5496 (X-1); 5572 (X-1); 5592 (XX-1); 5784 (VI-1); 5911 (XX-1); 5954 (XX-1); 6108 (XV-13); 7929 (XV-12); 7936 (XV-12); 8154 (XV-20); 8246 (XV-23); 9094 (XV-23); 9256 (XV-20); 9838 (IX-10e); 9858 (IX-10e); 9866 (III-33); 9926 (XX-2); 9929 (IX-10e); 10044 (XV-22); 10293 (IX-10e); 12579 (III-34); 12629 (XX-2); 12874 III-34); 12923 (V-2); 13036 (XV-13); 14497 (XX-1); 14503 (IX-10e); 14564 (III-34); 15094 (XX-2); 15602 (III-34).

Broadway, W. E., 539 (XV-10a); 542 (III-13a); 743 (II-2); 3088 (III-13b); 3165 (XV-10a); 3529 (III-13b); 3596 (XV-10a); 3800 (XV-10a); 4064 (III-13b); 4067 (III-13b); 4795 (III-13b); 5777 (III-13b); 6495 (XV-10a); 6759 (XV-10a); 7687 (XVIII-1); 10547 (XVIII-1).

Brooke, W. M. A., 5773 (VI-2).

Brown, S., 1830 (VI-1).

Bruijn, J. de, 1107 (XV-16a); 1333 (XV-21); 1685 (II-2); 1692 (II-2); 1700 (III-13b); 1722 (II-2).

Brumbach, W. C., 1381 (XII-2); 9387 (XX-1); 9413 (III-33); 9417 (XX-1).

Buchtíen, O., 88 (III-13a); 761 (III-13a); 1770 (II-2); 1772 (III-13a); 1774 (III-13a); 4202 (XV-4).

Bullock, S. H., 139 (X-3); 1138 (IV-1); 1876 (IV-1).

Bunch, P., 733 (VI-1).

Bunting, G. S., 3668 (XIV-9); 3926 (III-14b); 4048 (XIV-9); 5755 (V-1); 5760 (IX-10a); 6095 (V-1); 6243 (IX-10a); 6577 (XV-10b); 6938 (XV-16a); 6938b (XV-16a); 7718 (III-13a); 8779 (IX-10a); 8848 (IX-10a); 8849 (IX-10a); 9125 (XV-10b); 9143 (IX-10a); 9154 (V-1); 9808 (III-13a); 10711 (VI-1); 13201 (IX-10a); 13370 (XV-11).

Burch, D., 4332 (XX-1).

Burchell, W. J., 896 (XV-1); 898 (III-31); 944 (III-30); 1008A (I-2); 3414 (III-26); 4605 (III-31); 9427 (III-21); 9505-2 (III-21).

Bureau v. h. Boschwezen (including herb Acad. Rheno-traiect.), 20 (III-13b); 39 (II-2); 83 (III-13b); 88 (I-1); 107 (XIII-13b); 125 (II-2); 195 (III-13b); 237 (II-2); 239 (III-13b); 305 (XIV-7a); 312 (III-13b); 360 (I-2); 400 (III-13b); 417 (III-13b); 1105 (III-13b); 1188 (III-13b); 1222 (II-2); 1262 (II-2); 1333 (III-13b); 1462 (III-13b); 1529 (I-1); 1653 (III-13b); 1692 (III-13b); 1791 (II-2); 1951 (III-13b); 2718 (III-13b); 2815 (III-13b); 3127 (II-2); 3139 (II-2); 3195 (II-2); 3869 (III-13b); 3932 (II-2); 3996 (II-2); 4350 (I-1); 4357 (I-1); 4387 (II-2); 4391 (II-2); 4646 (II-2); 4800 (III-13b); 4868 (III-13b); 4895 (III-13b); 4991 (III-13b); 5099 (III-13b); 5520 (III-13b); 5561 (III-13b); 5972 (I-2); 5973 (I-1); 6056 (III-13b); 6196 (I-1); 6329 (III-13b); 6392 (III-13b); 6517 (III-13b); 6690 (III-13b); 6904 (II-2); 6952 (II-2); 8433 (XVI-10).

Burger, W., 4049 (V-1); 6698 (VI-1).

Burkart, A., 26977 (XV-19); 30482 (IX-7).

Buschbacher, R. J., 33 (III-13a).

Byrne, R., 253 (XX-2); 284 (XX-1).

Byron [probably Albuquerque, B.], 195 (I-3a); 554 (I-3a); 1039 (I-3a).

Byron & Elias, 67/47 (III-9).

Cabrera, E., 609 (XV-10a); 834 (XV-10a); 1215 (X-3); 1417 (XI-4); 1509 (X-3); 2442 (XV-21); 3381 (XX-1); 5584 (XV-21); 10723 (IX-10d); 11149 (IX-10e); 11548 (XI-4); 11599 (XI-4); 15995 (X-3).

Cabrera, I., 1977 (XIV-7b).

Calderón, C. E., 2888 (XIV-7b); 2918 (XIV-7b); 2929 (I-2).

Calderón, G. M., 1394 (XV-21).

Calderón, S., 307 (XV-9); 337 (VI-1); 338 (V-1); 550 (XV-21); 909 (XV-9); 1547 (V-1); 2042 (XV-12).

Caldwell, D. H., 8769 (XX-1); 8788 (XV-21).

Caletti, P. V., 135 (XVI-10).

Callejas, R. C., 1693 (III-28); 1914 (VI-2); 4034 (III-13a); 6103 (III22); 8267 (XV-12); 8590 (III-1); 8988 (III-22); 9813 (III-13a).

Calzada, J. I., 4513 (XV-21); 6487 (XI-4); 6632 (XI-4); 6795 (XV-21); 7144 (XI-4).

Camargo, J. C., 20 (III-27); 33 (III-27).

Campbell, D. G., et al., 9165 (III-13a); P21810 (III-19); P21830 (I-3a); P21922 (III-41); P22111 (III-20); P22438 (I-2).

Campbell, E., 6183 (IX-10a).

Campos Novaes, J. de, 246 (XV-2).

Campos Porto, P., 981 (III-30); 1663 (III-31); 2496 (XV-19).

Capegras, M. T., 7 (XIV-11).

Capucho, 593 (III-13a).

Capus, F. 148 (III-13a).

Carauta, J. P. P., 2368 (III-31); 2568 (III-31); 4642 (VI-1).

Cárdenas de Guevara, L., 304 (XV-10b); 1057 (XV-10a); 1342 (XVIII-1); 1357 (XV-8); 1387 (II-2); 2128 (XV-8); 2185 (XV-8); 2199 (XV-8); 2219 (XV-15); 2241 (XV-15); 2273 (XV-8); 2528 (XV-10a); 2587 (XV-10a); 2598 (XV-10a); 2633 (XV-10a); 2636 (XV-10a); 2673 (XV-10a); 2687 (XV-10a); 2689 (V-1); 2691 (XV-10a).

Cardona, F., 1140 (I-2); 1257 (XIV-7a); 1660 (III-13b); 2682 (III-4).

Carlos-Solano, S. & G. Morillo, 5078 (III-14b).

(VI-1); 210 (IX-10a); 221 (XV-19); 233 (XV-21); 300 (XV-15); 308 (XV-19); 323 (II-2); 331 (IX-8); 342 (III-30); 400 (XV-19); 452 (IX-10c); 483 (VI-1); 542 (X-3); 562 (IX-10c); 792 (VI-1); 850 (XV-10a); 914 (V-1); 1125 (XV-12); 1831 (I-3c).
Curtiss, A. H., 153 (XX-1); 278 (XX-2); 330 (III-34); 450 (VI-1); 475 (IX-10e); 716a (XX-1).

Daly, D. C., 766 (III-18); 914 (III-21); 977 (III-21); 1134 (III-21); 1312 (III-9); 1638 (III-13a); 1936 (III-21); 5935 (III-22); 5959 (III-14a); 6724 (XVI-10); 7797 (I-4).
Damasio, 8805 (III-26); 9448 (III-31); 10272 (III-31).
Damião, C., 624 (II-2); 2512 (I-3a); 2718 (XV-16a); 3055 (I-2).
Daniel, Bro., 37 (V-1); 3279 (III-23); 4040 (III-23).
D'Arcy, W. G., 6081 (XV-23).
Darwin, S., 2097 (XI-4).
Davidse, G., 4267 (X-3); 4411 (I-2); 4505 (III-13b); 4893A (I-1); 4982 (XV-21); 5332 (III-13b); 5434 (XIV-7b); 12709 (V-1); 12965 (XIV-7b); 13367 (XV-16a); 13835 (I-2); 13884 (XIV-7b); 14292 (XIV-7b); 14328 (XIV-7b); 14379 (I-2); 14457 (V-1); 14682 (XIV-7b); 14866 (VI-1); 16050 (XV-16a); 17003 (XIV-5b); 17067 (XIV-9); 17280 (XIV-9); 17608 (III-21); 17893 (III-21); 18586 (III-13a).
Davidson, C., 3561 (XIV-2); 9694 (I-2); 9810 (XIV-2).
Davis, D. H., 175 (XIV-7a); 189 (I-1); 810 (XV-16b).
Dawo, 167 (XV-15).
Dayton, W. A., 3097 (XV-9).
Delascio, C., F., 9481 (III-17); 13242 (XV-21).
Delgado, E., 97 (XV-10a).
Delgado S., A., 137 (X-2); 574 (XIII-2).
del Llano, M., 38 (I-2).
den Outer, R. W., 972 (III-13b).
Devoto, F., 1661 (IX-2); 2368 (IX-9).
Dias, A. C., 23 (III-31).
Díaz, C., 439 (XIV-2); 440 (XIV-2); 1156 (III-19).
Dick, C., 117 (III-18).
Diedrichs, E., 28 (III-13b).
Dimpl-von Lossau, E., 246 (XV-10a).
Dionizia, F., 153 (I-2).
Dixon, K., 400 (XII-2).
Dodson, C. H., 1716 (XV-17); 5150 (III-7); 6054 (III-7); 7480 (III-7); 8831 (III-7); 10062 (III-7); 12652 (XV-17); 13664 (XV-15); 13868 (XV-15).
Doesborg Jr., P. H. van, 648 (XIV-7a).
Donselaar, J. van., 215 (III-13b); 1120 (XIV-11); 1653 (III-13b); 1762 (III-13b).
Dorado, O., 1619 (IX-10d).
Dorantes, B., 2732 (XV-5); 3181 (XV-5).
Dorr, L. J., 7201 (III-13b).
Dreyer, D., 297 (X-3).
Duarte, A. P., 120 (III-30); 1078 (III-31); 3016 (VIII-2); 3191 (III-31); 4776 (III-30); 5620 (III-27); 5776 (IX-8); 6643 (III-28); 8566 (VII-1); 9202 (III-30).
Duarte, L. S. R., 10 (III-27); 77 (VIII-4).
Dubs, B., 506 (XV-10a); 1172a (VI-2).
Ducke, A., 35 (III-17a); 169 (III-13a); 202 (XIV-7b); 477 (I-3a); 493 (III-14b); 502 (III-21); 610 (III-21); 612 (III-13a); 631 (VI-2); 644 (IX-4); 653 (XIV-5a); 718 (I-2); 745 (XIV-6); 766 (XIV-7a); 805 (XVII-1); 821 (II-2); 847 (XVII-1); 983 (III-13a); 1008 (III-11b); 1097 (III-11b); 1378 (XV-1); 1412 (III-11b); 1428 (III-21); 1669 (IX-4); 1703 (III-15); 1952 (XVI-10); 1973 (III-1); 2050 (III-19); 2059 (III-10); 7528 (III-13a); 8175 (III-21); 8333 (XIV-6); 8484 (III-10); 8593 (III-21); 10159 (III-11c); 10163 (III-13a); 10770 (VI-2); 10771 (IX-4); 10827 (III-20); 10908 (XIV-6); 10946 (III-10); 10973 (III-31); 11599 (III-10); 11963 (XIV-6); 11971 (XIV-6); 12080 (III-9); 12675 (III-13a); 15248 (VI-3); 15627 (III-21); 15670 (XV-10a); 15713 (IX-4); 15938 (III-10); 15939 (III-10); 16034 (III-21); 16105 (IX-4); 16231 (IX-4); 16290 (III-13a); 16358 (IX-4); 16520 (III-21); 16584 (XV-10a); 16595 (III-10); 16622 (XIV-1); 16757 (III-20); 16758 (III-21); 16759 (III-11c); 16774 (XIV-6); 16775 (XIV-6); 16839 (IX-4); 23231 (III-17a); MG 4832 (I-2); MG 7191 (I-3b); MG 7343 (I-3a); MG 8475 (I-3b); MG 8667 (I-2); MG 10321 (III-20); MG 10881 (I-2); MG 11026 (I-2); MG 11800 (I-2); MG 15767 (I-2); MG 16224 (I-2); MG 16271 (II-3); MG 16470 (II-3); MG 16873 (III-20); MG 17000 (I-3a); MG 17276 (III-13a); MG 21325 (I-3a); RB 17496 (III-20); RB 20160 (III-10); RB 20161 (III-11c); RB 20165 (I-2); RB 23232 (III-20); RB 23233 (III-14b); RB 23236 (I-1); RB 23238 (III-15); RB 23288 (III-15); RB 24326 (III-1); RB 24328 (III-13a); RB 24330 (III-13a); RB 35527 (XIV-8); RB 35534 (III-41).
Duckett, F., 220 (XX-1).
Dugand, A., 309 (XV-15); 1143 (XV-10b); 2794 (V-1); 4185 (VI-1); 4620 (IX-10a); 6569 (IX-10a); 6638 (XV-10a).
Duivenvoorden, J., 1377 (III-14a).
Duke, J. A., 5113 (XV-9); 11601 (III-14a).
Dulmen, A. van, 40 (I-16a).
Dunn, D., 20169 (XII-2).
Duque-Jaramillo, J. M., 2313 (XV-16a); 2639 (XIX-1); 2943 (XIX-1); 4055A (VI-1);
Durán, R., 1022 (XI-4).
Dusén, P., 490 (IX-6); 8762 (III-31); 9943 (VIII-2); 9981 (VIII-2); 10383 (VIII-2); 11387 (III-27); 13572 (III-31); 13844 (III-27); 14447 (III-27); 14641 (III-27); 16831 (VIII-2); 17017 (III-27); 18023 (VIII-2).
Duss, A., 837 (IX-10c); 1163 (VI-1); 2629 (III-13a); 3036 (VI-1).
Dutra, J. 784 (XV-2).
Dwyer, J. D., 5001 (VI-1); 12673 (XX-1).

Edmond, Bro., 15 (III-34); 39 (III-34).
Edwall, G., 5608 (XV-2).
Edwards, M. T., 399 (XI-3); 542 (XI-3).
Eeburg, J., et al., 12631 (III-13b).
Eggers, H., 771 (III-13a); 1014 (III-13b); 1083 (III-13b); 1368 (III-13b); 1832 (XV-13); 5589 (III-13b); 5920 (XV-10a); 6786 (III-13a); 15791 (XV-17).
Egler, W. A., 769 (III-21); 882 (XIV-6); 1357 (III-20); 47635 (I-2).
Eiten, G., 3330 (III-27); 6143 (III-27).
Ekman, E. L., 786 (III-33); 787 (III-33); 1484 (XX-2); H 1845 (III-36); 2487 (III-37); 2698 (III-33); 4222 (III-33); 4919 (XX-2); 5119 (XX-1); 5265 (III-36); 5964 (IX-10d); 5998 (III-36); 6383 (III-32); 7237 (III-13a); 8247 (V-2); 8415 (V-2); 9373 (III-34); 9473 (V-2); 9551 (XX-2); 9816 (III-34); 9918 (III-32); 10020 (III-36); 10586 (III-34); 11161 (IX-10e); 11203 (XX-1); 11751 (III-34); 12280 (III-38); 12629 (V-2); 12929 (V-2); 13045 (XV-13); 14423 (III-33); 14483 (III-33); 14690 (III-34); 15518 (III-38); 15588 (X-1); 16035 (VI-1); 16145 (III-33); 17559 (IX-10e); 18401 (IX-10e); 18848 (III-33); 18859 (III-33); 19044 (III-33).

Gillespie, L. J., 797 (XIV-7a); 820 (III-13b); 1121 (XIV-7a); 1371 (I-1); 1374 (I-2); 2343 (XV-16a); 2444 (XV-16a); 2833 (I-1).
Ginzberger, A., 626 (XIV-2).
Giraldo-Cañas, D. A., 1012 (III-13a).
Glaziou, A., 627 (III-31); 1565 (IX-6); 2973 (III-31); 2974 (III-31); 5837 (VIII-2); 6188 (III-27); 6192 (III-31); 7589 (III-30); 7877 (XV-23); 8437 (XV-1); 8447 (VIII-2); 8630 (IX-6); 8654 (XIII-31); 8694 (III-31); 10595 (IX-6); 10601 (VI-2); 10602 (III-31); 10608 (IX-6); 10633 (VI-2); 10634 (III-31); 10680 (XV-19); 11925 (VI-2); 11927 (III-31); 11929 (II-2); 11930 (IX-6); 11931 (IX-6); 11933 (XV-1); 11939 (XVI-11); 12634 (IX-8); 12648 (XV-1); 12650 (IX-6); 13431 (III-26); 13772 (XV-20); 13788 (VII-1); 13789 (VIII-3); 13790 (VIII-3); 14646 (IX-6); 17607 (XVIII-1); 17609 (IX-6); 18202 (IX-8); 19109 (III-31); 19110 (III-27); 20292 (III-31).
Gleason, H. A., 819 (III-13b); 873 (XV-16a).
Glocker, 178b (XV-1).
Godli, A., MG 3931 (I-2).
Goeldi, A., 1094 (IX-4); 9659 (III-13a).
Goes, M. de, 16 (III-27); 32 (III-27).
Goés, O. C., 616 (III-31).
Goldman, E. A., 599 (XI-4).
Gomes, M., 204 (I-2).
Gomes da Silva, S. J., 100 (II-2).
Gómez J., J., 17 (I-2).
Gómez-Pompa, A., 1340 (XI-4).
Gonggrijp, J. W., 51 (XIV-7a); 79 (XIV-7a); 312 (III-13b); 2263 (XIV-7a); 4150 (XIV-11); 4211 (I-1).
González, A., 1406 (III-13b); 11951 (XV-10a); 11951bis (XV-10a).
González M., F., 10210 (XIII-2); 12312 (X-2); 17561 (XIII-2).
González O., J., 6566 (XI-1); 6699 (XI-1); 6799 (IX-10d).
González Q., L., 2370 (XI-3); 3528 (XIII-2).
Görts-van Rijn, A. R. A., 158 (II-13b).
Gottsberger, G. R., 213-23537 (IX-4).
Gould, F. W., 7691a (XV-22).
Goulding, M., 121 (XV-16a); 1102 (XV-16a); 1129 (I-2); 1130 (I-2); 1477 (XIV-6).
Grande, D. A. de, 27 (III-27).
Granville, J. J. de, 609 (I-2); 5287 (III-19); 5383 (III-13a); 9543 (III-13b); 10930 (III-19); B-4933-A (I-2); B-5108 (I-2).
Greenman, J. M., 339 (XI-4); 7021 (XI-3); 7022 (XII-2).
Grenand, 290 (III-13a); 1086 (III-13a); 1120 (I-2); 1319 (II-2); 1773 (I-2); 2462 (XIV-6); 2482 (XIV-6).
Grether, R., 1128 (X-3).
Grimes, J. W., 2786 (XIII-1); 2824 (III-5); 3113 (II-2); 3118 (III-15); 3119 (III-15); 3121 (III-15); 3123 (III-13a); 3128 (XIV-7a); 3134 (III-13a); 3135 (II-2); 3267 (XVI-9); 3287 (III-13a); 3310 (XVI-9); 3312 (III-6a); 3321 (III-18); 3332 (XVI-10); 3338 (III-3).
Guánchez, F., 1157 (III-17a); 3381 (I-2).
Gudiño, E., 627 (III-19); 1065 (III-13a).
Guedes, M., MG 2584 (III-21).
Guedes, T. N., 237 (III-13a); 326 (III-13a); 475 (III-13a); 476 (XV-1).
Guerrero C., B., 1004 (IV-1).
Guillemin, M., 45 (III-31); Cat.49 (III-31).
Gutiérrez, G. & R. E. Schultes, 764 (XIV-7b).
Gutiérrez, J., 70520 (II-34).
Gutiérrez R., A., 85 (II-2); 87 (XVII-1); 98 (XVII-1); 111 (II-2).
Hage, J. L., 101 (VI-3); 2180 (VI-1); 1851 (XV-21).
Hahn, L., 233 (XV-21); 1289 (IX-10c); 1430 (IX-10c).
Hahn, W., 3951 (III-13b); 4591 (I-2); 4873 (VI-1); 5104 (IX-10c); 5208 (XIV-7a); 5396 (III-14b); 5550 (I-2).
Hall, F., 17 (IX-10d).
Hammel, B., 12176 (XV-9); 17319 (II-3).
Handro, O., IAN-43810 (III-27); 2176 (III-31); 2301 (III-27); 2313 (VIII-2).
Hanson, H. C., 329 (XII-2); 455 (XI-3).
Hardmann, S., 1803 (III-28).
Harley, R. M., 10357 (XVI-10); 16440 (IX-9); 17408 (III-29); 17623 (III-28); 18493 (III-13a); 21421 (IX-9); 21458 (IX-9); 22007 (IX-9); 22142 (III-28); 22146 (III-29); 22210 (III-28); 22289 (II-2); 22516 (III-30); 22632 (II-2).
Harriman, N. A., 17046 (XX-1).
Harris, W., 3183 (III-35b); 8800 (III-35b); 8960 (III-35b); 8975 (III-35a); 9013 (IX-10a); 9077 (III-35b); 9182 (III-35a); 9231 (IX-10a); 9263 (III-35b); 9276 (VI-1); 9654 (XV-13); 9921 (III-35b); 9946 (XV-13); 9948 (III-35b); 10190 (XV-13); 10243 (III-35a); 10329 (III-35b).
Harrison, S. G., 610 (III-13b); 1085 (III-13b).
Hart, 6362 (XV-10a).
Hartshorn, G., 851 (XV-10a); 926 (I-2); 1517 (XV-9).
Hassler, E., 853 (IX-7); 1106 (IX-7); 2399 (IX-7); 3400 (IX-7); 3719 (XV-10a); 5406 (VI-2); 7217 (XV-19); 7468 (IX-7); 7527 (IX-7); 7616 (VI-2); 8006 (XV-10a); 11563 (IX-7); 12364 (XV-10a); 12389 (XV-19).
Hastings, J. R., 63-300 (XII-3); 64-183 (XII-3); 64-344 (IX-10d); 69-130 (XI-1).
Hatschbach, G., 1045 (III-27); 3261 (VIII-2); 3511 (XV-2); 8308 (III-31); 13089 (III-31); 15495 (XV-3); 15511 (XV-2); 15796 (XV-10a); 17165 (III-31); 17789 (XV-2); 17959 (III-31); 18060 (XV-2); 18293 (XV-2); 18430 (XV-1); 20050 (VIII-2); 21387 (III-27); 21948 (XV-19); 25304 (XV-10a); 25887 (XV-2); 26264 (III-27); 27100 (VIII-2); 30420 (VI-2); 33368 (XV-10a); 37172 (VIII-2); 39622 (III-30); 40206 (III-31); 40546 (XV-10a); 41710 (III-27); 41769 (XV-2); 42515 (VIII-2); 43300 (III-31); 43884 (III-27); 44195 (III-30); 45024 (VII-1); 47340 (XV-21); 47774 (III-28); 47789 (III-28); 48194 (VIII-1); 48345 (III-30); 48368 (IX-6); 50336 (XV-2); 50355 (III-31); 50724 (III-28); 52185 (VII-1); 52401 (XV-10a); 53377 (XVI-8); 55222 (XV-1); 58192 (VIII-2); 59387 (VIII-2); 59700 (VI-1); 59954 (XV-21); 60082 (III-27);
Haught, O., 270 (XV-17); 4061 (X-3); 4117 (IX-10a); 4366 (VI-1).
Havard, V., 112 (X-1).
Hayes, S., 656 (V-1).
Heithaus, E. R., 61 (VI-1).
Hekker, F., 10082 (III-1).
Heller, A. A., 1227 (VI-1).
Hellín, J. J., 6 (XV-9).
Henderson, A., 455 (I-3a); 688 (III-15).
Henrickson, J., 12427 (XI-3).
Heringer, E. P., 589 (III-27); 944 (XV-1); 2602 (III-27); 2718 (III-26); 3412 (III-13a); 5711 (XV-21); 6582 (IX-7); 6815 (XV-1); 7279 (XV-10a); 8982 (VIII-4); 9166 (VIII-4); 14106 (XV-10a); 15179 (III-28); 15274 (XV-10a); 16237 (IX-7).
Hermann, F. J., 11175 (III-19).
Hermann, H. A. van, 3959 (III-33).
Hernández, H. M., 157 (XI-3); 504 (X-3).
Hernández, L., 257 (I-1).

Hernández, N., 176 (XV-8).
Hernández C. & G. Pichon, 51 (IX-10f).
Hernández M., R., 4966 (XI-3); 6495 (XIII-2); 7929(XIII-1).
Herrera, G., 928 (XV-9); 6264 (XV-9).
Herter, W. G., 78972 (XVI-3); 94378 (XV-19).
Hess, W. E., 1997 (XV-20).
Heyligers, P.C., 103 (III-13b); 187 (III-13b).
Hill, S. R., 2266 (XX-1); 3408 (XX-2); 7317 (XII-2); 12937 (III-39); 13455 (XX-1).
Hinton, G. B., 928 (X-2); 4140 (X-2); 6046 (XV-5); 9980 (XV-5); 9984 (X-2); 10326 (XV-5); 11647 (IX-10d); 21422 (XI-3); 21578 (XI-3); 21832 ((XIII-2).
Hioram, Bro., 2072 (XX-2); 2174 (V-2); 2201 (XV-13); 9776 (X-1).
Hitchcock, A. S., 17173 (XIV-7a).
Hodge, W. H., 617 (III-13a); 1013 (III-13a); 1652 (III-13a); 2074 (III-13a); 2314 (III-13a); 2442 (III-13a); 2677 (III-13a); 3412 (III-13a).
Hoehne, F. C., 98 (III-30); 296 (III-31); 340 (VI-2); 498 (III-31); 1432 (III-30); 3156 (III-30); 3157 (III-31); 5662 (VI-2); 5891 (III-27); 6181 (III-31); 11348 (III-27); 12909 (III-27); 15563 (III-27); 23833 (III-31); 24850 (III-30); 29604 (III-27); 29842 (III-31); 30998 (VIII-2).
Hoehne, W., SPF 6225 (VIII-2); SPF 10498 (III-31); SPF 11320 (VIII-2); SPF 11432 (III-30); SPF 12253 (VIII-2); SPF 13156 (III-30); SPF 13157 (III-31).
Hoffman, B., 843 (VI-1); 853 (III-13b); 926 (XIV-7a); 930 (III-13b); 1225 (I-2); 2362 (I-2); 2719 (III-18); 3580 (III-12).
Holdridge, L. R., 520 (XV-13); 1112 (VI-1); 2316 (III-5).
Holmgren, I., 125 (V-1).
Holm-Nielsen, L., 27684 (VI-1); 27700 (VI-1); 27918 (VI-1); 27947 (VI-2).
Holst, B. K., 2188 (III-14b); 2690 (III-13b).
Holt, E. G. & W. Gehriger, 216 (I-3c).
Hopkins, H. C. & M. J. G., 310 (III-10).
Hopkins, M. J. G., 5 (III-2); 113 (IX-7).
Horton, 479 (III-13b).
Hostmann, F. W., 171 (XIV-7a); 214 (I-2) 259 (XV-21); 262 (I-2); 459 (XV-21); 479 (III-13b); 1190 (I-2); 1277 (XIV-7a).
Howard, R. A., 230 (XX-1); 5468 (XX-1); 6127 (III-37); 6594 (XX-1); 10562 (III-13a); 10939 (VI-1); 13690 (XV-13); 13857 (XV-13); 14666 (III-35b); 17698 (III-13a); 18755 (III-13a); 19057 (VI-1); 19178 (XV-10a); 19456 (XV-20); 19879 (III-13a).
Howell, J. T., 10406 (IX-10d).
Huashikat, V., 909 (III-19).
Huber, J., 764 (I-2); 1277 (III-13a).
Huber, O., 1593 (XIV-9); 1757 (XIV-9); 1783 (XIV-9); 1812 (XIV-9); 2172 (I-3c); 2934 (XIV-7a); 2971 (III-14b); 3655 (XIV-5b); 3684 (XIV-9); 5768 (XIV-5b); 5877 (XIV-9); 5979 (XIV-5b); 7403 (I-1); 9151 (III-13a); 10373 (III-13a); 11151 (III-14b); 16923 (XIV-5b); 17033 (XIV-5b).
Hughes, C. E., 1191 (XV-17); 1227 (IX-10d); 1231 (XV-10a); 1241 (VI-1); 1275 (X-3); 1284 (XII-2); 1292 (XV-5); 1296 (IV-1); 1297 (XI-2); 1307 (XV-5); 1334 (IV-1); 1335 (XV-5); 1339 (XV-9); 1342 (IV-1); 1343 (XV-10a); 1395 (XV-9); 1404 (XI-2); 1435 (XV-9); 1452 (IV-1); 1471 (XV-5); 1537 (IX-10d); 1539 (XII-3); 1543 (IV-1); 1576 (XV-6).
Humbert, H., 21115 (IX-7).
Humboldt, F. von & A. Bonpland, 603 (VI-1); 995 (XIV-5a); 3562 (XV-17); 3805 (V-1).
Ibarra Manríquez, G., 724 (XV-5); 1498 (XV-5); 1968 (XV-5).
Ibarrola, T., 827 (XV-19); 1727 (XV-19).
Idrobo, J. M. & R. E. Schultes, 932 (III-19).
Imray, J., 121 (III-13a).
INPA, 4114 (III-14b); 6225 (XIV-7b); 37797 (XIV-5a); 37835 (XIV-5a); 39493 (I-3a); 42115 (XIV-7a); 43842 (I-2); 48134 (XIV-7b); 50630 (III-1); 58162 (XV-16a); 58310 (II-2); 61607 (III-20); 81971 (XV-16a).
INPA/WWF, 1201.285 (III-21); 1301.338 (III-13a); 1302.2891 (XVII-1); 1302.4026 (III-45); 1302.4531 (III-45); 1302.41881 (III-18); 2206.619 (III-1); 2206.1515 (III-13a); 2206.1453 (III-45); 2303.982 (III-1); 2303.1476 (III-45); 2303.1519 (III-45); 2303.1822 (III-45); 2303.1824 (III-9); 2303.2010 (II-2); 3304.4574 (III-1).
Irwin, H. S., 127 (III-13b); 188 (XIV-7a); 377 (III-13b); 394 (XIV-7a); 667 (VI-1); 6815 (XV-1); 15672 (VIII-4); 30882 (III-30); 31255 (XV-1); 32460 (III-30); 32557 (III-30); 37204 (I-2); 47586 (I-2); 47849 (I-2); 48383 (I-2); 48832 (II-2); 54601 (III-13b); 55211 (I-2); 55802 (I-2).
Isely, D., 10697 (XII-2); 10713 (XII-2); 10977 (XX-1).

Jack, J. G., 3877 (VI-1); 5102 (XV-21); 5223 (VI-1); 5500 (XV-13); 6564 (XV-13); 6740 (XX-1); 6854 (III-36); 7251 (III-36); 7714 (IX-10e); 8293 (XV-23); 8647 (XX-2); 30312 (XV-12).
Jacquemin, H., 1844 (III-13a); 2097 (I-2).
Jalm, A., 1342 (III-13b).
Jangoux, J., 1144 (III-21).
Jansen-Jacobs, M. J., 1407 (I-2); 1721 (I-2); 1858 (II-2); 2643 (XV-14); 3031 (XVI-10).
Janzen, D., 502 (VI-1).
Jaramillo, J., 9036 (I-2).
Jardim, J. G., 126 (III-13a); 188 (II-2); 355 (II-2).
Játiva, C., 446 (V-1); 1009 (XV-17).
Jenman, G. S., 364 (XV-14); 528 (XIV-7a); 2112 (XV-21); 2429 (III-13b); 3904 (III-13b); 4266 (III-13b); 4964 (XIV-7a).
Jennings, O. E., 66 (IX-10e); 186 (XX-2); 209 (III-34); 519 (XX-1); 694 (III-34).
Jenssen S., E., 47 (XV-10a).
Jesús, J. A. de, 1786 (III-27); 2012 (III-27).
Jiménez, A., 336 (XV-9).
Jiménez, J. J., 5635 (XX-2).
Jiménez J., J. de, 1172 (III-13b); 3475 (III-34); 3666 (III-34).
Jiménez M., A., 1574 (V-1); 3445 (XV-12); 4126 (XV-12).
Jiménez-Saa, H., 1214 (XVI-10); 1249 (XV-16a); L.B.B. 14354 (II-2).
Job, M. M., 696 (XV-19); 958 (XV-19).
Jobert, 833 (XIV-7a).
Joergensen, P., 3210 (IX-7); 3608 (IX-7); 4189 (XV-10a).
Johnson, C. D., 118-76-2 (XI-1); 141-75 (IV-1); 165-76-3 (IX-10d); 341-78 (IX-10d); 1061-79 (XV-5); 2552-82 (IX-8).
Johnston, I. M., 1516 (III-7); 3059 (XII-3); 3082 (XI-1); 3110 (XI-1); 3440 (XII-3); 3442 (XII-3); 3494 (XII-3); 3498 (XII-3); 3565 (XII-3); 3917 (XII-3); 6081 (XII-2).
Johnston, J. R., 179 (VI-1).
Johnston, M. C., 5798 (XI-3); 6081 (XII-2); 6149 (XIII-2); 10324 (XI-3); 10469 (XIII-2); 11201B (XIII-2); 541178 (XI-3).
Joly, A. B., SPF 17763 (III-31).
Jones, G. C., et al., 3144 (III-5).
Jones, M. E., 17247 (IV-1); 17264 (XII-3); 23085 (XI-1); 23087 (IX-10d); 23088 (IX-10d); 29328 (XII-2).

José Eugenio, Padre, 578 (XV-1).
Joyal, E. 2056 ((XI-5).
Jultila, J., 767 (XV-21).
Junak, S. A., SC-1325 (XIX-1).

Kappler, A., 1341 (XIV-11).
Katsuro, J., 94D (XV-21).
Kawak, M. & L. Rossi, 47 (II-2).
Keel, S., 18 (XIV-5a); 283 (III-10).
Kegel, H. A. H., 706 (III-13b); 1022 (XV-16a); 1162 (XIV-7a); 1173 (XIV-11).
Keller, W. A., 7669 (V-1).
Kellerman, W. A., 7987 (IX-10d); 7993 (IX-10d).
Kelly, I., 958 (XV-21).
Kennedy, P. B., 7009 (XI-5).
Killeen, T., 1334 (VI-2).
Killip, E. P., 9043 (III-24); 14131 (XV-10b); 14345 (VI-1); 14494 (IX-10a); 14851 (III-13a); 21090 (X-3); 27033 (XIV-2); 27485 (III-19); 29746 (III-19); 29999 (XV-16a); 30079 (III-13a); 30559 (I-2); 37547 (I-2); 38182 (IX-10c); 38270 (VI-1); 38696 (III-14a); 43555 (III-34); 43566 (III-34); 44134 (III-34); 44520 (III-34).
Kimber, C., 1215 (IX-10c); 1352 (IX-10c).
King, R. M., 1399 (IX-10d); 1451 (XI-2); 1465 (XI-2); 1648 (XI-2); 2884 (IX-10d); 3839 (XV-21); 6170 (III-19).
King, S. R., 448 (III-19); 2884 (IX-10d).
Kirizawa, 1882 (III-27).
Klein, R., 58 (III-31); 151 (III-31); 168 (XII 31); 760 (III 31); 1759 (III-31); 2272 (III-31); 7983 (XV-2).
Klug, G., 658 (III-19); 794 (III-19); 955 (III-19); 1039 (III-19); 1063 (III-19); 1446 (I-2); 1534 (III-19); 1865 (III-19); 2504 (XIV-2); 2639 (III-19); 3079 (III-15); 3357 (III-13a); 3761 (VI-2); 3933 (III-19); 4173 (IX-10b).
Knapp, S., 3395 (VI-1); 4846 (IX-10a); 6699 (IX-10c).
Korning, J., 8591 (XVII-1); 8949 (XVII-1); 9271 (XVII-1).
Kral, R., 49115 (XV-21).
Kramer, K., 2652 (XIV-7a); 3186 (III-13b).
Kramer, K. V., 10764 (XV-19).
Krapovickas, A., 13233 (IX-7); 39098 (IX-2); 43791 (VI-2).
Krieger, L., 12533 (XIV-7b); 12613 (XIV-7b).
Krukoff, B. A., 805 (XVII-1); 1088 (XV-16a); 1541 (XV-16a); 1635 (VI-2); 5631 (I-4); 6339 (III-19); 6390 (XIV-8); 6457 (III-19); 6715 (I-3a); 6743 (I-2); 7239 (XIV-7a); 7938 (III-10); 8732 (XIV-7b); 8810 (XVII-1); 10299 (XV-4); 10886 (II-2).
Kubitzki, K., 78-104 (III-30); 79-76 (XIV-5a); 79-164 (I-2); 79-214 (I-2); 84/239 (XV-16a).
Kuhlmann, J. G., 7 (XV-1); 340 (XV-1); 445 (XV-1); 550 (XV-10a); 2035 (XIV-1); 3248 (I-2).
Kuhlmann, M., 442 (III-30); 444 (III-31); 1850 (III-20); 2187-b (III-26); 4054 (III-31); 17496 (III-20); 19073 (III-31); RB 13393 (III-30); RB 72791 (III-31); RB 78317 (III-26).
Kuntze, O., 684 (III-13b).

Labbiente, R., 55 (VI-1).
Lacerda, P., 104 (III-13a).
Lage, C., 268 (III-30).
Lamb, F. B., 155 (III-13a).
Landrum, L. R., 4180 (IX-8); 6309 (IV-1); 6467 (XX-1).
Langenheim, J. M., 3117 (XV-10a).
Langlassé, E., 107 (XV-5); 280 (XV-5); 687 (IX-10d); 687a (IX-10d).
Lanjouw, J., 413 (I-2); 417 (III-13b); 599 (III-13b); 957 (I-2); 1578 (III-13b); 2266 (III-13b).
Lanna & Castellanos, 1400 (III-28).
Lao, E. A., 26 (III-14a); 534 (III-5).
Lao Magin, R., 3 (XV-10a).
Larpin, D., 847 (III-14a); 979 (III-13b).
Lawrance, A. E., 319 (XV-12); 461 (III-13a).
Lavastre, Bro B., 924 (III-34); 1584 (XV-13).
Leavenworth, W. C., 228 (XI-3).
Ledingham, G. F., 4538 (VI-1)
Leeuenberg, A. J. M., 11691 (VI-1).
Legname et al., 8644 (IX-7).
Lehmann, B. T., 364 (III-23).
Leitão Filho, H. F., 359 (XV-12); 1515 (VIII-2); 9645 (III-31).
Leme F., H., 6294 (XV-12).
Lems, K., 5089 (II-2).
León, Bro., 2941 (III-33); 3626 (IX-10e); 3627 (IX-10e); 3646 (III-33); 3947 (XV-13); 4609 (III-34); 5394 (III-36); 5432 (XV-13); 5433 (XX-2); 8623 (III-33); 8624 (III-33); 9491 (IX-10e); 9638 (XV-13); 9693 (IX-10e); 9695 (XX-1); 9776 (X-1); 10203 (III-34); 10353 (XV-13); 10679 (III-33); 11673 (XX-1); 11677 (V-2); 11693 (XX-1); 11696 (XX-2); 11746 (III-34); 12095 (V-2); 12379 (IX-1); 13629 (III-33); 13705 (III-33); 13720 (V-2); 13807 (III-34); 14260 (IX-10e); 14370 (III-34); 14384 (III-33); 14997 (XX-2); 15503 (XX-1); 17976 (III 32); 19987 (III-37); 20058 (III-34); 20365 (III-37); 22603 (III-37).
Leonard, E. C., 4071 (XX-2); 4327 (III-36); 4919 (XX-2); 7490 (XV-7); 9479 (XX-2); 9481 (XX-1); 11292 (XX-1); 11371 (XX-1); 11733 (XV-13); 12570 (XV-13); 15423 (XX-1); 15575 (III-33); 15589 (XV-21).
Lepsch da Cunha, N. M., 286 (III-15).
Lescure, 616 (IX-4); 718 (III-13a); 877 (III-3); 1374 (III-13a); 2057 (XV-10a).
Level, J. S., 70 (XIV-5a); 73 (III-39); 88 (I-2); 108 (I-3b).
Lewis, G. P., 700 (III-28); 790 (III-28); 848 (II-2); 951 (II-2); 1061 (II-2); 1056 (III-28); 1167 (III-28); 1174 (III-29); 1422 (VI-3); 1505 (I-2); 1533 (XV-14 var.); 1534 (XV-16b); 1558 (I-2); 1590 (XV-14); 1613 (XV-1); 1627 (IX-8); 1645 (II-2); 1652 (XV-14 var.); 1707 (I-2); 1848 (XV-1); 1910 (XV-1); 1972 (IX-9); 1985 (III-30); 2020 (III-28); SPF 37144 (II-2).
Liesner, R., 3978 (XIV-7b); 5704 (I-2); 5903 (XVI-10); 6053 (IX-10c); 6408 (III-17a); 6596 (III-14b); 6782 (III-17a); 7004 (XIV-7b); 7143 (XIV-7a); 7330 (XIV-3); 7394 (III-13a); 7514 (III-19); 8477 (III-13a); 8538 (XIV-7b); 8572 (XIV-3); 8683 (XIV-7b); 8814 (XIV-8); 8816 (XIV-7b); 8856 (III-17a); 8861 (III-41); 8902 (XIV-7b); 8943 (XIV-7a); 8973 (XIV-7b); 9042 (XIV-7b); 9131 (XIV-9); 9486 (III-13b); 10571 (XVI-10); 11152 (XV-8); 11272 (VI-3); 16461 (III-17a); 18146 (III-14b); 19405 (III-13b); 19763 (III-13b); 20297 (III-13b); 20484 (I-1); 20697 (III-13b); 20729 (III-13b); 22113 (III-14b); 24242 (III-13b).
Lima, D. A., 49-319 (III-30).
Lima, D. P., 13235 (IX-9).
Lima, D. de A., 49-380 (II-2).
Lima, H. C. de, 1236 (III-27); 1606 (IX-8); 1787A (III-30); 1837 (II-2); 2018 (VIII-2); 2368 (III-31); 2585 (XV-1); 2593 (XV-1); 2743 (I-3a); 2745 (I-2); 2753 (I-3a); 2866 (XV-1); 2977 (III-28); 3875 (VIII-1); 3877 (VIII-1); 3894 (III-30); 3933 (IX-6) 3965 (III-30).

Mason, C. T., Jr., 1861 (IV-1).
Mason, H. L., 1828 (IX-10d); 1828a (IX-10d); 1869 (XII-3).
Mathews, A., 4183 (IX-10b).
Mattos, J., 9161 (III-27); 11784 (III-31); 11852a (III-31); 13817 (III-27).
Mattos Silva, L. A., 249 (VII-1); 395 (II-2); 444 (III-13a); 777 (III-28); 1499 (II-2); 1719 (III-28); 2300 (VI-3).
Matuda, E., 3267 (XV-5); 3486 (II-1); 3912 (XI-4); 3913 (XI-4); 4021 (V-1); 16725 (XV-9); 17060 (XV-9); 17328 (XV-21); 17361 (XV-10a); 17607 (V-1); 17697 (V-1); 17821 (XV-10a); 18377 (XV-9); 18682 (XV-5); 18782 (XV-5); 27961(X-2); 27962 (X-2); 27964 (X-2); 28158 (XIII-1).
Maxon, W. R., 312 (IX-10d).
May Nah, A., AM-292 (XII-2).
Mazuera, H., 27 (III-14a).
McClay, T. A., 41 (II-1).
McDaniel, S., 16112 (I-2); 16314 (XIV-2); 17985 (XV-16a); 20747 (I-2).
McDowell, T., et al., 1971 (I-2); 2059 (III-13b); 2070 (II-2); 2333 (III-13b); 2337 (III-13b); 3398 (I-2); 3611 (III-14a); 4073 (XIV-7a); 4149 (I-1).
McKenzie, C. L. H., 47 (XI-3).
McPherson, G., 7441 (III-14a).
McVaugh, R., 16834 (XIII-1); 18496 (X-2); 19084 (IV-1); 20725 (IV-1); 20852 (IV-1); 20854 (IV-1); 21994 (X-2); 22887 (X-2).
Medeiros, J., 62 (III-21); 65 (III-9).
Medrano, F. G., F-1150 (X-2).
Mejía, M., 7807 (XV-13); 8048 (XV-13); 8673 (XV-23); 9799 (VI-1); 11633 (VI-1); 17199 (XV-12).
Mélinon, 48 (III-13a); 119 (III-13a); 233 (XIV-7a); 356 (III-13a); 386 (III-13a).
Mell, C. D., 624 (XV-5); 2014 (XX-1).
Mello, F., 9 (XV-16a).
Mello Barreto, 1671 (XV-1); 2034 (VIII-2); 4525 (VIII-2); 6454 (XV-1); 6462 (III-26); 6464 (III-26); 6820 (VIII-2); 8297 (VIII-2).
Méndez Ton, A. 5870 (IV-1).
Meneces, E., 768 (XVI-10).
Meneses, E., 9 (XV-10a); 17 (XV-12); 718 (II-3).
Mennega, A. M. W., 20 (III-13b); 289 (XIV-7a); 336 (XIV-7a).
Mereles, F., 2419 (IX-7).
Mereno, S., 1 (VI-1).
Mesquita, A. L., 2608 (XV-1).
Mexia, Y., 5500 (XIII-27); 6007 (I-2); 6063 (XIV-7a); 6642 (III-7); 8931 (X-2).
Mexican Boundary Survey, 325a (XII-2).
Meyer, G., 111 (VI-2).
Meyer, T., 12631 (IX-7); 15876 (XV-19); 16085 (XV-19); 17107 (IX-9).
MG, 644 (IX-4); 1609 (IX-4); 1618 (XV-16a); 2148 (XV-19); 2296 (IX-4); 2579 (XV-16a); 4103 (XV-16a); 10614 (XV-16a); 11252 (XV-16a); 11511 (XIV-2); 11800 (XV-16a); 12268 (XIV-2); 12483 (XIV-2); 15704 (XVII-1); 15710 (XVII-1); 15724 (XV-18); 15773 (XV-19); 15881 (XV-16a); 15964 (XIV-2).
Miana de Faria, S., 542 (XIV-7b); 607 (XV-18); 616 (II-2); 617 (II-2); 626 (I-2).
Miers, J., 3835 (III-30).
Mileski, E., 150 (I-2).
Mille, L., 4 (XV-15); 6 (XV-15); 110 (XV-17); 808 (XV-17); 899 (XV-17).
Miller, O. O., 261 (XV-10a).
Milliken, W., 88 (XV-14); 548 (I-2).
Miranda, F. E., 972 (XIV-4); 6658 (XI-3); 7114 (IV-1).
Mizoguchi, K., 382 (III-27); 633 (XV-21); 820 (III-31).
Moldenke, H. N., 643 (XX-1); 3657 (XX-1); 25804 (XII-2); 28969 (XV-22); 29724 (XV-21); 31693 (XV-22).
Molina R., A., 1913 (III-5); 2945 (XV-9); 8638 (IX-10a); 11419 (V-1); 13566 (V-1); 13632 (IX-10d); 13921 (III40); 14301 (XI-2); 22863 (IX-10d); 26026 (V-1).
Monsalve B., M., 1373 (III-14a); 1425 (III-14a); 1430 (III-14a).
Montaldo, P., 3913 (XV-15).
Monteiro, O. P., 48 (III-20); 160 (III-19); 348 (III-1); 1221 (I-3b).
Monteiro da Costa, R., 180 (XIV-7a).
Montes, J. E., 27743 (IX-7).
Moore, H. E., 5001 (XI-3); 9617 (III-14b).
Moraes, H. de, 6455 (XV-12).
Moran, R., 2596 (XII-3); 7315 (XI-5).
Morawetz, W., 11-25888 (III-1).
Moreno, P., 164 (XX-1); 194 (XX-1).
Moreno, P. P., 10041 (XI-2); 22813 (XV-10a); 23350 (II-3); 25074 (XV-10a).
Moretti, C., 1135 (III-13a); 1251 (I-2).
Mori, S., 1738 (III-14a); 3008 (III-14a); 9041 (I-3a); 9374 (XV-1); 9431 (III-30); 9520 (IX-9); 9592 (III-28); 10001 (VII-1); 10044 (III-30); 10230 (III-28); 10340 (III-28); 10588 (III-28); 11086 (VII-1); 11314 (III-31); 11681 (III-28); 11852 (XV-1); 12191 (VI-3); 12205a (XV-1); 12247 (VII-1); 13006 (III-30); 13652 (III-28); 13689 (III-28); 13861 (III-13a); 13959 (II-2); 14943 (III-18); 14945 (III-18); 15047 (III-13a); 15351 (III-13a); 15432 (I-2); 15477 (III-18); 15635 (II-2); 15805 (II-2); 16066 (XVI-10); 16070 (XVI-10); 16087 (III-11c); 16097 (XVI-10); 16523 (XVI-10); 17455 (XIV-7a); 17620 (I-2); 19172 (III-6a); 20183 (III-1); 20327 (I-3a); 21300 (XIV-2); 21547 (XVII-1); 21791 (I-3a); 21919 (XIV-6); 21920 (I-3a); 21961a (XV-16); 22422 (I-3a).
Morillo, G., 5484 (III-14b); 5555 (III-13a); 8849 (I-2).
Morong, T., 801 (IX-7); 825 (XV-19).
Morton, C. V., 3119 (III-34).
Mosén, H., 2475 (VIII-2); 2836 (III-27); 3618 (III-31); 4112 (VIII-2).
Mota, C. D. A., 178 (III-14b).
Mueller, C. F., 531 (XI-3); 961 (XIII-2); 2633 (XI-3); 2969 (XI-3).
Muniz, C. F. S., 4668 (III-27).
Mutis, C., 2500 (XV-17); 3531 (III-14a); 4914 (V-1).

Nadruz, M., 721 (III-31).
Narvaez, M., 220 (XI-4).
Nascimento, O. C., 34 (III-17a); 81 (III-11a); 314 (VI-2); 368 (XIV-7a); 428 (I-2); 859 (III-21); 1040 (III-21).
Nascimento, P., 3 (III-21); 34 (III-17a); 81 (III-11a); 899 (III-21).
Nash, C. V., 1174 (XX-1); 1436 (XX-1).
Nealley, G. C., 153 (XII-2).
Nee, M., 9509 (XVI-10); 10770 (VI-1); 11084 (III-14a); 17171 (XV-21); 17683 (III-13b); 18234 (XII-2); 24430 (XI-3); 25811 (XII-2); 25813 (XII-2); 27119 (XII-2); 27835 (XV-9); 28127 (XV-9); 28682 (XV-21); 28718 (XI-3); 29627 (XV-21); 31644 (VI-2); 31894 (XV-16a); 32660 (XI-3); 32661 (XI-3); 32665 (XI-3); 32692 (XII-2); 33705 (VI-2); 35102 (VI-2); 35490 (IX-7); 35759 (IX-7); 35895 (IX-7); 36936 (XV-10a); 37021 (XV-10a);

(IX-10c); 11441 (XV-15); 11442 (V-1); 11450 (V-1); 11617 (IX-10c); 12163 (III-13b); 12184 (V-1); 12363 (IX-10c); 12467 (V-1); 14879 (XV-8).
Pl. Grisolanae, 26 (IX-10c).
Plowman, T., 3566 (X-3); 9429 (III-18); 9655 (III-21); 11210 (VI-2); 12394 (I-3a); 12487 (I-2); 12954 (XIV-7b); 13949 (III-28); 14135 (XX-1).
Poeppig, E. F., 2245 (III-19); D.2367 (III-19); 2728 (XV-16a); 2899 (XIV-7b); 2957 (III-13a).
Pohl, J. E., 201 (VIII-2); d. 1420 (XV-1); 2343 (IX-7).
Polak, M., 185 (III-13b); 287 (III-13b).
Pollard, C. L., 154 (XX-1).
Ponce, M., 130 (XV-5); 462 (XV-12).
Poole, J. M., 1765 (XIV-5a).
Porto, P. C., 7590 (III-30).
Prance, G. T., 1278 (III-21); 1508 (III-18); 1647 (III-13a); 1743 (I-2); 2220 (III-10); 2340 (XV-16a); 2475 (III-19); 2569 (XV-16a); 3191 (XIV-7b); 4057 (I-2); 4653 (III-10); 4729 (I-3a); 4952 (I-3a); 5309 (XIV-6); 5423 (XIV-6); 5662 (XIV-6); 6094 (I-2); 6191 (I-2); 6556 (III-13a); 8167 (XIV-6); 8189 (III-13a); 8435 (XV-16a); 8659 (XV-16a); 10442 (XIV-5a); 10676 (I-2); 10692 (I-2); 13854 (XVI-9); 14072 (III-19); 14090 (I-3a); 14576 (XIV-6); 14605 (III-13a); 14762 (XIV-6); 14934 (I-3a); 14978 (III-13a); 15206 (I-3a); 17366 (XIV-2); 17701 (XIV-7b); 17703 (I-2); 17873 (I-3a); 18025 (XIV-6); 19347 (VI-2); 19550 (XIV-7b); 20252 (XIV-7b); 20476 (III-1); 20616 (I-3a), 20959 (III-10); 21734 (XIV-6); 21752 (XIV-8); 21810 (III-19); 21935 (XIV-6); 22217 (III-15); 23016 (III-9); 23045 (III-15); 23591 (I-2); 23792 (III-19); 24504 (XV-16a); 25468 (III-21); 26292 (XV-10a); 26682 (XV-10a); 26683 (XV-10a); 26697 (VI-2); 28050 (III-15); 28391 (III-13b); 28521 (III-20); 28744 (XIV-10); 29407 (XIV-10); 29854 (XIV-10); 29929 (XV-21); 30200 (XVII-1); 30443 (XVII-1); 30475 (III-13a); 58262 (XV-10a); 59160 (XVI-10); 59273 (I-2).
Prévost, M.-F., 781 (XIV-11); 830 (III-13b); 1135 (IX-4); 1214 (IX-4); 1730 (I-2); 2003 (I-2); 2805 (III-13a).
Priest, J., 230 (V-1).
Pringle, C. G., 1461 (XIII-1); 1946 (XI-3); 1961 (XII-2); 2410 (XII-2); 2691 (XI-3); 5352 (IV-1); 6519 (XIII-1); 6790 (XIII-1); 8472 (X-2); 9770 (XI-3).
Proctor, G. R., 4699 (XIV-7a); 11211 (III-35b); 17056 (III-13a); 17242 (XV-10a); 17407 (XV-23); 17613 (III-13a); 26640 (III-35a); 35420 (III-35b); 39293 (III-33).
Proctor Cooper, G., 461 (III-5).
Pruski, J., 3264 (III-13a).
Pulle, A., 63 (XIV-7a); 65 (I-1); 530 (III-13b).
Purpus, C. A., 3869 (XIII-2); 4144 (X-2); 4849 (XI-3); 5176 (XI-3); 5177 (XIII-2); 6085 (XI-3); 8636 (IX-10c); 8717 (IX-10d); 8723 (XV-5); 8840 (IX-10c); 9086 (IX-10c); 10689 (XIII-1); 13043 (XV-5).
Pursell, R. A., 8160 (V-1); 8739 (V-1); 8904 (VI-1).

Questel, A. 2337 (III-13a).
Quevedo S., R. C., 16 (XVIII-1).

Rabelo, B. V., 589 (I-2); 909 (III-13a); 2716 (XIV-7a); 3016 (XIV-7a); 3045 (II-2); 3053 (III-11c).
Radford, A. E., 44916 (XV-22).
Ramalho, R. S., 176 (IX-9); 1069 (III-26); 1141 (III-27); 1212 (XV-1).
Ramcharan, E. K., 195 (III-13b).
Ramia, M., 8455 (V-1); 8471 (V-1); 8476 (VI-1); 8477 (V-1); 8490 (VI-1).
Ramírez, I., 361 (I-2).
Ramírez, N., 766 (IX-10c); 2464 (XV-21); 3082 (IX-10c); 3218 (IX-10c); 3482 (I-1).
Ramos, C. H., 320 (XI-3).
Ramos, J., INPA 62192 (I-2); INPA 62250 (I-3a).
Ramos, N. A., 764 (XVII-1).
Rangel, O., 2964 (V-1).
Ratter, J. A., 1432 (XVI-9); 2330 (XVI-10); 4455 (IX-4); 4495 (I-2); 4607 (XV-10a); 6263 (XV-14); 6309V (III-13b); 6403 (XV-10a).
Rawixtcher, F., SPF 17779 (III-27); RB 13393 (III-30).
RB 5277 (XV-1); 10166 (XV-1); 10167 (XV-10a); 10168 (XIV-6); 10169 (XIV-7a); 10174 (XV-18); 15246 (XV-1); 15248 (VI-3); 16774 (XIV-6); 16775 (XIV-6); 16791 (XV-19); 17497 (IX-10b); 20164 (XIV-7b); 20198 (XV-18); 24331 (XIV-2); 24335 (XVII-1); 35533 (XIV-9); 35537 (XIV-8); 71795 (IX-7); 7463 (XV-1).
Record, S. J., 7 (XV-10a); 20 (XV-9); 27 (XV-5); 51 (X-3); 59 (IX-10d); 75 (XV-10b); 7881 (III-14a); 10005 (IX-10d).
Reed, H. R., 449 (XV-21).
Regnell, A. F., II/96 (VIII-2).
Reitz, P. R., C134 (III-31); 1596 (III-31); 2190 (III-31); 2282 (III-31); 3221 (III-31); 3912 (III-31); 5647 (III-31); 7614 (XV-2); 8049 (III-31); 8342 (XV-2); 8649 (XV-2); 12190 (III 31); 14359 (XV-3); 17371 (III-31).
Renner, S., 97 (III-19); 1003 (XIV-6).
Restinga, 724 (III-30).
Restrepo, D., 167 (III-14a); 629 (I-2).
Revilla, J., 342 (XIV-2); 397 (XIV-2); 429 (XV-16a); 709 (I-2); 777 (I-2); 1218 (III-10); 1652 (III-19); 1971 (III-19); 2090 (XIV-2); 3478 (I-2); 3653 (XIV-2); 4140 (XIV-5a); 4194 (XIV-5a); 4195 (XIV-5a); 4483 (I-2); 7026 (III-10); 8405 (II-3); 8419 (III-9).
Ribeiro, B. G. S., 755 (III-14b); 1783 (III-21); 15323 (III-17a); 15328 (III-14b).
Riccobene, A., 008 (VI-1).
Richards, P. W., 497 (I-2).
Ricksecker, A. E., 283 (VI-1); 344 (XV-21); 352 (VI-1).
Rico, L., 725 (VI-1).
Riedel, L., 22 (VI-2); 24 (III-30); 26 (VIII-2); 42 (XV-16a); 118 (III-28); 204,205,206 (III-28); 372 (III-31); 443 (III-30); 444a (III-30); 444b (III-30); 529 (XIV-7a); 675 (VIII-2); 727 (IX-7); 2372 (III-30).
Rimachi Y., M., 1961 (XV-16a); 2482 (XV-16a); 2846 (III-19); 3267 (I-2); 4032 (III-1); 6108 (III-19); 8011 (XV-16a); 8530 (I-2); 8928 (XV-16a).
Rizzini, C. T., 1586 (III-30).
Roa, A., 264 (II-2); 517 (III-15).
Robert, A., 648 (VI-2); 802 (VI-2).
Robum, M. J., 288 (VIII-2).
Rodrigo, A. P., 969 (XV-19).
Rodrigues, W. A., 255 (XV-16a); 350 (XIV-4); 377 (XV-16a); 715 (XIV-5a); 1211 (III-14b); 1218 (III-10); 1325 (XIV-6); 1614 (XIV-7b); 1641 (III-13a); 1658 (I-3a); 1667 (III-14b); 1703 (III-14b); 1870 (III-10); 1872 (II-2); 2120 (III-10); 2145 (XIV-5a); 2150 (III-14b); 2267 (XIV-5a); 2349 (III-14b); 2447 (XIV-7b); 2549 (XIV-5a); 2910 (III-13a); 2911 (III-13a); 2993 (XIV-7b); 3012 (III-10); 3206 (III-10); 3285 (III-10); 3913 (III-10); 4005 (III-19); 4104 (III-10); 4220 (XIV-7a); 4455 (III-14b); 4873 (XIV-4); 5314 (III-9); 5359 (I-3a); 5475 (III-14b);

Shafer, J. A., 2 (XX-2); 4 (XX-1); 161 (III-33); 221 (VI-1); 413 (III-33); 390 (XX-2); 562 (XV-13); 835 (XX-2); 855 (XX-1); 980 (VI-1); 1194 (III-33); 1371 (XX-1); 1392 (XV-13); 1725 (III-34); 2577 (XX-2); 3096 (III-34); 3220 (III-37); 4048 (III-37); 8163 (III-37); 11117 (XX-1); 11226 (III-34); 12300 (XX-2); 13038 (XX-2).
Shakaim, S., RBAE0023 (II-2).
Sharp, A. J., 50376 (IX-10d).
Shattuck, O., 237 (III-14a).
Shepherd, J. D., 872 (III-14a).
Shillingford, C. A. & W. Ernst, 337 (III-33).
Shinners, L. H., 20445 (XV-22).
Shreve, F., 7280 (IX-10d).
Sieber (Fl. trinit.), 198 (III-13a).
Silva, A., 279 (II-2).
Silva, A. S., 163 (III-13a); 639 (III-19); 814 (III-15); 9169 (II-2).
Silva, A. V. V. da, et al., INPA/WWF 2206.2860.2 (III-45).
Silva, J. A., 389 (I-3a).
Silva, J. M., 296 (III-31).
Silva, J. M. da, INPA/WWF 1301.4431 (III-13a); INPA/WWF 3304.3041 (III-1); INPA/WWF 3304.5343 (III-13a).
Silva, J. S., 422 (III-31); 451 (III-27).
Silva, M., 137 (I-3a); 328 (IX-6); 837 (XV-16a); 964 (III-10); 985 (XIV-5a); 1396 (I-2); 1452 (III-10); 1787 (XV-19); 1856 (XIV-6); 1901 (XIV-7a); 2308 (III-18); 2338 (XVI-10); 2396 (III-13a); 2474 (XVI-10); 2531 (III-20).
Silva, M. de A., 095 (III-30).
Silva, M. F. da, 73 (III-1); 229 (I-3a); 239 (I-3a); 271 (III-21); 747 (III-10); 917 (III-1); 1061 (XIV-5a); 1143 (XIV-5a); 1179 (XIV-7b).
Silva, M. G., 2743 (III-18); 2850 (III-21); 3312 (VI-2); 3611 (XV-16a); 3861 (III-11c); 4053 (III-21); 4276 (XVII-1); 4781 (XVI-10); 5633 (III-13a); 5662 (III-21); 5669 (III-21); 6564 (III-13a).
Silva, N. T., 1174 (XVI-10); 1320 (XVI-10); 1379 (XVI-10); 1387 (XVI-10); 1612 (VI-2); 1696 (III-13a); 1818 (III-9); 1973 (III-9); 2217 (XVI-9); 2403 (III-9); 2409 (I-2); 2860 (III-13a); 3242 (II-3); 3372 (III-13a); 3648 (III-21); 3824 (III-13a); 3861 (III-11b); 3974 (III-13a); 4797 (I-2); 6357 (XIV-7a); 60677 (III-19).
Silveira, O. da, 17 (III-31).
Silverstone-Sopkin, A., 3079 (VI-1).
Simpson, D. R., 1 (XV-10a); 792 (XIV-2); 839 (XVII-1).
Sintenis, P., 6722 (VI-1).
Skutch, A. F., 2871 (XV-9).
Sloover, J. R. de., 178 (VI-2).
Small, J. K., 988 (XX-1); 1864 (XX-1); 2140 (XX-1); 2545 (XX-1); 2599 (XX-1); 3579 (XX-1); 3899 (XX-1); 3927 (XX-1); 4534 (XX-1); 5756 (XX-1); 6952 (XX-1); 12764 (XX-1).
Smith, A., P2589 (III-25).
Smith, A. C., 2156 (XV-16b); 2616 (XVI-10); 3316 (VI-3).
Smith, C. E., 3668 (X-2).
Smith, D. N., 5411 (XV-12); 13619 (VI-2); 13626 (III-13a); 13682 (II-2).
Smith, F. D., 102 (V-1); 172 (IX-10c).
Smith, H. H., 30 (X-3); 30a (X-3); 38 (XV-10b); 38a (XV-10b); 52 (X-3); 222 (III-13a); 295 (III-13a); 296 (IX-10a); 311 (VI-1); 902 (XV-10a); 2064 (IX-10a); 2065 (IX-10a); 2067 (XV-10b).
Smith, L. B., 1363 (III-30); 1643 (III-31); 5741 (III-31); 7304 (III-31); 13114 (XV-2); 13183 (XV-2); 13255 (XV-2).
Smith, L. D., 300 (XV-22).
Smith, R. D., 143 (XV-8).
Smith, R. F., 88810 (III-13b); M545 (XI-3).
Smith, S. F., 1434 (I-2).
Smith, S. G., 1263 (V-1).
Snethlage, E. M., 299 (III-21); MG 10042 (I-2).
Soares, E., 148 (I-2); 200 (I-2).
Soares, J., 2181 (III-30).
Sobel, G. L., 4603 (III-13a); 4676 (II-2); 4919 (III-13a).
Sobral, M., 5920 (XV-1).
Soejarto, D. D., 2522 (VI-1); 2960 (III-13a); 3471 (III-14a); 4495 (III-13a).
Sohns, E. R., 1665 (XX-1).
Solomon, J. C., 6404 (III-13a); 7862 (VI-2); 8181 (VI-2); 8571 (VI-2).
Soria, N., 3980 (IX-7).
Soto Nuñez, J. C., 17 (X-2); 23 (X-2); 55 (X-2); 662 (XV-5); 739 (XV-5); 751 (XV-5); 1576 (XV-5); 1688 (IV-1); 1714 (XI-5); 1905 (XV-5); 2108 (XV-5); 2779 (IX-10d); 3503 (IX-10d); 3504 (XV-5); 3879 (XV-5); 8024 (XV-5).
Soule, M., 4 (XII-3).
Sousa, M., 107 (XII-3); 224 (XII-3); 1983 (XV-5); 2800 (XV-5); 2871 (XV-5); 3119 (IV-1); 3174 (XV-5); 5265 (XV-9); 5291 (XV-5); 5377 (X-2); 5416 (X-2); 5450 (X-3); 5460 (XV-9); 5571 (XV-9); 5685 (XIII-2); 6046 (XV-5); 6363 (XIII-16); 6392 (XV-5); 7092 (XV-5); 7106 (IX-10d); 7158 (XV-9); 7169 (XIII-16); 7294 (XV-5); 7336 (XVI-10); 7349 (XV-10a); 7355 (X-3); 7394 (VI-1); 7395 (XV-10a); 7408 (X-3); 7432 (X-3); 7445 (IX-10d); 7472 (XI-2); 7517 (IX-10d); 7645 (III-16); 7654 (XV-5); 7671 (XV-5); 7707 (XIII-1); 7749 (IV-1); 7750 (X-2); 7949 (III-5); 7974 (XV-21); 7975 (VI-1); 8045 (IV-1); 8239 (XIII-2); 8579 (IX-10d); 8640 (X-3); 8718 (XV-10a); 8721 (IX-10d); 8835 (XV-5); 9097 (XV-10a); 9136 (V-1); 9141 (XV-10a); 9298 (XV-10a); 9306 (XIII-2); 9932 (XV-5); 9955 (III-16); 10080 (IX-10d); 10126 (X-3); 10141 (XV-10a); 10280 (XV-5); 10552 (III-16); 10565 (XV-5); 10571 (XV-5); 10579 (XV-5); 10673 (XV-5); 10915 (IX-10d); 10959 (XV-10a); 11323 (X-3); 11348 (XV-10a); 11597 (XV-10a); 11598 (XV-5); 11741 (V-1); 11826 (XV-9); 11938 (XI-2); 11944 (XI-2); 11947 (IX-10d); 12022 (XI-4); 12062 (III-16); 12074 (XV-5); 12134 (XI-4); 12150 (II-1); 12221 (XV-5); 12229 (XV-10a); 12245 (XI-4); 12560 (XV-5); 12584 (III-16); 12722 (XV-9); 12975 (XV-9); 13015 (XV-10a); 13055 (VI-1); 13065 (XV-5); 13094 (XV-5); 13201 (XV-5); 13211 (V-1).
Souza, L. A. G. de, 585 (XV-16a).
Souza, S. A. da M., 246 (XV-16a); 460 (IX-4).
SP 414 (XV-2); 1920 (XV-2); 13077 (XV-2); 14516 (VIII-2); 30998 (VIII-2).
Spada, J., 6/77 (III-30).
Sperling, C. R., 6111 (I-2); 6237 (XVI-10); 6269 (I-2).
Splitgerber, F. L., 917 (XV-16a).
Spruce, R., 245 (III-13a); 586 (III-20); 671 (VI-2); 686 (III-20); 771 (III-20); 812 (III-20); 882 (IX-4); 1088 (XV-10a); 1658 (II-3a); 1802 (III-10); 1810 (XIV-2); 1812 (III-10); 2288 (III-11a); 2351 (III-19); 2425 (II-3b); 2429 (I-2); 2471 (III-1); 2743 (III-11a); 2786 (III-17a); 2925 (XIV-7b); 3320 (II-3a); 3329 (XIV-9); 3364 (I-2); 4185 (IX-10b); 4405 (VI-2); 4407 (XV-10 affin.); 6330 (XV-15); 23233 (III-14b).
Stahel, G., 20 (III-13b); 107 (III-13b); 3195 (II-2); 6469 (II-2).
Standley, P. C., 20792 (VI-1); 21053 (VI-1); 21274 (VI-1); 22280 (XV-9); 22391 (XV-21); 26045 (XV-9); 29765 (IX-10a); 30383 (III-14a); 44209 (XV-9).

Stanford, Lauber & Taylor, 2155 (XI-3); 2317 (XI-3).
Stearn, W. L., 158 (IX-10a).
Steere, W. C., 1335 (XX-1).
Stehlé, H., 505 (VI-1); 617 (III-13a); 1743 (VI-1); 2618 (III-13a); 4756 (III-13a); 4954 (VI-1).
Steinbach, R., 538 (III-43); 6559 (IX-7); 6598 (VI-2); 8609 (IX-7).
Stergios, B., 3438 (XV-8); 4896 (VI-1); 6078 (I-2); 7351 (I 3b); 7741 (XIV-4); 8891 (I-2); 9042 (XIV-7a); 9263 (I-2); 9419 (I-3b); 9687 (I-2); 9690 (I-2); 9869 (XIV-7a); 9951 (XIV-7a); 10322 (I-2); 11366 (XIV-8); 11395 (I-2); 11593 (III-11a); 11680 (XIV-5a); 11857 (XVII-1); 12291 (I-2); 13023 (III-13b); 13087 (V-1); 15439 (XIV-7b).
Stern, W. L., 158 (IX-10a).
Stevens, W. D., 11042 (VI-1); 11284 (VI-1); 20294 (VI-2); 22769 (V-1); 22962 (XV-10a).
Stevenson, D. W. & J. F. Ramos, 1062 (III-17a); 1086 (III-17a).
Steward, W. C., 521 (XIV-5a); P13199 (III-19), P20252 (XIV-7b).
Steyermark, J. A., 384 (III-13b); 1060 (III-14b); 1374 (I-1); 17155 (V-1); 25785 (III-13b); 33295 (III-16); 37440 (III-16); 42078 (V-1); 53035 (XV-17); 54035 (XV-17); 59015 (III-14b); 59146 (I-1); 60511 (III-12); 60514 (III-14b); 86335 (II-2); 86394 (III-13b); 86467 (III-13b); 86528 (II-2); 86661 (III-13b); 86898 (VI-3); 87725 (XV-16a); 87793 (XIV-2); 88077 (II-2); 89317 (VI-3); 89485 (III-13b); 91368 (III-13b); 93316 (III-4); 95182 (III-13b); 96165 (III-13b); 101324 (III-19); 102129 (XV-10a); 102627 (XIV-7b); 102883 (III-17a); 102963 (XIV-9); 102996 (XIV-9); 106617 (I-1); 109167 (III-4); 110753 (IX-10c); 111389 (III-13b); 114780 (XIV-2); 115343 (XV-16a); 115528 (III-12); 116871 (III-13b); 116988 (III-13b); 117088 (XIV-7b); 117160 (XIV-7b); 117321 (III-14b); 119350 (III-13b); 120524 (III-13a); 122262 (XIV-7b); 127631 (III-2); 127960 (I-1); 128567 (III-14b); 130646 (III-13b); 131865 (I-1); 131954 (I-1).
Stimson, W. R., 597 (XX-1).
St. Hilaire, A. de, Cat. B′ no. 136 (IX-8); no. 282 (III-31); Cat. C′ no. 1089 (III-31); Cat. B^2 N. 2207 (III-31); Cat. B^2 N. 2208 (III-26).
Stoffers, A. L., 426 (XV-16b).
Strang, H. E., 328 (III-30).
Strudwick, J. J., 3104 (III-13a); 4425 (IX-4).
Stubblebine, W. H., 13226 (III-27); 13227 (II-2).
Sturrock, 455 (III-13a).
Sucre, D., 1212 (III-30); 9583 (III-27); 10611 (III-31); 10631 (III-31); 10702 (III-30).

Tamayo, F., 2038 (III-14a); 4007 (I-2).
Tate, G. H. H., 93 (VI-1); 863 (III-14b).
Tavakilian, G., 1141 (II-2).
Tavares, S., 1002 (III-30); 1055 (III-30).
Tawjoeran, J., LBB12580 (III-13b).
Taylor, A. A., 117 (III-34); 118 (III-34).
Taylor, J., 6107 (X-2).
Taylor, M., 233 (XIII-2).
Taylor, N., 231 (XX-2).
Teixeira, B. C., 300 (III-27).
Teixeira, G., 2967 (III-30).
Teixeira, L. O. A., 03 (III-21); 639 (XIV-6); 1145 (XIV-7a); 1231 (I-3a); 1239 (XIV-6).
Téllez, O., 398 (III-5); 571 (VI-1); 608 (V-1); 906 (XI-4); 939 (XI-4); 1018 (X-3); 1030 (XV-5); 1144 (X-3); 1151 (X-3); 1227 (X-3); 1520 (X-3); 1892 (XX-1); 1972 (XX-1); 2007 (VI-1); 2036 (XX-1); 2073 (X-3); 2261 (XX-1); 2465 (X-3); 2519 (XX-1); 3190 (XI-4); 3322 (X-2); 3342bis (XX-1); 3739 (XI-4); 3837 (XX-1).
Tenorio L., P., 202 (VI-1); 1025 (XI-1); 1305 (XII-3); 2817 (IX-10d); 3098 (XV-5); 4631 (XI-1); 5547 (XIII-2); 8329 bis (IX-10d); 8846 (X-2); 9147 (XI-3); 10404 (IV-1); 10475 (IV-1); 10477 (IX-10d); 11498 (XIII-1); 11563 (X-2); 12815 (XII-3).
Terceros, W., 768 (XVI-10).
Tessman, G., 3400 (XV-10a); 3678 (I-2); 4035 (XV-10a); 5257 (XIV-2); 5319 (I-2).
Teuhissen, M., 16394 (XIV-7a).
Thomas, W. W., 3442 (I-2); 3444 (I-2); 4067 (IX-4); 4594 (XV-19); 5115 (II-3); 8899 (III-13a); 9859 (III-28); 9965 (III-28); 10056 (III-28).
Thompson, S. A., 411 (XV-17); 413 (XV-17).
Thurber, G., 858 (XII-2).
Tillett, S. S., 45371 (I-1); 45554 (III-1); 45696 (I-1).
Tiwari, S., 610 (III-13b).
Tjon Lim Sang, R., LBB-16235 (XVI-10).
Toepffer, A., 771 (III-13a); 1014 (III-13b); 1083 (III-13b).
Tonduz, A., 8932 (XV-9); 9077 (XV-9); 13060 (III-25); 13531 (XV-9); 13854 (VI-1); 13885 (VI-1).
Torke, K., 395 (X-2).
Toro, R. A., 483 (XV-21); 522 (XV-12); 814 (XV-12).
Torres C., R., 80 (XVI-10); 233 (XV-9); 921 (XII-2); 986 (XI-3); 1693 (X-2); 2272 (XI-2); 3919 (XI-1); 5374 (X-3); 5872 (XI-2); 6673 (XIII-2); 8265 (XV-10a); 8376 (IX-10d); 11247 (V-1).
Toysffer, A., 771 (III-13a).
Tracy, J. & J. B. Alcorn, 450 (III-19).
Tresling, 211 (I-2).
Triana, J. J., 4450 (XV-12); 4481 (IX-10c).
Troncoso, N., 1346 (XV-19).
Trott, S., 100 (XI-3).
Trujillo, B., 13625 (III-14b).
Tucker, J. M., 917 (VI-1).
Tún Ortíz, R., 1210 (IX-10d).
Turner, B. L., 3972 (XI-3).
Turner, R. M., 71-22 (XI-5); 71-41 (XI-5); 75-45 (XI-5); 77-34 (IX-10d); 77-67 (IX-10d).
Tutin, T. G., 379 (III-13b).
Tyson, E., 2635 (XV-21); 3191 (III-14a); 5939 (XV-9).

Ucan, E., 516 (XI-4); 793 (X-3); 1146 (XII-2); 1203 (XI-4); 1374 (XI-4); 1414 (XI-4); 1523 (XI-4); 1833 (XI-4); 1910 (XI-4); 2238 (XI-4); 2329 (IX-10d).
Uchôa, A. M., 51 (III-30).
Ugent, D., 1375 (XV-22).
Ule, E., 5340 (XIV-7b); 5584 (III-19); 5648 (XV-16a); 5859 (XIV-7a); 6358 (IX-10b); 6394 (VI-2); 8130 (XV-1); 8874 (XIV-7b); 8876 (II-2); 9437 (XV-1).
Uribe-Uribe, L., 448 (VI-1); 867 (III-13b); 4102 (XV-12).
Utley, J. 4668 (VI-1).

Valenzuela, J. M., (III-34).
Valverde, F. M., 559 (VI-1).
Van Devender, T. R., 93-63 (XI-5); 93-982 (XV-6).
Vara, L. M., 290 (XX-1).
Vargas, I., 1172 (VI-2).
Vargas A., J. 26 (XV-17); 31 (XV-17).
Vasco, 27 (XV-1).
Vasconcellos, J. M., 481 (XV-1).
Vasques A., E., 13 (XV-10a).

Index to Scientific Names

Synonyms are in *italic*, new names and combinations in **boldface**. Page numbers in **boldface** indicate primary references. An asterisk (*) indicates an illustration or a map. The symbol = equates taxonomic synonyms; the symbol ≡, homotypic ones.